CAMBRIDGE STUDIES IN ADVANCED MATHEMATICS 206

THE GEOMETRY OF CUBIC HYPERSURFACES

Cubic hypersurfaces are described by almost the simplest possible polynomial equations, yet their behaviour is rich enough to demonstrate many of the central challenges in algebraic geometry. With exercises and detailed references to the wider literature, this thorough text introduces cubic hypersurfaces and all the techniques needed to study them.

The book starts by laying the foundations for the study of cubic hypersurfaces and of many other algebraic varieties, covering cohomology and Hodge theory of hypersurfaces, moduli spaces of those and Fano varieties of linear subspaces contained in hypersurfaces. The next three chapters examine the general machinery applied to cubic hypersurfaces of dimension two, three, and four. Finally, the author looks at cubic hypersurfaces from a categorical point of view and describes motivic features.

Based on the author's lecture courses, this is an ideal text for graduate students as well as an invaluable reference for researchers in algebraic geometry.

Daniel Huybrechts is Professor in the Mathematical Institute of the University of Bonn. He previously held positions at Université Denis Diderot Paris 7 and the University of Cologne. He has published five books, including *Lectures on K3 Surfaces* (2016) and *Fourier-Mukai Transforms in Algebraic Geometry* (2006).

CAMBRIDGE STUDIES IN ADVANCED MATHEMATICS

All the titles listed below can be obtained from good booksellers or from Cambridge University Press. For a complete series listing, visit www.cambridge.org/mathematics.

Already Published
166 D. Li & H. Queffelec *Introduction to Banach Spaces, I*
167 D. Li & H. Queffelec *Introduction to Banach Spaces, II*
168 J. Carlson, S. Müller-Stach & C. Peters *Period Mappings and Period Domains (2nd Edition)*
169 J. M. Landsberg *Geometry and Complexity Theory*
170 J. S. Milne *Algebraic Groups*
171 J. Gough & J. Kupsch *Quantum Fields and Processes*
172 T. Ceccherini-Silberstein, F. Scarabotti & F. Tolli *Discrete Harmonic Analysis*
173 P. Garrett *Modern Analysis of Automorphic Forms by Example, I*
174 P. Garrett *Modern Analysis of Automorphic Forms by Example, II*
175 G. Navarro *Character Theory and the McKay Conjecture*
176 P. Fleig, H. P. A. Gustafsson, A. Kleinschmidt & D. Persson *Eisenstein Series and Automorphic Representations*
177 E. Peterson *Formal Geometry and Bordism Operators*
178 A. Ogus *Lectures on Logarithmic Algebraic Geometry*
179 N. Nikolski *Hardy Spaces*
180 D.-C. Cisinski *Higher Categories and Homotopical Algebra*
181 A. Agrachev, D. Barilari & U. Boscain *A Comprehensive Introduction to Sub-Riemannian Geometry*
182 N. Nikolski *Toeplitz Matrices and Operators*
183 A. Yekutieli *Derived Categories*
184 C. Demeter *Fourier Restriction, Decoupling and Applications*
185 D. Barnes & C. Roitzheim *Foundations of Stable Homotopy Theory*
186 V. Vasyunin & A. Volberg *The Bellman Function Technique in Harmonic Analysis*
187 M. Geck & G. Malle *The Character Theory of Finite Groups of Lie Type*
188 B. Richter *Category Theory for Homotopy Theory*
189 R. Willett & G. Yu *Higher Index Theory*
190 A. Bobrowski *Generators of Markov Chains*
191 D. Cao, S. Peng & S. Yan *Singularly Perturbed Methods for Nonlinear Elliptic Problems*
192 E. Kowalski *An Introduction to Probabilistic Number Theory*
193 V. Gorin *Lectures on Random Lozenge Tilings*
194 E. Riehl & D. Verity *Elements of ∞-Category Theory*
195 H. Krause *Homological Theory of Representations*
196 F. Durand & D. Perrin *Dimension Groups and Dynamical Systems*
197 A. Sheffer *Polynomial Methods and Incidence Theory*
198 T. Dobson, A. Malnič & D. Marušič *Symmetry in Graphs*
199 K. S. Kedlaya *p-adic Differential Equations*
200 R. L. Frank, A. Laptev & T. Weidl *Schrödinger Operators:Eigenvalues and Lieb–Thirring Inequalities*
201 J. van Neerven *Functional Analysis*
202 A. Schmeding *An Introduction to Infinite-Dimensional Differential Geometry*
203 F. Cabello Sánchez & J. M. F. Castillo *Homological Methods in Banach Space Theory*
204 G. P. Paternain, M. Salo & G. Uhlmann *Geometric Inverse Problems*
205 V. Platonov, A. Rapinchuk & I. Rapinchuk *Algebraic Groups and Number Theory, I (2nd Edition)*

The Geometry of Cubic Hypersurfaces

DANIEL HUYBRECHTS
University of Bonn

Shaftesbury Road, Cambridge CB2 8EA, United Kingdom

One Liberty Plaza, 20th Floor, New York, NY 10006, USA

477 Williamstown Road, Port Melbourne, VIC 3207, Australia

314–321, 3rd Floor, Plot 3, Splendor Forum, Jasola District Centre,
New Delhi – 110025, India

103 Penang Road, #05–06/07, Visioncrest Commercial, Singapore 238467

Cambridge University Press is part of Cambridge University Press & Assessment,
a department of the University of Cambridge.

We share the University's mission to contribute to society through the pursuit of
education, learning and research at the highest international levels of excellence.

www.cambridge.org
Information on this title: www.cambridge.org/9781009280006

DOI: 10.1017/9781009280020

First published 2023

A catalogue record for this publication is available from the British Library.

Library of Congress Cataloging-in-Publication Data

Names: Huybrechts, Daniel, author.
Title: The geometry of cubic hypersurfaces / Daniel Huybrechts.
Description: Cambridge ; New York, NY : Cambridge University Press, 2023. |
Series: Cambridge studies in advanced mathematics ; 206. || Includes
bibliographical references and index.
Identifiers: LCCN 2022059690 (print) | LCCN 2022059691 (ebook) |
ISBN 9781009280006 (hardback) | ISBN 9781009280020 (epub)
Subjects: LCSH: Surfaces, Cubic. | Hypersurfaces. | Equations, Cubic. |
Geometry, Algebraic
Classification: LCC QA573 .H89 2023 (print) | LCC QA573 (ebook) |
DDC 516.3/52–dc23/eng20230328
LC record available at https://lccn.loc.gov/2022059690
LC ebook record available at https://lccn.loc.gov/2022059691
ISBN 978-1-009-28000-6 Hardback

Contents

	Preface		*page* xiii
	Notation		xv
1	**Basic Facts**		1
	1	Numerical and Cohomological Invariants	1
	2	Linear System and Lefschetz Pencils	16
	3	Automorphisms and Deformations	31
	4	Jacobian Ring	41
	5	Classical Constructions: Quadric Fibrations, Ramified Covers, etc.	58
2	**Fano Varieties of Lines**		75
	1	Construction and Infinitesimal Behaviour	75
	2	Lines of the First and Second Type	89
	3	Global Properties and a Geometric Torelli Theorem	101
	4	Cohomology and Motives	110
	5	Fano Correspondence	124
3	**Moduli Spaces**		134
	1	Quasi-projective Moduli Space and Moduli Stack	134
	2	Geometry of the Moduli Space	143
	3	Periods	148
4	**Cubic Surfaces**		156
	1	Picard Group	157
	2	Representing Cubic Surfaces	164
	3	Lines on Cubic Surfaces	175
	4	Moduli Space	183
5	**Cubic Threefolds**		193
	1	Lines on the Cubic and Curves on Its Fano Surface	196
	2	Albanese of the Fano Surface	208

3 Albanese, Picard, and Prym 217
4 Global Torelli Theorem and Irrationality 229
5 Nodal, Stable, and Other Special Cubic Threefolds 240
6 Appendix: Comparison of Cubic Threefolds and Cubic Fourfolds 246

6 Cubic Fourfolds 251
1 Geometry of Some Special Cubic Fourfolds 254
2 Pfaffian Cubic Fourfolds 273
3 The Fano Variety as a Hyperkähler Fourfold 292
4 Geometry of the Very General Fano Variety 310
5 Lattices and Hodge Theory of Cubics versus K3 Surfaces 330
6 Period Domains and Moduli Spaces 342

7 Derived Categories of Cubic Hypersurfaces 354
1 Kuznetsov's Component 356
2 Cubic Surfaces and Cubic Threefolds 377
3 Cubic Fourfolds 384
4 Chow Groups and Chow Motives 399
407

References 407
Index 434

Contents (Detailed)

Preface xiii
Notation xv

1 Basic Facts 1
1 Numerical and Cohomological Invariants 1
1.1 Lefschetz Hyperplane Theorem 1
1.2 Twisted Hodge Numbers and Picard Group 3
1.3 Euler and Betti Numbers 5
1.4 Hodge Numbers and χ_y-Genus 7
1.5 Intersection Form 12
1.6 Cubics over Other Fields 15
2 Linear System and Lefschetz Pencils 16
2.1 Universal Hypersurface 16
2.2 Discriminant Divisor 17
2.3 Resultant 20
2.4 Monodromy Group 22
2.5 Vanishing Classes 27
2.6 Diffeomorphisms 29
3 Automorphisms and Deformations 31
3.1 Infinitesimal Automorphisms 31
3.2 (Polarized) Automorphisms 33
3.3 (Polarized) Deformations 35
3.4 No Automorphisms Generically 37
4 Jacobian Ring 41
4.1 Hessian and Jacobian 41
4.2 Gorenstein and Poincaré 42
4.3 Mather–Yau and Donagi 46
4.4 Symmetrizer Lemma 49

4.5 Infinitesimal and Variational Torelli Theorem 52
5 Classical Constructions: Quadric Fibrations, Ramified Covers, etc. 58
5.1 Projection from a Linear Subspace 58
5.2 Quadric Fibrations 60
5.3 (Uni-)Rational Parametrizations I 63
5.4 Nodal Cubics 66
5.5 Hyperplane Sections 68
5.6 Triple Covers 71
2 Fano Varieties of Lines 75
1 Construction and Infinitesimal Behaviour 75
1.1 Representing the Fano Functor 75
1.2 Dimensions of the Fano Variety and Lines on Quadrics 80
1.3 Local Theory 81
1.4 Normal Bundle of a Line 84
1.5 (Uni-)Rational Parametrization II 87
2 Lines of the First and Second Type 89
2.1 Linear Spaces Tangent to a Line 89
2.2 Lines of the Second Type 94
2.3 Points Contained in Lines of the Second Type 96
3 Global Properties and a Geometric Torelli Theorem 101
3.1 Canonical Bundle and Picard Group 101
3.2 Connectedness and the Universal Line 103
3.3 Geometric Global Torelli Theorem 107
4 Cohomology and Motives 110
4.1 Grothendieck Ring of Varieties 110
4.2 Chow Motives 112
4.3 Degree and Euler Number 113
4.4 Hodge Theory via Motives 116
4.5 Integral Hodge Structures 118
4.6 χ_y-Genus and Low Dimensions 120
5 Fano Correspondence 124
5.1 Plücker Polarization and Fano Correspondence 125
5.2 Isometry 127
5.3 Quadratic Fano Correspondence 130
5.4 Dual Fano Correspondence 132
5.5 Fano Correspondences for Chow Groups 132
3 Moduli Spaces 134
1 Quasi-projective Moduli Space and Moduli Stack 134
1.1 Quotients 135

1.2 GIT Quotients 136
1.3 Stability of Hypersurfaces 137
1.4 Hilbert–Mumford 139
1.5 Moduli Quotient and Universal Family 140
1.6 Moduli Stacks 142
2 Geometry of the Moduli Space 143
2.1 Cohomology 144
2.2 Unirationality of the Moduli Space 147
3 Periods 148
3.1 Period Domain and Period Map 148
3.2 Infinitesimal and Local Torelli 152
3.3 Variational, General, and Generic Torelli 154

4 Cubic Surfaces 156
1 Picard Group 157
1.1 Intersection Form 157
1.2 Numerical Characterization of Lines 160
1.3 Effective Cone 162
1.4 Monodromy Group of 27 Lines 163
2 Representing Cubic Surfaces 164
2.1 Cubic Surfaces as Blow-Ups 164
2.2 Blowing Up $\mathbb{P}^2$ and $\mathbb{P}^1 \times \mathbb{P}^1$ 165
2.3 Cubic Surfaces As Double Covers 169
2.4 Conic Fibrations of Cubic Surfaces 171
2.5 Pfaffian Cubic Surfaces, Clebsch, and Cayley 172
3 Lines on Cubic Surfaces 175
3.1 Lines Are Exceptional 175
3.2 Two Disjoint Lines I 175
3.3 Ten Lines Intersecting a Given Line 176
3.4 Lines Generating the Picard Group 177
3.5 Two Disjoint Lines II 177
3.6 Configuration of Lines 178
3.7 Lines versus Bitangents for Double Covers 179
3.8 Eckardt Points 181
3.9 Cubic Surfaces and Lines over Other Fields 182
4 Moduli Space 183
4.1 GIT Description 183
4.2 Stable Cubic Surfaces 185
4.3 Period Description via Cubic Threefolds 186

5 Cubic Threefolds 193

0.1 Invariants of Cubic Threefolds 193
0.2 Invariants of Their Fano Variety 194
0.3 Chow Groups and Chow Motives 195

1 Lines on the Cubic and Curves on Its Fano Surface 196
1.1 Lines of the Second Type 196
1.2 Lines Intersecting a Given Line 199
1.3 Conic Fibration 204

2 Albanese of the Fano Surface 208
2.1 Tangent Bundle versus Universal Sub-bundle 209
2.2 Cohomology Ring of the Fano Surface 212
2.3 Albanese Morphism 213
2.4 Geometric Global Torelli Theorem for Threefolds 215

3 Albanese, Picard, and Prym 217
3.1 Albanese versus Intermediate Jacobian 217
3.2 Reminder on Prym Varieties 221
3.3 Theta Divisor of the Albanese 225
3.4 Chow Groups 228

4 Global Torelli Theorem and Irrationality 229
4.1 Torelli for Curves 230
4.2 Torelli for Cubic Threefolds 231
4.3 Andreotti's Proof 233
4.4 Singularity of the Theta Divisor 234
4.5 Cubic Threefolds Are Not Rational 236

5 Nodal, Stable, and Other Special Cubic Threefolds 240
5.1 Torelli for Nodal Cubic Threefolds 240
5.2 Semi-Stable Cubic Threefolds 243
5.3 Moduli Space 244
5.4 Pfaffian Cubics, Klein, and Segre 244

6 Appendix: Comparison of Cubic Threefolds and Cubic Fourfolds 246
6.1 Passing from Threefolds to Fourfolds 246
6.2 Summary 249

6 Cubic Fourfolds 251

0.1 Invariants of Cubic Fourfolds 251
0.2 Invariants of Their Fano Variety 252
0.3 Chow Groups and Chow Motives 254

1 Geometry of Some Special Cubic Fourfolds 254
1.1 Cubic Fourfolds Containing a Plane: Lattice Theory 255
1.2 Cubic Fourfolds Containing a Plane: Quadric Fibration 257

	1.3	Cubic Fourfolds Containing a Plane: Fano Correspondence	263
	1.4	Nodal Cubic Fourfolds: Blow-Up and Lattice Theory	265
	1.5	Nodal Cubic Fourfolds: Fano Variety	268
	1.6	Normal Scrolls and Veronese Surfaces	271
2	Pfaffian Cubic Fourfolds		273
	2.1	Universal Pfaffian	273
	2.2	Generic Linear Sections of the Universal Pfaffian	275
	2.3	Grassmannian Embeddings of X, $F(X)$, and $S^{[2]}$	277
	2.4	Fano Variety versus Hilbert Scheme	279
	2.5	Correspondence: Pfaffian Cubics versus K3 Surfaces	281
	2.6	Family of Quartic Normal Scrolls	285
	2.7	Pfaffian Cubic Fourfolds Are Rational	288
	2.8	Cohomology	290
3	The Fano Variety as a Hyperkähler Fourfold		292
	3.1	Hyperkähler Fourfolds	292
	3.2	Beauville–Donagi: Fano Variety versus Hilbert Scheme	296
	3.3	Global Torelli Theorem for Cubic Fourfolds	300
	3.4	Beauville–Donagi: Fano Correspondence	301
	3.5	Plücker Polarization	304
	3.6	Néron–Severi for Picard Rank Two	306
	3.7	Excluding $d = 2$ and $d = 6$	307
	3.8	Integral Cohomology of a Cubic and Its Fano Variety	309
4	Geometry of the Very General Fano Variety		310
	4.1	Algebraic Cohomology of the Very General Fano Variety	310
	4.2	Some Explicit Realizations	314
	4.3	Fano Variety of a Cubic Threefold Hyperplane Section	316
	4.4	Lines of the Second Type	318
	4.5	Lines Intersecting a Given Line	321
	4.6	Curves on the Fano Variety and Voisin's Endomorphism	327
5	Lattices and Hodge Theory of Cubics versus K3 Surfaces		330
	5.1	Hodge and Lattice Theory of K3 Surfaces	330
	5.2	Lattice Theory of Cubic Fourfolds	331
	5.3	Hassett's Rationality Conjecture	337
	5.4	Cubics versus K3 Surfaces via the Mukai Lattice	338
	5.5	Fano Variety: Hilbert Schemes and Lagrangian Fibrations	340
6	Period Domains and Moduli Spaces		342
	6.1	Baily–Borel	343
	6.2	Hassett Divisors of Special Fourfolds	344
	6.3	Reminder: K3 Surfaces	346

	6.4	Period Map for Cubic Fourfolds	347
	6.5	Laza and Looijenga	349
	6.6	Cubic Fourfolds and Polarized/Twisted K3 Surfaces	350
	6.7	Semi-Stable Cubic Fourfolds	352
7	**Derived Categories of Cubic Hypersurfaces**		354
	0.1	Orlov's Formulae	354
	0.2	Fourier–Mukai Functors	355
	1	Kuznetsov's Component	356
	1.1	Bondal–Orlov	356
	1.2	Admissible Subcategories	356
	1.3	Semi-Orthogonal Decompositions and Mutations	358
	1.4	Serre Functor	361
	1.5	Derived Categories of Hypersurfaces	363
	1.6	Serre Functor of Kuznetsov's Component	366
	1.7	Matrix Factorizations	369
	1.8	Derived Category of the Fano Variety	374
	2	Cubic Surfaces and Cubic Threefolds	377
	2.1	Cubic Surfaces	377
	2.2	Kuznetsov Component As a Category of Clifford Sheaves	378
	2.3	Categorical Global Torelli Theorem	380
	3	Cubic Fourfolds	384
	3.1	Fourfolds Containing a Plane	385
	3.2	Pfaffian Cubic Fourfolds	388
	3.3	Nodal Cubic Fourfolds	390
	3.4	Addington–Thomas	392
	3.5	Fourfolds with Equivalent Kuznetsov Components	393
	3.6	Fano Variety As a Moduli Space in $\mathcal{A}_X$	396
	3.7	Spherical Functor and the Hilbert Square $\mathcal{A}_X^{[2]}$	397
	4	Chow Groups and Chow Motives	399
	4.1	Chow Groups	399
	4.2	Decomposition of the Diagonal	401
	4.3	Finite-dimensional Motives	403
	4.4	Kuznetsov Component versus Motives	405
			407
	References		407
	Index		434

Preface

Algebraic geometry starts with cubic polynomial equations. Everything of smaller degree, like linear maps or quadratic forms, belongs to the realm of linear algebra. An important body of work, from the beginning of algebraic geometry to the present day, has been devoted to cubic equations. In fact, cubic hypersurfaces of dimension one, so elliptic curves, occupy a very special and central place in algebraic and arithmetic geometry and cubic surfaces with their 27 lines form one of the most studied classes of geometric objects.

Besides the intrinsic interest in cubic hypersurfaces, their study allows one to test key techniques in modern algebraic geometry. In fact, quite a few central notions were originally introduced to answer questions concerning cubic hypersurfaces before later developing into indispensable tools for a broad range of problems. Most of the material covered by these notes has been taught in classes and the guiding principle often was to first introduce a general concept and then to see how it works in practice for cubic hypersurfaces. As the title indicates, it is the geometry of these varieties that is centre stage. In particular, the many interesting arithmetic aspects of cubic hypersurfaces are barely touched upon. Moreover, as Hodge theory is one of the key technical tools, we often stick to hypersurfaces over complex numbers.

Chapters 1, 2, and 3 cover the general theory of cubic hypersurfaces, their Fano varieties, and their moduli spaces. Chapters 4, 5, and 6 are devoted to cubic hypersurfaces of dimension two, three, and four. The theory is less well developed beyond dimension four and we leave out completely the case of dimension one, i.e. of elliptic curves. Chapter 7 deals with some general categorical aspects of cubic hypersurfaces, again, mostly of dimension three and four. The (detailed) list of contents should give a fairly clear idea of the subjects and results that will be discussed.

These notes have their origin in a lecture course at the University of Bonn in the winter term 2017–2018. Other parts were presented during Lent term 2020 in Cambridge and in the winter term 2021–2022 again in Bonn. As the lectures, these notes assume a solid background in algebraic and complex geometry but are otherwise self-contained.

I am most grateful to colleagues and students who attended the lectures on which these notes are based. Their interest, questions, and comments have been extremely helpful and motivating. Many people have made valuable comments on these notes for which I am truly grateful. My sincere thanks go to Pieter Belmans, Frank Gounelas, Moritz Hartlieb, Emmanuel Kowalski, Alexander Kuznetsov, Robert Laterveer, Jia-Choon Lee, Andrés Rojas, Samuel Stark, Mauro Varesco, and Xiangsheng Wei.

Financial support and hospitality of the following institutions is gratefully acknowledged: The Hausdorff Center for Mathematics (Bonn), DPMMS and Gonville & Caius College (Cambridge), and the Institute for Theoretical Studies (ETH, Zurich). In the last phase, my research was partially funded by the ERC-Synergy Grant 854361 HyperK.

Cross references and proofs: Cross references of the type 'Theorem **1**.2.3' refer to Theorem 2.3 in another chapter, here in Chapter 1, whereas 'Proposition 2.3' refers to Proposition 2.3 within the same chapter.

Notation: In the first chapters we discuss cubic hypersurfaces of arbitrary dimensions. Later, when dealing with particular dimensions, we will usually denote cubic surfaces by $S \subset \mathbb{P}^3$, cubic threefolds by $Y \subset \mathbb{P}^4$, and cubic fourfolds by $X \subset \mathbb{P}^5$.

Notation

A	Albanese variety of a Fano surface: $\mathrm{Alb}(F)$.
A_2	A_2 lattice.
A_5	Moduli space of principally polarized abelian fivefolds.
$\mathcal{A}_X$	Kuznetsov component of a hypersurface.
$a\colon F \longrightarrow A$	Albanese map.
$\alpha_{\mathrm{P},X}$	Brauer class associated with a plane in a cubic fourfold.
$\mathrm{Aut}(X)$	Group of automorphisms.
$\mathrm{Aut}(X, \mathcal{O}_X(1))$	Group of polarized automorphisms.
$\mathrm{Aut}(\mathrm{D^b}(X))$, $\mathrm{Aut}_s(\mathrm{D^b}(X))$	Group of (symplectic) exact equivalences.
$\mathrm{Bs}(L)$	Base locus of a line bundle L.
$\mathrm{Bl}_L(X)$	Blow-up of a cubic in a line.
$\mathrm{Br}(X)$	Brauer group.
$\mathcal{C}$	Positive cone of a cubic surface.
$\mathcal{C}$	Arithmetic quotient of period domain for cubic fourfolds.
$\mathcal{C}_d$	Hassett divisor of special cubic fourfolds.
CH^*	Chow ring.
C_L	Curve of lines intersecting a given line $L \subset Y$.
C_x	Curve of lines through a point x in a cubic fourfold.
D_n	Period domain for cubics of dimension n.
$\mathrm{Coh}(X)$	Abelian category of coherent sheaves.
$\mathrm{D^b}(X)$	Bounded derived category of coherent sheaves.
$\mathrm{D^b}(X, \alpha)$	Bounded derived category of twisted coherent sheaves.
$\mathrm{Diff}(X)$	Diffeomorphism group.
$\mathrm{disc}(\Lambda)$	Discriminant of lattice Λ.
E_6	E_6-lattice.
E_8	E_8-lattice.
$F(X)$	Fano variety of lines in a cubic fourfold.
$F(Y)$	Fano variety of lines in a cubic threefold.
$F_2(X)$	Fano variety of lines of the second type.
$F_L(X)$	Fano variety of lines intersecting a given line.
F_{P}	Fano variety of lines intersecting a given plane.
g	Plücker polarization (as cohomology class).
Γ	Primitive cohomology of a cubic fourfold.
Γ_d	Transcendental lattice of a general special cubic fourfold.
$\Gamma(d, n)$	Monodromy group $\subset \mathrm{O}(H^n(X, \mathbb{Z}))$.
$\mathbb{G}$	Grassmann variety.

GL	General linear group.
h	Cohomology class of a hyperplane section.
$H^*(X)_{\mathrm{pr}}$	Primitive cohomology.
$\widetilde{H}(S,\mathbb{Z})$	Mukai lattice of a K3 surface.
$\widetilde{H}(X,\mathbb{Z})$, $\widetilde{H}(\mathcal{A}_X,\mathbb{Z})$	Mukai lattice of a cubic fourfold.
$\mathfrak{h}(X)$	Chow motive of a variety.
$\mathrm{HS}_{\mathbb{Z}}$	Category of polarizable integral pure Hodge structures.
$J(Y)$	Intermediate Jacobian of a cubic threefold.
$J(X)$	Jacobian ideal.
$\mathrm{Ker}(\omega)$	Kernel of a symplectic from: $\mathrm{Ker}(\omega\colon W \longrightarrow W^*)$.
$K(X) = K(\mathrm{Coh}(X))$	Grothendieck group.
K_d	Algebraic cohomology of a general special cubic fourfold.
$K_0(\mathrm{Var}_k)$	Grothendieck ring of varieties.
K_{top}	Topological K-theory.
L_d	Extended Picard lattice of a polarized K3 surface.
$\mathbb{L}$, $p\colon \mathbb{L} \longrightarrow F$	Universal line over the Fano variety.
$\mathbb{L}_{\mathbb{G}}$	Universal line over the Grassmann variety.
$q\colon \mathbb{L} \longrightarrow X$	Universal line over a cubic.
$\mathbb{L}^{[2]}$	Hilbert scheme of the universal line over the Fano variety.
$\mathbb{L}_{\mathcal{D}_0}$	Left mutation.
Λ	Lattice of a K3 surface: $H^2(S,\mathbb{Z})$.
Λ_d	Lattice of a polarized K3 of degree d: $H^2(S,\mathbb{Z})_{\mathrm{pr}}$.
$\widetilde{\Lambda}$	Mukai lattice: $\Lambda \oplus U$.
$\mathrm{MF}(F,\mathbb{Z})$	Category of graded matrix factorizations.
M_n	Moduli space of smooth cubics of dimension n.
$M_{d,n}$	Moduli space of smooth hypersurfaces: $\dim = n$ & $\deg = d$.
$\mathrm{Mot}(k)$	Category of rational Chow motives.
NS	Néron–Severi lattice.
$\mathcal{N}_{L/X}$	Normal bundle of a line in a cubic.
O	Orthogonal group of a lattice.
$\mathrm{O}^+ \subset \mathrm{O}$	Kernel of the spinor norm.
$\tilde{\mathrm{O}} \subset \mathrm{O}$	Orthogonal group fixing a class $h^{n/2}$.
$\mathbb{P}(V)$	Projective space of lines.
$\mathbb{P}(\mathcal{F})$	Projective bundle of lines in fibres of $\mathcal{F}$.
Pf	Pfaffian.
$\varphi\colon H^n(X) \longrightarrow H^{n-2}(F)$	Fano correspondence.
$\phi\colon \mathrm{Bl}_P \longrightarrow \mathbb{P}$	Quadric fibration.
$\mathrm{Pic}(X)$	Picard group.
$\mathcal{P}\colon S \longrightarrow D$	Period map.
P_L	Maximal linear space tangent to a line L.
$\mathrm{Prym}(C/D)$	Prym variety.
P^*	Dual plane of lines in $\mathrm{P} \subset X$.
$q(\alpha)$	Beauville–Bogomolov–Fujiki quadratic form.
$\mathcal{Q}$	Universal quotient bundle over the Grassmann variety.
$\mathcal{Q}_F$	Universal quotient bundle over the Fano variety.
$\mathbb{R}_{\mathcal{D}_0}$	Right mutation.
$R(Y)$	Curve of lines of the second type in cubic threefold.
$R(X)$	Jacobian ring.
$\rho(X)$	Picard number.
SL	Special linear group.
$\mathcal{S}$	Universal sub-bundle over the Grassmann variety.

$\mathcal{S}_F$	Universal sub-bundle over the Fano variety.
$\mathcal{S}_X$	Serre functor.
S	Cubic surface or K3 surface.
S_{P}	K3 surface associated with a plane in a cubic.
$S^{[2]}$	Hilbert scheme of a K3 surface.
Σ_P	Quartic normal scroll in Pfaffian cubic.
$\mathrm{sign}(\Lambda)$	Signature of lattice Λ.
$\mathbb{Z}(n)$	Tate twist or twist of the trivial rank one lattice.
$T_{[L]}F$	Tangent space of Fano variety at a line.
$\mathcal{T}_X$	Tangent sheaf of a variety.
$\mathbb{T}_x X$	Projective tangent space.
U	Hyperbolic plane with standard basis e, f.
V	Vector space underlying $\mathbb{P}^{n+1} = \mathbb{P}(V)$.
W	Weyl group.
Y	Cubic threefold.
X	Cubic fourfold.
Ξ	Theta divisor of Albanese variety.
X_V	Pfaffian cubic fourfold.
$X^{[2]}$	Hilbert scheme of subschemes of length two.
$(.)$	Intersection pairing.
$\chi(,)$	Euler pairing.
$(**)$	Hassett condition for special cubic fourfolds.
$(***)$	Addington–Hassett condition for special cubic fourfolds.

1

Basic Facts

This first chapter collects general results concerning smooth hypersurfaces, especially those of relevance to cubic hypersurfaces. Results that are particular to any special dimension – cubic surfaces, threefolds, etc., all behave very differently – will be dealt with in subsequent chapters in greater detail.

1 Numerical and Cohomological Invariants

The goal of this first section is to compute the standard invariants, numerical and cohomological, of smooth cubic hypersurfaces $X \subset \mathbb{P}^{n+1}$. Essentially all results and arguments are valid for arbitrary degree, but specializing to the case of cubics often simplifies the formulae. Most of the results hold for hypersurfaces over arbitrary (algebraically closed) fields. However, to keep the discussion as geometric as possible, we often provide arguments relying on the ground field being the complex numbers and only indicate how to reason in the general situation. See Section 1.6 for more specific comments.

1.1 Lefschetz Hyperplane Theorem Let us begin by recalling the Lefschetz hyperplane theorem, see e.g. [474, V.13] and, for the ℓ-adic versions over arbitrary fields, [210, Exp. XIII], [148, Exp. XI], or [146, IV]:

Assume $X \subset Y$ is a smooth ample hypersurface in a smooth complex projective variety Y of dimension $n + 1$. Pullback and push-forward define natural maps between (co)homology and homotopy groups. They satisfy:

(i) $H^k(Y, \mathbb{Z}) \twoheadrightarrow H^k(X, \mathbb{Z})$ is bijective for $k < n$ and injective for $k \leq n$.
(ii) $H_k(X, \mathbb{Z}) \twoheadrightarrow H_k(Y, \mathbb{Z})$ is bijective for $k < n$ and surjective for $k \leq n$.
(iii) $\pi_k(X) \twoheadrightarrow \pi_k(Y)$ is bijective for $k < n$ and surjective for $k \leq n$.

Combined with Poincaré duality $H^k(X, \mathbb{Z}) \simeq H_{2n-k}(X, \mathbb{Z})$, these results provide information about the cohomology groups of X in almost all degrees. For example, combining the Lefschetz hyperplane theorem with the usual ring isomorphism $H^*(\mathbb{P}^{n+1}, \mathbb{Z}) \simeq \mathbb{Z}[h_{\mathbb{P}}]/(h_{\mathbb{P}}^{n+2})$, where $h_{\mathbb{P}} := \mathrm{c}_1(\mathcal{O}(1)) \in H^2(\mathbb{P}^{n+1}, \mathbb{Z})$, implies the following result.

Corollary 1.1 *Let $X \subset \mathbb{P}^{n+1}$ be a smooth hypersurface of dimension $n > 1$ and degree d. Then X is simply connected and for $k \neq n$, one has*

$$H^k(X, \mathbb{Z}) \simeq \begin{cases} \mathbb{Z} & \text{if } k \text{ is even,} \\ 0 & \text{if } k \text{ is odd.} \end{cases}$$ □

For smooth cubic hypersurfaces X of dimension at least two, $\pi_1(X) = \{1\}$ can also be deduced from the unirationality of X, see Remark 5.13 or Remark **2**.1.23.

Exercise 1.2 To make the above more precise, prove for $h := h_{\mathbb{P}}|_X$ that

$$H^{2k}(X, \mathbb{Z}) = \begin{cases} \mathbb{Z} \cdot h^k & \text{if } 2k < n, \\ \mathbb{Z} \cdot (h^k/d) & \text{if } 2k > n. \end{cases}$$

Remark 1.3 At this point it is natural to wonder which of the classes $(1/d)\, h^k \in H^{2k}(X, \mathbb{Z})$, $2k > n$, are actually algebraic, i.e. can be written as an integral(!) linear combination $\sum n_i\, [Z_i]$ of fundamental classes $[Z_i]$ of subvarieties $Z_i \subset X$ of codimension k. This is not always possible and first examples of non-effective curve classes on hypersurfaces in $\mathbb{P}^4$ of large degree were constructed by Kollár [33].

However, every smooth cubic hypersurface $X \subset \mathbb{P}^{n+1}$, $n > 1$ contains a line $\mathbb{P}^1 \subset X$, see Proposition **2**.1.19 or Remark **2**.3.6. Hence, for $d = 3$, the generator $(1/3)\, h^{n-1}$ of $H^{2n-2}(X, \mathbb{Z})$ is algebraic. In Exercise 5.2 we will see that, for example, the generic cubic hypersurface $X \subset \mathbb{P}^7$ does not contain a linear $\mathbb{P}^2 \subset X$ and, therefore, there is no natural candidate for a cycle representing the generator $h^4/3$ of $H^8(X, \mathbb{Z})$.

Remark 1.4 According to the universal coefficient theorem, see e.g. [155, p. 186], there exist short exact sequences

$$0 \longrightarrow \mathrm{Ext}^1(H_{k-1}(X, \mathbb{Z}), \mathbb{Z}) \longrightarrow H^k(X, \mathbb{Z}) \longrightarrow \mathrm{Hom}(H_k(X, \mathbb{Z}), \mathbb{Z}) \longrightarrow 0.$$

We apply this to the hypersurface $X \subset \mathbb{P}^{n+1}$ and $k = n$. As $H_{n-1}(X, \mathbb{Z}) \simeq H_{n-1}(\mathbb{P}^{n+1}, \mathbb{Z})$ is trivial or isomorphic to $\mathbb{Z}$, one finds that

$$H^n(X, \mathbb{Z}) \simeq \mathrm{Hom}(H_n(X, \mathbb{Z}), \mathbb{Z}) \simeq \mathbb{Z}^{\oplus b_n(X)}.$$

In other words, $H^n(X, \mathbb{Z})$ is torsion free.

Exercise 1.5 Assume $X \subset \mathbb{P}^{n+1}$ is a smooth hypersurface of degree $d > 1$ and $\mathbb{P}^\ell \subset X$ is a linear subspace contained in X. Show then that $\ell \leq n/2$. The same result then holds for all smooth hypersurfaces over a field of characteristic zero. See Remark 3.3 for a geometric and more elementary argument which also works for $\mathrm{char}(k) > d$.

1.2 Twisted Hodge Numbers and Picard Group There is an algebraic proof of the Lefschetz hyperplane theorem, at least with coefficients in a field. The argument can be combined with Bott's vanishing results to gain control over more general twisted Hodge numbers (and not merely Betti numbers). As those will be frequently used in the sequel, we record them here.

We start with the classical *Bott vanishing* for $\mathbb{P} := \mathbb{P}(V) \simeq \mathbb{P}^{n+1}$, which can be deduced from the (dual of the) *Euler sequence*

$$0 \longrightarrow \Omega_{\mathbb{P}} \longrightarrow V^* \otimes \mathcal{O}(-1) \longrightarrow \mathcal{O} \longrightarrow 0 \tag{1.1}$$

and the short exact sequences obtained by taking exterior products

$$0 \longrightarrow \Omega_{\mathbb{P}}^p \longrightarrow \bigwedge^p (V^* \otimes \mathcal{O}(-1)) \longrightarrow \Omega_{\mathbb{P}}^{p-1} \longrightarrow 0.$$

A closer inspection of the associated long exact cohomology sequences reveals that

$$H^q(\mathbb{P},\ \Omega_{\mathbb{P}}^p(k)) = 0$$

for all p, q, and k with a short list of exceptions for which the dimensions $h^q(\mathbb{P}, \Omega_{\mathbb{P}}^p(k)) := \dim H^q(\mathbb{P}, \Omega_{\mathbb{P}}^p(k))$ are computed as follows:

(i) if $0 \leq p = q \leq n+1$, $k = 0$, then $h^{p,p}(\mathbb{P}) = h^p(\mathbb{P}, \Omega_{\mathbb{P}}^p) = 1$;

(ii) if $q = 0$, $k > p$, then $h^0(\mathbb{P}, \Omega_{\mathbb{P}}^p(k)) = \binom{n+1+k-p}{k} \cdot \binom{k-1}{p}$; or

(iii) if $q = n+1$, $k < p - (n+1)$, then $h^{n+1}(\mathbb{P}, \Omega_{\mathbb{P}}^p(k)) = \binom{-k+p}{-k} \cdot \binom{-k-1}{n+1-p}$.

The last two cases are Serre dual to each other and the well-known formula

$$h^0(\mathbb{P}, \mathcal{O}(k)) = \binom{n+1+k}{k} \tag{1.2}$$

is a special case of (ii).

To deduce vanishings for X, one then uses the standard short exact sequences

$$0 \longrightarrow \Omega_{\mathbb{P}}^p(-d) \longrightarrow \Omega_{\mathbb{P}}^p \longrightarrow \Omega_{\mathbb{P}}^p|_X \longrightarrow 0,$$

$$0 \longrightarrow \mathcal{O}_X(-d) \longrightarrow \Omega_{\mathbb{P}}|_X \longrightarrow \Omega_X \longrightarrow 0, \tag{1.3}$$

the dual of the normal bundle sequence, and the exterior powers of the latter

$$0 \longrightarrow \Omega_X^{p-1}(-d) \longrightarrow \Omega_{\mathbb{P}}^p|_X \longrightarrow \Omega_X^p \longrightarrow 0. \tag{1.4}$$

Note that as a special case of (1.4), one obtains the following *adjunction formula.*

Lemma 1.6 *The canonical bundle of a smooth hypersurface $X \subset \mathbb{P}^{n+1}$ of degree d is*

$$\omega_X \simeq \mathcal{O}_X(d - (n+2)). \tag{1.5}$$

It is ample for $d > n+2$, trivial for $d = n+2$, and anti-ample (i.e. its dual ω_X^* is ample) in all other cases. □

Applying cohomology and Bott vanishing to (1.3) and (1.4) then gives:

Corollary 1.7 *For $k < d$, the natural map*

$$H^q(\mathbb{P}, \Omega_{\mathbb{P}}^p(k)) \longrightarrow H^q(X, \Omega_X^p(k))$$

is bijective for $p+q < n$ and injective for $p+q \leq n$. □

Note that, in particular, *Kodaira vanishing* holds (over any field!):

$$H^q(X, \Omega_X^p(k)) = 0$$

for $k > 0$ and $p+q > n$, which is Serre dual to the vanishing for $p+q < n$ and $k < 0$.

Remark 1.8 For $d = 3$ and $n > 1$, the vanishing of $H^0(X, \Omega_X^p) = 0$, $p > 0$, can also be deduced (at least in characteristic zero) from the fact that cubic hypersurfaces are unirational, see Section **2**.1.5.

Corollary 1.9 *Let $X \subset \mathbb{P}^{n+1}$ be a smooth hypersurface of degree d. If $n > 2$, then*

$$\mathrm{Pic}(X) \simeq \mathbb{Z} \cdot \mathcal{O}_X(1).$$

For $n = 2$, $d \leq n+1 = 3$, and $k = \mathbb{C}$, one has $\mathrm{Pic}(X) \simeq H^2(X, \mathbb{Z})$.

Proof For $k = \mathbb{C}$, the proof is a consequence of the exponential sequence (in the analytic topology) $0 \longrightarrow \mathbb{Z} \longrightarrow \mathcal{O}_X \longrightarrow \mathcal{O}_X^* \longrightarrow 0$, which gives the exact sequence

$$H^1(X, \mathcal{O}_X) \longrightarrow H^1(X, \mathcal{O}_X^*) \xrightarrow{\sim} H^2(X, \mathbb{Z}) \longrightarrow H^2(X, \mathcal{O}_X).$$

Now, by the Lefschetz hyperplane theorem or Corollary 1.7, $H^1(X, \mathcal{O}_X) = 0$ for $n > 1$ and $H^2(X, \mathcal{O}_X) = 0$ for $n > 2$ and, using Serre duality, also for $d \leq n+1 = 3$.

See [210, XII. Cor 3.6] for a proof over arbitrary fields. The vanishing $H^2(X, \mathcal{O}_X) = 0$ is used there to extend any line bundle on X to a formal neighbourhood and then to $\mathbb{P}^{n+1}$ by algebraization. □

Exercise 1.10 Show that the cotangent bundle Ω_X, and hence the tangent bundle $\mathcal{T}_X$, of a smooth hypersurface $X \subset \mathbb{P}^{n+1}$ of degree $d \geq 3$ is stable. In other words, for any subsheaf $\mathcal{F} \subset \Omega_X$ of rank $0 < r < n$ with $\det(\mathcal{F}) \simeq \mathcal{O}_X(k)$, one has the slope inequality $k/r < (d-(n+2))/n$.

Remark 1.11 Let us rephrase the above results in the motivic setting, cf. [20, 367] for basic facts and standard notations. For the pure motive $\mathfrak{h}(X)$ of a smooth hypersurface $X \subset \mathbb{P}^{n+1}$ of degree d, there exists a decomposition

$$\mathfrak{h}(X) \simeq \mathfrak{h}(X)_{\mathrm{pr}} \oplus \bigoplus_{i=0}^{n} \mathbb{Q}(-i)$$

in the category of rational *Chow motives* Mot(k), cf. [387]. Here, $\mathbb{Q}(1)$ is the Tate motive (Spec(k), id, 1) and the primitive part $\mathfrak{h}(X)_{\rm pr}$ has cohomology concentrated in degree n. Moreover, $\mathrm{CH}^*(\mathfrak{h}(X)_{\rm pr})$ contains the homological trivial part of $\mathrm{CH}^*(X)$. The decomposition is known to be multiplicative, which is a result of Diaz [153] and Fu, Laterveer, and Vial [184].

Note that not so much more is known about the Chow ring of (cubic) hypersurfaces. However, according to Paranjape [384], see also [427], one knows $\mathrm{CH}^{n-1}(X)\otimes\mathbb{Q} \simeq \mathbb{Q}$ for smooth cubic hypersurfaces of dimension $n \geq 5$. See Section **7**.4. for further information on Chow groups and Chow motives and references.

1.3 Euler and Betti Numbers It remains to compute the Betti number $b_n(X) := \dim_{\mathbb{Q}} H^n(X, \mathbb{Q})$ of a smooth hypersurface $X \subset \mathbb{P} = \mathbb{P}^{n+1}$ and we approach this via the *Euler number*

$$e(X) := \sum_{i=0}^{2n} (-1)^i\, b_i(X) = \sum_{i=0,\neq n}^{2n} (-1)^i\, b_i(X) + (-1)^n\, b_n(X).$$

Using $b_i(X) = b_i(\mathbb{P})$ for $i = 0, \ldots, 2n$, but $i \neq n$, one finds

$$e(X) = \begin{cases} n + b_n(X) & \text{if } n \text{ is even,} \\ n + 1 - b_n(X) & \text{if } n \text{ is odd.} \end{cases}$$

Rephrasing this in terms of the *primitive Betti number* $b_n(X)_{\rm pr} := \dim_{\mathbb{Q}} H^n(X, \mathbb{Q})_{\rm pr}$, which equals $b_n(X) - 1$ for even $n > 0$ and $b_n(X)$ for n odd (use $b_{n-2}(X) = 1$ and 0, respectively), implies

$$b_n(X)_{\rm pr} = (-1)^n(e(X) - (n + 1)).$$

This reduces our task to computing $e(X) = \int_X \mathrm{c}_n(X)$. Now, the *total Chern class* of X can be computed by using the restriction of the Euler sequence (1.1) and the dual (1.3) of the normal bundle sequence:

$$\begin{aligned} \mathrm{c}(X) &:= \sum \mathrm{c}_i(X) = \mathrm{c}(\mathcal{T}_{\mathbb{P}}|_X) \cdot \mathrm{c}(\mathcal{O}_X(d))^{-1} = \mathrm{c}(\mathcal{O}_X(1))^{n+2} \cdot \mathrm{c}(\mathcal{O}_X(d))^{-1} \\ &= \frac{(1+h)^{n+2}}{(1+d\,h)} = \left(1 - d\,h + (d\,h)^2 \pm \cdots\right) \cdot \sum_{i=0}^{n} \binom{n+2}{i} h^i, \end{aligned}$$

where, as before, $h = \mathrm{c}_1(\mathcal{O}_X(1))$. Hence,

$$\begin{aligned} \mathrm{c}_n(X) &= \frac{1}{d^2} \cdot \left((-1)^{n+2} \cdot d^{n+2} \pm \cdots \pm \binom{n+2}{n} \cdot d^2\right) \cdot h^n \\ &= \frac{1}{d^2} \cdot \left((1-d)^{n+2} + d \cdot (n+2) - 1\right) \cdot h^n, \end{aligned}$$

which combined with $\int_X h^n = d$ leads to

$$e(X) = \frac{1}{d} \cdot \left((1-d)^{n+2} + d \cdot (n+2) - 1\right).$$

For $d = 3$, the right-hand side becomes

$$e(X) = \frac{1}{3}\left((-2)^{n+2} + 3 \cdot n + 5\right) \tag{1.6}$$

or, more instructively, for the Euler number e_n of the n-dimensional cubic hypersurface:

$$e_{2m} = 2m + 2 + \sum_{i=0}^{m} 4^i \quad \text{and} \quad e_{2m+1} = 2m + 2 - 2\sum_{i=0}^{m} 4^i.$$

Corollary 1.12 *The primitive middle Betti number of a smooth hypersurface $X \subset \mathbb{P}^{n+1}$ of degree d and dimension $n > 0$ is given by*

$$b_n(X)_{\mathrm{pr}} = \frac{(-1)^n}{d}\left(d - 1 + (1-d)^{n+2}\right),$$

which for $d = 3$, becomes $b_n(X)_{\mathrm{pr}} = (-1)^n \cdot (2/3) \cdot \left(1 + (-1)^n \cdot 2^{n+1}\right)$. □

We record the result for cubics and small dimensions in the following table. Further information about the intersection form, to be discussed in Section 1.5, is also included.

n	$e(X)$	$b_n(X)_{\mathrm{pr}}$	$b_n(X)$	$\tau(X)$	$(b_n^+(X), b_n^-(X))$
0	3	3	3	3	(3, 0)
1	0	2	2		
2	9	6	7	−5	(1, 6)
3	−6	10	10		
4	27	22	23	19	(21, 2)
5	−36	42	42		
6	93	86	87	−53	(17, 70)
7	−162	170	170		
8	351	342	343	163	(253, 90)
9	−672	682	682		
10	1 377	1 366	1 367	−485	(441, 926)

Exercise 1.13 Denote by $b_{n,\mathrm{pr}}$ the nth primitive Betti number of a smooth cubic hypersurface $X \subset \mathbb{P}^{n+1}$. Show then that

$$b_{n+1,\mathrm{pr}} = 2 \cdot b_{n-1,\mathrm{pr}} + b_{n,\mathrm{pr}}.$$

For a geometric reason behind this equality, see Remark 5.21.

Exercise 1.14 Let b_n be the nth Betti number of a smooth cubic hypersurface of dimension n. Show then that

$$b_n = \frac{1}{6}\left(2^{n+3} + 3 + (-1)^n \cdot 7\right)$$

or, alternatively,

$$b_{2m} = 2 + \sum_{i=0}^{m} 4^i, \quad b_{2m+1} = 2 \cdot \sum_{i=0}^{m} 4^i, \quad \text{and} \quad b_{2m+1} = 2\, b_{2m} - 4.$$

1.4 Hodge Numbers and χ_y-Genus After having computed all Betti numbers $b_i(X)$ of smooth hypersurfaces $X \subset \mathbb{P}^{n+1}$, we now aim at determining their *Hodge numbers* $h^{p,q}(X) := \dim H^q(X, \Omega_X^p)$. In principle, the Hodge numbers can be computed from the normal bundle sequence (1.3) and the Euler sequence (1.1) and it is a good exercise to do this in low dimensions. However, the information is more elegantly expressed via a universal formula that determines all the Hodge numbers for hypersurfaces of a fixed degree but of arbitrary dimension.

Exercise 1.15 Use the description of the canonical bundle ω_X in Lemma 1.6 and the sequences (1.4) to compute the remaining Hodge numbers for cubic hypersurfaces of dimension three and four. More precisely, show that for a cubic threefold $Y \subset \mathbb{P}^4$, one has

$$h^{3,0}(Y) = h^{0,3}(Y) = 0 \quad \text{and} \quad h^{2,1}(Y) = h^{1,2}(Y) = 5$$

and that for a cubic fourfold $X \subset \mathbb{P}^5$, the middle Hodge numbers are

$$h^{4,0}(X) = h^{0,4}(X) = 0, \quad h^{3,1}(X) = h^{1,3}(X) = 1, \quad \text{and} \quad h^{2,2}(X) = 21.$$

In general, the Hodge numbers are encoded by the *Hirzebruch χ_y-genus*, which for an arbitrary smooth projective variety X of dimension n is defined as the polynomial

$$\chi_y(X) := \sum_{p=0}^{n} \chi^p(X)\, y^p$$

with coefficients $\chi^p(X) := \chi(X, \Omega_X^p) = \sum_{q=0}^{n} (-1)^q\, h^{p,q}(X)$. For example, $\chi_y(\mathbb{P}^n) = 1 - y \pm \cdots + (-1)^n y^n$.

Corollary 1.16 *For a smooth hypersurface $X \subset \mathbb{P}^{n+1}$, one has*

$$\chi^p(X) = (-1)^{n-p}\, h^{p,n-p}(X) + \begin{cases} (-1)^p & \text{if } 2p \neq n, \\ 0 & \text{if } 2p = n \end{cases}$$

and, therefore, for $2p \neq n$,

$$h^{p,n-p}(X) \neq 0 \quad \textit{if and only if} \quad \chi^p(X) \neq (-1)^p$$

$$h^{p,n-p}(X) = 1 \quad \textit{if and only if} \quad \chi^p(X) = (-1)^{n-p} + (-1)^p. \qquad \square$$

This can be pictured by the Hodge diamond, which distinguishes the two cases: n even and n odd. The discussion prompts certain natural questions: for which d and n is $h^{n,0} \neq 0$? Or, how can one compute $\max\{p \mid h^{p,n-p} \neq 0\}$, which encodes the level of the Hodge structure of X? For example, by Corollary 1.7 one knows that $h^{n,0} = 0$ for cubic hypersurfaces of dimension $n > 1$.

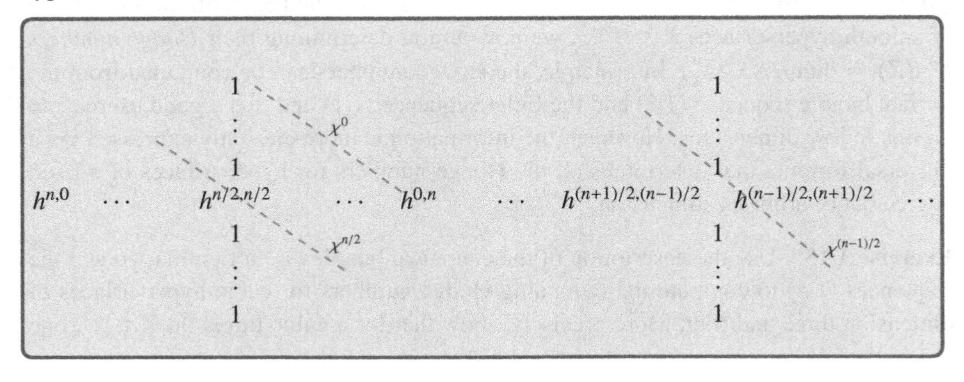

In principle, $\chi_y(X)$ can be computed by the Hirzebruch–Riemann–Roch formula

$$\chi^p(X) = \int_X \mathrm{ch}(\Omega_X^p) \cdot \mathrm{td}(X),$$

which expressed in terms of the Chern roots γ_i of $\mathcal{T}_X$ becomes

$$\chi_y(X) = \int_X \prod_{i=1}^{n} \frac{(1 - y\, e^{-\gamma_i})\,\gamma_i}{1 - e^{-\gamma_i}},$$

cf. [245, Cor. 5.1.4]. The characteristic classes of Ω_X^p and of $\mathcal{T}_X$ (the latter are needed for the computation of $\mathrm{td}(X)$) can all be explicitly determined by means of the Euler sequence for $\mathbb{P}$ and the normal bundle sequence for $X \subset \mathbb{P}$. However, the computation is not particularly enlightening until everything is put in a generating series, cf. [235, Thm. 22.1.1].

Theorem 1.17 (Hirzebruch) *For smooth hypersurfaces $X_n \subset \mathbb{P}^{n+1}$ of degree d, one has*

$$\sum_{n=0}^{\infty} \chi_y(X_n)\, z^{n+1} = \frac{1}{(1+yz)(1-z)} \cdot \frac{(1+yz)^d - (1-z)^d}{(1+yz)^d + y\,(1-z)^d}. \tag{1.7}$$

A variant of this formula for the primitive Hodge numbers

$$h^{p,q}(X)_{\rm pr} := \dim H^{p,q}(X)_{\rm pr} = h^{p,q}(X) - \delta_{p,q}$$

was worked out in [148, Exp. XI]:

$$\sum_{p,q\geq 0, n\geq 0} h^{p,q}(X_n)_{\rm pr}\, y^p z^q = \frac{1}{(1+y)(1+z)} \cdot \left[\frac{(1+y)^d - (1+z)^d}{(1+z)^d\, y + (1+y)^d\, z} - 1 \right].$$

We consider the usual specializations of the χ_y-genus for cubic hypersurfaces ($d = 3$):

(i) $y = 0$. So, we consider $\chi_{y=0}(X) = \chi^0(X) = \chi(X, \mathcal{O}_X)$. The left-hand side of (1.7) is readily computed as

$$\sum_{n=0}^{\infty} \chi(X_n, \mathcal{O}_{X_n})\, z^{n+1} = 3\,z + 0\,z^2 + z^3 + z^4 + \cdots.$$

Indeed, the first two coefficients are $\chi(X_0 = \{x_1, x_2, x_3\}, \mathcal{O}_{X_0}) = 3$ and $\chi(X_1 = E, \mathcal{O}_E) = 0$, where E is an elliptic curve. For $n > 1$, use Bott vanishing and the short exact sequence $0 \longrightarrow \mathcal{O}_{\mathbb{P}}(-3) \longrightarrow \mathcal{O}_{\mathbb{P}} \longrightarrow \mathcal{O}_X \longrightarrow 0$ to compute $\chi(X, \mathcal{O}_X) = \chi(\mathbb{P}, \mathcal{O}_{\mathbb{P}}) - \chi(\mathbb{P}, \mathcal{O}_{\mathbb{P}}(-3)) = 1$. Alternatively, simply use $h^{0,q}(X) = h^{q,0}(X) = \dim H^0(X, \Omega_X^q) = 0$ for $q > 0$.

To confirm (1.7) in this case, we compute its right-hand side and indeed find

$$\begin{aligned}
\frac{1}{1-z}\left(1 - (1-z)^3\right) &= \frac{1}{1-z} - (1-z)^2 \\
&= (1 + z + z^2 + \cdots) - (1 - 2z + z^2) \\
&= 3\,z + 0\,z^2 + z^3 + z^4 + \cdots.
\end{aligned}$$

(ii) $y = -1$. Observe that $\chi_{y=-1}(X) = e(X)$. In this case, (1.7) taken literally gives

$$\sum_{n=0}^{\infty} e(X_n)\, z^{n+1} = \frac{1}{(1-z)^2} \cdot \frac{(1-z)^3 - (1-z)^3}{(1-z)^3 - (1-z)^3},$$

which is of course not very instructive. Only when the right-hand side of (1.7) for $y = -1$ is computed as the limit for $y \to -1$ via L'Hôpital's rule, one obtains the useful formula

$$\begin{aligned}
\sum_{n=0}^{\infty} e(X_n)\, z^{n+1} &= \frac{3\,z}{(1-z)^2\,(1+2\,z)} \\
&= 3\,z \cdot (1 + z + z^2 + \cdots)^2 \cdot (1 - 2\,z + (2\,z)^2 - (2\,z)^3 \pm \cdots) \\
&= 3\,z + 0\,z^2 + 9\,z^3 + \cdots,
\end{aligned}$$

which sheds a new light on (1.6). The reader may want to check that one indeed gets the same answer.

(iii) $y = 1$. This is the most interesting case. According to the *Hirzebruch signature theorem* [235, Thm. 15.8.2], we have

$$\chi_{y=1}(X) = \tau(X).$$

To prove this, recall that for $n \equiv 0\,(2)$, the intersection pairing

$$H^n(X, \mathbb{R}) \times H^n(X, \mathbb{R}) \longrightarrow \mathbb{R}$$

is a non-degenerate symmetric bilinear form which, of course, can be diagonalized to become $\mathrm{diag}(+1, \ldots, +1, -1, \ldots, -1)$. Now, by definition,

$$\tau(X) := b_n^+(X) - b_n^-(X), \tag{1.8}$$

where $b_n^\pm(X)$ is the number of ± 1. Then the Hodge–Riemann bilinear relations imply $\tau(X) = \sum_{p,q}(-1)^p\, h^{p,q}(X) = \sum_{p,q}(-1)^q\, h^{p,q}(X)$, cf. [245, Cor. 3.3.18]. Note that, although the definition of the signature only involves the middle cohomology, indeed all Hodge numbers $h^{p,q}(X)$, also the ones for $p + q \neq n$, enter the sum.

As a side remark, observe that the right-hand side of (1.7) for $y = 1$ reads

$$\frac{1}{(1 - z^2)} \cdot \frac{(1 + z)^d - (1 - z)^d}{(1 + z)^d + (1 - z)^d},$$

which is anti-symmetric in z. Hence, only the X_n with $n \equiv 0\,(2)$ enter the computation, so that one need not worry about defining an analogue of the signature for alternating intersection forms. In any case, (1.7) implies for $d = 3$ the intriguing formula

$$\begin{aligned} \sum_{n=0}^{\infty} \tau(X_n)\, z^{n+1} &= \frac{6\,z + 2\,z^3}{(1 - z)^2\,(2 + 6\,z^2)} \\ &= z \cdot (3 + z^2) \cdot (1 + z^2 + z^4 + \cdots) \cdot (1 - 3\,z^2 + (3\,z^2)^2 - (3\,z^2)^3 \pm \cdots) \\ &= z \cdot (3 - 5\,z^2 + 19\,z^4 - 53\,z^6 + 163\,z^8 - 485\,z^{10} \pm \cdots). \end{aligned}$$

Maybe more instructive is the explicit formula for the signature of an even dimensional smooth cubic hypersurface $X_{2m} = X \subset \mathbb{P}^{2m+1}$:

$$\tau(X_{2m}) = (-1)^m \cdot 2 \cdot 3^m + 1. \tag{1.9}$$

Remark 1.18 Later, see Theorem 4.21 and (4.7) in Section 4.3, we will see that for a smooth cubic hypersurface $X \subset \mathbb{P}^{n+1}$ the middle Hodge numbers are given by

$$h^{p,n-p}(X)_{\mathrm{pr}} = \binom{n+2}{2n+1-3p}.$$

These numbers are reasonable in the sense that they satisfy complex conjugation $h^{p,n-p} = h^{n-p,p}$, but the combinatorial consequence of combining $\sum_{p=0}^{n} h^{p,n-p}(X)_{\mathrm{pr}} = b_n(X)_{\mathrm{pr}}$ with Corollary 1.12 seems less clear, see Exercise 4.13. From this description we will eventually be able to read off easily the properties of Hodge numbers. For example, one finds:

(i) $h^{p,n-p}(X)_{\mathrm{pr}} \neq 0$ if and only if $n-1 \leq 3p \leq 2n+1$.

(ii) $h^{p,n-p}(X)_{\mathrm{pr}} = 1$ if and only if $3p = 2n+1$ or $3p = n-1$.

(iii) The *level* of the Hodge structure

$$\ell = \ell(H^n(X)) := \max\{\, |p-q| \mid H^{p,q}(X) \neq 0 \,\}$$

satisfies $\ell > 1$ for $n > 5$ and $\ell > 2$ for $n > 8$. The first computations of this sort were done by Rapoport [397].

Note that the two cases in (ii) are Serre dual to each other.

Exercise 1.19 Show that only for $n = 3$ and $n = 5$, the middle cohomology $H^n(X, \mathbb{Z})$ is the Tate twist of the Hodge structure of a (principally polarized) abelian variety. In other words, only in these cases the intermediate Jacobian $J^n(X)$ is naturally a principally polarized abelian variety, see Section **2**.5.5 for a reminder of the definition.

In principle, we have now computed all Hodge numbers of smooth (cubic) hypersurfaces, but decoding (1.7) is not always easy. For later use, we record the middle Hodge numbers of smooth cubic hypersurfaces of dimension ≤10. See Section **3**.3.1 for a brief reminder of Hodge structures.

n	$b_n(X)_{\mathrm{pr}}$	H^n_{pr}	$h^{p,q}_{\mathrm{pr}}$
1	2	$H^{1,0} \oplus H^{0,1}$	1 1
2	6	$H^{1,1}_{\mathrm{pr}}$	6
3	10	$H^{2,1} \oplus H^{1,2}$	5 5
4	22	$H^{2,1} \oplus H^{2,2}_{\mathrm{pr}} \oplus H^{1,3}$	1 20 1
5	42	$H^{3,2} \oplus H^{2,3}$	21 21
6	86	$H^{4,2} \oplus H^{3,3}_{\mathrm{pr}} \oplus H^{2,4}$	8 70 8
7	170	$H^{5,2} \oplus H^{4,3} \oplus H^{3,4} \oplus H^{2,5}$	1 84 84 1
8	342	$H^{5,3} \oplus H^{4,4}_{\mathrm{pr}} \oplus H^{3,5}$	45 252 45
9	682	$H^{6,3} \oplus H^{5,4} \oplus H^{4,5} \oplus H^{3,6}$	11 330 330 11
10	1 366	$H^{7,3} \oplus H^{6,4} \oplus H^{5,5}_{\mathrm{pr}} \oplus H^{4,6} \oplus H^{3,7}$	1 220 924 220 1

1.5 Intersection Form Our next goal is to determine the intersection form on $H^n(X,\mathbb{Z})$ for a smooth cubic hypersurface $X \subset \mathbb{P} = \mathbb{P}^{n+1}$. Recall from Section 1.1 that $H^n(X,\mathbb{Z})$ is torsion free, i.e. $H^n(X,\mathbb{Z}) \simeq \mathbb{Z}^{\oplus b_n(X)}$. The non-degenerate, and in fact unimodular, intersection pairing

$$H^n(X,\mathbb{Z}) \times H^n(X,\mathbb{Z}) \longrightarrow H^{2n}(X,\mathbb{Z}) \simeq \mathbb{Z}$$

is symplectic for $n \equiv 1\,(2)$ and symmetric for $n \equiv 0\,(2)$. In the first case, $H^n(X,\mathbb{Z})$ admits a basis $\gamma_1,\ldots,\gamma_{b_n=2m}$ for which the intersection matrix has the standard form

$$\begin{pmatrix} 0 & & & 1 & & \\ & \ddots & & & \ddots & \\ & & 0 & & & 1 \\ -1 & & & 0 & & \\ & \ddots & & & \ddots & \\ & & -1 & & & 0 \end{pmatrix}.$$

For $n \equiv 0\,(2)$, the intersection pairing on $H^n(X,\mathbb{Z})$ defines a unimodular lattice. In other words, the determinant of the intersection matrix (with respect to any integral basis), i.e. the discriminant of the lattice, is ± 1. The classification of unimodular lattices is a classical topic. It distinguishes between *even lattices*, i.e. those for which $(\alpha.\alpha) \equiv 0\,(2)$ for all $\alpha \in \Lambda$, and *odd lattices*.

Assume that Λ is an odd lattice, i.e. that there exists $\alpha \in \Lambda$ with $(\alpha.\alpha) \equiv 1\,(2)$, which is unimodular and indefinite. Then, see [422, V. Thm. 4]

$$\Lambda \simeq \mathrm{I}_{r,s} := \mathbb{Z}(1)^{\oplus r} \oplus \mathbb{Z}(-1)^{\oplus s},$$

where $\mathbb{Z}(a)$ is the lattice of rank one with intersection form given by (1.1) $= a$. This can be applied to the middle cohomology of any even-dimensional, smooth hypersurface of odd degree, as $(h^{n/2}\,.\,h^{n/2}) = \int_X h^{n/2} \cdot h^{n/2} = d$.

That the intersection pairing on $H^n(X,\mathbb{Z})$ is indeed indefinite can be deduced easily (at least for cubic hypersurfaces) from a comparison of $\tau(X)$ and $b_n(X)$, cf. Corollary 1.12 and (1.9).

Corollary 1.20 *Let $X \subset \mathbb{P}^{n+1}$ be a smooth cubic hypersurface of even dimension. Then the intersection form on its middle cohomology describes a lattice isomorphic to*

$$H^n(X,\mathbb{Z}) \simeq \mathbb{Z}(1)^{\oplus b_n^+} \oplus \mathbb{Z}(-1)^{\oplus b_n^-} \simeq \mathrm{I}_{b_n^+,b_n^-}.$$

Here, $b_n^\pm := b_n^\pm(X)$ are uniquely determined by $b_n^+ + b_n^- = b_n(X) = (1/3)\,(2^{n+2}+5)$, see Corollary 1.12, *and $b_n^+ - b_n^- = \tau(X) = (-1)^{n/2}\cdot 2\cdot 3^{n/2}+1$, see* (1.9). □

More interesting, however, is the primitive cohomology $H^n(X,\mathbb{Z})_{\mathrm{pr}}$. The intersection form is still non-degenerate there, but not unimodular, and, as it turns out, not odd. By definition and using the fact that $b_{n-2} = 1$ for even $n > 0$, $H^n(X,\mathbb{Z})_{\mathrm{pr}}$ is the orthogonal complement $(h^{n/2})^\perp \subset H^n(X,\mathbb{Z})$.

It is important to note that

$$\mathbb{Z}\cdot h^{n/2}\oplus H^n(X,\mathbb{Z})_{\rm pr}\subset H^n(X,\mathbb{Z}) \tag{1.10}$$

is not an equality. It describes a subgroup of finite index. The square of the index is

$$\mathrm{ind}^2=\pm\mathrm{disc}\,(\mathbb{Z}\cdot h^{n/2})\cdot\mathrm{disc}\,\left(H^n(X,\mathbb{Z})_{\rm pr}\right)=\pm 3\cdot\mathrm{disc}\,\left(H^n(X,\mathbb{Z})_{\rm pr}\right),$$

where we use the fact that $H^n(X,\mathbb{Z})$ is unimodular and $\mathbb{Z}\cdot h^{n/2}\simeq\mathbb{Z}(3)$, see [249, Ch. 14.0.2] for the general statement and references. This also shows that the discriminant of the intersection form on $H^n(X,\mathbb{Z})_{\rm pr}$ is at least divisible by three and, therefore, $H^n(X,\mathbb{Z})_{\rm pr}$ is not unimodular. In fact, $\mathrm{disc}(H^n(X,\mathbb{Z})_{\rm pr})=3$, because the discriminant groups of $\mathbb{Z}\cdot h$ and $H^n(X,\mathbb{Z})_{\rm pr}$ are naturally isomorphic, cf. [249, Prop. 14.0.2]. This can also be deduced from the explicit description below.

The following is a folklore result for cubics (in dimension four, cf. [227] and Section **6**.5.2) and has been generalized to other degrees and complete intersections by Beauville [54]. For the definition of the lattices A_2, E_6, E_8, and U, see [249, Ch. 14] and the references therein.

Proposition 1.21 *Let $X\subset\mathbb{P}^{n+1}$ be a smooth cubic hypersurface of even, positive dimension. Then the intersection form on its middle primitive cohomology $H^n(X,\mathbb{Z})_{\rm pr}$ is described as follows:*

(i) *For $n=2$, one has $H^n(X,\mathbb{Z})_{\rm pr}\simeq E_6(-1)$.*
(ii) *For $n>2$, one has $H^n(X,\mathbb{Z})_{\rm pr}\simeq A_2\oplus E_8^{\oplus a}\oplus U^{\oplus b}$. Here, $b:=\min\{b_n^+(X)-3,b_n^-(X)\}$ and*

$$a:=\begin{cases}\frac{1}{8}(b_n^+-b_n^--3)=\frac{1}{8}(\tau(X)-3)=\frac{1}{4}(3^{n/2}-1) & \textit{if } n\equiv 0\,(4),\\[2mm] \frac{1}{8}(b_n^--b_n^++3)=\frac{1}{8}(3-\tau(X))=\frac{1}{4}(3^{n/2}+1) & \textit{if } n\equiv 2\,(4).\end{cases}$$

In particular, $\mathrm{disc}\left(H^n(X,\mathbb{Z})_{\rm pr}\right)=3$ *and the inclusion* (1.10) *has index three.*

Note that $n\equiv 0\,(4)$ if and only if $b_n^+\geq b_n^-$, see (1.9). Also, observe that $b\geq(1/3)(2^{n+1}-3^{n/2+1}-1)$, which is rather large for $n\geq 4$, i.e. $H^n(X,\mathbb{Z})_{\rm pr}$ contains many copies of the hyperbolic plane U. This often simplifies lattice theoretic arguments.

Proof Assume $n>2$, so that $b_n^+(X)>3$, and consider the odd, unimodular lattice

$$\Lambda:=\mathbb{Z}^{\oplus 3}\oplus E_8^{\oplus a}\oplus U^{\oplus b}.$$

It has rank $\mathrm{rk}(\Lambda) = b_n(X)$ and signature $\tau(\Lambda) = \tau(X)$.[1] Therefore, Λ and $H^n(X, \mathbb{Z})$ are odd, indefinite, unimodular lattices of the same rank and signature and hence isomorphic to each other (and to $\mathrm{I}_{b_n^+, b_n^-}$), cf. [422, V. Thm. 6].

Recall that a primitive vector $\alpha \in \Lambda$ in an odd unimodular lattice Λ is called *characteristic* if $(\alpha.\beta) \equiv (\beta.\beta)\,(2)$ for all $\beta \in \Lambda$. Obviously, the orthogonal complement $\alpha^\perp \subset \Lambda$ of a characteristic vector is always even. The converse also holds, cf. [322, Lem. 3.3]. Indeed, for any primitive $\alpha \in \Lambda$ in the unimodular lattice Λ there exists $\beta_0 \in \Lambda$ with $(\alpha.\beta_0) = 1$. Then, for all $\beta \in \Lambda$, the vector $\beta - (\alpha.\beta)\beta_0$ is contained in $\alpha^\perp$ and in particular of even square if $\alpha^\perp$ is assumed to be even. Hence, $(\beta.\beta) \equiv (\alpha.\beta)^2 \cdot (\beta_0.\beta_0)\,(2)$. As Λ is odd, there exists a β with $(\beta.\beta)$ odd and hence $(\beta_0)^2$ must be odd. Altogether this proves $(\beta.\beta) \equiv (\alpha.\beta)^2 \equiv (\alpha.\beta)\,(2)$ for all β, i.e. α is characteristic.

For example, $(1,1,1) \in \mathbb{Z}^{\oplus 3}$ is characteristic, for its orthogonal complement is A_2. In this case it can also be checked directly by observing that $((1,1,1).(x_1,x_2,x_3)) = x_1 + x_2 + x_3 \equiv x_1^2 + x_2^2 + x_3^2 = ((x_1,x_2,x_3).(x_1,x_2,x_3))\,(2)$. But then $(1,1,1) \in \mathbb{Z}^{\oplus 3} \subset \Lambda$ is also characteristic for Λ and its orthogonal complement is the lattice in (ii).

One now applies a general result for unimodular lattices from [487, Thm. 3]: two primitive vectors $\alpha, \beta \in \Lambda$ are in the same $\mathrm{O}(\Lambda)$-orbit if and only if $(\alpha.\alpha) = (\beta.\beta)$ and both are either characteristic or both are not.

Therefore, to prove the assertion, it suffices to show that $h^{n/2} \in H^n(X, \mathbb{Z})$ is characteristic or, equivalently, that $H^n(X, \mathbb{Z})_{\mathrm{pr}}$ is even. We postpone the proof of this statement to Corollary 2.14, where it fits more naturally in the discussion of Picard–Lefschetz reflections and of the monodromy action for the universal family of hypersurfaces. A more topological argument was given by Libgober and Wood [322].

It remains to deal with the case $n = 2$, where we have $H^2(X, \mathbb{Z}) \simeq \mathrm{I}_{1,6}$. It is easy to check that $\alpha := (3, 1, \ldots, 1) \in \mathrm{I}_{1,6}$ is characteristic with $(\alpha)^2 = 3$ and its orthogonal complement turns out to be $E_6(-1) \simeq \alpha^\perp \subset \mathrm{I}_{1,6}$. Indeed, a computation shows that $e_1 := (0,1,-1,0,0,0,0)$, $e_2 := (0,0,1,-1,0,0,0)$, $e_3 := (0,0,0,1,-1,0,0)$, $e_4 := (1,0,0,0,1,1,1)$, $e_5 := (0,0,0,0,1,-1,0)$, and $e_7 := (0,0,0,0,1,-1)$ span $\alpha^\perp$ and that their intersection matrix is indeed $E_6(-1)$. See also the discussion in Section **4**.3.4.

Now consider the class of the hyperplane section $h \in H^2(X, \mathbb{Z})$. As in this case $\mathrm{Pic}(X) \simeq H^2(X, \mathbb{Z})$, one can argue algebraically, using the Hirzebruch–Riemann–Roch formula, to prove that h is characteristic. Indeed,

$$\chi(X, L) = \frac{(L.L) + (L.h)}{2} + 1$$

implies $(L.L) \equiv (L.h) \equiv 0\,(2)$ and $(L.h) = 0$ for $L \in h^\perp$. Hence, using [487, Thm. 3] again, $H^2(X, \mathbb{Z})_{\mathrm{pr}} \simeq \alpha^\perp \simeq E_6(-1)$.

Later we will describe the isomorphisms $H^2(X, \mathbb{Z})_{\mathrm{pr}} \simeq E_6(-1)$ from a more geometric perspective and, in particular, write down bases of both lattices in terms of lines, see Sections **4**.1–3. □

[1] Note that $\tau \equiv 3\,(8)$ is a general fact for unimodular lattices containing a characteristic element α with $(\alpha.\alpha) \equiv 3\,(8)$, cf. [422, V. Thm. 2]. In our situation, $\tau(X) \equiv 3\,(8)$ can be deduced from (1.9).

Remark 1.22 Kulkarni and Wood [290, Thm. 11.1] show that the purely lattice theoretic description in Corollary 1.20 of the intersection product on $H^n(X,\mathbb{Z})$ can be realized geometrically in the following sense:

• For $n \equiv 0\,(4)$, a smooth cubic hypersurface $X \subset \mathbb{P}^{n+1}$ is diffeomorphic to a connected sum of the form $M\#k(S^n \times S^n)$ with $k = b_n^-(X)$ and, therefore, $b_n^+(M) = b_n(M) = \tau(M) = \tau(X)$.

• For $n \equiv 2\,(4)$, $n \geq 4$ the hypersurface is diffeomorphic to a connected sum of the form $M\#k(S^n{\times}S^n)$ with $k = b_n^+(X)-1$ and, therefore, $b_n(M) = b_n^-(M)+1 = -\tau(M)+2 = -\tau(X) + 2$.

• For $n \equiv 1\,(2)$, a smooth cubic hypersurface X is diffeomorphic to a connected sum $M\#k(S^n \times S^n)$, with $k = b_n(X)/2 - 1$ and, hence, $b_n(M) = 2$. For $n = 1, 3$, or 7, this can be improved to $k = b_n(X)/2$ and $b_n(M) = 0$.

• The remaining case of smooth cubic surfaces $X \subset \mathbb{P}^3$ is slightly different. Viewing X as the blow-up of $\mathbb{P}^2$ in six points, as in Section **4**.2.2, reveals that it is diffeomorphic to the connected sum $\mathbb{P}^2\#6\overline{\mathbb{P}^2}$.

1.6 Cubics over Other Fields We conclude with a number of comments on (cubic) hypersurfaces over arbitrary fields and notably over fields of positive characteristics. Most of the subtleties and pathologies that usually occur for varieties over fields of positive characteristic can safely be ignored for hypersurfaces. In the following, let $X \subset \mathbb{P}_k^{n+1}$ be a smooth hypersurface over an arbitrary field k.

(i) The *Hodge–de Rham spectral sequence*

$$E_1^{p,q} = H^q(X,\Omega_X^p) \Rightarrow H_{\mathrm{dR}}^{p+q}(X/k) \tag{1.11}$$

degenerates, cf. Section 4.5. For $\mathrm{char}(k) = 0$ or $\mathrm{char}(k) > \dim(X)$, this follows from results of Deligne and Illusie [147]. For the latter, use the fact that smooth hypersurfaces over fields of positive characteristic can of course be lifted to characteristic zero.

More directly and avoiding the assumption $\mathrm{char}(k) > \dim(X)$, one can argue as follows. The computations in Section 1.4 show in particular that the Hodge numbers $h^{p,q}(X) = \dim(E_1^{p,q})$ of smooth hypersurfaces only depend on d and n, but not on $\mathrm{char}(k)$. From (1.11) one deduces that $\sum_{p+q=m} h^{p,q}(X) \geq \dim H_{\mathrm{dR}}^m(X/k)$. Moreover, equality holds if and only if the spectral sequence degenerates. On the other hand, $\dim H_{\mathrm{dR}}^m(X/k)$ is upper semi-continuous. Hence, the degeneration of the spectral sequence in characteristic zero implies its degeneration in positive characteristic as well.

(ii) The *Kodaira vanishing* $H^q(X,\Omega_X^p \otimes L) = 0$ for $p + q > n$ and $L \in \mathrm{Pic}(X)$ ample holds. This can either be seen as a consequence of [147] for large enough characteristic or be read off from Corollaries 1.7 and 1.9. In particular, all numerical assertions on Hodge numbers remain valid over arbitrary fields. Also, for algebraically closed fields, the étale Betti numbers equal the ones computed in characteristic zero. The only case not covered by these comments is the case of cubic surfaces in characteristic two.

(iii) Assume $k = \mathbb{F}_q$. Then the *Weil conjectures* show that

$$Z(X, t) := \exp\left(\sum_{r=1}^{\infty} |X(\mathbb{F}_{q^r})| \frac{t^r}{r}\right) = \frac{P(t)^{(-1)^{n+1}}}{\prod_{i=0}^{n}(1 - q^i t)}$$

with $P(t) = \prod(1-\alpha_i t)$ of degree $b_n(X)_{\text{pr}}$ and α_i algebraic integers of absolute value $|\alpha_i| = q^{n/2}$. This was established by Bombieri and Swinnerton-Dyer [80] for cubic threefolds and by Dwork [166] for arbitrary hypersurfaces, prior to the proof of the general Weil conjectures by Deligne.

Of course, as cubic surfaces are rational, the Weil conjectures follow in this case from the Weil conjectures for $\mathbb{P}^2$. This was first noted by Weil himself [492].

2 Linear System and Lefschetz Pencils

This section discusses the linear system of (cubic) hypersurfaces. Basic facts concerning the discriminant divisor are reviewed and its degree is computed. We describe the monodromy group of the family of smooth hypersurfaces as a subgroup of the orthogonal group of the middle cohomology and complement the results with a comparison of the monodromy action with the action of the group of diffeomorphisms.

2.1 Universal Hypersurface Hypersurfaces $X \subset \mathbb{P} = \mathbb{P}^{n+1}$ of degree d are parametrized by the projective space

$$|\mathcal{O}_{\mathbb{P}}(d)| \simeq \mathbb{P}^{N(d,n)}, \; N(d,n) = \binom{n+1+d}{d} - 1.$$

We will often abbreviate $N(d,n) = h^0(\mathbb{P}^{n+1}, \mathcal{O}_{\mathbb{P}}(d)) - 1$ simply by N.

The *universal hypersurface* shall be denoted by

$$\mathcal{X} \subset \mathbb{P}^N \times \mathbb{P}. \tag{2.1}$$

It is a hypersurface of bidegree $(1, d)$, i.e. $\mathcal{X}$ is a divisor contained in the linear system $|\mathcal{O}_{\mathbb{P}^N}(1) \boxtimes \mathcal{O}_{\mathbb{P}}(d)|$, and the fibre of the (flat) first projection $\mathcal{X} \longrightarrow \mathbb{P}^N$ over the point corresponding to $X \subset \mathbb{P}$ is indeed just X.

More explicitly, $\mathcal{X}$ can be described as the zero set of the universal equation $G = \sum a_I x^I$, where $a_I \in H^0(\mathbb{P}^N, \mathcal{O}_{\mathbb{P}^N}(1))$ are the linear coordinates corresponding to the monomials $x^I \in H^0(\mathbb{P}, \mathcal{O}_{\mathbb{P}}(d))$. In other words, writing $\mathbb{P}$ as $\mathbb{P} = \mathbb{P}(V)$ for some vector space V of dimension $n + 2$, $H^0(\mathbb{P}, \mathcal{O}_{\mathbb{P}}(d)) = S^d(V^*)$, and $\mathbb{P}^N = \mathbb{P}(S^d(V^*))$, one has $H^0(\mathbb{P}^N, \mathcal{O}_{\mathbb{P}^N}(1)) = S^d(V)$ and G corresponds to

$$\text{id} \in \text{End}(S^d(V)) \simeq S^d(V) \otimes S^d(V^*) = H^0(\mathbb{P}^N \times \mathbb{P}, \mathcal{O}_{\mathbb{P}^N}(1) \boxtimes \mathcal{O}_{\mathbb{P}}(d)).$$

The universal hypersurface $\mathcal{X}$ is smooth. To prove this, observe that the second projection $\mathcal{X} \longrightarrow \mathbb{P}$ is the projective bundle $\mathbb{P}(\mathrm{Ker}(\mathrm{ev})) \longrightarrow \mathbb{P}$, where ev is the surjective(!) evaluation map

$$\mathrm{ev}\colon H^0(\mathbb{P}, \mathcal{O}_{\mathbb{P}}(d)) \otimes \mathcal{O} \twoheadrightarrow \mathcal{O}_{\mathbb{P}}(d).$$

The natural $\mathrm{SL}(n+2)$-action on $H^0(\mathbb{P}, \mathcal{O}_{\mathbb{P}}(d))$ descends to an action of $\mathrm{SL}(n+2)$ and $\mathrm{PGL}(n+2)$ on $|\mathcal{O}_{\mathbb{P}}(d)|$. Both are linearized in the sense that they are obtained by composing homomorphisms $\mathrm{SL}(n+2) \longrightarrow \mathrm{SL}(N+1)$ and $\mathrm{PGL}(n+2) \longrightarrow \mathrm{PGL}(N+1)$ with the natural actions of $\mathrm{SL}(N+1)$ and $\mathrm{PGL}(N+1)$ on $H^0(\mathbb{P}^N, \mathcal{O}_{\mathbb{P}^N}(1))$ and $|\mathcal{O}_{\mathbb{P}^N}(1)|$.

The following table records the dimensions of the linear system of cubic hypersurfaces of small dimensions. We also include information about the *moduli space*

$$M_n := |\mathcal{O}_{\mathbb{P}}(3)|_{\mathrm{sm}} /\!/ \mathrm{PGL}(n+2)$$

and the discriminant divisor $D(n) := D(3,n) := |\mathcal{O}_{\mathbb{P}}(3)| \setminus |\mathcal{O}_{\mathbb{P}}(3)|_{\mathrm{sm}}$, both to be discussed later, see Sections 2.3 and **3**.1.3. We write $N(n) := N(3,n)$.

n	$N(n)$	$\dim(\mathrm{PGL}(n+2))$	$\dim(M_n)$	$\deg(D(n))$
0	3	3	0	4
1	9	8	1	12
2	19	15	4	32
3	34	24	10	80
4	55	35	20	192
5	83	48	35	448

The closed formula for the dimension of the moduli space is

$$\dim(M_n) = \binom{n+2}{3} = \frac{n^3 + 3n^2 + 2n}{6}. \tag{2.2}$$

2.2 Discriminant Divisor We are mostly interested in smooth hypersurfaces. They are parametrized by a Zariski open subset which shall be denoted by

$$U(n,d) := |\mathcal{O}_{\mathbb{P}}(d)|_{\mathrm{sm}} := \{\, X \in |\mathcal{O}_{\mathbb{P}}(d)| \mid X \text{ smooth} \,\} \subset |\mathcal{O}_{\mathbb{P}}(d)|.$$

For an algebraically closed ground field k, Bertini's theorem shows that there exists a smooth hypersurface of the given degree d. Hence, $U(n,d)$ is non-empty and, therefore, dense. In fact, if $\mathrm{char}(k) = 0$ or at least $\mathrm{char}(k) \nmid d$, then the *Fermat hypersurface*

$$X := V\left(\sum_{i=0}^{n+1} x_i^d\right) \subset \mathbb{P}$$

is smooth, as it is easy to check using the Jacobian criterion. The following explicit equations for smooth hypersurfaces over arbitrary fields are taken from [270, p. 333]:

$$\begin{cases} \sum_{i=0}^{m-1} x_i\, x_{i+m} & \text{if } d = 2, n+2 = 2m, \\ \sum_{i=0}^{m-1} x_i\, x_{i+m} + x_{n+1}^2 & \text{if } d = 2, n+1 = 2m, \\ \sum_{i=0}^{n+1} x_i^d & \text{if } d \geq 3, \operatorname{char}(k) \nmid d, \\ \sum_{i=0}^{n} x_i\, x_{i+1}^{d-1} + x_0^d & \text{if } d \geq 3, \operatorname{char}(k) \mid d. \end{cases} \tag{2.3}$$

Hence, the set of k-rational points of $U(n,d) = |\mathcal{O}_{\mathbb{P}}(d)|_{\mathrm{sm}}$ is always non-empty.

Definition 2.1 The *discriminant divisor*

$$D(d,n) \subset |\mathcal{O}_{\mathbb{P}}(d)|$$

is the complement of the Zariski open (and dense) subset $U(d,n) \subset |\mathcal{O}_{\mathbb{P}}(d)|$ of smooth hypersurfaces. Thus, $D(d,n)$ is closed and, in a first step, it will be viewed with its reduced induced scheme structure. However, in Section 2.3 we observe that its natural scheme structure provided by its description as the zero set of the resultant is reduced.

Theorem 2.2 *The discriminant divisor $D(d,n) \subset |\mathcal{O}_{\mathbb{P}}(d)|$ is an irreducible divisor. Its degree is $(d-1)^{n+1} \cdot (n+2)$, which for $d = 3$ reads*

$$\deg(D(3,n)) = 2^{n+1} \cdot (n+2).$$

Proof Consider the universal hypersurface $\mathcal{X} \subset \mathbb{P}^N \times \mathbb{P}$ as above and define

$$\mathcal{X}_{\mathrm{sing}} := \mathcal{X} \cap \bigcap_{i=0}^{n+1} V_i,$$

where the $V_i := V(\partial_i G)$ are the hypersurfaces of bidegree $(1, d-1)$ defined by the derivatives of the equation of the universal hypersurface

$$\partial_i G := \sum a_I \frac{\partial x^I}{\partial x_i} \in H^0\left(\mathbb{P}^N \times \mathbb{P}, \mathcal{O}_{\mathbb{P}^N}(1) \boxtimes \mathcal{O}_{\mathbb{P}}(d-1)\right).$$

By the Jacobian criterion, $\mathcal{X}_{\mathrm{sing}} \subset \mathcal{X} \longrightarrow \mathbb{P}^N$ is the (non-flat) family of singular loci of the fibres $\mathcal{X}_t$, i.e. $(\mathcal{X}_{\mathrm{sing}})_t = (\mathcal{X}_t)_{\mathrm{sing}}$.

As the *Euler equation*, see [90, Ch. 4], holds in its universal form $\sum x_i\, \partial_i G = d \cdot G$, one has $\bigcap V_i \subset \mathcal{X}$ if $\operatorname{char}(k) \nmid d$ (which we will tacitly assume, but see Remark 2.3). Hence, $\mathcal{X}_{\mathrm{sing}} = \bigcap V_i$ and, therefore, $\operatorname{codim}(\mathcal{X}_{\mathrm{sing}}) \leq n+2$. To prove that equality holds, consider the other projection $\mathcal{X}_{\mathrm{sing}} \longrightarrow \mathbb{P}$, which we claim is a $\mathbb{P}^k$-bundle with $k = N - n - 2$. To see this, observe that the homomorphism of sheaves on $\mathbb{P}$

$$\varphi\colon H^0(\mathbb{P}, \mathcal{O}_{\mathbb{P}}(d)) \otimes \mathcal{O}_{\mathbb{P}} \longrightarrow \mathcal{O}_{\mathbb{P}}(d-1)^{\oplus n+2}, \ F \longmapsto (\partial_i F)$$

is surjective, which can be checked e.g. at the point $z = [1 : 1 : \cdots : 1]$ by using the fact that $(\partial_i x_j^d)(z) = d \cdot \delta_{ij}$. Then

$$\mathcal{X}_{\text{sing}} \simeq \mathbb{P}(\text{Ker}(\varphi)) \twoheadrightarrow \mathbb{P}.$$

This clearly proves $\text{codim}(\mathcal{X}_{\text{sing}}) = n + 2$, but also that $\mathcal{X}_{\text{sing}}$ is smooth and irreducible. To be precise, one needs to verify that $\mathcal{X}_{\text{sing}} \simeq \mathbb{P}(\text{Ker}(\varphi))$ as schemes and not only as sets, which is left to the reader.

Next, $D := D(d, n)$ is by definition the image of $\mathcal{X}_{\text{sing}}$ under the projection

$$\mathcal{X}_{\text{sing}} \subset \mathcal{X} \subset \mathbb{P}^N \times \mathbb{P} \longrightarrow \mathbb{P}^N.$$

Let us denote the pullbacks of the hyperplane sections on $\mathbb{P}^N$ and $\mathbb{P}$ (both denoted by h) to $\mathbb{P}^N \times \mathbb{P}$ by h_1 and h_2. Suppose D is of codimension > 1. Then $(h^{N-1} . D) = 0$, which, however, would contradict

$$(h_1^{N-1} . \mathcal{X}_{\text{sing}}) = (h_1^{N-1} . (h_1 + (d-1)\, h_2)^{n+2}) = (n+2) \cdot (d-1)^{n+1}.$$

Hence, $D \subset \mathbb{P}^N$ really is a divisor. The computation also shows that in order to prove the claimed degree formula for D, it suffices to prove that $\mathcal{X}_{\text{sing}} \twoheadrightarrow D$ is generically injective or, in other words, that the generic singular hypersurface $X \in |\mathcal{O}_{\mathbb{P}}(d)|$ has exactly one singular point (which is in fact an ordinary double point). (Note that one needs to assume $\text{char}(k) = 0$ for the set-theoretic injectivity to imply that the morphism is of degree one.) One way of doing this would be to write down examples of hypersurfaces in each degree with exactly one ordinary double point[2] or to argue geometrically (assuming $\text{char}(k) = 0$) by considering again the projective bundle $\mathcal{X}_{\text{sing}} \longrightarrow \mathbb{P}$. The fibre over a point z can be thought of as a linear system with z as its only base point. By Bertini's theorem with base points, see e.g. [223, III. Rem. 10.9.2], the generic element will then be singular exactly at z.

To see that generically it has to be an ordinary double point, just write down one hypersurface with such a singular point at z (but possibly other singular points), e.g. the union of $(d-2)$ generic hyperplanes $\mathbb{P}^n \subset \mathbb{P}$ and of a cone with vertex z over a quadric in some hyperplane. □

Remark 2.3 In [148, Exp. XVII] the discriminant divisor is viewed as the dual variety of the Veronese embedding $v_d \colon \mathbb{P} \hookrightarrow (\mathbb{P}^N)^*$, i.e. as the locus of hyperplanes (parametrized by $\mathbb{P}^N$) that are tangent to $v_d(\mathbb{P})$. It is also proved that the smooth locus of $D(d, n)$ is the maximal open subset over which $\mathcal{X}_{\text{sing}} \twoheadrightarrow D(d, n)$ is an isomorphism and that it coincides with the set of those singular hypersurfaces with one ordinary double point as only singularity.

[2] D. van Straten has provided me with examples in certain degrees. Note that writing down examples with just one (badly) singular point is easy, e.g. the cone over the smooth examples in (2.3) has only one singular point, which however is an ordinary double point only for $d = 2$. Since hypersurfaces with a singularity worse than an ordinary double point may deform generically to a hypersurface with more than one singular point, they cannot be used to prove injectivity in this fashion.

2.3 Resultant There is a classical and more algebraic approach to the discriminant divisor using resultants, cf. [101, 127, 150, 193]. Let us review some general facts. Consider homogeneous polynomials in $k[x_0, \dots, x_{n+1}]$ of degree $d_i > 0$, $i = 0, \dots, m$. Then there exists a unique polynomial, the *resultant*, $R(y_{i,I}) := R_{d,n}(y_{i,I}) \in k[y_{i,I}]$, $i = 0, \dots, m$, $|I| = d_i$, such that:

(i) For all choices of polynomials $f_i \in k[x_0, \dots, x_{n+1}]_{d_i}$, $i = 0, \dots, m$, the intersection $\bigcap V(f_i) \subset \mathbb{P}^{n+1}_{\bar{k}}$ is non-empty if and only if $R(f_0, \dots, f_m) = 0$.[3]
(ii) $R(x_0^{d_0}, \dots, x_m^{d_m}) = 1$ (normalization).
(iii) $R \in k[y_{i,I}]$ is irreducible.

Moreover, R is homogeneous of degree $\prod_{j \neq i} d_j$ in the variables $y_{i,I}$ for fixed i and so of total degree $\prod d_i \cdot \sum(1/d_i)$.

Consider a homogeneous polynomial $F \in k[x_0, \dots, x_{n+1}]$ of degree d and apply the above to $f_i = \partial_i F$, $i = 0, \dots, m = n+1$, which are all homogeneous of degree $d_i = d-1$. Then, $X = V(F)$ is singular, i.e. $\bigcap V(f_i) \neq \emptyset$, if and only if $X \in |\mathcal{O}_{\mathbb{P}}(d)|$ is in the zero locus of R.

Strictly speaking, R defines a hypersurface in $\mathbb{P}^{N'} = \mathrm{Proj}(k[y_{i,I}])$, $i = 0, \dots, n+1$, $|I| = d-1$, so $N' = (n+2)\binom{n+d}{d-1} - 1$. Its pullback via the linear embedding $\mathbb{P}^N \hookrightarrow \mathbb{P}^{N'}$ that maps x^I to $\left[i_j x^{I_j}\right]_{j=0,\dots,n+1}$, where for $I = (i_0, \dots, i_{n+1})$ one sets $I_j := (i_0, \dots, (i_j - 1), \dots, i_{n+1})$, describes the image of $\mathcal{X}_{\mathrm{sing}}$, i.e. the discriminant divisor. The irreducibility still holds, cf. [150, Sec. 5 & 6]. This leads to

$$D(d,n) = V(R_{d,n}(\partial_0 G, \dots, \partial_{n+1} G)) \subset \mathbb{P}^N = |\mathcal{O}_{\mathbb{P}}(d)|,$$

which also defines a natural scheme structure on $D(d,n)$.

Remark 2.4 The resultant is usually normalized to yield the *discriminant*

$$\Delta_{d,n} := d^{c_{d,n}} \cdot R_{d,n}(\partial_i G) \in H^0\left(\mathbb{P}^N, \mathcal{O}_{\mathbb{P}^N}((d-1)^{n+1} \cdot (n+2))\right), \tag{2.4}$$

where $c_{d,n} = (1/d)((-1)^{n+2} - (d-1)^{n+2})$. With this normalization, $\Delta_{d,n}$ becomes an irreducible polynomial in $\mathbb{Z}[y_I]$, which makes it unique up to a sign.

Example 2.5 The case $n = 0$ and $d = 3$, so three points in $\mathbb{P}^1$, leads to the classical discriminant for cubic polynomials $f(X)$. If $\alpha_1, \alpha_2, \alpha_3$ are the 0s of $f(X)$, then by definition $\Delta(f(X)) = ((\alpha_1 - \alpha_2)(\alpha_1 - \alpha_3)(\alpha_2 - \alpha_3))^2$. For $f(X) = X^3 + aX + b$, one has $\Delta(f(X)) = -4a^3 - 27b^2$.

The discriminant of a general polynomial $a_0 x_0^3 + a_1 x_0^2 x_1 + a_2 x_0 x_1^2 + a_3 x_1^3$ is the rather complicated polynomial of degree four

$$\Delta_{3,0} = a_1^2 a_2^2 - 4a_3 a_1^3 - 4a_2^3 a_0 - 27a_3^2 a_0^2 + 18a_0 a_1 a_2 a_3,$$

[3] Here, $R(f_0, \dots, f_m)$ is the shorthand for applying R to the coefficients of the polynomials f_i.

which thus defines the discriminant surface of degree four

$$D(3,0) = V(\Delta_{3,0}) \subset \mathbb{P}^3.$$

As an exercise, the reader may want to compare this with the normalization $R(x_0^2, x_1^2) = 1$, which in (2.4), using $c_{3,0} = -1$, implies $\Delta_{3,0}\big((1/3)(x_0^3 + x_1^3)\big) = (1/3)$.

To confirm Remark 2.3, one can verify that the singular set of $D(3,0) = V(\Delta_{3,0})$ is indeed the curve of triple points.

Example 2.6 (i) For $n = 1$ and $d = 3$, the discriminant divisor

$$D(3,1) \subset \mathbb{P}^9$$

is of degree 12 or, equivalently, the discriminant is an element of the vector space $H^0(\mathbb{P}^9, \mathcal{O}_{\mathbb{P}^9}(12))$, which is of dimension 293.930. Written as a linear combination of monomials, 12.894 of the coefficients are non-trivial, cf. [127, p. 99]. If the partial derivatives $\partial_i F$ are written as $\partial_0 F = a_{11}x_0^2 + a_{12}x_1^2 + a_{13}x_2^2 + a_{14}x_0x_1 + a_{15}x_0x_2 + a_{16}x_1x_2$, etc., and one defines $[\ell_1\ell_2\ell_3] := \det(a_{i,\ell_j}) \in H^0(\mathbb{P}^9, \mathcal{O}_{\mathbb{P}^9}(3))$, with pairwise distinct ℓ_i, then Δ is a polynomial of degree four in the $[\ell_1\ell_2\ell_3]$ that involves less, but still 68, terms. In short, the discriminant is complicated.

Maybe just one word on the comparison between the discriminant introduced here and the discriminant of a plane cubic $E \subset \mathbb{P}^2$ in Weierstrass form $y^2 = 4x^3 - g_2x - g_3$ which is classically defined as $\Delta(E) := g_2^3 - 27g_3^2$. This is a rather simple polynomial of degree three in the coefficients, whereas the full discriminant of cubic plane curves is a polynomial of degree 12. The reason for this is that bringing a cubic polynomial in the variables x_0, x_1, x_2 into Weierstrass form involves non-linear transformations. More concretely, the coefficients g_2 and g_3 of the Weierstrass form are of degree four and six, respectively, in the coefficients of the original cubic equation, see e.g. [278, Ch. 3].

(ii) For the case $n = 2$ and $d = 3$, the discriminant divisor $D(3,2) \subset \mathbb{P}^{19}$ is of degree 32. Its equation can be expressed in terms of certain fundamental invariants, see Section 4.4.1. This was first done by Salmon [415] with correction by Edge [171]. For more recent considerations, see [100].

As a consequence of Theorem 2.2 and the discussion in its proof, we deduce the following.

Corollary 2.7 *Assume k is algebraically closed. Then for the generic line $\mathbb{P}^1 \hookrightarrow \mathbb{P}^N$, the induced family $\mathcal{X}_{\mathbb{P}^1} \twoheadrightarrow \mathbb{P}^1$ has exactly $(d-1)^{n+1} \cdot (n+2)$ singular fibres $\mathcal{X}_1, \mathcal{X}_2, \ldots$, each with exactly one singular point $x_i \in \mathcal{X}_i$. Moreover, each x_i is an ordinary double point of $\mathcal{X}_i$ and they are all distinct as points in $\mathbb{P}^{n+1}$.* □

A pencil with these properties is called a *Lefschetz pencil*. Note that by Bertini's theorem [223, III. Cor. 10.9], at least when $\text{char}(k) = 0$, the total space $\mathcal{X}_{\mathbb{P}^1}$ is still smooth. See [148, Exp. XVII].

In more concrete terms, for the generic choice of polynomials $F_0, F_1 \in H^0(\mathbb{P}, \mathcal{O}_{\mathbb{P}}(d))$ for exactly $(d-1)^{n+1} \cdot (n+2)$ values $t = [t_0 : t_1]$ the hypersurface $\mathcal{X}_t = V(t_0 F_0 + t_1 F_1)$ is singular. Each singular fibre $\mathcal{X}_i$ has exactly one singular point x_i, which, moreover, is an ordinary double point. Note that $x_i \neq x_j$ for $i \neq j$, as otherwise x_i would be a singular point of all the fibres. Also observe that the projection $\mathcal{X} \longrightarrow \mathbb{P}$ is the blow-up of the base locus $V(F_0, F_1) \subset \mathbb{P}$.

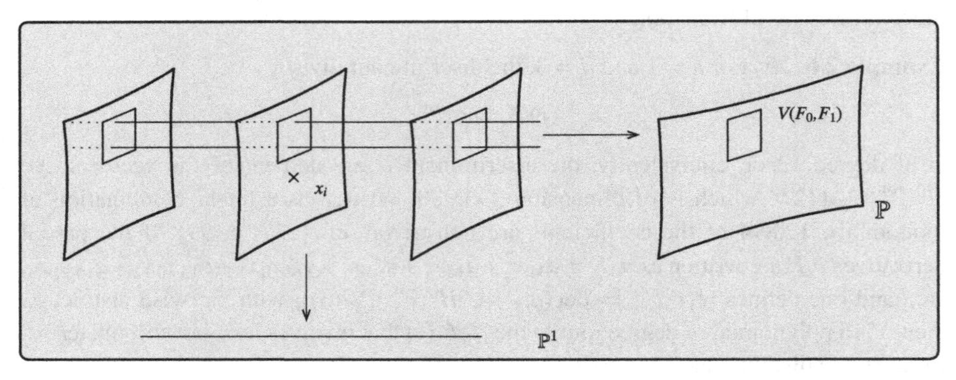

Example 2.8 There are, of course, pencils $\mathcal{X}_{\mathbb{P}^1} \longrightarrow \mathbb{P}^1 \hookrightarrow \mathbb{P}^N$ with more singular fibres, i.e. with more or worse singularities. The *Hesse pencil* of plane cubics $\mathcal{X}_t \subset \mathbb{P}^2$ given by

$$t_0\,(x_0^3 + x_1^3 + x_2^3) - t_1 3\, x_0 x_1 x_2$$

is such an example. Here, the fibre $\mathcal{X}_{[0:1]}$ consists of three lines yielding three singular points. The Hesse pencil is a special instance of the *Dwork pencil* (or *Fermat pencil*), see [68], defined by the equation

$$t_0 \left(\sum_{i=0}^{n+1} x_i^{n+2} \right) - t_1 d \prod_{i=0}^{n+1} x_i$$

of hypersurfaces of degree $d = n + 2$.

Clearly, the number of singular fibres of any pencil does not exceed $(d-1)^{n+1} \cdot (n+2)$, unless all fibres are singular. Note that for an arbitrary pencil the total space $\mathcal{X}_{\mathbb{P}^1}$ need not be smooth.

2.4 Monodromy Group We now assume $k = \mathbb{C}$ and consider the universal family

$$\pi \colon \mathcal{X} \longrightarrow U(d, n) \subset |\mathcal{O}_{\mathbb{P}^{n+1}}(d)|$$

of smooth hypersurfaces of degree d and dimension n. Note the change in notation. If needed later, we will denote the universal family of all hypersurfaces by $\bar{\mathcal{X}}$, which is smooth projective and contains $\mathcal{X}$ as a dense open subset. Fix a point $0 \in U(d,n)$ and denote the fibre over it by $X := \mathcal{X}_0$. The *monodromy representation*

$$\rho\colon \pi_1(U(d,n),0) \longrightarrow \mathrm{GL}(H^n(X,\mathbb{Z})) \tag{2.5}$$

is the homomorphism obtained by parallel transport with respect to the Gauss–Manin connection. Equivalently, $R^n\pi_*\mathbb{Z}$ is a locally constant system on $U(n,d)$ and (2.5) is the corresponding representation of the fundamental group. The *monodromy group* is by definition the image of the monodromy representation

$$\Gamma(d,n) := \mathrm{Im}\,(\rho\colon \pi_1(U(d,n),0) \longrightarrow \mathrm{GL}(H^n(X,\mathbb{Z})))\,.$$

It depends on the base point $0 \in U(d,n)$ only up to conjugation.

The monodromy group has been determined by Beauville [46] in complete generality. We discuss the result for $d = 3$ and use the shorthand

$$\Gamma_n := \Gamma(3,n) \subset \mathrm{GL}(H^n(X,\mathbb{Z})).$$

Theorem 2.9 *The monodromy group Γ_n of the universal smooth cubic hypersurface $\mathcal{X} \longrightarrow |\mathcal{O}_{\mathbb{P}^{n+1}}(3)|_{\mathrm{sm}}$ is the group*

$$\Gamma_n \simeq \begin{cases} \tilde{\mathrm{O}}^+(H^n(X,\mathbb{Z})) & \text{if } n \equiv 0\,(2), \\ \mathrm{SpO}(H^n(X,\mathbb{Z}),q) & \text{if } n \equiv 1\,(2). \end{cases}$$

In fact, Beauville shows that $\Gamma(d,n)$ for n even and arbitrary d and also for n odd and all odd d admits this description. If n is odd and d is even, then the monodromy group is the full symplectic group $\mathrm{Sp}(H^n(X,\mathbb{Z}))$.

Before sketching the main steps of the proof in Section 2.5, let us explain the notation and add a few related comments.

For n even, one defines $\tilde{\mathrm{O}}(H^n(X,\mathbb{Z})) \subset \mathrm{O}(H^n(X,\mathbb{Z}))$ as the subgroup of all orthogonal transformations $g\colon H^n(X,\mathbb{Z}) \xrightarrow{\sim} H^n(X,\mathbb{Z})$ with $g(h^{n/2}) = h^{n/2}$. Via the induced action on $H^n(X,\mathbb{Z})_{\mathrm{pr}}$ it can be identified with the subgroup, cf. [249, Prop. 14.2.6]:

$$\tilde{\mathrm{O}}(H^n(X,\mathbb{Z})) \simeq \left\{ g \in \mathrm{O}(H^n(X,\mathbb{Z})_{\mathrm{pr}}) \mid \mathrm{id} = \bar{g} \in \mathrm{O}(A_{H^n(X,\mathbb{Z})_{\mathrm{pr}}}) \right\}.$$

Here, $A_\Lambda := \Lambda^*/\Lambda$ is the discriminant group of a lattice Λ, which for the primitive cohomology $\Lambda = H^n(X,\mathbb{Z})_{\mathrm{pr}}$ of a smooth cubic hypersurface is $\mathbb{Z}/3\mathbb{Z}$, see Proposition 1.21.

Another natural group in this context is the subgroup

$$\mathrm{O}^+(H^n(X,\mathbb{Z})_{\mathrm{pr}}) := \mathrm{Ker}\left(\mathrm{sn}_n\colon \mathrm{O}(H^n(X,\mathbb{Z})_{\mathrm{pr}}) \longrightarrow \{\pm 1\}\right),$$

where the spinor norm sn_n is the group homomorphism that sends a reflection s_δ in a hyperplane $\delta^\perp$ to $(-1)^{n/2} \cdot (\delta)^2/|(\delta)^2|$. In other words, if by means of the Cartan–Dieudonné theorem g is written as a product $\prod s_{\delta_i}$ of reflections with $\delta_i \in H^n(X,\mathbb{R})_{\mathrm{pr}}$, then $\mathrm{sn}_n(g) = 1$ if for $n \equiv 0\,(4)$ the number of δ_i with $(\delta_i)^2 < 0$ is even and for $n \equiv 2\,(4)$ the number of δ_i with $(\delta_i)^2 > 0$ is even.

The orthogonal group in the theorem is the finite index subgroup of $\mathrm{O}(H^n(X,\mathbb{Z})_{\mathrm{pr}})$:

$$\tilde{\mathrm{O}}^+(H^n(X,\mathbb{Z})) := \tilde{\mathrm{O}}(H^n(X,\mathbb{Z})) \cap \mathrm{O}^+(H^n(X,\mathbb{Z})_{\mathrm{pr}}).$$

For n odd, the intersection product on $H^n(X,\mathbb{Z}) = H^n(X,\mathbb{Z})_{\mathrm{pr}}$ is alternating and can be put in the standard normal form. However, there exists an auxiliary rather subtle topological invariant, which is the quadratic form $q\colon H^n(X,\mathbb{Z}) \longrightarrow \mathbb{Z}/2\mathbb{Z}$, see [95, Sec. 1], that enters the definition of the group $\mathrm{SpO}(H^n(X,\mathbb{Z}),q)$ in the above theorem. Using the fact that any $\alpha \in H^n(X,\mathbb{Z})$ can be represented by an embedded sphere $S^n \hookrightarrow X$, one has for $n \neq 1,3,7$ that $q(\alpha) = 0$ if and only if the topological normal bundle of $S^n \hookrightarrow X$ is trivial.[4] The definition of q for $n = 1,3,7$ is more involved. In any case, for n odd $\mathrm{SpO}(H^n(X,\mathbb{Z}),q)$ is defined as the group of all isomorphisms $g\colon H^n(X,\mathbb{Z}) \xrightarrow{\sim} H^n(X,\mathbb{Z})$ that are compatible with the alternating intersection form $(\,.\,)$ and the quadratic form q.

Remark 2.10 The occurrence of the primitive cohomology in Theorem 2.9 is not a surprise. Indeed, the restriction $h^{n/2}$ of $h_{\mathbb{P}}^{n/2} = \mathrm{c}_1(\mathcal{O}_{\mathbb{P}}(1))^{n/2} \in H^n(\mathbb{P}^{n+1},\mathbb{Z})$ to any of the fibres $\mathcal{X}_t$ defines a section of the locally constant system $R^n\pi_*\mathbb{Z}$. Hence, the primitive cohomology groups $H^n(\mathcal{X}_t,\mathbb{Z})_{\mathrm{pr}}$ of the fibres glue to a locally constant subsheaf $R^n_{\mathrm{pr}}\pi_*\mathbb{Z} \subset R^n\pi_*\mathbb{Z}$. Equivalently, the monodromy representation (2.5) satisfies $\rho(\gamma)(h^{n/2}) = h^{n/2}$ for all $\gamma \in \pi_1(U(d,n))$, i.e. $h^{n/2}$ is monodromy invariant.

In fact, $h^{n/2}$ is the only monodromy invariant class up to scaling. Indeed, Deligne's invariant cycle theorem [474, V. Thm. 16.24] shows that the monodromy invariant part $H^n(X,\mathbb{Q})^\rho$ of $H^n(X,\mathbb{Q})$ is the image of the restriction $H^n(\bar{\mathcal{X}},\mathbb{Q}) \longrightarrow H^n(X,\mathbb{Q})$, where $\bar{\mathcal{X}} \subset \mathbb{P}^N \times \mathbb{P}$ denotes the universal family of all hypersurfaces. Now writing $\bar{\mathcal{X}}$ as a projective bundle over $\mathbb{P}$ shows that

$$H^n(\bar{\mathcal{X}},\mathbb{Q}) \simeq \bigoplus H^{n-2i}(\mathbb{P},\mathbb{Q}) \cdot \mathrm{c}_1(\mathcal{O}_{\mathbb{P}^N}(1))^i.$$

As $\mathrm{c}_1(\mathcal{O}_{\mathbb{P}^N}(1))$ restricts trivially to the fibres of the first projection $\bar{\mathcal{X}} \longrightarrow \mathbb{P}^N$, only $H^n(\mathbb{P},\mathbb{Q})$ survives the map $H^n(\bar{\mathcal{X}},\mathbb{Q}) \longrightarrow H^n(X,\mathbb{Q})$ and, therefore, its image is spanned by $h^{n/2}$.

Similarly, the monodromy representation preserves the intersection form on $H^n(X,\mathbb{Z})$. Therefore, $\mathrm{Im}(\rho) \subset \mathrm{O}(H^n(X,\mathbb{Z}))$ for n even and $\mathrm{Im}(\rho) \subset \mathrm{Sp}(H^n(X,\mathbb{Z}))$ for n odd.

Note that one can deduce from the theorem the well-known fact [474, V. Thm. 15.27] that $H^n(X,\mathbb{Q})_{\mathrm{pr}}$ is an irreducible $\Gamma(d,n)$-module or, equivalently, that $R^n_{\mathrm{pr}}\pi_*\mathbb{Q}$ cannot

[4] The *Arf invariant* of q, also called the *Kervaire invariant* of X, is often viewed as the analogue of the discriminant of the symmetric intersection form for n even. Recall that the Arf invariant $A(q) \in \mathbb{F}_2$ of the binary quadratic form $q = ax^2 + xy + by^2$ is ab. For arbitrary q, which can be written as a direct sum of those, it is defined by additive extension, cf. [94, Ch. III]. According to a result of Kulkarni and Wood [290, Prop. 12.1], the Kervaire invariant is non-trivial for cubic hypersurfaces.

be written as a direct sum of non-trivial locally constant systems. Remark 2.11 and Corollary 2.12 below are related to this.

Remark 2.11 In the general setting, where instead of working over $\mathbb{C}$ the ground field can be an arbitrary algebraically closed field k, the geometric ℓ-adic monodromy group is the Zariski closure G of the image of the representation $\pi_1^{\text{ét}}(U) \longrightarrow \text{GL}(H^n(X, \bar{\mathbb{Q}}_\ell))$. It has been determined in [146, Sec. 4.4]:

$$G = \begin{cases} \text{finite} & \text{if } n = 2, \\ \text{O}(H^n(X, \bar{\mathbb{Q}}_\ell)) & \text{if } 2 < n \equiv 0\,(2), \\ \text{Sp}(H^n(X, \bar{\mathbb{Q}}_\ell)) & \text{if } n \equiv 1\,(2). \end{cases}$$

Instead of working with coefficients in $\bar{\mathbb{Q}}_\ell$ one can also use coefficients in $\mathbb{C}$. For n, even the proof comes down to the fact that an algebraic subgroup $G \subset \text{O}(V)$ of a complex vector space V with a non-degenerate symmetric bilinear form is $\text{O}(V)$ as soon as there exists a G-orbit of classes δ with $(\delta)^2 = 2$ generating V and such that G contains all reflections s_δ induced by classes in that orbit, cf. [146, Lem. 4.4.2] or the account in [389].

When working over $\mathbb{C}$, one defines the algebraic monodromy group as the algebraic group $G \subset \text{GL}(H^n(X, \mathbb{Q})_{\text{pr}})$ over $\mathbb{Q}$ obtained as the Zariski closure of the monodromy group $\Gamma_n \subset \text{GL}(H^n(X, \mathbb{Z})_{\text{pr}}) \subset \text{GL}(H^n(X, \mathbb{Q})_{\text{pr}})$.

For the following result, which for n even strengthens Remark 2.10, recall the following two notions:

(i) A very general point in $|\mathcal{O}_{\mathbb{P}}(3)|$ is a point in the complement of a countable union of proper Zariski closed subsets.
(ii) The Hodge structure $H^n(X, \mathbb{Q})_{\text{pr}}$ is irreducible if it cannot be written as a direct sum of non-trivial sub-Hodge structures or, equivalently, if it does not contain any non-trivial proper sub-Hodge structure.

The equivalence of the two characterizations in (ii) uses the existence of a polarization.

Corollary 2.12 *Assume $X \in |\mathcal{O}_{\mathbb{P}}(3)|$ is a very general cubic hypersurface of (even) dimension $n > 2$. Then $H^n(X, \mathbb{Q})_{\text{pr}}$ is an irreducible Hodge structure. In particular, the (rational) Hodge conjecture holds for the very general cubic hypersurface.*

Proof We follow the exposition of Peters and Steenbrink [389, Sec. 7]. The main input is that for $n > 2$, the identity component $G^0 \subset G$ of the geometric monodromy group (thought of as algebraic groups over $\mathbb{Q}$) acts irreducibly on $H^n(X, \mathbb{Q})_{\text{pr}}$, see Remark 2.11.

We write the primitive cohomology of the very general cubic hypersurface X as direct sum $H^n(X, \mathbb{Q})_{\text{pr}} \simeq \bigoplus_{i=1}^k V_i$ of irreducible Hodge structures. Then the projection $H^n(X, \mathbb{Q})_{\text{pr}} \twoheadrightarrow V_1$ can be extended to a multivalued flat section of the local system

$\mathrm{End}(R^n_{\mathrm{pr}}\pi_*\mathbb{Q})$ which is everywhere of type $(0,0)$. As any polarizable variation of Hodge structures of type $(0,0)$ has finite monodromy, this section becomes univalued after passing to a finite étale cover of $|\mathcal{O}_{\mathbb{P}}(3)|_{\mathrm{sm}}$. In the process, the group G^0 does not change, so that it still acts irreducibly. The image of the flat section of $\mathrm{End}(R^n_{\mathrm{pr}}\pi_*\mathbb{Q})$ defines a locally constant subsystem of $R^n_{\mathrm{pr}}\pi_*\mathbb{Q}$ with image V_1 at the point corresponding to X. However, as G^0 acts irreducibly this shows that $k = 1$, i.e. $H^n(X,\mathbb{Q})_{\mathrm{pr}}$ is irreducible.

Concerning the Hodge conjecture, note that for an arbitrary smooth cubic hypersurface X and $2p \neq n$, we have $H^{p,p}(X,\mathbb{Q}) = H^{2p}(X,\mathbb{Q}) = \mathbb{Q}\cdot h^p$. For a very general cubic hypersurface X of even dimension $n = 2p$, the irreducibility of the Hodge structures $H^n(X,\mathbb{Q})_{\mathrm{pr}}$ in particular says that $H^{p,p}(X,\mathbb{Q})_{\mathrm{pr}} = 0$. Therefore, Hodge classes in $H^n(X,\mathbb{Q})$ are again just multiples of h^p. □

Note that for $n = 2$, the identity component G^0 is trivial and indeed $H^2(S,\mathbb{Q})_{\mathrm{pr}}$ is of type $(1,1)$ for any cubic surface. As it is of dimension six, it is certainly not irreducible.

Remark 2.13 (i) The identity component $G^0 \subset G$ of the geometric monodromy group is contained in the *Mumford–Tate group* of the Hodge structure $H^n(X,\mathbb{Q})_{\mathrm{pr}}$ of the very general X, see [145, Prop. 7.5] or [19, Lem. 4]. By definition, the Mumford–Tate group of a Hodge structure determines the space of all Hodge classes in tensor products of the Hodge structure and of its dual. Thus, whenever the monodromy group is big, the Mumford–Tate group also is and, therefore, the various tensor products have only few Hodge classes. The proof makes this philosophy explicit for $H^n(X,\mathbb{Q})\otimes H^n(X,\mathbb{Q})$.

So, the arguments in the above proof actually show that for the very general cubic hypersurface X any endomorphism of the Hodge structure $H^n(X,\mathbb{Q})_{\mathrm{pr}}$ is a multiple of the identity, cf. [483, Lem. 5.1]:

$$\mathrm{End}_{\mathrm{Hdg}}(H^n(X,\mathbb{Q})_{\mathrm{pr}}) \simeq \mathbb{Q}.$$

(ii) Arguments similar to the ones above also show that for the very general cubic X of dimension $n > 2$ the Hodge structures $S^2H^n(X,\mathbb{Q})_{\mathrm{pr}}$ for n even and $\bigwedge^2 H^n(X,\mathbb{Q})$ for n odd split into the direct sum of two irreducible Hodge structures, i.e.

$$S^2H^n(X,\mathbb{Q})_{\mathrm{pr}} \simeq \mathbb{Q}\cdot q_X \oplus q_X^\perp \text{ resp. } \bigwedge^2 H^n(X,\mathbb{Q})_{\mathrm{pr}} \simeq \mathbb{Q}\cdot q_X \oplus q_X^\perp.$$

Here, q_X denotes the symmetric resp. alternating bilinear intersection form on the primitive cohomology $H^n(X,\mathbb{Q})_{\mathrm{pr}}$.

To see this, one uses the classical fact that the orthogonal group $\mathrm{O}(V,q)$ respectively the symplectic group $\mathrm{Sp}(V,q)$ of a vector space V with a non-degenerate symmetric or alternating bilinear form q acts irreducibly on the orthogonal complement $q^\perp$ in S^2V respectively $\bigwedge^2 V$. For example, for a symmetric forms the orthogonal complement $q^\perp$ is the space of harmonic polynomials, i.e. those contained in the kernel of the Laplacian, which is an irreducible representation of $\mathrm{O}(V)$, see [197, Ch. 10].

2.5 Vanishing Classes The computation of the monodromy group Γ_n, or more generally of $\Gamma(d, n)$, proceeds in three steps.

(i) Show that $\Gamma(d, n)$ equals the monodromy group of the smooth part of a Lefschetz pencil $\mathcal{X}_{\mathbb{P}^1} \twoheadrightarrow \mathbb{P}^1$.

(ii) Assume $\mathcal{X} \twoheadrightarrow \Delta$ is a family of hypersurfaces over a disc with $\mathcal{X}$ and $\mathcal{X}_{t\neq 0}$ smooth and such that the central fibre $\mathcal{X}_0$ has one ordinary double point as its only singularity. Let γ be the simple loop around $0 \in \Delta$. Describe the induced monodromy operation $\rho(\gamma)\colon H^n(X, \mathbb{Z}) \xrightarrow{\sim} H^n(X, \mathbb{Z})$ as a reflection s_δ. Here, $X = \mathcal{X}_\varepsilon$ is a distinguished smooth fibre.

(iii) Let $\mathcal{X}_{\mathbb{P}^1} \twoheadrightarrow \mathbb{P}^1$ be a Lefschetz pencil with nodal singular fibres over $t_1, \ldots, t_\ell \in \mathbb{P}^1 \setminus \infty$. Describe the subgroup $\langle s_{\delta_i} \rangle \subset \mathrm{GL}(H^n(X, \mathbb{Z}))$ generated by the monodromy operations around all the nodal fibres $\mathcal{X}_{t_1}, \ldots, \mathcal{X}_{t_\ell}$.

To have at least a rough idea, let us give a few more details for all three steps. For details of the statements and of the proofs we have to refer to the literature, cf. [474].

(i) Similar to the Lefschetz hyperplane theorem for smooth hyperplane sections of smooth projective varieties, cf. Section 1.1, a result of Zariski, see [217] or [474, V. Thm. 15.22], shows that for a very general line $\mathbb{P}^1 \subset \mathbb{P}^N = |\mathcal{O}_{\mathbb{P}}(d)|$ the induced map

$$\pi_1(\mathbb{P}^1 \setminus D) \twoheadrightarrow \pi_1(\mathbb{P}^N \setminus D) \tag{2.6}$$

is surjective. (From now on we omit mentioning the base point in $\mathbb{P}^1 \subset \mathbb{P}^N$ in the notation.) The restriction of $R^n\pi_*\mathbb{Z}$ to $\mathbb{P}^1 \setminus D$, which is isomorphic to the higher direct image for the restriction of the family to $\mathbb{P}^1 \setminus D$, corresponds to the representation

$$\pi_1(\mathbb{P}^1 \setminus D) \twoheadrightarrow \pi_1(\mathbb{P}^N \setminus D) \longrightarrow \mathrm{GL}(H^n(X, \mathbb{Z}))$$

obtained by composing (2.5) with (2.6). Hence, $\Gamma(d, n)$ can be computed as the monodromy group of an arbitrary Lefschetz pencil $\mathcal{X}_{\mathbb{P}^1} \twoheadrightarrow \mathbb{P}^1$, i.e. as the image of

$$\rho_{\mathbb{P}^1} : \pi_1(\mathbb{P}^1 \setminus D) \longrightarrow \mathrm{GL}(H^n(X, \mathbb{Z})). \tag{2.7}$$

By Theorem 2.2, $\mathbb{P}^1 \setminus D \simeq \mathbb{P}^1 \setminus \{t_1, \ldots, t_\ell\}$ with $\ell = (d-1)^{n+1} \cdot (n+2)$. Therefore, $\pi_1(\mathbb{P}^1 \setminus D)$ is isomorphic to a quotient of the free group $\pi_1(\mathbb{C} \setminus \{t_1, \ldots, t_\ell\}) \simeq \mathbb{Z}^{*\ell}$ with free generators given by the simple loops γ_i around the points $t_i \in \mathbb{C}$.

Thus, in order to describe the image of (2.7), we need to compute the monodromy operators $\rho_{\mathbb{P}^1}(\gamma_i)$ and the group they generate. (In our discussion the details concerning the base point and the dependence on the path connecting it to circles around the critical values are suppressed.)

(ii) Let $x \in \mathcal{X}_0$ be the ordinary double point of the central fibre of the family $\mathcal{X} \twoheadrightarrow \Delta$ obtained from the Lefschetz pencil above by restriction to a small disk $\Delta \hookrightarrow \mathbb{P}^1$, $0 \longmapsto t_i$. The intersection of a ball $B(x) \subset \mathcal{X}$ around x with the nearby smooth fibre

$X = \mathcal{X}_\varepsilon$ retracts to a sphere $S^n \subset B(x) \cap X \subset X$. It is called the *vanishing sphere* and its cohomology class $\delta = [S^n] \in H^n(X, \mathbb{Z})$ is the *vanishing class*. Its main property, responsible for the name and verified by a local computation, is that it generates the kernel of the push-forward map, cf. [474, V. Cor. 14.17]:

$$H^n(X, \mathbb{Z}) \simeq H_n(X, \mathbb{Z}) \longrightarrow H_n(\mathcal{X}, \mathbb{Z}).$$

The self intersection $(\delta.\delta)$, determined by the normal bundle of $S^n \subset X$, is given by

$$(\delta.\delta) = \begin{cases} 0 & \text{if } n \equiv 1\,(2), \\ -2 & \text{if } n \equiv 2\,(4), \\ 2 & \text{if } n \equiv 0\,(4). \end{cases} \tag{2.8}$$

Of course, the vanishing for odd n follows from the fact that in this case the intersection pairing on the middle cohomology is alternating. The other two cases are obtained by an explicit computation, see [474, IV.15.2].

The crucial input is the description of the monodromy operation $\rho(\gamma)$ induced by a simple loop around $0 \in \Delta$. It is described by the *Picard–Lefschetz formula*:

$$\rho(\gamma) = s_\delta \colon \alpha \longmapsto \alpha + \varepsilon_n \cdot (\alpha.\delta) \cdot \delta, \;\text{ with }\; \varepsilon_n = \begin{cases} 1 & \text{if } n \equiv 2, 3\,(4), \\ -1 & \text{if } n \equiv 0, 1\,(4). \end{cases} \tag{2.9}$$

Note that the sign is such that for n even, s_δ is a reflection in $\delta^\perp$ and so, in particular, $s_\delta(\delta) = -\delta$ and $s_\delta^2 = \text{id}$. For n odd, the monodromy is not of finite order, as $s_\delta^k(\alpha) = \alpha + \varepsilon_n \cdot k \cdot (\alpha.\delta) \cdot \delta$.[5]

(iii) We have computed the images $\rho_{\mathbb{P}^1}(\gamma_i)$ of the free loops around the singular fibres $\mathcal{X}_{t_i}$, $i = 1, \ldots, \ell = \deg D(d, n)$, as the operators s_{δ_i} associated with the corresponding classes δ_i. They are reflections for even n and of infinite order for odd n.

Consider now families $\mathcal{X}^i \longrightarrow \Delta^i$ around each $t_i \in \mathbb{P}^1$ as in (ii). We may assume that the smooth reference fibre is X for all of them. We observe that all vanishing classes $\delta_i \in H^n(X, \mathbb{Z})$ are contained in the primitive cohomology. This follows from describing the composition

$$H^n(X, \mathbb{Z}) \simeq H_n(X, \mathbb{Z}) \longrightarrow H_n(\mathcal{X}^i, \mathbb{Z}) \longrightarrow H_n(\mathbb{P}, \mathbb{Z}) \simeq H^{n+2}(\mathbb{P}, \mathbb{Z}) \longrightarrow H^{n+2}(X, \mathbb{Z})$$

as the product with the hyperplane class. In fact, the *vanishing cohomology*

$$H^n(X, \mathbb{Z})_{\text{van}} := \text{Ker}(H^n(X, \mathbb{Z}) \longrightarrow H^{n+2}(\mathbb{P}, \mathbb{Z})),$$

which in our situation coincides with the primitive cohomology, is generated over $\mathbb{Z}$ by the vanishing classes, see [474, V. Lem. 14.26] or for the algebraic treatment [146, Sec. 4.3]. This has the following consequence.

[5] Note that in [46] the sign of the intersection form is changed for $n \equiv 3\,(4)$, so that in this case $s_\delta(\alpha) = \alpha + (\alpha.\delta) \cdot \delta$ also.

Corollary 2.14 *The primitive cohomology $H^n(X,\mathbb{Z})_{\rm pr}$ of a smooth hypersurface $X \subset \mathbb{P}^{n+1}$ is generated by classes δ with $(\delta.\delta)$ even as in* (2.8). *In particular, for n even, the lattice $H^n(X,\mathbb{Z})_{\rm pr}$ is even.* □

Note that the fact that $H^n(X,\mathbb{Z})_{\rm pr}$ is generated by the vanishing classes δ_i in particular shows that $b_n(X)_{\rm pr} \leq \deg D(d,n)$, which is confirmed by a quick comparison of Corollary 1.12 with Theorem 2.2.

For even n, the *Weyl group* is the subgroup

$$W \subset {\rm O}(H^n(X,\mathbb{Z})_{\rm pr})$$

generated by the reflections s_{δ_i}. For n odd, the Weyl group $W \subset {\rm Sp}(H^n(X,\mathbb{Z}))$ is defined analogously. In both cases W acts transitively on the set of vanishing classes $\Delta := \{\delta_i\}$, cf. [326, Prop. 7.5] or [474, Prop. 15.23].

A lattice Λ (symmetric or alternating) with a class of vectors $\Delta \subset \Lambda$ generating Λ and with the associated Weyl group acting transitively on Δ is called a *vanishing lattice*, see [168, 261]. By our discussion so far, we have $\Gamma(d,n) = {\rm Im}(\rho) = {\rm Im}(\rho_{\mathbb{P}^1}) = W$.

The proof of Theorem 2.9 for even $n > 2$ is in [46] reduced to a purely lattice theoretic result by Ebeling [168] describing the Weyl group of a *complete* vanishing lattice as this particular subgroup of the orthogonal group of the lattice. The lattice $H^n(X,\mathbb{Z})_{\rm pr}$ is complete, which by definition means that Δ contains a certain configuration of six vanishing classes. The fact that for $n > 2$, in accordance with Proposition 1.21, the lattice contains $A_2 \oplus U^{\oplus 2}$ is part of the picture.

The case of cubic surfaces is well known classically and is usually stated as

$$\Gamma(3,2) \simeq W(E_6).$$

This is the only case in which the monodromy group of cubic hypersurfaces is actually finite. Indeed, it is an index two subgroup of the finite orthogonal group ${\rm O}(H^2(X,\mathbb{Z})_{\rm pr})$ of the definite lattice $H^2(X,\mathbb{Z})_{\rm pr}$. We shall come back to it in Section **4**.1.4. For n odd, the result is deduced from [261].

Exercise 2.15 Consider the case $n = 0$ for which the universal smooth cubic hypersurface $\mathcal{X} \twoheadrightarrow \mathbb{P}^3 \setminus S$, defined over the complement of a quartic surface S with the explicit equation given in Example 2.5, is an étale cover of degree three. Show that the monodromy group $\Gamma(0,3)$ is in fact $\mathfrak{S}_3$. It equals the Galois group of the field extension $K(\mathbb{P}^3) \subset K(\mathcal{X})$. See [219] for further information.

2.6 Diffeomorphisms Clearly, any monodromy transformation is induced by a diffeomorphism. Hence, the monodromy group Γ_n is a subgroup of the image of the natural representation

$$\tau\colon {\rm Diff}^+(X) \twoheadrightarrow {\rm O}(H^n(X,\mathbb{Z}))$$

of the group of orientation preserving diffeomorphisms. It turns out that $\mathrm{Im}(\tau)$ is slightly larger than Γ_n. Details have been worked out by Beauville [46]. Here are the main steps.

Let us first consider the case that n is even and $n > 2$. Clearly, $\mathrm{Diff}^+(X)$ also acts on $H^2(X,\mathbb{Z}) \simeq \mathbb{Z} \cdot h$ and, therefore, sends h to h or to $-h$. The latter is realized by complex conjugation defined on any X defined by an equation with coefficients in $\mathbb{R}$. As a consequence, $\mathrm{Diff}^+(X)$ respects the direct sum decomposition $H^n(X,\mathbb{Q}) = H^n(X,\mathbb{Q})_{\mathrm{pr}} \oplus \mathbb{Q} \cdot h^{n/2}$. This eventually leads to

$$\mathrm{Im}(\tau) \simeq \begin{cases} \tilde{\mathrm{O}}(H^n(X,\mathbb{Z})) & \text{if } n \equiv 0\,(4), \\ \mathrm{O}(H^n(X,\mathbb{Z})_{\mathrm{pr}}) & \text{if } n \equiv 2\,(4). \end{cases}$$

To prove this note that $\tilde{\mathrm{O}}^+(H^n(X,\mathbb{Z})) \subset \mathrm{Im}(\tau)$ and that for $n \equiv 2\,(4)$, complex conjugation induces an element in $\mathrm{Im}(\tau)$ the restriction of which to $H^n(X,\mathbb{Z})_{\mathrm{pr}}$ acts non-trivially on the discriminant $A_{H^n_{\mathrm{pr}}} \simeq A_{\mathbb{Z}\cdot h^{n/2}} \simeq \mathbb{Z}/3\mathbb{Z}$.

Hence, it is enough to find an orientation preserving diffeomorphism g which acts with spinor norm $\mathrm{sn}_n(\tau(g)) = -1$ on $H^n(X,\mathbb{Z})$ and fixes $h^{n/2}$. For this, one uses the connected sum decomposition of X as $M'\#(S^n \times S^n)$, cf. Remark 1.22, and the diffeomorphism g obtained by gluing the identity on M' with the product $\iota \times \iota$ of the diffeomorphism $\iota\colon S^n \longrightarrow S^n$, $(x_1,\ldots,x_{n+1}) \longmapsto (x_1,\ldots,x_n,-x_{n+1})$. It acts on the induced orthogonal decomposition $H^n(X,\mathbb{Z}) \simeq H^n(M',\mathbb{Z}) \oplus H^n(S^n \times S^n,\mathbb{Z})$, for which we may assume that $h^{n/2} \in H^n(M',\mathbb{Z})$, by $-\mathrm{id}$ on $H^n(S^n \times S^n,\mathbb{Z}) \simeq U$, and by id on $U^\perp = H^n(M',\mathbb{Z})$. Write $-\mathrm{id}_U = s_{e-f} \circ s_{e+f}$, with $e, f \in U$ the standard basis, to see that indeed $\mathrm{sn}_n(\tau(g)) = -1$ and $\tau(g)(h^{n/2}) = h^{n/2}$.[6]

For cubic surfaces, there is no reason for a diffeomorphism to respect the hyperplane class (up to sign) and indeed $\mathrm{Im}(\tau) = \mathrm{O}(H^2(X,\mathbb{Z}))$, cf. [488] and Section **4**.1.4.

For n odd, Beauville's result reads

$$\mathrm{Im}(\tau) = \begin{cases} \mathrm{SpO}(H^n(X,\mathbb{Z}),q) & \text{if } n \neq 1,3,7, \\ \mathrm{Sp}(H^n(X,\mathbb{Z})) & \text{if } n = 1,3,7. \end{cases}$$

Indeed, the description of $q([S^n])$ for $n \neq 1,3,7$ in terms of the topological normal bundle of $S^n \subset X$ is invariant under diffeomorphisms. In the other cases one proves that s_δ is realized by a diffeomorphism for any primitive $\delta \in H^n(X,\mathbb{Z})$. As those generate the symplectic group, this is enough to prove the claim for $n = 1,3,7$. Concretely, for a given δ, there exist δ' with $(\delta.\delta') = 1$ and a decomposition $X \simeq M'\#(S^n \times S^n)$ with $H^n(S^n \times S^n,\mathbb{Z})$ spanned by δ,δ'. In [46] it is then observed that the reflection s_δ is

[6] In [46] the result for $n \equiv 2\,(4)$ is stated as $\mathrm{Im}(\tau) = \tilde{\mathrm{O}}(H^n(X,\mathbb{Z})) \times \{\pm 1\}$. Indeed, complex conjugation defines an element of order two in $\mathrm{Im}(\tau)$ that acts non-trivially on the discriminant of $H^n(X,\mathbb{Z})_{\mathrm{pr}}$. Moreover, it commutes with the index two subgroup $\Gamma_n = \tilde{\mathrm{O}}^+(H^n(X,\mathbb{Z}))$, as the universal family is defined over $\mathbb{R}$ and hence monodromy commutes with complex conjugation. However, that complex conjugation also commutes with the additional diffeomorphism g would seem to need an additional argument.

realized by gluing the identity on M' with the diffeomorphism $(x,y)\longmapsto(x,x\cdot y)$, where $x\cdot y$ is the multiplication in $\mathbb{C}$, $\mathbb{H}$, or $\mathbb{O}$ for the three cases $n=1,3,7$.

3 Automorphisms and Deformations

Smooth hypersurfaces behave nicely in many respects. For example, for most of them the deformation theory is easy to understand, not showing any of the pathological features to be reckoned with for arbitrary smooth projective varieties. Similarly, their groups of automorphisms are usually finite and generically even trivial. We will assume $d\geq 3$ throughout this section. The only slightly exotic cases that need special care are $(n,d)=(1,3)$ and $(n,d)=(2,4)$, i.e. plane cubic (elliptic) curves and quartic K3 surfaces.

3.1 Infinitesimal Automorphisms First order information about the group of automorphisms of a smooth hypersurface $X\subset\mathbb{P}^{n+1}$ and about its deformations are encoded by the cohomology groups $H^0(X,\mathcal{T}_X)$ and $H^1(X,\mathcal{T}_X)$, respectively. Those can be computed in terms of the standard exact sequences. We begin, however, with the following well-known fact.

Lemma 3.1 *Assume* char$(k)\nmid d$. *Then a hypersurface $X\subset\mathbb{P}^{n+1}$ of degree d defined by $F\in k[x_0,\ldots,x_{n+1}]_d$ is smooth if and only if the partial derivatives ∂_iF form a regular sequence in $k[x_0,\ldots,x_{n+1}]$.*

Proof A standard result in commutative algebra shows that a sequence $a_i\in A$, $i=1,\ldots,\dim(A)$, in a regular local ring A is regular if and only if $\mathrm{ht}((a_i))=\dim(A)$, cf. [343, Thm. 16.B]. Hence, the partial derivatives (∂_iF) form a regular sequence in the polynomial ring $k[x_0,\ldots,x_{n+1}]$ if and only if the affine intersection $V((\partial_iF))=\bigcap V(\partial_iF)\subset\mathbb{A}^{n+2}$ is zero-dimensional. However, as the polynomials ∂_iF are homogeneous, $V((\partial_iF))$ is $\mathbb{G}_m$-invariant. Hence, (∂_iF) is a regular sequence if and only if the projective intersection $V((\partial_iF))\subset\mathbb{P}^{n+1}$ is empty. This implies that also $X\cap V((\partial_iF))$ is empty and, by the Jacobian criterion, that X is smooth.

Conversely, if X is smooth and char$(k)\nmid d$, the Euler equation $d\cdot F=\sum_{i=0}^{n+1}x_i\,\partial_iF$ shows that $V((\partial_iF))=X_{\mathrm{sing}}=\emptyset$, i.e. (∂_iF) is a regular sequence. □

Example 3.2 The assumption on the characteristic is needed, as shown by the example $F=x_0^2x_1-x_0x_1^2$ with char$(k)=3$. Indeed, in this case $X=\{0,\infty,[1:1]\}$ is smooth, but $\partial_0F=-x_1\,(x_0+x_1)$ and $\partial_1F=x_0\,(x_0+x_1)$ have a common 0 in $[1:-1]$.

Remark 3.3 The smoothness of a hypersurface X expressed in terms of the partial derivatives of its defining equation has concrete geometric consequences: For example, it can be used to prove that a smooth hypersurface $X=V(F)\subset\mathbb{P}^{n+1}$ of degree $d\geq 2$

cannot contain a linear subspace $\mathbb{P}^k \subset \mathbb{P}^{n+1}$ of dimension $k > n/2$. This was the content of Exercise 1.5, which used cohomology. The following argument is even easier.

Assume $\mathbb{P}^k \subset \mathbb{P}^{n+1}$ is the linear subspace $V(x_{k+1}, \dots, x_{n+1})$. If $\mathbb{P}^k \subset X$, then the equation of X can be written as $F = \sum_{j=k+1}^{n+1} x_j G_j$ with $\deg(G_j) \geq 1$. Hence, for a point $x \in \mathbb{P}^k \subset X$, one has $\partial_j F(x) = G_j(x)$ for $j \geq k+1$ and $\partial_j F(x) = 0$ for $j \leq k$. If $k \geq (n+1)/2$, then $V(G_{k+1}, \dots, G_{n+1}) \cap \mathbb{P}^k \neq \emptyset$ and, therefore, there exists a point $x \in \mathbb{P}^k \subset X$ such that $\partial_j F(x) = 0$ for all $j = 0, \dots, n+1$, which contradicts the smoothness of X at x.

Exercise 3.4 Show that a hypersurface $X = V(F) \subset \mathbb{P}^{n+1}$ of degree d is singular if for some i the degree of F as a polynomial in x_i is $\deg_{x_i}(F) \leq d-2$, see [196, Lem. 1.2].

According to a classical observation of Kodaira and Spencer [279, Lem. 14.2], one has the following.

Corollary 3.5 *Let $X \subset \mathbb{P} = \mathbb{P}^{n+1}$ be a smooth hypersurface of degree d.*

(i) *If $n \geq 0$ and $d \geq 3$ but $(n,d) \neq (1,3)$, then $H^0(X, \mathcal{T}_X) = 0$.*
(ii) *If $n > 2$ or $d \leq 3$, then $H^1(X, \mathcal{T}_{\mathbb{P}}|_X) = 0$ and the normal bundle sequence induces a surjection*

$$H^0(X, \mathcal{O}_X(d)) \twoheadrightarrow H^1(X, \mathcal{T}_X).$$

Proof We shall give a proof under the additional assumption that $\mathrm{char}(k) \nmid d$ and refer to [270, Sec. 11.7] for the general case.

Combining the Euler sequence $0 \to \mathcal{O}_{\mathbb{P}} \to \mathcal{O}_{\mathbb{P}}(1)^{\oplus n+2} \to \mathcal{T}_{\mathbb{P}} \to 0$ and the normal bundle sequence $0 \to \mathcal{T}_X \to \mathcal{T}_{\mathbb{P}}|_X \to \mathcal{O}_X(d) \to 0$, we obtain a diagram

$$\begin{array}{ccccccccc} H^0(\mathcal{T}_X) & \longrightarrow & H^0(\mathcal{T}_{\mathbb{P}}|_X) & \longrightarrow & H^0(\mathcal{O}_X(d)) & \longrightarrow & H^1(\mathcal{T}_X) & \longrightarrow & H^1(\mathcal{T}_{\mathbb{P}}|_X). \\ & & \uparrow & \nearrow_{(\partial_i F)} & & & & & \\ & & H^0(\mathcal{O}_X(1))^{\oplus n+2} & & & & & & \end{array} \tag{3.1}$$

Here, as before, $\partial_i F \in H^0(\mathbb{P}, \mathcal{O}_{\mathbb{P}}(d-1))$ are the $n+2$ partial derivatives of the homogeneous polynomial $F \in H^0(\mathbb{P}, \mathcal{O}_{\mathbb{P}}(d))$ defining X. The cokernel of the vertical map is contained in $H^1(\mathcal{O}_X)$, which is trivial for $n \neq 1$.

Now, for the first assertion in the case $n > 1$, observe that $H^0(X, \mathcal{T}_X) = 0$ if and only if the kernel of the composition

$$(\partial_i F) \colon H^0(X, \mathcal{O}_X(1))^{\oplus n+2} \longrightarrow H^0(X, \mathcal{O}_X(d))$$

is spanned by the vector $(x_0, \dots, x_{n+1})$. Assume $\sum h_i\, \partial_i F$ vanishes on X for some $h_i \in H^0(\mathbb{P}, \mathcal{O}_{\mathbb{P}}(1))$. Then, after rescaling, $\sum h_i\, \partial_i F = d \cdot F = \sum x_i\, \partial_i F$ and, therefore, $\sum (h_i - x_i)\, \partial_i F = 0$. Using that $(\partial_i F)$ is a regular sequence, see Lemma 3.1, and $d \geq 3$, this implies $h_i = x_i$. There is nothing to prove for $n = 0$, so the remaining case is $n = 1$, $d > 3$, for which the assertion follows from the fact that $H^0(C, \omega_C^*) = 0$ for a smooth curve of genus $g(C) > 1$.

For the second assertion observe that $H^1(X, \mathcal{T}_{\mathbb{P}}|_X) = 0$, whenever $H^1(X, \mathcal{O}_X(1)) = 0 = H^2(X, \mathcal{O}_X)$, which holds as soon as $n > 2$, see Corollary 1.7. We leave it to the reader to complete the argument in the cases $n = 1, 2$ and $d = 1, 2, 3$. □

3.2 (Polarized) Automorphisms Let X be a smooth projective variety and assume $\mathcal{O}_X(1)$ is an ample line bundle. We are interested in the two groups:

$$\mathrm{Aut}(X, \mathcal{O}_X(1)) \subset \mathrm{Aut}(X).$$

Here, $\mathrm{Aut}(X)$ is the group of all automorphisms $g\colon X \xrightarrow{\sim} X$ over k. The subgroup $\mathrm{Aut}(X, \mathcal{O}_X(1))$ is the group of all such automorphisms with the additional property that $g^*\mathcal{O}_X(1) \simeq \mathcal{O}_X(1)$. These groups are in fact the groups of k-rational points of group schemes over k, which we shall also denote by $\mathrm{Aut}(X, \mathcal{O}_X(1))$ and $\mathrm{Aut}(X)$.

Remark 3.6 Standard Hilbert scheme theory, see e.g. [178] or [253], ensures that the group scheme $\mathrm{Aut}(X, \mathcal{O}_X(1))$ is a quasi-projective variety and that $\mathrm{Aut}(X)$ is at least locally of finite type. Indeed, there exists an open embedding

$$\mathrm{Aut}(X) \hookrightarrow \mathrm{Hilb}(X \times X),\ g \longmapsto \Gamma_g,$$

mapping an automorphism to its graph. The *Hilbert scheme* $\mathrm{Hilb}(X \times X)$ of $X \times X$ is locally of finite type. More precisely, it is the disjoint union $\coprod \mathrm{Hilb}^P(X \times X)$, $P \in \mathbb{Q}[T]$, of projective varieties $\mathrm{Hilb}^P(X \times X)$ parametrizing closed subschemes $Z \subset X \times X$ with *Hilbert polynomial* $\chi(Z, (\mathcal{O}_X(m) \boxtimes \mathcal{O}_X(m))|_Z) = P(m)$.

The Hilbert polynomial of the graph Γ_g of an arbitrary isomorphism is $\chi(X, \mathcal{O}_X(m) \otimes g^*\mathcal{O}_X(m))$. Thus, for $P(m) := \chi(X, \mathcal{O}_X(2m))$, one has a locally closed embedding

$$\mathrm{Aut}(X,\ \mathcal{O}_X(1)) \hookrightarrow \mathrm{Hilb}^P(X \times X).$$

Note that it may fail to be open in general, as $\chi(X, \mathcal{O}_X(m) \otimes g^*\mathcal{O}_X(m)) = \chi(X, \mathcal{O}_X(2m))$ may not necessarily imply that $g^*\mathcal{O}_X(1) \simeq \mathcal{O}_X(1)$.

Proposition 3.7 *The Zariski tangent spaces of* $\mathrm{Aut}(X)$ *and* $\mathrm{Aut}(X, \mathcal{O}_X(1))$ *at the identity satisfy*

$$T_{\mathrm{id}}\mathrm{Aut}(X, \mathcal{O}_X(1)) \subset T_{\mathrm{id}}\mathrm{Aut}(X) \simeq H^0(X, \mathcal{T}_X). \tag{3.2}$$

The inclusion is an equality if $H^1(X, \mathcal{O}_X) = 0$.

Proof This follows from the description of the tangent space of the Hilbert scheme of closed subschemes of Y at the point $[Z] \in \mathrm{Hilb}(Y)$ corresponding to $Z \subset Y$ as

$$T_{[Z]}\mathrm{Hilb}(Y) \simeq \mathrm{Hom}(\mathcal{I}_Z, \mathcal{O}_Z),$$

cf. [178, Thm. 6.4.9] or [253, Sec. 2.2] and Section **2**.1.3. For $Z := \Gamma_{\rm id} \subset Y := X \times X$, this becomes

$$T_{\rm id}{\rm Aut}(X) \simeq {\rm Hom}(\mathcal{I}_\Delta, \mathcal{O}_\Delta) \simeq H^0(\Delta, \mathcal{N}_{\Delta/X\times X}) \simeq H^0(X, \mathcal{T}_X).$$

As for our purposes the inclusion in (3.2) is all we need, we leave the second assertion to the reader. Hint: use $H^1(X, \mathcal{O}_X) \simeq T_{[\mathcal{O}_X]}{\rm Pic}(X)$. □

Exercise 3.8 Refine the proposition by showing that $T_{\rm id}{\rm Aut}(X, \mathcal{O}_X(1))$ is the kernel of the map $H^0(X, \mathcal{T}_X) \twoheadrightarrow H^1(X, \mathcal{O}_X)$ induced by the contraction with the first Chern class $c_1(L)$.

For a smooth hypersurface of dimension $n \geq 2$ and degree $d \geq 3$, the result immediately gives

$$T_{\rm id}{\rm Aut}(X, \mathcal{O}_X(1)) = T_{\rm id}{\rm Aut}(X) \simeq H^0(X, \mathcal{T}_X) \simeq 0,$$

which allows one to prove the following general finiteness result. The original proof by Matsumura–Monsky [344] is different, it avoids cohomological methods and relies on techniques from commutative algebra. See [375, Rem. 6] for historical remarks.

Corollary 3.9 *Let $X \subset \mathbb{P}^{n+1}$ be a smooth hypersurface of dimension $n \geq 0$ and degree $d \geq 3$, but $(n, d) \neq (1, 3)$. Then* ${\rm Aut}(X, \mathcal{O}_X(1))$ *is finite and* ${\rm Aut}(X)$ *is discrete. In fact, if $(n, d) \neq (1, 3), (2, 4)$, then* ${\rm Aut}(X, \mathcal{O}_X(1)) = {\rm Aut}(X)$*, and then both groups are finite.*

Proof As ${\rm Aut}(X)$ and ${\rm Aut}(X, \mathcal{O}_X(1))$ are group schemes, all tangent spaces are isomorphic and in our case trivial by Corollary 3.5. Hence, ${\rm Aut}(X)$, which is locally of finite type, is a countable set of reduced isolated points. As ${\rm Aut}(X, \mathcal{O}_X(1))$ is quasi-projective, it must be a finite set of reduced isolated points.

The equality ${\rm Aut}(X, \mathcal{O}_X(1)) = {\rm Aut}(X)$ for $n > 2$ follows from Corollary 1.9. For $n = 2$ and $d \neq 4$, use that $\omega_X \simeq \mathcal{O}(d - (n + 2))$ is preserved by all automorphisms and that ${\rm Pic}(X)$ is torsion free, see Remark 1.4. □

Similarly, if $(n, d) \neq (1, 3), (2, 4)$, any isomorphism $X \simeq X'$ between two smooth hypersurfaces $X, X' \subset \mathbb{P}^{n+1}$ of degree d is induced by a coordinate change of $\mathbb{P}^{n+1}$.

Remark 3.10 (i) For $n = 1$ and $d = 3$, the result really fails, but not too badly. For a smooth plane cubic curve $E \subset \mathbb{P}^2$ and ${\rm char}(k) \neq 3$, one has

$$0 = T_{\rm id}{\rm Aut}(E, \mathcal{O}_E(1)) \subset T_{\rm id}{\rm Aut}(E) \simeq H^0(E, \mathcal{T}_E) \simeq H^0(E, \mathcal{O}_E) \simeq k,$$

see [270, Sec. 11.7.5]. So, even in this case, the group ${\rm Aut}(E, \mathcal{O}_E(1))$ is in fact finite, although the bigger group ${\rm Aut}(E)$ certainly is not.

(ii) The finiteness of ${\rm Aut}(X)$ also fails for $n = 2$ and $d = 4$ in general. Indeed, there exist quartic K3 surfaces with infinite automorphism groups, see [249, Sec. 15.2.5] for examples and references.

The groups of automorphisms of the universal smooth hypersurface of degree d, denoted $\mathcal{X} \longrightarrow U = |\mathcal{O}(d)|_{\text{sm}}$, form a quasi-projective family

$$\mathbf{Aut} := \text{Aut}(\mathcal{X}/U,\ \mathcal{O}_{\mathcal{X}}(1)) \longrightarrow U = |\mathcal{O}_{\mathbb{P}}(d)|_{\text{sm}}. \tag{3.3}$$

More precisely, there exist functorial bijections between $\text{Mor}_U(T, \mathbf{Aut})$ and the set of automorphisms $g\colon \mathcal{X}_T \xrightarrow{\sim} \mathcal{X}_T$ over T with $g^*\mathcal{O}_{\mathcal{X}_T}(1) \simeq \mathcal{O}_{\mathcal{X}_T}(1)$ modulo $\text{Pic}(T)$. As in the absolute case, mapping g to its graph, describes a locally closed embedding $\mathbf{Aut} \subset \text{Hilb}(\mathcal{X} \times_U \mathcal{X}/U)$ into the relative Hilbert scheme.

According to Corollary 3.9, the fibres of $\mathbf{Aut} \longrightarrow U$, i.e. the groups $\text{Aut}(X, \mathcal{O}_X(1))$, are finite and, therefore, $\mathbf{Aut} \longrightarrow U$ is a quasi-finite morphism. In fact, it turns out to be finite, cf. Remark **3**.1.7, which is a consequence of the GIT stability of smooth hypersurfaces. Note that the general result of [346] proving properness for families of non-ruled varieties is not applicable to cubic hypersurfaces of dimension at least two.

3.3 (Polarized) Deformations The description of the first order deformations of a smooth projective variety X is similar. Firstly, there is a bijection between $H^1(X, \mathcal{T}_X)$ and the set of flat morphisms $\mathcal{X} \longrightarrow \text{Spec}(k[\varepsilon])$ with closed fibre $\mathcal{X}_0 = X$, cf. [223, II. Ex. 9.13.2]. This can be extended to the following picture, cf. [178, Ch. 6]: If $H^0(X, \mathcal{T}_X) = 0$, then the functor

$$F_X\colon (Art/k) \longrightarrow (Set),$$

mapping a local Artinian k-algebra A to the set of flat morphisms $\mathcal{X} \longrightarrow \text{Spec}(A)$ with the choice of an isomorphism $\mathcal{X}_0 \simeq X$ for the closed fibre $\mathcal{X}_0$ has a pro-representable hull, see [178, Def. 6.3.1]. This means that there exist a complete local k-algebra R and a 'versal' flat family $\mathcal{X} \longrightarrow \text{Spf}(R)$, $\mathcal{X}_0 \simeq X$, for which the induced transformation

$$h_R = \text{Mor}_{k\text{-}alg}(R,\ ) \longrightarrow F_X \tag{3.4}$$

is bijective for $A = k[\varepsilon]$. We shall write $\text{Def}(X) := \text{Spf}(R)$ with the distinguished closed point $0 \in \text{Def}(X)$ and the Zariski tangent space $T_0\text{Def}(X) \simeq H^1(X, \mathcal{T}_X)$.

Similarly, one considers the polarized version

$$F_{X,\mathcal{O}_X(1)}\colon (Art/k) \longrightarrow (Set)$$

mapping A to the set of flat polarized families $(\mathcal{X}, \mathcal{O}_{\mathcal{X}}(1)) \longrightarrow \text{Spec}(A)$ with closed fibre $(\mathcal{X}, \mathcal{O}_{\mathcal{X}}(1)) \simeq (X, \mathcal{O}_X(1))$. Again, the functor $F_{X,\mathcal{O}_X(1)}$ has a pro-representable hull R' with a 'versal' flat family $(\mathcal{X}, \mathcal{O}_{\mathcal{X}}(1)) \longrightarrow \text{Def}(X, \mathcal{O}_X(1)) := \text{Spf}(R')$.

Only if $\text{Aut}(X)$ is trivial, one can expect a universal family to exist, i.e. (3.4) to be an isomorphism. Then F_X is said to be pro-representable (and similarly for $F_{X,\mathcal{O}_X(1)}$). This is the difference between a universal and a versal family.

The natural forgetful transformation $F_{X,\mathcal{O}_X(1)} \longrightarrow F_X$ describes a morphism

$$\text{Def}(X, \mathcal{O}_X(1)) \longrightarrow \text{Def}(X), \tag{3.5}$$

which in general is neither injective nor surjective. The first Chern class

$$\mathrm{c}_1(\mathcal{O}_X(1)) \in H^1(X, \Omega_X) \simeq \mathrm{Ext}^1(\mathcal{T}_X, \mathcal{O}_X)$$

interpreted as an extension class defines an exact sequence

$$0 \longrightarrow \mathcal{O}_X \longrightarrow \mathcal{D}(\mathcal{O}_X(1)) \longrightarrow \mathcal{T}_X \longrightarrow 0.$$

Here, the sheaf $\mathcal{D}(\mathcal{O}_X(1))$ can be thought of as the sheaf of differential operators of $\mathcal{O}_X(1)$ of order ≤ 1.

Then $T_0\mathrm{Def}(X, \mathcal{O}_X(1)) \simeq H^1(X, \mathcal{D}(\mathcal{O}_X(1)))$ and the tangent map of (3.5) is part of a long exact sequence, see [421, Sec. 3.3] for more details:

$$\begin{array}{ccccc} \cdots \to H^1(X, \mathcal{O}_X) \to & H^1(X, \mathcal{D}(\mathcal{O}_X(1))) & \longrightarrow & H^1(X, \mathcal{T}_X) & \to H^2(X, \mathcal{O}_X) \to \cdots . \\ & \simeq T_0\mathrm{Def}(X, \mathcal{O}_X(1)) & & \simeq T_0\mathrm{Def}(X). & \end{array}$$

In fact, for most hypersurfaces the outer terms are trivial.

Remark 3.11 Over $\mathbb{C}$, the formal spaces $\mathrm{Def}(X)$ and $\mathrm{Def}(X, \mathcal{O}_X(1))$ can alternatively be thought of as germs of complex spaces. Standard deformation theory ensures that the universal families $\mathcal{X} \longrightarrow \mathrm{Def}(X)$ and $(\mathcal{X}, \mathcal{O}_{\mathcal{X}}(1)) \longrightarrow \mathrm{Def}(X, \mathcal{O}_X(1))$ can in fact be extended from families over formal bases to families over some small complex spaces. While this remains true in the algebraic setting for $(\mathcal{X}, \mathcal{O}_{\mathcal{X}}(1)) \longrightarrow \mathrm{Def}(X, \mathcal{O}_X(1))$, cf. [178, Thm. 8.4.10], it fails for the unpolarized situation.

The universal family of smooth hypersurfaces $\mathcal{X} \longrightarrow U = |\mathcal{O}_{\mathbb{P}}(d)|_{\mathrm{sm}}$ induces a morphism $(U, 0) \longrightarrow \mathrm{Def}(X, \mathcal{O}_X(1))$ from the formal neighbourhood of $0 := [X] \in U$. We think of $|\mathcal{O}_{\mathbb{P}}(d)|$ as a component of the Hilbert scheme $\mathrm{Hilb}(\mathbb{P}^{n+1})$ and of $\mathcal{O}_X(d)$ as the normal bundle $\mathcal{N}_{X/\mathbb{P}^{n+1}}$. Then $T_0U \simeq H^0(X, \mathcal{O}_X(d))$ and the tangent map of the composition

$$(U, 0) \longrightarrow \mathrm{Def}(X, \mathcal{O}_X(1)) \longrightarrow \mathrm{Def}(X)$$

is the boundary map of the normal bundle sequence $H^0(X, \mathcal{O}_X(d)) \longrightarrow H^1(X, \mathcal{T}_X)$. Conversely, for an arbitrary deformation $(\mathcal{X}, \mathcal{O}_{\mathcal{X}}(1)) \longrightarrow \mathrm{Spec}(A)$ of X over a local ring A there exists a relative embedding $\mathcal{X} \hookrightarrow \mathbb{P}^{n+1}_A$ extending the given one $X \subset \mathbb{P}^{n+1}$. Here one uses the fact that $H^1(X, \mathcal{O}_X(1)) = 0$, which ensures that all sections of $\mathcal{O}_X(1)$ on X extend to sections of $\mathcal{O}_{\mathcal{X}}(1)$, see [421, Sec. 3.3].

Proposition 3.12 *Let $X \subset \mathbb{P}^{n+1}$ be a smooth hypersurface of degree d. Assume $n > 2$ or $n = 2$, $d \leq 3$. Then the natural map*

$$H^0(X, \mathcal{O}_X(d)) \simeq T_0|\mathcal{O}_{\mathbb{P}}(d)| \twoheadrightarrow T_0\mathrm{Def}(X, \mathcal{O}_X(1)) \xrightarrow{\sim} T_0\mathrm{Def}(X) \simeq H^1(X, \mathcal{T}_X)$$

is surjective. Furthermore, the forgetful morphism (3.5) *is an isomorphism*

$$\mathrm{Def}(X, \mathcal{O}_X(1)) \xrightarrow{\sim} \mathrm{Def}(X)$$

between smooth germs.

Proof Most of the proposition is an immediate consequence of the preceding discussion and the vanishings $H^1(X, \mathcal{O}_X) = 0 = H^2(X, \mathcal{O}_X)$. In order to see that the isomorphism $T_0\mathrm{Def}(X, \mathcal{O}_X(1)) \simeq T_0\mathrm{Def}(X)$ between the tangent spaces is induced by an isomorphism $\mathrm{Def}(X, \mathcal{O}_X(1)) \simeq \mathrm{Def}(X)$ it suffices to observe that both spaces are smooth and so isomorphic to $\mathrm{Spf}(k[[z_1, \ldots, z_m]])$ with $m = \dim T_0$. This could either be deduced from the vanishing $H^2(X, \mathcal{D}(\mathcal{O}_X(1))) = H^2(X, \mathcal{T}_X) = 0$ for $n > 3$ [421, Thm. 3.3.11] or, simply, from the fact that $|\mathcal{O}_{\mathbb{P}}(d)|$ is smooth. □

Remark 3.13 The kernel of $H^0(X, \mathcal{O}_X(d)) \twoheadrightarrow H^1(X, \mathcal{T}_X)$ is a quotient of $H^0(X, \mathcal{T}_{\mathbb{P}}|_X)$ (and in fact equals it for $n \geq 2$ and $d \geq 3$). The latter should be thought of as the tangent space of the orbit through $[X] \in |\mathcal{O}_{\mathbb{P}}(d)|$ of the natural $\mathrm{GL}(n+2)$-action on $|\mathcal{O}_{\mathbb{P}}(d)|$, see Section **3**.1.3. This leads, for cubic hypersurfaces, to the dimension formula $\dim H^1(X, \mathcal{T}_X) = \binom{n+2}{3}$, for which an alternative proof will be given in Example 4.15.

For later reference, we also note that $H^2(X, \mathcal{T}_X) = 0$ for all smooth cubic hypersurfaces $X \subset \mathbb{P}^{n+1}$. For $n \geq 3$, this can be deduced from the Kodaira vanishing $H^i(X, \mathcal{T}_X) \simeq H^i(X, \Omega_X^{n-1} \otimes \omega_X^*) = 0$ for $i > 1$, as ω_X^* is ample, and for $n = 2$, it follows from Serre duality.

It may be worth pointing out the following consequence, which we will only state for cubic hypersurfaces.

Corollary 3.14 *Any local deformation of a smooth cubic hypersurface $X \subset \mathbb{P}^{n+1}$ as a variety over k is again a cubic hypersurface.* □

For $n = 2$, so for cubic surfaces, it is easy to construct smooth projective global deformations that are not cubic surfaces any longer, see Remark **4**.2.8. However, for $n > 2$, the fact that $\rho(X) = 1$ allows one to prove that global smooth projective deformations of cubic hypersurfaces are again cubic hypersurfaces, cf. [259, Thm. 3.2.5].[7] The situation is more complicated when one is interested in non-projective or, equivalently, non-Kählerian global deformations.

3.4 No Automorphisms Generically It turns out that for generic hypersurfaces the automorphism group is trivial and this has been generalized to complete intersections.

Theorem 3.15 *Assume $n > 0$, $d \geq 3$, and $(n, d) \neq (1, 3)$. Then there exists a dense open subset $V \subset |\mathcal{O}_{\mathbb{P}}(d)|_{\mathrm{sm}}$ such that for all geometric points $[X] \in V$ one has*

$$\mathrm{Aut}(X) = \mathrm{Aut}(X, \mathcal{O}_X(1)) = \{\mathrm{id}\}.$$

[7] Thanks to J. Ottem for the reference. Compare this to the well-known fact that $\mathrm{Def}(\mathbb{P}^1 \times \mathbb{P}^1)$ is a reduced point but yet $\mathbb{P}^1 \times \mathbb{P}^1$ can be deformed to any other Hirzebruch surface $\mathbb{F}_n = \mathbb{P}(\mathcal{O} \oplus \mathcal{O}(n))$ with n even.

There are three proofs in the literature. The original one by Matsumura and Monsky [344] and two more recent ones by Poonen [393] and Chen, Pan, and Zhang [115]. Benoist [63] discusses the more general situation of complete intersections. For $(n,d) = (1,3)$, i.e. for the generic cubic curve, one still has $\mathrm{Aut}(X, \mathcal{O}_X(1)) = \{\pm \mathrm{id}\}$ as long as $\mathrm{char}(k) \neq 3$. For simplicity, we shall assume that $(n,d) \neq (1,3),(2,4)$ and so $\mathrm{Aut}(X) = \mathrm{Aut}(X,\ \mathcal{O}_X(1))$, see Corollary 3.9.

In [393] the result is proved by writing down an explicit equation of one smooth hypersurface without any non-trivial polarized automorphisms, cf. Remark 3.19. We will follow [115] adapting the arguments to our situation.[8] We shall begin with the following result which is of independent interest.

Proposition 3.16 *Assume $X \subset \mathbb{P}^{n+1}$ is a smooth hypersurface of degree d over a field of characteristic zero with $n > 0$, $d \geq 3$, and $(n,d) \neq (1,3)$. Then* $\mathrm{Aut}(X)$ *acts faithfully on $H^1(X, \mathcal{T}_X)$.*

Proof We may assume that k is algebraically closed. Suppose $g \in \mathrm{Aut}(X)$ acts trivially on $H^1(X, \mathcal{T}_X)$. As an element $g \in \mathrm{Aut}(X) \subset \mathrm{PGL}(n+2)$ it can be lifted to an element in $\mathrm{SL}(n+2)$ which we shall also call g. It is still of finite order and, after a linear coordinate change, can be assumed to act by $g(x_i) = \lambda_i\, x_i$ for some roots of unities λ_i. This is where one needs $\mathrm{char}(k) = 0$.

Let $F \in H^0(\mathbb{P}, \mathcal{O}_{\mathbb{P}}(d))$ be a homogeneous polynomial defining X. As $g(X) = X$, the induced action of g on $H^0(\mathbb{P}, \mathcal{O}_{\mathbb{P}}(d))$ satisfies $g(F) = \mu F$ for some root of unity μ. Hence, changing g by $\mu^{-1/d}$ we may assume that $\mu = 1$ (but possibly g is now only a finite order element in $\mathrm{GL}(n+2)$). For greater clarity, we rewrite (3.1) as the short exact sequence

$$W := (V \otimes V^*)/k \cdot \mathrm{id} \simeq H^0(X, \mathcal{T}_{\mathbb{P}}|_X) \hookrightarrow H^0(X, \mathcal{O}_X(d)) \twoheadrightarrow H^1(X, \mathcal{T}_X),$$

with $V = \langle x_0, \ldots, x_{n+1} \rangle$ and using $H^0(X, \mathcal{T}_X) = 0 = H^1(X, \mathcal{T}_{\mathbb{P}}|_X)$ observed earlier. All maps are compatible with the action of g and also the isomorphism is GL-equivariant. Note that $H^0(X, \mathcal{O}_X(d))$ is endowed with the action of g by interpreting $\mathcal{O}_X(d)$ as the normal bundle $\mathcal{N}_{X/\mathbb{P}}$. The induced action is compatible with the natural one on $H^0(\mathbb{P}, \mathcal{O}(d))$ under the isomorphism $H^0(X, \mathcal{O}_X(d)) \simeq H^0(\mathbb{P}, \mathcal{O}(d))/\langle F \rangle$.

As g has finite order, any class $v \in H^1(X, \mathcal{T}_X)$ fixed by g can be lifted to a g-invariant section $(1/|g|) \sum g^i s \in H^0(X, \mathcal{O}_X(d))$, where s is an arbitrary pre-image of v. Thus, in order to arrive at a contradiction, it suffices to show that the g-invariant part $H^0(X, \mathcal{O}_X(d))^g$ cannot map onto $H^1(X, \mathcal{T}_X)$. So, it is enough to show that its dimension $h^0(\mathcal{O}_X(d))^g$ satisfies

$$h^0(\mathcal{O}_X(d))^g < h^1(X, \mathcal{T}_X) + \dim(W^g).$$

[8] Thanks to O. Benoist for the reference.

As $h^1(X, \mathcal{T}_X) = h^0(\mathcal{O}_X(d)) - \dim(W)$, this reduces the task to proving

$$\dim(W) - \dim(W^g) < h^0(\mathcal{O}_X(d)) - h^0(\mathcal{O}_X(d))^g. \tag{3.6}$$

Note that the left-hand side of (3.6) equals $\dim(V \otimes V^*) - \dim(V \otimes V^*)^g$ and, as $g(F) = F$, the right-hand side is nothing but $h^0(\mathbb{P}, \mathcal{O}(d)) - h^0(\mathbb{P}, \mathcal{O}(d))^g$. The weak inequality in (3.6) follows from the obvious equality $W^g = W \cap H^0(X, \mathcal{O}_X(d))^g$.

The strict equality follows from purely combinatorial considerations for which we refer to [115]. The idea is to write $V = \bigoplus V_\lambda$ with $V_\lambda := \langle x_i \mid \lambda_i = \lambda \rangle$. Then, the left-hand side is $\dim(W) - \dim(W^g) = (n+2)^2 - \sum \dim(V_\lambda)^2$. To compute the right-hand side, one decomposes $S^d(V) = S^d(\bigoplus V_\lambda)$ and shows that for $d \geq 3$, the dimension of the non-invariant part on the left exceeds the one on the right.[9] □

Exercise 3.17 In order to gain a concrete understanding of the combinatorial part of the proof, consider the situation $V = V_{\lambda_1} \oplus V_{\lambda_2}$, $\lambda_1 \neq \lambda_2$ and $d = 3$. Then $S^3V = S^3V_{\lambda_1} \oplus (S^2V_{\lambda_1} \otimes V_{\lambda_2}) \oplus (V_{\lambda_1} \otimes S^2V_{\lambda_2}) \oplus S^3V_{\lambda_2}$ and $\dim(S^3V)^g$ is maximal when $\lambda_1^3 = \lambda_2^3 = 1$. Show that in this case (3.6) holds, i.e. $\binom{n_1+2}{n_1-1} + \binom{n_2+2}{n_2-1} + (n+2)^2 < \binom{n+4}{n+1} + n_1^2 + n_2^2$, where $n_i = \dim V_{\lambda_i}$ and $n_1 + n_2 = n + 2 > 2$.

Corollary 3.18 *Let X be a smooth complex hypersurface of dimension $n > 0$ and degree $d \geq 3$ with $(n, d) \neq (1, 3)$. Then the action of the group* $\mathrm{Aut}(X)$ *on the middle cohomology $H^n(X, \mathbb{Z})$ is faithful.*

Proof The assertion follows from the proposition by using that the contraction map $H^1(X, \mathcal{T}_X) \longrightarrow \mathrm{End}(H^n(X, \mathbb{C}))$ is equivariant and injective, cf. Corollary 4.25. □

For $(n, d) = (1, 3)$, so plane cubic curves, the subgroup $\mathrm{Aut}(X, \mathcal{O}_X(1))$ still acts faithfully on $H^1(X, \mathbb{Z})$. An alternative proof of the corollary relying on the Lefschetz trace formula, applicable to hypersurfaces in positive characteristic, was worked out by Pan [383] and Javanpeykar and Loughran [262].

Proof of Theorem 3.15 As $H^0(X, \mathcal{T}_X) = 0$, the morphism $\mathbf{Aut} \longrightarrow U$ in (3.3) is unramified. After passing to a dense open subset $V \subset U$, we may assume it to be étale. Fix $[X] \in V$ and assume there exists $\mathrm{id} \neq g \in \mathrm{Aut}(X)$. After base change to an open neighbourhood of $[g] \in \mathbf{Aut}$, considered as an étale open neighbourhood of $[X] \in V$, there exists a relative automorphism $\mathbf{g}\colon \mathcal{X} \xrightarrow{\sim} \mathcal{X}$, so $\pi \circ \mathbf{g} = \mathbf{g}$, with $\mathbf{g}|_X = g$. The base change is suppressed in the notation.

The relative tangent sequence $0 \longrightarrow \mathcal{T}_X \longrightarrow \mathcal{T}_{\mathcal{X}}|_X \longrightarrow T_{[X]}V \otimes \mathcal{O}_X \longrightarrow 0$ induces an exact sequence $H^0(X, \mathcal{T}_{\mathcal{X}}|_X) \longrightarrow T_{[X]}V \twoheadrightarrow H^1(X, \mathcal{T}_X)$. The surjectivity follows from

[9] It is interesting to observe that the argument breaks down at this point for $n = 0$. And, indeed, the automorphism group of a cubic $X \subset \mathbb{P}^1$ is never trivial. For $n = 1$ and $d = 3$, the arguments still show that $\mathrm{Aut}(X, \mathcal{O}_X(1))$ acts faithfully on $H^1(X, \mathcal{T}_X)$.

Proposition 3.12 and all maps are compatible with the action of **g**. However, as π is **g**-invariant, the action on $T_{[X]}V$ is trivial. Therefore, the action of g on $H^1(X, \mathcal{T}_X)$ is trivial as well, which contradicts Proposition 3.16. □

Remark 3.19 Poonen [393] provides equations for smooth hypersurfaces X defined over the prime field k, so $k = \mathbb{F}_p$ or $k = \mathbb{Q}$, such that $\mathrm{Aut}(\bar{X}, \mathcal{O}_{\bar{X}}(1)) = \{\mathrm{id}\}$ for $\bar{X} := X \times_k \bar{k}$. For cubic hypersurfaces, the equations are of the form $c\, x_0^3 + \sum_{i=0}^n x_i\, x_{i+1}^2 + x_{n+1}^3$, where $n > 2$ and $\mathrm{char}(k) \neq 3$. The hard part of this approach is then the verification that the hypersurface given by this equation has really no polarized automorphisms.

Remark 3.20 (i) Assume $n \geq 2$ and $d \geq 3$, as before. Then $|\mathrm{Aut}(X)|$ is universally bounded, i.e. there is a constant $C(d, n)$ such that for all smooth $X \in |\mathcal{O}(d)|$

$$|\mathrm{Aut}(X)| < C(d, n).$$

This follows again from the fact that $\mathrm{Aut}(\mathcal{X}/U, \mathcal{O}_{\mathcal{X}}(1)) \longrightarrow U$ is a finite morphism, see Remark **3**.1.7.

The bound $C(d, n)$ can be made effective. Howard and Sommese [237] show that $C(d, n)$ can be chosen of the form $C(d, n) = C(n) \cdot d^n$. The bound is unlikely to be optimal. See also Remark 4.7.

(ii) A more recent result of González-Aguilera and Liendo [196, Thm. 1.3] determines the possible orders of automorphisms of a smooth hypersurface. For cubic hypersurfaces of dimension at least two, their result says that there exists a smooth cubic hypersurface $X \subset \mathbb{P}^{n+1}$ with an automorphism of order m not divisible by 2 or 3 if and only there exists an $\ell \in \{1, \ldots, n+2\}$ such that $(-2)^\ell \equiv 1\ (m)$.

For example, for an automorphism g of a smooth cubic surface $S \subset \mathbb{P}^3$ the result says that $|g| \leq 5$ if $|g|$ is not divisible by 2 or 3. A complete classification is known in this case, see [158, Table 9.5] or Hosoh's work [236], and shows that $|g| \leq 12$ without any divisibility condition.

In fact, the case of cubic hypersurfaces was treated earlier already. It was observed [195, Thm. 2.6 & Cor. 2.8] that if a prime $p > 3$ can be realized as the order of an automorphism of a smooth cubic hypersurface $X \subset \mathbb{P}^{n+1}$, then $p < 2^{n+1}$.

(iii) For smooth cubic threefolds, maximal groups of automorphisms have been classified by Wei and Yu [490, Thm. 1.1]. There are exactly six of them, including groups like $\mathbb{Z}/3\mathbb{Z} \times \mathfrak{S}_5$ and $\mathrm{PGL}_2(\mathbb{F}_{11})$, see Section **5**.5.4. For smooth cubic fourfolds, an essentially complete classification of groups of symplectic automorphisms, i.e. those acting trivially on $H^{3,1}(X)$, has been obtained by Fu [182] and Laza and Zheng [315]. Also, for a smooth cubic fourfold X, one has $|\mathrm{Aut}(X)| \leq 174\,960 = 2^4 \cdot 3^7 \cdot 5$ and the upper bound is attained by the Fermat cubic.

4 Jacobian Ring

The Jacobian ring is a finite-dimensional quotient of the coordinate ring of a smooth hypersurface obtained by dividing by the partial derivatives of the defining equation. At first glance, it looks like a rather coarse invariant but it turns out to encode the isomorphism type of the hypersurface as an abstract variety. There are purely algebraic aspects of the Jacobian ring as well as Hodge theoretic ones, which shall be explained or at least sketched in this section.

4.1 Hessian and Jacobian We shall assume that the characteristic of k is 0 or, at least, prime to d and $d-1$, where d is the degree d of the hypersurfaces under consideration. The polynomial ring

$$S := k[x_0, \dots, x_{n+1}] \simeq \bigoplus_{i \geq 0} S_i$$

is naturally graded. Here, S_i is the subspace of all homogeneous polynomials of degree i. For $F \in S_d$, we write $\partial_i F \in S_{d-1}$ for the partial derivatives $\partial_i F := \partial F / \partial x_i$. The *Hessian* of F is the matrix of homogeneous polynomials of degree $d-2$

$$H(F) := \left(\frac{\partial^2 F}{\partial x_i \, \partial x_j} \right)_{i,j}.$$

For its determinant one has $\det H(F) \in S_\sigma$, where from now on we use the shorthand

$$\sigma := (n+2) \cdot (d-2). \tag{4.1}$$

The reader may want to compare this number to the much larger degree of the discriminant divisor $\deg D(d,n) = (d-1)^{n+1}(n+2)$. Also, for the case of interest to us, $d = 3$, one simply has

$$\sigma = n + 2.$$

Recall that a polynomial $F \in S_d$ can be recovered from its partial derivatives by means of the Euler equation

$$d \cdot F = \sum_{i=0}^{n+1} x_i \, \partial_i F. \tag{4.2}$$

Definition 4.1 The *Jacobian ideal* of a homogeneous polynomial $F \in S_d$ of degree d is the homogeneous ideal

$$J(F) := (\partial_i F) \subset S = k[x_0, \dots, x_{n+1}]$$

generated by its partial derivatives,. The *Jacobian ring* or *Milnor ring* of F is the quotient

$$S \twoheadrightarrow R(F) := S/J(F),$$

which is naturally graded as well.

An immediate consequence of (4.2) is that the quotient map factors through the coordinate ring of X:

$$S \twoheadrightarrow S/(F) \twoheadrightarrow R(F).$$

If $X \subset \mathbb{P} := \mathbb{P}^{n+1}$ is the hypersurface defined by F, then we shall also write $J(X)$ and $R(X)$ instead of $J(F)$ and $R(F)$. As F is determined by X up to scaling, there is no ambiguity. If F or X are understood, we will abbreviate further to $J = J(F)$ and $R = R(F)$.

As an immediate consequence of Lemma 3.1, one obtains the following result.

Corollary 4.2 *For a smooth hypersurface $X \subset \mathbb{P}$ defined by a homogeneous polynomial F, the Jacobian ring $R(X) = R(F)$ is a zero-dimensional local ring and a finite-dimensional k-algebra.* □

4.2 Gorenstein and Poincaré We next want to show that the Jacobian ring R is Gorenstein with its (one-dimensional) socle in degree $\sigma = (n+2) \cdot (d-2)$. We will also compute the dimensions of its graded pieces.

Proposition 4.3 *Assume that the homogeneous polynomial $F \in S_d$ defines a smooth hypersurface $X \subset \mathbb{P} = \mathbb{P}^{n+1}$. Then the Jacobian ring $R := R(X) = R(F)$ has the following properties:*

(i) *The ring R is an Artinian graded ring with $R_i = 0$ for $i > \sigma$ and $R_\sigma \simeq k$. Moreover, R_σ is generated by the class of* $\det H(F)$.
(ii) *Multiplication defines a perfect pairing*

$$R_i \times R_{\sigma-i} \longrightarrow R_\sigma \simeq k.$$

(iii) *The Poincaré polynomial of R is given by*

$$P(R) := \sum_{i=0}^{\sigma} \dim(R_i)\, t^i = \left(\frac{1-t^{d-1}}{1-t}\right)^{n+2}. \tag{4.3}$$

For $d = 3$, the dimensions of the graded pieces of the Jacobian ring $R(F)$ are simply

$$\dim(R_i) = \binom{n+2}{i}. \tag{4.4}$$

Proof We write $f_i := \partial_i F$. Then, by Lemma 3.1, $f_0, \dots, f_{n+1} \in S$ is a regular sequence of homogeneous polynomials of degree $d-1$. This is in fact all we need for the proof.

Let us begin by recalling basic facts about the *Koszul complex* of a regular sequence $f_0, \dots, f_{n+1} \in S$. As always, $\mathbb{P}^{n+1} = \mathbb{P}(V)$ and so $V^* = \langle x_0, \dots, x_{n+1} \rangle$. Then the Koszul complex is the complex (concentrated in (homological) degree $n+2, \dots, 0$)

$$K_\bullet(f_i)\colon \left(\bigwedge^{n+2} V^* \longrightarrow \cdots \longrightarrow \bigwedge^k V^* \longrightarrow \cdots \longrightarrow \bigwedge^2 V^* \longrightarrow V^* \longrightarrow k \right) \otimes_k S$$

with differentials

$$\partial_p(x_{i_1} \wedge \cdots \wedge x_{i_p}) = \sum (-1)^j f_{i_j} \cdot x_{i_1} \wedge \cdots \wedge \widehat{x_{i_j}} \wedge \cdots \wedge x_{i_p}.$$

Now, for a regular sequence (f_i), the Koszul complex is exact in degree $\neq 0$ with:

$$H_0(K_\bullet(f_i)) \simeq \mathrm{Coker}\,(V^* \otimes S \longrightarrow S) \simeq R := S/(f_i),$$

see [424]. The exactness of the complex $K_\bullet(f_i) \longrightarrow R$ and the fact that the differentials in the Koszul complex are homogeneous of degree $d-1$ shows

$$\begin{aligned} \dim R_i &= \dim(S_i) - (n+2)\dim(S_{i-(d-1)}) \pm \cdots \\ &= \sum_{j=0}^{n+2} (-1)^j \binom{n+2}{j} \dim(S_{i-j(d-1)}). \end{aligned}$$

Of course, $\dim(S_{i-j(d-1)}) = h^0(\mathbb{P}, \mathcal{O}(i - j(d-1))) = \binom{n+1+i-j(d-1)}{i-j(d-1)}$, see (1.2). This in principle allows one to compute the right-hand side.

The argument can be made more explicit by observing that in $K_\bullet(f_i)$ only the differentials depend on the sequence (f_i). Hence, $\dim R_i$ can be computed by choosing a particular sequence, e.g. $f_i = x_i^{d-1}$. In this case, if a monomial $x^I = x_0^{i_0} \cdots x_{n+1}^{i_{n+1}}$ is not contained in the Jacobian ideal $(f_i = x_i^{d-1})$, then all $i_j \leq d-2$ and hence $|I| \leq (n+2)\cdot(d-2) = \sigma$. In other words, $R_i = 0$ for $i > \sigma$, which is not quite so obvious from the above dimension formula. Moreover, if $x^I \notin (f_i)$ for $|I| = \sigma$, then $x^I = \prod x_i^{d-2}$, i.e. R_σ is one-dimensional and generated by the Hessian determinant of $F = \sum x_i^d$. For the computation of the Poincaré polynomial, observe that

$$R\left(\sum x_i^d\right) \simeq k[x_0]/(x_0^{d-1}) \otimes \cdots \otimes k[x_{n+1}]/(x_{n+1}^{d-1})$$

and hence

$$P\left(R\left(\sum x_i^d\right)\right) = P\left(k[x]/(x^{d-1})\right)^{n+2} = (1 + t + \cdots + t^{d-2})^{n+2} = \left(\frac{1 - t^{d-1}}{1-t}\right)^{n+2}.$$

One can also argue without specializing to the case of a Fermat (or any other) hypersurface and without relying on the Koszul complex as follows. For an exact sequence $0 \longrightarrow M^m \longrightarrow \cdots \longrightarrow M^0 \longrightarrow 0$ of graded S-modules, the additivity of the Poincaré polynomial implies $\sum (-1)^j P(M^j) = 0$. Now, define $R^i := S/(f_0, \ldots, f_i)$ and consider the sequences $0 \longrightarrow R^{i-1} \xrightarrow{\cdot f_i} R^{i-1} \longrightarrow R^i \longrightarrow 0$. Since the sequence (f_i) is regular, they are exact. Then

$$P(R^i) = P(R^{i-1}) - t^{d-1}P(R^{i-1}) = (1 - t^{d-1})P(R^{i-1})$$

and by induction,

$$P(R) = (1 - t^{d-1})^{n+2}\, P(S).$$

As $P(S) = 1/(1-t)^{n+2}$, this implies (iii) and, thus, $R_i = 0$ for $i > \sigma$ and $R_\sigma \simeq k$.

Let us next show that the Hessian determinant $\det(\partial_i f_j)$ is not contained in the ideal (f_i) and thus generates R_σ. For this, consider the dual Koszul complex $K^\bullet(f_i) = \mathrm{Hom}_S(K_\bullet(f_i), S)$, which quite generally satisfies the duality

$$H^p(K^\bullet(f_i)) \simeq H_{n+2-p}(K_\bullet(f_i)),$$

see [291, Ch. 4]. So, for a regular sequence (f_i) the dual Koszul complex $K^\bullet(f_i)$ is exact in degree $\neq n+2$ and $H^{n+2}(K^\bullet) \simeq R$. This can also be checked directly, for example, by using the fact that $H^i(K^\bullet(f_i)) \simeq \mathrm{Ext}^i_S(R, S)$. Suppose now that $H := (h_{ij})$ is a matrix of homogeneous polynomials of degree $d-2$ such that $H \cdot (x_j)_j = (f_i)_i$, i.e. $\sum h_{ij}\, x_j = f_i$. Then H induces a morphism of complexes $\bigwedge^\bullet H\colon K_\bullet(f_i) \longrightarrow K_\bullet(x_i)$, the dual of which is a morphism $K^\bullet(x_i) \longrightarrow K^\bullet(f_i)$. The latter induces in degree $n+2$ the map

$$k \simeq S/(x_i) \simeq H^{n+2}(K^\bullet(x_i)) \longrightarrow H^{n+2}(K^\bullet(f_i)) \simeq R,\ 1 \longmapsto \det(H),$$

which can also be interpreted as the map $\eta\colon \mathrm{Ext}^{n+2}_S(k, S) \longrightarrow \mathrm{Ext}^{n+2}_S(R, S)$ induced by the short exact sequence $0 \longrightarrow (x_i)/(f_i) \longrightarrow R \longrightarrow k \longrightarrow 0$. As $(x_i)/(f_i)$ has zero-dimensional support and, thus, $\mathrm{Ext}^{n+1}_S((x_i)/(f_i), S) = 0$, the map η is injective. Therefore, $\det(H) \neq 0$ in R. To relate this to our assertion, observe that the Euler equation (4.2) implies $H(F) \cdot (x_j)_j = (d-1)\,(\partial_i F)_i$ and set $H := (1/(d-1)) \cdot H(F)$.

It remains to prove (ii), i.e. that the pairing defined by multiplication is perfect. Evidence comes from the equation $t^\sigma \cdot P(1/t) = P(t)$ for the Poincaré polynomial computed above. This already shows that $\dim R_i = \dim R_{\sigma-i}$. Thus, to verify that the pairing is non-degenerate, it suffices to prove that for any homogeneous $g \notin (f_i)$ there exists a homogeneous polynomial h with $0 \neq \bar{g} \cdot \bar{h} \in R_\sigma$ or, equivalently, that the degree σ part $(\bar{g})_\sigma$ of the homogeneous ideal $(\bar{g})$ in R is not trivial. Let i be maximal with $(\bar{g})_i \neq 0$ and pick $0 \neq \bar{G} \in (\bar{g})_i$. Suppose $i < \sigma$. Then $G \cdot (x_i) \subset (f_i)$, which induces a non-trivial homomorphism of S-modules $k \longrightarrow R$, $1 \longmapsto \bar{G}$. Hence, $\dim_k \mathrm{Hom}_S(k, R) > 1$, but this is impossible. Indeed, splitting the Koszul complex $K_\bullet(f_i)$ into short exact sequences and using that $\mathrm{Ext}^i_S(k, \bigwedge^p V^* \otimes S) = 0$ for $i < n+2$, one finds a sequence of inclusions

$$\mathrm{Hom}_S(k, R) \hookrightarrow \mathrm{Ext}^1_S(k, \mathrm{Ker}(\partial_0)) \hookrightarrow \quad \cdots \quad \hookrightarrow \mathrm{Ext}^{n+2}_S(k, \textstyle\bigwedge^{n+2} V^* \otimes S) \simeq k.$$

This concludes the proof of (ii) and of the proposition. □

Remark 4.4 Let us add a more analytic argument for the fact that the Hessian determinant generates the socle, cf. [205]. For this, we assume $k = \mathbb{C}$ and define the residue of $g \in \mathbb{C}[x_0, \ldots, x_{n+1}]$ with respect to F as

$$\mathrm{Res}(g) := \left(\frac{1}{2\pi i}\right)^{n+2} \int_\Gamma \frac{g\, dx_0 \wedge \cdots \wedge dx_{n+1}}{f_0 \cdots f_{n+1}},$$

where, as before, $f_i = \partial_i F$ and $\Gamma := \{x \in \mathbb{C}^{n+2} \mid |f_i(x)| = \varepsilon_i\}$ with $0 < \varepsilon_i \ll 1$. Then one checks the following two assertions:

(i) If $g \in (f_i)$, then $\mathrm{Res}(g) = 0$. This follows from Stokes's theorem. Indeed, for example, for $g = h \cdot f_0$ one has

$$(2\pi i)^{n+2}\,\mathrm{Res}(g) = \int_\Gamma \frac{h\, dx_0 \wedge \cdots \wedge dx_{n+1}}{f_1 \cdots f_{n+1}} = \int_{\Gamma_0} d\left(\frac{h}{f_1 \cdots f_{n+1}}\right) \wedge dx_0 \wedge \cdots \wedge dx_{n+1} = 0,$$

as $h/(f_1 \cdots f_{n+1})$ is holomorphic around $\Gamma_0 := \{z \in \mathbb{C}^{n+2} \mid |f_0(z)| < \varepsilon_0,\ |f_{i>0}(z)| = \varepsilon_i\}$.

(ii) The residue of $g = \det H(F)$ is non-zero. More precisely, $\mathrm{Res}(\det H(F)) = \deg(f)$. Here, $f \colon \mathbb{C}^{n+2} \longrightarrow \mathbb{C}^{n+2}$ is the map $x = (x_i) \longmapsto (f_i(x))$, which is of degree $\deg(f) = \dim \mathcal{O}_{\mathbb{C}^{n+2},0}/(f_i)$. Indeed,

$$\left(\frac{1}{2\pi i}\right)^{n+2} \int_\Gamma \frac{\det H(F)\, dx_0 \wedge \cdots \wedge dx_{n+1}}{f_0 \cdots f_{n+1}} = \left(\frac{1}{2\pi i}\right)^{n+2} \int_\Gamma \frac{df_0 \wedge \cdots \wedge df_{n+1}}{f_0 \cdots f_{n+1}}$$

$$= \left(\frac{1}{2\pi i}\right)^{n+2} \int_\Gamma f^*\left(\frac{dz_0}{z_0} \wedge \cdots \wedge \frac{dz_{n+1}}{z_{n+1}}\right) = \deg(f) \cdot \prod_{j=0}^{n+1} \frac{1}{2\pi i} \int_{|z_j| = \varepsilon_j} \frac{dz_j}{z_j}$$

$$= \deg(f).$$

Clearly, (i) and (ii) together imply $\det H(F) \notin J(F)$.

Saito [413] proves the above proposition by reducing the assertion to statements in local duality theory as in [222]. Voisin [474, Ch. 18] deduces the result from global Serre duality on $\mathbb{P}^{n+1}$.

Here is an immediate consequence of the perfectness of the pairing $R_i \times R_{\sigma-i} \longrightarrow R_\sigma$.

Corollary 4.5 *Assume $i + j \leq \sigma$. Then the natural map*

$$R_i \hookrightarrow \mathrm{Hom}(R_j, R_{i+j})$$

induced by multiplication is injective. □

In Remark 4.26 we will explain how the injectivity can be interpreted in more geometric terms.

Remark 4.6 The Jacobian rings of a smooth cubic surface $S \subset \mathbb{P}^3$ and of a smooth cubic threefold $Y \subset \mathbb{P}^4$ enjoy the Lefschetz property. More precisely, Dimca, Gondim, and Ilardin [156, Prop. 2.22] show that for the generic element $x \in R_1(S)$ multiplication defines an isomorphism $x^2\cdot\colon R_1(S) \xrightarrow{\sim} R_3(S)$ and Bricalli, Favale, and Pirola [92, Thm. C] prove that for the generic element $x \in R_1(Y)$ the two multiplication maps $x^3\cdot\colon R_1(Y) \xrightarrow{\sim} R_4(Y)$ and $x\cdot\colon R_2(Y) \xrightarrow{\sim} R_3(Y)$ are isomorphisms.

Remark 4.7 Let $X = V(F) \subset \mathbb{P}^{n+1}$ be a smooth hypersurface of degree d. Then its group of polarized automorphisms $\mathrm{Aut}(X, \mathcal{O}_X(1))$, which according to Corollary 3.9 is essentially always finite, acts on the finite-dimensional Jacobian ring $R = R(X)$. The action is graded and faithful. As a generalization of the Poincaré polynomial $P(R)$ one considers for any $g \in \mathrm{Aut}(X, \mathcal{O}_X(1))$ the polynomial

$$P(R, g) := \sum \mathrm{tr}\,(g \,|\, R_i)\, t^i.$$

Then the Poincaré polynomial is recovered as $P(R) = P(R, \mathrm{id})$. The equation (4.3) has been generalized by Bott–Tate and Orlik–Solomon [375] to

$$P(R, g) = \frac{\det(1 - g\, t^{d-1} \,|\, V)}{\det(1 - g\, t \,|\, V)}, \tag{4.5}$$

where $V^* = S_1 = \langle x_0, \ldots, x_{n+1} \rangle$. This can then be used to see that $|\mathrm{Aut}(X, \mathcal{O}_X(1))|$ is bounded by a function only depending on d and n, see [375, Cor. 2.7], which we have hinted at already in Remark 3.20.

4.3 Mather–Yau and Donagi As a graded version of a result of Mather and Yau [342], Donagi [161] showed that the Jacobian ring of a hypersurface determines the hypersurface up to projective equivalence.

Example 4.8 To motivate Donagi's result, let us discuss the case of smooth cubic curves $E = X \subset \mathbb{P} = \mathbb{P}^2$. The interesting information encoded by the Jacobian ring

$$R = R_0 \oplus R_1 \oplus R_2 \oplus R_3$$

is the perfect pairing $R_1 \times R_2 \longrightarrow R_3 \simeq k$. We shall describe this for a plane cubic in Weierstraß form $y^2 = 4\,x^3 - g_2\,x - g_3$, i.e. with $F = x_1^2 x_2 - 4\,x_0^3 + g_2\,x_0 x_2^2 + g_3\,x_2^3$. The partial derivatives are

$$\partial_0 F = -12\,x_0^2 + g_2\,x_2^2, \quad \partial_1 F = 2\,x_1 x_2, \quad \text{and} \quad \partial_2 F = x_1^2 + 2g_2\,x_0 x_2 + 3g_3\,x_2^2.$$

From this one deduces bases for R_1, R_2, and R_3, namely:

$$R_1 = \langle \bar{x}_0, \bar{x}_1, \bar{x}_2 \rangle, \quad R_2 = \langle \bar{x}_2^2, \bar{x}_0\bar{x}_1, \bar{x}_0\bar{x}_2 \rangle, \quad \text{and} \quad R_3 = \langle \bar{x}_2^3 \rangle.$$

With respect to theses bases, the multiplication $R_1 \times R_2 \longrightarrow R_3$ is described by the matrix

$$\begin{pmatrix} \bar{x}_0\bar{x}_2^2 & \bar{x}_0^2\bar{x}_1 & \bar{x}_0^2\bar{x}_2 \\ \bar{x}_1\bar{x}_2^2 & \bar{x}_0\bar{x}_1^2 & \bar{x}_0\bar{x}_1\bar{x}_2 \\ \bar{x}_2^3 & \bar{x}_0\bar{x}_1\bar{x}_2 & \bar{x}_0\bar{x}_2^2 \end{pmatrix} = \begin{pmatrix} -3g_3/(2g_2) & 0 & g_2/12 \\ 0 & (27g_3^2 - g_2^3)/(6g_2) & 0 \\ 1 & 0 & -3g_3/(2g_2) \end{pmatrix}.$$

Recall that the discriminant of an elliptic curve in Weierstraß form is by definition $\Delta(E) = g_2^3 - 27\, g_3^2$, see Example 2.6, and its j-function is $j(E) = 1\,728\, \frac{g_2^3}{\Delta(E)}$, cf. [223, Sec. IV.4]. Hence, the perfect pairing $R_1 \times R_2 \longrightarrow R_3 \simeq k$ determines $j(E)$ and, therefore (at least for k algebraically closed), the isomorphism type of E. Note that already the determinant

$$\frac{\Delta(E)^2}{72\, g_2^3} = 24 \cdot 1\,728 \cdot g_2^3 \cdot j(E)^{-2}$$

of the above matrix almost remembers the isomorphism type of E.

Proposition 4.9 *Let $X, X' \subset \mathbb{P} = \mathbb{P}^{n+1}$ be two smooth hypersurfaces such that there exists an isomorphism $R(X) \simeq R(X')$ of graded rings. Then the two hypersurfaces are equivalent, i.e. there exists an automorphism $g \in \mathrm{PGL}(n+2)$ of the ambient $\mathbb{P}$ with $g(X) = X'$.*

Proof We follow the proof in [474, Ch. 18]. First, we may assume that the polynomials F and F' defining X and X' are of the same degree $d > 2$. Then the given graded isomorphism $R(F) \xrightarrow{\sim} R(F')$ can be uniquely lifted to an isomorphism $g \colon S \xrightarrow{\sim} S$ with $g(J(F)) = J(F')$ which reduces the proof to the case $g = \mathrm{id}$ and $J(F) = J(F')$.

Next, consider the path $F_t := t \cdot F' + (1-t) \cdot F$ connecting F and F'. On the one hand, we have $J(F_t)_d = J(F)_d$ for essentially all t, which by deriving with respect to t gives $dF_t/dt = F' - F$. The latter is contained in the ideal $(F) + (F') \subset J(F) = J(F')$. On the other hand, the tangent space of the $\mathrm{GL}(n+2)$-orbit at F_t is just $J(F_t)_d = J(F)_d$, which can be seen by computing for $A = (a_{ij}) \in M(n+2, \mathbb{C})$

$$\frac{d}{ds} F_t\left((\mathrm{id} + s \cdot A)x\right)|_{s=0} = \sum_i \partial_i F_t \sum_j a_{ij} x_j.$$

Hence, the path F_t is tangent to all intersecting orbits and, therefore, stays inside the $\mathrm{GL}(n+2)$-orbit through F, cf. [161, Lem. 1.2]. This proves the proposition. □

Remark 4.10 There exist examples of smooth projective varieties X that can be embedded as hypersurfaces $X \hookrightarrow \mathbb{P}$ in non-equivalent ways. For example, the Fermat quartic $X \subset \mathbb{P}^3$ is known to admit exactly three equivalence classes of degree four polarizations [142]. The three Jacobian rings are therefore non-isomorphic. However, for cubics of dimension at least two this does not occur, as the hyperplane bundle is determined by the canonical bundle.

Remark 4.11 In Proposition 4.9 it is enough to assume that there is a ring isomorphism $R(X) \simeq R(X')$, not necessarily graded. Indeed, any ring isomorphism induces a graded isomorphism $\bigoplus \mathfrak{m}_R^i/\mathfrak{m}_R^{i+1} \simeq \bigoplus \mathfrak{m}_{R'}^i/\mathfrak{m}_{R'}^{i+1}$, where $\mathfrak{m}_R \subset R(X)$ and $\mathfrak{m}_{R'} \subset R(X')$ are the maximal ideals. Then use that $R \simeq \bigoplus \mathfrak{m}_R^i/\mathfrak{m}_R^{i+1}$ as graded k-algebras.[10]

For reasons that will become clear later (see Section 4.5), we are interested in a certain subspace of the Jacobian ring $R(X)$ which only takes into account the degrees

$$t(p) := (n - p + 1) \cdot d - (n + 2).$$

Observe that these indices enjoy the symmetry

$$t(p) + t(n - p) = (n + 2) \cdot (d - 2) = \sigma.$$

Therefore, multiplication describes perfect pairings

$$R_{t(p)} \times R_{t(n-p)} \longrightarrow R_\sigma \simeq k.$$

Let us first check for which p one finds a non-trivial $R_{t(p)}$. This is the case if and only if $0 \leq t(p) \leq \sigma = (n + 2) \cdot (d - 2)$, i.e. for

$$\frac{n + 2 - d}{d} \leq p \leq \frac{(n + 1) \cdot (d - 1) - 1}{d}.$$

For $d = 3$, this becomes

$$\frac{n - 1}{3} \leq p \leq \frac{2n + 1}{3}. \tag{4.6}$$

Observe that $t(p) = \sigma$ if and only if $n - p + 1 = (1/d) \cdot (n + 2) \cdot (d - 1)$, which leads to the next result.

Lemma 4.12 *For a given n and d, the following conditions are equivalent:*

(i) $d \mid (n + 2)$.
(ii) *There exists* $p \in \mathbb{Z}$ *with* $t(p) = 0$.
(iii) *There exists* $p \in \mathbb{Z}$ *with* $t(p) = \sigma$.
(iv) *There exists* $p \in \mathbb{Z}$ *with* $t(p) = d$.
(v) *There exists* $p \in \mathbb{Z}$ *with* $\dim R_{t(p)} = 1$.
(vi) $\bigoplus R_{t(p)} \simeq \bigoplus R_{md}$. □

We also record that for $d = 3$, cf. (4.4):

$$\dim(R_{t(p)}) = \binom{n + 2}{3\,(n - p + 1) - (n + 2)} = \binom{n + 2}{2\,n + 1 - 3p}. \tag{4.7}$$

[10] Thanks to J. Rennemo for explaining this to me.

Exercise 4.13 Let $X \subset \mathbb{P}^{n+1}$ be a smooth hypersurface of degree d. Show that

$$\sum \dim(R_{t(p)}) = b_n(X)_{\text{pr}},$$

where $b_n(X)_{\text{pr}}$ was computed in Section 1.3. For $d = 3$, this becomes the mysterious combinatorial formula

$$\sum_p \binom{n+2}{2n+1-3p} = (-1)^n \cdot (2/3) \cdot \left(1 + (-1)^n \cdot 2^{n+1}\right), \tag{4.8}$$

cf. Remark 1.18. A geometric explanation will be given below, see Theorem 4.21.

4.4 Symmetrizer Lemma There is a beautiful technique going back to Donagi [161] that, under certain numerical conditions, allows one to recover the full Jacobian ring $R := R(X)$ from just the mutiplications $R_d \times R_{t(p)} \longrightarrow R_{t(p)+d}$. This is useful as R_d and the various $R_{t(p)}$ can be described geometrically. We start with the geometric description of $R_d(X)$. We recommend [126] for an instructive brief discussion and [474] for a more detailed one. See also [470, Lem. 1.8] for generalizations to cohomology of polyvector fields.[11]

Lemma 4.14 *Let $X \subset \mathbb{P}^{n+1}$ be a smooth hypersurface of degree d. Assume* $\dim(X) > 2$ *or $d \leq 3$. Then there exists a natural isomorphism*

$$R_d(X) \simeq H^1(X, \mathcal{T}_X).$$

Proof This follows from (3.1) in the proof of Corollary 3.5 and $H^1(X, \mathcal{T}_{\mathbb{P}}|_X) = 0$, which holds under the present assumptions. The case $n = 1, d = 3$ needs an additional argument which is left to the reader. □

Example 4.15 Observe that the isomorphism confirms the dimension formula (2.2):

$$\dim(\mathcal{M}_{3,n}) = \dim H^1(X, \mathcal{T}_X) = \dim R_3(X) = \binom{n+2}{3},$$

where $\mathcal{M}_{3,n}$ is the Deligne–Mumford moduli stack of all smooth cubic hypersurfaces, cf. Remark **3**.1.16.

Before turning to the geometric interpretation of the spaces $R_{t(p)}(X)$, we present the following purely algebraic result. It was proved by Donagi [161] for generic polynomials and by Donagi–Green [162] in general.

Proposition 4.16 (Symmetrizer lemma) *Assume the following inequalities:*

$$\text{(i) } i < j, \text{ (ii) } i + j \leq \sigma - 1, \textit{ and } \text{(iii) } d + j \leq \sigma + 3.$$

[11] With thanks to P. Belmans for the reference.

Then the image of the injection

$$R_{j-i} \hookrightarrow \mathrm{Hom}(R_i, R_j),$$

see Corollary 4.5, is the subspace of all linear maps $\varphi\colon R_i \longrightarrow R_j$ *such that for all* $g, h \in R_i$ *one has* $g \cdot \varphi(h) = h \cdot \varphi(g) \in R_{i+j}$.

Proof The subspace described by the symmetry condition is the kernel of

$$\mathrm{Hom}(R_i, R_j) \longrightarrow \mathrm{Hom}(\textstyle\bigwedge^2 R_i, R_{i+j}), \qquad \varphi \longmapsto (g \wedge h \longmapsto g \cdot \varphi(h) - h \cdot \varphi(g)).$$

As R_{i-j} is obviously contained in it, one has to prove the exactness of the sequence

$$R_{j-i} \longrightarrow \mathrm{Hom}(R_i, R_j) \longrightarrow \mathrm{Hom}(\textstyle\bigwedge^2 R_i, R_{i+j}).$$

This is done by comparing it to a certain Koszul complex on $\mathbb{P}^{n+1}$. See [474, Prop. 18.21] for details.

Note that for cubic hypersurfaces, the discussion below makes use of the symmetrizer lemma only for $i = 1, 2$, but the proof does not seem to be any easier in these cases. □

To recover large portions of $R(X)$, the proposition is applied repeatedly. Suppose $R_i \times R_j \longrightarrow R_{i+j}$ is known. Then one recovers $R_i \times R_{j-i} \longrightarrow R_j$, for which in addition (ii) and (iii) still hold. However, it may happen that (i) ceases to hold, i.e. that $i \geq j - i$, but this can be remedied by swapping the factors, which does not effect the conditions (ii) and (iii). The procedure stops at some $R_\ell \times R_\ell \longrightarrow R_\ell$ and a moment's thought reveals that $\ell = \gcd(i, j)$. Applied to $i = d$ and $j = t(p)$ (or, if necessary, with reversed order), this procedure eventually implies the following result, see Remark 4.19 which makes the procedure more explicit for $d = 3$.

Proposition 4.17 *Assume* $(2n + 1)/n \leq d$. *Fix* p *such that* $0 < t(p) \leq \sigma - d - 1$ *and let* $\ell := \gcd(d, n + 2) = \gcd(d, t(p))$. *Then the multiplication*

$$R_d \times R_{t(p)} \longrightarrow R_{d+t(p)} = R_{t(p-1)}$$

determines the multiplication $R_\ell \times R_\ell \longrightarrow R_{2\ell}$. □

Note that in general ℓ is not of the form $t(p)$.

The next corollary is a special case of a more general result, which beyond the cubic case is known for all smooth hypersurfaces except when $(d, n) = (4, 4m)$ or $d \mid (n + 2)$, see [161, 162, 474]. The argument is easier for cubic hypersurfaces and so we restrict to this case.

Corollary 4.18 *Assume* $X = V(F) \subset \mathbb{P}^{n+1}$ *is a smooth cubic hypersurface of dimension* $n > 2$ *with* $3 \nmid (n + 2)$. *Then there exist integers* p *with* $0 < t(p) \leq n - 2$ *and for each such* p *the multiplication*

$$R_3 \times R_{t(p)} \longrightarrow R_{3+t(p)}$$

determines the graded algebra $R = R(X)$, *and hence by Proposition 4.9 also* X, *uniquely.*

Proof For $d = 3$, the condition $0 < t(p) \leq \sigma - d - 1$ in Proposition 4.17 turns into, cf. (4.6):

$$n + 3 \leq 3p < 2n + 1, \tag{4.9}$$

which has integral solutions for all $n > 2$. For any such p, multiplication $R_3(X) \times R_{t(p)} \longrightarrow R_{3+t(p)}$ determines $R_1 \times R_1 \longrightarrow R_2$, as $3 \nmid (n+2)$.

More precisely, if for two smooth cubic hypersurfaces $X, X' \subset \mathbb{P}^{n+1}$ and their Jacobian rings R and R' there exists an isomorphism $[R_3 \times R_{t(p)} \longrightarrow R_{3+t(p)}] \simeq [R'_3 \times R'_{t(p)} \longrightarrow R'_{3+t(p)}]$, then one also has an isomorphism $[R_1 \times R_1 \longrightarrow R_2] \simeq [R'_1 \times R'_1 \longrightarrow R'_2]$. In particular, the isomorphism $R_1 \simeq R'_1$ corresponds to a linear coordinate change $g\colon \langle x_0, \ldots, x_{n+1} \rangle \xrightarrow{\sim} \langle x_0, \ldots, x_{n+1} \rangle$ and the compatibility with the multiplication can be interpreted as an isomorphism

$$\left[k[x_0, \ldots, x_{n+1}]_2 \simeq S^2(R_1) \longrightarrow R_2\right] \simeq \left[k[x_0, \ldots, x_{n+1}]_2 \simeq S^2(R'_1) \longrightarrow R'_2\right].$$

Hence, g identifies their kernels, which are spanned by the partial derivatives $\partial_i F$ and $\partial_i F'$ of the defining equations. Thus, g induces a ring isomorphism $k[x_0, \ldots, x_{n+1}] \xrightarrow{\sim} k[x_0, \ldots, x_{n+1}]$ that restricts to $J(X) \xrightarrow{\sim} J(X')$ and, hence, $R(X) \simeq R(X')$. □

Remark 4.19 For $d = 3$ and $3 \nmid (n+2)$, as in Corollary 4.18, there always exists a p with $t(p) = 1$ or $t(p) = 2$. The result covers, for example, cubics of dimension $n = 2, 3, 5, 6, 8, 9$, for which we list the admissible $t(p)$:

n	$0 < t(p) \leq \sigma - 4 = n - 2$
3	$\mathbf{t(2) = 1}$
5	$t(2) = 5$, $\mathbf{t(3) = 2}$
6	$t(3) = 4$, $\mathbf{t(4) = 1}$
8	$t(3) = 8$, $t(4) = 5$, $\mathbf{t(5) = 2}$
9	$t(4) = 7$, $t(5) = 4$, $\mathbf{t(6) = 1}$

Let us spell out how to recover the multiplication $R_1 \times R_1 \longrightarrow R_2$ from the multiplication $R_3 \times R_{t(p)} \longrightarrow R_{t(p)+3}$ or, equivalently, the map $R_3 \longrightarrow \mathrm{Hom}(R_{t(p)}, R_{t(p)+3})$ for the case $t(p) = 1$. We leave the case $t(p) = 2$ as an exercise. Given $R_3 \times R_1 \longrightarrow R_4$ allows one to write down the condition $g \cdot \varphi(h) = h \cdot \varphi(g)$ in R_4, where $g, h \in R_1$ and $\varphi \in \mathrm{Hom}(R_1, R_3)$. Since the assumptions of the symmetrizer lemma Proposition 4.16 are met, the subspace of all such φ is the image of the injection $R_2 \hookrightarrow \mathrm{Hom}(R_1, R_3)$, i.e. $R_2 \times R_1 \longrightarrow R_3$ is recovered. Applying the same procedure again, one reconstructs the multiplication $R_1 \times R_1 \longrightarrow R_2$.

Remark 4.20 The result is sharp. For example, for $n = 4$, the only $0 \leq t(p) \leq \sigma = 6$ are $t(1) = 6$, $t(2) = 3$, and $t(3) = 0$. But the pairing $R_3 \times R_3 \longrightarrow R_6 \simeq k$ certainly does not determine the cubic nor does the multiplication by scalars $R_3 \times R_0 \longrightarrow R_3$.

4.5 Infinitesimal and Variational Torelli Theorem The next step is to describe the parts $R_{t(p)}(X)$ geometrically. This is the following celebrated result of Carlson and Griffiths [103].

Theorem 4.21 (Carlson–Griffiths) *Let $X \subset \mathbb{P}^{n+1}$ be a smooth hypersurface of degree d. Assume $n > 2$ or $d \leq 3$. Then for all integers p there exists an isomorphism*

$$H^{p,n-p}(X)_{\rm pr} \simeq R_{t(p)}(X), \tag{4.10}$$

with $t(p) = (n - p + 1) \cdot d - (n + 2)$, compatible with the natural pairings on both sides, i.e. there exist commutative diagrams

$$\begin{array}{ccccc} H^{p,n-p}(X)_{\rm pr} & \times & H^{n-p,p}(X)_{\rm pr} & \longrightarrow & H^{n,n}(X)_{\rm pr} \\ \downarrow \simeq & & \downarrow \simeq & & \downarrow \simeq \\ R_{t(p)}(X) & \times & R_{t(n-p)}(X) & \longrightarrow & R_\sigma(X). \end{array} \tag{4.11}$$

Moreover, using the isomorphism $H^1(X, \mathcal{T}_X) \simeq R_d(X)$, cf. Lemma 4.14, *and the pairing $\mathcal{T}_X \times \Omega_X^p \longrightarrow \Omega_X^{p-1}$, one obtains commutative diagrams*

$$\begin{array}{ccccc} H^1(X, \mathcal{T}_X) & \times & H^{p,n-p}(X)_{\rm pr} & \longrightarrow & H^{p-1,n-p+1}(X)_{\rm pr} \\ \downarrow \simeq & & \downarrow \simeq & & \downarrow \simeq \\ R_d(X) & \times & R_{t(p)}(X) & \longrightarrow & R_{t(p-1)}(X). \end{array} \tag{4.12}$$

The proof of the theorem is involved and we will not attempt to present it in full. However, we will outline the most important parts of the general theory that enter the proof and, in particular, explain how to establish a link between the Jacobian ring and the primitive cohomology at all. As we will restrict to the case of hypersurfaces in $\mathbb{P}^{n+1}$ throughout, certain aspects simplify. We refer to [104] and [474] for more details and some of the crucial computations. In Section **3**.3 the results will be interpreted more geometrically in terms of moduli spaces.

(i) The de Rham complex of a (smooth) k-variety X of dimension n is the complex

$$\Omega_X^\bullet : \mathcal{O}_X \longrightarrow \Omega_X \longrightarrow \Omega_X^2 \longrightarrow \cdots \longrightarrow \Omega_X^n.$$

The sheaves $\Omega_X^i := \bigwedge^i \Omega_X$ are coherent (here in the Zariski topology), but the differentials $d \colon \Omega_X^i \longrightarrow \Omega_X^{i+1}$ are only k-linear. The *de Rham cohomology* of X is then defined as the hypercohomology of this complex:

$$H^*_{\mathrm{dR}}(X/k) := \mathbb{H}^*(X, \Omega^\bullet_X),$$

which can be computed via the Hodge–de Rham spectral sequence

$$E_1^{p,q} = H^q(X, \Omega^p_X) \Rightarrow H^{p+q}_{\mathrm{dR}}(X/k). \tag{4.13}$$

Note that the E_1-terms are just cohomology groups of coherent sheaves. The spectral sequence is associated with the *Hodge filtration*, which is induced by the complexes

$$F^p\Omega^\bullet_X : \Omega^p_X \longrightarrow \cdots \longrightarrow \Omega^n_X$$

concentrated in degrees $p, \ldots, n$ and the natural morphism $F^p\Omega^\bullet_X \longrightarrow \Omega^\bullet_X$. Then one defines

$$F^p H^*_{\mathrm{dR}}(X/k) := \mathrm{Im}\left(\mathbb{H}^*(F^p\Omega^\bullet_X) \longrightarrow \mathbb{H}^*(\Omega^\bullet_X) = H^*_{\mathrm{dR}}(X/k)\right). \tag{4.14}$$

Remark 4.22 If X is smooth and projective over a field k satisfying $\mathrm{char}(k) = 0$ or $\mathrm{char}(k) = p > \dim(X)$ and X is liftable to $W_2(k)$, then (4.13) degenerates [147]. This applies to smooth hypersurfaces $X \subset \mathbb{P}^{n+1}$, for which the assumption on the characteristic of k can be avoided, cf. Section 1.6. Once (4.13) is known to degenerate, the map in (4.14) is injective.

(ii) For open varieties, the Hodge–de Rham spectral sequence does not necessarily degenerate, but a replacement is available. For example, let us consider the open complement $j\colon U := \mathbb{P} \setminus X \hookrightarrow \mathbb{P} = \mathbb{P}^{n+1}$ of a smooth hypersurface $X \subset \mathbb{P}$. There are quasi-isomorphisms (see the discussion following (4.22) below)

$$\Omega^\bullet_{\mathbb{P}}(\log(X)) \xrightarrow{\sim} \Omega^\bullet_{\mathbb{P}}(*X) = j_*\Omega^\bullet_U. \tag{4.15}$$

Here, $\Omega^p_{\mathbb{P}}(*X) := j_*\Omega^p_U$ (in the Zariski topology!) is the sheaf of meromorphic p-forms on $\mathbb{P}$ with poles (of arbitrary order) along X. Furthermore, $\Omega^1_{\mathbb{P}}(\log(X)) \subset \Omega^1_{\mathbb{P}}(*X)$ is defined as the subsheaf locally generated by $d\log(f) = \frac{df}{f}$, where f is the local equation for X, and $\Omega^p_{\mathbb{P}}(\log(X)) := \bigwedge^p \left(\Omega^1_X(\log(X))\right)$. In our case, $X = V(F)$ and

$$\Omega^1_{\mathbb{P}}(\log(X))|_{U_j} = d\log(F_j)\,\mathcal{O}_{U_j} = \frac{dF_j}{F_j}\,\mathcal{O}_{U_j}$$

on the standard open subset $U_j := \mathbb{P} \setminus V(x_j)$ with $F_j := F(x_0/x_j, \ldots, x_{n+1}/x_j)$. The differentials in both complexes in (4.15) are the usual ones. By construction, $\Omega^\bullet_{\mathbb{P}}(\log(X))$ is the subcomplex of forms α with α and $d\alpha$ having at most simple poles along X. The Hodge filtration in the open case is defined by

$$F^p\Omega^\bullet_{\mathbb{P}}(\log(X)) : \Omega^p_{\mathbb{P}}(\log(X)) \longrightarrow \cdots \longrightarrow \Omega^{n+1}_{\mathbb{P}}(\log(X))$$

(and not as the direct image of $F^p\Omega_U^\bullet$) in degrees $p, \dots, n+1$. It induces the spectral sequence

$$E_1^{p,q} = H^q(\mathbb{P}, \Omega_{\mathbb{P}}^p(\log(X))) \Rightarrow \mathbb{H}^{p+q}(\mathbb{P}, \Omega_{\mathbb{P}}^\bullet(\log(X))), \tag{4.16}$$

where the right-hand side is isomorphic to

$$\mathbb{H}^\bullet(\mathbb{P}, \Omega_{\mathbb{P}}^\bullet(\log(X))) \simeq \mathbb{H}^*(\mathbb{P}, \Omega_{\mathbb{P}}^\bullet(*X)) \simeq H_{\mathrm{dR}}^*(U/k).$$

Again by [147], this spectral sequence degenerates under the assumptions of Remark 4.22 and so in particular for smooth hypersurfaces.

Observe that the residue map

$$\mathrm{res}\colon \frac{df}{f}h \longmapsto h|_X$$

leads to a short exact sequence $0 \longrightarrow \Omega_{\mathbb{P}}^1 \longrightarrow \Omega_{\mathbb{P}}^1(\log(X)) \longrightarrow i_*\mathcal{O}_X \longrightarrow 0$. Taking exterior powers, one obtains an exact sequence of complexes

$$\begin{array}{c} 0 \longrightarrow \Omega_{\mathbb{P}}^\bullet \longrightarrow \Omega_{\mathbb{P}}^\bullet(\log(X)) \xrightarrow{\mathrm{res}} (i_*\Omega_X^\bullet)[-1] \longrightarrow 0 \\ \simeq \Omega_{\mathbb{P}}^\bullet(*X) \end{array}$$

with $\mathrm{res}\left(\frac{df}{f} \wedge \alpha\right) = \alpha|_X$. Additionally, the sequence is compatible with the Hodge filtrations of all three complexes, which provides us with exact sequences

$$0 \longrightarrow F^p\Omega_{\mathbb{P}}^\bullet \longrightarrow F^p\Omega_{\mathbb{P}}^\bullet(\log(X)) \xrightarrow{\mathrm{res}} (i_*F^{p-1}\Omega_X^\bullet)[-1] \longrightarrow 0.$$

The induced long exact cohomology sequences read

$$\cdots \to H_{\mathrm{dR}}^i(\mathbb{P}/k) \to H_{\mathrm{dR}}^i(U/k) \to H_{\mathrm{dR}}^{i-1}(X/k) \to H_{\mathrm{dR}}^{i+1}(\mathbb{P}/k) \to \cdots \tag{4.17}$$

and

$$\cdots \to F^pH_{\mathrm{dR}}^i(\mathbb{P}/k) \to F^pH_{\mathrm{dR}}^i(U/k) \to F^{p-1}H_{\mathrm{dR}}^{i-1}(X/k) \to F^pH_{\mathrm{dR}}^{i+1}(\mathbb{P}/k) \to . \tag{4.18}$$

Note that the Hodge filtration $F^pH_{\mathrm{dR}}^*(U/k)$ is defined as the image of

$$\mathbb{H}^*(\mathbb{P}, F^p\Omega_{\mathbb{P}}^\bullet(\log(X))) \longrightarrow \mathbb{H}^*(\mathbb{P}, \Omega_{\mathbb{P}}^\bullet(\log(X))) \simeq H_{\mathrm{dR}}^*(U/k) \tag{4.19}$$

and not via the Hodge filtration of $\Omega_U^\bullet$. The exactness of (4.18) relies on the injectivity of (4.14) and (4.19), which is equivalent to the degeneration of the spectral sequences (4.13) and (4.16).

If X is defined over $k = \mathbb{C}$, there is an analytic version of the above for the associated complex manifold X^{an}. The de Rham complex $\Omega_{X^{\mathrm{an}}}^\bullet$ is defined similarly (now in the analytic topology) and so is the Hodge–de Rham spectral sequence

$$E_1^{p,q} = H^q(X^{\mathrm{an}}, \Omega_{X^{\mathrm{an}}}^p) \Rightarrow H_{\mathrm{dR}}^{p+q}(X^{\mathrm{an}}). \tag{4.20}$$

The Poincaré lemma shows that in the analytic topology the inclusion $\mathbb{C} \hookrightarrow \mathcal{O}_{X^{\mathrm{an}}}$ leads to a quasi-isomorphism $\mathbb{C} \xrightarrow{\sim} \Omega^\bullet_{X^{\mathrm{an}}}$ and hence an isomorphism

$$H^*(X^{\mathrm{an}}, \mathbb{C}) \xrightarrow{\sim} H^*_{\mathrm{dR}}(X^{\mathrm{an}}) = \mathbb{H}^*(\Omega^\bullet_{X^{\mathrm{an}}}).$$

The natural morphism $X^{\mathrm{an}} \longrightarrow X$ of ringed spaces provides a comparison map from the algebraic to the analytic de Rham cohomology. For X smooth and projective, GAGA shows that $H^q(X, \Omega^p_X) \xrightarrow{\sim} H^q(X^{\mathrm{an}}, \Omega^p_{X^{\mathrm{an}}})$. Hence, the left-hand sides of (4.13) and (4.20) coincide and, therefore, also the right-hand sides do, i.e. there exists an isomorphism

$$H^*_{\mathrm{dR}}(X/\mathbb{C}) \xrightarrow{\sim} H^*_{\mathrm{dR}}(X^{\mathrm{an}}), \tag{4.21}$$

which is compatible with the Hodge filtration. In fact, (4.21) continues to hold for arbitrary smooth varieties without any projectivity assumption, see [213, Thm. 1'].

Also, the open case can be cast in the analytic setting, where (4.15) is replaced by

$$\Omega^\bullet_{\mathbb{P}^{\mathrm{an}}}(\log(X^{\mathrm{an}})) \xrightarrow{\sim} \Omega^\bullet_{\mathbb{P}^{\mathrm{an}}}(*X^{\mathrm{an}}) \xrightarrow{\sim} j_*\mathcal{A}^\bullet_U \xrightarrow{\sim} Rj_*\mathbb{C}_U. \tag{4.22}$$

Here, the complex $\mathcal{A}^\bullet_U$ is the standard C^∞-de Rham complex.

The verification of the quasi-isomorphisms in (4.22), and, similarly in (4.15), is readily reduced to the case of $U = \mathbb{C} \setminus \{0\} \hookrightarrow \mathbb{C}$. In this case,

$$\begin{array}{rcccccc} \Omega^\bullet_{\mathbb{C}}(\log(\{0\}))(\mathbb{C}): & & \mathcal{O}_{\mathbb{C}} & \longrightarrow & \frac{dz}{z}\mathcal{O}_{\mathbb{C}}, & & \\ \Omega^\bullet_{\mathbb{C}}(*\{0\})(\mathbb{C}): & & \sum z^n\,\mathcal{O}_{\mathbb{C}} & \longrightarrow & \sum z^n dz\,\mathcal{O}_{\mathbb{C}}, & & \\ j_*\mathcal{A}^\bullet(\mathbb{C}): & & C^\infty_U & \longrightarrow & dx\,C^\infty_U + dy\,C^\infty_U & \longrightarrow & (dx \wedge dy)\,C^\infty_U. \end{array}$$

The cohomology of all three satisfies $H^i \simeq \mathbb{C}$ for $i = 0, 1$ and $H^i = 0$ otherwise.

Also, there is an analytic version of (4.16) and there exists a natural isomorphism

$$\mathbb{H}^*(\mathbb{P}, \Omega^\bullet_{\mathbb{P}}(\log(X))) \simeq \mathbb{H}^*(\mathbb{P}^{\mathrm{an}}, \Omega^\bullet_{\mathbb{P}^{\mathrm{an}}}(\log(X^{\mathrm{an}}))).$$

To simplify the notation, we will from now on also write X, U, and $\mathbb{P}$ for the associated analytic varieties. Then the exact sequence (4.17) becomes the classical Gysin sequence

$$\cdots \longrightarrow H^i(\mathbb{P}, \mathbb{C}) \longrightarrow H^i(U, \mathbb{C}) \longrightarrow H^{i-1}(X, \mathbb{C}) \longrightarrow H^{i+1}(\mathbb{P}, \mathbb{C}) \longrightarrow \cdots.$$

This is interesting only for $i - 1 = n$. As the map $H^k(X, \mathbb{C}) \longrightarrow H^{k+2}(\mathbb{P}, \mathbb{C})$ is surjective for $k = n - 1$ and $k = n$, simply because $H^k(\mathbb{P}) \longrightarrow H^k(X) \longrightarrow H^{k+2}(\mathbb{P})$ is multiplication with $[X] \in H^2(\mathbb{P})$, one finds

$$\begin{array}{ccc} H^{n+1}(U, \mathbb{C}) & \xrightarrow{\quad\sim\quad} & H^n(X, \mathbb{C})_{\mathrm{pr}} \\ \cup & & \cup \\ F^p H^{n+1}(U, \mathbb{C}) & \xrightarrow{\sim} & F^{p-1} H^n(X, \mathbb{C})_{\mathrm{pr}}. \end{array}$$

However, the Hodge filtration is difficult to compute and it is preferable to replace it by the *pole filtration* $F^\bullet_{\rm pol}$ of the complex $\Omega^\bullet_{\mathbb{P}}(*X)$. Under the quasi-isomorphisms (4.15) and (4.22) the two compare as follows:

$$\begin{array}{ccccccccc} F^p\Omega^\bullet_{\mathbb{P}}(\log(X)) & : & \Omega^p_{\mathbb{P}}(\log(X)) & \longrightarrow & \Omega^{p+1}_{\mathbb{P}}(\log(X)) & \longrightarrow & \cdots & \longrightarrow & \Omega^{n+1}_{\mathbb{P}}(\log(X)) \\ \cap & & \cap & & \cap & & & & \cap \\ F^p_{\rm pol}\Omega^\bullet_{\mathbb{P}}(*X) & : & \Omega^p_{\mathbb{P}}(X) & \longrightarrow & \Omega^{p+1}_{\mathbb{P}}(2X) & \longrightarrow & \cdots & \longrightarrow & \Omega^{n+1}_{\mathbb{P}}((n-p+2)X). \end{array}$$

This is usually not a quasi-isomorphism. However, using that $H^*(\mathbb{P},\mathbb{C})_{\rm pr}=0$ and applying Bott vanishing, see Section 1.2, one finds

$$F^p_{\rm pol}H^{n+1}(U,\mathbb{C})\simeq F^pH^{n+1}(U,\mathbb{C}).$$

The advantage of using the pole filtration stems from the following result.

Lemma 4.23 (Griffiths) *Let $X\subset\mathbb{P}=\mathbb{P}^{n+1}$ be a smooth hypersurface. Then $F^pH^n(X,\mathbb{C})_{\rm pr}\simeq F^{p+1}H^{n+1}(U,\mathbb{C})$ is isomorphic to*

$$F^{p+1}_{\rm pol}H^{n+1}(U,\mathbb{C})\simeq\frac{H^0(\mathbb{P},\Omega^{n+1}_{\mathbb{P}}((n-p+1)X))}{dH^0(\mathbb{P},\Omega^n_{\mathbb{P}}((n-p)X))}\tag{4.23}$$

and $H^{p,n-p}(X)_{\rm pr}\simeq F^{p+1}H^{n+1}(U,\mathbb{C})/F^{p+2}H^{n+1}(U,\mathbb{C})$ is isomorphic to

$$\frac{F^{p+1}_{\rm pol}H^{n+1}(U,\mathbb{C})}{F^{p+2}_{\rm pol}H^{n+1}(U,\mathbb{C})}\simeq\frac{H^0(\mathbb{P},\Omega^{n+1}_{\mathbb{P}}((n-p+1)X))}{H^0(\mathbb{P},\Omega^{n+1}_{\mathbb{P}}((n-p)X))+dH^0(\mathbb{P},\Omega^n_{\mathbb{P}}((n-p)X))}.\tag{4.24}$$

Proof By definition, the left-hand side in (4.23) is the image of the map

$$\mathbb{H}^{n+1}(\mathbb{P},F^{p+1}_{\rm pol}\Omega^\bullet_{\mathbb{P}}(*X))\longrightarrow\mathbb{H}^{n+1}(\mathbb{P},\Omega^\bullet_{\mathbb{P}}(*X)).$$

The natural map $\Omega^{n+1}_{\mathbb{P}}((n-p+1)X)[-(n+1)]\longrightarrow F^{p+1}_{\rm pol}\Omega^\bullet_{\mathbb{P}}(*X)$ induces

$$H^0(\mathbb{P},\Omega^{n+1}_{\mathbb{P}}((n-p+1)X))\longrightarrow F^{p+1}_{\rm pol}\mathbb{H}^{n+1}(\mathbb{P},\Omega^\bullet_{\mathbb{P}}(*X))\simeq F^{p+1}_{\rm pol}H^{n+1}(U,\mathbb{C}).$$

It is rather straightforward to show that the map is surjective and that its kernel is the image of $d\colon H^0(\mathbb{P},\Omega^n_{\mathbb{P}}((n-p)X))\longrightarrow H^0(\mathbb{P},\Omega^{n+1}_{\mathbb{P}}((n-p+1)X))$. The isomorphism in (4.24) follows. □

Let us come back to the discussion of Theorem 4.21. Observe that

$$\begin{aligned}H^0(\mathbb{P},\Omega^{n+1}_{\mathbb{P}}((n-p+1)X))&\simeq H^0(\mathbb{P},\mathcal{O}((n-p+1)\cdot d-(n+2)))\\&\simeq H^0(\mathbb{P},\mathcal{O}(t(p))).\end{aligned}$$

Thus, in order to prove (4.10), it suffices to show that the image of

$$(\partial_iF)\colon H^0(\mathbb{P},\mathcal{O}((n-p)\cdot d-(n+1)))^{\oplus n+2}\longrightarrow H^0(\mathbb{P},\mathcal{O}(t(p)))$$

equals $H^0(\mathbb{P}, \Omega_{\mathbb{P}}^{n+1}((n-p)X)) + dH^0(\mathbb{P}, \Omega_{\mathbb{P}}^n((n-p)X))$. This is a rather unpleasant computation in terms of rational differential forms on $\mathbb{P}$ and $\mathbb{C}^{n+2}$. We omit this here and refer to [474, Thm. 18.10] or [104, Ch. 3.2].[12]

To prove the commutativity of (4.11), one first needs to fix an appropriate isomorphism $H^{n,n}(X)_{\rm pr} \simeq R_\sigma(X)$ which again involves rational forms. Also, the commutativity of (4.12) is not straightforward. One has to argue that the multiplication

$$H^0(\mathbb{P}, \mathcal{O}(d)) \times H^0(\mathbb{P}, \mathcal{O}(t(p))) \longrightarrow H^0(\mathbb{P}, \mathcal{O}(t(p-1)))$$

is related to the contraction

$$H^1(X, \mathcal{T}_X) \times H^{n-p}(X, \Omega_X^p) \longrightarrow H^{n-p+1}(X, \Omega_X^{p-1})$$

via the surjection $H^0(\mathbb{P}, \mathcal{O}(d)) \longrightarrow H^0(X, \mathcal{O}_X(d)) \longrightarrow H^1(X, \mathcal{T}_X)$. Note that the multiplication takes place in the top degree $n+1$ of $F_{\rm pol}^{p+1}\Omega_{\mathbb{P}}^\bullet(*X)$, whereas the contraction applies to the lowest degree (from degree p to $p-1$).

This finishes our discussion of the main ideas that go into the proof of Theorem 4.21.

Example 4.24 For later use, we spell out the example of a smooth cubic threefold $Y \subset \mathbb{P}^4$ and its complement $U := \mathbb{P}^4 \setminus Y$. Then the residue defines an isomorphism $H^4(U, \mathbb{C}) \simeq H^3(Y, \mathbb{C})$ which restricts to $F^3H^4(U, \mathbb{C}) \simeq F^2H^3(Y, \mathbb{C}) = H^{2,1}(Y)$. According to the lemma, the space is naturally identified with the five-dimensional space

$$H^{2,1}(Y) \simeq H^0(\mathbb{P}^4, \Omega_{\mathbb{P}^4}^4((3-2+1)\cdot 3)) \simeq H^0(\mathbb{P}^4, \mathcal{O}(1)).$$

We conclude this section by a result that will later be used to prove that the period map is unramified, cf. Section **3**.3.2. We state the result for cubic hypersurfaces only.

Corollary 4.25 (Infinitesimal Torelli theorem) *Let $X \subset \mathbb{P}^{n+1}$ be a smooth cubic hypersurface of dimension $n > 2$. Then the contraction $\mathcal{T}_X \times \Omega_X^p \longrightarrow \Omega_X^{p-1}$ defines an injection*

$$H^1(X, \mathcal{T}_X) \hookrightarrow \mathrm{Hom}\left(\bigoplus_{p+q=n} H^{p,q}(X)_{\rm pr}, \bigoplus_{p+q=n} H^{p-1,q+1}(X)_{\rm pr}\right). \tag{4.25}$$

Proof There exists a p such that $0 \leq t(p) \leq \sigma - 3$. Then, by Corollary 4.5, multiplication in the Jacobian ring $R(X)$ gives an injection

$$R_3 \hookrightarrow \mathrm{Hom}(R_{t(p)}, R_{t(p)+3}).$$

Conclude by using $R_{t(p)} \simeq H^{p,n-p}(X)_{\rm pr}$ and the compatibility of the multiplication in $R(X)$ with the contraction map, cf. Theorem 4.21. □

[12] One could try an alternative argument: apply $\bigwedge^{n+1}$ to the Euler sequence to obtain $0 \longrightarrow \mathcal{O}(-(n+2)) \longrightarrow \mathcal{O}(-(n+1))^{\oplus n+2} \longrightarrow \Omega_{\mathbb{P}}^n \longrightarrow 0$. Tensor with $\mathcal{O}((n-p)\cdot d)$ and consider the composition $H^0(\mathbb{P}, \mathcal{O}((n-p)\cdot d - (n+1)))^{\oplus n+2} \twoheadrightarrow H^0(\mathbb{P}, \Omega_{\mathbb{P}}^n((n-p)\cdot d)) \xrightarrow{d} H^0(\mathbb{P}, \Omega_{\mathbb{P}}^{n+1}((n-p+1)\cdot d))$, which should be compared to the map given by $(\partial_i F)$. However, this attempt becomes quickly as technical as the standard approach.

Remark 4.26 If $X \subset \mathbb{P}^{n+1}$ is smooth cubic hypersurface of even dimension $n = 2m$, then the injectivity $R_3 \hookrightarrow \mathrm{Hom}(R_{t(m)}, R_{t(m)+3})$ translates into the injectivity of

$$H^1(X, \mathcal{T}_X) \hookrightarrow \mathrm{Hom}(H^{m,m}(X)_{\mathrm{pr}}, H^{m-1,m+1}(X)_{\mathrm{pr}}),$$

which geometrically can be interpreted as saying that for any first order deformation there exists a primitive class of type (m, m) that does not stay of type (m, m).

In fact, our discussion and, more precisely, Corollary 4.18 imply a stronger result, at least for cubic hypersurfaces in two thirds of all dimensions.

Corollary 4.27 (Variational Torelli theorem) *Let $X \subset \mathbb{P}^{n+1}$ be a smooth cubic hypersurface of dimension $n > 2$ such that $3 \nmid (n+2)$. Then* (4.25) *determines X uniquely.* □

It is tempting to try to reduce the information needed to recover X further. For example, one could try to apply the symmetrizer lemma Proposition 4.16 to determine the image of (4.25) in terms the cup product $H^{p,n-p} \times H^{n-p,p} \longrightarrow H^{n,n}$. For this to determine the contraction with classes in $H^1(X, \mathcal{T}_X)$, one would need $n - p = p - 1$. However, then the condition (ii) $t(p) + t(p) + 3 \leq \sigma - 1 = n + 1$ in Proposition 4.16 is not met.

In Section 3.3.2 the last two corollaries will be reformulated as variants of the Torelli theorem.

5 Classical Constructions: Quadric Fibrations, Ramified Covers, etc.

This section presents standard constructions for cubic hypersurfaces. We will explain how linear projections turn cubic hypersurfaces into quadric fibrations, see Section 5.1, and how triple covers of projective spaces ramified along cubic hypersurfaces provide cubic hypersurfaces of higher dimensions, see Section 5.5. There is also a discussion of nodal cubics in Section 5.4 and of (uni)rational parametrizations in Section 5.3.

5.1 Projection from a Linear Subspace To get a feeling how many smooth cubic hypersurfaces $X \subset \mathbb{P}^{n+1}$ contain a linear subspace $\mathbb{P}^{k-1} \subset \mathbb{P}^{n+1}$, let us first look at a special case and then describe the global picture.

Remark 5.1 Consider the Fermat cubic hypersurface $X = V(\sum x_i^3) \subset \mathbb{P}^{n+1}$. Then, for n even, $V(x_0 + x_1, x_2 + x_3, \ldots, x_n + x_{n+1})$ describes a linear subspace $\mathbb{P}^{n/2} \subset \mathbb{P}^{n+1}$ contained in X. Analogously, for n odd, $V(x_0 + x_1, x_2 + x_3, \ldots, x_{n-1} + x_n, x_{n+1})$ describes a linear subspace $\mathbb{P}^{(n-1)/2} \subset \mathbb{P}^{n+1}$ contained in X.

Clearly, this implies that the Fermat cubic contains linear subspaces $\mathbb{P}^\ell$ of any dimensions $\ell \leq n/2$. Recall that a cubic smooth hypersurface cannot contain linear subspaces of higher dimension, cf. Exercise 1.5.

Exercise 5.2 Let $\mathrm{P} := \mathbb{P}^{k-1} \subset \mathbb{P}^{n+1}$, $n > 0$, be a linear subspace of dimension $k-1 > 0$.

(i) Compute the dimension of the linear system $|\mathcal{O}_{\mathbb{P}}(3)\otimes\mathcal{I}_{\mathrm{P}}|$ of all cubic hypersurfaces containing P.

(ii) Show that $\dim|\mathcal{O}_{\mathbb{P}}(3)\otimes\mathcal{I}_{P}| < \dim(M_n)$, where M_n is the moduli space of all smooth cubic hypersurfaces of dimension n, if and only if

$$n + 2 < \binom{k+2}{3}. \tag{5.1}$$

Conclude from this that the generic smooth cubic $X \subset \mathbb{P}^{n+1}$ does not contain a linear subspace $\mathbb{P}^{k-1}$ if k satisfies (5.1).

(iii) For example, neither the generic cubic fourfold $X \subset \mathbb{P}^5$ nor the generic cubic sixfold $X \subset \mathbb{P}^7$ contains a plane $\mathbb{P}^2$. In the first case, one has $\dim(\mathbb{P}^2) = (1/2)\dim(X)$, while in the second case $\dim(\mathbb{P}^2) < (1/2)\dim(X)$, cf. Remark 1.3.

(iv) For any $n > 0$ and $1 < k \leq (n+1)/2$, there exists a family $\mathcal{X} \longrightarrow S$ of smooth cubic hypersurfaces of dimension n over a connected base S and an S-smooth subscheme $\mathcal{P} \subset \mathcal{X}$ such that each fibre $\mathcal{P}_t \subset \mathcal{X}_t$ is isomorphic to a linear $\mathbb{P}^{k-1} \subset \mathcal{X}_t \subset \mathbb{P}^{n+1}$ and every such pair $\mathbb{P}^{k-1} \subset X$ occurs as one of the fibres, cf. [472, §1 Lem. 1].

We now write a linear subspace as $\mathrm{P} := \mathbb{P}(W) \subset \mathbb{P} := \mathbb{P}(V)$ with $\dim(V) = n+2$ and $\dim(W) = k$. Additionally, we pick a generic linear subspace $\mathbb{P}(U) \subset \mathbb{P}(V)$ of codimension k. Here, generic means that the composition $U \subset V \twoheadrightarrow V/W$ is an isomorphism or, equivalently, that $U + W = V$ or, still equivalently, $\mathbb{P}(W) \cap \mathbb{P}(U) = \emptyset$.

The linear projection $\mathbb{P} \dashrightarrow \mathbb{P}(U) \simeq \mathbb{P}(V/W)$ from P is the rational map that sends $x \in \mathbb{P} \setminus \mathrm{P}$ to the unique point of intersection of the linear subspace $\overline{x\mathrm{P}} \simeq \mathbb{P}^k$ with $\mathbb{P}(U) \simeq \mathbb{P}^{n+1-k}$. It is the rational map associated with the linear system $|\mathcal{I}_{\mathrm{P}} \otimes \mathcal{O}(1)| \subset |\mathcal{O}(1)|$ with base locus $\mathrm{P} \subset \mathbb{P}$, which is resolved by a simple blow-up. The resulting morphism $\phi\colon \mathrm{Bl}_{\mathrm{P}}(\mathbb{P}) \longrightarrow \mathbb{P}(V/W)$ is then associated with the complete linear system $|\tau^*\mathcal{O}(1) \otimes \mathcal{O}(-E)|$:

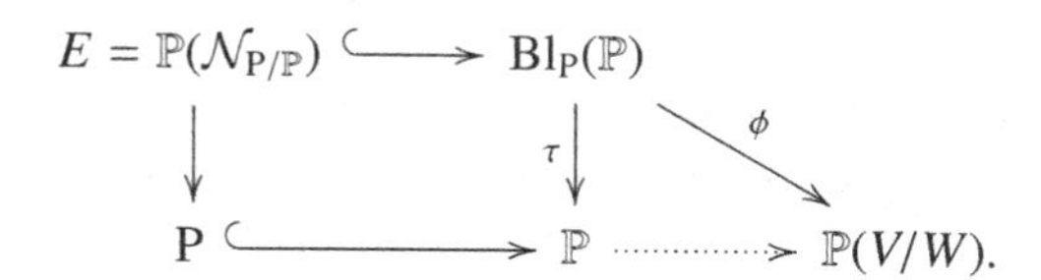

The fibre $\phi^{-1}(y)$, $y \in \mathbb{P}(U) \simeq \mathbb{P}(V/W)$, is the strict transform of $\mathbb{P}^k \simeq \overline{y\mathrm{P}} \subset \mathbb{P}$ in $\mathrm{Bl}_{\mathrm{P}}(\mathbb{P})$ which for dimension reasons is isomorphic to $\mathbb{P}^k$. To visualize the situation observe that $E \cap \phi^{-1}(y)$ is a section of $\tau|_E\colon E \twoheadrightarrow \mathrm{P}$ which over a point $x \in \mathrm{P}$ picks out the normal direction $v \in \mathbb{P}(\mathcal{N}_{\mathrm{P}/\mathbb{P}}(x))$ given by the line $\overline{xy}$.

All fibres of $\phi\colon \mathrm{Bl}_{\mathrm{P}}(\mathbb{P}) \twoheadrightarrow \mathbb{P}(V/W)$ are projective spaces $\mathbb{P}^k$ and, indeed, it is the projective bundle $\mathrm{Bl}_{\mathrm{P}}(\mathbb{P}) \simeq \mathbb{P}(\mathcal{F}^*)$ with the locally free sheaf

$$\mathcal{F} := \phi_*\tau^*\mathcal{O}(1)$$

on $\mathbb{P}(V/W)$, which is of rank $k+1$. To determine $\mathcal{F}$ explicitly, tensor the structure sequence of the exceptional divisor $E \subset \mathrm{Bl}_{\mathrm{P}}(\mathbb{P})$ with $\tau^*\mathcal{O}(1)$ to get the short exact sequence

$$\begin{array}{ccccccccc} 0 & \longrightarrow & \tau^*\mathcal{O}(1)\otimes\mathcal{O}(-E) & \longrightarrow & \tau^*\mathcal{O}(1) & \longrightarrow & \tau^*\mathcal{O}(1)|_E & \longrightarrow & 0. \\ & & \simeq \phi^*\mathcal{O}(1) & & \simeq \mathcal{O}_\phi(1) & & \simeq \mathcal{O}(1,0) & & \end{array} \tag{5.2}$$

Here, we use the fact that $\mathcal{N}_{\mathrm{P}/\mathbb{P}} \simeq V/W \otimes \mathcal{O}(1)$, from which we deduce an isomorphism

$$E = \mathbb{P}(\mathcal{N}_{\mathrm{P}/\mathbb{P}}) \simeq \mathrm{P} \times \mathbb{P}(V/W),$$

compatible with the natural projections, and that $\mathcal{O}(E)|_E \simeq \mathcal{O}(1,-1)$ on $E \simeq \mathrm{P}\times\mathbb{P}(V/W)$. In particular, $\phi^*\mathcal{O}(1)|_E \simeq (\tau^*\mathcal{O}(1)\otimes\mathcal{O}(-E))|_E \simeq \mathcal{O}(0,1)$. Therefore, ϕ restricted to E is the projection onto $\mathbb{P}(V/W)$. Thus, $\mathcal{F}$ is described by the direct image under ϕ of (5.2), which reads

$$0 \longrightarrow \mathcal{O}(1) \longrightarrow \mathcal{F} \longrightarrow H^0(\mathrm{P},\mathcal{O}(1))\otimes\mathcal{O}_{\mathbb{P}(V/W)} \longrightarrow 0.$$

The sequence splits, which gives a non-canonical isomorphism

$$\mathcal{F} \simeq \mathcal{O}(1)\oplus(W^*\otimes\mathcal{O}_{\mathbb{P}(V/W)}) \simeq \mathcal{O}(1)\oplus\mathcal{O}^{\oplus k}$$

and, hence, $\det(\mathcal{F}) \simeq \mathcal{O}(1)$, which is all we shall use for the moment.

Now let $X \subset \mathbb{P} = \mathbb{P}(V)$ be a cubic hypersurface with equation $F \in H^0(\mathbb{P},\mathcal{O}(3))$. The pullback τ^*F is a section of $\tau^*\mathcal{O}(3)$, whose zero divisor $V(\tau^*F)$ is the total transform of X. If $\mathrm{P} \subset \mathbb{P}$ is contained in X, cf. Exercise **2.3.5**, then the total transform has two components, the exceptional divisor E and the strict transform of $X \subset \mathbb{P}$. The latter is the blow-up $\mathrm{Bl}_{\mathrm{P}}(X)$ of X in $\mathrm{P} \subset X$. More precisely, in this case F is contained in $H^0(\mathbb{P},\mathcal{O}(3)\otimes\mathcal{I}_{\mathrm{P}}) \subset H^0(\mathbb{P},\mathcal{O}(3))$ and τ^*F lies in $H^0(\mathrm{Bl}_{\mathrm{P}}(\mathbb{P}),\tau^*\mathcal{O}(3)\otimes\mathcal{O}(-E)) \subset H^0(\mathrm{Bl}_{\mathrm{P}}(\mathbb{P}),\tau^*\mathcal{O}(3))$. Therefore, as a subvariety of $\mathrm{Bl}_{\mathrm{P}}(\mathbb{P})$, the blow-up of X is

$$\mathrm{Bl}_{\mathrm{P}}(X) = V(\tau^*F) \in |\tau^*\mathcal{O}(3)\otimes\mathcal{O}(-E)|.$$

5.2 Quadric Fibrations The next goal is to describe $\mathrm{Bl}_{\mathrm{P}}(X)$ as a quadric fibration over $\mathbb{P}(V/W)$. To this end, we compute the direct image of τ^*F under ϕ. First, observe that

$$\tau^*\mathcal{O}(3)\otimes\mathcal{O}(-E) \simeq \tau^*\mathcal{O}(2)\otimes(\tau^*\mathcal{O}(1)\otimes\mathcal{O}(-E)) \simeq \mathcal{O}_\phi(2)\otimes\phi^*\mathcal{O}(1).$$

Here, the relative tautological line bundle $\mathcal{O}_\phi(1)$ correspond to writing $\mathrm{Bl}_{\mathrm{P}}(\mathbb{P}) \simeq \mathbb{P}(\mathcal{F}^*)$ with $\mathcal{F} = \phi_*\tau^*\mathcal{O}(1)$, as before. Thus,

$$\phi_*(\tau^*\mathcal{O}(3)\otimes\mathcal{O}(-E)) \simeq \phi_*(\mathcal{O}_\phi(2)\otimes\phi^*\mathcal{O}(1)) \simeq S^2(\mathcal{F})\otimes\mathcal{O}(1)$$

and, therefore, τ^*F can be thought of as a section $q \in H^0(\mathbb{P}(V/W), S^2(\mathcal{F}) \otimes \mathcal{O}(1))$ or as a symmetric homomorphism

$$q\colon \mathcal{F}^* \longrightarrow \mathcal{F} \otimes \mathcal{O}(1).$$

Hence, the fibre of $\mathrm{Bl}_{\mathrm{P}}(X) \subset \mathrm{Bl}_{\mathrm{P}}(\mathbb{P}) \simeq \mathbb{P}(\mathcal{F}^*)$ over $y \in \mathbb{P}(V/W)$, i.e. the residual quadric Q_y of the intersection $\mathrm{P} \subset \overline{y\mathrm{P}} \cap X$, is the quadric defined by $q_y \in S^2(\mathcal{F}(y))$. In particular, the fibre is smooth if and only if this quadric is non-degenerate. Thus, the discriminant divisor D_{P} of $\phi\colon \mathrm{Bl}_{\mathrm{P}}(X) \longrightarrow \mathbb{P}(V/W)$ is

$$D_{\mathrm{P}} = V(\det(q)) \subset \mathbb{P}(V/W).$$

Here, $\det(q)\colon \det(\mathcal{F})^* \longrightarrow \det(\mathcal{F}) \otimes \mathcal{O}(k+1)$ is viewed as a section of the line bundle $\det(\mathcal{F})^2 \otimes \mathcal{O}(k+1) \simeq \mathcal{O}(k+3)$.

The discussion is summarized by the following classical fact, see e.g. [80, Lem. 2] for $n = 3$ and [41, Ch. 1]. See also Corollary **5**.1.23 for a discussion in dimension three.

Proposition 5.3 *Assume that the smooth cubic hypersurface $X \subset \mathbb{P}^{n+1}$ contains a linear subspace $\mathrm{P} = \mathbb{P}^{k-1}$ such that there exists no linear subspace $\mathbb{P}^k \subset X$ containing* P. *Then the linear projection from* P *defines a morphism*

$$\phi\colon \mathrm{Bl}_{\mathrm{P}}(X) \twoheadrightarrow \mathbb{P}^{n+1-k} \tag{5.3}$$

with the following properties:

(i) *The fibre over $y \in \mathbb{P}^{n+1-k}$ is the residual quadric Q_y of $\mathrm{P} \subset \overline{y\mathrm{P}} \cap X$, i.e.*

$$\overline{y\mathrm{P}} \cap X = \mathrm{P} \cup Q_y.$$

(ii) *The fibres are singular exactly over the discriminant divisor $D_{\mathrm{P}} \in |\mathcal{O}(k+3)|$.*
(iii) *The morphism $\phi\colon \mathrm{Bl}_{\mathrm{P}}(X) \twoheadrightarrow \mathbb{P}^{n+1-k}$ is flat.*

Proof The first two assertions follow from the preceding discussion. For (iii) use 'miracle flatness' which asserts that the smoothness of $\mathrm{Bl}_{\mathrm{P}}(X)$ and of $\mathbb{P}^{n+1-k}$ together with the fact that all fibres are of dimension $k-1$ imply flatness of ϕ. □

Example 5.4 As we shall see in Proposition **2**.1.19 and Remark **2**.3.6, every smooth cubic hypersurface of dimension $n \geq 2$ contains a line, which corresponds to the case $k = 2$ above. Thus, we can always project from a line $L \subset X$, which defines a morphism $\mathrm{Bl}_L(X) \twoheadrightarrow \mathbb{P}^{n-1}$, the fibre of which are conics. The discriminant hypersurface $D_L \subset \mathbb{P}^{n-1}$ is contained in the linear system $|\mathcal{O}(5)|$ and the fibres Q_y over points $y \in D_L$ consist either of two intersecting lines or of a plane double line. We will come back to this in Remark **2**.5.2.

Exercise 5.5 Let $E \subset \mathrm{Bl}_L(X)$ be the exceptional divisor of the blow-up of a line $L \subset X$ contained in a smooth cubic hypersurface $X \subset \mathbb{P}^{n+1}$. For a generic point $y \in \mathbb{P}^{n-1}$ of a complementary linear subspace, the residual conic $Q_y \subset \overline{yL} \cap X$ intersects L in two points $x_1, x_2 \in L$. Observe that the line $\overline{x_iy} \subset \mathbb{P}^{n+1}$ is tangent to X at x_i. Show that the two induced normal directions of $L \subset X$ at x_1 and x_2 correspond to the two points contained in the fibre of $\phi|_E\colon E \twoheadrightarrow \mathbb{P}^{n-1}$ over y. In particular, the restriction $\phi|_E$ is generically finite of degree two.

Remark 5.6 If there exists a linear subspace of bigger dimension contained in X, then the fibre dimension of ϕ is not constant anymore. For example, if

$$\mathbb{P}^{k-1} \simeq \mathrm{P} \subset \mathbb{P}^k \simeq \mathrm{P}' \subset X,$$

then the fibre $\phi^{-1}(y)$ of $\phi\colon \mathrm{Bl}_{\mathrm{P}}(X) \twoheadrightarrow \mathbb{P}^{n+1-k}$ over the point of intersection of $\mathbb{P}^k \cap \mathbb{P}^{n+1-k} = \{y\}$ will be $\mathrm{P}' \simeq \mathbb{P}^k$. The description of the discriminant divisor as an element $D_{\mathrm{P}} \in |\mathcal{O}(k+3)|$ remains unchanged.

Example 5.7 Consider the Fermat cubic $X = V(\sum x_i^3) \subset \mathbb{P}^{n+1}$ of even dimension and let $\mathrm{P} = \mathbb{P}^{n/2} \subset X$ be the linear subspace $V(x_0 + x_1, \ldots, x_n + x_{n+1})$, see Remark 5.1. Show that then $D_{\mathrm{P}} \subset \mathbb{P}^{n+1-k} = V(x_0 - x_1, \ldots, x_n - x_{n+1})$ is the union of $n/2 + 1$ hyperplanes and the cubic $X \cap \mathbb{P}^{n+1-k}$.

Remark 5.8 In this lengthy remark we shall discuss various questions related to the smoothness of the discriminant divisor D_{P}.

(i) In order to understand the local structure of D_{P} we first have a look at the space of symmetric matrices

$$M := \{A \in M(r \times r) \mid A^t = A\}$$

over an algebraically closed field. It is stratified by the closed subvarieties

$$M_\ell := \{A \in M \mid \mathrm{rk}(A) \leq \ell\}.$$

Using normal forms of symmetric matrices, one shows that $M_\ell \setminus M_{\ell-1}$ comes with a transitive action of $\mathrm{GL}(r)$ and is therefore smooth. For our purposes, only the two strata $M_{r-2} \subset M_{r-1} \subset M_r = M$ are relevant. Clearly, $M_{r-1} \subset M_r$ is a divisor cut out by the determinant $\det(A)$ and, using normal forms, one proves that $M_{r-2} \subset M_{r-1}$ is of codimension two. Furthermore, a local calculation reveals that M_{r-2} is the singular set of M_{r-1} and that M_{r-1} has ordinary quadratic singularities along $M_{r-2} \setminus M_{r-3}$, i.e. the generic three-dimensional section of $M_{r-2} \subset M_{r-1} \subset M$ through a point $A \in M_{r-2} \setminus M_{r-3}$ produces a surface with an ordinary double point, see [36, Lem. 2].

The stratification can be put into families. For the locally free sheaf $\mathcal{F}$ of rank $r = k+1$ and the twist $S^2(\mathcal{F})(1)$ it defines a stratification

$$M_{k-1}(S^2(\mathcal{F})(1)) \subset M_k(S^2(\mathcal{F})(1)) \subset M_{k+1}(S^2(\mathcal{F})(1)) = |S^2(\mathcal{F})(1)|.$$

Now, by construction, D_{P} is the degeneracy locus

$$D_{\mathrm{P}} = M_k(q) = \{\, y \mid \mathrm{rk}(q_y) \leq k \,\} = q^{-1}M_k(S^2(\mathcal{F})(1)),$$

where q is viewed as a section of the vector bundle $|S^2(\mathcal{F})(1)| \longrightarrow \mathbb{P}^{n+1-k}$. As $\mathcal{F} \simeq \mathcal{O}(1) \oplus \mathcal{O}^{\oplus k}$, the sheaf $S^2(\mathcal{F})(1)$ is globally generated, which allows us to apply the following Bertini type argument, see [36, Lem. 4]. Namely, for the generic section $q \in H^0(\mathbb{P}(U), S^2(\mathcal{F})(1))$ the pre-image of the deeper stratum

$$q^{-1}M_{k-1}(S^2(\mathcal{F})(1)) \subset D_{\mathrm{P}} = q^{-1}M_k(S^2(\mathcal{F})(1)) \subset \mathbb{P}(U) \simeq \mathbb{P}^{n+1-k}$$

is of codimension three in $\mathbb{P}^{n+1-k}$ and D_{P} has ordinary quadratic singularities along the subset $q^{-1}M_{k-1}(S^2(\mathcal{F})(1)) \setminus q^{-1}M_{k-2}(S^2(\mathcal{F})(1))$.

However, a priori there is no reason that the section q induced by the equation defining $X \subset \mathbb{P}^{n+1}$ is generic in this sense and often it is not.

(ii) Note that the discussion suggests that D_{P} should not be expected to be smooth for $n + 1 - k \geq 3$. More precisely, the only cases when it definitely is are the cases of lines in cubic surfaces, of lines in cubic threefolds and of planes in cubic fourfolds.

For a smooth cubic threefold $Y \subset \mathbb{P}^4$ and for the generic (but not for every!) line $L \subset Y$, the discriminant $D_L \subset \mathbb{P}^2$ is a smooth curve. This follows from (iii) below, see Corollary **5**.1.9 and also Corollary **5**.1.23. However, for planes $\mathbb{P}^2 \subset X \subset \mathbb{P}^5$ in a cubic fourfold, the discriminant curve can indeed be singular (and planes in X are rigid and, thus, have all to be considered generic), see Exercise 5.7. If X is also allowed to vary, more flexibility is gained, see Remark **6**.1.5.

Note that, in general, whenever there exists a smooth cubic $X_0 \subset \mathbb{P}^{n+1}$ containing a linear $\mathbb{P}^{k-1} \simeq \mathrm{P}_0 \subset X_0$ with D_{P_0} smooth, then by Exercise 5.2 the smoothness of D_{P} holds for the generic pair $\mathbb{P}^{k-1} \simeq \mathrm{P} \subset X$.

When the smoothness of D_{P} cannot be achieved, the next best would be for D_{P} to have mild singularities, e.g. ordinary double points. This is the case for the generic line $L \subset X$ in a smooth cubic fourfold, see Lemma **6**.4.12, but as our section q may a priori not be generic even for the generic line, it does not follow for general reasons.

(iii) Using the quadric fibration $\mathrm{Bl}_{\mathrm{P}}(X) \longrightarrow \mathbb{P}(U)$ and the description of D_{P} as its discriminant divisor, one sometimes obtains a more concrete understanding of the singularities of D_{P}. More precisely, D_{P} is smooth at $y \in D_{\mathrm{P}}$ if and only if the fibre $\phi^{-1}(y)$ is a quadric cone with an isolated singularity, cf. [41, Prop. 1.2] or [27, Prop. 1.2.5] for the statement and for references.

5.3 (Uni-)Rational Parametrizations I The larger the integer k or, equivalently, the smaller the dimension of the target space $\mathbb{P}(U)$ is, the more special X is. According to Exercise 1.5, the dimension of a linear subspace contained in a smooth cubic hypersurface $X \subset \mathbb{P}^{n+1}$ cannot exceed $n/2$, i.e. $k \leq n/2 + 1$. Thus, the most special and, hence, the geometrically most revealing case is $\mathbb{P}(U) \simeq \mathbb{P}^{n/2}$ for n even and $\mathbb{P}(U) \simeq \mathbb{P}^{(n+1)/2}$ for n odd.

Corollary 5.9 *Assume $X \subset \mathbb{P}^{n+1}$ is a smooth cubic hypersurface of even dimension containing a linear $\mathrm{P} = \mathbb{P}^{n/2} \subset X \subset \mathbb{P}^{n+1}$. The linear projection from P defines a quadric fibration $\mathrm{Bl}_{\mathrm{P}}(X) \longrightarrow \mathbb{P}^{n/2}$ with discriminant divisor $D_{\mathrm{P}} \in |\mathcal{O}((n/2)+3)|$. A similar result holds for odd n.* □

This construction can be very useful. As an example, we explain how the existence of a quadric fibration in dimension four can be used to prove unirationality (of degree two). But we emphasize that unirationality (of degree two) in fact holds true for all cubics of dimension $n > 1$, cf. Corollary **2**.1.21.

Example 5.10 Assume $\mathbb{P}^2 \simeq \mathrm{P} \subset X \subset \mathbb{P} = \mathbb{P}^5$. Pick a generic $\mathbb{P}^3 \subset \mathbb{P}$ and let $\tilde{S}$ be the intersection of $\tau^{-1}(\mathbb{P}^3) \subset \mathrm{Bl}_{\mathrm{P}}(\mathbb{P})$ and $\mathrm{Bl}_{\mathrm{P}}(X) \subset \mathrm{Bl}_{\mathrm{P}}(\mathbb{P})$. Observe that $\tau^{-1}(\mathbb{P}^3)$ is the blow-up of $\mathbb{P}^3$ in the point of intersection x of $\mathbb{P}^3$ and P. Then $\tilde{S}$ is the blow-up of the cubic surface $S := X \cap \mathbb{P}^3$ in x. The generic fibre of

$$\phi|_{\tilde{S}} \colon \tilde{S} \longrightarrow \mathbb{P}^2$$

over $y \in \mathbb{P}^2$ is the intersection of the quadric $\phi^{-1}(y) \subset \overline{y\mathrm{P}}$ with the line $\mathbb{P}^1 \simeq \mathbb{P}^3 \cap \overline{y\mathrm{P}}$ and, therefore, consists of two points, i.e. $\tilde{S} \longrightarrow \mathbb{P}^2$ is of degree two, cf. the discussion in Section **4**.2.3. The base change $\mathrm{Bl}_{\mathrm{P}}(X) \times_{\mathbb{P}^2} \tilde{S} \longrightarrow \tilde{S}$ is a quadric fibration with a section over the rational surface $\tilde{S}$ and hence rational. Thus, $\mathrm{Bl}_{\mathrm{P}}(X)$ and, therefore, X are unirational.

One can try to run the same argument for any linear subspace $\mathrm{P} \subset X$. For example, for a line $\mathbb{P}^1 \simeq L \subset X$, which always exists as we will see, one picks a generic $\mathbb{P}^n$ and lets $S := \mathbb{P}^n \cap X$, which is a cubic hypersurface of dimension $n-1$. The base change $\mathrm{Bl}_{\mathrm{P}}(X) \times_{\mathbb{P}^n} \tilde{S} \longrightarrow \tilde{S}$ is a conic fibration with a section but now, by induction, one only knows that $\tilde{S}$ is unirational of degree two.

If the existence of a second, complementary linear space contained in the cubic hypersurface X is assumed, not only unirationality but in fact rationality of X can be deduced.

Corollary 5.11 *Assume that a smooth cubic hypersurface $X \subset \mathbb{P}^{n+1}$ of even dimension $n = 2m$ contains two complementary linear subspaces $\mathbb{P}^m \simeq \mathbb{P}(W) \subset X$ and $\mathbb{P}^m \simeq \mathbb{P}(W') \subset X$, i.e. such that $W \oplus W' = V$ and, in particular, $\mathbb{P}(W) \cap \mathbb{P}(W') = \emptyset$. Then the quadric fibration* (5.3) *admits a section and X is rational.*

Proof The section is of course given by the inclusion $\mathbb{P}^m \simeq \mathbb{P}(W') \subset X$, for which the linear projection induces an isomorphism $\mathbb{P}(W') \xrightarrow{\sim} \mathbb{P}(U)$. As any quadric admitting a rational point is rational, the scheme-theoretic generic fibre $\phi^{-1}(\eta)$ is a rational quadric over $K(\mathbb{P}^m)$. Hence, $\mathrm{Bl}_{\mathrm{P}}(X)$ is rational and, therefore, X itself is. □

In principle, the argument would work for complementary linear subspaces $\mathbb{P}^{k-1} \simeq \mathbb{P}(W) \subset X$ and $\mathbb{P}^{n+1-k} \simeq \mathbb{P}(W') \subset X$. However, unless $n = 2(k-1)$, the dimension of one of the two subspaces will exceed $n/2$, which is excluded by Exercise 1.5.

Example 5.12 In fact, assuming the existence of two complementary linear subspaces $\mathrm{P} := \mathbb{P}(W), \mathrm{P}' := \mathbb{P}(W') \subset X$, i.e. such that $W \oplus W' = V$, the rationality of X can also be deduced from the following, in fact easier, construction. Consider the rational map

$$\psi \colon \mathrm{P} \times \mathrm{P}' \dashrightarrow X$$

that sends a pair (x, x') to the residual point of intersection y of $\{x, x'\} \subset \overline{xx'} \cap X$. The map is well defined for all (x, x') for which the line $\overline{xx'}$ is not contained in X. This describes a non-empty open subset of $\mathrm{P} \times \mathrm{P}'$.

Observe that any point in the complement of $\mathrm{P} \cup \mathrm{P}'$ is contained in the image of ψ. More concretely, $y \in X \setminus (\mathrm{P} \cup \mathrm{P}')$ is the image of (x, x'), where x and x' are determined by $\{x\} = \mathrm{P} \cap \overline{y\mathrm{P}'}$ and $\{x'\} = \overline{y\mathrm{P}} \cap \mathrm{P}'$. As any line through a point $y \in X \setminus (\mathrm{P} \cup \mathrm{P}')$ that intersects both P and P′ meets P and P′ exactly in the points of intersection x and x', the argument also shows that ψ is generically injective. Thus, X is rational.

As mentioned before, Exercise 1.5 shows that for smooth X the situation can only occur if $\dim(X) = 2m$ and $\dim(\mathrm{P}) = \dim(\mathrm{P}') = m$. In this case, the construction leads to a rational parametrization $\mathbb{P}^m \times \mathbb{P}^m \dashrightarrow X \subset \mathbb{P}^{2m+1}$. Here are two concrete examples:

(i) The Fermat cubic fourfold $X = V(x_0^3 + \cdots + x_5^3) \subset \mathbb{P}^5$ contains the two disjoint planes $V(x_0 - \lambda x_1 x_2 - \mu x_3, x_4 - \nu x_5)$ and $V(x_0 - \mu x_1, x_2 - \nu x_3, x_4 - \lambda x_5)$. Here, λ, μ, and ν are the three 3rd roots of unity. In particular, the Fermat cubic fourfold is rational.

(ii) The cubic fourfold $X = V(F) \subset \mathbb{P}^5$ with $F = x_0^2 x_1 - x_0 x_1^2 + x_2^2 x_3 - x_2 x_3^2 + x_4^2 x_5 - x_4 x_5^2$ which contains the two disjoint planes $\mathrm{P} = V(x_0, x_2, x_4)$ and $\mathrm{P}' = V(x_1, x_3, x_5)$, cf. [240, Sec. 5] or [228, Sec. 1].

The rationality of X, for example in dimension four, can also be deduced from the existence of other types of surfaces. We recommend [228] for further information, see also Remark **5**.4.15, Section **6**.1.2, Conjecture **6**.5.15, Conjecture **7**.3.1, and Theorem **7**.3.14.

Remark 5.13 Unirationality of cubic hypersurfaces of dimension at least two can be shown alternatively by the following crude trick which, however, does not give any information about the degree of unirationality.

Fix a smooth cubic hypersurface $X \subset \mathbb{P}^{n+1}$ of dimension $n > 2$. Then consider two generic hyperplane sections $Y_1, Y_2 \subset X$ and the associated rational map

$$Y_1 \times Y_2 \dashrightarrow X, \;\; (y_1, y_2) \longmapsto x.$$

Here, x is the residual point of intersection of $\{y_1, y_2\} \subset \overline{y_1 y_2} \cap X$. The map is dominant as for the generic point $x \in X$ and any point $y_1 \in Y_1$ the line $\overline{xy_1}$ intersects Y_2 in a (unique) point y_2. Since by induction we may assume that Y_1 and Y_2 are unirational, then X is also. Note that cubic surfaces over an algebraically closed field are in fact rational, see Proposition **4**.2.4.

One can always project a smooth cubic hypersurface from a point $x_0 \in X$. This defines a morphism $\phi\colon \mathrm{Bl}_{x_0}(X) \twoheadrightarrow \mathbb{P}^n$ which is generically finite of degree two. Thus, there exists a dominant rational map of degree two

$$X \dashrightarrow\!\!\!\!\!\rightarrow \mathbb{P}^n$$

or, in other words, the *degree of irrationality* is bounded $\mathrm{irr}(X) \leq 2$.

Indeed, for generic $y \in \mathbb{P}^n$, the fibre $\phi^{-1}(y)$ consists of the two residual points x_1, x_2 of the intersection $x_0 \in \overline{yx_0} \cap X$, i.e. $\overline{yx_0} \cap X = \{x_0, x_1, x_2\}$, where possibly $x_0 = x_1$, $x_0 = x_2$, or $x_1 = x_2$. The discriminant divisor in this case is contained in the linear system $|\mathcal{O}(4)|$, but since there exists a line through every point, cf. Proposition **2**.1.19, projection from a point is only generically finite but not finite. The set of points in $\mathbb{P}^n$ with positive-dimensional fibre will be described in Remark **2**.3.6.

In Corollary **2**.1.21 we will explain that there also exists a dominant rational map of degree two with the role of X and $\mathbb{P}^n$ reversed:

$$\mathbb{P}^n \dashrightarrow\!\!\!\!\!\rightarrow X.$$

Remark 5.14 Of course, one could also consider the linear projection of a cubic hypersurface $X \subset \mathbb{P}^{n+1}$ onto a generic $\mathbb{P}^n \subset \mathbb{P}^{n+1}$ from a point $x \in \mathbb{P}^{n+1} \setminus X$. The resulting morphism $\rho\colon X \longrightarrow \mathbb{P}^n$ is of degree three and its Tschirnhaus bundle is $\rho_*\mathcal{O}_X/\mathcal{O}_{\mathbb{P}^n} \simeq \mathcal{O}(-1) \oplus \mathcal{O}(-1)$, see [353, Cor. 8.5]. The branch locus of the projection is usually singular, possibly reducible and non-reduced. For $n = 2$, it is a curve of degree six in $\mathbb{P}^2$ with six cusps, see [353, Lem. 10.1].

5.4 Nodal Cubics Projecting from a point $x_0 \in X$ becomes more interesting when X is singular at x_0. The simplest case is that of a nodal cubic hypersurface with exactly one ordinary double point x_0 as its only singularity.

Recall that an isolated singularity $x_0 \in X$ is an ordinary double point (or a node) if the exceptional fibre $E_{x_0} \subset \mathrm{Bl}_{x_0}(X)$ is a non-degenerate quadric in the exceptional divisor $\mathbb{P}^n \simeq E \subset \mathrm{Bl}_{x_0}(\mathbb{P}^{n+1})$. Then for any line $x_0 \in \mathbb{P}^1 \subset \mathbb{P}$, the intersection $\mathbb{P}^1 \cap X$ has multiplicity at least two at x_0.

Exercise 5.15 Assume $X \subset \mathbb{P}^{n+1}$ is a *nodal hypersurface* with exactly one node at the point $[0 : \cdots : 0 : 1] \in \mathbb{P}^{n+1}$. Show then that X is defined by an equation of the form

$$F(x_0, \ldots, x_n) + x_{n+1} \cdot G(x_0, \ldots, x_n),$$

where F and G define hypersurfaces of degree three and two in $V(x_{n+1}) \simeq \mathbb{P}^n$.

As in the smooth case, the blow-up $\mathrm{Bl}_{x_0}(X)$, which is smooth, can be described as the strict transform of X in the blow-up of $\mathbb{P}$:

$$\begin{array}{ccccccc} E_{x_0} & \hookrightarrow & \mathrm{Bl}_{x_0}(X) & \hookrightarrow & \mathrm{Bl}_{x_0}(\mathbb{P}) & \twoheadrightarrow & \mathbb{P}^n \\ \downarrow & & \downarrow & & \downarrow{\scriptstyle\tau} & & \\ \{x_0\} & \hookrightarrow & X & \hookrightarrow & \mathbb{P} & & \end{array}$$

with $E_{x_0} \subset E = \mathbb{P}(T_{x_0}\mathbb{P}) \simeq \mathbb{P}^n$ a smooth quadric hypersurface. The difference to the smooth case is that the pullback τ^*F of the defining equation for X now vanishes along E to order two so that $\mathrm{Bl}_{x_0}(X) \in |\tau^*\mathcal{O}(3)\otimes\mathcal{O}(-2E)|$. By a similar argument as in the smooth case, τ^*F can then be viewed as an element in $H^0(\mathbb{P}^n, \mathcal{F}\otimes\mathcal{O}(2))$, for $\tau^*\mathcal{O}(3)\otimes\mathcal{O}(-2E) \simeq \tau^*\mathcal{O}(1) \otimes \phi^*\mathcal{O}(2)$.

As in this situation $\mathcal{F} \simeq \mathcal{O}(1) \oplus \mathcal{O}$, the blow-up is realized as a closed subscheme $\mathrm{Bl}_{x_0}(X) \subset \mathrm{Bl}_{x_0}(\mathbb{P}) \simeq \mathbb{P}(\mathcal{F}^*)$ of the $\mathbb{P}^1$-bundle $\mathbb{P}(\mathcal{F}^*)$ given by τ^*F viewed as a section (t_1, t_2) of $\mathcal{O}(3) \oplus \mathcal{O}(2)$. Here, the zero locus of t_1 can be thought of as the intersection of X with $\mathbb{P}^n$, while the zero locus of t_2 is the intersection of the non-degenerate quadric E with $\mathbb{P}^n$. Thus, for $y \notin V(t_1) \cap V(t_2)$, the fibre $\phi^{-1}(y)$ consists of the residual point of $x_0 \in \overline{yx_0} \cap X$ and for $y \in V(t_1) \cap V(t_2)$ one has $\phi^{-1}(y) \simeq \mathbb{P}^1$. In other words,

$$\phi\colon \mathrm{Bl}_{x_0}(X) \simeq \mathrm{Bl}_Z(\mathbb{P}^n) \twoheadrightarrow \mathbb{P}^n$$

is the blow-up of $\mathbb{P}^n$ in the complete intersection $Z := V(t_1) \cap V(t_2) \subset \mathbb{P}^n$ of type $(3, 2)$ or, alternatively, ϕ contracts every line that passes through x_0. Compare this discussion with Remark **2**.3.6, where an interpretation in terms of Fano varieties is provided.

The birational correspondence can alternatively be described as an isomorphism

$$X \setminus \bigcup_{x_0 \in L} L \simeq \mathbb{P}^n \setminus Z.$$

Corollary 5.16 *A cubic hypersurface X with an ordinary double point x_0 as its only singularity is rational. The blow-up $\mathrm{Bl}_{x_0}(X)$ is isomorphic to a blow-up $\mathrm{Bl}_Z(\mathbb{P}^n)$ with $Z \subset \mathbb{P}^n$ a smooth, complete intersection of type* $(3, 2)$. □

Remark 5.17 It is an interesting and often intriguing question to determine the maximal number $\mu_n(d)$ of ordinary double points an otherwise smooth hypersurface $X \subset \mathbb{P}^{n+1}$ of degree d can acquire, even for surfaces the question is not fully understood. However, for cubic hypersurfaces, the maximal number of ordinary double points of an otherwise smooth cubic hypersurface $X \subset \mathbb{P}^{n+1}$ is known, namely

$$\mu_n := \mu_n(3) = \binom{n+2}{\left[\frac{n+1}{2}\right]}.$$

For example, $\mu_2 = 4$, $\mu_3 = 10$, and $\mu_4 = 15$. In dimension two and three, the maximum is realized by a unique cubic: the *Cayley surface* described by the equation

$$x_0\, x_1\, x_2\, x_3 \cdot \left(\frac{1}{x_0} + \frac{1}{x_1} + \frac{1}{x_2} + \frac{1}{x_3} \right) = 0,$$

see also Remark **4**.2.16, and by the *Segre cubic threefold*

$$\sum_{i=0}^{5} x_i^3 = \sum_{i=0}^{5} x_i = 0,$$

see [158, Thm. 9.4.14]. Note that the Cayley cubic surface is the hyperplane section of the Segre cubic threefold with $x_4 = x_5$. Uniqueness fails in higher dimension. In fact, there is a positive-dimensional family of cubic fourfolds with the maximal number 15 of nodes. Goryunov [199] and Kalker [265] give explicit equations for cubics attaining the maximum.[13]

5.5 Hyperplane Sections Let $X \subset \mathbb{P}^{n+1} = \mathbb{P}(V)$ be a smooth cubic hypersurface of dimension n. Then the intersection with a generic hyperplane $V(h) \subset \mathbb{P}^{n+1}$, $h \in V^* = H^0(\mathbb{P}(V), \mathcal{O}(1))$, defines a smooth cubic hypersurface $Y := X \cap V(h) \subset V(h) \simeq \mathbb{P}^n$ of dimension $n - 1$. We say that the family of hyperplane sections of X has maximal variation if the rational map

$$\Phi_X \colon \mathbb{P}(V^*) \dashrightarrow M_n = |\mathcal{O}_{\mathbb{P}^n}(3)|_{\rm sm}/\!/\mathrm{PGL}(n+1),\ h \longmapsto X \cap V(h)$$

is generically finite, i.e. its image is of dimension $n + 1$, and we say it has zero variation if the map is constant. For simplicity, we restrict to $\mathrm{char}(k) = 0$ and then the variation is measured by the derivative

$$d\Phi_{X,Y} \colon H^0(Y, \mathcal{O}_Y(1)) \longrightarrow H^1(Y, \mathcal{T}_Y) \tag{5.4}$$

of Φ_X at the generic hyperplane section $Y = X \cap V(h)$. The variation is maximal if $d\Phi_{X,Y}$ is injective and it is 0 if $d\Phi_{X,Y} = 0$.

The situation has been first studied by Beauville [40]. We restrict to the case of cubic hypersurfaces but the arguments below easily generalize to higher degrees.

Proposition 5.18 *Consider smooth cubic hypersurfaces of dimension $n \geq 2$.*

(i) *For the generic cubic hypersurface $X \subset \mathbb{P}^{n+1}$, the family of hyperplane sections has maximal variation.*
(ii) *For no smooth cubic hypersurface $X \subset \mathbb{P}^{n+1}$, the variation of the family of hyperplane sections is* 0.

[13] I am grateful to S. Stark for the reference.

Proof For the proof of the first assertion, we follow Opstall and Veliche [465], who in turn rely heavily on [40], and it naturally splits into two steps. First, the property of having a maximal varying family of hyperplane sections is a Zariski open condition in $|\mathcal{O}_{\mathbb{P}(V)}(3)|$. Indeed, the condition $d\Phi_{X,Y}$ being injective describes an open subset in the incidence variety of all pairs (X, Y) consisting of a smooth cubic hypersurface $X \subset \mathbb{P}(V)$ together with smooth hyperplane section $Y = X \cap V(h)$. Thus, it suffices to exhibit one pair $Y = X \cap V(h) \subset X \subset \mathbb{P}(V)$ with $d\Phi_{X,Y}$ injective. It turns out that the generic hyperplane section of the Fermat cubic $X := V\left(\sum_{i=0}^{n+1} x_i^3\right) \subset \mathbb{P}^{n+1} \simeq \mathbb{P}(V)$ has this property. For instance, the hyperplane section defined by $h = x_{n+1} - \sum_{i=0}^{n} a_i x_i$ is $Y = V(F) \subset \mathbb{P}^n$ with

$$F(x_0, \ldots, x_n) := \sum_{i=0}^{n} x_i^3 + \left(\sum_{i=0}^{n} a_i x_i\right)^3.$$

First order deformations of $Y = X \cap V(h)$ inside X can be written as $Y_\varepsilon = V(F_\varepsilon)$ with

$$F_\varepsilon(x_0, \ldots, x_n) := \sum_{i=0}^{n} x_i^3 + \left(\sum_{i=0}^{n} a_i x_i + \varepsilon \sum_{i=0}^{n} b_i x_i\right)^3.$$

Then the induced class in $R_3(Y) \simeq H^1(Y, \mathcal{T}_Y)$, see Lemma 4.14, is (up to the factor 3)

$$\left(\sum_{i=0}^{n} a_i x_i\right)^2 \cdot \left(\sum_{i=0}^{n} b_i x_i\right) \tag{5.5}$$

modulo the Jacobian ideal, which is generated by the derivatives $(1/3)\,\partial_j F = x_j^2 + a_j\,(\sum a_i x_i)^2$. However, the attempt to write (5.5) as a linear combination $\sum_j g_j \cdot (x_j^2 + a_j\,(\sum_i a_i x_i)^2)$, with linear polynomials $g_j = g_j(x_0, \ldots, x_n)$, leads to the equation

$$\left(\sum_{i=0}^{n} a_i x_i\right)^2 \cdot \left(\sum_{i=0}^{n} (b_i x_i - a_i g_i)\right) = \sum_{j=0}^{n} g_j \cdot x_j^2.$$

As the right-hand side is a linear combination of monomials $x_i\, x_j^2$, involving only two variables, while on the left-hand side several monomials in three different variables occur for the generic choice of a_i, this leads to a contradiction. Hence, for the generic choice of a hyperplane section $Y = X \cap V\left(x_{n+1} - \sum_{i=0}^{n} a_i x_i\right)$, no first order deformation corresponding to $\sum b_i x_i$ vanishes in $H^1(Y, \mathcal{T}_Y)$.

For the proof of the second assertion, we follow [40]. The first step consists of showing that $d\Phi_{X,Y} = 0$ implies that Y is contained in the *polar variety* $P_u X = V\left(\sum_{i=0}^{n+1} u_i\, \partial_i F\right)$ of $X = V(F)$ for some point $u \in \mathbb{P}^{n+1} \setminus Y$. For this, we may assume $Y = X \cap V(x_{n+1}) = V(F(x_0, \ldots, x_n, 0)) \subset \mathbb{P}^n$, so that first order deformations of Y in X are of the form $Y_\varepsilon = V(F_\varepsilon) \subset \mathbb{P}^n$ with

$$F_\varepsilon = F\left(x_0, \ldots, x_n, \varepsilon \sum_{i=0}^n b_i x_i\right) = F(x_0, \ldots, x_n, 0) + \varepsilon\,(\partial_{n+1}F)(x_0, \ldots, x_n, 0) \sum_{i=0}^n b_i x_i.$$

Those correspond to a trivial first order deformation of Y, if the corresponding class in $R_3(Y) \simeq H^1(Y, \mathcal{T}_Y)$ is trivial, i.e.

$$(\partial_{n+1}F)(x_0, \ldots, x_n, 0) \sum_{i=0}^n b_i x_i = \sum_{i=0}^n g_i \cdot (\partial_i F)(x_0, \ldots, x_n, 0)$$

for certain linear polynomials $g_i = g_i(x_0, \ldots, x_n, 0)$. By virtue of Proposition 4.3, this holds for all choices of b_i if and only if $(\partial_{n+1}F)(x_0, \ldots, x_n, 0) = 0$ in $R_2(Y)$, i.e.

$$(\partial_{n+1}F)(x_0, \ldots, x_n, 0) = \sum_{i=0}^n u_i\,(\partial_i F)(x_0, \ldots, x_n, 0)$$

for suitable scalars $u_0, \ldots, u_n$. However, then for $u := [u_0 : \cdots : u_n : -1] \in \mathbb{P}^{n+1} \setminus Y$, the polar $P_u X$ contains Y.

The second step of the argument exploits the Gauss map

$$\gamma\colon X \longrightarrow X^* \subset \mathbb{P}^*, \qquad x \longmapsto [\partial_0 F(x) : \cdots : \partial_{n+1}F(x)],$$

which describes the normalization of the *dual variety* X^*, see [174, Ch. 10] and Section 2.2.2. The pullback of a hyperplane section is the intersection $P_u X \cap X$ of some polar variety with X. Hence, by Bertini's theorem $P_u X \cap X$ is smooth for generic u and, therefore, cannot contain any hyperplane section Y. Thus, $d\Phi_{X,Y} \neq 0$ for generic Y. □

Whether the family of hyperplane sections of a particular smooth cubic hypersurface has maximal variation seems to be an open question. For threefolds, see [357, Pbl. 4.2].

As a consequence of this result, or rather of its proof, and using the fact that the moduli space of cubic surfaces is of dimension four, one finds the following well-known result in the theory of cubic surfaces, which is a classical result by Sylvester, see [158, Cor. 9.4.2].

Corollary 5.19 *The generic cubic surface $S \subset \mathbb{P}^3$ over an algebraically closed field k of* $\mathrm{char}(k) \neq 2, 3$ *is isomorphic to a hyperplane section of the Fermat cubic threefold $Y := V\left(\sum_{i=0}^4 x_i^3\right) \subset \mathbb{P}^4$, i.e.*

$$S \simeq Y \cap \mathbb{P}^3 \simeq V\left(\sum_{i=0}^4 x_i^3, \sum_{i=0}^4 a_i x_i\right) \subset \mathbb{P}^4.$$

Equivalently, $S \simeq V\left(\sum_{i=0}^4 b_i x_i^3, \sum_{i=0}^4 x_i\right) \subset \mathbb{P}^4$. □

Clearly, in higher dimensions cubic hypersurfaces of dimension n obtained by intersecting a single fixed cubic hypersurface of dimension $n+1$ make up for only a small subset of the whole moduli space.

Remark 5.20 At this point it is natural to consider Lefschetz pencils of hyperplane sections of a fixed smooth cubic hypersurface $X \subset \mathbb{P}^{n+1}$. Similar to the discussion in Section 2.3, where we considered Lefschetz pencils of cubic hypersurfaces in $\mathbb{P}^{n+1}$, the choice of a generic linear $\mathbb{P}^1 \hookrightarrow |\mathcal{O}(1)|$ describes a flat projective family

$$\mathcal{Y} \twoheadrightarrow \mathbb{P}^1, \tag{5.6}$$

where the fibre $\mathcal{Y}_t$ are the hyperplane sections $X \cap H_t$, with $t \in \mathbb{P}^1$. Alternatively, the inclusions $\mathcal{Y}_t = X \cap H_t \hookrightarrow X$ describe $\mathcal{Y}$ as the the blow-up $\mathcal{Y} \simeq \mathrm{Bl}_Y(X) \twoheadrightarrow X$ in the base locus of the pencil which is a cubic hypersurface in $H_{t_1} \cap H_{t_2} \simeq \mathbb{P}^{n-1}$ and so of dimension $n-2$.

The generic fibre of (5.6) is smooth and each singular fibre has exactly one ordinary double point. The latter follows from the *Gauss map* $X \twoheadrightarrow X^*$ being generically injective and the classical fact that the generic singular hyperplane section has exactly one singular point, see [148, Exp. XVII, Prop. 4.2], [174, Cor. 10.21], and the discussion in Section 2.2.

The number m of singular fibres $\mathcal{Y}_{t_1}, \ldots, \mathcal{Y}_{t_m} \subset \mathcal{Y}$ is the degree of the dual variety which is also known classically, namely

$$m = \left|\{\, t \in \mathbb{P}^1 \mid \mathcal{Y}_t \text{ singular } \}\right| = \deg(X^*) = 3 \cdot 2^n.$$

In [148, Exp. XVIII, (3.2.4)] this number is computed as

$$\deg(X^*) = (-1)^n \left(e(X_n) + e(X_{n-2}) - 2e(X_{n-1})\right),$$

where $e(X_n)$ is the Euler number of a smooth (cubic) hypersurface of dimension n and fixed degree. Then use (1.6) in Section 1.3. For a more direct approach, see [174, Prop. 2.9].

5.6 Triple Covers There is a way to link cubics of dimension n to cubics of dimension $n+1$ in an almost canonical way. In the end, one finds that every smooth cubic hypersurface of dimension n is a hyperplane section of some smooth cubic hypersurface of dimension $n+1$. In fact, the construction works for hypersurfaces of any degree d.

Let $X = V(F) \subset \mathbb{P}^{n+1}$ be an arbitrary hypersurface of degree d given by a polynomial $F = F(x_0, \ldots, x_{n+1}) \in H^0(\mathbb{P}^{n+1}, \mathcal{O}(d))$. Then

$$\tilde{F} := F - x_{n+2}^d \in H^0(\mathbb{P}^{n+2}, \mathcal{O}(d))$$

describes a hypersurface $\tilde{X} := V(\tilde{F}) \subset \mathbb{P}^{n+2}$. Clearly, X is isomorphic to a hyperplane section of $\tilde{X}$, namely

$$X = \tilde{X} \cap V(x_{n+2}).$$

Observe that X is smooth if and only if $\tilde{X}$ is smooth.

Note that $X \subset \mathbb{P}^{n+1}$ determines its defining equation F only up to a scaling factor, i.e. $V(F) = V(\lambda F)$ for all $\lambda \in k^*$, and this does effect the equation $\tilde{F}$. However, at least if k contains dth roots of unity, the two hypersurfaces $V(F - x_{n+2}^d)$ and $V(\lambda F - x_{n+2}^d)$ only differ by a linear coordinate change $[x_0 : \cdots : x_{n+1} : x_{n+2}] \longmapsto [x_0 : \cdots : x_{n+1} : \lambda^{1/d} x_{n+2}]$.

The rational map given by the linear projection $\mathbb{P}^{n+2} \dashrightarrow \mathbb{P}^{n+1}$ that drops the last coordinate is regular along $\tilde{X} \subset \mathbb{P}^{n+2}$. It defines a finite morphism, and in fact a cyclic cover, of degree d

$$\pi \colon \tilde{X} \twoheadrightarrow \mathbb{P}^{n+1}$$

branched over $X \subset \mathbb{P}^{n+1}$. More precisely, $\pi^{-1}(X) = d\,X$ as divisors in $\tilde{X}$ and

$$\pi \colon X = \tilde{X} \cap V(x_{n+2}) \xrightarrow{\sim} X.$$

The Galois group of the covering is generated by $[x_0 : \cdots : x_{n+2}] \longmapsto [x_0 : \cdots : \rho x_{n+2}]$ with ρ a dth primitive root of unity.

In short, up to finite quotients by μ_d one obtains an inclusion

$$|\mathcal{O}_{\mathbb{P}^1}(d)|_{\rm sm} \subset |\mathcal{O}_{\mathbb{P}^2}(d)|_{\rm sm} \subset |\mathcal{O}_{\mathbb{P}^3}(d)|_{\rm sm} \subset \cdots,$$

which is compatible with the linear actions $\mathrm{PGL}(2) \subset \mathrm{PGL}(3) \subset \mathrm{PGL}(4) \subset \cdots$.

This basic construction has been successfully used to relate moduli spaces of cubic hypersurfaces of different dimensions. We shall come back to this in Sections **4**.4.3 and **5**.6.1.

Remark 5.21 The construction can be carried out multiple times. We restrict to the case of cubics, i.e. $d = 3$. In this fashion, by applying the construction twice, one obtains

$$\begin{array}{ccccc} & & X_{n+1} & \hookrightarrow & \mathbb{P}^{n+2} \\ & & \downarrow & & \\ X_n & \hookrightarrow & \mathbb{P}^{n+1} & & \\ \downarrow & & & & \\ X_{n-1} \hookrightarrow \mathbb{P}^n, & & & & \end{array}$$

where X_{n-1} is an arbitrary smooth cubic in $\mathbb{P}^n$ and X_n and X_{n+1} are obtained by the above procedure. Then there exists an algebraic isomorphism of Hodge structures

$$H^{n+1}(X_{n+1}, \mathbb{Q})_{\rm pr} \simeq H^{n-1}(X_{n-1}, \mathbb{Q})_{\rm pr}(-1)^{\oplus 2} \oplus (H^n(X_n, \mathbb{Q})_{\rm pr} \otimes H^1(E, \mathbb{Q}))^{\rm fix}. \tag{5.7}$$

The isomorphism was established by van Geemen and Izadi [464, Prop. 3.5], based on work of Shioda and Katsura [436, 437]. Here, $E \subset \mathbb{P}^2$ is the Fermat cubic, which is thought of as X_1 obtained as the triple cover of $\mathbb{P}^1$ branched over three points $X_0 \subset \mathbb{P}^1$, and the fixed part on the right-hand side is taken with respect to the action of a primitive root of unity. Observe that (5.7) explains the formula in Exercise 1.13 relating the primitive Betti numbers of cubic hypersurfaces of three consecutive dimensions.

The reason for (5.7) is the following geometric correspondence originally described by Katsura and Shioda for Fermat hypersurfaces: write

$$\begin{aligned} X_0 &= V(u_0^3 + u_1^3) \subset \mathbb{P}^1,\ X_1 = V(y_0^3 + y_1^3 + y_2^3) \subset \mathbb{P}^2, \\ X_n &= V(F(x_0, \ldots, x_n) + x_{n+1}^3) \subset \mathbb{P}^{n+1}, \\ \text{and}\quad X_{n+1} &= V(F(z_0, \ldots, z_n) + z_{n+1}^3 + z_{n+2}^3) \subset \mathbb{P}^{n+2}. \end{aligned}$$

Then consider the rational map

$$X_n \times X_1 \dashrightarrow X_{n+1}$$

given by $z_i = x_i y_2$, $i = 0, \ldots, n$ and $z_{n+1+j} = \varepsilon\, x_{n+1} y_j$, $j = 0, 1$, where $\varepsilon^3 = -1$, i.e.

$$([x_0 : \cdots : x_n : x_{n+1}], [y_0 : y_1 : y_2]) \longmapsto [x_0 y_2 : \cdots : x_n y_2 : \varepsilon\, x_{n+1} y_0 : \varepsilon\, x_{n+1} y_1].$$

The indeterminacy locus is $Z := V(x_{n+1}, y_2) \simeq X_{n-1} \times X_0$. A simple blow-up resolves the indeterminacies and leads to

$$\begin{array}{ccc} \mathrm{Bl}_Z(X_n \times X_1) & \longrightarrow & \mathrm{Bl}_Z(X_n \times X_1)/\mu_3 \\ \downarrow & & \downarrow \pi \\ X_n \times X_1 & \dashrightarrow & X_{n+1}. \end{array}$$

Here, μ_3 acts by

$$([x_0 : \cdots : x_n : x_{n+1}], [y_0 : y_1 : y_2]) \longmapsto [\varepsilon\, x_0 : \cdots : \varepsilon\, x_n : -x_{n+1}], [\varepsilon\, y_0 : \varepsilon\, y_1 : -y_2])$$

and π describes contractions

$$\mathbb{P}^n \times X_0 \longrightarrow X_0 \simeq V(z_0, \ldots, z_n) \cap X_{n+1} \quad \text{and} \quad X_{n-1} \times \mathbb{P}^1 \longrightarrow X_{n-1} \simeq V(z_{n+1}, z_{n+2}) \cap X_{n+1}.$$

The construction sketched above not only works for cubic hypersurfaces. For hypersurfaces of degree d, however, (5.7) involves $d-1$ copies of the cohomology of X_{n-1}.

Remark 5.22 Assume $X \subset \mathbb{P}^{n+1}$ is a (smooth) cubic hypersurface and consider the triple cover $\tilde{X} \twoheadrightarrow \mathbb{P}^{n+2}$, as above. The isomorphism class of $\tilde{X}$ only depends on the isomorphism class of X and not on its embedding $X \subset \mathbb{P}^{n+1}$. Is the converse also true, i.e. does $\tilde{X}$ determine X? It turns out that for the generic cubic $X \subset \mathbb{P}^{n+1}$ of dimension $n \geq 2$, this is indeed true. In other words, using the notation in Section **3**.1.5, mapping a smooth cubic hypersurface X of dimension n to the naturally associated triple cover $\tilde{X} \twoheadrightarrow \mathbb{P}^{n+1}$ branched over X defines a generically injective morphism

$$M_{3,n} \longrightarrow M_{3,n+1}, \ X \longmapsto \tilde{X} \tag{5.8}$$

between the moduli spaces of smooth cubic hypersurface of dimension n and $n+1$.

Indeed, for generic $X = V(F) \subset \mathbb{P}^{n+1}$, the determinant of the Hessian $\det H(f)$ is irreducible, for example, this is the case for $F(x_0, x_1, x_2) = x_0^3 + x_1^3 + x_2^3 + x_0 x_1 x_2$. As the determinant of the Hessian of the equation $\tilde{F} = F(x_0, \ldots, x_{n+1}) - x_{n+2}^3$ is

$$\det H(\tilde{F}) = 6 \cdot \det H(F) \cdot x_{n+2},$$

x_{n+2} is its only linear factor. Since $X \simeq \tilde{X} \cap V(x_{n+2})$ and since the intersection does not depend on the choice of the embedding $X \subset \mathbb{P}^{n+1}$, this shows that the isomorphism class of the generic cubic X of dimension n is uniquely determined by the isomorphism class of the cubic $\tilde{X}$ of dimension $n+1$.

Note that the Jacobian rings of X and $\tilde{X}$ are related by $R(\tilde{X}) \simeq R(X) \otimes k[x_{n+2}]/x_{n+2}^2$. Are there techniques to reconstruct $R(X)$ from $R(\tilde{X})$? In such a case, the triple cover $\tilde{X}$ again determines the original cubic X.

The fact that (5.8) is generically injective is at the heart of an approach of Allcock, Carlson, and Toledo [13, 14] to link cubic surfaces to cubic threefolds, see Section **4**.4.3, and cubic threefolds to cubic fourfolds, see Section **5**.6.1.

2

Fano Varieties of Lines

With any (cubic) hypersurface $X \subset \mathbb{P}^{n+1}$ one associates its Fano variety of lines $F(X)$ or, more generally, of m-planes, contained in X. For a smooth cubic surface $S \subset \mathbb{P}^3$, the Fano variety $F(S)$ consists of 27 reduced points corresponding to the 27 lines contained in S. In higher dimensions, Fano varieties are even more interesting and have become a central topic of study in the theory of cubic hypersurfaces, especially in dimension three and four.

The classical references for Fano varieties of lines and planes are the articles by Altman and Kleiman [15] and by Barth and Van de Ven [37]. For cubic hypersurfaces many arguments simplify, and we will restrict to cubics whenever this is the case. For enumerative aspects, we recommend [15, 174].

This chapter covers the general theory of Fano varieties of lines in Section 1, provides a detailed discussion of lines of the first and second type on cubic hypersurfaces in Section 2, and presents global properties of the Fano variety of lines in Section 3. The remaining two Sections 4 and 5 are devoted to numerical and motivic properties.

1 Construction and Infinitesimal Behaviour

We shall begin with an outline of the techniques that go into the construction of the Fano variety of linear subspaces $\mathbb{P}^m \subset \mathbb{P}^{n+1}$ contained in a given projective variety $X \subset \mathbb{P}^{n+1}$. The main tool is Grothendieck's Quot-scheme, which also provides information on the tangent space of the Fano variety at a point corresponding to a linear subspace $\mathbb{P}^m \subset X$.

1.1 Representing the Fano Functor We work over an arbitrary field k and use the shorthand $\mathbb{P} := \mathbb{P}^{n+1}_k$. Often it is preferable to think of $\mathbb{P}$ as $\mathbb{P}(V) = \mathrm{Proj}(S^*(V^*))$ for some fixed k-vector space V of dimension $n+2$. We consider an arbitrary subvariety $X \subset \mathbb{P}$ and fix an integer $0 \leq m \leq n+1$. Then the *Fano functor* of m-planes is the functor

$$\underline{F}(X,m)\colon (Sch/k)^o \longrightarrow (Set), \tag{1.1}$$

which sends a k-scheme T (of finite type) to the set of all T-flat closed subschemes $L \subset T \times X$ such that all fibres $L_t \subset X_{k(t)} \subset \mathbb{P}_{k(t)}$ are linear subspaces of dimension m. We shall mostly be interested in the case of lines, i.e. $m = 1$, and will write $\underline{F}(X) := \underline{F}(X, 1)$ in this case.

Remark 1.1 Here are a few examples and easy observations.

(i) For $X = \mathbb{P}$, one obtains the *Grassmann functor*

$$\underline{F}(\mathbb{P}, m) = \underline{\mathbb{G}}(m, \mathbb{P}).$$

(ii) For $m = 0$, the functor $\underline{F}(X, 0)$ is the functor of points h_X.

(iii) For nested closed subschemes $X \subset X' \subset \mathbb{P}$, there are natural inclusions

$$\underline{F}(X,m) \subset \underline{F}(X',m) \subset \underline{F}(\mathbb{P},m) = \underline{\mathbb{G}}(m,\mathbb{P}).$$

(iv) Set $P_m(\ell) := \binom{m+\ell}{\ell}$ and let $\underline{\mathrm{Hilb}}^{P_m}(X)$ be the Hilbert functor that sends a k-scheme T to the set of all T-flat closed subschemes $Z \subset T \times X$ with fibrewise Hilbert polynomial $\chi(Z_t, \mathcal{O}_{Z_t}(\ell)) = P_m(\ell)$. Then

$$\underline{F}(X,m) = \underline{\mathrm{Hilb}}^{P_m}(X).$$

Here, we leave it as an exercise to show that any closed subvariety $Z \subset \mathbb{P}$ with Hilbert polynomial P_m is indeed a linear subspace $\mathbb{P}^m \subset \mathbb{P}$.

Theorem 1.2 *The Fano functor $\underline{F}(X,m)$ of m-planes is represented by a projective k-scheme $F(X,m)$, the* Fano variety of m-planes *in $X \subset \mathbb{P}$.*

There are various ways to argue. However, in the end, the proof always comes down to the representability of the Grassmann functor.

• Use (iv) above and the representability of $\underline{\mathrm{Hilb}}^P(X)$ (for arbitrary projective X and Hilbert polynomial P) by the Hilbert scheme $\mathrm{Hilb}^P(X)$. This in turn is a special case of the representability of the Grothendieck Quot-functor $\underline{\mathrm{Quot}}^P_{X/\mathcal{E}}$ of quotients $\mathcal{E} \twoheadrightarrow \mathcal{F}$ with Hilbert polynomial P. Indeed, $\underline{\mathrm{Hilb}}^P(X) \simeq \underline{\mathrm{Quot}}^P_{X/\mathcal{O}_X}$. Recall that the existence of the Quot-scheme is eventually reduced to the existence of the Grassmann variety, cf. [253, Ch. 2.2] or the original [212] or [178, Part 2].

• The inclusion $\underline{F}(X,m) \subset \underline{\mathbb{G}}(m,\mathbb{P})$ in (iii) above describes a closed subfunctor. As the Grassmann functor $\underline{\mathbb{G}}(m,\mathbb{P})$ is representable by the Grassmann variety $\mathbb{G}(m,\mathbb{P})$, $\underline{F}(X,m)$ is represented by a closed subscheme $F(X,m) \subset \mathbb{G}(m,\mathbb{P})$.

Let us spell out the second approach a bit further. But first recall that

$$\mathbb{G}(m, \mathbb{P}^{n+1}) \simeq \mathrm{Gr}(m+1, n+2),$$

where $\mathrm{Gr}(m+1, n+2)$ is the Grassmann variety of linear subspaces of $k^{\oplus n+2}$ of dimension $m+1$ or, in other words,

$$\mathbb{G}(m, \mathbb{P}^{n+1}) \simeq \mathrm{Gr}(m+1, n+2) \simeq \mathrm{Quot}^{P\equiv n+1-m}_{\mathrm{Spec}(k)/V}.$$

The isomorphisms between the corresponding functors

$$\underline{\mathrm{Gr}}(m+1, n+2) \xrightarrow{\sim} \underline{\mathbb{G}}(m, \mathbb{P}^{n+1}) \quad \text{and} \quad \underline{\mathrm{Gr}}(m+1, n+2) \xrightarrow{\sim} \underline{\mathrm{Quot}}^{P}_{\mathrm{Spec}(k)/V}$$

are given by

$$[\mathcal{G} \subset V \otimes \mathcal{O}_T] \longmapsto \mathbb{P}(\mathcal{G}) \quad \text{and} \quad [\mathcal{G} \subset V \otimes \mathcal{O}_T] \longmapsto [V \otimes \mathcal{O}_T \twoheadrightarrow V \otimes \mathcal{O}_T/\mathcal{G}].$$

On k-rational points this gives

$$[W \subset V] \longmapsto \mathbb{P}(W) \subset \mathbb{P}(V) \quad \text{and} \quad [W \subset V] \longmapsto [V \twoheadrightarrow V/W].$$

Also recall that $\mathbb{G}(m, \mathbb{P})$ is an irreducible, smooth, projective variety of dimension

$$\dim(\mathbb{G}(m, \mathbb{P})) = (m+1) \cdot (n+1-m).$$

It is naturally embedded into $\mathbb{P}(\bigwedge^{m+1} V)$ via the Plücker embedding

$$\mathbb{G} := \mathbb{G}(m, \mathbb{P}) \hookrightarrow \mathbb{P}\left(\bigwedge^{m+1} V\right), \qquad L = \mathbb{P}(W) \longmapsto [\det(W)]. \tag{1.2}$$

Under this embedding, $\mathcal{O}(1)|_{\mathbb{G}} \simeq \bigwedge^{m+1}(\mathcal{S}^*)$. Here, $\mathcal{S}$ is the universal sub-bundle of rank $m+1$, which is part of the universal exact sequence

$$0 \longrightarrow \mathcal{S} \longrightarrow V \otimes \mathcal{O}_{\mathbb{G}} \longrightarrow \mathcal{Q} \longrightarrow 0. \tag{1.3}$$

The universal family of m-planes over $\mathbb{G}(m, \mathbb{P})$ is the $\mathbb{P}^m$-bundle associated with $\mathcal{S}$:

$$p\colon \mathbb{L}_{\mathbb{G}} := \mathbb{P}(\mathcal{S}) \longrightarrow \mathbb{G}(m, \mathbb{P}),$$

which comes with its relative tautological line bundle $\mathcal{O}_p(1)$.[1] The inclusion $\mathcal{S} \subset V \otimes \mathcal{O}_{\mathbb{G}}$ corresponds to the natural embedding

$$\mathbb{L}_{\mathbb{G}} \subset \mathbb{G}(m, \mathbb{P}) \times \mathbb{P}^{n+1}$$

and the induced projection $q\colon \mathbb{L}_{\mathbb{G}} \longrightarrow \mathbb{P}^{n+1}$ satisfies

$$q^*\mathcal{O}_{\mathbb{P}}(1) \simeq \mathcal{O}_p(1).$$

Assume now that $X \subset \mathbb{P}$ is a hypersurface defined by the homogenous polynomial $F \in k[x_0, \ldots, x_{n+1}]_d = H^0(\mathbb{P}, \mathcal{O}(d)) \simeq S^d(V^*)$. Dualizing (1.3) and taking symmetric powers provide us with a natural surjection

$$S^d(V^*) \otimes \mathcal{O}_{\mathbb{G}} \twoheadrightarrow S^d(\mathcal{S}^*)$$

[1] Throughout, we will work with the geometric, i.e. non-Grothendieck, convention for projective bundles, so $p_*\mathcal{O}_p(1) \simeq \mathcal{S}^*$.

and hence a map $S^d(V^*) \longrightarrow H^0(\mathbb{G}, S^d(\mathcal{S}^*))$. Let $s_F \in H^0(\mathbb{G}, S^d(\mathcal{S}^*))$ denote the image of $F \in S^d(V^*)$ under this map, so

$$S^d(V^*) \longrightarrow H^0(\mathbb{G}, S^d(\mathcal{S}^*)),\ F \longmapsto s_F.$$

Then the Fano variety of m-planes on X is the closed subvariety of the Grassmann variety defined as the zero locus of s_F, i.e.

$$F(X,m) = V(s_F) \subset \mathbb{G}(m,\mathbb{P}). \tag{1.4}$$

In particular, whenever $F(X,m)$ is non-empty, then

$$\begin{aligned}\dim(F(X,m)) &\geq \dim(\mathbb{G}(m,\mathbb{P})) - \mathrm{rk}(S^d(\mathcal{S}^*)) \\ &= (m+1)\cdot(n+1-m) - \binom{m+d}{d}.\end{aligned} \tag{1.5}$$

Moreover, in the case of equality the class of $F(X,m)$, in the Chow ring or just in the cohomology of $\mathbb{G}(m,\mathbb{P})$, can be expressed as the rth Chern class of $S^d(\mathcal{S}_F^*)$:

$$[F(X,m)] = \mathrm{c}_r(S^d(\mathcal{S}^*)), \tag{1.6}$$

where $r = \mathrm{rk}(S^d(\mathcal{S}^*)) = \binom{m+d}{d}$. We will come back to this later, see Section 4.3.

For more general subvarieties $X \subset \mathbb{P}$, the argument is similar: if $X = \bigcap V(F_i)$, then $F(X,m) = \bigcap V(s_{F_i})$, where $s_{F_i} \in H^0(\mathbb{G}, S^{d_i}(\mathcal{S}^*))$, $d_i = \deg(F_i)$. But unless X is a complete intersection, it is more complicated to compute the class of its Fano scheme.

We shall denote the universal family of m-planes over $F(X,m)$ by $p\colon \mathbb{L} \longrightarrow F(X,m)$, which is nothing but the restriction

$$\mathbb{L} = \mathbb{L}_{\mathbb{G}}|_{F(X,m)} = \mathbb{P}(\mathcal{S}_F := \mathcal{S}|_{F(X,m)})$$

of $\mathbb{L}_{\mathbb{G}}$ to $F(X,m) \subset \mathbb{G}(m,\mathbb{P})$. We also think of $\mathbb{L}$ as the universal family

$$\mathbb{L} = \{\, (L,x) \mid x \in L \subset X \,\}$$

of pairs (L,x) consisting of an m-plane $\mathbb{P}^m \simeq L \subset \mathbb{P}$ contained in X and a point $x \in L$. With its two projections, one has the following diagram, to which we later refer as the (geometric) *Fano correspondence*, see Section 5:

$$\begin{array}{ccc} \mathbb{L} & \xrightarrow{\ q\ } & X \\ {\scriptstyle p}\downarrow & & \\ F(X,m). & & \end{array} \tag{1.7}$$

The composition of the natural inclusion $F(X,m) \subset \mathbb{G}(m,\mathbb{P})$ with the Plücker embedding (1.2) of $\mathbb{G}(m,\mathbb{P})$ defines the Plücker embedding of the Fano variety of X

$$F(X,m) \hookrightarrow \mathbb{G}(m,\mathbb{P}) \hookrightarrow \mathbb{P}\left(\textstyle\bigwedge^{m+1} V\right).$$

The restriction of the hyperplane line bundle $\mathcal{O}(1)|_{F(X,m)} \simeq \bigwedge^{m+1}(\mathcal{S}^*|_{F(X,m)})$ is called the *Plücker polarization* and its first Chern class will be denoted

$$g = c_1(\mathcal{S}^*|_{F(X,m)}) \in \mathrm{CH}^1(F(X,m)), \tag{1.8}$$

often also considered as a cohomology class

$$g \in H^2(F(X,m),\mathbb{Z})(1).$$

The following universal variant will be useful. Consider the universal hypersurface $\mathcal{X} \longrightarrow |\mathcal{O}(d)| = \mathbb{P}^N$, cf. Section **1.2**. Then denote by

$$\underline{F}(\mathcal{X},m)\colon (Sch/|\mathcal{O}(d)|)^o \longrightarrow (Set) \tag{1.9}$$

the functor that sends a morphism $T \longrightarrow |\mathcal{O}(d)|$ to the set of all T-flat closed subschemes $L \subset \mathcal{X}_T \subset T \times \mathbb{P}$ parametrizing m-planes $\mathbb{P}^m \subset \mathbb{P}$ in the fibres of the pullback $\mathcal{X}_T \longrightarrow T$. Using the relative version of the Quot-scheme or of the Grassmann variety, one finds that $\underline{F}(\mathcal{X},m)$ as a scheme over $|\mathcal{O}(d)|$ is represented by a projective morphism

$$F(\mathcal{X},m) \longrightarrow |\mathcal{O}(d)|.$$

By functoriality, the fibre over $[X] \in |\mathcal{O}(d)|$ is $F(X,m)$ and one should think of $F(\mathcal{X},m)$ as parametrizing pairs $(L \subset X)$ of m-planes contained in hypersurfaces of degree d. As in the absolute case, $F(\mathcal{X},m)$ can be realized as a closed subscheme of the relative Grassmannian

$$F(\mathcal{X},m) = V(s_G) \subset |\mathcal{O}(d)| \times \mathbb{G}(m,\mathbb{P}),$$

where $s_G \in H^0(|\mathcal{O}(d)| \times \mathbb{G}(m,\mathbb{P}), \mathcal{O}(1) \boxtimes S^d(\mathcal{S}^*))$ is the image of the universal equation $G \in H^0(|\mathcal{O}(d)| \times \mathbb{P}, \mathcal{O}(1) \boxtimes \mathcal{O}_{\mathbb{P}}(d)) = H^0(|\mathcal{O}(d)|, \mathcal{O}(1)) \otimes S^d(V^*)$, see Section **1.2.1**, under

$$\mathcal{O}(1) \boxtimes (S^d(V^*) \otimes \mathcal{O}_{\mathbb{G}}) \twoheadrightarrow \mathcal{O}(1) \boxtimes S^d(\mathcal{S}^*). \tag{1.10}$$

Let us now look at the other projection $\pi\colon F(\mathcal{X},m) \longrightarrow \mathbb{G}(m,\mathbb{P})$. From the description of $F(\mathcal{X},m)$ as $V(s_G) \subset |\mathcal{O}(d)| \times \mathbb{G}(m,\mathbb{P})$ one deduces an isomorphism

$$F(\mathcal{X},m) \simeq \mathbb{P}(\mathcal{K}) \longrightarrow \mathbb{G}(m,\mathbb{P}),$$

where $\mathcal{K} := \mathrm{Ker}(S^d(V^*) \otimes \mathcal{O}_{\mathbb{G}} \twoheadrightarrow S^d(\mathcal{S}^*))$. In more concrete terms, the fibre of $\mathcal{K}$ at the point $L \in \mathbb{G}(m,\mathbb{P})$ is the vector space $H^0(\mathbb{P}, \mathcal{I}_L \otimes \mathcal{O}(d))$.

Remark 1.3 Instead of introducing the two moduli functors (1.1) and (1.9) and arguing that they can be represented by projective schemes, one could alternatively just use this description of $F(\mathcal{X},m)$ as a projective bundle over $\mathbb{G}(m,\mathbb{P})$ and define $F(X,m)$ as its fibre over X under the other projection $F(\mathcal{X},m) \longrightarrow |\mathcal{O}_{\mathbb{P}}(d)|$. However, as soon as the local structure of these Fano schemes is needed, a functorial approach is preferable.

1.2 Dimensions of the Fano Variety and Lines on Quadrics The above description of the universal Fano scheme allows one to compute its dimension.

Proposition 1.4 *The relative Fano variety $F(\mathcal{X}, m)$ of m-planes in hypersurfaces of degree d in $\mathbb{P}^{n+1}$ is an irreducible, smooth, projective variety of dimension*

$$\dim(F(\mathcal{X}, m)) = (m+1)\cdot(n+1-m) + \binom{n+1+d}{d} - \binom{m+d}{d} - 1. \qquad \square$$

For example, for $m = 1$ and $d = 3$, the formula reads

$$\dim(F(\mathcal{X}, m)) = (2n-4) + \binom{n+4}{3} - 1.$$

The first part of the following immediate consequence confirms (1.5).

Corollary 1.5 *If for an arbitrary hypersurface $X \subset \mathbb{P}^{n+1}$ of degree d the Fano variety $F(X, m)$ is not empty, then*

$$\begin{aligned} \dim(F(X, m)) &\geq \dim(F(\mathcal{X}, m)) - \dim|\mathcal{O}(d)| \qquad (1.11) \\ &= (m+1)\cdot(n+1-m) - \binom{m+d}{d}. \end{aligned}$$

Moreover, equality holds in (1.11) for generic $X \in |\mathcal{O}(d)|$ unless $F(X, m)$ is empty. $\square$

Example 1.6 (i) For a hyperplane $X \simeq \mathbb{P}^n \subset \mathbb{P}^{n+1}$, the Fano variety $F(X, m)$ is simply the Grassmann variety $\mathbb{G}(m, \mathbb{P}^n) \simeq \mathrm{Gr}(m+1, n+1)$. It is of dimension $(m+1)\cdot(n-m)$, as predicted by (1.11).

(ii) For a smooth quadric $X = Q \subset \mathbb{P}^{n+1}$, it is interesting to consider the Fano variety $F(Q, m) \subset \mathbb{G}(m, \mathbb{P}^{n+1})$ of linear subspaces of maximal possible dimension m, which is $m = \lfloor n/2 \rfloor$. In this case, $F(Q, m)$ parametrizes linear subspaces that are isotropic with respect to the quadratic form q defining Q. It is also called the orthogonal or *isotropic Grassmann variety* and sometimes denoted by $\mathrm{OGr}(q, V)$. The following results are classical, see [205, p. 735] or [445, Sec. 2.1]:

• If $n = 2m$, then $F(Q, m)$ consists of two isomorphic, smooth, irreducible components of dimension $\binom{m+1}{2}$. Furthermore, two linear subspaces $\mathbb{P}(W_1), \mathbb{P}(W_2) \subset Q \subset \mathbb{P}^{n+1}$ are contained in the same connected component if and only if $\dim(W \cap W')$ is even. As an example, consider a two-dimensional quadric $Q \simeq \mathbb{P}^1 \times \mathbb{P}^1 \subset \mathbb{P}^3$. In this case, $F(Q, 1) = \mathbb{P}^1 \sqcup \mathbb{P}^1$, where the two components are the factors of $Q \simeq \mathbb{P}^1 \times \mathbb{P}^1$ parametrizing the fibres of the corresponding projections. Note that fibres of the same projection indeed intersect in a $\mathbb{P}^1$ or not at all.

• If $n = 2m + 1$, then $F(Q, m)$ is smooth and irreducible of dimension $\binom{m+2}{2}$. For example, for $Q \simeq \mathbb{P}^1 \subset \mathbb{P}^2$ the Fano variety $F(Q, 0)$ is isomorphic to Q.

The cases of even and odd dimensions are related as follows: consider a smooth quadric $Q \subset \mathbb{P}^{n+1} = \mathbb{P}(V)$ with $n = 2m$ and pick a hyperplane $\mathbb{P}^n \simeq \mathbb{P}(V') \subset \mathbb{P}(V)$ such that $Q' := Q \cap \mathbb{P}(V')$ is smooth. Then $\mathbb{P}(W) \longmapsto \mathbb{P}(W \cap V')$ defines an isomorphism of each of the two components S_m of $F(Q, m) \subset \mathbb{G}(m, \mathbb{P}(V))$ with $F(Q', m-1) \subset \mathbb{G}(m-1, \mathbb{P}(V'))$. The varieties $S_m \simeq F(Q' \subset \mathbb{P}^{2m}, m-1)$ are called *spinor varieties.* For example, $S_1 \simeq \mathbb{P}^1$ and one also knows that $S_2 \simeq \mathbb{P}^3$.

The spinor varieties are homogenous under the action of $\mathrm{SO}(V)$ (and also of $\mathrm{Spin}(V)$) with $\mathrm{Pic}(S_m) \simeq \mathbb{Z}$. The (very) ample generator $\mathcal{O}(1/2)$ is a square root of the Plücker polarization. It is given by a closed embedding

$$S_m \hookrightarrow \mathbb{P}(\textstyle\bigwedge^+ U),$$

where $U \oplus U' \simeq V \simeq k^{2n+2}$ is a fixed decomposition into isotropic subspaces. Here, the even part $\bigwedge^+ U \subset \bigwedge^* U$ is viewed as the half-spinor representation of $\mathrm{Spin}(V)$ and the embedding is obtained by observing that both actions of $\mathrm{Spin}(V)$ have the same stabilizer. In particular, $h^0(S_m, \mathcal{O}(1/2)) = 2^m$.

The case of interest to us is $d = 3$ and $m = 1$. In this case, (1.11) becomes

$$\dim(F(X)) \geq 2n - 4$$

for non-empty $F(X)$. Using deformation theory, we shall see that $F(X)$ really is non-empty of dimension $2n - 4$ for all smooth cubic hypersurfaces of dimension at least two. This shall be explained next.

Remark 1.7 Also relevant for us is the case $d = 3$ and $m = 2$. Then $\dim(F(X, 2)) \geq 3n - 13$ as soon as $F(X, 2)$ is not empty. The right-hand side is non-negative for $n \geq 5$. For $n < 5$, one can conclude that $F(X, 2)$ is empty for generic $X \in |\mathcal{O}(3)|$. So, for example, the generic cubic fourfold does not contain planes. We know already that a smooth cubic threefold cannot contain a plane, see Exercise **1**.1.5 and Remark **1**.3.3.

Remark 1.8 If there exists one smooth cubic hypersurface $X_0 \subset \mathbb{P}^{n+1}$ containing a linear $\mathbb{P}^m$, i.e. $F(X_0, m) \neq \emptyset$, and $(m+1)^2 + 9(m+1) + 2 \leq 6(n+2)$, then $F(X, m) \neq \emptyset$ for all smooth cubic hypersurfaces $X \subset \mathbb{P}^{n+1}$, see Exercise **1**.5.2.

1.3 Local Theory Any further study of the Fano variety of m-planes needs at least some amount of deformation theory. Let us begin with a recollection of some classical facts and a reminder of the main arguments. Most of the following can be found in standard textbooks, e.g. [178, 224, 253, 281, 421]. As (non-)smoothness is preserved under base change, we may assume for simplicity that k is algebraically closed.

As the Fano variety of m-planes is a special case of the Hilbert scheme which in turn is a special case of the Quot-scheme, let us start with the latter.

Let $q := [\mathcal{E} \twoheadrightarrow \mathcal{F}_0] \in \text{Quot} = \text{Quot}_{X/\mathcal{E}}$ be a k-rational point in the Quot-scheme of quotients of a given sheaf $\mathcal{E}$ on X. We denote the kernel by $\mathcal{K}_0 := \text{Ker}\,(\mathcal{E} \twoheadrightarrow \mathcal{F}_0)$. Then there exists a natural isomorphism

$$T_q\text{Quot} \simeq \text{Hom}(\mathcal{K}_0, \mathcal{F}_0),$$

see [212, Exp. 221, Sec. 5]. Moreover, if $\text{Ext}^1(\mathcal{K}_0, \mathcal{F}_0) \simeq 0$, then Quot is smooth at q.

Let us quickly recall the main arguments for both statements. See [178, Ch. 6.4] or [253, Ch. 2.2] for technical details. By the functorial property of the Quot-scheme, the tangent space $T_q\text{Quot}$ parametrizes quotients $\mathcal{E}_{k[\varepsilon]} \twoheadrightarrow \mathcal{F}$ of $\mathcal{E}_{k[\varepsilon]} = \mathcal{E} \boxtimes k[\varepsilon]$ on $X_\varepsilon := X \times \text{Spec}(k[\varepsilon])$ which are flat over $k[\varepsilon]$ and the restriction of which to $X \subset X_{k[\varepsilon]}$ gives back q. It is convenient to study the following more general situation. Let A be a local Artinian k-algebra with residue field k and assume an extension $q_A = [\mathcal{E}_A \twoheadrightarrow \mathcal{F}]$ of $q = [\mathcal{E} \twoheadrightarrow \mathcal{F}_0]$ to $X_A = X \times \text{Spec}(A)$ has been found already. Consider a small extension $A' \twoheadrightarrow A = A'/I$, i.e. a local Artinian k-algebra A' with maximal ideal $\mathfrak{m}_{A'}$ such that $I \cdot \mathfrak{m}_{A'} = 0$. Any further extension of q_A to $q_{A'} = [\mathcal{E}_{A'} \twoheadrightarrow \mathcal{F}']$ leads to a commutative diagram of vertical and horizontal short exact sequences of the form

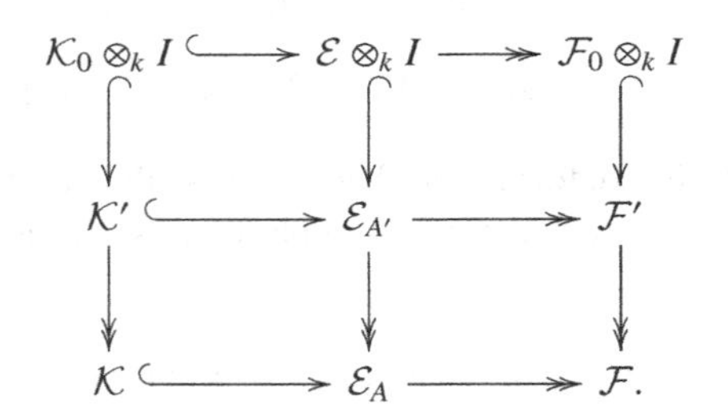

Here, one uses that $\mathcal{F}' \otimes_{A'} I \simeq \mathcal{F}_0 \otimes_k I$, etc. Next observe that

$$\mathcal{F}' \simeq \text{Coker}\,(\psi\colon \mathcal{K} \longrightarrow \mathcal{E}_{A'}/(\mathcal{K}_0 \otimes_k I)),$$

where ψ is the obvious map. Furthermore, the composition of ψ with the projection $\varphi\colon \mathcal{E}_{A'}/(\mathcal{K}_0 \otimes_k I) \twoheadrightarrow \mathcal{E}_A$ is the given inclusion $\mathcal{K} \hookrightarrow \mathcal{E}_A$. Conversely, one can define an extension $\mathcal{F}'$ in this way if the short exact sequence of coherent sheaves on X_A

$$0 \longrightarrow \mathcal{F}_0 \otimes_k I \longrightarrow \varphi^{-1}(\mathcal{K}) \longrightarrow \mathcal{K} \longrightarrow 0 \tag{1.12}$$

is split. The class of (1.12) is an element

$$\mathfrak{o} \in \text{Ext}^1_{X_A}(\mathcal{K}, \mathcal{F}_0 \otimes_k I) \simeq \text{Ext}^1_X(\mathcal{K}_0, \mathcal{F}_0) \otimes_k I,$$

where we use flatness of $\mathcal{K}$ for the isomorphism. If this class is 0 and a split has been chosen, then all other extensions differ by elements in

$$\text{Hom}_{X_{A'}}(\mathcal{K}, \mathcal{F}_0 \otimes_k I) \simeq \text{Hom}_X(\mathcal{K}_0, \mathcal{F}_0) \otimes_k I.$$

Hence, Quot is (formally) smooth at $q = [\mathcal{E} \twoheadrightarrow \mathcal{F}_0]$ if $\mathrm{Ext}^1(\mathcal{K}_0, \mathcal{F}_0) = 0$ and the possible extensions of q to $X \times \mathrm{Spec}(k[\varepsilon])$ are parametrized by $\mathrm{Hom}(\mathcal{K}_0, \mathcal{F}_0)$.

Applied to the Hilbert scheme $\mathrm{Hilb}(X) \simeq \mathrm{Quot}_{X/\mathcal{O}_X}$, one finds that the tangent space at the point $Z \in \mathrm{Hilb}(X)$ is given by

$$T_Z\mathrm{Hilb}(X) \simeq \mathrm{Hom}(\mathcal{I}_Z, \mathcal{O}_Z)$$

and that $\mathrm{Hilb}(X)$ is smooth at the point $Z \in \mathrm{Hilb}(X)$ if $\mathrm{Ext}^1(\mathcal{I}_Z, \mathcal{O}_Z) = 0$. Now assume that $Z \subset X$ is a regular embedding with normal bundle $\mathcal{N}_{Z/X}$. Then, $\mathrm{Hom}(\mathcal{I}_Z, \mathcal{O}_Z) \simeq H^0(Z, \mathcal{N}_{Z/X})$ and the local to global spectral sequence, cf. [246, Ch. 3],

$$E_2^{p,q} = H^p(Z, \mathcal{E}xt^q_X(\mathcal{I}_Z, \mathcal{O}_Z)) \Rightarrow \mathrm{Ext}^{p+q}_X(\mathcal{I}_Z, \mathcal{O}_Z)$$

provides us with an exact sequence

$$H^1(Z, \mathcal{N}_{Z/X}) \hookrightarrow \mathrm{Ext}^1(\mathcal{I}_Z, \mathcal{O}_Z) \xrightarrow{\delta} H^0(Z, \mathcal{E}xt^1_X(\mathcal{I}_Z, \mathcal{O}_Z)) \longrightarrow H^2(Z, \mathcal{N}_{Z/X}).$$

Furthermore, the local obstructions in $\mathcal{E}xt^1_X(\mathcal{I}_Z, \mathcal{O}_Z)$ to deform a smooth subvariety are all trivial. Hence, any obstruction in $\mathrm{Ext}^1(\mathcal{I}_Z, \mathcal{O}_Z)$ maps to 0 under δ and, therefore, is in fact contained in $H^1(Z, \mathcal{N}_{Z/X})$.

Example 1.9 Let us test this in the case of $\mathbb{G}(m, \mathbb{P}) \simeq \mathrm{Hilb}^{P_m}(\mathbb{P})$. On the one hand, we know that at $L = \mathbb{P}(W) \in \mathbb{G} = \mathbb{G}(m, \mathbb{P})$ the tangent space $T_L\mathbb{G}$ is isomorphic to $\mathrm{Hom}(W, V/W)$ or, more globally, that $\mathcal{T}_{\mathbb{G}} \simeq \mathcal{H}om(\mathcal{S}, \mathcal{Q})$ with $\mathcal{S}$ and $\mathcal{Q}$ as in (1.3). On the other hand, $T_L\mathrm{Hilb}(\mathbb{P}) \simeq \mathrm{Hom}(\mathcal{I}_L, \mathcal{O}_L)$. Indeed, there is a natural isomorphism

$$\mathrm{Hom}(W, V/W) \simeq \mathrm{Hom}(\mathcal{I}_L, \mathcal{O}_L)$$

between the two descriptions obtained by applying $\mathrm{Hom}(\ \ , \mathcal{O}_L)$ to the Koszul complex

$$\cdots \longrightarrow \bigwedge^2(V/W)^* \otimes \mathcal{O}(-2) \longrightarrow (V/W)^* \otimes \mathcal{O}(-1) \longrightarrow \mathcal{I}_L \longrightarrow 0,$$

associated with the equations $(V/W)^* \hookrightarrow V^*$ for $L = \mathbb{P}(W)$, and by using the natural isomorphisms $\mathrm{Hom}((V/W)^* \otimes \mathcal{O}(-1), \mathcal{O}_L) \simeq (V/W) \otimes H^0(L, \mathcal{O}_L(1)) \simeq \mathrm{Hom}(W, V/W)$.

Applied to the case $L \in F(X, m) = \mathrm{Hilb}^{P_m}(X)$ for an m-plane $L \subset X$ in a variety $X \subset \mathbb{P}$ that is assumed to be smooth (along L), one obtains the following result.

Proposition 1.10 *Let $L \subset X$ be an m-plane contained in a variety $X \subset \mathbb{P}^{n+1}$ which is smooth along L. Then the tangent space $T_LF(X, m)$ of the Fano variety $F(X, m)$ at the point $L \in F(X, m)$ corresponding to L is naturally isomorphic to $H^0(L, \mathcal{N}_{L/X})$, so*

$$T_LF(X, m) \simeq H^0(L, \mathcal{N}_{L/X}).$$

Furthermore, if $H^1(L, \mathcal{N}_{L/X}) = 0$, then $F(X, m)$ at the point $L \in F(X, m)$ is smooth of dimension $h^0(L, \mathcal{N}_{L/X})$. □

Remark 1.11 Using the description $\mathbb{L} = \{(L, x) \mid x \in L \subset X\}$, we see that the fibres of the projection $q\colon \mathbb{L} \longrightarrow X$ parametrize all m-planes in X passing through a fixed point. On an infinitesimal level, this is expressed by

$$H^0(L, \mathcal{N}_{L/X} \otimes \mathcal{I}_x) \simeq \operatorname{Ker}(dq\colon T_{(L,x)}\mathbb{L} \longrightarrow T_xX),$$

cf. [281, Thm. II.1.7]. The infinitesimal version of (1.7) takes the form

$$\begin{array}{ccccc} & & T_xL & & \\ & & \big\downarrow & & \\ H^0(L,\mathcal{N}_{L/X}\otimes\mathcal{I}_x) & \hookrightarrow & T_{(L,x)}\mathbb{L} & \xrightarrow{dq} & T_xX \\ \big\downarrow & & \big\downarrow{\scriptstyle dp} & & \\ H^0(L,\mathcal{N}_{L/X}) & \xrightarrow{\sim} & T_LF(X,m). & & \end{array}$$

Remark 1.12 There exists a relative version of Proposition 1.10. Assume $\mathcal{X} \longrightarrow S$ is a projective morphism over a locally Noetherian base S and $\mathcal{E}$ is a coherent sheaf on $\mathcal{X}$. Then the relative Quot-scheme $\pi\colon \mathrm{Quot}_{\mathcal{X}/S/\mathcal{E}} \longrightarrow S$ parametrizes T-flat quotients $\mathcal{E}_T \twoheadrightarrow \mathcal{F}_0$ on $\mathcal{X} \times_S T$ for all S-schemes T. It is a locally projective S-scheme with fibres $\pi^{-1}(s) = \mathrm{Quot}_{X/\mathcal{E}|_X}$, where $X := \mathcal{X}_s$. In particular, the relative tangent space at a $k(s)$-rational point $q = [\mathcal{E}|_X \twoheadrightarrow \mathcal{F}_0] \in \mathrm{Quot}_{X/\mathcal{E}|_X} \subset \mathrm{Quot}_{\mathcal{X}/S/\mathcal{E}} \longrightarrow S$ is the tangent space of the fibre $\pi^{-1}(s) = \mathrm{Quot}_{X/\mathcal{E}|_X}$, i.e.

$$T_q\pi^{-1}(s) \simeq T_q\mathrm{Quot}_{X/\mathcal{E}|_X} \simeq \mathrm{Hom}_X(\mathcal{K}_0, \mathcal{F}_0),$$

where $\mathcal{K}_0 = \operatorname{Ker}(\mathcal{E}|_X \twoheadrightarrow \mathcal{F}_0)$. More interestingly, if locally in X there are no obstructions to deform $\mathcal{E}|_X \twoheadrightarrow \mathcal{F}_0$ and $H^1(X_s, \mathcal{H}om(\mathcal{K}_0, \mathcal{F}_0)) = 0$, then π is smooth at q.

This applies to our situation. Consider the universal family $\mathcal{X} \longrightarrow |\mathcal{O}(d)|$ of hypersurfaces of degree d and let $F(\mathcal{X}, m) \longrightarrow |\mathcal{O}(d)|$ be the associated family of Fano varieties of m-planes in the fibres. Then the morphism $F(\mathcal{X}, m) \longrightarrow |\mathcal{O}(d)|$ is smooth at a point L corresponding to an m-plane $L \subset X$ in a smooth (along L) fibre $X = \mathcal{X}_s$ if $H^1(X, \mathcal{N}_{L/X}) = 0$.

1.4 Normal Bundle of a Line To compute the normal bundle $\mathcal{N}_{L/X}$ of an m-plane $\mathbb{P}^m \simeq L \subset X$ we use the short exact sequence

$$0 \longrightarrow \mathcal{N}_{L/X} \longrightarrow \mathcal{N}_{L/\mathbb{P}} \longrightarrow \mathcal{N}_{X/\mathbb{P}}|_L \longrightarrow 0 \tag{1.13}$$

of locally free sheaves on $L \simeq \mathbb{P}^m$, where we again assume that X is smooth (along L).

The normal bundle $\mathcal{N}_{L/\mathbb{P}}$ can be readily computed by comparing the Euler sequences for $L \simeq \mathbb{P}^m$ and for $\mathbb{P} = \mathbb{P}^{n+1}$. One finds $\mathcal{N}_{L/\mathbb{P}} \simeq \mathcal{O}(1)^{\oplus n+1-m}$. More precisely, if $L = \mathbb{P}(W) \subset \mathbb{P} = \mathbb{P}(V)$, then there is a natural isomorphism $\mathcal{N}_{L/\mathbb{P}} \simeq \mathcal{O}_L(1) \otimes (V/W)$.

If now $X \subset \mathbb{P}$ is a smooth (at least along L) hypersurface of degree d, then the exact sequence (1.13) becomes

$$0 \longrightarrow \mathcal{N}_{L/X} \longrightarrow \mathcal{O}_L(1)^{\oplus n+1-m} \longrightarrow \mathcal{O}_L(d) \longrightarrow 0. \tag{1.14}$$

After a coordinate change, the surjection is given by $\partial_i F$, $i = m+1, \ldots, n+1$. Indeed, assume that $L = V(x_{m+1}, \ldots, x_{n+1})$. Then

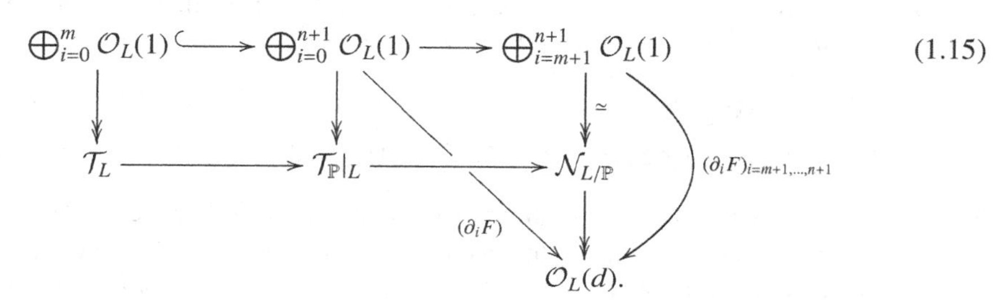

$$\begin{array}{ccccc} \bigoplus_{i=0}^{m} \mathcal{O}_L(1) & \hookrightarrow & \bigoplus_{i=0}^{n+1} \mathcal{O}_L(1) & \longrightarrow & \bigoplus_{i=m+1}^{n+1} \mathcal{O}_L(1) \\ \downarrow & & \downarrow & & \downarrow \simeq \\ \mathcal{T}_L & \longrightarrow & \mathcal{T}_{\mathbb{P}}|_L & \longrightarrow & \mathcal{N}_{L/\mathbb{P}} \\ & & & \searrow^{(\partial_i F)} & \downarrow \\ & & & & \mathcal{O}_L(d). \end{array} \tag{1.15}$$

with the curved arrow $\bigoplus_{i=m+1}^{n+1} \mathcal{O}_L(1) \to \mathcal{O}_L(d)$ labelled $(\partial_i F)_{i=m+1,\ldots,n+1}$.

Compare the proof of Corollary 2.6 for a more invariant version of the diagram.

Observe that (1.14) has the following numerical consequences:

$$\det(\mathcal{N}_{L/X}) = \mathcal{O}_L((n+1-m)-d), \quad \mathrm{rk}(\mathcal{N}_{L/X}) = n-m, \quad \text{and}$$

$$\begin{aligned} \chi(\mathcal{N}_{L/X}) &= \chi(\mathcal{O}_L(1)) \cdot (n+1-m) - \chi(\mathcal{O}_L(d)) \\ &= (m+1) \cdot (n+1-m) - \binom{m+d}{d}, \end{aligned}$$

which equals the right-hand side of (1.11).

For $m = 1$ and $d = 3$, these observations allow us to classify all normal bundles.

Lemma 1.13 *Let $L \subset X$ be a line in a smooth (along L) cubic hypersurface $X \subset \mathbb{P}^{n+1}$. Then $\mathcal{N}_{L/X} \simeq \mathcal{O}_L(a_1) \oplus \cdots \oplus \mathcal{O}_L(a_{n-1})$, $a_1 \geq \cdots \geq a_{n-1}$, with*

$$(a_1, \ldots, a_{n-1}) = \begin{cases} (1, \ldots, 1, 0, 0) & \text{or} \\ (1, \ldots, 1, 1, -1). \end{cases}$$

Proof As $L \simeq \mathbb{P}^1$, any locally free sheaf on L is isomorphic to a direct sum of invertible sheaves, so $\mathcal{N}_{L/X} \simeq \bigoplus \mathcal{O}_L(a_i)$ with $\sum a_i = n-3$. However, the inclusion $\mathcal{N}_{L/X} \subset \mathcal{O}_L(1)^{\oplus n}$ implies $a_i \leq 1$, which is enough to conclude. □

Together with Proposition 1.10 and Remark 1.12, the lemma implies the following.

Corollary 1.14 *The Fano variety of lines $F(X)$ of a smooth cubic hypersurface $X \subset \mathbb{P}^{n+1}$ is smooth and of dimension $2n-4$ if not empty.*

Furthermore, the universal Fano variety $F(\mathcal{X}) \longrightarrow |\mathcal{O}_{\mathbb{P}}(3)|$ is smooth over $|\mathcal{O}_{\mathbb{P}}(3)|_{\mathrm{sm}}$.

Proof Indeed, for both cases, $(a_i) = (1, \ldots, 1, 0, 0)$ and $(a_i) = (1, \ldots, 1, 1, -1)$, we have $H^1(L, \mathcal{N}_{L/X}) = 0$ and hence $F(X)$ is smooth. The dimension formula follows from $h^0(L, \bigoplus \mathcal{O}(a_i)) = 2n - 4$ in the two cases. □

Definition 1.15 Lines with $(a_i) = (1, \ldots, 1, 0, 0)$ and $(a_i) = (1, \ldots, 1, 1, -1)$ are called *lines of the first type* and *of the second type*, respectively.

Remark 1.16 It will come in handy, see e.g. Remark 2.20, to write the normal bundle of lines of both types in a more invariant form as

$$\mathcal{N}_{L/X} \simeq \begin{cases} \left(H^0(L, \mathcal{N}_{L/X}(-1)) \otimes \mathcal{O}_L(1)\right) \oplus \left(H^1(L, \mathcal{N}_{L/X}(-2)) \otimes \det(W^*) \otimes \mathcal{O}_L\right), & \text{respectively} \\ \left(H^0(L, \mathcal{N}_{L/X}(-1)) \otimes \mathcal{O}_L(1)\right) \oplus \left(H^1(L, \mathcal{N}_{L/X}(-1)) \otimes \det(W^*) \otimes \mathcal{O}_L(-1)\right). & \end{cases}$$

The factor $\det(W^*)$ in the second summands comes in since $H^1(L, \mathcal{O}_L(-2))$ is one-dimensional, but it does not come with a natural basis. Indeed, from the Euler sequence one has the natural isomorphism $\omega_L \simeq \mathcal{O}_L(-2) \otimes \det(W^*)$, which combined with the natural trivialization of $H^1(L, \omega_L)$ implies $H^1(L, \mathcal{O}_L(-2)) \simeq \det(W)$.

Hence, for a line $L \subset X$ of the second type, there exists a natural isomorphism

$$\det(\mathcal{N}_{L/X}) \simeq \det H^0(L, \mathcal{N}_{L/X}(-1)) \otimes H^1(L, \mathcal{N}_{L/X}(-1)) \otimes \det(W^*) \otimes \mathcal{O}_L(n-3).$$

On the other hand, the normal bundle sequence for $L \subset X$ leads to a natural isomorphism $\det(\mathcal{N}_{L/X}) \simeq \omega_X^*|_L \otimes \omega_L \simeq \det(W^*) \otimes \mathcal{O}_L(n-3)$. Combining the two isomorphisms, one obtains a natural (in L and for fixed X) isomorphism

$$\det H^0(L, \mathcal{N}_{L/X}(-1)) \simeq H^1(L, \mathcal{N}_{L/X}(-1))^*. \tag{1.16}$$

Exercise 1.17 Show that for any line $L \subset X$ in a smooth cubic hypersurface, the normal bundle sequence splits and, therefore, $\mathcal{T}_X|_L \simeq \mathcal{O}(2) \oplus \mathcal{N}_{L/X}$. Thus, the property of being of the first or of the second type can also be read off from the shape of $\mathcal{T}_X|_L$.

Exercise 1.18 Consider the normal bundle $\mathcal{N}_{\mathbb{L}/F(X)\times X}$ of the natural inclusion $\mathbb{L} \subset F(X) \times X$ and globalize Proposition 1.10 to the isomorphism

$$\mathcal{T}_{F(X)} \simeq p_* \mathcal{N}_{\mathbb{L}/F(X)\times X}$$

of the tangent bundle of the Fano variety.

Proposition 1.19 *Let $X \subset \mathbb{P}^{n+1}$ be a smooth cubic hypersurface, $n \geq 2$. Then the Fano variety of lines $F(X)$ is smooth, projective, and of dimension*

$$\dim(F(X)) = 2n - 4.$$

Proof The preceding discussion essentially proves the claim. It only remains to show that $F(X)$ is non-empty. For this, consider the Fermat cubic $X_0 = V(x_0^3 + \cdots + x_{n+1}^3)$ which is smooth for $\mathrm{char}(k) \neq 3$. Then clearly the line $L_0 := V(x_0 + x_1, x_2 + x_3, x_4, \ldots, x_{n+1})$

is contained in X_0 and hence $F(X_0) \neq \emptyset$. See Remark 1.5.1 for a similar observation. For char$(k) = 3$ with $\xi = \sqrt{-1} \in k$, one may take $X_0 := V(\sum_0^n x_i x_{i+1}^2 + x_0^3)$, see Section 1.2.2, and the line $L_0 := V(x_0 - \xi x_1, x_2, x_4, \ldots, x_{n+1})$. Of course, for the assertion one may assume $k = \bar{k}$.

According to Remark 1.12, the vanishing $H^1(L_0, \mathcal{N}_{L_0/X}) = 0$ not only proves that the fibre $F(X_0)$ of $F(\mathcal{X}) \twoheadrightarrow |\mathcal{O}(3)|$ over the point $X_0 \in |\mathcal{O}(d)|$ is smooth at the point $L_0 \in F(X_0)$ but that in fact the morphism is smooth at the point L_0. In particular, the projective morphism $F(\mathcal{X}) \twoheadrightarrow |\mathcal{O}(d)|$ is surjective which proves $F(X) \neq \emptyset$ for all cubics. Alternatively, one can combine $\dim(F(X_0)) = 2n-4$ with Corollary 1.5 to conclude that the generic non-empty fibre is of dimension exactly $2n - 4$. Hence, again by Corollary 1.5, $F(\mathcal{X}) \twoheadrightarrow |\mathcal{O}(3)|$ has to be surjective, i.e. $F(X)$ is non-empty for all X. Another, more direct argument will be given in Remark 3.6. □

Exercise 1.20 Assume that a cubic hypersurface $X \subset \mathbb{P}^{n+1}$ contains two distinct, but intersecting, lines $L_1, L_2 \subset X$. Show that there then exists a plane $\mathbb{P}^2 \subset \mathbb{P}^{n+1}$ such that either $L_1 \cup L_2 \subset \mathbb{P}^2 \subset X$ or the intersection $\mathbb{P}^2 \cap X$ is a union $L_1 \cup L_2 \cup L$ of three lines. This applies to all smooth cubic hypersurfaces of dimension at least two.

1.5 (Uni-)Rational Parametrization II The existence of lines in cubic hypersurfaces has the following immediate consequence.

Corollary 1.21 *For a smooth cubic hypersurface $X \subset \mathbb{P}^{n+1}$ of dimension $n > 1$ defined over an algebraically closed field, there exists a rational dominant map of degree two*

$$\mathbb{P}^n \dashrightarrow X.$$

Thus, cubic hypersurfaces of dimension at least two are unirational, cf. Example **1.5.10**.

The assumption on the field can be weakened, cf. [283].[2] One only needs the existence of one line contained in X.

As observed at the end of Section **1.5.3**, there also exists a dominant rational map of degree two

$$X \dashrightarrow \mathbb{P}^n.$$

Proof Pick a line $L \subset X$ and consider the projectivization of the restricted tangent bundle $\mathbb{P}(\mathcal{T}_X|_L) \twoheadrightarrow L$. A point in $\mathbb{P}(\mathcal{T}_X|_L)$ is represented by a tangent vector $0 \neq v \in T_xX$, which then defines a unique line $L_v \subset \mathbb{P}$ passing through x with $T_xL_v \subset T_xX$ spanned by v. Then, either the line L_v is contained in X or, and this is the generic case, it is not. In the latter case, L_v intersects X in a unique point $y_v \in X$ with the property that the scheme theoretic intersection $L_v \cap X$ is $2x + y_v$. Note that $y_v = x$ can occur.

[2] Thanks to X. Wei for the reference.

Unless X contains a hyperplane, one defines in this way a rational map

$$\mathbb{P}(\mathcal{T}_X|_L) \dashrightarrow X, \quad v \longmapsto y_v, \tag{1.17}$$

which is regular on a dense open subset intersecting each fibre $\mathbb{P}(T_xX)$, $x \in L$.

Now, pick a point $y \in X \setminus L$ in the image and consider the cubic curve $C_y := \overline{yL} \cap X$. If the residual conic Q of $L \subset C_y = L \cup Q$ does not contain L, then there exist at most two lines $y \in L_i$, $i = 1, 2$, that intersect L and, at the same time, are tangent to X at their intersection points. They correspond to the (at most) two points of the intersection $L \cap Q$. In other words, under the assumption on C_y the non-empty fibre of $v \longmapsto y_v$ over y would consists of at most two points. Furthermore, for dimension reasons, $\mathbb{P}(\mathcal{T}_X|_L) \dashrightarrow X$ would be dominant. To show that $C_y = L \cup Q$ satisfies $L \not\subset Q$ for at least one (and then for the generic) y, take any tangent vector $v \in T_x\mathbb{P}^{n+1} \setminus T_xX$, $x \in L$. Then the line L_v is not tangent to X at x and, therefore, $L \not\subset Q$ for the residual conic of $L \subset \overline{L_vL} \cap X$. But then for any point $y \in Q$ the cubic C_y has the desired property. Hence, the rational map (1.17) is generically of degree two and dominant. □

Remark 1.22 The restriction of the map (1.17) to the fibre over one point $x \in L$ defines a generically injective rational map

$$\mathbb{P}^{n-1} \simeq \mathbb{P}(T_xX) \hookrightarrow X. \tag{1.18}$$

The indeterminacies of this map are contained in the set of tangent directions v for which the line L_v is contained in X. Since the line L_v is determined by the two points x, y_v as soon as $x \neq y_v$, the map is injective on the open subset of tangent directions with $L_v \cap X \neq 3x$. Note that this open set is not empty. Indeed, it contains the tangent direction of any line $x \in L \subset \mathbb{P}$ tangent to X at x and going through a point y in $X \cap \mathbb{P}^n$, where $\mathbb{P}^n \subset \mathbb{P}$ is a generic linear subspace not containing x, cf. Remark 3.6. Warning: the rational subvariety (1.18) is not linear.

Remark 1.23 The *Lüroth problem* is a classical question in algebraic geometry that asks whether a unirational variety is automatically rational. This holds in dimension one and in dimension two over algebraically closed fields of characteristic zero. Much of the work on cubic hypersurfaces has been triggered by the Lüroth problem. Cubic surfaces are in fact rational. However, smooth cubic threefolds are never rational, see Section **5**.4.5. The general cubic fourfold is not expected to be rational, but special ones are, see Conjectures **6**.5.15 and **7**.3.1.

Sometimes rationality can be excluded by topological properties not necessarily satisfied by unirational varieties. For example, the celebrated article by Artin and Mumford [24] uses torsion in $H^3(X, \mathbb{Z})$ to exhibit a unirational threefold that is not rational. However, the easiest topological invariant, the fundamental group $\pi_1(X)$, does not distinguish between rational and unirational varieties. Indeed, as proved by Serre [423], unirational varieties in characteristic zero are simply connected and, more generally, rationally connected varieties are simply connected by a result of Kollár [282]. For cubic hypersurfaces, this provides us with an alternative proof of Corollary **1**.1.1.

Note that in positive characteristic, already for surfaces, unirationality neither implies rationality nor simply connectedness, see [433, 434].

Remark 1.24 There is a general expectation (conjecture of Debarre–de Jong) that for $d \leq n+1$ and $\text{char}(k) = 0$ or at least $\text{char}(k) \geq d$, the Fano variety of lines $F(X)$ is smooth of the expected dimension $2n - d - 1$. For $d \leq 6$, this has been proved in characteristic zero by Beheshti [60] to which we also refer for further references. See also [174, Prop. 6.40], where the claim is reduced to the case $d = n + 1$.

Remark 1.25 Note that for $m \geq 2$ and an m-plane $L \subset X$ contained in a smooth (along L) hypersurface X of degree d such that $\mathcal{N}_{L/X} \simeq \bigoplus \mathcal{O}(a_i)$, the Fano variety $F(X, m)$ at the point $L \in F(X, m)$ is smooth of dimension $\sum h^0(\mathbb{P}^m, \mathcal{O}(a_i))$. However, in contrast to locally free sheaves on $\mathbb{P}^1$, there is a priori no reason why $\mathcal{N}_{L/X}$ on $L \simeq \mathbb{P}^m$ should be a direct sum of invertible sheaves. Also the dimension of $F(X, m)$ is harder to control as other cohomology groups $H^i(L, \mathcal{N}_{L/X})$ enter the picture.

2 Lines of the First and Second Type

We come back to the difference between lines of the first and of the second type. Various characterizations are available, e.g. via the Gauss map or via linear subspaces tangent to the line. We introduce the Fano variety $F_2(X)$ of lines of the second type and determine its dimension, similar to the computation for $F(X)$ itself.

2.1 Linear Spaces Tangent to a Line As a warm-up we propose the following.

Exercise 2.1 Prove that for a line $L \subset X$ in a smooth cubic hypersurface the following conditions are equivalent:

(i) L is of the first type;
(ii) $H^1(L, \mathcal{N}_{L/X}(-1)) = 0$; and
(iii) $h^0(L, \mathcal{N}_{L/X}(-1)) = n - 3$.

Find similar descriptions in terms of the restriction of the tangent bundle $\mathcal{T}_X|_L$, cf. Exercise 1.17. Assume now that L is contained in a smooth hyperplane section $Y = X \cap H$ and show that if L is of the first type as a line in Y then it is so as a line in X. The converse does not hold in general.

Remark 2.2 From Exercise 2.1 and (1.14) we deduce that a line $L \subset X$ in a smooth cubic hypersurface $X = V(F)$ is of the first type if and only if the partial derivatives $\partial_i F|_L \in H^0(L, \mathcal{O}_L(2))$ span the three-dimensional space $H^0(L, \mathcal{O}_L(2))$. Hence, L is of the second type if and only if $\langle \partial_i F|_L \rangle \subset H^0(L, \mathcal{O}_L(2))$ is of dimension two. Note that $\dim(\langle \partial_i F|_L \rangle) \geq 2$ holds for all lines, as otherwise the $\partial_i F$ would have a common 0 in at least one point of L, contradicting the smoothness of X.

For $L = \mathbb{P}(W)$, the surjection $\mathcal{N}_{L/\mathbb{P}} \twoheadrightarrow \mathcal{N}_{X/\mathbb{P}}|_L$ can be viewed as a map $V/W \otimes \mathcal{O}_L(1) \longrightarrow \mathcal{O}_L(3)$, cf. (1.15). After twisting and taking global sections, it gives the short exact sequence

$$0 \longrightarrow \mathcal{N}_{L/X}(-1) \longrightarrow \underset{\simeq V/W \otimes \mathcal{O}_L}{\mathcal{N}_{L/\mathbb{P}}(-1)} \longrightarrow \mathcal{O}_L(2) \longrightarrow 0$$

with its associated long exact cohomology sequence

$$0 \longrightarrow H^0(L, \mathcal{N}_{L/X}(-1)) \longrightarrow V/W \longrightarrow \underset{= S^2(W^*).}{H^0(L, \mathcal{O}_L(2))} \longrightarrow H^1(L, \mathcal{N}_{L/X}(-1)) \longrightarrow 0$$

The map in the middle

$$\psi_L \colon V/W \longrightarrow H^0(L, \mathcal{O}_L(2)) = S^2(W^*) \tag{2.1}$$

is of rank at least two. It is of rank three, i.e. surjective, if and only if L is of the first type.

The map (2.1) can be described more abstractly: the cubic polynomial defining X is an element $F \in S^3(V^*)$, which by contraction defines a map $i_F \colon V \longrightarrow S^2(V^*)$. In other words, i_F maps a vector $x \in V$ to the partial derivative $\partial_x F$. Composing with the natural projection $S^2(V^*) \twoheadrightarrow S^2(W^*)$ gives

$$\begin{array}{ccccc} V & \xrightarrow{i_F} & S^2(V^*) & \twoheadrightarrow & S^2(W^*) \\ & \searrow & & \nearrow_{\psi_L} & \\ & & V/W. & & \end{array}$$

Here, the dotted arrow exists as $L = \mathbb{P}(W) \subset X$.

Example 2.3 Consider the Fermat cubic $X = V(F = \sum x_i^3) \subset \mathbb{P}^{n+1}$ over a field k with $\mathrm{char}(k) \neq 3$. Then $L := V(x_0 + x_1, x_2 + x_3, x_4, \ldots, x_{n+1})$ is a line of the second type contained in X, cf. the proof of Proposition 1.19.

Indeed, under $\mathbb{P}^1 \xrightarrow{\sim} L$, $[t : s] \longmapsto [t : -t : s : -s : 0 : \cdots : 0]$ the partial derivatives $\partial_i F$, $i = 0, \ldots, n+1$, pull back to $3\,t^2, 3\,t^2, 3\,s^2, 3\,s^2, 0, \ldots, 0$ and, therefore, span a subspace of $H^0(\mathbb{P}^1, \mathcal{O}_{\mathbb{P}^1}(2))$ of dimension two only. We leave it to the reader to work out an example in the case $\mathrm{char}(k) = 3$.

Lines L of the first and the second type in X can also be distinguished by the existence of linear subspaces containing L that are tangent to X at every point of L. We need to recall some facts from classical algebraic geometry.

Definition 2.4 The *projective tangent space* of a hypersurface $X = V(F)$ at a point $y \in X$ is the linear space

$$\mathbb{P}^n \simeq \mathbb{T}_y X := V\left(\sum x_i\, \partial_i F(y)\right) \subset \mathbb{P}^{n+1}.$$

The projective tangent space is independent of the choice of the equation F and of the linear coordinates $x_0, \ldots, x_{n+1}$. The Euler equation implies $y \in \mathbb{T}_y X$.

Lemma 2.5 *Consider a hypersurface $X \subset \mathbb{P}^{n+1}$ and a linear subspace $\mathbb{P}^m \simeq H \subset \mathbb{P}^{n+1}$ not contained in X. Then for $y \in H \cap X$, the following conditions are equivalent:*

(i) *$H \subset \mathbb{T}_y X$, i.e. H is tangent to X at the point y.*
(ii) *y is a singular point of $X \cap H$.*

Proof Choose linear coordinates such that $H = V(x_{m+1}, \ldots, x_{n+1})$. Then $H \subset \mathbb{T}_y X$ if and only if $x_0\, \partial_0 F(y) + \cdots + x_m\, \partial_m F(y) = 0$ for all $[x_0 : \cdots : x_m]$, which of course is equivalent to $\partial_0 F(y) = \cdots = \partial_m F(y) = 0$. As the $\partial_i F(y)$, $i = 0, \ldots, m$, are the partial derivatives of the restriction $F|_H$, which defines the intersection $X \cap H$, this proves the assertion. □

In Section **5**.1.1, especially Lemma **5**.1.8, and Section **6**.0.2, the next result will be discussed again and in more detail for cubic hypersurfaces of dimension three and four.

Corollary 2.6 *Assume $L \subset X \subset \mathbb{P}^{n+1}$ is a line contained in a smooth cubic.*

(i) *The line L is of the first type if and only if there exists a unique linear subspace $L \subset P_L \simeq \mathbb{P}^{n-2}$ that is tangent to X at every point $y \in L$.*
(ii) *The line L is of the second type if and only if there exists a linear subspace $L \subset P_L \simeq \mathbb{P}^{n-1}$ that is tangent to X at every point $y \in L$. In this case, P_L is unique.*

Furthermore, the linear subspace P_L can be described as the intersection of all tangent spaces at points of L, i.e.

$$P_L = \bigcap_{y \in L} \mathbb{T}_y X.$$

Proof We start with a direct argument to prove (ii).

If L is of the second type, then by Remark 2.2 all derivatives $\partial_i F|_L$ are linear combinations of two quadratic forms $Q, Q' \in H^0(L, \mathcal{O}_L(2))$, i.e. $\partial_i F(y) = a_i\, Q(y) + a'_i\, Q'(y)$ for all $y \in L$. Hence, we can write $\sum x_i\, \partial_i F(y) = (\sum a_i\, x_i) Q(y) + (\sum a'_i x_i) Q'(y)$ and define

$$P_L := V\left(\sum a_i\, x_i, \sum a'_i\, x_i\right) \subset \mathbb{P}^{n+1}.$$

Then $P_L \simeq \mathbb{P}^{n-1}$ satisfies $P_L \subset \mathbb{T}_y X$ for all $y \in L$.

Conversely, assume for a line $L \subset X$ that there exists $L \subset P_L \simeq \mathbb{P}^{n-1}$ with $P_L \subset \mathbb{T}_y X$ for all $y \in L$. If, to simplify the notation, $L = V(x_2, \ldots, x_{n+1}) \subset P_L = V(x_n, x_{n+1})$, this is equivalent to $\partial_0 F(y) = \cdots = \partial_{n-1} F(y) = 0$ for all $y \in L$. Thus, the derivatives $\partial_i F|_L$ span only a two-dimensional space in $H^0(L, \mathcal{O}_L(2))$ and, therefore, L is of the second type.

An alternative argument, which is more invariant and also proves (i), goes as follows: if $L = \mathbb{P}(W)$ is contained in a linear subspace $\mathbb{P}(U)$, then rewriting (1.15) as

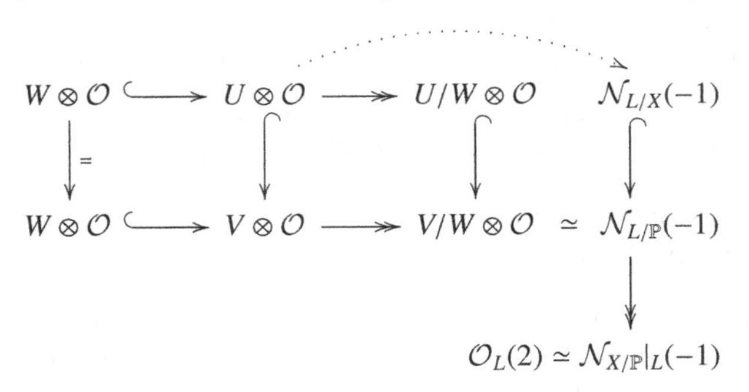

shows that $\mathbb{P}(U)$ is tangent to X at every point of L if and only if the subspace $U \subset V$ maps to $\mathcal{N}_{L/X}(-1) \subset \mathcal{N}_{L/\mathbb{P}}(-1)$ under the surjection $V \otimes \mathcal{O} \twoheadrightarrow \mathcal{N}_{L/\mathbb{P}}(-1)$.

In (i) and (ii), we let $P_L = \mathbb{P}(U)$ with $U \subset V$ defined as the pre-image under the map $V \otimes \mathcal{O} \twoheadrightarrow \mathcal{N}_{L/\mathbb{P}}(-1)$ of the maximal trivial sub-bundle $\mathcal{O}_L^{\oplus k} \subset \mathcal{N}_{L/X}(-1) \subset \mathcal{N}_{L/\mathbb{P}}(-1) \simeq V/W \otimes \mathcal{O}$ with $k = n-3$ and $k = n-2$, respectively.

To prove the last assertion, use that by virtue of Lemma 2.5, $P_L \subset \mathbb{T}_y X$ if and only if $X \cap P_L$ is singular at y. Since by definition, P_L is tangent to X at every point $y \in L$, i.e. $X \cap P_L$ is singular at every point $y \in L$, this proves the inclusion $P_L \subset \bigcap \mathbb{T}_y X$. As the intersection is linear, the same argument proves equality. □

Remark 2.7 For later use, we emphasize that P_L can be identified as

$$P_L = \mathbb{P}(U) \subset \mathbb{P}(V) \quad \text{with } W \subset U \twoheadrightarrow H^0(L, \mathcal{N}_{L/X}(-1)) \subset V/W,$$

i.e. $U \subset V$ is the pre-image of $H^0(L, \mathcal{N}_{L/X}(-1)) \subset V/W$ under the natural projection.

Remark 2.8 If L is of the second type, then the linear subspaces $L \subset \mathbb{P}^{n-2} \subset \mathbb{P}^{n+1}$ that are tangent to X at all points of L are exactly the linear subspaces $L \subset \mathbb{P}^{n-2} \subset P_L$. They are parametrized by $\mathbb{P}((U/W)^*) \simeq \mathbb{P}^{n-3}$, where we write $L = \mathbb{P}(W)$ and $P_L = \mathbb{P}(U)$.

Exercise 2.9 Fix a line $L = \mathbb{P}(W) \subset X$ in a smooth cubic hypersurface $X \subset \mathbb{P}^{n+1}$.

(i) Consider a plane $L = \mathbb{P}(W) \subset \mathbb{P}^2 \subset \mathbb{P}^{n+1}$ not contained in X and let $Q \subset \mathbb{P}^2 \cap X$ be the residual conic of $L \subset \mathbb{P}^2 \cap X$. Show that $L \subset Q$ if and only if $\mathbb{P}^2 \subset P_L$.

(ii) Deduce from (i) that the line $W_y \subset V/W$ corresponding to a point $y \in \mathbb{P}^{n+1} \setminus L$ is contained in the kernel of $\psi_L\colon V/W \longrightarrow S^2(W^*) = H^0(L, \mathcal{O}(2))$, see Remark 2.2, if and only if $y \in P_L$.

(iii) Show that the zero set in L of $\psi_L(W_y) \subset H^0(L, \mathcal{O}(2))$ is the intersection $L \cap Q_y$ of L with the residual conic Q_y of $L \subset \overline{yL} \cap X$.

Note that by (ii) for $y \in P_L$, the image $\psi(W_y)$ is trivial and, thus, its zero set is all of L, which fits with (i) saying $L \subset Q_y$ in this case.

Another way of rephrasing the difference between lines of the first and of the second type in more geometric terms uses the *Gauss map*.[3] Recall that the Gauss map for a hypersurface $X = V(F) \subset \mathbb{P} := \mathbb{P}^{n+1}$ is the map

$$\gamma_X \colon \mathbb{P} \longrightarrow \mathbb{P}^*, \quad x \longmapsto [\partial_0 F(x) : \cdots : \partial_{n+1} F(x)],$$

which is regular for smooth X, see Lemma **1**.3.1. For $d > 1$, the morphism is not constant and hence finite. Then, also its restriction

$$\gamma_X \colon X \twoheadrightarrow X^* := \gamma_X(X)$$

onto the dual variety X^* is finite. Geometrically, γ_X maps $y \in X$ to the projective tangent space $\mathbb{T}_y X = V(\sum x_i\, \partial_i F(y))$. Hence, the fibre of γ_X over a point $[H] \in X^*$ corresponding to a hyperplane $H \subset \mathbb{P}$ is the set of all points $x \in X$ at which H is tangent to X. Moreover, γ_X is generically injective as $\gamma_{X^*} \circ \gamma_X = \mathrm{id}$, cf. [174, Ch. 10]. In other words, $\gamma_X \colon X \twoheadrightarrow X^*$ is the normalization of X^*.

As an immediate consequence of Remark 2.2, one then finds the following.

Exercise 2.10 Let $L \subset X$ be a line contained in a smooth cubic hypersurface X. Prove the following assertions:

(i) The line L is of the first type if and only if $\gamma_X \colon L \xrightarrow{\sim} \gamma_X(L)$ is an isomorphism onto a smooth plane conic.

(ii) The line L is of the second type if $\gamma_X \colon L \longrightarrow \gamma_X(L)$ is a degree two covering of a line.

Note that Lemma 1.13 applied to the case $n = 2$ leaves only one possibility, namely $\mathcal{N}_{L/X} \simeq \mathcal{O}_L(-1)$, which counts as a line of the first type. For $n > 2$, there are two cases and both can be geometrically realized on any smooth cubic hypersurface, see [174, Prop. 6.30]. Indeed, for lines of the second type, this follows from Example 2.3 and Proposition 2.13. For lines of the first type, combine the dimension formulae in Lemma 2.12 and Proposition 1.19.

[3] Often, when the Gauss map is involved, some assumptions on the characteristic $\mathrm{char}(k)$ of the ground field k have to be made. As we will usually only consider hypersurfaces of degree three, it will suffice to assume that $\mathrm{char}(k) \neq 2, 3$.

2.2 Lines of the Second Type Assume $n > 2$. Lines of the first type are generic or, equivalently, the set of lines of the second type

$$F_2(X) := \{ L \mid h^0(L, \mathcal{N}_{L/X}(-1)) \geq n-2 \} \subset F(X) \tag{2.2}$$

is a proper closed subscheme.

Remark 2.11 The characterization of lines of the second type given in Remark 2.2 in terms of the map ψ_L in (2.1) globalizes to the following description of $F_2(X)$: writing $\mathcal{S}_F$ and $\mathcal{Q}_F$ for the restriction of the universal sub-bundle and the universal quotient bundle to $F = F(X) \hookrightarrow \mathbb{G}(1, \mathbb{P})$, there exists a canonical sheaf homomorphism

$$\psi\colon \mathcal{Q}_F \longrightarrow S^2(\mathcal{S}_F^*)$$

for which $F_2(X)$ is the degeneracy locus

$$F_2(X) = M_2(\psi) := \{ L \in F(X) \mid \mathrm{rk}(\psi_L) \leq 2 \},$$

As $\mathrm{rk}(S^2(\mathcal{S}_F^*)) = 3$, the locus $F_2(X)$ is isomorphic to the zero locus of a global section of a locally free sheaf. Indeed, consider the projective bundle $\pi\colon \mathbb{P}(S^2(\mathcal{S}_F)) \longrightarrow F$ with the tautological injection $\mathcal{O}_\pi(-1) \longrightarrow \pi^* S^2(\mathcal{S}_F)$. The composition with the pullback of the dual of ψ leads to

$$\mathcal{O}_\pi(-1) \longrightarrow \pi^* S^2(\mathcal{S}_F) \xrightarrow{\pi^*\psi^*} \pi^* \mathcal{Q}_F^*,$$

which can also be considered as a global section $\tilde{\psi} \in H^0(\mathbb{P}(S^2(\mathcal{S}_F)), \pi^* \mathcal{Q}_F^* \otimes \mathcal{O}_\pi(1))$. Then, the projection π induces an isomorphism

$$\pi\colon V(\tilde{\psi}) \xrightarrow{\sim} F_2(X).$$

This observation immediately implies a dimension formula for $F_2(X)$.

Lemma 2.12 *If not empty, the locus $F_2(X) \subset F(X)$ of lines of the second type contained in a smooth cubic hypersurface X of dimension n is of dimension*

$$\dim(F_2(X)) = n - 2 = (1/2)\dim(F(X)).$$

Furthermore, the closed set of all $x \in X$ contained in a line of the second type is of dimension at most $n-1$. See Corollary 2.15 *for a stronger statement.*

Proof The proof has two parts. First, the usual dimension formula for degeneracy loci shows $\mathrm{codim}(M_2(\psi)) \leq (\mathrm{rk}(\mathcal{Q}_F) - 2) \cdot (\mathrm{rk}(S^2(\mathcal{S}_F^*)) - 2) = n - 2$, cf. [23, 174, 189]. Therefore, $\dim(F_2(X)) \geq n-2$. The second part consists of proving that $n-2$ is an upper bound for the dimension which was first observed by Clemens and Griffiths [119, Cor. 7.6].

As $\gamma_X\colon X \longrightarrow X^*$ is generically injective, the image of $q\colon \mathbb{L}_2 := p^{-1}(F_2) \longrightarrow X$ is a proper closed subscheme and, therefore, of dimension at most $n-1$. Hence, as the first projection $p\colon \mathbb{L}_2 \twoheadrightarrow F_2$ is of relative dimension one, it suffices to show that the map $q_2 := q|_{\mathbb{L}_2}\colon \mathbb{L}_2 \twoheadrightarrow q(\mathbb{L}_2) \subset X$ is generically finite, which will also prove the last assertion.

For $x \in q_2(\mathbb{L}_2)$, consider $\iota_x\colon q_2^{-1}(x) \longrightarrow X$, $L \longmapsto \iota_L(x)$, where $\iota_L\colon L \longrightarrow L$ is the covering involution for the restriction of the Gauss map $\gamma_X|_L\colon L \longrightarrow X^*$, see Exercise 2.10:

$$\begin{array}{ccc} \iota_L \circlearrowright L & \hookrightarrow & X \\ {\scriptstyle 2:1}\downarrow & & \downarrow{\scriptstyle \gamma_X} \\ \gamma_X(L) & \hookrightarrow & X^*. \end{array}$$

Thus, $\gamma_X(\iota_x(L)) = \gamma_X(x)$ for all $L \in q_2^{-1}(x)$ and, therefore, the image of ι_x is finite. If $x \neq \iota_L(x)$, the line $L \in q_2^{-1}(x)$ is the unique line through x and $\iota_L(x)$. Hence, ι_x is injective on the open subset of lines $L \in q_2^{-1}(x)$ satisfying $x \neq \iota_L(x)$, which implies that this set is finite. It remains to prove that the lines with $x = \iota_L(x)$ do not affect our dimension count. For this, observe that the set of points $(L, x) \in \mathbb{L}_2$ with $x = \iota_L(x)$ is (fibrewise with respect to p) of codimension one. As for the assertion, it suffices to prove finiteness of q_2 restricted to its complement, this concludes the proof. □

Proposition 2.13 *For a generic smooth cubic hypersurface $X \subset \mathbb{P}^{n+1}$ of dimension $n > 2$, the locus $F_2(X) \subset F(X)$ of lines of the second type is non-empty, smooth, and of dimension*

$$\dim(F_2(X)) = n - 2 = (1/2)\dim(F(X)).$$

Proof The dimension formula has been established already for all smooth X already by Lemma 2.12. So, we only have to prove the non-emptiness and the smoothness for generic X. However, the following arguments, adapted from the four-dimensional case treated by Amerik [17], also provide an alternative proof for the dimension formula.

Consider the universal Fano variety of lines of the second type

$$F_2(\mathcal{X}) \subset F(\mathcal{X}) \subset |\mathcal{O}_{\mathbb{P}}(3)| \times \mathbb{G}(1, \mathbb{P}),$$

cf. Proposition 1.5. So, the fibre of $F_2(\mathcal{X}) \longrightarrow |\mathcal{O}_{\mathbb{P}}(3)|$ over the point corresponding to a smooth cubic X is $F_2(X)$. The fibres of the other projection $\pi\colon F_2(\mathcal{X}) \longrightarrow \mathbb{G}(1, \mathbb{P})$ are all isomorphic to, say, the fibre over the line $L = V(x_0, \ldots, x_{n-1}) \subset \mathbb{P}$. It is a closed subscheme of the fibre of $F(\mathcal{X})$, which is the projective space $\mathbb{P}_L := |\mathcal{I}_L \otimes \mathcal{O}_{\mathbb{P}}(3)|$. Recall from Remark 2.2 and Exercise 2.10 that $\pi^{-1}(L) \subset \mathbb{P}_L$ in points corresponding to smooth cubics is characterized by the property that $\partial_i F|_L \in H^0(L, \mathcal{O}_L(2))$, $i = 0, \ldots, n-1$, span a two-dimensional subspace. Thus, $\pi^{-1}(L)$ is the degeneracy locus $M_2(\psi) \subset \mathbb{P}_L$ of

$$\psi\colon \langle x_0, \ldots, x_{n-1} \rangle \otimes \mathcal{O}_{\mathbb{P}_L}(-1) \longrightarrow H^0(L, \mathcal{O}_L(2)) \otimes \mathcal{O}_{\mathbb{P}_L},$$

which at the point $[F] \in \mathbb{P}_L$ is $\psi_F(x_i) = \partial_i F|_L$, cf. Remark 2.11. The coefficients of ψ are the $3n$ coordinates corresponding to the monomials

$$x_i \cdot x_n^2,\ x_i \cdot x_n x_{n+1}, \quad \text{and} \quad x_i \cdot x_{n+1}^2,$$

$i = 0, \ldots, n-1$, among all the linear coordinates of $\mathbb{P}_L = |\mathcal{I}_L \otimes \mathcal{O}_{\mathbb{P}}(3)|$.

Thus, with respect to the monomial basis of $\mathbb{P}_L$ and the basis of $H^0(L, \mathcal{O}_L(2))$ given by the restrictions of x_n^2, $x_n x_{n+1}$, and x_{n+1}^2, the situation is described by the matrix

$$\psi = \begin{pmatrix} y_0 & \cdots & y_{n-1} \\ y_n & \cdots & y_{2n-1} \\ y_{2n} & \cdots & y_{3n-1} \end{pmatrix}$$

on a projective space $\mathbb{P}^N$ with coordinates $y_0, \ldots, y_{3n-1}, y_{3n}, \ldots, y_N$. Hence, $M_k(\psi)$ is the pre-image under the linear projection $\mathbb{P}^N \dashrightarrow \mathbb{P}(M(3,n))$ of the universal determinantal variety in $\mathbb{P}(M(3,n))$ (with the $3n$ coordinates $y_0, \ldots, y_{3n-1}$). Hence, the classical formulae apply, see [23, 174], and show that $\operatorname{codim}(M_k(\psi)) = (n-k)\cdot(3-k)$. In particular, the fibre $\pi^{-1}(L)$ is of codimension $n-2$ in $|\mathcal{I}_L \otimes \mathcal{O}_{\mathbb{P}}(3)|$ and the singularities of the fibre over L are contained in $M_1(\psi)$. This proves that

$$\dim(F_2(\mathcal{X})) = \dim|\mathcal{O}_{\mathbb{P}}(3)| + n - 2.$$

As the image of $F_2(\mathcal{X}) \longrightarrow |\mathcal{O}_{\mathbb{P}}(3)|$ meets the smooth locus, cf. Example 2.3, and the fibre over any smooth $X \in |\mathcal{O}_{\mathbb{P}}(3)|$ is of dimension at most $n-2$, cf. Lemma 2.12, $F_2(X)$ is indeed of dimension $n-2$ for all smooth X.

Using the fact that the $\partial_i F|_L \in H^0(L, \mathcal{O}_L(2))$ for a smooth cubic $X = V(F)$ always span at least a two-dimensional space, we conclude that the image of $M_1(\psi)$ in $|\mathcal{O}_{\mathbb{P}}(3)|$ does not meet the open subset of smooth cubics. In other words, the open subset of $F_2(\mathcal{X})$ lying over $|\mathcal{O}_{\mathbb{P}}(3)|_{\text{sm}} \subset |\mathcal{O}_{\mathbb{P}}(3)|$ is smooth. Hence, $F_2(X) \subset F(X)$ is smooth of dimension $n-2$ for the generic smooth cubic hypersurface X. □

Exercise 2.14 In the above proof, we observed that $F_2(X)$ is non-empty of dimension $n-2$ for all smooth cubic hypersurfaces. Alternatively, one can deduce the non-emptyness using semi-continuity and (2.2). We stress that for special smooth cubic hypersurfaces $F_2(X)$ may well be singular, see Remark **5**.1.7.

2.3 Points Contained in Lines of the Second Type As a consequence we show that the generic point $x \in X$ is not contained in a line of the second type and that the generic point in the locus of points that are contained in a line of the second type is contained in only finitely many such lines. Implicitly, we have seen this already in the proof of Lemma 2.12.

Corollary 2.15 *Consider the restriction of* (1.7) *to* $F_2 := F_2(X)$

$$\begin{array}{ccccc} \mathbb{L}_2 := p^{-1}(F_2) & \hookrightarrow & \mathbb{L} & \xrightarrow{q} & X \\ \downarrow & & \downarrow p & & \\ F_2 & \hookrightarrow & F. & & \end{array}$$

The subvariety $\mathbb{L}_2 \subset \mathbb{L}$ *is the non-smooth locus of* $q\colon \mathbb{L} \twoheadrightarrow X$, *i.e.*

$$\mathbb{L}_2 = \{\, (L, x) \in \mathbb{L} \mid dq\colon T_{(L,x)}\mathbb{L} \longrightarrow T_xX \text{ is not surjective} \,\},$$

and its image $q(\mathbb{L}_2) \subset X$ *is a divisor in* X.

Proof The $\mathbb{P}^1$-bundle $\mathbb{L}_2 \twoheadrightarrow F_2(X)$ was introduced already in the proof of Lemma 2.12. There, we showed that $q\colon \mathbb{L}_2 \twoheadrightarrow X$ is finite on the dense open subset of points $(L, x) \in \mathbb{L}_2$ for which the restriction of the Gauss map $\gamma_X|_L$ is not ramified in x. This proves the second assertion.

For the first assertion, we evoke Remark 1.11, which in our situation shows that the kernel of $dq\colon T_{(L,x)}\mathbb{L} \twoheadrightarrow T_xX$ is naturally isomorphic to $H^0(L, \mathcal{N}_{L/X}(-1))$. As for lines of the first type, one has $h^0(L, \mathcal{N}_{L/X}(-1)) = n - 3$ and for those of the second type, $h^0(L, \mathcal{N}_{L/X}(-1)) = n - 2$, this immediately proves the assertion. □

Note that purity of branch loci predicts the non-smooth locus to be of codimension at most $1 + \dim \mathbb{L} - \dim(X) = n - 2$, i.e. of dimension at least $n - 1$. This fits the above description of the branch locus as $\mathbb{L}_2$ which, as a $\mathbb{P}^1$-bundle over $F_2(X)$, is of dimension $\dim F_2(X) + 1 = n - 2 + 1 = n - 1$, see Lemma 2.12.

Remark 2.16 The description

$$\operatorname{Ker}(dq\colon T_{(L,x)}\mathbb{L} \twoheadrightarrow T_xX) \simeq H^0(L, \mathcal{N}_{L/X}(-1)),$$

as used in the above proof, also shows that dq drops rank at most by one, i.e.

$$n - 1 \leq \operatorname{rk}(dq\colon T_{(L,x)}\mathbb{L} \twoheadrightarrow T_xX) \leq n$$

and, in particular, for all $x \in X$,

$$n - 3 \leq \dim(q^{-1}(x)) \leq n - 2.$$

While for the generic cubic, all fibres $q^{-1}(x)$ are of dimension $n - 3$, the differential dq has never constant rank, i.e. the projection q is not smooth, for $n \geq 3$.

Gounelas and Kouvidakis [200, Rem. 3.8] show that the projection $q\colon \mathbb{L}_2 \twoheadrightarrow X$ is birational onto its image, at least for the generic cubic $X \subset \mathbb{P}^{n+1}$, $n \geq 3$. In other words, the generic point in X that is contained in a line of the second type is contained in exactly one such line. For a direct argument for cubic threefolds, see Remark **5**.1.17.

Remark 2.17 There is a link between Corollary 2.6 and Corollary 2.15. For $L \in F(X)$, consider $P_L = \mathbb{P}(U)$ as in Corollary 2.6 and pick $x \in L \subset P_L$. Now compare the deformations of $L \subset X$ through x and deformations of $L \subset P_L$ through x. Using the diagram in the proof of Corollary 2.6 and combining it with the commutative diagram

$$\begin{array}{ccccccccc} 0 & \longrightarrow & \mathcal{N}_{L/X\cap P_L}(-1) & \longrightarrow & \mathcal{N}_{L/P_L}(-1) & \longrightarrow & \mathcal{N}_{X\cap P_L/P_L}|_L(-1) & \longrightarrow & 0 \\ & & \downarrow & & \downarrow & & \downarrow \simeq & & \\ 0 & \longrightarrow & \mathcal{N}_{L/X}(-1) & \longrightarrow & \mathcal{N}_{L/\mathbb{P}}(-1) & \longrightarrow & \mathcal{O}_L(2) & \longrightarrow & 0, \end{array}$$

one finds that to first order these deformations are the same, i.e.

$$H^0(L, \mathcal{N}_{L/X}(-1)) \simeq H^0(L, \mathcal{N}_{L/X\cap P_L}(-1)) \simeq H^0(L, \mathcal{N}_{L/P_L}(-1)).$$

In particular, as $P_L \simeq \mathbb{P}^{n-1}$ for lines of the second type, there are more deformations of L inside P_L through a fixed point x than for lines of the first type for which $P_L \simeq \mathbb{P}^{n-2}$.

Later we will see that, for specific n, the locus $F_2(X)$ often has a concrete geometric meaning, providing a different proof for $\dim(F_2(X)) = n - 2$, For example, for $n = 3$, so smooth cubic threefolds $Y \subset \mathbb{P}^4$, $F_2(Y)$ is a curve in the Fano surface $F(Y)$, see Section **5**.1.1. Note that $F_2(Y)$ can be singular for specific smooth cubic threefolds and $q\colon \mathbb{L}_2 \longrightarrow Y$ might have positive-dimensional fibres, cf. Remark **5**.1.7. Hence, for the smoothness of $F_2(X)$, the assumption in Proposition 2.13 that X is generic is essential.

Remark 2.18 The description of $F_2(X)$ as a degeneracy locus of the expected dimension allows one to compute its fundamental class

$$[F_2(X)] \in H^{n-2}(F(X), \mathbb{Z}),$$

which is the middle cohomology of the Fano variety $F(X)$, in terms of Chern classes of $\mathcal{S}_F$ or, alternatively, of $\mathcal{Q}_F$ (Porteous formula). We will not do this in general, but see Proposition **5**.1.1 for cubic threefolds and Proposition **6**.4.1 for the computation in the case of cubic fourfolds.

Remark 2.19 Recall from Corollary 2.6 that for lines of the second type, there exists a unique linear subspace $L \subset P_L \simeq \mathbb{P}^{n-1} \subset \mathbb{P}^{n+1}$ that is tangent to X at every point of L. Unlike the case of lines of the first type, linear subspaces $L \subset \mathbb{P}^{n-2}$ that are tangent to X along L are not unique. They are all contained in P_L and parametrized by a subspace $\mathbb{P}^{n-3} \subset P_L$ complementary to $L \subset P_L$.

This leads one to consider the incidence variety $\widetilde{F}(X)$ of all pairs $(L, \mathbb{P}(U)) \in F(X) \times \mathbb{G}(n-2, \mathbb{P})$ consisting of a line $L \subset X$ and a linear subspace $\mathbb{P}^{n-2} \simeq \mathbb{P}(U) \subset \mathbb{P}$ such that L is contained in $\mathbb{P}(U)$ and $\mathbb{P}(U)$ is tangent to X along L. Then

$$\tau\colon \widetilde{F}(X) \longrightarrow F(X),\ (L, \mathbb{P}(U)) \longmapsto L$$

is an isomorphism over $F(X) \setminus F_2(X)$ and a $\mathbb{P}^{n-3}$-bundle over $F_2(X) \subset F(X)$. So, at least over the open complement of the set $F_2(X)_{\rm sing} \subset F(X)$ of singular points of $F_2(X)$, it looks like the blow-up $\sigma\colon \mathrm{Bl}_{F_2(X)}(F(X)) \twoheadrightarrow F(X)$.[4]

To make the identification $\mathrm{Bl}_{F_2(X)}(F(X)) \simeq \widetilde{F}(X)$ rigorous (while assuming for simplicity that $F_2(X)$ is smooth), we have to find a natural identification between the fibres of τ and σ over a point $L \in F_2(X)$. By the discussion in the proof of Corollary 2.6, see also Remark 2.20 below, we know

$$\tau^{-1}(L) = \mathbb{P}(H^0(L, \mathcal{N}_{L/X}(-1))^*).$$

The description of $\sigma^{-1}(L)$ is more involved. As for any blow-up, the fibre over a point in the centre of the blow-up is the projectivization of the normal bundle at that point, so

$$\sigma^{-1}(L) = \mathbb{P}(T_L F(X)/T_L F_2(X)).$$

For a global description of the normal bundle for cubic fourfolds, see Proposition **6**.4.8. We know that $T_L F(X) \simeq H^0(L, \mathcal{N}_{L/X})$ and that the subspace $T_L F_2(X) \subset T_L F(X)$ is the subspace of all first order deformations of L in X such that L stays of the second type, i.e. such that $h^0(L, \mathcal{N}_{L/X}(-1)) = n-2$ is preserved. Assume $L_\varepsilon \subset X_\varepsilon := X \times \mathrm{Spec}(k[\varepsilon])$ corresponds to $v \in T_L F(X) \simeq H^0(L, \mathcal{N}_{L/X})$. The boundary map of the associated short exact sequence

$$0 \longrightarrow \varepsilon \cdot \mathcal{N}_{L/X} \longrightarrow \mathcal{N}_{L_\varepsilon/X_\varepsilon} \longrightarrow \mathcal{N}_{L/X} \longrightarrow 0$$

twisted by $\mathcal{O}_L(-1)$ is a map

$$\partial_v\colon H^0(L, \mathcal{N}_{L/X}(-1)) \longrightarrow H^1(L, \mathcal{N}_{L/X}(-1)) \simeq k$$

that sends a section s to the obstruction to deform it sideways to a section of $\mathcal{N}_{L_\varepsilon/X_\varepsilon}(-1)$.[5] Altogether, we obtain a map

$$\begin{aligned} T_L F(X) \simeq H^0(L, \mathcal{N}_{L/X}) &\longrightarrow \mathrm{Hom}(H^0(L, \mathcal{N}_{L/X}(-1)), H^1(L, \mathcal{N}_{L/X}(-1))) \\ &\simeq H^0(L, \mathcal{N}_{L/X}(-1))^* \otimes H^1(L, \mathcal{N}_{L/X}(-1)), \end{aligned}$$

the kernel of which is $T_L F_2(X)$. Thus, for smooth points $L \in F_2(X)$, there exists a natural isomorphism $T_L F(X)/T_L F_2(X) \simeq H^0(L, \mathcal{N}_{L/X}(-1))^*$, up to tensoring with the line $H^1(L, \mathcal{N}_{L/X}(-1))$, and hence a natural identification $\sigma^{-1}(L) \simeq \tau^{-1}(L)$.

Recall from Remark 1.16 that there exists a natural isomorphism $H^1(L, \mathcal{N}_{L/X}(-1)) \simeq \det H^0(L, \mathcal{N}_{L/X}(-1))^*$.

Remark 2.20 This remark is rather lengthy and a little technical. We recommend skipping it at first reading and coming back to it when it is later used, see Section **6**.4.4.

[4] Has the following any chance of being true for $n > 3$: if there is no $\mathbb{P}^{n-2}$ contained in X, then $\widetilde{F}$ is smooth. If $\widetilde{F}$ is smooth, then F_2 is smooth.

[5] The isomorphisms here and in the next displays are not canonical.

Let us look at the other projection

$$\alpha\colon \widetilde{F}(X) \longrightarrow \mathbb{G}(n-2,\mathbb{P}),\ (L,\mathbb{P}(U)) \longmapsto \mathbb{P}(U).$$

On the open part $F(X)\setminus F_2(X) \subset \widetilde{F}(X)$ this is just

$$\alpha_0\colon F(X)\setminus F_2(X) \longrightarrow \mathbb{G}(n-2,\mathbb{P}),\ L \longmapsto P_L.$$

Similarly, one can consider

$$\beta\colon F_2(X) \longrightarrow \mathbb{G}(n-1,\mathbb{P}),\ L \longmapsto P_L.$$

We wish to describe the pullbacks of the Plücker polarizations on $\mathbb{G}(n-2,\mathbb{P})$ and $\mathbb{G}(n-1,\mathbb{P})$ under α and β. They can be compared to the standard Plücker polarizations $\mathcal{O}_{F_2}(1)$ and $\mathcal{O}_F(1)$ on $F_2(X) \subset F(X) \subset \mathbb{G}(1,\mathbb{P})$ as follows:

$$\alpha^*\mathcal{O}(1) \simeq \tau^*\mathcal{O}_F(3)(-E) \quad \text{and} \quad \beta^*\mathcal{O}(2) \simeq \mathcal{O}_{F_2}(4).$$

Here, $E \subset \widetilde{F}(X)$ is the exceptional $\mathbb{P}^{n-3}$-bundle over $F_2(X)$. Note that in particular $\alpha_0^*\mathcal{O}(1) \simeq \mathcal{O}_F(3)|_{F\setminus F_2}$, which is of particular interest for $n > 3$, as then $F_2(X) \subset F(X)$ is of codimension at least two.

Let us sketch the argument assuming $n \geq 4$ and $\widetilde{F}(X)$ smooth. In both cases, we will establish isomorphisms between the fibres of the involved line bundles that are natural and, therefore, glue to isomorphisms between the line bundles themselves. The actual gluing is left as an exercise, but see Proposition **5**.2.2 for similar arguments where we do also explain the gluing.

Let $L = \mathbb{P}(W) \subset X$ be of the first type. Then $P_L = \mathbb{P}(U) \simeq \mathbb{P}^{n-2}$, where U is given as an extension

$$0 \longrightarrow W \longrightarrow U \longrightarrow H^0(L,\mathcal{N}_{L/X}(-1)) \longrightarrow 0, \tag{2.3}$$

see the proof of Corollary 2.6. Furthermore, according to Remark 2.2, there exists a short exact sequence

$$0 \longrightarrow H^0(L,\mathcal{N}_{L/X}(-1)) \longrightarrow V/W \longrightarrow S^2(W^*) \longrightarrow 0. \tag{2.4}$$

After fixing $\det(V) \simeq k$, the fibre of $\alpha_0^*\mathcal{O}(-1)$ at the point L is naturally isomorphic to

$$\begin{aligned}\det(U) &\simeq \det(W)\otimes\det H^0(\mathcal{N}_{L/X}(-1)) \simeq \det(W)\otimes\det(V/W)\otimes\det(S^2(W))\\ &\simeq \det(W)\otimes\det(W)^*\otimes\det(W)^3 \simeq \det(W)^3,\end{aligned}$$

which is naturally identified with the fibre of $\mathcal{O}_F(-3)$ at L. Globalizing the argument, one obtains the isomorphism $\alpha_0^*\mathcal{O}(1) \simeq \mathcal{O}(3)|_{F\setminus F_2}$. To conclude the description of the pullback $\alpha^*\mathcal{O}(1)$ it suffices to show that it restricts to $\mathcal{O}(1)$ on the fibres of $E \longrightarrow F_2(X)$. The fibre $\sigma^{-1}(L) \simeq \tau^{-1}(L) \simeq \mathbb{P}(H^0(L,\mathcal{N}_{L/X}(-1)))$ over $L = \mathbb{P}(W) \in F_2(X)$ parametrizes all extensions

$$0 \longrightarrow W \longrightarrow U_0 \longrightarrow \bar{U}_0 \longrightarrow 0,$$

where $\bar{U}_0 \subset H^0(L, \mathcal{N}_{L/X}(-1))$ is a hyperplane. At such a point and with W fixed, the fibre of $\alpha^*\mathcal{O}(-1)$ is naturally identified with

$$\det(U_0) \simeq \det(W) \otimes \det(\bar{U}_0) \simeq \det(H^0(L, \mathcal{N}_{L/X}(-1))/\bar{U}_0)^*,$$

which is canonically isomorphic to the fibre of $\mathcal{O}(-1) \simeq \mathcal{O}(E)|_{\tau^{-1}(L)}$. Here, the second isomorphism relies on the natural identification of $\det(W)^*$ and $\det(H^0(L, \mathcal{N}_{L/X}(-1)))$ (or rather of the squares, which is enough for our purpose), which in turn is a consequence of Remark 2.2 and (1.16) in Remark 1.16, see also the argument below.

To prove $\beta^*\mathcal{O}(2) \simeq \mathcal{O}_{F_2}(4)$ we have to establish a natural isomorphism $\det(U)^2 \simeq \det(W)^4$ for which we again use (2.3), only that now U is of dimension n:

$$\begin{aligned}\det(U)^2 &\simeq \det(W)^2 \otimes \det H^0(L, \mathcal{N}_{L/X}(-1)) \otimes \det H^0(L, \mathcal{N}_{L/X}(-1)) \\ &\simeq \det(W)^2 \otimes \det H^0(L, \mathcal{N}_{L/X}(-1)) \otimes H^1(L, \mathcal{N}_{L/X}(-1))^* \\ &\simeq \det(W)^2 \otimes \det(V/W) \otimes \det(S^2(W^*))^* \simeq \det(W)^4.\end{aligned}$$

Here, the second isomorphism follows from (1.16) in Remark 1.16 and the third one from the exact sequence

$$0 \longrightarrow H^0(L, \mathcal{N}_{L/X}(-1)) \longrightarrow V/W \longrightarrow S^2(W^*) \longrightarrow H^1(L, \mathcal{N}_{L/X}(-1)) \longrightarrow 0,$$

which is the analogue of (2.4) for lines of the second type, see Remark 2.2. The isomorphism $\beta^*\mathcal{O}(2) \simeq \mathcal{O}_{F_2}(4)$ suggest that maybe in fact $\beta^*\mathcal{O}(1) \simeq \mathcal{O}_{F_2}(2)$, but it turns out that this is not true, see Remark **6**.4.9.

3 Global Properties and a Geometric Torelli Theorem

No information is lost when passing from a smooth cubic hypersurface of dimension at least three to its Fano variety of lines. For cubic fourfolds, the Plücker polarization of the Fano variety has to be taken into account and, of course, for smooth cubic surfaces, where the Fano variety consists of just 27 reduced points, the result fails.

3.1 Canonical Bundle and Picard Group Recall the isomorphism $\det(\mathcal{S}^*) \simeq \mathcal{O}(1)|_{\mathbb{G}}$ for the Plücker embedding $\mathbb{G} \hookrightarrow \mathbb{P}(\bigwedge^{m+1} V)$, see Section 1.1. The following result is [15, Prop. 1.8].

Lemma 3.1 *For a smooth cubic hypersurface $X \subset \mathbb{P}^{n+1}$, the canonical bundle ω_F of the Fano variety of lines $F = F(X) \subset \mathbb{G}(1, \mathbb{P}) \hookrightarrow \mathbb{P}^N$, $N = \binom{n+2}{2} - 1$, is*

$$\omega_F \simeq \mathcal{O}(4 - n)|_F.$$

Proof As the Fano variety is the zero set $F(X) = V(s_F) \subset \mathbb{G}(1, \mathbb{P})$ of a regular section $s_F \in H^0(\mathbb{G}, S^3(\mathcal{S}^*))$, see Section 1.1, the normal bundle sequence for $F = F(X) \subset \mathbb{G}$ takes the form

$$0 \longrightarrow \mathcal{T}_F \longrightarrow \mathcal{T}_{\mathbb{G}}|_F \longrightarrow S^3(\mathcal{S}_F^*) \longrightarrow 0.$$

The adjunction formula then implies

$$\omega_F = \det(\mathcal{T}_F^*) \simeq \omega_{\mathbb{G}}|_F \otimes \det(S^3(\mathcal{S}_F^*)).$$

As $\mathcal{T}_{\mathbb{G}} \simeq \mathcal{H}om(\mathcal{S}, \mathcal{Q})$, one has $\omega_{\mathbb{G}} \simeq \det(\mathcal{S} \otimes \mathcal{Q}^*) \simeq \det(\mathcal{S})^n \otimes \det(\mathcal{Q}^*)^2 \simeq \mathcal{O}(-n-2)|_{\mathbb{G}}$. Thus, it remains to prove that $\det(S^3(\mathcal{S})) \simeq \det(\mathcal{S})^6$, which one deduces from the splitting principle and the following computation: write formally $\mathcal{S} = M_0 \oplus M_1$ and observe $S^3(M_0 \oplus M_1) \simeq M_0^3 \oplus (M_0^2 \otimes M_1) \oplus (M_0 \otimes M_1^2) \oplus M_1^3$. □

Thus, for smooth cubic threefolds $Y \subset \mathbb{P}^4$, the Fano variety of lines $F(Y)$ is a smooth projective surface with very ample canonical bundle, in particular, $F(Y)$ is of general type. For smooth cubic fourfolds $X \subset \mathbb{P}^5$, the Fano variety $F(X)$ has trivial canonical bundle $\omega_F \simeq \mathcal{O}_F$ and we will later see that $F(X)$ is a four-dimensional hyperkähler manifold, see Theorem **6**.3.10. Eventually, for $n > 4$, the Fano variety becomes a Fano variety in the sense that its anti-canonical bundle $\omega_{F(X)}^*$ is ample, and in fact very ample. In short:

$$\omega_{F(X)} = \begin{cases} \text{ample} & \text{if } n = 3, \\ \text{trivial} & \text{if } n = 4, \\ \text{anti-ample} & \text{if } n > 4. \end{cases}$$

Exercise 3.2 Use the arguments in the proof above to compute the Chern character $\mathrm{ch}(F) := \mathrm{ch}(\mathcal{T}_F)$. More precisely, if formally we write $\mathrm{c}(\mathcal{S}_F^*) = (1 + \ell_0) \cdot (1 + \ell_1)$, so that $\mathrm{c}_1(\mathcal{S}_F^*) = \ell_0 + \ell_1$ and $\mathrm{c}_2(\mathcal{S}_F) = \ell_0 \cdot \ell_1$, then for $x_i := \exp(\ell_i)$,

$$\mathrm{ch}(F) = (x_0 + x_1) \cdot \left(n + 2 - (1/x_0) - (1/x_1) - x_0^2 - x_1^2\right).$$

This gives back the above result $\mathrm{c}_1(F) = (n - 4) \cdot g$, where $g = \mathrm{c}_1(\mathcal{O}(1)|_F) = -\mathrm{c}_1(\mathcal{S}_F) = \ell_0 + \ell_1$, and

$$\mathrm{ch}_2(F) = (n/2 - 7) \cdot g^2 + (12 - n) \cdot \mathrm{c}_2(\mathcal{S}_F),$$

which, for $n = 3$ and $n = 4$, becomes $\mathrm{c}_2(F) = 6 \cdot g^2 - 9 \cdot \mathrm{c}_2(\mathcal{S}_F)$ and $\mathrm{c}_2(F) = 5 \cdot g^2 - 8 \cdot \mathrm{c}_2(\mathcal{S}_F)$, see Section **5**.2.1 and Proposition **6**.4.1. For $n = 4$, this was computed by Diamond [151, Sec. 4.2.1].

We will later determine the (rational) cohomology of $F(X)$ for any smooth cubic hypersurface $X \subset \mathbb{P}^{n+1}$, cf. Section 4.6. At least in characteristic zero, the positivity property of the canonical bundle $\omega_{F(X)}$ already implies certain vanishings, e.g.

$$H^q(F(X), \mathcal{O}) = 0$$

for $n > 4$ and $q > 0$. See also Corollary 4.17 for an alternative approach. This allows one to prove the following result, although strictly speaking only for $n > 4$, but see Corollaries 4.18 and **6**.3.11.

Corollary 3.3 *Let $F = F(X)$ be the Fano variety of lines contained in a smooth cubic hypersurface $X \subset \mathbb{P}^{n+1}$ over $\mathbb{C}$. Then $H^1(F, \mathbb{Z}) = 0$ for $n \geq 4$ and, in particular,* $\mathrm{Pic}^0(F) = 0$. *For $n \geq 5$, one has* $\mathrm{Pic}(F) \simeq H^2(F, \mathbb{Z})(1)$. □

In Corollary 4.18 we will see that $\mathrm{Pic}(F) \simeq \mathbb{Z}$ for $n \geq 5$ and in Remark 4.19 we explain that $F(X)$ is in fact simply connected.

All these assertions remain valid over arbitrary fields. For example, Kodaira vanishing holds for liftable varieties [147], see the comments in Section **1**.1.6, and implies the vanishing of $\mathrm{Pic}^0(F)$.

3.2 Connectedness and the Universal Line Apart from the Fano variety of lines on a cubic surface, all others are connected, see [15, Thm. 1.16] and [37, Thm. 6]. This leads to the following strengthening of Proposition 1.19.[6]

Proposition 3.4 *Let $X \subset \mathbb{P}^{n+1}$ be a smooth cubic hypersurface of dimension $n > 2$. Then $F(X)$ is an irreducible, smooth, projective variety of dimension $2n - 4$.*

Proof More generally, Barth and Van de Ven [37] prove that the Fano variety of lines $F(X)$ on any, not necessarily smooth, hypersurface $X \subset \mathbb{P}^{n+1}$ of degree d is connected if $d < 2(n-1)$. They argue by bounding the dimension of the ramification locus (of the Stein factorization) of $\mathbb{L} \longrightarrow X$. Altman and Kleiman [15] use instead the Koszul complex

$$\cdots \longrightarrow \bigwedge^2 (S^3(\mathcal{S})) \longrightarrow S^3(\mathcal{S}) \longrightarrow \mathcal{O}_{\mathbb{G}} \longrightarrow \mathcal{O}_{F(X)} \longrightarrow 0. \tag{3.1}$$

The induced spectral sequence

$$E_1^{p,q} = H^q(\mathbb{G}, \bigwedge^{-p} (S^3(\mathcal{S}))) \Rightarrow H^{p+q}(F(X), \mathcal{O}_{F(X)})$$

combined with generalized Bott vanishing results for Grassmann varieties:

$$H^p(\mathbb{G}, \bigwedge^p (S^3(\mathcal{S}))) = 0 \quad \text{for all } p \neq 0$$

shows $H^0(\mathbb{G}, \mathcal{O}_{\mathbb{G}}) \xrightarrow{\sim} H^0(F(X), \mathcal{O}_{F(X)})$. Hence, $F(X)$ is connected. Together with the smoothness of $F(X)$, this shows that $F(X)$ is irreducible.

For alternative arguments, see Exercise 3.7, for $n \geq 4$, and Example 4.21. □

Exploiting similar techniques, Borcea [85] proved connectedness of certain Fano varieties $F(X, m)$ of m-planes in complete intersections.

[6] One could think of applying the Fulton–Lazarsfeld connectivity [316, Thm. 7.2.1], to prove the connectivity of $F(X, m)$ whenever $\dim(F(X, m)) > 0$. However, this would need the ampleness of $S^d(\mathcal{S}^*)$, which is just wrong.

Exercise 3.5 Let $X \subset \mathbb{P}^{n+1}$ be a cubic hypersurface and assume that m is an integer with $m^2 + 11m \leq 6n$. Show that there exists a linear subspace $\mathbb{P}^m \subset X \subset \mathbb{P}^{n+1}$ of dimension m contained in X.

Remark 3.6 Assume $n > 2$. Then, the projection

$$q\colon \mathbb{L} \twoheadrightarrow X$$

of the universal line $\mathbb{L} = \mathbb{P}(\mathcal{S}|_{F(X)}) \longrightarrow F(X)$ is surjective or, equivalently, through every point $y \in X$ there exists at least one line $y \in L \subset X$, possibly defined only over a finite extension of the residue field of y. To prove this claim, we may assume that k is algebraically closed. In fact, there are many ways to go about it; here are a few.

(i) We have computed that $\dim(F(X)) = 2n - 4$, which implies that $\dim(\mathbb{L}) = 2n - 3$, and $\dim(F_2(X)) = n - 2$. Hence, $q\colon \mathbb{L} \longrightarrow X$ over the complement of $\mathbb{L}|_{F_2(X)} \subset \mathbb{L}$ has fibre dimension $n - 3$, see Remark 2.16, and, therefore, the image of q is of dimension n, i.e. q is surjective.

(ii) Avoiding any prior dimension computations, one can argue as follows, cf. [209]. For a fixed point $y \in X$, let $\mathbb{P}^n \subset \mathbb{P}^{n+1}$ be a hyperplane not containing y. We may assume $y = [1 : 0 : \cdots : 0]$ and $\mathbb{P}^n = V(x_0)$. If $X = V(F)$, then the projective tangent space at the point $y \in X$ is the hyperplane $\mathbb{T}_yX = V(\sum x_i\,\partial_iF(y)) \simeq \mathbb{P}^n$, see Definition 2.4, and any line $y \in L \subset \mathbb{T}_yX$ has intersection multiplicity $\mathrm{mult}_y(X, L) \geq 2$. For dimension reasons, there exists a point

$$z \in \mathbb{P}^n \cap \mathbb{T}_yX \cap X \cap V(\partial_0F).$$

Then let $L := \overline{yz}$ be the line connecting the two points. We may choose coordinates such that $z = [0 : 1 : 0 : \cdots : 0]$, in which case $F|_L$ is the polynomial $F(x_0, x_1, 0, \ldots, 0)$. By definition $\partial_0F(z) = 0$ and by the Euler equation also $\partial_1F(z) = 0$. Therefore, $\mathrm{mult}_z(X, L) \geq 2$. However, a line $L \subset \mathbb{P}^{n+1}$ intersecting a cubic hypersurface $X \subset \mathbb{P}^{n+1}$ in two distinct points with multiplicity at least two at each of them is contained in X.

(iii) Another possibility to argue is to first take hyperplane sections to reduce to the case of smooth cubic threefolds $Y \subset \mathbb{P}^4$. In this case the assertion was first observed by Clemens and Griffiths [119, Cor. 8.2]. Another more direct argument was given by Coskun and Starr [125, Lem. 2.1].

Note that the argument in (ii) also shows that the fibre of $\mathbb{L} \longrightarrow X$ over y has the expected dimension $n - 3$ if and only if the intersection $\mathbb{P}^n \cap \mathbb{T}_yX \cap X \cap V(\partial_0F)$ is of the expected dimension $n - 3$. It is known [125, Cor. 2.2] that the fibre can be of dimension bigger than $n - 3$ for at most finitely many points $x \in X$.

Exercise 3.7 Assume $n \geq 4$ and observe then that $\mathbb{P}^n \cap \mathbb{T}_yX \cap X \cap V(\partial_0F)$ in Remark 3.6 is connected by Bertini's theorem. Deduce from this that $F(X)$ is connected, thus proving Proposition 3.4 again for cubic hypersurfaces of dimension $n \geq 4$.

Remark 3.8 Consider the morphism $\varpi\colon \mathbb{L} \longrightarrow \mathbb{P}(\mathcal{T}_X)$ that is induced by the inclusion $\mathcal{T}_{\mathbb{L}/F(X)} \hookrightarrow q^*\mathcal{T}_X$ and the isomorphism $\mathbb{L} \simeq \mathbb{P}(\mathcal{T}_{\mathbb{L}/F(X)})$. Concretely, ϖ maps a point $(L, y) \in \mathbb{L}$ to the tangent direction of L at $y \in L$. On each fibre of the projection $p\colon \mathbb{L} \longrightarrow F(X)$ the morphism ϖ is described by the natural embedding $L \simeq \mathbb{P}(\mathcal{T}_L) \hookrightarrow \mathbb{P}(\mathcal{T}_X)$. It is injective, as the line L is uniquely determined by the point y and the tangent line T_yL. One checks that ϖ also separates tangent directions and that, therefore, it is in fact a closed embedding.

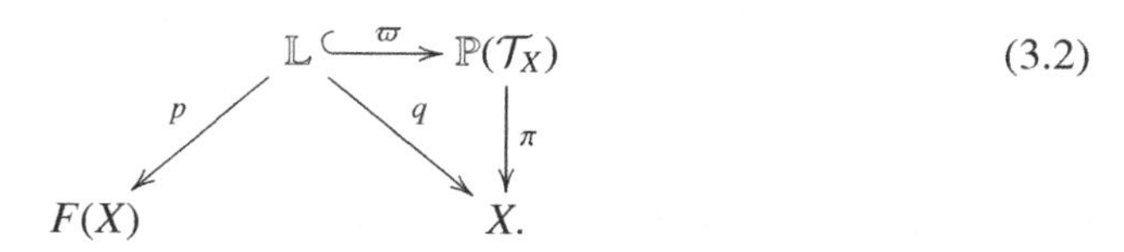

Note that the various line bundles enjoy the following compatibilities:

$$q^*\mathcal{O}_X(1) \simeq \mathcal{O}_p(1) \quad \text{and} \quad \varpi^*\mathcal{O}_\pi(1) \simeq \mathcal{O}_p(-2) \otimes p^*\mathcal{O}_F(1).$$

For the latter, combine $\varpi^*\mathcal{O}_\pi(-1) \simeq \mathcal{T}_{\mathbb{L}/F(X)}$, which holds by definition of ϖ, and the relative Euler sequence for p, see the proof of Proposition 3.10 below for more details.

Note that $\mathbb{L} \subset \mathbb{P}(\mathcal{T}_X)$ is of codimension two and all fibres of $q\colon \mathbb{L} \longrightarrow X$ satisfy

$$\dim(q^{-1}(y)) \leq n - 1.$$

In fact, the upper bound can be improved to $n - 2$, see Remark 2.16.

Using Remark 3.6 one sees that the fibre $q^{-1}(y) \subset \mathbb{P}(T_yX) \simeq \mathbb{P}^n \cap \mathbb{T}_yX$ is an intersection of a cubic and a quadric. So, if it is of the expected dimension $n-3$, it is a $(2, 3)$ complete intersection and, in particular, of degree six, cf. Lemma 5.11.

Exercise 3.9 Show that the normal bundle of the natural embedding $\mathbb{L} \subset F(X) \times X$ is isomorphic to

$$\mathcal{N}_{\mathbb{L}/F(X)\times X} \simeq \varpi^*(\mathcal{T}_\pi \otimes \mathcal{O}_\pi(-1)) \simeq \varpi^*\mathcal{T}_\pi \otimes \mathcal{O}_p(2) \otimes p^*\mathcal{O}_F(-1).$$

We conclude this section by a description of $\mathbb{L} \subset \mathbb{P}(\mathcal{T}_X)$ as the zero set of a section of a rank two bundle. This is a variation of an argument of Shen [426, Prop. 5.1] for cubic fourfolds.

Proposition 3.10 *There exists a locally free sheaf E of rank two on $\mathbb{P}(\mathcal{T}_X)$ given as an extension*

$$0 \longrightarrow \pi^*\mathcal{O}(3) \otimes \mathcal{O}_\pi(3) \longrightarrow E \longrightarrow \pi^*\mathcal{O}(3) \otimes \mathcal{O}_\pi(2) \longrightarrow 0$$

and a section $s \in H^0(\mathbb{P}(\mathcal{T}_X), E)$ with $V(s) = \mathbb{L} \subset \mathbb{P}(\mathcal{T}_X)$. Hence,

$$\mathbb{L} = V(s_2) \subset V(s_1) \subset \mathbb{P}(\mathcal{T}_X)$$

with $s_1 \in H^0(\mathbb{P}(\mathcal{T}_X), \pi^*\mathcal{O}(3) \otimes \mathcal{O}_\pi(2))$ *and* $s_2 \in H^0(V(s_1), (\pi^*\mathcal{O}(3) \otimes \mathcal{O}_\pi(3))|_{V(s_1)})$.[7]

Proof We use the two descriptions of the universal line over the Grassmannian $\mathbb{G} = \mathbb{G}(1, \mathbb{P})$ as

$$p\colon \mathbb{L}_\mathbb{G} \simeq \mathbb{P}(\mathcal{S}) \longrightarrow \mathbb{G} \quad \text{and} \quad \pi\colon \mathbb{L}_\mathbb{G} \simeq \mathbb{P}(\mathcal{T}_\mathbb{P}) \longrightarrow \mathbb{P}.$$

From the two relative Euler sequences

$$0 \longrightarrow \mathcal{O} \longrightarrow p^*\mathcal{S} \otimes \mathcal{O}_p(1) \longrightarrow \mathcal{T}_p \longrightarrow 0 \tag{3.3}$$

$$\text{and} \quad 0 \longrightarrow \mathcal{O} \longrightarrow \pi^*\mathcal{T}_\mathbb{P} \otimes \mathcal{O}_\pi(1) \longrightarrow \mathcal{T}_\pi \longrightarrow 0$$

we deduce $\omega_p \simeq p^*\mathcal{O}_\mathbb{G}(1) \otimes \mathcal{O}_p(-2)$ and $\omega_\pi \simeq \pi^*\mathcal{O}(-n-2) \otimes \mathcal{O}_\pi(-n-1)$. Inserted into

$$p^*\omega_\mathbb{G} \otimes \omega_p \simeq \omega_{\mathbb{L}_\mathbb{G}} \simeq \pi^*\omega_\mathbb{P} \otimes \omega_\pi,$$

while using $\mathcal{O}_p(1) \simeq \pi^*\mathcal{O}(1)$ and $\omega_\mathbb{G} \simeq \det(\mathcal{S} \otimes \mathcal{Q}^*) \simeq \mathcal{O}_\mathbb{G}(-n-2)$, this leads to

$$p^*\mathcal{O}_\mathbb{G}(1) \simeq \pi^*\mathcal{O}(2) \otimes \mathcal{O}_\pi(1). \tag{3.4}$$

Tensoring (3.3) with $\mathcal{O}_p(-1)$ provides us with the short exact sequence

$$0 \longrightarrow \mathcal{O}_p(-1) \longrightarrow p^*\mathcal{S} \longrightarrow p^*\mathcal{O}_\mathbb{G}(-1) \otimes \mathcal{O}_p(1) \longrightarrow 0,$$

which shows that $p^*S^3(\mathcal{S}^*)$ admits a natural filtration

$$0 = E_0 \subset E_1 \subset E_2 \subset E_3 \subset E_4 = p^*S^3(\mathcal{S}^*)$$

with quotients $E_{i+1}/E_i \simeq p^*\mathcal{O}_\mathbb{G}(3-i) \otimes \mathcal{O}_p(2i-3) \simeq \pi^*\mathcal{O}(3) \otimes \mathcal{O}_\pi(3-i)$. Now consider the pullback $p^*s_F \in H^0(\mathbb{L}_\mathbb{G}, p^*S^3(\mathcal{S}^*))$ of the global section $s_F \in H^0(\mathbb{G}, S^3(\mathcal{S}^*))$ with $F(X) = V(s_F) \subset \mathbb{G}$, see (1.4). Its projection s_3 to $E_4/E_3 \simeq \pi^*\mathcal{O}(3)$ is nothing but the equation $F \in H^0(\mathbb{P}, \mathcal{O}(3))$ of X. Thus, $V(s_3) = \mathbb{P}(\mathcal{T}_\mathbb{P}|_X)$ and the restriction of p^*s_F to $V(s_3)$ projects to a section s_2 of $(E_3/E_2)|_{V(s_3)} \simeq \pi^*\mathcal{O}(3) \otimes \mathcal{O}_\pi(1)$ which cuts out the fibrewise linear divisor $\mathbb{P}(\mathcal{T}_X) \subset \mathbb{P}(\mathcal{T}_\mathbb{P}|_X)$. Hence, p^*s_F restricted to $\mathbb{P}(\mathcal{T}_X)$ is a section of the bundle $E := E_2 \subset S^3(\mathcal{S}^*)$ which is an extension of $E_2/E_1 \simeq \pi^*\mathcal{O}(3) \otimes \mathcal{O}_\pi(2)$ by $E_1 \simeq \pi^*\mathcal{O}(3) \otimes \mathcal{O}_\pi(3)$. □

Note that on each fibre of $\pi\colon \mathbb{P}(\mathcal{T}_X) \longrightarrow X$ the result above confirms the description of $q^{-1}(y)$ in Remark 3.6.

Exercise 3.11 Use the description of $\mathbb{L} \subset \mathbb{P}(\mathcal{T}_X)$ as the zero section of an extension of $\pi^*\mathcal{O}(3) \otimes \mathcal{O}_\pi(2)$ by $\pi^*\mathcal{O}(3) \otimes \mathcal{O}_\pi(3)$ to reprove Lemma 3.1.

[7] In [426] it is claimed that the extension describing E splits. However, Ottem [244, App.] shows that this fails in dimension four.

3.3 Geometric Global Torelli Theorem The following *geometric global Torelli theorem* generalizes a well-known result for $n = 3$, which we shall explain in Section 5.2.4. It turns out that the general proof is less geometric, but in the end much easier. We follow Charles [114], where a more general version is proved allowing the cubic hypersurfaces to have isolated singularities.

Proposition 3.12 (Geometric global Torelli theorem) *Assume $X, X' \subset \mathbb{P}^{n+1}$ are smooth cubic hypersurfaces of dimension $n > 2$ and let $F(X)$ and $F(X')$ be their Fano varieties of lines endowed with the natural Plücker polarizations $\mathcal{O}_F(1)$ and $\mathcal{O}_{F'}(1)$.*

Then $X \simeq X'$ if and only if $(F(X), \mathcal{O}_F(1)) \simeq (F(X'), \mathcal{O}_{F'}(1))$ as polarized varieties. For $n \neq 4$, this is equivalent to $F(X) \simeq F(X')$ as unpolarized varieties.

Proof Any isomorphism $X \simeq X'$ is induced by an automorphism of the ambient projective space, cf. Corollary 1.3.9. Therefore, it naturally induces an isomorphism between the Fano varieties of lines, which in addition is automatically polarized.

Before proving the converse, let us show that for $n \neq 4$, any isomorphism $F(X) \simeq F(X')$ is automatically polarized. Indeed, for $n \neq 4$, the canonical bundle $\omega_F \simeq \mathcal{O}_F(4-n)$ is a non-trivial, possibly negative, multiple of the Plücker polarization $\mathcal{O}_F(1)$ and any isomorphism $F(X) \simeq F(X')$ respects the canonical bundle. Hence, if $\mathrm{Pic}(F(X))$ is torsion free for $n > 4$, then any isomorphism $F(X) \simeq F(X')$ is automatically polarized.

To prove that $\mathrm{Pic}(F(X))$ is torsion free, cf. Corollary 3.3, observe that for a torsion line bundle L one has $\chi(F(X), L) = \chi(F(X), \mathcal{O}) = 1$. As $F(X)$ is a Fano variety for $n > 4$, we have $H^i(F(X), L) \simeq H^i(F(X), L \otimes \omega_F^* \otimes \omega_F) = 0$ for $i > 0$ by Kodaira vanishing (which holds also in positive characteristic, as with X also $F(X)$ is liftable, see Section 1.1.6). Therefore, $H^0(F(X), L) \neq 0$ and, hence, $L \simeq \mathcal{O}$.[8] For $n = 3$, the fact that $\omega_F \simeq \mathcal{O}(1)|_F$ suffices to conclude.

Now, to prove the converse, let us assume that we are given a polarized isomorphism between two Fano varieties $(F(X), \mathcal{O}_F(1)) \simeq (F(X'), \mathcal{O}_{F'}(1))$. The restriction under the Plücker embedding $F(X) \subset \mathbb{G} \subset \mathbb{P}(\bigwedge^2 V)$ leads to the isomorphisms

$$H^0(\mathbb{P}(\textstyle\bigwedge^2 V), \mathcal{O}(1)) \xrightarrow{\sim} H^0(\mathbb{G}, \mathcal{O}_{\mathbb{G}}(1)) \xrightarrow{\sim} H^0(F(X), \mathcal{O}_F(1))$$

and similarly for $F(X')$. The first isomorphism is classical, e.g. a standard Bott formula for Grassmann varieties can be used to show that *Plücker coordinates* form a complete linear system. For the second one, use the Koszul complex (3.1) twisted by $\mathcal{O}(1)$, the

[8] Thanks to S. Stark for the argument. Alternatively, one can use [141]. As an aside, $\mathrm{Pic}(F(X))$ is also torsion free for $n = 4$, as then $F(X)$ is a hyperkähler manifold, see Theorem 6.3.10 and, therefore, $\mathrm{Pic}(F(X)) \simeq \mathrm{NS}(F(X))$. In Remark 4.19, we give an argument that for $n > 4$, the Fano variety $F(X)$ is a Fano variety and, hence, (algebraically) simply connected.

associated spectral sequence, and $H^i(\mathbb{G}, \bigwedge^i S^3(\mathcal{S})(1)) = 0$ for all $i > 0$, see [15, Thm. 5.1]. Thus, any given polarized isomorphism sits in a commutative diagram

$$\begin{array}{ccccc} F(X) & \hookrightarrow & \mathbb{G} & \hookrightarrow & \mathbb{P}(\bigwedge^2 V) \\ \downarrow\simeq & & \vdots & & \downarrow\simeq \\ F(X') & \hookrightarrow & \mathbb{G} & \hookrightarrow & \mathbb{P}(\bigwedge^2 V). \end{array} \tag{3.5}$$

The next step consists of completing the diagram by an automorphism of $\mathbb{G}$. For this use again (3.1), now twisted by $\mathcal{O}(2)$. On the one hand, according to [15, Thm. 1.15], restriction defines an injection $H^0(\mathbb{G}, \mathcal{O}_{\mathbb{G}}(2)) \hookrightarrow H^0(F(X), \mathcal{O}_F(2))$, while on the other hand, it is known classically, see [198, Ex. 8.12], that $\mathbb{G} \subset \mathbb{P}(\bigwedge^2 V)$ is cut out by quadrics, i.e. by the kernel of the restriction map $H^0(\mathbb{P}(\bigwedge^2 V), \mathcal{O}(2)) \longrightarrow H^0(\mathbb{G}, \mathcal{O}_{\mathbb{G}}(2))$. As this kernel coincides with the kernel of the restriction map to $F(X)$, the automorphism of $\mathbb{P}(\bigwedge^2 V)$ in (3.5) restricts to an automorphism of $\mathbb{G}$.

However, automorphisms of $\mathbb{G}$ are classified. In our situation, they are all induced by automorphisms of V, see [116] or [220, Thm. 10.19], which leads to an automorphism of the whole correspondence $\mathbb{G} \longleftarrow \mathbb{P}(\mathcal{S}) \longrightarrow \mathbb{P}(V)$. The final result is the commutative diagram

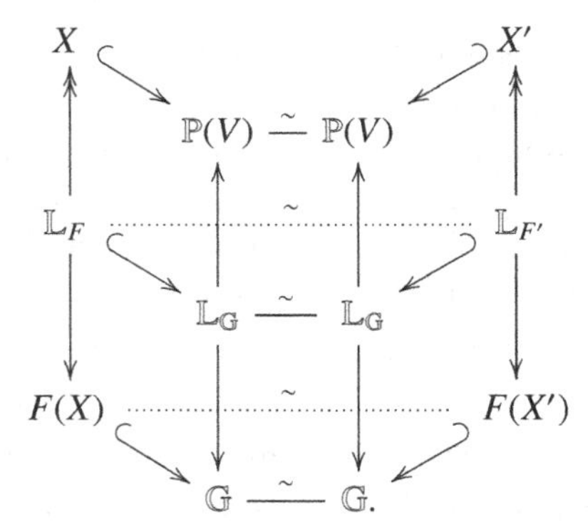

The surjectivity of $\mathbb{L} \twoheadrightarrow X$, i.e. the fact that there exists a line through every point, cf. Remark 3.6, eventually implies that the automorphism of $\mathbb{P}(V)$ restricts to an isomorphism $X \simeq X'$. □

The proof above shows more, namely that any (polarized) isomorphism $F(X) \simeq F(X')$ is induced by a unique isomorphism $X \simeq X'$. For $n = 3$, we will provide a different proof which relies on an isomorphism between the restriction of the tautological bundle $\mathcal{S}_F$ and the tangent sheaf $\mathcal{T}_F$, see Proposition **5**.2.12.

Corollary 3.13 *For a smooth cubic hypersurface $X \subset \mathbb{P}^{n+1}$ of dimension $n \neq 4$, the group of automorphisms* $\mathrm{Aut}(F(X))$ *of its Fano variety of lines $F(X)$ is finite and* $H^0(F(X), \mathcal{T}_{F(X)}) = 0$. *The vanishing holds for $n = 4$ also.*

Proof The assertions are obvious for $n \leq 2$. For $2 < n \neq 4$, we know by the arguments in the proof above that any automorphism $F(X) \simeq F(X)$ is induced by an automorphism of X. As by Corollary **1**.3.9 the group $\mathrm{Aut}(X)$ is finite for all cubic hypersurfaces of dimension at least two, this proves the assertion. The tangent space of the smooth group scheme $\mathrm{Aut}(F(X))$ is $H^0(F(X), \mathcal{T}_{F(X)})$, which therefore has to vanish.

For $n = 4$, one argues similarly using polarized automorphisms. Alternatively, it follows from $F(X)$ being a hyperkähler manifold, see Theorem **6**.3.10. □

For future reference, let us make the observation in the above proof more explicit as follows. The natural map

$$\mathrm{Aut}(X) \hookrightarrow \mathrm{Aut}(F(X)) \tag{3.6}$$

is injective for any smooth cubic hypersurface of dimension $2 < n$ and it is in fact an isomorphism for all $2 < n \neq 4$. For cubic fourfolds, one finds $\mathrm{Aut}(X) \simeq \mathrm{Aut}(F(X), \mathcal{O}(1))$.

In the same vain,

$$H^1(X, \mathcal{T}_X) \hookrightarrow H^1(F(X), \mathcal{T}_{F(X)}) \tag{3.7}$$

that associates to a first order deformation of X a first order deformation of $F(X)$ is injective for $n > 2$. We give a Hodge theoretic proof of this fact later, see Corollary 5.10. For cubic threefolds, (3.7) is in fact bijective, see Proposition **5**.2.14, but not for cubic fourfolds, see Corollary **6**.3.12.

Remark 3.14 Stark [441] shows that for $n > 4$, the natural map (3.7) is also surjective and hence bijective. By the previous corollary $H^0(F(X), \mathcal{T}_{F(X)}) = 0$ and Kodaira vanishing proves

$$H^i(F(X), \mathcal{T}_{F(X)}) \simeq H^i(F(X), \Omega_{F(X)}^{\dim F(X)-1} \otimes \omega_{F(X)}^*) = 0$$

for $i > 1$, since $\omega_{F(X)}$ is anti-ample. Thus, at least in principal, a Riemann–Roch computation can be used to show

$$\begin{aligned} - h^1(F(X), \mathcal{T}_{F(X)}) &= \chi(F(X), \mathcal{T}_{F(X)}) \\ &= \int_{\mathbb{G}} c_4(S^3(\mathcal{S}^*)) \cdot (\mathrm{ch}(\mathcal{S}^* \otimes \mathcal{Q}) - \mathrm{ch}(S^3(\mathcal{S}^*))) \cdot \mathrm{td}(\mathcal{S}^* \otimes \mathcal{Q}) \cdot \mathrm{td}(S^3(\mathcal{S}^*))^{-1} \\ &= -\binom{n+2}{3} = -h^1(X, \mathcal{T}_X), \end{aligned}$$

see the proof of Proposition 4.6 for similar computations. The arguments in [441] use Borel–Bott–Weil theory. Certain vanishing results entering the proof had earlier been observed by Borcea [85].

4 Cohomology and Motives

According to Proposition 3.12, the Fano variety $F(X)$ of lines contained in a smooth cubic hypersurface X determines the hypersurface. Therefore, essentially all and, in particular, all cohomological and motivic information about X should be encoded by $F(X)$. In this section we study the cohomology of $F(X)$ and we do this by first looking at the motive of $F(X)$.

4.1 Grothendieck Ring of Varieties As the Fano variety $F(X)$ itself, its motive is an interesting object to study. Now, the motive of $F(X)$ may mean various things. Here, we are interested in the class $[F(X)]$ of $F(X)$ in the *Grothendieck ring of varieties* $K_0(\mathrm{Var}_k)$ and in its motive $\mathfrak{h}(F(X))$ in the category $\mathrm{Mot}(k)$ of rational Chow motives.

We begin with the Grothendieck ring $K_0(\mathrm{Var}_k)$ of varieties over a field k. Recall that by definition it is the abelian group generated by classes $[Y]$ of quasi-projective varieties modulo the relations

$$[Y] = [Z] + [U],$$

the *scissor relation*. Here, $Z \subset Y$ is a closed subset and $U = Y \backslash Z$ is its open complement. The abelian group $K_0(\mathrm{Var}_k)$ becomes a ring with multiplication defined by the formula $[Y] \cdot [Y'] = [Y \times Y']$.

The *Lefschetz motive* is the class $\ell := [\mathbb{A}^1]$ of the affine line. An important consequence of the scissor relation is the fact that $[Y] = [F] \cdot [Z]$ for any Zariski locally trivial fibration $Y \twoheadrightarrow Z$ with fibre F. See, for example, [20, Ch. 13] or [113, Ch. 2] for more details.

Exercise 4.1 Note that, for the last assertion, it is not enough to assume that the fibration $Y \twoheadrightarrow Z$ is étale locally trivial. Show that otherwise one would have $\ell = 0$ in $K_0(\mathrm{Var}_k)$.

Galkin and Shinder [190] relate the class $[F(X)] \in K_0(\mathrm{Var}_k)$ to the class $[X^{[2]}] \in K_0(\mathrm{Var}_k)$ of the *Hilbert scheme* $X^{[2]}$ of subschemes of X of length two.[9] In general, the Hilbert square of a smooth variety can be obtained as the blow-up of the symmetric product $X^{(2)} := (X \times X)/\mathfrak{S}_2$ along the diagonal $X \simeq \Delta \subset X^{(2)}$. Hence, in $K_0(\mathrm{Var}_k)$ one has

$$[X^{[2]}] - [\mathbb{P}^{n-1}] \cdot [X] = [X^{(2)}] - [X]. \tag{4.1}$$

Proposition 4.2 (Galkin–Shinder) *Let $X \subset \mathbb{P}^{n+1}$ be a smooth cubic hypersurface. Then in $K_0(\mathrm{Var}_k)$ the following equations hold:*

$$[X^{[2]}] = [\mathbb{P}^n] \cdot [X] + \ell^2 \cdot [F(X)] \tag{4.2}$$

[9] As an aside, it is known that the Hilbert scheme of a smooth cubic hypersurface of dimension $n \geq 3$ is a Fano variety, see [62, Thm. C].

and

$$[X^{(2)}] = (1+\ell^n)\cdot[X] + \ell^2\cdot[F(X)]. \tag{4.3}$$

Proof We follow closely the argument in [190], where one also finds a version for singular cubics.

Consider the universal family $F(X) \longleftarrow \mathbb{L} \longrightarrow X$ of lines contained in X. As $\mathbb{L} \longrightarrow F(X)$ is the $\mathbb{P}^1$-bundle $\mathbb{P}(\mathcal{S}|_{F(X)}) \longrightarrow F(X)$, its class in $K_0(\mathrm{Var}_k)$ is given by

$$[\mathbb{L}] = [\mathbb{P}^1]\cdot[F(X)]. \tag{4.4}$$

Similarly, we denote by $\mathbb{G} = \mathbb{G}(1,\mathbb{P}) \longleftarrow \mathbb{L}_{\mathbb{G}} \longrightarrow \mathbb{P}$ the universal family of lines in $\mathbb{P} = \mathbb{P}^{n+1}$ and let $\mathbb{L}_{\mathbb{G}}|_X$ be the pre-image of X under the second projection. Then $\mathbb{L}_{\mathbb{G}}|_X$ parametrizes pairs (x, L) consisting of a line $L \subset \mathbb{P}$ and a point $x \in X \cap L$. It can also be described as the $\mathbb{P}^n$-bundle $\mathbb{P}(\mathcal{T}_{\mathbb{P}}|_X) \longrightarrow X$, cf. the construction in the proof of Corollary 1.21. This shows that in $K_0(\mathrm{Var}_k)$ one has

$$[\mathbb{L}_{\mathbb{G}}|_X] = [\mathbb{P}^n]\cdot[X]. \tag{4.5}$$

Next, consider the morphism $\mathbb{L}_{\mathbb{G}}|_X \setminus \mathbb{L} \longrightarrow X^{[2]}$ that sends (x, L) to the residual intersection $[(L\cap X)\setminus\{x\}] \in X^{[2]}$. It is inverse to the morphism $X^{[2]}\setminus\mathbb{L}^{[2]} \longrightarrow \mathbb{L}_{\mathbb{G}}|_X$ that sends $Z \in X^{[2]}$ to the pair (x, L_Z). Here, $L_Z \subset \mathbb{P}$ is the unique line containing the length two-subscheme $Z \subset \mathbb{P}$ and x is the residual point of the inclusion $Z \subset L_Z \cap X$. By definition, $\mathbb{L}^{[2]}$ is the relative symmetric product of the universal line $p\colon \mathbb{L} \longrightarrow F(X)$, which equivalently can be described as the relative Hilbert scheme of subschemes of length two in the fibres of p or, still equivalently, as the $\mathbb{P}^2$-bundle $\mathbb{L}^{[2]} \simeq \mathbb{P}(S^2(\mathcal{S}^*|_{F(X)})) \longrightarrow F(X)$.

Now, since

$$[\mathbb{L}^{[2]}] = [\mathbb{P}^2]\cdot[F(X)] \tag{4.6}$$

in $K_0(\mathrm{Var}_k)$, the isomorphism

$$\mathbb{L}_{\mathbb{G}}|_X \setminus \mathbb{L} \simeq X^{[2]} \setminus \mathbb{L}^{[2]}$$

together with (4.4), (4.5), and (4.6) proves the first equation (4.2). The second equation (4.3) follows from (4.1). □

Remark 4.3 The discussion shows that $X^{[2]}$ and $\mathbb{L}_{\mathbb{G}}|_X$ are birational. As the latter is simply $\mathbb{P}(\mathcal{T}_{\mathbb{P}}|_X)$, which is birational to $X\times\mathbb{P}^n$, one concludes that $X^{[2]}$ and X are stably birational. This can also be deduced from reducing (4.2) modulo ℓ, at least when $\mathrm{char}(k) = 0$. Indeed, by a result of Larsen and Lunts [306], the quotient $K_0(\mathrm{Var}_k) \twoheadrightarrow K_0(\mathrm{Var}_k)/(\ell)$ is isomorphic to the monoid ring $\mathbb{Z}[\mathrm{SB}_k]$, see also [113]. Here, SB_k is the monoid of equivalence classes of smooth projective varieties modulo *stable birationality*. Using the fact that $[\mathbb{P}^n] = 1 + \cdots + \ell^n \equiv 1$ modulo ℓ, (4.2) then shows

$$[X^{[2]}] = [X] \text{ in } \mathbb{Z}[\mathrm{SB}],$$

i.e. $X^{[2]}$ and X are stably birational.

Remark 4.4 As observed by Voisin [483, Prop. 2.9], a closer inspection reveals that the construction in the proof above leads to the following picture:

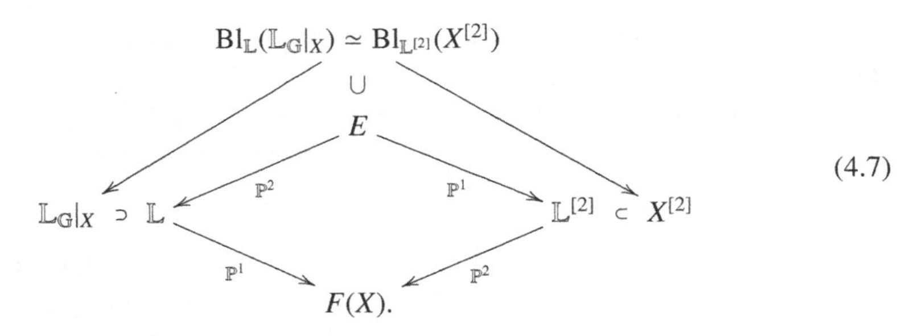

(4.7)

Here, E is the exceptional divisor of both blow-ups. In fact, $\mathrm{Bl}_{\mathbb{L}}(\mathbb{L}_{\mathbb{G}}|_X) \simeq \mathrm{Bl}_{\mathbb{L}^{[2]}}(X^{[2]})$ is an irreducible component of the incidence variety

$$\{ (x, L, Z) \mid x \in L \cap X,\ Z \subset L \cap X \} \subset X \times \mathbb{L}_{\mathbb{G}}|_X \times X^{[2]}$$

and $E \simeq \mathbb{L} \times_{F(X)} \mathbb{L}^{[2]}$.

4.2 Chow Motives Let us apply the standard formulae for cohomology and motives of smooth blow-ups and projective bundles to (4.7). For example, using $\operatorname{codim}(\mathbb{L} \subset \mathbb{L}_{\mathbb{G}}|_X) = 3$ and $\operatorname{codim}(\mathbb{L}^{[2]} \subset X^{[2]}) = 2$, the following isomorphisms hold in the category of rational *Chow motives* $\mathrm{Mot}(k)$

$$\mathfrak{h}(\mathrm{Bl}_{\mathbb{L}}(\mathbb{L}_{\mathbb{G}}|_X)) \oplus \mathfrak{h}(\mathbb{L})(-3) \simeq \mathfrak{h}(\mathbb{L}_{\mathbb{G}}|_X) \oplus \mathfrak{h}(E)(-1)$$

and

$$\mathfrak{h}(\mathrm{Bl}_{\mathbb{L}^{[2]}}(X^{[2]})) \oplus \mathfrak{h}(\mathbb{L}^{[2]})(-2) \simeq \mathfrak{h}(X^{[2]}) \oplus \mathfrak{h}(E)(-1),$$

see [20, 367]. Here, $\mathfrak{h}(Y)(-i) := \mathfrak{h}(Y) \otimes (\mathbb{P}^1, [\mathbb{P}^1 \times x])^{\otimes i}$ is the twist with the ith power of the Lefschetz motive. This can be combined with the standard formula for projective bundles, which in our situation gives

$$\begin{aligned} \mathfrak{h}(\mathbb{L}) &\simeq \mathfrak{h}(F(X)) \oplus \mathfrak{h}(F(X))(-1), \\ \mathfrak{h}(\mathbb{L}_{\mathbb{G}}|_X) &\simeq \mathfrak{h}(X) \oplus \cdots \oplus \mathfrak{h}(X)(-n), \text{ and} \\ \mathfrak{h}(\mathbb{L}^{[2]}) &\simeq \mathfrak{h}(F(X)) \oplus \mathfrak{h}(F(X))(-1) \oplus \mathfrak{h}(F(X))(-2). \end{aligned}$$

The isomorphism $\mathrm{Bl}_{\mathbb{L}}(\mathbb{L}_{\mathbb{G}}|_X) \simeq \mathrm{Bl}_{\mathbb{L}^{[2]}}(X^{[2]})$ then implies a formula which, assuming cancellation holds in $\mathrm{Mot}(k)$, would look like this:

$$\mathfrak{h}(F(X))(-2) \oplus \bigoplus_{i=0}^{n} \mathfrak{h}(X)(-i) \simeq \mathfrak{h}(X^{[2]}). \tag{4.8}$$

That cancellation in our case does hold was proved by Laterveer [307] and independent proof of the isomorphism was given by Diaz [153]. The decomposition was refined incorporating the primitive decomposition by Fu, Laterveer, and Vial [183]. The formula can be combined with the isomorphism

$$\mathfrak{h}(X^{[2]}) \simeq S^2\mathfrak{h}(X) \oplus \bigoplus_{i=1}^{n-1} \mathfrak{h}(X)(-i), \tag{4.9}$$

which shows that finite-dimensionality of $\mathfrak{h}(X)$ in the sense of Kimura and O'Sullivan implies finite-dimensionality of $\mathfrak{h}(F(X))$, cf. [307, Thm. 4].[10]

Thus, assuming cancellation holds, one has:

$$\mathfrak{h}(F(X))(-2) \oplus \mathfrak{h}(X) \oplus \mathfrak{h}(X)(-n) \simeq S^2\mathfrak{h}(X). \tag{4.10}$$

Using the decomposition $\mathfrak{h}(X) \simeq \mathfrak{h}(X)_{\mathrm{pr}} \oplus \bigoplus_{i=0}^{n} \mathbb{Q}(-i)$, cf. Remark **1**.1.11, and assuming cancellation holds, this then becomes

$$\mathfrak{h}(F(X))(-2) \oplus \mathbb{Q}(-n) \simeq S^2\mathfrak{h}(X)_{\mathrm{pr}} \oplus \bigoplus_{i=1}^{n-1} \mathfrak{h}(X)_{\mathrm{pr}}(-i) \oplus \bigoplus_{0<i\leq j<n} \mathbb{Q}(-i-j), \tag{4.11}$$

see [183]. Here, $\mathbb{Q}(1)$ is the Tate motive $(\mathrm{Spec}(k), \mathrm{id}, 1)$, the dual of the Lefschetz motive.

Remark 4.5 As a consequence of Bittner's result [70], see also (4.15) below, assigning the Chow motive $\mathfrak{h}(X)$ to a smooth projective variety X defines a linear map

$$K_0(\mathrm{Var}_k) \longrightarrow K_0(\mathrm{Mot}_k).$$

The Grothendieck group on the right-hand side is by definition the abelian group generated by all rational Chow motives $\mathfrak{h}$ subject to the relation $[\mathfrak{h}] + [\mathfrak{h}'] = [\mathfrak{h} \oplus \mathfrak{h}']$. This allows one to deduce the above isomorphisms as equalities in $K_0(\mathrm{Mot}_k)$ without assuming cancellation.

For a categorical version of this result, also just using (4.7), see Section **7**.1.8.

4.3 Degree and Euler Number Combining the information obtained from the description of $F(X) \subset \mathbb{G}(1, \mathbb{P})$ as the zero set $V(s_F)$ and the description of its class $[F(X)] \in K_0(\mathrm{Var}_k)$, one can deduce the numerical information in Proposition 4.6 below, see [15, Prop. 1.6] and [190, Cor. 5.2]. The case $n = 3$ goes back to Bombieri and Swinnerton-Dyer [80].

Proposition 4.6 (Altman–Kleiman, Galkin–Shinder) *Let $X \subset \mathbb{P}^{n+1}$ be a smooth cubic hypersurface and let $F(X)$ be its Fano variety of lines considered with its Plücker embedding*

$$F(X) \hookrightarrow \mathbb{G}(1, \mathbb{P}) \hookrightarrow \mathbb{P}^N,$$

[10] Finite-dimensionality of $\mathfrak{h}(X)$ is known for $n = 3$, Section **5**.3, and $n = 5$, see Remark **7**.4.7. It is not known whether smooth cubic hypersurfaces of dimension $5 \neq n > 3$ are Kimura finite-dimensional.

where $N = \binom{n+2}{2} - 1$. *Then the degree and the Euler number of* $F(X)$ *are given by the following formulae*

$$\deg(F(X)) = 27 \cdot \frac{(2n-4)!}{n! \cdot (n-1)!} \cdot (3n^2 - 7n + 4) \tag{4.12}$$

and

$$e(F(X)) = \frac{e(X) \cdot (e(X) - 3)}{2} = \frac{2^{2n+4} + (-2)^{n+2} \cdot (6n+1) + 3n \cdot (3n+1) - 20}{18}. \tag{4.13}$$

In Corollary 4.16 below, the formula for the Euler number is generalized to a formula for the χ_y-genus.

Proof As a special case of (1.6), we know that $[F(X)] = \mathrm{c}_4(S^3(\mathcal{S}^*))$ in the cohomology ring or in the Chow ring of $\mathbb{G}$. Writing formally $\mathcal{S}^* = L_0 \oplus L_1$ and $S^3(\mathcal{S}^*) = L_0^3 \oplus (L_0^2 \otimes L_1) \oplus (L_0 \otimes L_1^2) \oplus L_1^3$, allows one to compute, cf. Exercise 3.2:

$$\begin{aligned} \mathrm{c}_4(S^3(\mathcal{S}^*)) &= 9 \cdot (5\,\ell_0^2\ell_1^2 + 2\,(\ell_0^3\ell_1 + \ell_0\ell_1^3)) \\ &= 9 \cdot \left(2\,\mathrm{c}_1(\mathcal{S}^*)^2 + \mathrm{c}_2(\mathcal{S}^*)\right) \cdot \mathrm{c}_2(\mathcal{S}^*), \end{aligned}$$

where $\ell_i = \mathrm{c}_1(L_i)$. Hence,

$$\deg(F(X)) = 9 \int_{\mathbb{G}} \mathrm{c}_1(\mathcal{S}^*)^{2n-4} \cdot \left(2\,\mathrm{c}_1(\mathcal{S}^*)^2 + \mathrm{c}_2(\mathcal{S}^*)\right) \cdot \mathrm{c}_2(\mathcal{S}^*).$$

Using standard Schubert calculus (Pieri's and Gambelli's formulae), this is turned in [15] into a rather complicated formula which then can be simplified to the above. Compare Remark 4.7 below for an alternative approach.

In principle, the second assertion can also be deduced by Schubert calculus, as

$$e(F(X)) = \int_{F(X)} \mathrm{c}_{2n-4}(\mathcal{T}_{F(X)}) = \int_{\mathbb{G}} \left(\frac{c(\mathcal{T}_{\mathbb{G}})}{c(S^3(\mathcal{S}^*))}\right)_{2n-4} \cdot \mathrm{c}_4(S^3(\mathcal{S}^*)).$$

But here is a more illuminating way of doing this. Taking Euler numbers of (4.3) in Proposition 4.2 shows

$$e(X^{(2)}) = 2 \cdot e(X) + e(F(X)),$$

where we use the additivity, respectively, the multiplicativity, of the Euler number and $e(\ell^n) = 1$, cf. [70]. Taking cohomology commutes with taking symmetric products, in other words, $H^*(X^{(n)}) = S^n H^*(X)$ (say with coefficients in a field of characteristic zero), cf. [211, Prop. 5.2.3] or [331]. Hence, $e(X^{(2)}) = \binom{e(X)+1}{2}$.[11] This proves the first equality in (4.13) and the second follows from (**1**.1.6). □

[11] The closed formula proved by MacDonald says

$$\sum_{n=0}^{\infty} e(X^{(n)})\, z^n = (1-z)^{-e(X)} = \exp\left(e(X) \cdot \sum_{r=1}^{\infty} z^r / r\right), \tag{4.14}$$

Remark 4.7 The above classical computation of the degree relies on the representation of $F(X)$ as a subvariety of the Grassmannian $\mathbb{G}$. As the degree can also be computed as

$$\deg(F(X)) = \int_{\mathbb{L}} \mathrm{c}_1(p^*\mathcal{O}(1))^{2n-4} \cdot \mathrm{c}_1(\mathcal{O}_p(1))$$

on the universal line $\mathbb{L}$, we can change perspective and instead exploit the projection $q\colon \mathbb{L} \twoheadrightarrow X$ and the description of $\mathbb{L}$ as a subscheme of $\mathbb{P}(\mathcal{T}_X)$ of codimension two provided by Proposition 3.10. More precisely, using (3.4) in its proof, allows one to compute the degree as

$$\deg(F(X)) = \int_{\mathbb{P}(\mathcal{T}_X)} (2\pi^*h + u)^{2n-4} \cdot \pi^*h \cdot (3\pi^*h + 3u) \cdot (3\pi^*h + 2u),$$

where $h = \mathrm{c}_1(\mathcal{O}_X(1))$ and $u = \mathrm{c}_1(\mathcal{O}_\pi(1))$.

Exercise 4.8 Use the projection $\pi\colon \mathbb{P}(\mathcal{T}_X) \twoheadrightarrow X$ and $\sum u^i \cdot \pi^*\mathrm{c}_{n-i}(\mathcal{T}_X) = 0$ to conclude the computation in the previous remark and to confirm (4.12).

Remark 4.9 Recall from Section **1**.1.4 that the Euler numbers $e(X_n)$ of smooth cubic hypersurfaces $X_n \subset \mathbb{P}^{n+1}$ of arbitrary dimensions are encoded by the generating series

$$\sum_{n=0}^{\infty} e(X_n)\, z^{n+1} = \frac{3\,z}{(1-z)^2\,(1+2\,z)}.$$

A formal computation using Mathematica[12] reveals

$$\sum_{n=2}^{\infty} e(F(X_n))\, z^{n+1} = \frac{27\,(1-2\,z)\,z^3}{(-1+z)^3\,(1+2\,z)^2\,(-1+4\,z)},$$

but a conceptual understanding in the sense of Theorem **1**.1.17 is not known.[13]

n	$\omega_{F(X)}$	$\dim(F(X))$	$\deg(F(X))$	$e(F(X))$
2	$\mathcal{O}$	0	27	27
3	$\mathcal{O}(1)$	2	45	27
4	$\mathcal{O}$	4	108	324
5	$\mathcal{O}(-1)$	6	297	702

which is the geometric analogue of the well-known equality $\sum_{n=0}^{\infty} |X^{(n)}(\mathbb{F}_q)|\, z^n = \exp\left(\sum |X(F_{q^r})|\, z^r/r\right)$ for the Zeta function of a variety over a finite field $\mathbb{F}_q$ and which generalizes to an equality for the Poincaré polynomial

$$\sum_{n=0}^{\infty} \left(\sum_i (-1)^i b_i(X^{(n)})\, y^i \right) z^n = \frac{(1-y^1z)^{b_1(X)} \cdot (1-y^3z)^{b_3(X)} \dots}{(1-z)^{b_0(X)} \cdot (1-y^2z)^{b_2(X)} \dots}.$$

[12] With thanks to P. Magni.

[13] Mathematica did not come up with a generating series for $\deg(F(X_n))$.

4.4 Hodge Theory via Motives The computation of the Euler number is only a faint shadow of the full cohomological information available. There are various ways to unpack the information encoded by the above motivic approach. We shall focus on the Hodge theoretic content and so assume from now on that $k = \mathbb{C}$.

To start, let $\mathrm{HS}_{\mathbb{Z},n}$ be the additive category of polarizable, pure Hodge structures of weight n, see [172] for more on this and the notation. Recall that the Tate twist defines an equivalence

$$\mathrm{HS}_{\mathbb{Z},n} \xrightarrow{\sim} \mathrm{HS}_{\mathbb{Z},n-2},\ H \longmapsto H(1) := H \otimes \mathbb{Z}(1),$$

where $\mathbb{Z}(1) = (2\pi i)\mathbb{Z}$ is the pure Hodge structure of weight $(-1,-1)$ geometrically realized by the dual of $H^2(\mathbb{P}^1, \mathbb{Z})$. Let $\mathrm{HS}_{\mathbb{Z}} = \bigoplus \mathrm{HS}_{\mathbb{Z},n}$ be the additive category of graded pure, polarizable integral Hodge structures and denote its Grothendieck group by $K_0(\mathrm{HS}_{\mathbb{Z}})$. By definition, this is the group generated by isomorphism classes of integral polarizable Hodge structures with the condition that $[H]+[H'] = [H \oplus H']$. In particular, two Hodge structures H and H' define the same class $[H] = [H']$ in $K_0(\mathrm{HS}_{\mathbb{Z}})$ if and only if there exists a Hodge structure H_0 such that $H \oplus H_0 \simeq H' \oplus H_0$. Note that the tensor product defines a natural ring structure on $K_0(\mathrm{HS}_{\mathbb{Z}})$.

According to Bittner [70], $K_0(\mathrm{Var}_{\mathbb{C}})$ can also be described as the quotient of the free abelian group generated by isomorphism classes of smooth projective varieties by the relation

$$[\mathrm{Bl}_Z(Y)] + [Z] = [Y] + [E]$$

for every blow-up $\mathrm{Bl}_Z(Y) \longrightarrow Y$ of a smooth projective variety Y along a smooth projective subvariety $Z \subset Y$ of codimension r with exceptional divisor E. Using the fact that for each such smooth blow-up there exists a graded isomorphism of polarizable integral Hodge structures, cf. [474, Ch. 7]:

$$H^*(\mathrm{Bl}_Z(Y), \mathbb{Z}) \oplus H^*(Z, \mathbb{Z})(-r) \simeq H^*(Y, \mathbb{Z}) \oplus H^*(E, \mathbb{Z})(-1), \tag{4.15}$$

one finds that there exists a ring homomorphism

$$K_0(\mathrm{Var}_{\mathbb{C}}) \longrightarrow K_0(\mathrm{HS}_{\mathbb{Z}}),\ [X] \longmapsto [H^*(X, \mathbb{Z})(\dim X)], \tag{4.16}$$

where X is smooth and projective. Under this map, $\ell = [\mathbb{A}^1] = [\mathbb{P}^1] - [\mathrm{pt}]$ is sent to the class of the Hodge structure $\mathbb{Z}(-1)$.

Exercise 4.10 Show that $X \longmapsto (-y)^{-\dim X} \cdot \chi_y(X)$ defines a ring homomorphism

$$\tilde{\chi}_y \colon K_0(\mathrm{Var}_{\mathbb{C}}) \longrightarrow \mathbb{Z}[y, y^{-1}],$$

which factors through (4.16).

Corollary 4.11 *Let $X \subset \mathbb{P}^{n+1}$ be a smooth cubic hypersurface. Then in $K_0(\mathrm{HS}_{\mathbb{Z}})$ the following equality holds*

$$[H^*(X^{[2]},\mathbb{Z})(2)] = [H^*(\mathbb{P}^n,\mathbb{Z})(2)] \cdot [H^*(X,\mathbb{Z})] + [H^*(F(X),\mathbb{Z})]. \tag{4.17}$$

As we shall see shortly, there is an isomorphism in $\mathrm{HS}_{\mathbb{Z}}$ behind (4.17), see Corollary 4.12 below.

Proof Apply (4.16) to Proposition 4.2 and twist by the Hodge structure $\mathbb{Z}(2)$. □

Next we consider the natural functor

$$\mathrm{HS}_{\mathbb{Z}} \longrightarrow \mathrm{HS}_{\mathbb{Q}},\ H \longmapsto H \otimes_{\mathbb{Z}} \mathbb{Q}$$

to the category of graded pure, polarizable, rational Hodge structures $\mathrm{HS}_{\mathbb{Q}}$. It induces a linear map

$$K_0(\mathrm{Var}_{\mathbb{C}}) \longrightarrow K_0(\mathrm{HS}_{\mathbb{Z}}) \longrightarrow K_0(\mathrm{HS}_{\mathbb{Q}}). \tag{4.18}$$

As the category $\mathrm{HS}_{\mathbb{Q}}$ is semi-simple, cf. [390, Cor. 2.12], the natural map

$$\mathrm{HS}_{\mathbb{Q}} \hookrightarrow K_0(\mathrm{HS}_{\mathbb{Q}})$$

is injective, i.e. two rational Hodge structures H and H' are isomorphic if and only if $[H] = [H']$ in $K_0(\mathrm{HS}_{\mathbb{Q}})$.[14] Thus, (4.17) becomes a graded isomorphism of rational Hodge structures

$$H^*(X^{[2]},\mathbb{Q})(2) \simeq (H^*(\mathbb{P}^n,\mathbb{Q}) \otimes H^*(X,\mathbb{Q}))\,(2) \oplus H^*(F(X),\mathbb{Q}). \tag{4.19}$$

There is a shortcut to arrive at the isomorphism (4.19) by just applying cohomology to (4.8), i.e. using the commutativity of the diagram

$$\begin{array}{ccccc}
(\mathrm{SmProj}(\mathbb{C})) & & \longrightarrow & & \mathrm{Mot}(\mathbb{C}) \\
\downarrow & & & & \downarrow \\
K_0(\mathrm{Var}_{\mathbb{C}}) & \longrightarrow & K_0(\mathrm{HS}_{\mathbb{Z}}) & \longrightarrow & K_0(\mathrm{HS}_{\mathbb{Q}}).
\end{array}$$

Similarly, either by applying cohomology to (4.9) or by using (4.18), one obtains an isomorphism of Hodge structures

$$H^*(X^{[2]},\mathbb{Q}) \simeq S^2H^*(X,\mathbb{Q}) \oplus \bigoplus_{i=1}^{n-1} H^*(X,\mathbb{Q})(-i).$$

[14] Injectivity does not hold for integral Hodge structures. Indeed, there exist elliptic curves E, E', and E_0 such that E and E' are non-isomorphic but $E \times E_0 \simeq E' \times E_0$, see [435]. In this case, $[H^1(E,\mathbb{Z})] = [H^1(E',\mathbb{Z})]$ in $K_0(\mathrm{HS}_{\mathbb{Z}})$, but $H^1(E,\mathbb{Z})$ and $H^1(E',\mathbb{Z})$ are non-isomorphic Hodge structures. Thanks to B. Moonen for the reference.

Altogether this leads to the isomorphism of Hodge structures

$$(S^2H^*(X,\mathbb{Q}))(2) \simeq H^*(X,\mathbb{Q})(2) \oplus H^*(X,\mathbb{Q})(2-n) \oplus H^*(F(X),\mathbb{Q}) \tag{4.20}$$

and, after decomposing into primitive parts, to

$$\begin{aligned} H^*(F(X),\mathbb{Q}) \oplus \mathbb{Q}(2-n) \simeq & \bigoplus_{i=1}^{n-1} H^n(X,\mathbb{Q})_{\mathrm{pr}}(2-i) \oplus \bigoplus_{0<i\leq j<n} \mathbb{Q}(2-i-j) \\ & \oplus \begin{cases} (S^2H^n(X,\mathbb{Q})_{\mathrm{pr}})(2) & \text{for } n \equiv 0\,(2), \\ (\bigwedge^2 H^n(X,\mathbb{Q})_{\mathrm{pr}})(2) & \text{for } n \equiv 1\,(2), \end{cases} \end{aligned} \tag{4.21}$$

cf. [190]. Of course, here and below $H^n(X,\mathbb{Q})_{\mathrm{pr}} = H^n(X,\mathbb{Q})$ for n odd. Note that (4.21) can also be obtained by taking cohomology of (4.11). In particular, for the middle cohomology of the Fano variety, the formula proves

$$H^{2n-4}(F(X),\mathbb{Q}) \simeq \mathbb{Q}(2-n)^{\oplus m} \oplus \begin{cases} (S^2H^n(X,\mathbb{Q})_{\mathrm{pr}})(2) \oplus H^n(X,\mathbb{Q})_{\mathrm{pr}}(1-\frac{n}{2}) & \text{for } n \equiv 0\,(2), \\ (\bigwedge^2 H^n(X,\mathbb{Q})_{\mathrm{pr}})(2) & \text{for } n \equiv 1\,(2) \end{cases}$$

with $m = [n/2] - 1$.

4.5 Integral Hodge Structures Instead of using the abstract language of motives, it is possible to work entirely on the level of cohomology. In fact, working directly with cohomology makes some of the results more concrete and more precise, e.g. (4.19) is valid for cohomology with integral coefficients, and shows that the isomorphisms in (4.21) are in fact algebraic.

Start with the diagram (4.7) and apply the blow-up formula for cohomology, cf. [474, Ch. 7], to $\sigma_1\colon B := \mathrm{Bl}_{\mathbb{L}}(\mathbb{L}_{\mathbb{G}}|_X) \longrightarrow \mathbb{L}_{\mathbb{G}}|_X$. Note that its exceptional divisor $\tau_1\colon E \longrightarrow \mathbb{L}$ is a $\mathbb{P}^2$-bundle. We obtain isomorphisms

$$\begin{aligned} H^*(B,\mathbb{Z}) & \simeq H^*(\mathbb{L}_{\mathbb{G}}|_X,\mathbb{Z}) \oplus H^*(\mathbb{L},\mathbb{Z})(-1) \oplus H^*(\mathbb{L},\mathbb{Z})(-2) \\ & \simeq (H^*(\mathbb{P}^n,\mathbb{Z}) \otimes H^*(X,\mathbb{Z})) \oplus \left\{ \begin{matrix} H^*(F,\mathbb{Z})(-1) \\ \oplus \\ H^*(F,\mathbb{Z})(-2) \end{matrix} \right\} \oplus \left\{ \begin{matrix} H^*(F,\mathbb{Z})(-2) \\ \oplus \\ H^*(F,\mathbb{Z})(-3) \end{matrix} \right\}. \end{aligned} \tag{4.22}$$

The inverse of the first isomorphism is up to the sign given by

$$(\alpha,\alpha_1,\alpha_2) \longmapsto \sigma_1^*\alpha + j_*\tau_1^*\alpha_1 + j_*\tau_1^*\alpha_2 \cdot [E].$$

Here, $j\colon E \hookrightarrow B$ is the inclusion. For the second isomorphism, apply the Leray–Hirsch formula for the cohomology of a projective bundle to the $\mathbb{P}^n$-bundle $\mathbb{L}_{\mathbb{G}}|_X \longrightarrow X$ and to the $\mathbb{P}^1$-bundle $p\colon \mathbb{L} \longrightarrow F = F(X)$.

Next we use the blow-up formula for $\sigma_2\colon B \simeq \mathrm{Bl}_{\mathbb{L}^{[2]}}(X^{[2]}) \longrightarrow X^{[2]}$. This time the exceptional divisor $\tau_2\colon E \longrightarrow \mathbb{L}^{[2]}$ is a $\mathbb{P}^1$-bundle and the Leray–Hirsch formula applied to the $\mathbb{P}^2$-bundle $p^{[2]}\colon \mathbb{L}^{[2]} \longrightarrow F$, shows

$$\begin{aligned} H^*(B,\mathbb{Z}) &\simeq H^*(X^{[2]},\mathbb{Z}) \oplus H^*(\mathbb{L}^{[2]},\mathbb{Z})(-1) \\ &\simeq H^*(X^{[2]},\mathbb{Z}) \oplus \left\{ H^*(F,\mathbb{Z})(-1) \oplus H^*(F,\mathbb{Z})(-2) \oplus H^*(F,\mathbb{Z})(-3) \right\}. \end{aligned}$$

Combining the two descriptions of $H^*(B,\mathbb{Z})$ gives the following isomorphism of Hodge structures which gives back Corollary 4.11.

Corollary 4.12 *Let $X \subset \mathbb{P}^{n+1}$ be a smooth cubic hypersurface. Then there exists an isomorphism*

$$\begin{aligned} H^*(X^{[2]},\mathbb{Z}) &\simeq (H^*(\mathbb{P}^n,\mathbb{Z}) \otimes H^*(X,\mathbb{Z})) \oplus H^*(F,\mathbb{Z})(-2) \\ \oplus\, H^*(\mathbb{L}^{[2]},\mathbb{Z})(-1) &\qquad \oplus\, H^*(\mathbb{L}^{[2]},\mathbb{Z})(-1) \end{aligned} \tag{4.23}$$

of integral Hodge structures. □

Exercise 4.13 Show that the cohomology $H^*(F(X),\mathbb{Z})$ of the Fano variety of a smooth cubic hypersurface is torsion free.

Use the fact that the integral cohomology $H^*(X^{[2]},\mathbb{Z})$ of a complex manifold X with torsion free cohomology $H^*(X,\mathbb{Z})$ is again torsion free, see [452]. An argument based on the weaker version of (4.23) provided by the equality (4.17) in $K_0(\mathrm{Hdg}_{\mathbb{Z}})$ was given by Shinder [432]. See Remark **5**.1.21 for comments on the case of cubic threefolds.[15]

Note that apart from the obvious copy of $H^*(F,\mathbb{Z})(-2)$ on the right-hand side of (4.23) there is another one hidden in $H^*(\mathbb{L}^{[2]},\mathbb{Z})(-1)$ on both sides. The two copies arise from the natural inclusions of direct summands

$$H^*(F,\mathbb{Z})(-2) \hookrightarrow H^*(\mathbb{L},\mathbb{Z})(-1) \quad \text{and} \quad H^*(F,\mathbb{Z})(-2) \hookrightarrow H^*(\mathbb{L},\mathbb{Z})(-2)$$

in (4.22). Composing the inclusion $H^*(X^{[2]},\mathbb{Z}) \hookrightarrow H^*(B,\mathbb{Z})$ with the projection

$$H^*(B,\mathbb{Z}) \longrightarrow H^*(\mathbb{L},\mathbb{Z})(-1) \oplus H^*(\mathbb{L},\mathbb{Z})(-2) \longrightarrow H^*(F,\mathbb{Z})(-2) \oplus H^*(F,\mathbb{Z})(-2)$$

defines a map

$$(f_1,f_2)\colon H^*(X^{[2]},\mathbb{Z}) \longrightarrow H^*(F,\mathbb{Z})(-2) \oplus H^*(F,\mathbb{Z})(-2).$$

Lemma 4.14 *With the above notation, $f_1 = 0$ and $f_2(\alpha) = p^{[2]}_*(\alpha|_{\mathbb{L}^{[2]}})$.*

[15] Thanks to S. Stark for various discussions related to this. His question on MathOverflow prompted the above argument and Shinder's proof.

Proof The computation of $f_2(\alpha)$ is a consequence of the commutativity of the diagram

$$\begin{array}{ccc} H^*(X^{[2]}) & \xrightarrow{\sigma_2^*} & H^*(B) \\ {\scriptstyle ()|_{\mathbb{L}^{[2]}}}\downarrow & & \downarrow{\scriptstyle ()|_E} \\ H^*(\mathbb{L}^{[2]}) & \xrightarrow{\tau_2^*} & H^*(E) \\ {\scriptstyle p_*^{[2]}}\downarrow & & \downarrow{\scriptstyle \tau_{1*}} \\ H^{*-4}(F) & \xrightarrow{p^*} & H^{*-4}(\mathbb{L}). \end{array}$$

To compute f_1, we write $[E]|_E = \tau_1^* e_1 + \tau_2^* e_2$, possibly up to classes on F which will not effect the following. Here, e_1 and e_2 are the relative tautological classes of p and $p^{[2]}$. Then use $\tau_{1*}(\sigma_2^*\delta|_E \cdot [E]|_E) = \tau_{1*}(\tau_2^*(\delta|_{\mathbb{L}^{[2]}}) \cdot [E]|_E)$ and $\delta|_{\mathbb{L}^{[2]}} = p^{[2]*}\delta_1 \oplus (p^{[2]*}\delta_2 \cdot e_2)$, which leads to $\tau_2^*(\delta|_{\mathbb{L}^{[2]}}) = \tau_1^* p^*\delta_1 \oplus (\tau_1^* p^*\delta_2 \cdot \tau_2^* e_2)$ for some classes $\delta_1 \in H^*(F)$ and $\delta_2 \in H^{*-2}(F)$. Hence, $\tau_{1*}(\tau_2^*(\delta|_{\mathbb{L}^{[2]}}) \cdot [E]|_E)$ is the sum of $\tau_{1*}\big(\tau_1^*(p^*\delta_1 \cdot e_1) \oplus (\tau_1^* p^*\delta_1 \cdot \tau_2^* e_2)\big)$ and $\tau_{1*}\big((\tau_1^* p^*(\delta_2 \cdot e_1) \cdot \tau_2^* e_2) \oplus \tau_1^* p^*\delta_2 \cdot \tau_2^* e_2^2)\big)$. The two parts of the first summand are trivial, because $\tau_{1*}\tau_1^* = 0$ and $\tau_{1*}\tau_2^* e_2 = 0$, for τ_1 is a $\mathbb{P}^2$-bundle. Similarly, the first part of the second summand is trivial. Therefore, $\tau_{1*}(\tau_2^*(\delta|_{\mathbb{L}^{[2]}}) \cdot [E]|_E) = p^*\delta_2 \cdot \tau_{1*}\tau_2^* e_2^2$, the projection of which to $H^{*-4}(F, \mathbb{Z}) \subset H^{*-2}(\mathbb{L}, \mathbb{Z})$ is trivial. □

Corollary 4.15 *For the general smooth cubic hypersurface $X \subset \mathbb{P}^{n+1}$, the rational Hodge conjecture holds for $F(X)$ in the middle degree $2n-4$. The space of Hodge classes in this degree satisfies* $\dim H^{n-2,n-2}(F(X), \mathbb{Q}) = [n/2]$.

Proof Consider the isomorphism (4.21) which by the preceding discussion is algebraic. On the right-hand side, the direct sum $\mathbb{Q}(2-n)^{\oplus m}$ is spanned by Hodge classes which are all obviously algebraic. According to Remark **1**.2.13, (ii), up to scalars the only Hodge class in the summand $S^2(H^n(X, \mathbb{Q})_{\rm pr})$ for n even and in $\bigwedge^2(H^n(X, \mathbb{Q})_{\rm pr})$ for n odd is the class that corresponds to the intersection form q which is algebraic. □

4.6 χ_y-Genus and Low Dimensions Both Corollary 4.11 or alternatively (4.21), allow one to compute the Betti and Hodge numbers of $F(X)$. The formula below evaluated at $y = -1$ gives back (4.13).

Corollary 4.16 *The χ_y-genus of the Fano variety of lines $F(X)$ of a smooth cubic hypersurface $X \subset \mathbb{P}^{n+1}$ is given by*

$$\chi_y(F(X)) = \frac{\chi_y(X) - 2(-1)^n y^n - 1}{2y^2} \cdot \chi_y(X).$$

Proof We use $\chi_y(\ell) = -y$, $\chi_y(\ell^n) = (-y)^n$, $\chi_y(\mathbb{P}^n) = 1 - y \pm \cdots + (-1)^n y^n$, and

$$\chi_y(X^{(2)}) = \binom{\chi_y(X) + 1}{2}.$$

The latter is again a special case of a general *MacDonald formula* for the χ_y-genus analogous to the one for the Euler number (4.14), see [89, 347, 500]. This gives

$$y^2\cdot\chi_y(F)=\binom{\chi_y(X)+1}{2}-(1+(-y)^n)\cdot\chi_y(X)$$
$$=\frac{\chi_y(X)-1-2(-y)^n}{2}\cdot\chi_y(X)$$

by applying the homomorphism $\tilde{\chi}_y\colon K_0(\mathrm{Var}_k)\longrightarrow K_0(\mathrm{HS}_{\mathbb{Q}})\longrightarrow\mathbb{Z}[y,y^{-1}]$, see Exercise 4.10, to (4.3) in Proposition 4.2 or using Corollary 4.11 and only the second map. □

In principle, it should be possible to combine this with Hirzebruch's formula for the generating series $\sum_{n=0}^{\infty}\chi_y(X_n)\,z^{n+1}$ of all χ_y-genera of cubic hypersurfaces, see Theorem 1.1.17, to express $\sum_{n=0}^{\infty}\chi_y(F(X_n))\,z^{n+1}$ as a rational function, cf. Remarks 4.9 and 4.22.

Before making some of the computations explicit in low-dimensional cases, we shall draw a few further consequences from (4.21).

Corollary 4.17 *Let X be a smooth cubic hypersurface of dimension n.*

(i) *If n is even, then $H^*(X,\mathbb{Q})=H^{\mathrm{ev}}(X,\mathbb{Q})$ and $H^*(F(X),\mathbb{Q})=H^{\mathrm{ev}}(F(X),\mathbb{Q})$.*
(ii) *If n is odd, then $H^*(X,\mathbb{Q})=H^{\mathrm{ev}}(X,\mathbb{Q})\oplus H^n(X,\mathbb{Q})$ and*

$$H^*(F(X),\mathbb{Q})\simeq H^{\mathrm{ev}}(F(X),\mathbb{Q})\oplus\bigoplus_{i=1}^{n-1}H^n(X,\mathbb{Q})(2-i).\qquad\square$$

The description of the first cohomology provides us with an alternative proof of Corollary 3.3, which we state again in the following improved form.

Corollary 4.18 *Let $X\subset\mathbb{P}^{n+1}$ be a smooth cubic hypersurface and let $F(X)$ be its Fano variety of lines.*

(i) *If $n\geq 4$, then the Picard variety* $\mathrm{Pic}^0(F(X))$ *is trivial.*
(ii) *If $n\geq 5$, then the Picard group is of rank one, i.e.* $\mathrm{Pic}(F(X))\simeq\mathbb{Z}$.

Proof Indeed, as $H^{\mathrm{odd}}(F(X),\mathbb{Q})$ is trivial for even n and otherwise $H^{\mathrm{odd}}(F(X),\mathbb{Q})\simeq\bigoplus_{i=1}^{n-1}H^n(X,\mathbb{Q})(2-i)$, one finds $H^1(F(X),\mathbb{Q})=0$ for $n\geq 4$. Hence, $H^{0,1}(F(X))=0$ which implies that $\mathrm{Pic}^0(F(X))$ is trivial.

To prove the second assertion, observe that (4.21) implies $H^2(F(X),\mathbb{Z})(1)\simeq\mathbb{Z}$. □

Debarre and Manivel [141, Prop. 1 & Ex. 3.3] prove a more precise form of (ii), namely $\mathrm{Pic}(F(X))\simeq\mathbb{Z}\cdot\mathcal{O}_F(1)$ for $n\geq 5$.

Remark 4.19 In fact, $F(X)$ is known to be simply connected for $n \geq 4$.[16] For $n = 4$, this is a result of Beauville and Donagi [58], see Section **6**.3.2, and for $n > 4$, the result follows from a result by Sommese [440].

Alternatively, one can use the fact that for $n > 4$, the Fano variety of lines $F(X)$ is a Fano variety, i.e. it has negative canonical bundle, see Lemma 3.1. Quite generally, according to the results of Campana and Kollár, rationally connected varieties are simply connected, cf. [138, 139]. That the algebraic fundamental group is trivial follows from the observation that any finite étale cover $\pi\colon \tilde{Z} \twoheadrightarrow Z$ of a Fano variety Z would again be Fano and, therefore, $1 = \chi(\tilde{Z}, \mathcal{O}) = \chi(Z, \mathcal{O}) \cdot \deg(\pi) = \deg(\pi)$, see [138, Cor. 4.18] for the rest of the argument.

The middle cohomology $H^n(X, \mathbb{Q})$ of the cubic hypersurface X carries most of the information. As we will see again and again, for the Fano variety of lines it is the cohomology in degree $n - 2$, which is below the middle for all $n > 2$. And indeed, the next result says that the two are intimately related. In Section 5.1, yet another approach to computing the cohomology of the Fano variety is explained. Instead of the birational correspondence between $\mathbb{L}_{\mathbb{G}}|_X$ and $X^{[2]}$ it relies more directly on the Fano correspondence between $F(X)$ and X provided by the universal line $\mathbb{L}$.

Corollary 4.20 *Let X be a smooth cubic hypersurface of dimension $n > 2$.*

(i) *If n is even, then there exists an isomorphism of Hodge structures*

$$H^{n-2}(F(X), \mathbb{Q}) \simeq H^n(X, \mathbb{Q})_{\rm pr}(1) \oplus \bigoplus \mathbb{Q}(2 - i - j),$$

where the direct sum is over all $0 < i \leq j < n$ such that $2(i + j) = n + 2$.

(ii) *If n is odd, then $H^{{\rm odd}<n-2}(F(X), \mathbb{Q}) = 0$ and there exists an isomorphism of Hodge structures*

$$H^{n-2}(F(X), \mathbb{Q})_{\rm pr} = H^{n-2}(F(X), \mathbb{Q}) \simeq H^n(X, \mathbb{Q})(1) \simeq H^n(X, \mathbb{Q})_{\rm pr}(1). \qquad \square$$

Example 4.21 Let us start by computing $H^0(F(X), \mathbb{Q})$. For this, compare the proof of (4.21). We distinguish the two cases:

(i) For $n = 2$, we obtain the isomorphism of vector spaces

$$H^0(F(X), \mathbb{Q}) \oplus \mathbb{Q} \simeq S^2(H^2(X, \mathbb{Q})_{\rm pr}) \oplus H^2(X, \mathbb{Q})_{\rm pr} \oplus \mathbb{Q}.$$

Taking dimensions while using $b_2(X)_{\rm pr} = 6$, shows

$$b_0(F(X)) + 1 = 21 + 6 + 1.$$

Hence, $b_0(F(X)) = 27$, i.e. $F(X)$ consists of 27 isolated points. We stress that using étale cohomology, the same conclusion can be drawn for smooth cubic surfaces over arbitrary algebraically closed fields.

[16] I am indebted to R. Laterveer and S. Stark for pointing this out to me and for providing the references.

(ii) For $n > 2$, one finds $H^0(F(X), \mathbb{Q}) \simeq \mathbb{Q}$, where the right-hand side comes from $\mathbb{Q}(2 - 1 - 1)$. This proves again that $F(X)$ is connected, cf. Proposition 3.4 and Exercise 3.7.

We shall exploit (4.21) to compute the Hodge diamond of $F(X)$ for smooth cubic hypersurfaces of dimensions $n \leq 5$, cf. [190]. For the computation of the Hodge diamonds for the corresponding cubic hypersurface, see Section **1**.1.4.

***n* = 3**: Here, the formulae lead to the following isomorphisms of Hodge structures

$$H^1(F(X), \mathbb{Q}) \simeq H^3(X, \mathbb{Q})(1),$$

where we use that $H^3(X, \mathbb{Q})$ is primitive, and

$$H^2(F(X), \mathbb{Q}) \simeq \left(\bigwedge^2 H^3(X, \mathbb{Q})\right)(2).$$

Note that a priori the formula involves a direct summand $\mathbb{Q}(1)$ on both sides, which then cancels out. Combining the two isomorphisms defines an isomorphism

$$\bigwedge^2 H^1(F(X), \mathbb{Q}) \simeq H^2(F(X), \mathbb{Q}).$$

At this point, we do not yet know that the isomorphism is given by exterior product in cohomology, but see Section 5.3 and Section **5**.2.2. For the Hodge diamond this leads to:

$$\begin{array}{lccccc} b_0(F(X)) = 1 & & & 1 & & \\ b_1(F(X)) = 10 & & 5 & & 5 & \\ b_2(F(X)) = 45 & 10 & & 25 & & 10. \end{array}$$

***n* = 4**: In this case, $\dim(F(X)) = 4$ and the cohomology of $F(X)$ is concentrated in even degree.

$$H^2(F(X), \mathbb{Q}) \simeq H^4(X, \mathbb{Q})_{\mathrm{pr}}(1) \oplus \mathbb{Q}(-1)$$

and

$$H^4(F(X), \mathbb{Q}) \simeq S^2(H^4(X, \mathbb{Q})_{\mathrm{pr}})(2) \oplus H^4(X, \mathbb{Q})_{\mathrm{pr}} \oplus \mathbb{Q}(-2).$$

For the Betti and Hodge numbers this implies

$$\begin{array}{lccccc} b_0(F(X)) = 1 & & & 1 & & \\ b_2(F(X)) = 23 & & 1 & 21 & 1 & \\ b_4(F(X)) = 276 & 1 & 21 & 232 & 21 & 1. \end{array}$$

***n* = 5**: Here, $\dim(F(X)) = 6$ and for the rational cohomology of $F(X)$ we have

$$H^1(F(X), \mathbb{Q}) = 0, \quad H^2(F(X), \mathbb{Q}) \simeq \mathbb{Q}(-1), \quad H^3(F(X), \mathbb{Q}) \simeq H^5(X, \mathbb{Q})(1) \simeq H^5(X, \mathbb{Q})_{\mathrm{pr}}(1),$$

$$H^4(F(X),\mathbb{Q}) \simeq \mathbb{Q}(-2)^{\oplus 2}, \quad H^5(F(X),\mathbb{Q}) \simeq H^5(X,\mathbb{Q}) \simeq H^5(X,\mathbb{Q})_{\rm pr},$$

and

$$H^6(F(X),\mathbb{Q}) \simeq \left(\bigwedge^2 H^5(X,\mathbb{Q})\right)(2) \oplus \mathbb{Q}(-3).$$

Thus, the non-trivial part of the Hodge diamond below the middle looks like this:

$b_0(F(X)) = 1$			1			
$b_1(F(X)) = 0$						
$b_2(F(X)) = 1$			1			
$b_3(F(X)) = 42$		21		21		
$b_4(F(X)) = 2$			2			
$b_5(F(X)) = 42$		21		21		
$b_6(F(X)) = 862$	210		442		210.	

Remark 4.22 Instead of considering Hodge structures of hypersurfaces over $\mathbb{C}$ and of their Fano varieties, it is also interesting to study hypersurfaces over finite fields. Galkin and Shinder [190] give the following formula

$$|F(X)(\mathbb{F}_q)| = \frac{|X(\mathbb{F}_q)|^2 - 2\,(1+q^n)\,|X(\mathbb{F}_q)| + |X(\mathbb{F}_{q^2})|}{2\,q^2},$$

which is a direct consequence of Proposition 4.2 or its version (4.9) in Mot(k), applying arguments similar to the ones to compute $e(F(X))$ in the proof of Proposition 4.6.

One way to see this is uses that the Zeta function $Z(X,z) = \exp\left(\sum_{r=1}^{\infty} |X(\mathbb{F}_{q^r})| \frac{z^r}{r}\right)$ can also be written as $Z(X,z) = \sum z^{\deg(Z)}$, with the sum running over all zero cycles, and so $|X^{(2)}(\mathbb{F}_q)| = (1/2)\,(|X(\mathbb{F}_q)|^2 + |X(\mathbb{F}_{q^2})|)$. In principle, this allows one to write $Z(F(X),z)$ in terms of $Z(X,z)$, which according to the Weil conjectures has a rather special form for cubic hypersurfaces, cf. Section **1**.1.6. This has been explained in detail by Debarre, Laface, and Roulleau [140] who in particular discuss the existence of lines defined over any finite base field.

5 Fano Correspondence

The way we related $H^n(X,\mathbb{Q})$ and $H^{n-2}(F(X),\mathbb{Q})$ was rather abstract and we shall now explain a more direct and canonical way. This makes use of the *Fano correspondence* (5.1) and the *quadratic Fano correspondence* (5.7).

5.1 Plücker Polarization and Fano Correspondence The Fano correspondence is the diagram

$$\begin{array}{ccc} \mathbb{L} & \xrightarrow{q} & X \\ {\scriptstyle p}\big\downarrow & & \\ F(X), & & \end{array} \tag{5.1}$$

which induces for any m a homomorphism of integral Hodge structures

$$\varphi := p_* \circ q^* : H^m(X,\mathbb{Z}) \longrightarrow H^{m-2}(F(X),\mathbb{Z})(-1). \tag{5.2}$$

Depending on the context, it may also be useful to consider the correspondence on the level of Chow groups

$$\varphi\colon \mathrm{CH}^i(X) \longrightarrow \mathrm{CH}^{i-1}(F(X)) \tag{5.3}$$

or to use other types of cohomology theories.

We apply the Leray–Hirsch decomposition to the $\mathbb{P}^1$-bundle $\mathbb{L} \simeq \mathbb{P}(\mathcal{S}_F) \longrightarrow F(X)$ and write

$$H^*(\mathbb{L},\mathbb{Z}) \simeq p^*H^*(F(X),\mathbb{Z}) \oplus u \cdot p^*H^{*-2}(F(X),\mathbb{Z}).$$

Here, $u := \mathrm{c}_1(\mathcal{O}_p(1))$ is the relative tautological class. Note that the pullback map $p^*: H^*(F(X),\mathbb{Z}) \longrightarrow H^*(\mathbb{L},\mathbb{Z})$ is injective. Moreover,

$$u^2 + u \cdot p^*\mathrm{c}_1(\mathcal{S}_F) + p^*\mathrm{c}_2(\mathcal{S}_F) = 0 \quad \text{and} \quad p_*(p^*\gamma + u \cdot p^*\gamma') = \gamma'.$$

Similar formulae hold for Chow groups.

Lemma 5.1 *The twisted correspondence* $\varphi(2)\colon H^4(X,\mathbb{Z})(2) \longrightarrow H^2(F(X),\mathbb{Z})(1)$ *maps the square of the hyperplane class* h^2 *to the Plücker polarization* g, *cf.* (1.8)*:*

$$\varphi(h^2) = g.$$

Similarly, $h^2 \in \mathrm{CH}^2(X)$ *is mapped to* $\mathrm{c}_1(\mathcal{O}_F(1)) \in \mathrm{CH}^1(F(X))$ *under* (5.3).

In particular, for $2 < n$, *we have* $\varphi(h^2) \neq 0$ *and, more generally, for all* $0 < k \leq n$

$$0 \neq \varphi(h^k) \in H^{2k-2}(F(X),\mathbb{Q}).$$

Proof Recall that $\mathbb{L} \simeq \mathbb{P}(\mathcal{S}_F) \subset \mathbb{P}(V \otimes \mathcal{O}_F) \simeq F \times \mathbb{P}(V)$ is induced by the natural inclusion $\mathcal{S}_F \subset V \otimes \mathcal{O}_F$ and, thus,

$$\mathcal{O}_p(1) \simeq q^*\mathcal{O}(1).$$

Hence, $p_*q^*h^2 = p_*(q^*\mathrm{c}_1(\mathcal{O}(1))^2) = p_*(u^2) = -\mathrm{c}_1(\mathcal{S}_F) = g$.

The argument for the second assertion is geometric. Fix a generic point $(L,x) \in \mathbb{L}$, so $x \in L \subset X$, and consider generic hyperplane sections $Z_k := H_1 \cap \cdots \cap H_k \cap X$ through x. Then $q^*(h^k) \in H^{2k}(\mathbb{L},\mathbb{Q})$ is the fundamental class of the subvariety $q^{-1}(Z_k)$.

As all hyperplanes H_i contain x and were otherwise chosen generically, the fibre of $p|_{q^{-1}(Z_k)}\colon q^{-1}(Z_k) \longrightarrow F(X)$ over $L \in F(X)$ consists of the one point (L, x). Hence, the restriction $p|_{q^{-1}(Z_k)}$ is generically finite on at least one irreducible component and, therefore, $\varphi(h^k) = p_*[q^{-1}(Z_k)] \neq 0$. □

Remark 5.2 (i) Arguing as in the proof above, we find that $\varphi(h^k)$ is represented by the subvariety

$$F_{Z_k} := \{ L \mid L \cap Z_k \neq \emptyset \},$$

where $Z_k \subset \mathbb{P}^{n+1-k}$ is the cubic obtained as a generic linear section $Z_k := \mathbb{P}^{n+1-k} \cap X$ of codimension k in X. In particular, the Plücker polarization is represented by the divisor $F_{\mathbb{P}^{n-1} \cap X}$ of all lines $L \in F(X)$ intersecting a generically chosen linear section $\mathbb{P}^{n-1} \cap X$.

(ii) We can also work with singular linear intersections. For example, for a generic line $L \subset X$ and a generic line $L \neq L' \subset X$ that intersects L, the intersection of the plane $\mathbb{P}^2 \simeq \overline{LL'}$ with X consists of L, L' and a third residual line L'', see Exercise 1.20. Then the closure F_L of the locus of all lines $L' \in F(X)$ such that $L \neq L'$ and $L \cap L' \neq \emptyset$ is of dimension $n-2$ and its middle cohomology class $[F_L] \in H^{2n-4}(F(X), \mathbb{Z})$ satisfies

$$3 \cdot [F_L] = [F_L] + [F_{L'}] + [F_{L''}] = \varphi(h^{n-1}).$$

Note that $L' \longmapsto L''$ defines an involution of F_L and its quotient is the projection

$$F_L \twoheadrightarrow D_L \subset \mathbb{P}^{n-1},$$

that maps a line $L' \in F_L$ to its intersection point with a generically chosen linear subspace $\mathbb{P}^{n-1} \subset \mathbb{P}^{n+1}$. Here, $D_L \in |\mathcal{O}(5)]$ is the discriminant hypersurface of the linear projection $\mathrm{Bl}_L(X) \longrightarrow \mathbb{P}^{n-1}$ from $L \subset X$, see Example **1**.5.4. The situation will be studied in detail for $n = 3$ and $n = 4$ in Chapters 5 and 6.

Exercise 5.3 Show that more generally $\varphi(h^m) \in H^{2m-2}(F(X), \mathbb{Z})$ can be expressed as a polynomial in the two Chern classes $\mathrm{c}_1(\mathcal{S}_F)$ and $\mathrm{c}_2(\mathcal{S}_F)$ of the universal sub-bundle $\mathcal{S}_F$. Concretely, for example, in $H^4(F(X), \mathbb{Z})$ or even in $\mathrm{CH}^2(F(X))$ one has

$$\varphi(h^3) = \mathrm{c}_1^2(\mathcal{S}_F) - \mathrm{c}_2(\mathcal{S}_F).$$

Remark 5.4 The formula in the last exercise can be interpreted geometrically as follows. Fix generic hyperplanes H, H_1, H_2 and let $S_i := H_i \cap H \cap X$. Then $F_i := F_{S_i} := \{L \mid L \cap S_i \neq \emptyset\}$, $i = 1, 2$, both represent the Plücker polarization $g = \mathrm{c}_1(\mathcal{S}_F^*)$. The intersection $F_1 \cap F_2$, which represents the class $g^2 = \mathrm{c}_1^2(\mathcal{S}_F)$, consists of all lines intersecting S_1 and S_2. Hence,

$$F_1 \cap F_2 = F(Z) \cup F_{S_1 \cap S_2}.$$

Here, $Z = H \cap X$ and the Fano variety $F(Z) \subset F(X)$ of lines in Z is viewed as the zero set of the associated canonical section of $\mathcal{S}_F^*$. Then use $[F(Z)] = \mathrm{c}_2(\mathcal{S}_F)$ and $[F_{S_1 \cap S_2}] = \varphi(h^3)$.

5.2 Isometry We next generalize results by Clemens–Griffiths [119] and Beauville–Donagi [58] in the case of $n = 3$ and $n = 4$ to higher dimensions. A purely topological proof for the first part was given by Shimada [431].

Proposition 5.5 *Assume $n \geq 3$. The Fano correspondence defines an injective map*

$$\varphi\colon H^n(X,\mathbb{Z}) \hookrightarrow H^{n-2}(F(X),\mathbb{Z})(-1)$$

and satisfies

$$(\alpha.\beta) = -\frac{1}{6}\int_{F(X)} \varphi(\alpha)\cdot\varphi(\beta)\cdot g^{n-2} \tag{5.4}$$

for all primitive classes $\alpha,\beta \in H^n(X,\mathbb{Z})_{\mathrm{pr}}$.

The pairing on the left-hand side of (5.4) is the standard intersection pairing on the middle cohomology $H^n(X,\mathbb{Z})$. On the right-hand side, the pairing is up to the scalar factor $-1/6$ of the Hodge–Riemann pairing associated with the Plücker polarization g.

Proof The injectivity of the map $\varphi\colon H^n(X,\mathbb{Z})_{\mathrm{pr}} \hookrightarrow H^{n-2}(F(X),\mathbb{Z})(-1)$ follows from (5.4) which in turn is proved by the following computation. The pullback of $\alpha \in H^n(X)$ can be written uniquely as

$$q^*\alpha = p^*\varphi(\alpha)' + u\cdot p^*\varphi(\alpha). \tag{5.5}$$

If α is primitive, then $h\cdot\alpha = 0$ and hence $u\cdot q^*\alpha = 0$. Using $u^2 = -p^*c_2(\mathcal{S}_F)+u\cdot p^*g$, this becomes $-p^*(\varphi(\alpha)\cdot c_2(\mathcal{S}_F))+u\cdot p^*(\varphi(\alpha)'+g\cdot\varphi(\alpha)) = 0$, which implies (i) $\varphi(\alpha)'+g\cdot\varphi(\alpha) = 0$ and $\varphi(\alpha)\cdot c_2(\mathcal{S}_F) = 0$. The latter then implies (ii) $u^2\cdot p^*\varphi(\alpha) = u\cdot p^*(g\cdot\varphi(\alpha))$.

Taking the product of (5.5) with the corresponding equation for another primitive class β, one obtains

$$q^*(\alpha\cdot\beta) = p^*(\varphi(\alpha)'\cdot\varphi(\beta)') + u\cdot p^*(\varphi(\alpha)\cdot\varphi(\beta)' + \varphi(\alpha)'\cdot\varphi(\beta)) + u^2\cdot p^*(\varphi(\alpha)\cdot\varphi(\beta)).$$

The first summand on the right-hand side becomes trivial under p_*. By (i), the direct image p_* of the second can be written as $-2(g\cdot\varphi(\alpha)\cdot\varphi(\beta))$ and, according to (ii), the last summand equals $u\cdot p^*(g\cdot\varphi(\alpha)\cdot\varphi(\beta))$. Altogether, one obtains the equation

$$p_*q^*(\alpha\cdot\beta) = -g\cdot\varphi(\alpha)\cdot\varphi(\beta),$$

the left-hand side of which can also be written as $(\alpha.\beta)\cdot p_*q^*[\mathrm{pt}]$. Taking the product with g^{n-3}, for which we have to assume $n \geq 3$, and integrating proves

$$(\alpha.\beta)\cdot\deg(p(q^{-1}(z))) = -\int_{F(X)} \varphi(\alpha)\cdot\varphi(\beta)\cdot g^{n-2}$$

for generic $z \in X$. The claim then follows from Lemma 5.11 below.

It remains to prove that φ is not only injective on $H^n(X,\mathbb{Z})_{\mathrm{pr}}$ but on all of $H^n(X,\mathbb{Z})$. Of course, the two are different only for n even, in which case we may write $H^n(X,\mathbb{Q}) =$

$H^n(X, \mathbb{Q})_{\mathrm{pr}} \oplus \mathbb{Q} \cdot h^{n/2}$. As $H^n(X, \mathbb{Z})$ is torsion free, it suffices to prove injectivity with rational coefficients which amounts to prove that $\varphi(h^{n/2})$ is not contained in $\varphi(H^n(X, \mathbb{Q})_{\mathrm{pr}})$. For this, we may assume that X is very general, for φ and $h^{n/2}$ are constant in families. However, for very general X, the Hodge structure $H^n(X, \mathbb{Q})_{\mathrm{pr}}$ is irreducible, cf. Corollary **1**.2.12. Therefore, neither $H^n(X, \mathbb{Q})_{\mathrm{pr}}$ nor its isomorphic image under φ can contain the non-trivial Hodge class $\varphi(h^{n/2})$. □

Exercise 5.6 Assume $n = 3$. Then h^3 is represented by three generic points in X and, therefore, $\varphi(h^3)$ by 18 lines. Show that this confirms Exercise 5.3.

Remark 5.7 (i) Note that for n odd, $H^{n-2}(F(X), \mathbb{Q}) = H^{n-2}(F(X), \mathbb{Q})_{\mathrm{pr}}$, cf. Corollary 4.20, and so φ maps $H^n(X, \mathbb{Q}) = H^n(X, \mathbb{Q})_{\mathrm{pr}}$ to $H^{n-2}(F(X), \mathbb{Q})_{\mathrm{pr}}(-1)$. This also holds true for $n = 4$, but the argument is more involved: one may assume that X is general, in which case $H^n(X, \mathbb{Q})_{\mathrm{pr}}$ is an irreducible Hodge structure, see Corollary **1**.2.12. As the Fano correspondence φ sends $H^{3,1}(X)$ to $H^{2,0}(F(X))$, the whole primitive cohomology $H^4(X, \mathbb{Q})_{\mathrm{pr}}$ is mapped into the minimal sub-Hodge structure of $H^2(F(X), \mathbb{Q})$ containing the one-dimensional $H^{2,0}(F(X))$, hence into $H^2(F(X), \mathbb{Q})_{\mathrm{pr}}$ and, for dimension reasons, isomorphically onto it.

(ii) In [260, Thm. 4] it is claimed in full generality that the composition of the restriction of φ to the primitive part with the projection onto the primitive cohomology describes an isomorphism of integral Hodge structures. However, the projection does usually not map into integral cohomology and, therefore, one needs to at least invert some integers. For n odd or $n = 4$, there are injections $H^n(X, \mathbb{Z})_{\mathrm{pr}} \hookrightarrow H^{n-2}(F(X), \mathbb{Z})_{\mathrm{pr}}(-1)$, which we shall see to be an isomorphism for $n = 3$ and $n = 4$, see Corollary **5**.3.3 and Proposition **6**.3.19. The following result is the key observation.

Corollary 5.8 *Let n be odd and assume that for all $\gamma_1, \gamma_2 \in H^{n-2}(F(X), \mathbb{Z})$ one has $\int_{F(X)} \gamma_1 \cdot \gamma_2 \cdot g^{n-2} \equiv 0\,(6)$. Then the Fano correspondence determines an isomorphism of Hodge structures*

$$\varphi\colon H^n(X, \mathbb{Z}) \xrightarrow{\sim} H^{n-2}(F(X), \mathbb{Z})(-1).$$

Proof Under the assumptions on n, the two cohomologies $H^n(X, \mathbb{Z})$ and $H^{n-2}(F(X), \mathbb{Z})$ are torsion free modules of the same rank, cf. Exercise 4.13. According to Proposition 5.5, the Fano correspondence is injective and compatible with the (alternating) intersection product on $H^n(X, \mathbb{Z})$ and the pairing $(-1/6) \int_F \gamma_1 \cdot \gamma_2 \cdot g^{n-2}$ on $H^{n-2}(F, \mathbb{Z})$, which by assumption is integral. As the former is unimodular, this suffices to conclude. □

It seems that for n odd only in the case $n = 3$, the assumption on the divisibility of the Hodge–Riemann pairing has been proved. Note that for n even, the case $n = 4$ is the only one in which the Fano correspondence $\varphi\colon H^n(X, \mathbb{Z}) \hookrightarrow H^{n-2}(F(X), \mathbb{Z})(-1)$ is a morphism of integral Hodge structures of the same rank. Once again, it is indeed an isomorphism, which will be discussed in Section **6**.3.4.

Remark 5.9 For $n = 3$ and $n = 4$, we will see, cf. Section **5**.3.1 and Corollary **6**.4.3, that for primitive classes $\gamma_1, \gamma_2 \in H^{n-2}(F(X), \mathbb{Z})_{\mathrm{pr}}$, one has

$$\int_{F(X)} \gamma_1 \cdot \gamma_2 \cdot g^{n-2} = 3 \int_{F(X)} \gamma_1 \cdot \gamma_2 \cdot [F_L] = \int_{F(X)} \gamma_1 \cdot \gamma_2 \cdot \varphi(h^{n-1}),$$

cf. Remark 5.2. Is this true in higher dimensions, say for classes $\gamma_i = \varphi(\alpha_i)$ with $\alpha_i \in H^n(X, \mathbb{Z})_{\mathrm{pr}}$?

Any deformation of a smooth cubic hypersurface $X \subset \mathbb{P}^{n+1}$ induces a deformation of the associated Fano variety $F(X)$. Using Hodge theory, the induced linear map between the spaces of deformations of first order, see (3.7), is shown to be injective.

Corollary 5.10 *For $n \geq 3$, any non-trivial first order deformation of a smooth cubic hypersurface $X \subset \mathbb{P}^{n+1}$ induces a non-trivial first order deformation of its Fano variety $F(X)$, i.e.*

$$H^1(X, \mathcal{T}_X) \hookrightarrow H^1(F(X), \mathcal{T}_{F(X)}).$$

Proof According to the infinitesimal Torelli theorem, see Corollary **1**.4.25, the map $H^1(X, \mathcal{T}_X) \longrightarrow \mathrm{Hom}(H^n(X, \mathbb{C})_{\mathrm{pr}}, H^n(X, \mathbb{C})_{\mathrm{pr}})$, measuring the first order variation of the Hodge structure $H^n(X, \mathbb{C})_{\mathrm{pr}}$, is injective.

Similarly, one considers the map $H^1(F, \mathcal{T}_F) \longrightarrow \mathrm{Hom}(H^{n-2}(F, \mathbb{C}), H^{n-2}(F, \mathbb{C}))$ about which we do not know anything a priori. However, if a class $v \in H^1(X, \mathcal{T}_X)$ is mapped to 0 in $H^1(F, \mathcal{T}_F)$ then the induced infinitesimal variation of the Hodge structure $H^{n-2}(F, \mathbb{C})$ is trivial. However, by Proposition 5.5, in this case also the variation of the Hodge structure $H^n(X, \mathbb{C})_{\mathrm{pr}}$ is trivial and hence $v = 0$. □

Lemma 5.11 *For $n \geq 3$, the generic fibre of the morphism $q \colon \mathbb{L} \longrightarrow X$ is of dimension $n - 3$ and degree six with respect to the Plücker polarization g, i.e.*

$$\int_{q^{-1}(z)} g^{n-3} = 6.$$

Proof Fix a generic point $z \in X$ and pick a hyperplane $\mathbb{P}^n \subset \mathbb{P}^{n+1}$ not containing z. Then the linear embedding

$$\mathbb{P}^n \hookrightarrow \mathbb{G}(1, \mathbb{P}) \hookrightarrow \mathbb{P}(\textstyle\bigwedge^2 V), \quad y \longmapsto \overline{yz} \tag{5.6}$$

induces an isomorphism $\{y \in \mathbb{P}^n \mid \overline{yz} \subset X\} \simeq p(q^{-1}(z))$. As in Remark 3.6, this proves

$$\{\, y \in \mathbb{P}^n \mid \overline{yz} \subset X \,\} \simeq \mathbb{P}^n \cap \mathbb{T}_z X \cap X \cap P_z X.$$

Here, $\mathbb{T}_z X = V(\sum x_i \partial_i F(z))$ is the projective tangent space of $X = V(F)$ at $z \in X$ and $P_z X = V(\sum z_i \partial_i F)$ is its polar, cf. also Section **4**.2.3. For generic choices of $z \in X$ and $\mathbb{P}^n \subset \mathbb{P}^{n+1}$, this is a transversal intersection of the cubic X, the quadric $P_z X$, and the two hyperplanes $\mathbb{P}^n$ and $\mathbb{T}_z X$ and, therefore, of degree six. Here we use that the pullback of the Plücker polarization on $\mathbb{G}(1, \mathbb{P})$ under (5.6) is $\mathcal{O}(1)$ on $\mathbb{P}^n$, which can be checked by a direct computation. □

Example 5.12 For a smooth cubic threefold $Y \subset \mathbb{P}^4$, so $n = 3$, the result says that there are exactly six lines passing through every point in a Zariski dense, open subset of Y. We shall come back to this in Section **5**.1.

Remark 5.13 Barth and Van de Ven [37] verify that, for $n \geq 4$, the fibres of $q \colon \mathbb{L} \longrightarrow X$ are connected by proving that the codimension of the ramification locus is at least of codimension two, cf. the proof of Proposition 3.4. The description of the fibres as the intersection of two hypersurfaces in $\mathbb{P}^{n-1}$ as in the proof above shows this more directly. Also, observe that the connectedness of the fibres implies, once again, that $F(X)$ is connected for $n > 3$, cf. Proposition 3.4, Exercise 3.7, and Example 4.21.

5.3 Quadratic Fano Correspondence Let us now turn to the quadratic version of the Fano correspondence (5.1):

$$\begin{array}{ccc} \mathbb{L}^{[2]} & \overset{q^{[2]}}{\hookrightarrow} & X^{[2]} \\ {\scriptstyle p^{[2]}}\downarrow & & \\ F(X). & & \end{array} \tag{5.7}$$

Here, $q^{[2]} \colon \mathbb{L}^{[2]} \hookrightarrow X^{[2]}$ is the natural inclusion and $p^{[2]} \colon \mathbb{L}^{[2]} \longrightarrow F(X)$ is the projection, see Remark 4.4. The quadratic Fano correspondence defines a homomorphism of integral Hodge structures

$$\varphi^{[2]} := p^{[2]}_* \circ q^{[2]*} \colon H^m(X^{[2]}, \mathbb{Z}) \longrightarrow H^{m-4}(F(X), \mathbb{Z})(-2). \tag{5.8}$$

Lemma 5.14 *Assume n is even. Then the homomorphism* (5.8) *for $m = 2n$ composed with the natural map $S^2H^n(X, \mathbb{Z}) \longrightarrow H^{2n}(X^{[2]}, \mathbb{Z})$ equals the composition*

$$S^2H^n(X, \mathbb{Z}) \xrightarrow{S^2(\varphi)} S^2(H^{n-2}(F, \mathbb{Z})(-1)) \xrightarrow{\wedge} H^{2n-4}(F(X), \mathbb{Z})(-2). \tag{5.9}$$

A similar statement holds for n odd with S^2H^n replaced by $\bigwedge^2 H^n$.

Proof The assertion follows from the commutativity of the diagram

$$\begin{array}{ccccc} H^n(X) & \times & H^n(X) & \longrightarrow & H^{2n}(X^{[2]}) \\ \downarrow & & \downarrow & & \downarrow \\ H^n(\mathbb{L}) & \times & H^n(\mathbb{L}) & \longrightarrow & H^{2n}(\mathbb{L}^{[2]}) \\ \downarrow & & \downarrow & & \downarrow \\ H^{n-2}(F) & \times & H^{n-2}(F) & \longrightarrow & H^{2n-4}(F). \end{array}$$

The commutativity of the upper square is obvious and for the lower square it follows from the commutative diagram

$$\begin{array}{ccccc} \mathbb{L}\times\mathbb{L} & \xhookleftarrow{\Delta_{\mathbb{L}/F}} & \mathbb{L}\times_F\mathbb{L} & \longrightarrow & \mathbb{L}^{[2]} \\ \downarrow & & \downarrow & \swarrow & \\ F\times F & \xhookleftarrow{\Delta_F} & F, & & \end{array}$$

where the square is a fibre product. □

Recall from Lemma 4.14 that $(p_*^{[2]}\circ q^{[2]*})(\alpha) = p_*^{[2]}(\alpha|_{\mathbb{L}^{[2]}}) = f_2(\alpha)$, where

$$f_2\colon H^m(X^{[2]},\mathbb{Z}) \longrightarrow H^{m-4}(F(X),\mathbb{Z})$$

is the projection to the second copy of $H^{2n-4}(F(X),\mathbb{Z})$ on the right-hand side of (4.22).

Corollary 5.15 *Let $n>2$ be even. Then the square $S^2(\varphi)$ of the Fano correspondence φ or, equivalently, the restriction of $\varphi^{[2]}$ to $S^2H^n(X,\mathbb{Z})_{\rm pr}\subset H^{2n}(X^{[2]},\mathbb{Z})$ is an injective homomorphism of integral Hodge structures*

$$S^2(\varphi)\colon S^2H^n(X,\mathbb{Z})_{\rm pr} \hookrightarrow H^{2n-4}(F(X),\mathbb{Z})(-2). \tag{5.10}$$

A similar statement holds for n odd with S^2H^n replaced by $\bigwedge^2 H^n$.

Proof We restrict to the case that n is even, the odd case is similar. Also, as $H^n(X,\mathbb{Z})$ is torsion free, the assertion is equivalent to the corresponding one for rational Hodge structures, so we may work with rational coefficients. Finally, as $S^2(\varphi)$ does not change under deformations, we may assume that X is general.

Now split $S^2H^n(X,\mathbb{Q})_{\rm pr}\simeq \mathbb{Q}\cdot q_X\oplus q_X^\perp$, where q_X denotes the class corresponding to the intersection form. By Proposition 5.5, $q_X\in S^2H^n(X,\mathbb{Q})_{\rm pr}$ is mapped to a non-trivial Hodge class on $F(X)$. According to Remark **1**.2.13, the Hodge structure $q_X^\perp$ is irreducible and, in particular, there are no non-trivial Hodge classes neither in $q_X^\perp$ nor in its image under (5.10). Thus, it suffices to verify the injectivity of the restriction of (5.10) to $q_X^\perp\subset S^2H^n(X,\mathbb{Q})_{\rm pr}$, which, again by the irreducibility of $q_X^\perp$, would follow from $q_X^\perp\longrightarrow H^{2n-4}(F(X),\mathbb{Z})$ being non-trivial.

Clearly, $q_X^\perp$ maps injectively into the direct sum on the right-hand side of (4.22). By Lemma 4.14, the component f_1 to the first copy of $H^{2n-4}(F(X),\mathbb{Z})$ is trivial and by Lemma 5.14 the component f_2 to the second component is $S^2(\varphi)$. Thus, it suffices to show that all other components of $q_X^\perp\longrightarrow H^{2n}(B,\mathbb{Z})$ vanish. However, the remaining part on the right-hand side of (4.22) decomposes into Hodge structures of dimension $<\dim q_X^\perp$. Thus, none of the projections into one of those can be injective on $q_X^\perp$ and, therefore, they all have to be trivial. □

5.4 Dual Fano Correspondence Let us study a few more formal aspects of the correspondence (5.1). On the level of cohomology, we are interested in the two maps:

$$\varphi := p_* \circ q^* \colon H^n(X,\mathbb{Z}) \longrightarrow H^{n-2}(F(X),\mathbb{Z})(-1)$$

and

$$\psi := q_* \circ p^* \colon H^{3n-6}(F(X),\mathbb{Z}) \longrightarrow H^n(X,\mathbb{Z})(3-n).$$

The degree shift for the map ψ is caused by $q\colon \mathbb{L} \longrightarrow X$ having generic fibre of dimension $n-3$. Note that Poincaré duality for X and $F(X)$ defines natural isomorphisms

$$H^n(X,\mathbb{Z})^* \simeq H^n(X,\mathbb{Z}) \quad \text{and} \quad H^{n-2}(F(X),\mathbb{Z})^* \simeq H^{3n-6}(F(X),\mathbb{Z}),$$

where we use the fact that the cohomology of X and $F(X)$ is torsion free, see Remark 1.1.4 and Exercise 4.13. The projection formula shows that φ and ψ are dual to each other, i.e.

$$(\varphi(\alpha).\gamma)_F = (\alpha.\psi(\gamma))_X$$

for all $\alpha \in H^n(X,\mathbb{Z})$ and $\gamma \in H^{3n-6}(F(X),\mathbb{Z})$. Here, $(\,.\,)_X$ and $(\,.\,)_F$ denote the intersection pairings on X and $F(X)$.

Shimada [431] considers the correspondence ψ as a map $H_{n-2}(F(X),\mathbb{Z}) \longrightarrow H_n(X,\mathbb{Z})$ and shows its surjectivity, which gives an alternative proof of Proposition 5.5. Then, for n odd, it is automatically an isomorphism up to torsion, which follows from a comparison of Betti numbers, cf. Corollary 5.8.

5.5 Fano Correspondences for Chow Groups The same formalism works on the level of Chow groups, but one has to distinguish between cubics of even and odd dimension.

Assume $\mathbf{n} \equiv \mathbf{0}\,(\mathbf{2})$ and write $n = 2m$. Then (5.1) induces maps

$$\mathrm{CH}^{3m-3}(F(X)) \xrightarrow{\ \psi\ } \mathrm{CH}^m(X) \xrightarrow{\ \varphi\ } \mathrm{CH}^{m-1}(F(X)).$$

Using the compatibility with the cycle class maps, one obtains the commutative diagram

$$\begin{array}{ccccc}
\mathrm{CH}^{3m-3}(F(X)) & \xrightarrow{\ \psi\ } & \mathrm{CH}^m(X) & \xrightarrow{\ \varphi\ } & \mathrm{CH}^{m-1}(F(X)) \\
\downarrow & & \downarrow & & \downarrow \\
H^{6m-6}(F(X),\mathbb{Z})(3m-3) & \xrightarrow{\ \psi\ } & H^{2m}(X,\mathbb{Z})(m) & \xrightarrow{\ \varphi\ } & H^{2m-2}(F(X),\mathbb{Z})(m-1).
\end{array}$$

To avoid potential confusion, let us stress that the diagram is not supposed to suggest that the rows are exact or even that the compositions are 0.

For $\mathbf{n} \equiv \mathbf{1}\,(\mathbf{2})$, we write $n = 2m-1$ and consider, as above,

$$\mathrm{CH}^{3m-4}(F(X)) \xrightarrow{\ \psi\ } \mathrm{CH}^m(X) \xrightarrow{\ \varphi\ } \mathrm{CH}^{m-1}(F(X)).$$

However, in this case the cycle map does not relate this to the middle cohomology of X. Instead, one has to restrict to the homologically trivial parts and use the Abel–Jacobi maps to intermediate Jacobians, which for a smooth projective variety Z of dimension N are the complex tori, cf. [474] for the general theory:

$$J^{2k-1}(Z) := \frac{H^{2k-1}(Z,\mathbb{C})}{F^kH^{2k-1}(Z) + H^{2k-1}(Z,\mathbb{Z})} \simeq \frac{F^{N-k+1}H^{2N-2k+1}(Z)^*}{H_{2N-2k+1}(Z,\mathbb{Z})}.$$

Both descriptions are used in the following commutative diagram

$$\begin{array}{ccccc}
\mathrm{CH}^{3m-4}(F(X))_{\mathrm{hom}} & \xrightarrow{\psi} & \mathrm{CH}^m(X)_{\mathrm{hom}} & \xrightarrow{\varphi} & \mathrm{CH}^{m-1}(F(X))_{\mathrm{hom}} \\
\downarrow{\scriptstyle \mathrm{AJ}_F} & & \downarrow{\scriptstyle \mathrm{AJ}_X} & & \downarrow{\scriptstyle \mathrm{AJ}_F} \\
J^{3n-6}(F(X)) & \xrightarrow{\psi} & J^n(X) & \xrightarrow{\varphi} & J^{n-2}(F(X)) \\
\simeq \frac{F^{(n-1)/2}H^{n-2}(F(X))^*}{H^{n-2}(F(X),\mathbb{Z})} & & \simeq \frac{F^{(n+1)/2}H^n(X)^*}{H_n(X,\mathbb{Z})} & & \simeq \frac{H^{3n-6}(F(X),\mathbb{C})}{F^{3m-3}H^{n-2}(F(X))+H^{3n-6}(F(X),\mathbb{Z})}.
\end{array}$$

Note that the intermediate Jacobian $J^n(X)$ is self-dual and the two maps in the bottom row are naturally dual to each other.

3

Moduli Spaces

To study the geometry of a particular hypersurface $X \subset \mathbb{P} = \mathbb{P}^{n+1}$ or to understand how a certain feature changes when X is deformed, the actual embedding of X into the projective space $\mathbb{P}$ is often of no importance. This viewpoint leads to the notion of moduli spaces of varieties isomorphic to hypersurfaces of fixed degree and fixed dimension. There are various ways to construct these moduli spaces and we will discuss the most fundamental ones.

Further details on moduli spaces of cubic hypersurfaces of dimension two, three, and four can be found in Chapters 4, 5, and 6.

1 Quasi-projective Moduli Space and Moduli Stack

The embeddings of a fixed X into $\mathbb{P}$ are parametrized by the choice of a basis of $H^0(X, \mathcal{O}_X(1))$ up to scaling.[1] So, instead of the linear system $|\mathcal{O}_{\mathbb{P}}(d)|$ one is really interested in the quotient $|\mathcal{O}_{\mathbb{P}}(d)|/\mathrm{GL}(n+2)$. Ideally, one would like this quotient to exist in the category of varieties or schemes and to come with a universal family. However, as it turns out, this is too much to ask for.

Example 1.1 Consider the easiest case of interest to us: $d = 3$ and $n = 0$, i.e. three points in $\mathbb{P}^1$. Up to a linear coordinate change, there are only three possibilities: $\{x_1, x_2, x_3\}$ (three distinct points), $\{2 \cdot x_1, x_2\}$ (two distinct points, one with multiplicity two), or $\{3x\}$ (a triple point). Thus, the moduli space parametrizing all varieties isomorphic to hypersurfaces $X \subset \mathbb{P}^1$ of degree three should consists of three points. However, together with all possible embeddings they are parametrized by the projective space $|\mathcal{O}_{\mathbb{P}^1}(3)|$, which is connected and, therefore, does not admit a morphism onto a disconnected space.

[1] For $n \geq 2$ and $d \neq n+2$, the line bundle $\mathcal{O}_X(1)$ itself does not depend on the embedding, as it is determined by the property that $\mathcal{O}_X(d-(n+2)) \simeq \omega_X$. For $n > 2$, one can alternatively use that $\mathcal{O}_X(1)$ is the ample generator of $\mathrm{Pic}(X)$, see Corollary 1.1.9.

The same phenomenon can be described in terms of orbit closures. For example, the limit of the one-parameter subgroup $\mathrm{diag}(t, 1/t)$ applied to the set $V(x_0^2 x_1 - x_0 x_1^2) = \{0 = [0:1], \infty = [1:0], [1:1]\}$ viewed as a point in $|\mathcal{O}_{\mathbb{P}^1}(3)|$ is

$$\lim \{0, \infty, [1:1]\} = \begin{cases} \{2 \cdot 0, \infty\} & \text{for } t \longrightarrow 0, \\ \{0, 2 \cdot \infty\} & \text{for } t \longrightarrow \infty. \end{cases}$$

Hence, all these points should be identified under the quotient map to any moduli space that has a reasonable geometric structure.

Similar phenomena occur in higher dimensions and for all $d > 1$. The way out is to allow only *stable* hypersurfaces. Those are parametrized by an open subset of $|\mathcal{O}_{\mathbb{P}}(d)|$ and include all smooth hypersurfaces. This then leads to a quasi-projective moduli space (without a universal family in general) parametrizing orbits of hypersurfaces. To obtain a projective moduli space one has to add *semi-stable* hypersurfaces. This, however, leads to a moduli space that identifies certain orbits.

We briefly review the main features of GIT needed to understand moduli spaces of (smooth, cubic) hypersurfaces. We recommend [317, Ch. 6] for a quick introduction and Mumford's classic [362] or the textbooks [157, 356] for more details and references. Although we definitely want the moduli spaces to be defined over arbitrary fields, we usually assume that k is algebraically closed, just to keep the discussion geometric.

1.1 Quotients Let A be a finite type (say integral) k-algebra and G a linear algebraic group over k with an action on $X = \mathrm{Spec}(A)$ or, equivalently, an action on A. If a quotient $X \longrightarrow X/G$ in the geometric sense exists, then $X/G = \mathrm{Spec}(A^G)$, where $A^G \subset A$ is the invariant ring. In order for X/G to be a variety, the ring A^G needs to be again of finite type. This is Hilbert's 14th problem which has been answered by Hilbert himself in characteristic zero for $G = \mathrm{SL}$ and in general by Nagata and Harboush, see [362] or the entertaining [360] for a historic account, references, and proofs:

If G is reductive, then A^G is again a finite type k-algebra.

This seems to settle the question in the affine case by just defining $X/G := \mathrm{Spec}(A^G)$ with the quotient morphism $X \longrightarrow X/G$ induced by the inclusion $A^G \subset A$. However, this is, in general, a quotient only in a weaker sense.

Definition 1.2 A morphism $\pi\colon X \longrightarrow Y$ is a *categorical quotient* for the action of a group G on X if

(i) π is G-invariant[2] and

[2] So, pre-composing π with either of the two natural morphisms $G \times X \longrightarrow X$, the second projection or the group action, gives the same morphism.

(ii) any other G-invariant morphism $\pi' : X \longrightarrow Y'$ factors uniquely through a morphism $Y \longrightarrow Y'$.

A G-invariant morphism $\pi : X \longrightarrow Y$ is a *good quotient* if the following conditions hold:

(i) π is affine and surjective,
(ii) $\pi(Z)$ of any closed G-invariant subset $Z \subset X$ is closed,
(iii) $\pi(Z_1) \cap \pi(Z_2) = \pi(Z_1 \cap Z_2)$ for all closed G-invariant sets $Z_1, Z_2 \subset X$, and
(iv) $\mathcal{O}_Y$ is the sheaf of G-invariant sections of $\mathcal{O}_X$, i.e. $\mathcal{O}_Y \simeq (\pi_*\mathcal{O}_X)^G$ or, in other words, $\pi^* : \mathcal{O}_Y(U) \xrightarrow{\sim} \mathcal{O}_X(\pi^{-1}(U))^G$ for all open subsets $U \subset Y$.

A good quotient is *geometric* if in addition the pre-image of any closed point is an orbit.[3]

By definition, any geometric quotient is a good quotient and, as proved in [362, Prop. 0.1], any good quotient is also a categorical quotient:

$$\text{geometric} \Rightarrow \text{good} \Rightarrow \text{categorical}.$$

Note that a good quotient is equipped with the quotient topology and parametrizes the closed orbit of the action. Hence, a good quotient is geometric exactly when all orbits are closed, see [317, Prop. 6.1.7]. The main result on affine quotients is the following, cf. [362, Thm. 1.1] or [317, Prop. 6.3.1]:

> *Assume A is a finite type k-algebra and G is a reductive group acting on $X = \mathrm{Spec}(A)$. Then $X \longrightarrow X/\!/G := \mathrm{Spec}(A^G)$ is a good quotient.*

In particular, it is a categorical quotient, but usually not a geometric one.

1.2 GIT Quotients With certain modifications, the same recipe can be applied to projective varieties. Assume $A = \bigoplus_{i \geq 0} A_i$ is a graded k-algebra of finite type generated by A_1 and assume that the projective variety $X = \mathrm{Proj}(A)$ is endowed with the action of a reductive linear algebraic group G. Note that, in contrast to the affine case, the action is not necessarily induced by an action of G on A. However, we shall assume it is, in which case it is induced by a G-action on A_1. This is called a *linearization*. Geometrically it is realized by an embedding $X \hookrightarrow \mathrm{Proj}(S^*(A_1)) \simeq \mathbb{P}^m$ such that the action of G on X is the restriction of an action of G on $\mathbb{P}^m$ induced by a linear representation $G \longrightarrow \mathrm{GL}(A_1)$.

One is tempted to imitate the affine case and define the quotient simply as $\mathrm{Proj}(A^G)$. Note that A^G is naturally graded and again of finite type, but possibly not generated by elements of degree one. This can be easily remedied by passing to $\bigoplus_{i \geq 0} A_{mi}$ for an appropriate $m > 0$. However, the graded inclusion $A^G \subset A$ does not define a morphism between the associated projective schemes. Indeed, a homogeneous prime or maximal

[3] The exact definition of these notions varies from source to source. The subtle differences will be of no importance in our situation.

ideal in A may intersect A^G in its inessential ideal $(A^G)_+ := \bigoplus_{i>0}(A^G)_i$. In other words, there exists a morphism

$$X^{\mathrm{ss}} := X \setminus V((A^G)_+) \longrightarrow X^{\mathrm{ss}}/\!/G := \mathrm{Proj}(A^G)$$

only on the open set $X^{\mathrm{ss}} \subset X$. This naturally leads to the central definition of GIT.

Definition 1.3 A point $x \in X$ is *semi-stable* if it is contained in the open subset $X^{\mathrm{ss}} \subset X$, i.e. if there exists a homogeneous G-invariant $f \in A_i$, for some $i > 0$, with $f(x) \neq 0$.

A point $x \in X$ is *stable* if x is semi-stable and the induced orbit morphism $G \longrightarrow X^{\mathrm{ss}}$ is proper, i.e. the orbit $G \cdot x$ is closed in X^{ss} and the stabilizer G_x is finite. The set X^{s} of stable points is an open subset of X^{ss}.

Exercise 1.4 For a linearized action of a linear algebraic reductive group G on $\mathbb{P}(V)$, a point $[x] \in \mathbb{P}(V)$ is semi-stable if and only if $0 \notin \overline{G \cdot x} \subset V$. A point $[x] \in \mathbb{P}(V)$ is stable if and only if the morphism $G \longrightarrow V$, $g \longmapsto g \cdot x$ is proper.

Using open affine covers, the problem is reduced to the affine case which eventually leads to the following key result in GIT [362, Thm. 1.10].

Theorem 1.5 (Mumford) *Assume that a linearization of the action of a reductive linear algebraic group G on $X = \mathrm{Proj}(A)$ has been fixed. Then the natural morphism $X^{\mathrm{ss}} \longrightarrow X^{\mathrm{ss}}/\!/G$ is a good quotient and the restriction $X^{\mathrm{s}} \longrightarrow X^{\mathrm{s}}/\!/G$ is a geometric quotient.*

1.3 Stability of Hypersurfaces Let us turn to the concrete GIT problem that concerns us. Consider $G := \mathrm{SL}(n+2)$ with its natural action on $\mathbb{P}^{n+1}$ and the induced action on all complete linear systems $|\mathcal{O}(d)|$. Instead of $\mathrm{SL}(n+2)$ one often considers $\mathrm{PGL}(n+2)$. Both groups are reductive and the orbits of their actions on $|\mathcal{O}(d)|$ are of course the same. The advantage of working with SL is that its action on $|\mathcal{O}(d)|$ comes with a natural linearization. The relevant result for us is the following, see [270, Sec. 11.8] for the arithmetic version over $\mathrm{Spec}(\mathbb{Z})$.

Corollary 1.6 *Every smooth hypersurface $X \subset \mathbb{P}$ of degree $d \geq 3$ defines a stable point $[X] \in |\mathcal{O}(d)|$ for the action of $G = \mathrm{SL}(n+2)$, i.e.*

$$U(d,n) = |\mathcal{O}(d)|_{\mathrm{sm}} \subset |\mathcal{O}(d)|^{\mathrm{s}}.$$

Proof The semi-stability is an immediate consequence of Theorem **1**.2.2 and holds in fact for $d > 1$. Indeed, the complement of $U(d,n) \subset \mathbb{P}^N = |\mathcal{O}(d)|$ is the discriminant divisor $D = D(d,n)$, which is the zero set $V(\Delta)$ of the discriminant $\Delta = \Delta_{d,n} \in H^0(\mathbb{P}^N, \mathcal{O}(\ell))$, $\ell = (d-1)^{n+1} \cdot (n+2)$. As the smoothness of a hypersurface $X \subset \mathbb{P}$ does not depend on the embedding, the discriminant divisor D is invariant under the action of GL. Hence, for all $g \in \mathrm{GL}$, the induced action on $H^0(\mathbb{P}^N, \mathcal{O}(\ell))$, sending Δ to $g^*\Delta$, satisfies

$D = V(\Delta) = V(g^*\Delta)$. Therefore, $g^*\Delta = \lambda_g \cdot \Delta$ for some $\lambda_g \in \mathbb{G}_m$. This in fact defines a morphism of algebraic groups $\mathrm{GL} \longrightarrow \mathbb{G}_m$, $g \longmapsto \lambda_g$. However, the only characters of GL are powers of the determinant, which by definition is trivial on $G = \mathrm{SL}$. Hence, Δ is a G-invariant homogeneous polynomial that does not vanish at any point $[X] \in |\mathcal{O}(d)|$ corresponding to a smooth hypersurface. In other words, $U \subset |\mathcal{O}(d)|^{\mathrm{ss}}$.

In order to show stability, one has to prove that for X the morphism

$$G \longrightarrow |\mathcal{O}(d)|^{\mathrm{ss}},\ g \longmapsto g[X]$$

is proper. Let us first prove that the stabilizer $G_{[X]}$ is finite. Clearly, any $g \in G_{[X]}$ induces an automorphism of the polarized variety $(X, \mathcal{O}_X(1))$. This defines a morphism $G_{[X]} \longrightarrow \mathrm{Aut}(X, \mathcal{O}_X(1))$, the fibre of which is contained in the finite subgroup $\mu_{n+2} = \mathrm{Ker}(\mathrm{SL}(n+2) \longrightarrow \mathrm{PGL}(n+2))$. Now use Corollary **1**.3.9 and Remark **1**.3.10.

To conclude, one needs to show that the orbit $G \cdot [X]$ is closed in $|\mathcal{O}(d)|^{\mathrm{ss}}$. Let us first show it is closed in the open subset $U = |\mathcal{O}(d)|_{\mathrm{sm}} \subset |\mathcal{O}(d)|^{\mathrm{ss}}$. Consider its closure $\overline{G \cdot [X]}$ in U and suppose there exists a point $[X'] \in \overline{G \cdot [X]} \setminus G \cdot [X]$. Then $G \cdot [X'] \subset \overline{G \cdot [X]} \setminus G \cdot [X]$ and hence $\dim(G \cdot [X']) < \dim(G \cdot [X])$ which would imply $\dim(G_{[X']}) > 0$ contradicting the above discussion. Now, consider the morphism

$$\pi\colon |\mathcal{O}(d)|^{\mathrm{ss}} \longrightarrow |\mathcal{O}(d)|^{\mathrm{ss}}/\!/G = \mathrm{Proj}\left(k[H^0(\mathbb{P}^N, \mathcal{O}(1))]^G\right).$$

Clearly, U is the pre-image of the open non-vanishing locus of $\Delta \in H^0(\mathbb{P}^N, \mathcal{O}(\ell))^G \subset k[H^0(\mathbb{P}^N, \mathcal{O}(1))]^G$ and, therefore, $\pi^{-1}(\pi([X])) \subset U$ for all smooth X. As the subset $G \cdot [X]$ of $\pi^{-1}(\pi(x))$ is closed in the bigger set U, it is also closed in $\pi^{-1}(\pi([X]))$. However, the fibre $\pi^{-1}(\pi([X]))$ as the pre-image of a closed point is closed in $|\mathcal{O}(d)|^{\mathrm{ss}}$. Altogether this proves that $G \cdot [X] \subset |\mathcal{O}(d)|^{\mathrm{ss}}$ is closed. □

Remark 1.7 The techniques of the proof show that the morphism

$$\mathrm{PGL}(n+2) \times U \longrightarrow U \times U, \quad (g, [X]) \longmapsto ([X], g[X])$$

is proper. Now, the pre-image of the diagonal $\Delta \subset U \times U$ can be interpreted as the scheme **Aut** $= \mathrm{Aut}(\mathcal{X}/U, \mathcal{O}_{\mathcal{X}}(1)) \longrightarrow U$ of polarized automorphisms of the universal family of smooth hypersurfaces $\mathcal{X} \longrightarrow U \subset |\mathcal{O}(d)|$, cf. Section **1**.3.2. So, in particular, the fibre over $[X] \in U$ is the finite group $\mathrm{Aut}(X, \mathcal{O}_X(1))$. Note that as a consequence one finds that $\mathrm{Aut}(\mathcal{X}/U, \mathcal{O}_{\mathcal{X}}(1)) \longrightarrow U$ is a finite morphism, cf. [270, Cor. 11.8.4].

Example 1.8 For $d = 1$, i.e. for hyperplanes, no $[X] \in |\mathcal{O}(1)|$ is semi-stable. Indeed, in this case, $U(1, n) = |\mathcal{O}(1)| \simeq \mathbb{P}^*$ and $k[x_0, \ldots, x_{n+1}]^{\mathrm{SL}} = k$.

In contrast, smooth quadrics, so $d = 2$, are semi-stable by the above, but they are not stable. Indeed, the stabilizer of a quadric, say of $\sum x_i^2$ and in fact of every smooth quadric is of this form after a linear coordinate change, is the special orthogonal group $\mathrm{SO}(n+2) \subset \mathrm{SL}(n+2)$, which is not finite.

Exercise 1.9 The above proof did not cover the case $n = 0, 1$.

(i) Verify that stability still holds in these cases. The only problematic case is $n = 1$ and $d = 3$.
(ii) Show that for $n = 0$ and $d = 3$, (semi-)stability is equivalent to smoothness.

The next questions one should ask are: is the inclusion $|\mathcal{O}(d)|_{\rm sm} \subset |\mathcal{O}(d)|^{\rm s}$ is strict? How can one interpret its complement geometrically? How big is $|\mathcal{O}(d)|^{\rm ss} \setminus |\mathcal{O}(d)|^{\rm s}$?

1.4 Hilbert–Mumford The Hilbert–Mumford criterion is a powerful tool to decide whether a point is stable or semi-stable. It roughly says that it suffices to check one-parameter subgroups and gives a numerical criterion for those.

A one-parameter subgroup of a (reductive) group G is a non-constant morphism $\lambda\colon \mathbb{G}_m \longrightarrow G$ of algebraic groups. If a linear action $\rho\colon G \longrightarrow \mathrm{GL}(V)$ is given, then the induced action $\rho \circ \lambda\colon \mathbb{G}_m \longrightarrow \mathrm{GL}(V)$ can be diagonalized, i.e. there exists a basis (e_i) of V such that $\lambda(t)(e_i) = t^{r_i}\, e_i$, $r_i \in \mathbb{Z}$. The *Hilbert–Mumford weight* of a point $x = \sum x_i\, e_i \in V$ with respect to this one-parameter subgroup is defined as

$$\mu(x, \lambda) := -\min\{\, r_i \mid x_i \neq 0 \,\}.$$

Theorem 1.10 (Hilbert–Mumford criterion) *For a linearized action of a reductive group G on $\mathbb{P}(V)$, a point $[x] \in \mathbb{P}(V)$ is semi-stable if and only if $\mu(x, \lambda) \geq 0$ for all one-parameter subgroups $\lambda\colon \mathbb{G}_m \longrightarrow G$. The point $[x]$ is stable if and only if strict inequality holds for all non-trivial λ.*

Using Exercise 1.4, one direction is easy to prove. The difficulty lies in checking that it suffices to test one-parameter subgroups.

Example 1.11 A plane cubic curve $E \subset \mathbb{P}^2$ is stable if and only if it is smooth. It is semi-stable if and only if it has at most ordinary double points as singularities, cf. [356, Exa. 7.2] or [362]. See also Section 2.2 below for a description of the moduli space of all semi-stable plane cubic curves.

In later chapters we will discuss stability of cubic hypersurfaces in dimension $n \leq 4$, see Sections **4**.4.2, **5**.5.2, and **6**.6.7. But applying the Hilbert–Mumford criterion is typically quite tricky. For example, the proof of the stability of smooth hypersurfaces in Corollary 1.6 did not make use of it, but uses the discriminant instead. Also, the argument to prove stability of smooth cubic surfaces does not easily generalize to dimension three or higher.

Remark 1.12 Fedorchuk [181] proves that a hypersurface $X \subset \mathbb{P}^{n+1}$ defines a semi-stable point in $|\mathcal{O}_{\mathbb{P}^{n+1}}(d)|$ if and only if the subspace $\langle \partial_i F \rangle \subset k[x_0, \ldots, x_{n+1}]_d$ defines a semi-stable point in $\mathrm{Gr}(n+2, k[x_0, \ldots, x_{n+1}]_d)$ with respect to the natural $\mathrm{SL}(n+2)$-action on the Grassmann variety.

1.5 Moduli Quotient and Universal Family Ideally, one would like the universal family $\mathcal{X} \longrightarrow |\mathcal{O}(d)|_{\rm sm}$ of smooth hypersurfaces to descend to a universal family over the quotient $\bar{\mathcal{X}} \longrightarrow |\mathcal{O}(d)|_{\rm sm}//G$ (of varieties isomorphic to smooth hypersurfaces). The natural (and only) choice for such a family would be the quotient $\bar{\mathcal{X}} := \mathcal{X}//G$, where the action of $G = \mathrm{SL}(n+2)$ on $|\mathcal{O}(d)|_{\rm sm}$ is lifted to the natural action on $\mathcal{X} \subset |\mathcal{O}(d)|_{\rm sm} \times \mathbb{P}^{n+1}$. However, over a point $[X] \in |\mathcal{O}(d)|_{\rm sm}//G$ the fibre of this family would be the quotient $X/\mathrm{Aut}(X, \mathcal{O}_X(1))$ of X by the finite group $\mathrm{Aut}(X, \mathcal{O}_X(1))$ and not X itself. This is the reason why the quotient

$$M_{d,n} := |\mathcal{O}(d)|_{\rm sm}//G \tag{1.1}$$

typically does not represent the *moduli functor*

$$\mathcal{M}_{d,n}\colon (Sch/k)^o \longrightarrow (Set). \tag{1.2}$$

Here, by definition, $\mathcal{M}_{d,n}$ sends a k-scheme T to the set $\mathcal{M}_{d,n}(T)$ of all equivalence classes of polarized smooth projective families $(\mathcal{X}, \mathcal{O}_{\mathcal{X}}(1)) \longrightarrow T$, $\mathcal{O}_{\mathcal{X}}(1) \in \mathrm{Pic}_{\mathcal{X}/T}(T)$, such that all geometric fibres are isomorphic to some smooth hypersurface $X \subset \mathbb{P}^{n+1}$ (over the appropriate field) of degree d with the polarization $\mathcal{O}_{\mathcal{X}}(1)|_X$ given by the restriction $\mathcal{O}_{\mathbb{P}^{n+1}}(1)|_X$. Here, two such families are equivalent if there exists an isomorphism between the T-schemes that respects the two polarizations up to the twist by an invertible sheaf on T.

However, $M_{d,n}$ is still a *coarse moduli space* which means the following.

Corollary 1.13 *For $d \geq 3$, there exists a natural transformation $\phi\colon \mathcal{M}_{d,n} \longrightarrow \underline{M}_{d,n}$ that satisfies the following conditions:*

(i) *The induced map $\phi(k')\colon \mathcal{M}_{d,n}(k') \xrightarrow{\sim} \underline{M}_{d,n}(k')$ is bijective for any algebraically closed field extension k'/k.*
(ii) *Any natural transformation $\mathcal{M}_{d,n} \longrightarrow \underline{N}$ to a k-scheme factorizes uniquely through a morphism $M_{d,n} \longrightarrow N$ over k.*

The second condition is essentially a consequence of the fact that $|\mathcal{O}(d)|_{\rm sm} \longrightarrow M_{d,n}$ is a categorical quotient. The inclusion $|\mathcal{O}(d)|_{\rm sm} \subset |\mathcal{O}(d)|^{\rm s}$ together with the fact that $|\mathcal{O}(d)|^{\rm s} \longrightarrow |\mathcal{O}(d)|^{\rm s}//G$ is a geometric quotient implies the first one. For an outline of the details of the arguments, see e.g. the discussion in [249, Sec. 5.2].

Remark 1.14 As some geometric arguments make use of actual families, one often has to find substitutes for it. The following techniques are the most frequent ones:

(i) Instead of working with a universal family over $M_{d,n}$, which does not exist, one uses the universal family $\mathcal{X} \longrightarrow |\mathcal{O}(d)|_{\rm sm}$ and the fact that $|\mathcal{O}(d)|_{\rm sm} \longrightarrow M_{d,n}$ is a geometric quotient.

(ii) Assume $n > 0$, $d \geq 3$, and $(n, d) \neq (1, 3)$. Then, according to Theorem **1**.3.15, there exists an open and dense subset $V \subset |\mathcal{O}(d)|_{\rm sm}$ such that $\mathrm{Aut}(X, \mathcal{O}_X(1)) = \{\mathrm{id}\}$ for all $[X] \in V$. We may choose V to be invariant under G. Then there exists a universal family

$$\bar{\mathcal{X}} \longrightarrow \bar{V} \subset M_{d,n}$$

over the dense open subset $\bar{V} := V//G \subset M_{d,n}$. Explicitly, set $\bar{\mathcal{X}} := \mathcal{X}|_V/G$. It would be useful to have control over the closed set $M_{d,n} \setminus \bar{V}$, e.g. to know its codimension. The locus of smooth cubic surfaces with a non-trivial automorphism is of codimension one, see Remark **4**.3.10 and Section **4**.4.1.

(iii) *Luna's étale slice theorem* can be applied and gives the following: for any point $x := [X] \in |\mathcal{O}(d)|_{\rm sm}$ there exists a G_x-invariant smooth locally closed subscheme $x \in S \subset |\mathcal{O}(d)|_{\rm sm}$, the *slice* through x, such that both natural morphisms

$$S \times^{G_x} G \longrightarrow |\mathcal{O}(d)|_{\rm sm} \quad \text{and} \quad S/G_x \longrightarrow M_{d,n}$$

are étale (and automatically quasi-finite). The morphism $S \longrightarrow S/G_x$ is finite and a 'universal' family exists over S, namely the pullback of $\mathcal{X} \longrightarrow |\mathcal{O}(d)|_{\rm sm}$. In this sense, universal families exist étale locally over appropriate finite covers. See, for example, [286] for more on Luna's étale slice theorem.

(iv) A universal family may not even exist in a formal neighbourhood of a point $[X] \in M_{d,n}$. Using the notation in Section **1**.3.3, for any X the restriction of the moduli functor $\mathcal{M}_{d,n}$ to $(Art/k) \hookrightarrow (Sch/k)^o$, $A \longmapsto {\rm Spec}(A)$, is the union of all

$$F_X \simeq F_{X,\mathcal{O}_X(1)}$$

(under the numerical assumptions of Proposition **1**.3.12). This defines a finite morphism ${\rm Def}(X, \mathcal{O}_X(1)) \simeq {\rm Def}(X) \longrightarrow M_{d,n}$ onto the formal neighbourhood of $[X] \in M_{d,n}$, which is in fact the quotient by the natural action of ${\rm Aut}(X)$ on ${\rm Def}(X)$. Over ${\rm Def}(X, \mathcal{O}_X(1)) \simeq {\rm Def}(X)$ there does exist a 'universal' family, which is a formal variant of (iii).

(v) Finally, using finite *level structures*, there exists a finite morphism $\tilde{M}_{d,n} \longrightarrow M_{d,n}$ with a 'universal' family $\tilde{\mathcal{X}} \longrightarrow \tilde{M}_{d,n}$, cf. [249, §5.4.2] for a discussion in the case of K3 surfaces. The key input to this approach is the fact that, for $(n, d) \neq (1, 3)$ and $n > 0$, the action of ${\rm Aut}(X, \mathcal{O}_X(1))$ on the middle cohomology is faithful, see Corollary **1**.3.18.

Remark 1.15 The non-existence of a universal family or, equivalently, the possibility of non-trivial automorphisms, is also responsible for the difference between the field of moduli and the (or, rather, a) field of definition. This is expressed by saying that for non-closed fields k the map $\phi(k)\colon \mathcal{M}_{d,n}(k) \longrightarrow M_{d,n}(k)$ is usually not bijective.

To make this precise, let $X \subset \mathbb{P}^{n+1}_{\bar{k}}$ be a hypersurface of degree d and $[X] \in M_{d,n}(\bar{k})$ the corresponding closed point in the moduli space. The moduli space $M_{d,n}$ is defined over the ground field k and so the point $[X] \in M_{d,n}$ has a residue field $k \subset k_{[X]} \subset \bar{k}$, the *field of moduli* of X, which is finite over k. However, X may not be defined over its field of moduli $k_{[X]}$, but only over some finite extension $k_{[X]} \subset k_X$ of it (which is not necessarily unique), i.e. there exists a variety X_o over $k_X \subset \bar{k}$ such that $X \simeq X_o \times_{k_X} \bar{k}$. That X may not be defined over $k_{[X]}$ is the reason for $\phi(k_{[X]})$ not to be necessarily surjective. Also, since $M_{d,n}(k_X) \subset M_{d,n}(\bar{k})$, the potential non-uniqueness of X_o causes the non-injectivity of $\phi(k_X)$.

Moreover, for all field automorphisms $\sigma \in \mathrm{Aut}(k_X/k_{[X]})$, there exists a polarized automorphism $\varphi_\sigma \colon X_o^\sigma \xrightarrow{\sim} X_o$ over k_X. In fact, $k_{[X]}$ is the fixed field of all $\sigma \in \mathrm{Aut}(\bar{k}/k)$ with $X^\sigma \simeq X$. The isomorphisms φ_σ do not necessarily define a descent datum, as X_o may have non-trivial automorphisms. However, if $\mathrm{Aut}(X)$ is trivial, then indeed $k_{[X]}$ is a field of definition, which in this case is unique, and so $k_{[X]} = k_X$. As a consequence, one finds that for all $[X]$ in the open subset $\bar{V} \subset M_{d,n}$ of hypersurfaces without automorphisms the field of definition and the field of moduli coincide, i.e. X is defined over the residue field $k_{[X]}$ of $[X] \in M_{d.n}$.

1.6 Moduli Stacks We change perspective and replace the moduli functor $\mathcal{M}_{d,n}$ as in (1.2) by the category $\mathcal{M}_{d,n} \longrightarrow (Sch/k)$ fibred in groupoids (CFG). By definition, the fibre $\mathcal{M}_{d,n}(T)$ over a k-scheme T is the category of all polarized smooth projective families $(\mathcal{X}, \mathcal{O}_\mathcal{X}(1)) \longrightarrow T$ of polarized varieties isomorphic to hypersurfaces of degree d and dimension n with isomorphisms of polarized families as morphisms in the category.

• The CFG $\mathcal{M}_{d,n} \longrightarrow (Sch/k)$ is a stack. This entails two assertions:

(i) For two families $(\mathcal{X}_1, \mathcal{O}_{\mathcal{X}_1}(1)) \longrightarrow T$ and $(\mathcal{X}_2, \mathcal{O}_{\mathcal{X}_2}(1)) \longrightarrow T$, as above, the functor $\mathrm{Isom}(\mathcal{X}_1, \mathcal{X}_2) \colon (Sch/T)^o \longrightarrow (Set)$ that sends $T' \longrightarrow T$ to the set of isomorphisms of the T'-families $(\mathcal{X}_1, \mathcal{O}_{\mathcal{X}_1}(1))_{T'}$ and $(\mathcal{X}_2, \mathcal{O}_{\mathcal{X}_2}(1))_{T'}$ is a sheaf in the étale topology.
(ii) Every descent datum in $\mathcal{M}_{d,n}$ is effective, i.e. for an étale covering $T' \longrightarrow T$ and $(\mathcal{X}', \mathcal{O}_{\mathcal{X}'}(1)) \in \mathcal{M}_{d,n}(T')$ together with an isomorphisms of the two pullbacks to $T' \times_T T'$ satisfying a natural cocycle condition over $T' \times_T T' \times_T T'$ there always exists a family $(\mathcal{X}, \mathcal{O}_\mathcal{X}(1)) \in \mathcal{M}_{d,n}(T)$ the pullback of which to T' is isomorphic to $(\mathcal{X}', \mathcal{O}_{\mathcal{X}'}(1))$.

The proofs are by now standard and valid in broad generality. For an account of the analogous statements for polarized K3 surfaces and further references, see [249, Sec. 5.4.1].

By the very definition, $\mathcal{M}_{d,n}$ is isomorphic to the quotient stack $[U_{d,n}/\mathrm{SL}(n+2)]$ of the open set $U_{d,n} = |\mathcal{O}_{\mathbb{P}^{n+1}}(d)|_{\mathrm{sm}} \subset |\mathcal{O}_{\mathbb{P}^{n+1}}(d)|$ of smooth hypersurfaces.

• The stack $\mathcal{M} = \mathcal{M}_{d,n}$ of hypersurfaces of degree d and dimension n is a *Deligne–Mumford stack*. In other words, one has the following:

(i) The diagonal morphism $\Delta \colon \mathcal{M} \longrightarrow \mathcal{M} \times \mathcal{M}$ is representable, quasi-compact, and separated, i.e. it is quasi-separated.

(ii) There exists an étale covering $\underline{U} \longrightarrow \mathcal{M}$ by a scheme.

The representability of the diagonal is the assertion that the functor $\mathrm{Isom}(\mathcal{X}_1, \mathcal{X}_2)$ is representable, cf. Section **1**.3.2. The diagonal is finite, which can be seen as a consequence of the stability of smooth hypersurfaces, see Remark 1.7. The existence of an étale covering by a scheme follows from the diagonal being unramified and the existence of a smooth covering by a scheme. The arguments are again well known, cf. [249, Sec. 5.4.2] for the case of polarized K3 surfaces.

Remark 1.16 In Remark 1.14 we explained that locally the coarse moduli space $M_{d,n}$ looks like the quotient of $\mathrm{Def}(X)$ by the natural action of $\mathrm{Aut}(X)$. In this sense, we can think of the local analytic stacks $[\mathrm{Def}(X)/\mathrm{Aut}(X)]$ as covering the Deligne–Mumford stack $\mathcal{M}_{d,n}$.

As smooth cubic hypersurfaces are unobstructed, more precisely $H^2(X, \mathcal{T}_X) = 0$, see Remark **1**.3.13, these local charts provided by $\mathrm{Def}(X)$ are smooth, cf. Proposition **1**.3.12 and its proof. We conclude that the moduli stack of smooth cubic hypersurfaces $\mathcal{M}_{3,n}$ is a smooth Deligne–Mumford stack of dimension $\binom{n+2}{3}$, see Section **1**.2.1. The coarse moduli space $M_{d,n}$ is singular, but all singularities are finite quotient singularities.

The difference between $\mathcal{M}_{3,n}$ and its coarse moduli space $M_{3,n}$ is also detected on the level of tangent spaces. The tangent space of the stack at a point corresponding to a smooth cubic $X \subset \mathbb{P}^{n+1}$ is $T_{[X]}\mathcal{M}_{3,n} \simeq H^1(X, \mathcal{T}_X)$, but it can be bigger for the coarse moduli space, which looks locally like the finite quotient germ $\mathrm{Def}(X)/\mathrm{Aut}(X)$. While the stack $\mathcal{M}_{3,n}$ is smooth, the smooth locus of the moduli space $M_{3,n}$ is the strictly smaller but still Zariski dense open subset of all smooth cubics with trivial automorphism group.

Remark 1.17 Naturally, one would like to compactify $M_{d,n}$ to a projective variety that is well behaved and in particular not too singular. The obvious GIT compactification $M_{d,n} \subset |\mathcal{O}(d)|^{ss}/\!/G$ has neither a modular interpretation, for characterizing semistable surfaces geometrically is complicated, nor is it a particularly nice variety, as its singularities are typically rather bad. But $M_{d,n}$ also admits a GIT compactification with finite quotient singularities. This is provided by the Kirwan blow-up which is obtained by successively blowing up $|\mathcal{O}(d)|^{ss}$ and then take the GIT compactification, see [273, 274, 275].

2 Geometry of the Moduli Space

The quasi-projective moduli space $M_{d,n}$ of polarized varieties isomorphic to smooth hypersurfaces of degree d and dimension n has been introduced as the quotient

$$U_{d,n} \longrightarrow M_{d,n} = U_{d,n}/\!/\mathrm{PGL}(n+2) = U_{d,n}/\!/\mathrm{SL}(n+2)$$

of the open subset $U_{d,n} = |\mathcal{O}_{\mathbb{P}^{n+1}}(d)|_{\rm sm} \subset |\mathcal{O}_{\mathbb{P}^{n+1}}(d)|$ of all smooth hypersurfaces by the natural action of $\mathrm{SL}(n+2)$. As such, $M_{d,n}$ is an open subscheme of the GIT quotient $|\mathcal{O}_{\mathbb{P}^{n+1}}(d)|^{\rm ss}//\mathrm{SL}(n+2)$, which is a projective scheme.

But the moduli space $M_{d,n}$ can also be described as an affine quotient. To make this precise, observe that the open set $\widetilde{U}_{d,n} := H^0(\mathbb{P}^{n+1}, \mathcal{O}(d))_{\rm sm}$ of all homogenous polynomials of degree d defining smooth hypersurfaces is the affine variety $\mathrm{Spec}(A)$. Here, A is the homogeneous localization of the polynomial ring $k[a_I] = k[H^0(\mathbb{P}^{n+1}, \mathcal{O}(d))^*]$ with respect to the discriminant $\Delta_{d,n} \in k[a_I]_e$, $e = (d-1)^{n+1} \cdot (n+2)$, see Section **1**.2.2 and Section **1**.2.3. Clearly, $M_{d,n}$ is then the affine quotient

$$\widetilde{U}_{d,n} = \mathrm{Spec}(A) \longrightarrow M_{d,n} = \mathrm{Spec}(A^{\mathrm{GL}(n+2)}) = \widetilde{U}_{d,n}/\mathrm{GL}(n+2) = U_{d,n}/\mathrm{SL}(n+2)$$

and, in particular, $M_{d,n}$ is an affine variety. What else can we say about its geometric structure? Note that $M_{d,n}$ is a normal variety with at most finite quotient singularities.

2.1 Cohomology The first step in understanding the topology of the moduli space $M_{d,n}$ is to relate its cohomology to the cohomology of $U_{d,n}$, see [388].[4]

Theorem 2.1 (Peters–Steenbrink) *Assume $d \geq 3$. Then there exists an isomorphism of graded $\mathbb{Q}$-vector spaces*

$$H^*(U_{d,n}, \mathbb{Q}) \simeq H^*(M_{d,n}, \mathbb{Q}) \otimes H^*(\mathrm{SL}(n+2), \mathbb{Q}) \tag{2.1}$$

and the Leray spectral sequence

$$E_2^{p,q} = H^p(M_{d,n}, R^q\pi_*\mathbb{Q}) \Rightarrow H^{p+q}(U_{d,n}, \mathbb{Q}) \tag{2.2}$$

for the quotient morphism $\pi \colon U_{d,n} \longrightarrow M_{d,n}$ degenerates.

Remark 2.2 (i) In fact, the isomorphism (2.1) is an isomorphism of mixed Hodge structures. Indeed the mixed Hodge structure $H^*(\mathrm{SL}(n+2), \mathbb{Q})$ is of Hodge type and the pullback π^* is a morphism of mixed Hodge structures.

(ii) In [388] the moduli space $M_{d,n}$ is rather considered as the quotient of $\widetilde{U}_{d,n}$ by $\mathrm{GL}(n+2)$. The result is fundamentally the same, namely there exists an isomorphism

$$H^*(\widetilde{U}_{d,n}, \mathbb{Q}) \simeq H^*(M_{d,n}, \mathbb{Q}) \otimes H^*(\mathrm{GL}(n+2), \mathbb{Q}). \tag{2.3}$$

(iii) The cohomology of $\mathrm{SL}(N, \mathbb{C})$ and $\mathrm{GL}(N, \mathbb{C})$ is well known. Indeed, the inclusions of their compact real forms $\mathrm{SU}(N) \subset \mathrm{SL}(N, \mathbb{C})$ and $\mathrm{U}(N) \subset \mathrm{GL}(N, \mathbb{C})$ are homotopy equivalences and according to a result of Borel [86, Thm. 8.2] this then gives

$$H^*(\mathrm{GL}(N), \mathbb{Q}) \simeq \textstyle\bigwedge^*\langle \eta_1, \ldots, \eta_N \rangle \quad \text{and} \quad H^*(\mathrm{SL}(N), \mathbb{Q}) \simeq \textstyle\bigwedge^*\langle \eta_2, \ldots, \eta_N \rangle,$$

[4] Thanks to O. Banerjee for discussions related to this section.

with η_k of degree $2k-1$. An explicit realization of the classes η_k is described in [388, Sec. 5], which is crucial for Lemma 2.3 below. Namely, η_k can be realized by the locus of those matrices for which the first $N+1-k$ columns are linearly dependent.

Note that the mixed Hodge structure of $H^*(\mathrm{GL}(N), \mathbb{Q})$ is such that the classes $\eta_k \in H^{2k-1}(\mathrm{GL}(N), \mathbb{Q})$ are of type (k, k).

(iv) Finally, the cohomology $H^*(M_{d,n}, \mathbb{Q})$ is naturally isomorphic to the equivariant cohomology $H^*_{\mathrm{GL}(N+2)}(\widetilde{U}_{d,n}, \mathbb{Q})$ and $H^*_{\mathrm{SL}(N+2)}(U_{d,n}, \mathbb{Q})$, cf. [93, Sec. 1].

The main step in the proof of Theorem 2.1 is the following.

Lemma 2.3 *The orbit map*

$$\varphi_F \colon \mathrm{GL}(n+2) \longrightarrow \widetilde{U}_{d,n},\ g \longmapsto g \cdot F$$

through a polynomial $F \in \widetilde{U}_{d,n}$ induces a surjection

$$\varphi_F^* \colon H^*(\widetilde{U}_{d,n}, \mathbb{Q}) \twoheadrightarrow H^*(\mathrm{GL}(n+2), \mathbb{Q}).$$

Proof We only sketch the main ideas and refer to [388, Sec. 6] for the technical details. Let $\widetilde{D} \subset W := H^0(\mathbb{P}^{n+1}, \mathcal{O}(d))$ be the lift of the discriminant divisor, so that its complement is $\widetilde{U} := \widetilde{U}_{d,n} = W \setminus \widetilde{D}$. Then, for a subvariety $Y_k \subset \widetilde{D}$ of codimension k in W, there exists a natural map

$$\mathbb{Q} \simeq H^0(Y_k, \mathbb{Q}) \simeq H^{2k}_{Y_k}(W, \mathbb{Q}) \longrightarrow H^{2k}_{\widetilde{D}}(W, \mathbb{Q}) \simeq H^{2k-1}(\widetilde{U}, \mathbb{Q}).$$

Now, the orbit map φ_F extends naturally to a map $\bar{\varphi}_F \colon M_{n+2} \longrightarrow W$ from the space of $(n+2) \times (n+2)$ matrices M_{n+2} to the vector space W. The extension has the property that the inverse image of the discriminant divisor $\widetilde{D} \subset W$ is the divisor $D_{n+2} := M_{n+2} \setminus \mathrm{GL}(n+2)$ of all non-invertible matrices. This leads to the commutative diagram

$$\begin{array}{ccc} H^{2k}_{\widetilde{D}}(W, \mathbb{Q}) & \xrightarrow{\sim} & H^{2k-1}(\widetilde{U}, \mathbb{Q}) \\ \big\downarrow{\scriptstyle \bar{\varphi}_F^*} & & \big\downarrow{\scriptstyle \varphi_F^*} \\ H^{2k}_{D_{n+2}}(M_{n+2}, \mathbb{Q}) & \xrightarrow{\sim} & H^{2k-1}(\mathrm{GL}(n+2), \mathbb{Q}). \end{array}$$

The last step of the proof is to show that the subvariety $Y_i \subset \widetilde{D} \subset W$ of those polynomials F for which the hypersurface $V(F) \subset \mathbb{P}^{n+1}$ has singularities contained in the linear section $V(F) \cap V(x_0, \ldots, x_{n+2-k})$ pulls back to the subvarieties defining the classes η_k as in Remark 2.2 above. □

Proof of Theorem 2.1 Equipped with this lemma, a minor modification of the standard proof of the Leray–Hirsch theorem is enough to prove the GL-version (2.3). First, for a section $s \colon H^*(\mathrm{GL}(n+2), \mathbb{Q}) \hookrightarrow H^*(\widetilde{U}_{d,n}, \mathbb{Q})$ of φ_F^*, i.e. $\varphi_F^* \circ s = \mathrm{id}$, the map

$$\begin{array}{ccccc} H^*(\mathrm{GL}(n+2), \mathbb{Q}) & \otimes & H^*(M_{d,n}, \mathbb{Q}) & \longrightarrow & H^*(\widetilde{U}_{d,n}, \mathbb{Q}) \\ \alpha & \otimes & \beta & \longmapsto & s(\alpha) \cdot \pi^*\beta \end{array}$$

is an isomorphism, see [388, Sec. 2] for a few more details. The proof of the SL-version (2.1) is similar.

Next, the higher direct image sheaves $R^q\pi_*\mathbb{Q}$ can be trivialized by means of the section s and then the above isomorphism implies

$$\sum_{p+q=k} \dim H^p(M_{d,n}, R^q\pi_*\mathbb{Q}) = b_k(\widetilde{U}_{d,n}),$$

which suffices to deduce that the Leray spectral sequence degenerates. □

Example 2.4 (i) The moduli space of smooth cubic surfaces $M_{3,2}$ has the cohomology of a point, i.e. $H^*(M_{3,2}, \mathbb{Q}) \simeq H^0(M_{3,2}, \mathbb{Q}) \simeq \mathbb{Q}$. Indeed, $M_{3,2} \simeq \mathbb{A}^4/\mu_4$, see Section **4**.4.1. So in this case, the theorem says that

$$H^*(U_{3,2}, \mathbb{Q}) \simeq H^*(\mathrm{SL}(4), \mathbb{Q}),$$

which is a result first proved by Vasil'ev [466].

(ii) Consider the universal hypersurface of smooth cubic surfaces $\mathcal{S} \longrightarrow U = U_{3,2}$. The relative Fano variety of lines $F := F(\mathcal{S}/U) \longrightarrow U$ is an étale morphism of degree 27, see Section **4**.1.4 for the computation of its monodromy group. According to Das [134, Thm. 1.1], passing to the étale cover does not change the cohomology, i.e.

$$H^*(F(\mathcal{S}/U), \mathbb{Q}) \simeq H^*(U, \mathbb{Q}) \simeq H^*(\mathrm{SL}(4), \mathbb{Q}).$$

(iii) Also the cohomology of the total space of the universal family $\mathcal{S} \longrightarrow U = U_{3,2}$ of smooth cubic surfaces has been computed. Das [133, Thm. 1.1] shows that the natural inclusion $\mathcal{S} \subset U \times \mathbb{P}^3$ induces isomorphisms

$$H^*(\mathcal{S}, \mathbb{Q}) \simeq H^*(U \times \mathbb{P}^3, \mathbb{Q})/h^3 \simeq H^*(U, \mathbb{Q}) \otimes H^*(\mathbb{P}^2, \mathbb{Q}) \simeq H^*(\mathrm{SL}(4), \mathbb{Q}) \otimes H^*(\mathbb{P}^2, \mathbb{Q}).$$

Remark 2.5 Instead of fixing the degree, for us $d = 3$, it is interesting to let it grow. One then asks whether the natural inclusions obtained from (2.1)

$$H^k(\mathrm{SL}(n+2), \mathbb{Q}) \hookrightarrow H^k(U_{d,n}, \mathbb{Q}) \tag{2.4}$$

stabilize the cohomology of $U_{d,n}$ for k and n fixed and large d. This has first been studied by Vakil and Wood [457] as a problem in the Grothendieck ring of varieties and subsequently by Tommasi [451] who proves that (2.4) is an isomorphism in degree $k < (d+1)/2$. Conversely, combined with the above result it proves the vanishing of the cohomology $H^k(M_{d,n}, \mathbb{Q})$ for $0 < k < (d+1)/2$.

For the moduli space of smooth cubic hypersurfaces, these results only give

$$H^1(U_{3,n}, \mathbb{Q}) = 0 \quad \text{and} \quad H^1(M_{3,n}, \mathbb{Q}) = 0.$$

Note that the first vanishing (for arbitrary d) can be deduced from the Gysin sequence for the open embedding $U_{d,n} \subset |\mathcal{O}_{\mathbb{P}^{n+1}}(d)|$ and the irreducibility of the discriminant divisor

$D(d, n) = |\mathcal{O}_{\mathbb{P}^{n+1}}(d)| \setminus U_{d,n}$, see Theorem **1**.2.2. Note, however, that the fundamental group of $U_{d,n}$ and $M_{d,n}$ is quite an intricate object that has been studied intensively, see [325] for general results and references and [321] for the case of cubic surfaces.

Remark 2.6 There are other interesting and meaning full questions concerning the action of $\mathrm{GL}(n+1)$ on $|\mathcal{O}_{\mathbb{P}^{n+1}}(d)|$.

For example, one may want to know the degree of the orbit closure $\overline{\mathrm{GL}(n+1)\cdot[X]} \subset |\mathcal{O}_{\mathbb{P}^{n+1}}(d)|$. This turns out to be a difficult problem which has been studied in detail only for plane curves, see work by Aluffi and Faber [16].

Also, it is natural to study the (intersection) cohomology of the GIT compactification of $M_{d,n}$ or of the Kirwan blow-up [275]. Apart from low-dimensional cases very little is known. Again, for d large, the cohomology of $M_{d,n}$ and its GIT compactification stabilizes in small degree.

2.2 Unirationality of the Moduli Space As the GIT quotient of $|\mathcal{O}_{\mathbb{P}^{n+1}}(d)|$, the moduli space $M_{d,n}$ is unirational, but is it also rational or stably rational? This is a classical problem with a vast literature, especially the case $n = 1$ of plane curves which has attracted a lot of attention. We shall here restrict to the case $d = 3$:

- For $n = 0$, the moduli space $M_{3,0}$ consists of just one point and hence is rational.
- For $n = 1$, the moduli space $M_{3,1}$ is isomorphic to the affine line $\mathbb{A}^1$ and, therefore, is rational, cf. [157, Ch. 10.3].

In fact, its compactification provided by the GIT quotient $|\mathcal{O}_{\mathbb{P}^2}(3)|^{\mathrm{ss}}/\!/\mathrm{SL}(3)$ is by construction $\mathrm{Proj}(S)$, where $S = k[a_I]^{\mathrm{SL}(3)}$, $I = (i_0, i_1, i_2)$ with $i_0+i_1+i_2 = 3$, is the invariant ring. It is known classically that S is generated by two algebraically independent polynomials $g_2 \in k[a_I]_4$ and $g_3 \in k[a_I]_6$, cf. Example **1**.2.6 and, therefore,

$$|\mathcal{O}_{\mathbb{P}^2}(3)|^{\mathrm{ss}}/\!/\mathrm{SL}(3) \simeq \mathrm{Proj}(S) \simeq \mathrm{Proj}(S^{(12)}) \simeq \mathrm{Proj}(k[g_2^3, g_3^2]) \simeq \mathbb{P}^1.$$

The points in $\mathbb{A}^1 \subset \mathbb{P}^1$ are in bijection to smooth cubic curves and the closed orbit through the union $V(x_0 \cdot x_1 \cdot x_2)$ of three lines corresponds to $\infty \in \mathbb{P}^1$. Other singular plane nodal cubic curves, like the union $V(x_0 \cdot (x_1^2 - x_2^2))$ of a conic and a line, correspond to non-closed orbits mapping to ∞ also. There is ample literature on this particular case, see the surveys [361, 369] or the textbooks [157, 356].

- For $n = 2$, the GIT quotient $|\mathcal{O}_{\mathbb{P}^3}(3)|^{\mathrm{ss}}/\!/\mathrm{PGL}(4)$ is the weighted projective space $\mathbb{P}(1, 2, 3, 4, 5)$, see Section **4**.4.1, and the moduli space $M_{3,2}$ is the open subset,

$$M_{3,2} = \mathbb{A}^4/\mu_4 \subset \mathbb{P}(1, 2, 3, 4, 5),$$

cf. Corollary **4**.4.3. Since weighted projective spaces are toric varieties and, therefore, rational, also $M_{3,2}$ is rational. See Section **4**.4.2 for the description of singular stable and semi-stable cubic surfaces.

• For $n \geq 3$, it seems an open question as to whether the moduli space $M_{3,n}$ is rational. In fact, I am not even sure it is known to be stably rational.

Remark 2.7 The moduli space $M = M_{3,4}$ of smooth cubic fourfolds is of dimension 20. It contains distinguished Noether–Lefschetz or Hassett divisors $M \cap C_d \subset M$ for all $6 < d \equiv 0, 2\,(6)$, see Section **6**.6.4. These divisors are known to be of general type for large enough d, see Remark **6**.6.16.

So, instead of asking for (uni-)rationality or uniruledness of these divisors, which only holds for small values of d, e.g. for $d = 42$ [304, Thm. 0.2], other birational properties are of interest. In [11] one finds upper bounds for the degree of irrationality, i.e. the minimal degree of a dominant rational map $C_d \dashrightarrow \mathbb{P}^{10}$.

3 Periods

Periods provide a transcendental approach to studying smooth complex projective varieties. Typically, for X of dimension n, one considers the primitive middle cohomology $H^n(X, \mathbb{Z})_{\rm pr}$ with its intersection pairing and its Hodge structure. The period of X is then by definition this linear algebra datum naturally associated with X. The two principal goals of introducing periods are an alternative approach to the construction of moduli spaces of varieties of a particular type, complementing the constructions sketched in the previous paragraphs, and, not unrelated, the formulation of a global Torelli theorem, ideally proving that the period of a variety uniquely determines it.

3.1 Period Domain and Period Map Let us briefly recall the abstract notion of a Hodge structure, see [47, 144, 204] or [249, Sec. 3.1]. Consider a free $\mathbb{Z}$-module Γ of finite rank. A *Hodge structure* of weight n on Γ consists of a decomposition of the complex vector space $\Gamma_{\mathbb{C}} := \Gamma \otimes \mathbb{C}$ as

$$\Gamma_{\mathbb{C}} \simeq \bigoplus_{p+q=n} H^{p,q}$$

satisfying the condition $\overline{H^{p,q}} = H^{p,q}$. A Hodge structure can alternatively be encoded by the associated *Hodge filtration*

$$0 \subset \cdots \subset F^n \subset \cdots \subset F^0 \subset \cdots \subset \Gamma_{\mathbb{C}}.$$

Here, $F^p := \bigoplus_{i \geq p} H^{i,j}$, for which we have $F^p \oplus \overline{F^q} = \Gamma_{\mathbb{C}}$ for all $p + q = n + 1$. Conversely, starting with a Hodge filtration one reconstructs the Hodge structure by defining $H^{p,q} := F^p \cap \overline{F^q}$.

Example 3.1 The prime example of a Hodge structure of weight n is of course the cohomology $H^n(X, \mathbb{Z})$ (modulo torsion) of a smooth, complex projective variety X with the Hodge decomposition $H^n(X,\mathbb{C}) = \bigoplus_{p+q=n} H^{p,q}(X)$. In this case, $F^{n+1} = 0$ and $F^0 = H^n(X,\mathbb{C})$.

If an ample class $c_1(L) \in H^2(X,\mathbb{Z})$ is fixed, then the primitive integral cohomology $H^n(X,\mathbb{Z})_{L\text{-pr}}$ (modulo torsion) also has a natural Hodge structure and, in addition, comes with a natural *polarization* $(\alpha,\beta) \longmapsto (-1)^{n(n-1)/2} \int \alpha \wedge \beta \wedge c_1(L)^{\dim(X)-n}$, which for $\dim(X) = n$ is just the intersection pairing up to a sign.

The abstract notion of a polarization for an arbitrary Hodge structure of weight n is a morphism of Hodge structures $\psi\colon \Gamma \otimes \Gamma \longrightarrow \mathbb{Z}(-n)$ such that its $\mathbb{R}$-linear extension leads to a positive symmetric symmetric form $(\alpha,\beta) \longmapsto \psi(\alpha, C\beta)$ on the real vector space $(H^{p,q} \oplus H^{q,p}) \cap \Gamma_{\mathbb{R}}$. Here, C is the *Weil operator* which acts by i^{p-q} on $H^{p,q}$.

Note that the assumption that ψ is a morphism of Hodge structures means in practice that $\psi(\alpha_1,\alpha_2) = 0$ for $\alpha_i \in H^{p_i,q_i}$ unless $(p_1,q_1) + (p_2,q_2) = (n,n)$, which can also be expressed as the orthogonality condition $F^p \perp F^{n-p+1}$ with respect to the $\mathbb{C}$-linear extension of ψ.

Example 3.2 Let us spell out these notions for smooth cubic hypersurfaces of dimension two, three, and four.

(i) For a smooth cubic surface $S \subset \mathbb{P}^3$, the Hodge decomposition is trivial, i.e. $H^2(S,\mathbb{C}) = H^{1,1}(S)$ and, accordingly, the Hodge filtration collapses to

$$F^2 = 0 \subset F^1 = H^{1,1}(S) = F^0 = H^2(S,\mathbb{C})$$

and similarly for the primitive cohomology.

(ii) For a smooth cubic threefold $Y \subset \mathbb{P}^4$, the situation is already more interesting. Here, we have $H^3(Y,\mathbb{C})_{\text{pr}} = H^3(Y,\mathbb{C}) = H^{2,1}(Y) \oplus H^{1,2}(Y)$ and hence

$$F^3 = 0 \subset F^2 = H^{2,1}(Y) \subset F^1 = F^0 = H^3(Y,\mathbb{C})$$

of dimensions $f^2 = 5$ and $f^1 = f^0 = 10$.

(iii) Eventually, for a smooth cubic fourfold $X \subset \mathbb{P}^5$, the Hodge decomposition of the primitive cohomology gives rise to a filtration

$$F^4 = 0 \subset F^3 = H^{3,1}(X) \subset F^2 = H^{3,1}(X) \oplus H^{2,2}(X)_{\text{pr}} \subset F^1 = F^0 = H^4(X,\mathbb{C})_{\text{pr}}$$

of dimensions $f^3 = 1$, $f^2 = 21$, and $f^1 = f^0 = 22$.

For more details for the following discussion, we recommend the classic [204, Ch. I] or the more recent textbook [104, Ch. 4.4].

Fix a free $\mathbb{Z}$-module $\Gamma \simeq \mathbb{Z}^b$, a $\mathbb{Z}$-linear map $\psi\colon \Gamma \otimes \Gamma \longrightarrow \mathbb{Z}$ and a collection of Hodge numbers $h^{p,q}$, $p+q=n$, with $\sum h^{p,q} = b$. Then the set D of ψ-polarized Hodge structures on Γ of weight n with $\dim H^{p,q} = h^{p,q}$ is naturally a complex manifold, the *period domain*. More precisely, for $f^p = \sum_{i\geq p} h^{i,j}$, it is an open subset of the algebraic variety, called the *compact dual* of D,

$$D^\vee \subset \prod \mathrm{Gr}(f^p, \Gamma_{\mathbb{C}})$$

of all flags $(F^p) \in \prod \mathrm{Gr}(f^p, \Gamma_{\mathbb{C}})$ satisfying the orthogonality condition $F^p \perp F^{n-p+1}$ with respect to the $\mathbb{C}$-linear extension of ψ.

The compact dual $D^\vee$ is a smooth projective variety and $D \subset D^\vee$ is an open in the classical topology, cut out by the condition that $\psi(\ ,C\ )$ be positive definite. The compact dual $D^\vee$ comes with a transitive action of $G_{\mathbb{C}} = \mathrm{O}(\Gamma_{\mathbb{C}}, \psi)$ and, therefore, can be written as the homogenous space $D^\vee \simeq G_{\mathbb{C}}/B$, where B is the stabilizer of a flag in $D^\vee$. Similarly, D can be described as a homogenous space

$$D \simeq G_{\mathbb{R}}/(B \cap G_{\mathbb{R}}) \subset D^\vee \simeq G_{\mathbb{C}}/B.$$

We will be interested in the quotient $\mathrm{O}' \setminus D$ of D by the natural action of a discrete group $\mathrm{O}' \subset G_{\mathbb{R}}$ commensurable with the full orthogonal group $\mathrm{O}(\Gamma, \psi)$. Since $B \cap G_{\mathbb{R}}$ is a compact group, the action of $\mathrm{O}(\Gamma, \psi)$ is properly discontinuous and any such quotient $\mathrm{O}' \setminus D$ is a normal complex space with finite quotient singularities.

Consider a smooth projective family $\pi\colon \mathcal{X} \twoheadrightarrow S$ of varieties of dimension n with a relative ample line bundle $\mathcal{L}$. The families of cohomologies $H^n(\mathcal{X}_t, \mathbb{Z})$ and $H^n(\mathcal{X}_t, \mathbb{Z})_{\mathcal{L}_t\text{-pr}}$ form locally constant systems $R^n\pi_*\mathbb{Z}$ and $R^n_{\mathrm{pr}}\pi_*\mathbb{Z}$. We assume that S is connected, which implies that the Hodge numbers $h^{p,q} := h^{p,q}(\mathcal{X}_t)$ are constant. If S is simply connected, then the two local systems are canonically isomorphic to the two constant systems $\underline{H}^n(X, \mathbb{Z})$ and $\underline{H}^n(X, \mathbb{Z})_{\mathrm{pr}}$ for a fixed fibre $X = \mathcal{X}_0$ with respect to the ample line bundle $\mathcal{L}_0$. In this situation, the *period map*

$$\mathcal{P}\colon S \longrightarrow D,\ t \longmapsto (F^pH^n(\mathcal{X}_t, \mathbb{C})_{\mathcal{L}_t\text{-pr}})$$

is well defined and, by a classical result of Griffiths, holomorphic. Here, the period domain $D \subset D^\vee \subset \prod \mathrm{Gr}(f^p, H^n(X, \mathbb{C})_{\mathrm{pr}})$ is defined in terms of the polarization of the primitive cohomology $H^n(X, \mathbb{Z})_{\mathrm{pr}}$. The construction applies to the universal deformation $\mathcal{X} \twoheadrightarrow \mathrm{Def}(X) = \mathrm{Def}(X, \mathcal{O}_X(1))$ of a cubic hypersurface of dimension $n \geq 2$, see Section **1**.3.2, and describes the local period map

$$\mathcal{P}\colon \mathrm{Def}(X) \longrightarrow D \subset D^\vee \subset \prod \mathrm{Gr}(f^p, H^n(X, \mathbb{C})_{\mathrm{pr}}). \tag{3.1}$$

If S is not simply connected, one can still consider the map $\mathcal{P}\colon S \longrightarrow \mathrm{O}' \setminus D$, where $\mathrm{O}' = \mathrm{Im}(\pi_1(S) \longrightarrow \mathrm{O}(H^n(X, \mathbb{Z})_{\mathrm{pr}}))$ is the monodromy group of the family.

Let us apply the discussion to the universal family of smooth cubic hypersurfaces

$$\mathcal{X} \longrightarrow U(n) = |\mathcal{O}_{\mathbb{P}^{n+1}}(3)|_{\mathrm{sm}} \subset |\mathcal{O}_{\mathbb{P}^{n+1}}(3)|,$$

see Section **1**.2.4. The monodromy group $\Gamma_n := \mathrm{Im}(\pi_1(U(n)) \longrightarrow \mathrm{GL}(H^n(X, \mathbb{Z})_{\mathrm{pr}}))$, for $X = \mathcal{X}_0$ a fixed distinguished fibre, has been computed in Theorem **1**.2.9. For even n, we have found $\Gamma_n \simeq \tilde{\mathrm{O}}^+(H^n(X,\mathbb{Z}))$, which is a finite index subgroup of $\mathrm{O}(H^n(X,\mathbb{Z})_{\mathrm{pr}})$. Similarly, $\Gamma_n \simeq \mathrm{SpO}(H^n(X,\mathbb{Z}),q)$ for n odd. In any case, the period map is then a holomorphic map

$$\mathcal{P}\colon U(n) \longrightarrow \Gamma_n \setminus D_n,$$

where $D_n \subset D_n^\vee \subset \prod \mathrm{Gr}(f_n^p, H^n(X,\mathbb{C})_{\mathrm{pr}})$ with $f_n^p = \sum_{i\geq p} h^{i,j}(X)_{\mathrm{pr}}$. Locally around every point in $U(d)$ the period map $\mathcal{P}$ admits a holomorphic lift $\Delta \longrightarrow D_n$. In order to descend the period map to the moduli space of cubics, one might have to divide out by a slightly larger discrete group, but since an isomorphism always acts by an isometry on the middle cohomology, we obtain a period map

$$\mathcal{P}\colon M_n = U(n)//G \longrightarrow \mathrm{O} \setminus D_n. \tag{3.2}$$

Here, $\mathrm{O} = \mathrm{O}(H^n(X,\mathbb{Z})_{\mathrm{pr}})$ for n even and $\mathrm{O} = \mathrm{Sp}(H^n(X,\mathbb{Z}))$ for n odd. Alternatively, one can use the existence of a universal family locally, see Remark 1.14, to glue local period maps to the global period map (3.2). Yet another alternative is to introduce the moduli space of marked cubics parametrizing smooth cubics X together with an isometry of $H^n(X,\mathbb{Z})_{\mathrm{pr}}$ with the abstract lattice described in Proposition **1**.1.21 for n even or with the standard symplectic lattice for n odd.

Remark 3.3 Instead of working with the singular spaces M_n and $\mathrm{O} \setminus D_n$, one can view the period map as a map between smooth analytic Deligne–Mumford stacks:

$$\mathcal{P}\colon \mathcal{M}_n \longrightarrow [\mathrm{O} \setminus D_n].$$

This has the advantage of keeping track of the finite groups of automorphisms $\mathrm{Aut}(X)$. Furthermore, the existence of the universal family over $\mathcal{M}_n$ often simplifies arguments.

Example 3.4 Typically, the quotients $\mathrm{O} \setminus D_n$ will not be Kähler or algebraic [105, 206]. However, for cubic hypersurfaces of dimension three and four, they are.

(i) For smooth cubic threefolds, the dimensions of the Hodge filtrations are $f^3 = 0$, $f^2 = 5$, and $f^1 = f^0 = b_3 = 10$. The orthogonality conditions $F^p \perp F^{n-p+1}$ reduce to $F^2 \perp F^2$, which cuts out $D_3^\vee \subset \mathrm{Gr}(5,10)$. In other words, $D^\vee$ can be identified with the Lagrangian Grassmann variety $\mathrm{LG}(5,10)$ of all Lagrangian subspaces of a ten-dimensional symplectic vector space. Thus, its dimension is $\dim D_3^\vee = \dim \mathrm{LG}(5,10) = 15$, which is also the dimension of the moduli space A_5 of principally polarized abelian varieties of dimension five.

In our situation, the symplectic vector space is the complex vector space associated with the standard symplectic structure on $\mathbb{Z}^{\oplus 10}$ and thus comes with the action of the symplectic group $\mathrm{Sp}(10,\mathbb{Z})$. The quotient $\mathrm{Sp}(10,\mathbb{Z}) \setminus D_3$ is naturally endowed with the structure of a normal algebraic variety with finite quotient singularities. It can be thought of as the moduli space of principally polarized abelian varieties of dimension five:

$$A_5 \simeq \mathrm{Sp}(10,\mathbb{Z}) \setminus D_3,$$

which is more commonly written as the quotient $A_5 \simeq \mathrm{Sp}(10,\mathbb{Z})\setminus\mathbb{H}_5$ of the Siegel upper half space $\mathbb{H}_5$ of all symmetric 5×5 matrices τ with $\mathrm{Im}(\tau)$ positive definite.

The period map for cubic threefolds is thus the map

$$\mathcal{P}\colon M_3 \longrightarrow \mathrm{Sp}(10,\mathbb{Z}) \setminus D_3 \simeq A_5,$$

which later will be shown to be a locally closed embedding.[5] Geometrically, it can be interpreted as the map that sends a smooth cubic threefold $Y \subset \mathbb{P}^4$ to its intermediate Jacobian $J(Y)$. For more on this case, see Section **5**.5.3.

(ii) For smooth cubic fourfolds, the dimensions of the Hodge filtration are $f^4 = 0$, $f^3 = 1$, $f^2 = 21$, and $f^1 = f^0 = b_2(X)_{\mathrm{pr}} = 22$. The orthogonality conditions $F^p \perp F^{n-p+1}$ come down to a single condition $F^3 = H^{3,1} \perp F^2 = H^{3,1} \oplus H^{2,2}_{\mathrm{pr}}$, which can also be expressed by saying that the line $H^{3,1}(X) \subset H^4(X,\mathbb{C})_{\mathrm{pr}}$ determines also $H^{2,2}(X)_{\mathrm{pr}}$. Note that this condition implies that the projection of $D_4^\vee \subset \mathrm{Gr}(1,22) \times \mathrm{Gr}(21,22)$ to the first factor is still injective. Thus, it is more natural to view the period domain as the analytically open subset of a non-degenerate quadric hypersurface of dimension 20:

$$D_4 \subset D_4^\vee \subset \mathrm{Gr}(1,22) \simeq \mathbb{P}^{21} \simeq \mathbb{P}(\Gamma \otimes \mathbb{C}).$$

Here, $\Gamma := E_8(-1)^{\oplus 2} \oplus U^{\oplus 2} \oplus A_2(-1)$. The analytically open subset $D_4 \subset D_4^\vee$ is cut out by the positivity condition $(x.\bar{x}) > 0$ for $[x] \in D^\vee \subset \mathbb{P}(\Gamma \otimes \mathbb{C})$. Also in this case, the quotient $\mathrm{O}(\Gamma) \setminus D_4$ is indeed a quasi-projective variety. See Remark **6**.6.13 for a discussion why dividing by $\mathrm{O}(\Gamma)$ and $\tilde{\mathrm{O}}^+(\Gamma)$ amounts to essentially the same. In this case, the period map is a map display, which is later shown to be an open immersion,

$$\mathcal{P}\colon M_4 \longrightarrow \mathrm{O}(\Gamma) \setminus D_4,$$

see Section **6**.6 for a detailed discussion.

Although $\mathrm{O}(\Gamma) \setminus D$ is in general not algebraic, it has been conjectured by Griffiths and recently proved by Bakker, Brunebarbe, and Tsimerman [32] that the image of the period map is always contained in a quasi-projective variety.

3.2 Infinitesimal and Local Torelli For a smooth cubic hypersurface X of dimension $n \geq 2$, the Zariski tangent space of its universal deformation space $\mathrm{Def}(X) = \mathrm{Def}(X, \mathcal{O}_X(1))$ is naturally identified with $T_0\mathrm{Def}(X) \simeq H^1(X, \mathcal{T}_X)$, see Section **1**.3.3. Thus, the differential of the local period map (3.1) can be viewed as a map

[5] It might seem more natural to divide out by the monodromy group $\Gamma_3 = \mathrm{SpO}(H^3(Y,\mathbb{Z}), q) \subset \mathrm{Sp}(10,\mathbb{Z})$. However, there are automorphisms of cubic threefolds that are not contained in the monodromy group Γ_3, see Remark **6**.6.13.

$$d\mathcal{P}\colon H^1(X,\mathcal{T}_X)\longrightarrow T_{\mathcal{P}(0)}D_n\subset \operatorname{Hom}\left(\bigoplus_{p+q=n} H^{p,q}(X)_{\rm pr},\ \bigoplus_{p+q=n} H^{p-1,q+1}(X)_{\rm pr}\right), \qquad (3.3)$$

which is in fact described by contraction, see [474, Ch. 17.1]. The injectivity of the map (3.3), see Corollary **1**.4.25, Remark **1**.4.26, and Remark 1.16, can be rephrased as the following statement.

Corollary 3.5 (Infinitesimal Torelli) *The period map*

$$\mathcal{P}\colon \mathcal{M}_n\longrightarrow [\mathrm{O}\setminus D_n]$$

from the moduli stack of all smooth cubics of dimension $n>2$ is unramified. □

Recall from Remark **1**.4.26 that for cubic hypersurfaces of even dimension $n=2m$ already $H^{m,m}(X)_{\rm pr}$ detects first order deformations of X.

Classically, the result is stated in terms of the moduli space $\widetilde{M}_n$ of marked cubics parametrizing pairs (X,φ) consisting of a smooth cubic hypersurface of dimension n and an isometry $H^n(X,\mathbb{Z})_{\rm pr}\simeq\Gamma$. The period map in this setting is then a holomorphic map

$$\widetilde{\mathcal{P}}\colon \widetilde{M}_n\longrightarrow D_n$$

and the infinitesimal Torelli asserts that it is unramified.

Remark 3.6 Often, the infinitesimal Torelli theorem implies the so-called local Torelli theorem, see Section 3.3. Concretely, if Hodge isometries $H^n(X,\mathbb{Z})_{\rm pr}\simeq H^n(X,\mathbb{Z})_{\rm pr}$ are induced by automorphisms of X, then the period map

$$\mathcal{P}\colon M_n\longrightarrow \mathrm{O}\setminus D_n$$

is also unramified. Otherwise, Hodge isometries not induced by an automorphism might cause problems.

Recall that the converse does not hold, i.e. the local (or even global) Torelli does not necessarily imply the infinitesimal Torelli theorem. Indeed, for smooth curves of genus $g>2$, the corresponding map $\widetilde{\mathcal{P}}$ (or its stack version) is not unramified (exactly over the hyperelliptic locus) but the period map $\mathcal{P}$ that associates with a smooth curve the polarized abelian variety provided by its Jacobian defines a closed embedding and is in particular unramified.

Example 3.7 The hypothesis in the above remark that would build upon the infinitesimal Torelli theorem to prove a local Torelli theorem is not easy to ensure, not even for cubic hypersurfaces. The only known cases are cubics of dimension three and four:

(i) As for smooth projective curves, any Hodge isometry $H^3(Y,\mathbb{Z})\simeq H^3(Y',\mathbb{Z})$ for two smooth cubic threefolds is up to a sign induced by a unique isomorphism $Y\simeq Y'$, see Remark **5**.4.9.

(ii) The assumption for the local Torelli theorem is satisfied for cubic fourfolds, which we will prove as part of the global Torelli theorem in Section **6**.3.3. More precisely, every Hodge isometry $H^4(X,\mathbb{Z}) \simeq H^4(X,\mathbb{Z})$ preserving h_X^2 is induced by an automorphism of X.

3.3 Variational, General, and Generic Torelli Let us consider the following possible Torelli statements for smooth cubic hypersurfaces. We leave it to the reader to write down appropriate versions for other types of varieties, e.g. curves, K3 surfaces, abelian varieties, hypersurfaces of other degrees, etc. We recommend Beauville's Bourbaki talk [47] for further information and references.

(i) Variational Torelli theorem: a smooth cubic hypersurface X of dimension n can be reconstructed from the (real) Hodge structure $H^n(X,\mathbb{C})_{\rm pr} \simeq \bigoplus H^{p,q}(X)_{\rm pr}$ together with the map (3.3).
(ii) General Torelli theorem: assume X is a very general smooth cubic hypersurface of dimension n. Any other smooth cubic hypersurface X' for which there exists an isomorphism(!) of Hodge structures $H^n(X,\mathbb{Q})_{\rm pr} \simeq H^n(X',\mathbb{Q})_{\rm pr}$ is isomorphic to X:

$$H^n(X,\mathbb{Q})_{\rm pr} \simeq H^n(X',\mathbb{Q})_{\rm pr} \;\Rightarrow\; X \simeq X'.$$

(iii) Generic Torelli theorem: the period map $\mathcal{P}\colon M_n \longrightarrow {\rm O}' \setminus D$ is of degree one, i.e. generically injective, for any discrete group ${\rm O} \subset {\rm O}' \subset {\rm O}_{\mathbb{R}}$.
(iv) Global Torelli theorem: two smooth cubic hypersurfaces X and X' of dimension n are isomorphic if and only if there exists a Hodge isometry(!) between their middle primitive integral Hodge structures:

$$(H^n(X,\mathbb{Z})_{\rm pr}, (\ .\ )) \simeq (H^n(X',\mathbb{Z})_{\rm pr}, (\ .\ )) \;\Leftrightarrow\; X \simeq X'.$$

(v) Infinitesimal Torelli theorem: the stacky period map $\mathcal{P}\colon \mathcal{M}_n \longrightarrow [{\rm O} \setminus D_n]$ is unramified or, equivalently, (3.3) is injective, see Corollary 3.5.
(vi) Local Torelli theorem: the period map $\mathcal{P}\colon M_n \longrightarrow {\rm O}\backslash D_n$ is unramified, see Remark 3.6.

These different versions of the Torelli theorem are interlinked.[6] For example, by the work of Carlson–Griffiths [103], Donagi [161], Voisin [485], and many others, one knows:

$$\text{variational} \;\Rightarrow\; \text{general} \;\Rightarrow\; \text{generic}.$$

See [104, Ch. 8.2] for a detailed discussion of the second implication. The freedom in choosing any discrete group ${\rm O}'$ in the generic Torelli theorem (iii) is surprising at first,

[6] And sometimes confused, especially infinitesimal with local.

especially when compared with the case of K3 surfaces. For example, for cubic threefolds, $M_3 \hookrightarrow A_5 \simeq \mathrm{Sp}(10,\mathbb{Z})\backslash D_3$ is a closed embedding and, somewhat unexpectedly, even dividing out by a bigger discrete subgroup $\mathrm{Sp}(10,\mathbb{Z}) \subset \mathrm{O}' \subset \mathrm{Sp}(10,\mathbb{R})$ still leads to a generically injective composition $M_3 \hookrightarrow A_5 \simeq \mathrm{Sp}(10,\mathbb{Z}) \setminus D_3 \twoheadrightarrow \mathrm{O}' \setminus D_3$.

Although, the variational Hodge conjecture only implies a general or a generic version of the Torelli theorem, it is in a certain sense stronger than the global Torelli theorem, but neither one of the two implies the other:

$$\text{variational} \not\Rightarrow \text{global} \not\Rightarrow \text{variational or generic.}$$

For example, the global Torelli theorem holds for K3 surfaces, see Section **6**.6.3, but in general a K3 surface can neither be reconstructed from its infinitesimal variation, so the variational Torelli does not hold, nor does the period map inject the moduli space of polarized K3 surfaces into $\mathrm{O}' \setminus D$ for a discrete group O' bigger than $\mathrm{O}(H^2(S,\mathbb{Z})_{\mathrm{pr}})$. Also, neither the local nor even the global Torelli theorem implies the infinitesimal Torelli theorem, cf. Remark 3.6. Nevertheless, an infinitesimal or local Torelli theorem is often seen as good first evidence for a global Torelli theorem.

As we have noted already in Section **1**.4.5, see Corollary **1**.4.27 in particular, the Torelli theorems (i)–(iii) hold for two-thirds of all smooth cubic hypersurfaces.

Corollary 3.8 (Variational, general, and generic Torelli) *For smooth cubic hypersurfaces of dimension $n > 2$ with $3 \nmid (n+2)$, the variational, general, and generic Torelli theorems hold true.* □

Thus, the first case for which a generic Torelli does not follow from general results is the case of cubic fourfolds. Indeed, in this case the generic Torelli does not hold, and consequently, neither does the variational or the general. But a global Torelli theorem holds. This is the content of Section **6**.3.3.

Remark 3.9 (i) At this point, for smooth cubic hypersurfaces of dimension $n = 3m - 2 = 7, 10, 13, \ldots$, neither the global nor the generic (and hence neither the variational nor the general) Torelli theorem is known to hold.

(ii) For cubic hypersurfaces covered by Corollary 3.8, the global Torelli theorem is only known in dimension three. So, the first case for which the generic but not the global Torelli theorem is known to hold is the case of cubic fivefolds.

(iii) By Corollary 3.5, the infinitesimal Torelli theorem holds for all smooth cubic hypersurfaces of dimension $n > 2$, but the local one in the formulation of (vi) above is not known beyond dimension four.

4

Cubic Surfaces

The general theory presented in Chapters 1–3 applied to the case of smooth cubic surfaces $S \subset \mathbb{P}^3$ provides us with some crucial information.

On the purely numerical side, we have seen that the Hodge diamond is only non-trivial in bidegree (p, p), i.e.

$$H^1(S, \mathcal{O}_S) = H^0(S, \Omega_S) = 0 \quad \text{and} \quad H^2(S, \mathcal{O}_S) = H^0(S, \Omega_S^2) = 0,$$

and, moreover, see Sections **1**.1.2 and **1**.1.3:

$$H^{1,1}(S) = H^1(S, \Omega_S) \simeq k^{\oplus 7}.$$

The linear system of all cubic surfaces $|\mathcal{O}(3)| \simeq \mathbb{P}^{19}$ comes with a natural action of PGL(4) and its GIT quotient, the moduli space of semi-stable cubic surfaces, is four-dimensional, see Sections **1**.2.1 and **3**.1.3

We have also seen that the Fano variety $F(S)$ of lines contained in S is non-empty, smooth, and zero-dimensional of degree 27, see Proposition **2**.4.6 and Example **2**.4.21. Hence, over an algebraically closed field k, the Fano variety $F(S)$ consists of 27 reduced k-rational points. So, any smooth cubic surface $S \subset \mathbb{P}^3$ defined over an algebraically closed field contains exactly 27 lines. In this chapter we denote them by $\ell_1, \ldots, \ell_{27}$, so

$$F(S) = \{ \ell_1, \ldots, \ell_{27} \}$$

or, viewing S as a blow-up of $\mathbb{P}^2$, as $E_1, \ldots, E_6, L_1, \ldots, L_6, L_{12}, \ldots, L_{56}$, see below. There are more classical arguments to deduce this result and we will touch upon some of the techniques in this chapter. However, we will have to resist the temptation to dive into the classical theory too much and instead refer to the rich literature on the subject, see, for example, [49, 158, 223, 229, 284, 335, 462]. Also, there is a vast literature on the arithmetic of cubic surfaces over non-algebraically closed fields which will not be mentioned at all, see e.g. the recent [269] for a modern approach and references.

1 Picard Group

Let $S \subset \mathbb{P}^3$ be a smooth cubic surface over an arbitrary field k. We will see that its Picard group $\mathrm{Pic}(S)$ coincides with the numerical Picard group $\mathrm{Num}(S)$ and the Néron–Severi group $\mathrm{NS}(S) = \mathrm{Pic}(S)/\mathrm{Pic}^0(S)$. So, it is endowed with the intersection pairing $(\mathcal{L}.\mathcal{L}')$ which satisfies the Hodge index theorem. In particular, the inequality $(\mathcal{L}.\mathcal{L}')^2 \geq (\mathcal{L}.\mathcal{L}) \cdot (\mathcal{L}'.\mathcal{L}')$ for all line bundles $\mathcal{L}, \mathcal{L}'$ with $(\mathcal{L}.\mathcal{L}) \geq 0$.

1.1 Intersection Form The only line bundles that come for free on any smooth cubic surface are $\mathcal{O}_S(1) := \mathcal{O}(1)|_S$ and its powers $\mathcal{O}_S(a)$. For example, the canonical bundle is described by the adjunction formula, see Lemma **1**.1.6, as

$$\omega_S \simeq \mathcal{O}_S(-1),$$

with the very ample dual $\omega_S^* \simeq \mathcal{O}_S(1)$. The Hirzebruch–Riemann–Roch formula for a line bundle $\mathcal{L}$ on S takes the form

$$\chi(S, \mathcal{L}) = \frac{(\mathcal{L}.\mathcal{L}) + (\mathcal{L}.\mathcal{O}_S(1))}{2} + 1, \tag{1.1}$$

where we use $\chi(S, \mathcal{O}_S) = 1$, see Section **1**.1.4.

Lemma 1.1 *Any numerically trivial line bundle $\mathcal{L}$ on a smooth cubic surface S is trivial. In particular,* $\mathrm{Pic}(S)$ *is torsion free of finite rank and* $\mathrm{Pic}^0(S) = 0$.

Proof Indeed, if a numerically trivial line bundle $\mathcal{L}$ is not trivial, then $(\mathcal{L}.\mathcal{O}_S(1)) = 0$ implies $H^0(S, \mathcal{L}) = 0$ and $H^2(S, \mathcal{L}) \simeq H^0(S, \mathcal{L}^* \otimes \omega_S)^* = 0$. Hence, $\chi(S, \mathcal{L}) \leq 0$, which contradicts (1.1) showing $\chi(S, \mathcal{L}) = 1$. □

Corollary 1.2 *For a smooth cubic surface $S \subset \mathbb{P}^3$ over an arbitrary field k, one has*

$$\mathrm{Pic}(S) \simeq \mathrm{NS}(S) \simeq \mathrm{Num}(S) \simeq \mathbb{Z}^{\oplus \rho(S)}$$

with $1 \leq \rho(S) \leq 7$. *For a field extension $k \subset k'$, the base change map*

$$\mathrm{Pic}(S) \hookrightarrow \mathrm{Pic}(S_{k'}) \tag{1.2}$$

is injective. Moreover, if k is algebraically closed, then $\rho(S) = 7$ and for any further base change (1.2) *is an isomorphism.*

Proof Recall that an invertible sheaf $\mathcal{L}$ on S is trivial if and only if $H^0(S, \mathcal{L}) \neq 0$ and $H^0(S, \mathcal{L}^*) \neq 0$. As $H^0(S_{k'}, \mathcal{L}_{k'}) \simeq H^0(S, \mathcal{L}) \otimes_k k'$, this shows the injectivity of (1.2).

For $k = \mathbb{C}$, the exponential sequence gives

$$\mathrm{Pic}(S) \simeq H^2(S, \mathbb{Z}) \simeq \mathbb{Z}^{\oplus 7},$$

while for an arbitrary algebraically closed field k the Kummer sequence

$$0 \longrightarrow \mu_n \longrightarrow \mathbb{G}_m \xrightarrow{(\ )^n} \mathbb{G}_m \longrightarrow 0,$$

with $n = \ell^m$ prime to char(k), provides injections $\mathrm{Pic}(S) \otimes \mathbb{Z}/\ell^m\mathbb{Z} \hookrightarrow H^2_{\acute{e}t}(S, \mu_{\ell^m})$ with a cokernel contained in $H^2(S, \mathbb{G}_m)$. Taking limits, one obtains

$$\mathrm{Pic}(S) \otimes \mathbb{Z}_\ell \hookrightarrow H^2_{\acute{e}t}(S, \mathbb{Z}_\ell(1)) \simeq \mathbb{Z}_\ell(1)^{\oplus 7}, \tag{1.3}$$

as $b_2(S) = 7$, cf. Section **1**.1.6. Together with (1.2), this proves $\rho(S) \leq 7$ for arbitrary base field k.

For k algebraically closed, the Brauer group is trivial, i.e. $\mathrm{Br}(S) = H^2(S, \mathbb{G}_m) = 0$. This is analogous to $H^2(S, \mathcal{O}_S^*) = H^2(S, \mathcal{O}_S)/H^2(S, \mathbb{Z}) = 0$ for $k = \mathbb{C}$. Hence, (1.3) is an isomorphism and, therefore, $\rho(S) = 7$. The last assertion follows from a standard 'spreading out' argument and the fact that for $k = \bar{k}$ the Picard variety Pic_S consists of isolated, reduced, k-rational points, cf. [249, Lem. 17.2.2]. □

Example 1.3 Examples of smooth cubics with $\rho(S) < 7$ can be produced easily.

(i) For example, if $\mathcal{S} \longrightarrow |\mathcal{O}(3)|$ is the universal cubic surface, then the scheme-theoretic generic fibre $\mathcal{S}_\eta$ satisfies $\mathrm{Pic}(\mathcal{S}_\eta) \simeq \mathbb{Z} \cdot \mathcal{O}(1)|_{\mathcal{S}_\eta}$. Here, $\mathcal{S}_\eta$ is a smooth cubic surface over the (non-algebraically closed) function field $k(\eta) \simeq k(t_1, \ldots, t_{19})$.

(ii) Similarly, if $\mathcal{S}_{\mathbb{P}^1} \longrightarrow \mathbb{P}^1 \hookrightarrow |\mathcal{O}(3)|$ is a Lefschetz pencil, then the other projection $\tau\colon \mathcal{S}_{\mathbb{P}^1} \longrightarrow \mathbb{P}^3$ is the blow-up of $\mathbb{P}^3$ in the smooth intersection $S_1 \cap S_2 \subset \mathbb{P}^3$ of two smooth cubics. Hence, $\mathrm{Pic}(\mathcal{S}_{\mathbb{P}^1}) \simeq \mathbb{Z} \cdot \mathcal{O}(1)|_{\mathcal{S}_{\mathbb{P}^1}} \oplus \mathbb{Z} \cdot \mathcal{O}(E)$ by the blow-up formula, where $E = \mathbb{P}(\mathcal{N}_{S_1 \cap S_2/\mathbb{P}^3})$ is the exceptional divisor of τ. Therefore, the fibre $\mathcal{S}_\eta$ over the generic point $\eta \in \mathbb{P}^1$, with residue field $k(\eta) \simeq k(t)$, satisfies $\mathrm{Pic}(\mathcal{S}_\eta) \simeq \mathbb{Z} \cdot \mathcal{O}_S(1)$.

Note that according to Corollary **1**.2.7, a Lefschetz pencil of cubic surfaces has exactly 32 singular fibres, each with one ordinary double point as the only singularity.

Remark 1.4 At this point, one could mention two famous results concerning cubic surfaces over non-algebraically closed fields going back to Segre and Manin, cf. [335, Thm. 33.1 & 33.2] and also [284, Ch. 3]:

- A smooth cubic surface of Picard number one is not rational.
- Birational smooth cubic surfaces of Picard number one are actually isomorphic.

Both results follow from the same principle that, for a birational correspondence that is not an isomorphism, the ample line bundles on both sides of the correspondence force the Picard group to be of rank at least two.

Remark 1.5 (i) Similarly to Example 1.3, it should be possible to construct examples of smooth cubic surfaces with an arbitrary prescribed Picard number $1 \leq \rho \leq 7$. But it is an entirely different matter to produce cubic surfaces with a prescribed Picard number over special types of fields, like number fields or finite fields. As the Picard number of a cubic surface over a finite field $\mathbb{F}_q$ can be read off its Zeta function (as the multiplicity of q^{-1} as a root of the denominator), computing $\rho(S)$ and $|S(\mathbb{F}_q)|$ are essentially equivalent. For the latter we refer to [34] and the references therein.

(ii) The Weil conjectures had been verified for cubic surfaces over finite fields early on by Weil himself [492], cf. [335, Thm. 27.1]. A finer analysis of the possibilities for the Zeta function was attempted in [335, 410, 442]. The complete classification was eventually given by Banwait, Fité, and Loughran [34] and a topological approach to the average number was described by Das [133].

By the Hodge index theorem, the Picard group $\mathrm{Pic}(S) \simeq \mathrm{NS}(S) \simeq \mathbb{Z}^{\oplus \rho(S)}$ together with the non-degenerate intersection pairing defines a lattice of signature $(1, \rho(S) - 1)$. It is an odd lattice, because $(\mathcal{O}_S(1).\mathcal{O}_S(1)) = 3$. The orthogonal complement $\mathcal{O}_S(1)^\perp \subset \mathrm{Pic}(S)$ is negative definite of rank ≤ 6.

For $k = \mathbb{C}$, the exponential sequence leads to an isomorphism of lattices

$$\mathrm{Pic}(S) \simeq H^2(S, \mathbb{Z}).$$

As $H^2(S, \mathbb{Z})$ is unimodular and odd, it is isomorphic to $\mathrm{I}_{1,6}$ and $\mathcal{O}_S(1)^\perp \simeq H^2(S, \mathbb{Z})_{\mathrm{pr}} \simeq E_6(-1)$, cf. Corollary **1**.1.20 and Proposition **1**.1.21. The same conclusions hold over an arbitrary algebraically closed field, as we will show next.

Corollary 1.6 *Let $S \subset \mathbb{P}^3$ be a smooth cubic surface over an algebraically closed field. Then*

$$\mathrm{Pic}(S) \simeq \mathrm{I}_{1,6} \quad \text{and} \quad \mathcal{O}_S(1)^\perp \simeq E_6(-1). \tag{1.4}$$

For an explicit basis of both lattices in terms of lines, see Section 3.4.

Proof Completely geometric arguments for this description exist. For example, one can use the fact that S is a blow-up of $\mathbb{P}^2$ in six points, which, however, we will deduce later from (1.4), or that S admits a conic fibration $S \twoheadrightarrow \mathbb{P}^1$ with five singular fibres, see Section 2.4 and [412, IV.2.5]. Here, we shall derive the claim from the description of the intersection pairing $H^2(S, \mathbb{Z})$ of a smooth cubic surface over $\mathbb{C}$.

Indeed, in characteristic zero, the assertion follows from the complex case and the standard Lefschetz principle. In positive characteristic, the assertion is proved by means of the specialization map

$$\mathrm{Pic}(S_{\bar{\eta}}) \hookrightarrow \mathrm{Pic}(S_{\bar{t}}).$$

Here, $S \twoheadrightarrow \mathrm{Spec}(R)$ is a smooth family of cubic surfaces over a DVR and t and η are the closed and generic points with residue fields $k(t)$ and $k(\eta)$ of positive and zero characteristic, respectively.

Specialization is injective, because it is compatible with the intersection form. However, $\mathrm{Pic}(S_{\bar{\eta}}) \simeq \mathrm{I}_{1,6}$ is a unimodular lattice and any isometric embedding of finite index of a unimodular lattice is an isomorphism. Once $\mathrm{Pic}(S)$ is determined, its primitive part is described as in the proof of Proposition **1**.1.21. □

Remark 1.7 The Galois group $\mathrm{Gal}(\bar{k}/k)$ naturally acts on $\mathrm{Pic}(S_{\bar{k}})$ and on the sublattice $\mathcal{O}_S(1)^\perp \simeq E_6(-1)$. It therefore defines a subgroup $G \subset \mathrm{O}(E_6)$, which is in fact contained in the Weyl group $W(E_6) \subset \mathrm{O}(E_6)$, cf. Section **1**.2.5. Alternatively, the Galois group acts on the configuration of lines $\mathcal{L}(S)$, see Remark 1.3 and Section 3.6, whose automorphism group is $W(E_6)$. Which subgroups can be realized in this way? It is a classical fact that the scheme theoretic generic cubic surface, which lives over the function field of $|\mathcal{O}(3)|$, leads to $G = W(E_6)$, see Corollary 1.14. For information concerning the case of finite fields, in which case G is a cyclic group, see [34].

1.2 Numerical Characterization of Lines We next aim at a purely numerical characterization of lines contained in smooth cubic surfaces.

Remark 1.8 (i) Observe that any $\mathbb{P}^1 \simeq L \subset S$ with $(L.L) = -1$ is in fact a line, i.e. the degree of L as a subvariety of the ambient $\mathbb{P}^3$ is $\deg(L) = 1$ or, still equivalently, $(\mathcal{O}_S(1).\mathcal{O}(L)) = \deg(\mathcal{O}_S(1)|_L) = 1$. Indeed, by adjunction $\mathcal{O}_{\mathbb{P}^1}(-2) \simeq \omega_L \simeq (\omega_S \otimes \mathcal{O}(L))|_L = \mathcal{O}_S(-1)|_L \otimes \mathcal{O}_{\mathbb{P}^1}(-1)$.

(ii) For a geometrically integral curve $C \subset S$, we deduce from (1.1) that

$$1 \geq 1 - h^1(C, \mathcal{O}_C) = \chi(C, \mathcal{O}_C) = \chi(S, \mathcal{O}_S) - \chi(S, \mathcal{O}_S(-C)) = -\frac{(C.C) - \deg(C)}{2}$$

and, therefore,

$$(C.C) \geq \deg(C) - 2 \geq -1. \tag{1.5}$$

If in addition $(C.C) = -1$ holds, which implies geometrically integral, then automatically $\deg(C) = 1$ and $h^1(C, \mathcal{O}_C) = 0$. Hence, again, $L := C \simeq \mathbb{P}^1$ is a line.

So, combining (i) and (ii), we find that a (-1)-curve, i.e. a (geometrically integral) curve with $(C.C) = -1$, on a smooth cubic surface is the same thing as a line.

(iii) Similarly, if $\mathcal{L} \in \mathrm{Pic}(S)$ with $(\mathcal{L}.\mathcal{O}_S(1)) = 1$ and $(\mathcal{L}.\mathcal{L}) = -1$, then $\chi(S, \mathcal{L}) = 1$ by (1.1) and, therefore, $H^0(S, \mathcal{L}) \neq 0$. Hence, $\mathcal{L} \simeq \mathcal{O}_S(L)$ for some curve $L \subset S$ which, using $\deg(L) = (\mathcal{L}.\mathcal{O}_S(1)) = 1$, implies that L is geometrically integral and hence a line.

Note that these arguments only use the numerical properties of the polarized surface $(S, \mathcal{O}_S(1))$ and the fact that $\omega_S \simeq \mathcal{O}_S(-1)$. This will be useful later on, see, for example, the proof of Proposition 2.7.

Thus, if $\mathrm{Pic}(S) \simeq \mathrm{I}_{1,6}$ and $\alpha \in \mathrm{I}_{1,6}$ is a characteristic vector of square $(\alpha.\alpha) = 3$, cf. proof of Proposition **1**.1.21, then there are natural bijections

$$\begin{aligned} \{\, \mathbb{P}^1 \simeq L \subset S \mid \text{line} \,\} &\simeq \{\, C \subset S \mid \text{integral, } (C.C) = -1 \,\} \\ &\simeq \{\, \beta \in \mathrm{I}_{1,6} \mid (\beta.\beta) = -1,\ (\alpha.\beta) = 1 \,\}, \end{aligned}$$

see also the proof of Corollary 1.9 below.

We draw two immediate but crucial consequences from this. The first one is usually deduced from a concrete geometric reasoning, which is avoided in the present approach.

Corollary 1.9 *Assume that $S \subset \mathbb{P}^3$ is a smooth cubic surface over an algebraically closed field. Then S contains six pairwise disjoint lines $\ell_1, \ldots, \ell_6 \subset S$.*

Proof By Corollary 1.6, the Picard lattice is $\text{Pic}(S) \simeq \text{I}_{1,6}$, and this is all that is needed in the following. In particular, the assumption on k to be algebraically closed can be weakened.

As argued in the proof of Proposition **1**.1.21, the class $\alpha = (3, -1, \ldots, -1) \in \text{I}_{1,6}$ (with the harmless but convenient sign change), written in the standard basis $v_0, \ldots, v_6$, and the hyperplane section h_S are both characteristic classes of the same square $(\alpha.\alpha) = (h_S.h_S) = 3$. Hence, after applying an appropriate orthogonal transformation, they coincide. But then the classes v_i, $i = 1, \ldots, 6$ correspond to line bundles $\mathcal{L}_i$ with $(\mathcal{L}_i.\mathcal{L}_i) = -1$ and $(\mathcal{L}_i.\mathcal{O}_S(1)) = 1$. According to the above remark, $\mathcal{L}_i \simeq \mathcal{O}(\ell_i)$, where the curves $\ell_i \subset S$ are lines. As $(\mathcal{L}_i.\mathcal{L}_j) = (v_i.v_j) = 0$ for $i \neq j$, they are pairwise disjoint. □

Note that the existence of two disjoint lines already implies that S is rational, see Corollary **1**.5.11.

Remark 1.10 It is curious to observe that one can reverse the flow of information and deduce from the geometry of a cubic surface information about the lattices $\text{I}_{1,6}$ and E_6. For example, the fact that the Fano variety $F(S)$ of lines on a smooth cubic surface over an algebraically closed field consists of 27 isolated, smooth k-rational points translates to the fact that in the lattice $\text{I}_{1,6}$ there exist exactly 27 classes ℓ with $(\ell.(3, 1, \ldots, 1)) = 1$ and $(\ell.\ell) = -1$.

Corollary 1.11 *Assume $S \subset \mathbb{P}^3$ is a smooth cubic surface over an arbitrary field k. Then, an invertible sheaf $\mathcal{L}$ is ample if and only if $(\mathcal{L}.\mathcal{L}) > 0$ and $(\mathcal{L}.L) > 0$ for every line $L \subset S_{\bar{k}}$.*

Proof Only the 'if-direction' requires a proof. For this, let us first recall the *Nakai–Moishezon criterion* for smooth projective surfaces over arbitrary fields, cf. [30]: an invertible sheaf $\mathcal{L}$ is ample if and only if $(\mathcal{L}.\mathcal{L}) > 0$ and $(\mathcal{L}.C) > 0$ for every curve $C \subset S$. It is of course enough to test integral curves C, but we may not necessarily be able to reduce to geometrically integral ones. For this reason, one has to take all lines in the base change $S_{\bar{k}}$ into account.

As $\mathcal{L}$ is ample if and only if its base change to $S_{\bar{k}}$ is ample, one can reduce to the case $k = \bar{k}$. Then any integral curve C is geometrically integral and by (1.5) either $(C.C) = -1$, in which case $\mathbb{P}^1 \simeq C$ is a line, or $(C.C) \geq 0$. To prove $(\mathcal{L}.C) > 0$ in the latter case we shall apply the Hodge index theorem. First note that there exists a hyperplane $\mathbb{P}^2 \subset \mathbb{P}^3$ such that the intersection $S \cap \mathbb{P}^2$ consists of three lines $\ell_1 \cup \ell_2 \cup \ell_3$. We postpone the proof of this fact, cf. Sections 2.4 and 3.3. As $(\mathcal{L}.\ell_i) > 0$, also $(\mathcal{L}.\mathcal{O}_S(1)) > 0$. Hence, $\mathcal{L}$ and $\mathcal{O}_S(1)$ are contained in the same connected component $\mathcal{C}^\circ$ of the positive cone

$$\mathcal{C} := \{ x \in \text{NS}(S) \otimes \mathbb{R} \mid (x.x) > 0 \} = \mathcal{C}^\circ \sqcup (-\mathcal{C}^\circ).$$

Similarly, $(\mathcal{O}_S(1).C) > 0$ implies $[C] \in \overline{\mathcal{C}^\circ} \setminus \{0\}$ and, therefore, also $(\mathcal{L}.C) > 0$.

The same remark as the one at the end of Remark 1.8 applies: only the numerical properties of $(S, \mathcal{O}_S(1))$, the isomorphism $\omega_S \simeq \mathcal{O}_S(-1)$, and the inequality $(\mathcal{L}.\mathcal{O}_S(1)) > 0$ have been used in the proof. □

1.3 Effective Cone We summarize the situation by a description of the ample cone and the effective cone. By definition, the *effective cone* is the cone of all finite, non-negative real linear combinations of curves

$$\mathrm{NE}(S) := \left\{ \sum a_i [C_i] \mid a_i \in \mathbb{R}_{\geq 0} \right\},$$

where $C_i \subset S$ are arbitrary irreducible (or integral) curves. The dual $\mathrm{NE}(S)^*$ of $\mathrm{NE}(S)$ is the nef cone which can also be described as the closure of the (open) *ample cone*

$$\mathrm{Amp}(S) := \left\{ \sum a_i \mathcal{L}_i \mid a_i \in \mathbb{R}_{>0},\ \mathcal{L}_i \text{ ample} \right\} \subset \mathrm{Pic}(S) \otimes \mathbb{R}.$$

In the following description of the effective cone we use the fact that there exist exactly 27 lines on a cubic surface over an algebraically closed field. This has been deduced by cohomological methods in Example **2**.4.21 already and the existence of at least 27 lines will be shown again in Remark 2.5.

Proposition 1.12 *Let S be a smooth cubic surface over an algebraically closed field. Then the effective cone is*

$$\mathrm{NE}(S) = \left\{ \sum_{i=1}^{27} a_i [\ell_i] \mid a_i \in \mathbb{R}_{\geq 0} \right\},$$

the closed rational polyhedron spanned by the 27 lines $\ell_1, \ldots, \ell_{27} \subset S$. The ample cone is the interior of its dual $\mathrm{NE}(S)^*$*, which is again rationally polyhedral:*

$$\mathrm{Amp}(S) = \mathrm{Int}\,(\mathrm{NE}(S)^*).$$

Proof As above, $\mathcal{C}^o$ denotes the connected component of the positive cone that contains $\mathcal{O}_S(1)$. By (1.1) all integral classes in $\mathcal{C}^o$ are contained in $\mathrm{NE}(S)$. Furthermore, any integral curve C with $[C]$ not contained in the closure of $\mathcal{C}^o$ is a line, see Remark 1.8. Hence, the closure of $\mathrm{NE}(S)$ is spanned by the closure of $\mathcal{C}^o$ and $K := \sum_{i=1}^{27} \mathbb{R} \cdot [\ell_i]$.

Now, $K \cap \mathcal{C}^o \neq \emptyset$. Indeed, as used before, the class $\mathcal{O}_S(1)$ can be written as the sum of three lines. In order to show that $\mathcal{C}^o \subset K$, it therefore suffices to argue that no class in $\mathcal{C}^o$ can be written as $a\ell_i + b\ell_j$ with $a, b \geq 0$. As two distinct lines are either disjoint or intersect transversally in exactly one point, $(\ell_i.\ell_j) = 0$ or $= 1$. Hence, $(a\ell_i + b\ell_j.a\ell_i + b\ell_j) = -(a^2 + b^2)$ or $= -(a^2 + b^2 - 2ab)$, which are both not positive. □

This result in particular shows that the ample cones of smooth cubic surfaces over algebraically closed fields all look the same. This is in stark contrast to other types of surfaces, for example K3 surfaces, cf. [249, Ch. 8].

For non-algebraically closed fields, these cones can be described via the inclusion $\text{Pic}(S) \hookrightarrow \text{Pic}(S_{\bar{k}})$ as $\text{Amp}(S) = \text{Amp}(S_{\bar{k}}) \cap (\text{Pic}(S) \otimes \mathbb{R})$ and, dually, $\text{NE}(S) = \text{NE}(S_{\bar{k}}) \cap (\text{Pic}(S) \otimes \mathbb{R})$. Hence, rephrasing Corollary 1.11, $\mathcal{L}$ is ample if and only if $(\mathcal{L}.C) > 0$ for all curves C which after base change to the algebraic closure are unions (with multiplicities) of lines.

Remark 1.13 It is not difficult to prove that an ample invertible sheaf on a cubic surface is automatically very ample, see [223, V. Thm. 4.11].

1.4 Monodromy Group of 27 Lines Consider the family of all smooth cubic surfaces $\mathcal{S} \longrightarrow U := |\mathcal{O}(3)|_{\text{sm}}$. In Section **1**.2.5 we discussed the monodromy group of this family, i.e. for $k = \mathbb{C}$ the image of the natural representation

$$\rho_{\mathcal{S}} : \pi_1(U) \longrightarrow \text{O}(H^2(S, \mathbb{Z})),$$

where $S = \mathcal{S}_0$ is a distinguished smooth fibre. According to Theorem **1**.2.9, this is the group $\tilde{\text{O}}^+(H^2(S, \mathbb{Z}))$ of all orthogonal transformations of the lattice $H^2(S, \mathbb{Z})$ with trivial spinor norm that fix the hyperplane class. In fact, in the discussion there we argued that the monodromy group, as a subgroup of the orthogonal group of the lattice $H^2(S, \mathbb{Z})_{\text{pr}} \simeq E_6(-1)$, is the Weyl group $W(E_6)$. Recall that its order is

$$|W(E_6)| = 51.840 = 2^7 \cdot 3^4 \cdot 5$$

and that $W(E_6)$ is a subgroup of index two of $\text{O}(E_6)$, only the coset of the orthogonal transformation given by a global sign change is missing, cf. [129, Sec. 15].

Let us rephrase this in terms of the family of 27 lines. Recall from Corollary **2**.1.14 that the relative Fano variety of lines of the family $\mathcal{S} \longrightarrow U$ is an étale morphism $F := F(\mathcal{S}/U) \longrightarrow U$ of degree 27. Furthermore, by Proposition **2**.1.4, F is connected. Using the same choice of a point $[S] \in U$ and identifying $\text{Bij}(F(S)) \simeq \mathfrak{S}_{27}$, we obtain the map

$$\rho_F : \pi_1(U) \longrightarrow \mathfrak{S}_{27}.$$

Its image is the monodromy group of the family of lines, only well defined up to conjugacy. The image is isomorphic to the Galois group of the covering, cf. [219, Sec. 1]. Compare the following classical fact also with the discussion in Section 3.6 and Remark 3.8.

Corollary 1.14 *The Galois group or, equivalently, the monodromy group* $\text{Im}(\rho_F) \subset \mathfrak{S}_{27}$ *of the universal family* $F \longrightarrow U$ *of the 27 lines contained in smooth cubic surfaces* $S \subset \mathbb{P}^3$ *is isomorphic to the Weyl group* $W(E_6)$.

Proof The image of the monodromy representation $\rho_{\mathcal{S}} : \pi_1(U) \longrightarrow \text{O}(H^2(S, \mathbb{Z}))$ is the Weyl group $W(E_6)$. An element in its kernel fixes every line and thus is also contained in the kernel of $\pi_1(U) \longrightarrow \mathfrak{S}_{27}$. The induced map $W(E_6) = \text{Im}(\rho_{\mathcal{S}}) \twoheadrightarrow \text{Im}(\rho_F) \subset \mathfrak{S}_{27}$ is injective, which proves the claim. □

Remark 1.15 Harris [219, Sec. III.3] describes the monodromy group as $\mathrm{O}^-(6, \mathbb{F}_2)$, the orthogonal group of $\mathbb{F}_2^{\oplus 6}$ endowed with the quadratic form $q := \sum_{i<j} x_i x_j$. Indeed, with every line $\ell \subset S$ one associates the class $\bar{\ell} \in H^2(S, \mathbb{F}_2)_{\rm pr}$ given as the image of $h - \ell$ which satisfies $(h - \ell.h - \ell) \equiv (h.h - \ell) \equiv 0\,(2)$. The intersection form on $H^2(S, \mathbb{F}_2)_{\rm pr}$ is seen to be isomorphic to q when expressed with respect to $\bar{\ell}_i$ for $\ell_1, \ldots, \ell_6$ the exceptional lines of a representation of S as a blow-up of $\mathbb{P}^2$, see Proposition 2.4. The quadratic form q on $\mathbb{F}_2^{\oplus 6}$ has exactly 27 non-trivial 0s which via $\ell \longmapsto \bar{\ell}$ are in bijection to the lines contained in S. This leads to an identification of $\mathrm{O}^-(6, \mathbb{F}_2) := \mathrm{O}(H^2(S, \mathbb{F}_2)_{\rm pr}, q)$ with the group of automorphisms of the line configuration on S. To conclude, one argues that every such automorphism is induced by a monodromy operation which uses elementary automorphisms of cubic surfaces.

Remark 1.16 It has been mentioned already in Section **1**.2.6 that apart from those diffeomorphisms of a cubic surface that can be described by monodromy (and, which, therefore preserve the hyperplane class and only define elements in the Weyl group), there are others such that in fact all orthogonal transformations of $H^2(S, \mathbb{Z})$ are realized by diffeomorphisms.

Note that from the perspective of lines on a cubic surface, it is clear that there must be many diffeomorphisms that do not preserve the hyperplane class. For example, use the fact that any two disjoint lines are alike, see Section 3.2, and that a cubic surface is a blow-up of $\mathbb{P}^2$, see Proposition 2.4. So, even without using any topological argument, just the investigation of lines on a smooth cubic surfaces shows that $\tau(\mathrm{Diff}(S))$ is much bigger than $W(E_6)$.

2 Representing Cubic Surfaces

Cubic surfaces can be viewed from different angles and can be described geometrically in various ways. Each representation highlights particular features. We will briefly describe the most common ones.

2.1 Cubic Surfaces as Blow-Ups To start, let us try to realize cubic surfaces as blow-ups of simpler surfaces.

Let $S \subset \mathbb{P}^3$ be a smooth cubic surface over an arbitrary field k and let $\mathbb{P}^1 \simeq E \subset S$ be a smooth, integral, rational curve. Assume that E is a (-1)-curve, i.e. $(E.E) = -1$ or, equivalently, that E is a line, cf. Remark 1.8. Then S is the blow-up

$$\tau \colon S \longrightarrow \bar{S}$$

of a smooth projective surface $\bar{S}$ in a point $x \in \bar{S}$ with exceptional line E. This is a special case of Castelnuovo's theorem [30, 49, 223]. Alternatively, one may use the linear system $|\mathcal{O}_S(1) \otimes \mathcal{O}(E)|$, which is indeed base point free and contracts the curve E.

More generally, one proves the following.

Lemma 2.1 *Assume $E_1, \ldots, E_m \subset S$ are m pairwise disjoint (-1)-curves. Then S is isomorphic to the blow-up $\tau\colon S \simeq \mathrm{Bl}_{\{x_i\}}(\bar{S}) \longrightarrow \bar{S}$ of a smooth, projective surface $\bar{S}$ with $E_i = \tau^{-1}(x_i)$ as exceptional lines. Furthermore, the following assertions hold.*

(i) *The Picard number of S satisfies $m \leq \rho(S) - 1 \leq 6$.*
(ii) *If $m = 6$, then $\bar{S} \simeq \mathbb{P}^2$.*
(iii) *If $m = 5$, then $\bar{S} \simeq \mathrm{Bl}_x(\mathbb{P}^2)$ or $\bar{S} \simeq \mathbb{P}^1 \times \mathbb{P}^1$.*

Proof Indeed, the blow-up of a smooth surface in one point increases the Picard number by one. As $\bar{S}$ is projective, $\rho(\bar{S}) \geq 1$. This proves the lower bound on $\rho(S)$ in (i). For the upper bound, use Corollary 1.2.

If $m = 6$, then $\bar{S}$ is minimal and its canonical bundle $\omega_{\bar{S}}$ satisfies $\omega_S \simeq \tau^*\omega_{\bar{S}} \otimes \mathcal{O}(\sum E_i)$, where E_i, $i = 1, \ldots, 6$, are the exceptional lines. Thus, $(\omega_{\bar{S}}.\omega_{\bar{S}}) = 9$. Hence, the classification theory of minimal surfaces of Kodaira dimension $-\infty$ proves that $\bar{S} \simeq \mathbb{P}^2$. Ruled surfaces over curves of positive genus can be ruled out, since $H^1(S, \mathcal{O}_S) = 0$.

If $m = 5$, then, similarly, $(\omega_{\bar{S}}.\omega_{\bar{S}}) = 8$. Now, if $\bar{S}$ is not minimal, it can be blown down once more and the resulting surface will then have to be $\mathbb{P}^2$. If $\bar{S}$ is minimal, then by classification theory S is a Hirzebruch surface, i.e. $\bar{S} \simeq \mathbb{F}_n := \mathbb{P}(\mathcal{O} \oplus \mathcal{O}(n))$ over $\mathbb{P}^1$ with $0 \leq n \neq 1$. We need to exclude all the cases $0 < n$. To this end, use the fact that $C_n = \mathbb{P}(\mathcal{O}(n)) \subset \mathbb{F}_n$ is a smooth rational curve with $(C_n.C_n) = -n$. Its strict transform in S is thus a smooth rational curve $\tilde{C}_n$ with self-intersection $(\tilde{C}_n.\tilde{C}_n) \leq -n$. Hence, according to Remark 1.8, we have $n = 0$ or $n = 1$. □

Exercise 2.2 Show that in the case $\bar{S} = \mathrm{Bl}_x(\mathbb{P}^2)$, none of the lines E_i, $i = 1, \ldots, 5$, is mapped to a point in the exceptional line over x.

Remark 2.3 Thus, eventually the situation reduces to the two cases $\tau\colon S \longrightarrow \mathbb{P}^2$ and $\tau\colon S \longrightarrow \mathbb{P}^1 \times \mathbb{P}^1$. They are given by the linear systems $\mathcal{O}_S(1) \otimes \mathcal{O}(\sum_{i=1}^m E_i)$ with $m = 6$ and $m = 5$, respectively. Hence, for degree reasons,

$$\mathcal{O}_S(1) \simeq \tau^*\mathcal{O}_{\mathbb{P}^2}(3) \otimes \mathcal{O}\left(-\sum_{i=1}^{6} E_i\right) \text{ resp. } \mathcal{O}_S(1) \simeq \tau^*\mathcal{O}(2,2) \otimes \mathcal{O}\left(-\sum_{i=1}^{5} E_i\right).$$

Here, $\mathcal{O}(2,2) := \mathcal{O}_{\mathbb{P}^1}(2) \boxtimes \mathcal{O}_{\mathbb{P}^1}(2)$ on $\mathbb{P}^1 \times \mathbb{P}^1$. Numerically, in the second case $\mathcal{O}_S(1)$ could a priori also be, for example, $\tau^*\mathcal{O}(4,1) \otimes \mathcal{O}(-\sum_{i=1}^5 E_i)$. However, in this case the first ruling would lead to a family of lines on S, which we know does not exist.

2.2 Blowing Up $\mathbb{P}^2$ and $\mathbb{P}^1 \times \mathbb{P}^1$ Assume a smooth cubic surface $S \subset \mathbb{P}^3$ contains six pairwise disjoint lines $E_1, \ldots, E_6 \subset S$. The induced classes $[E_i] \in \mathrm{Pic}(S_{\bar{k}}) \simeq \mathrm{I}_{1,6}$ generate a sublattice $\mathrm{I}_{0,6} \subset \mathrm{I}_{1,6}$. In fact, together with $\tau^*\mathcal{O}_{\mathbb{P}^2}(1) \in \mathrm{Pic}(S)$ they form a standard basis of $\mathrm{I}_{1,6}$ and so in particular $\mathrm{Pic}(S) \xrightarrow{\sim} \mathrm{Pic}(S_{\bar{k}}) = \mathrm{I}_{1,6}$. Note that $\mathcal{O}_S(1) = \mathcal{O}_{\mathbb{P}^3}(1)|_S$ and the classes of $E_1, \ldots, E_6$ span a proper sublattice of $\mathrm{Pic}(S)$ of index three.

Thus, as a consequence of Corollary 1.9, we obtain the following classical description of cubic surfaces as blow-ups of $\mathbb{P}^2$. The assumption on k being algebraically closed can be weakened to $\mathrm{Pic}(S)$ being a unimodular lattice of rank seven or, equivalently, $\mathrm{Pic}(S) \simeq \mathrm{I}_{1,6}$.

Proposition 2.4 *Let $S \subset \mathbb{P}^3$ be a smooth cubic surface over an algebraically closed field. Then S is isomorphic to the blow-up $\mathrm{Bl}_{\{x_i\}}(\mathbb{P}^2)$ of $\mathbb{P}^2$ in six distinct points $x_i \in \mathbb{P}^2$, $i = 1, \ldots, 6$.* □

Of course, for a given cubic surface $S \subset \mathbb{P}^3$ described by an explicit polynomial F, it is typically not easy to find the four cubic polynomials $f_i(y_0, y_1, y_2)$, $i = 0, \ldots, 3$, for which the (closure of the) image of the rational map

$$\mathbb{P}^2 \dashrightarrow \mathbb{P}^3,\ [y_0 : y_1 : y_2] \longmapsto [f_0(y_0, y_1, y_2) : \cdots : f_3(y_0, y_1, y_2)]$$

is S. For concrete aspects of this problem, see [392].

Remark 2.5 Assume S is presented as $\mathrm{Bl}_{\{x_i\}}(\mathbb{P}^2)$ as in the proposition. Then there are three sets of curves readily visible that will turn out to be lines:

(i) The exceptional lines $E_1, \ldots, E_6$.
(ii) The strict transforms L_{ij}, $i \neq j$, of the lines $\bar{L}_{ij} \subset \mathbb{P}^2$ passing through $x_i \neq x_j \in \mathbb{P}^2$.
(iii) The strict transforms L_i of any smooth conic $\bar{L}_i \subset \mathbb{P}^2$ passing through the five points $x_j \in \mathbb{P}^2$, $j \neq i$.

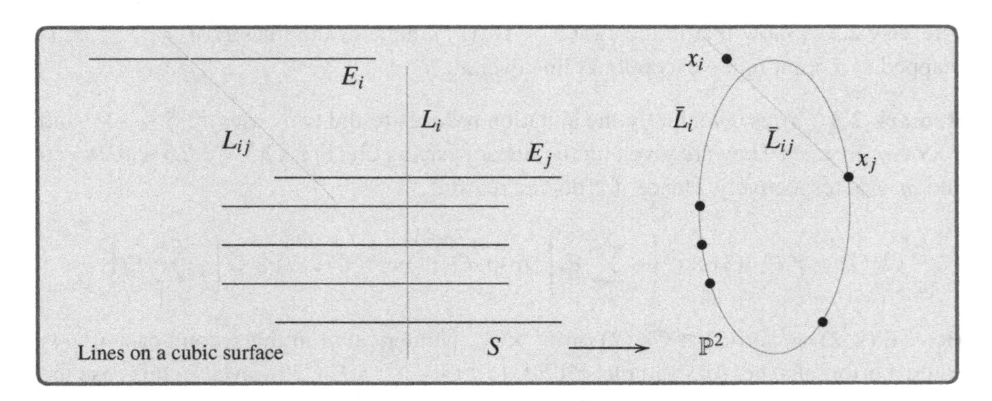

Lines on a cubic surface

Let us count them. There are six curves of type (i) and there are 15 curves of type (ii) assuming that no three points are collinear, i.e. that no $x_k \in \bar{L}_{ij}$ for any k distinct from i and j. To count the curves of type (iii), observe that $|\mathcal{O}_{\mathbb{P}^2}(2)|$ is of dimension five. Hence, for arbitrary five points, there exists a conic C containing them all. This conic C

is either smooth, the union of two distinct lines, or a double line. Hence, again under the assumption that no three of the points $x_1, \dots, x_6$ are collinear, the conic must be smooth and, therefore, there exists a unique L_i for every i. This gives another six curves exactly when the six points are not all contained in one conic.

As it turns out, these conditions are automatically satisfied, see Remark 2.8. Moreover, under these conditions the L_{ij} and the L_i are indeed lines, i.e. $(L_{ij}.L_{ij}) = (L_i.L_i) = -1$, cf. Remark 1.8. Hence, starting with a smooth cubic surface S, the lines of type (i), (ii), and (iii) account for 27, and hence all, lines.

Remark 2.6 One word on the intersection behaviour of these lines. First observe that $E_i \cap E_j = \emptyset$ for $i \neq j$. Similarly, $L_{ij} \cap L_{k\ell} = \emptyset$ for $\{i, j\} \neq \{k, \ell\}$. Also, for $i \neq j$ one has $L_i \cap L_j = \emptyset$ and $(L_i.E_j) = 1$, but $L_i \cap E_i = \emptyset$. Furthermore, $L_{ij} \cap E_k = \emptyset$ for $k \notin \{i, j\}$, but $(L_{ij}.E_i) = 1 = (L_{ij}.E_j)$. Finally, $(L_i.L_{ij}) = 1 = (L_j.L_{ij}) = 1$, cf. Remark 3.1.

Let us now address the converse and consider a blow-up $\tau\colon \mathrm{Bl}_{\{x_i\}}(\mathbb{P}^2) \twoheadrightarrow \mathbb{P}^2$ in six distinct points $x_1, \dots, x_6 \in \mathbb{P}^2$. Is this blow-up then automatically a cubic surface? It turns out that the same conditions on the points $\{x_i\}$, as above, need to be imposed.

Proposition 2.7 *Assume $x_1, \dots, x_6 \in \mathbb{P}^2$ are general in the sense that no three of them are collinear and there is no conic that contains them all. Then the blow-up $\mathrm{Bl}_{\{x_i\}}(\mathbb{P}^2)$ is isomorphic to a cubic surface $S \subset \mathbb{P}^3$.*

Proof More precisely, one shows that the invertible sheaf

$$\mathcal{L} := \tau^*\mathcal{O}(3) \otimes \mathcal{O}\left(-\sum E_i\right)$$

is very ample and that the image of the induced closed embedding

$$\phi_{\mathcal{L}}\colon \mathrm{Bl}_{\{x_i\}}(\mathbb{P}^2) \xrightarrow{\sim} S \subset \mathbb{P}^3$$

is a cubic surface. Here, as before, $E_1, \dots, E_6$ denote the exceptional lines.

Classically, the assertion is proved by showing that $\mathcal{L}$ separates points and tangent directions, cf. [49, 223]. We shall instead give an argument that uses the general Nakai–Moishezon criterion and some of our earlier considerations.

First, note that numerically $(\mathrm{Bl}_{\{x_i\}}(\mathbb{P}^2), \mathcal{L})$ indeed behaves like a cubic surface. By the blow-up formula, its Néron–Severi lattice is isomorphic to $\mathrm{I}_{1,6}$ with $\mathcal{L}$ corresponding to the characteristic vector $(3, -1, \dots, -1)$ and, in particular, $(\mathcal{L}.\mathcal{L}) = 3$. Hence, Corollary 1.11 is valid, see the comment at the end of its proof. In fact, only $(\mathcal{L}.\mathcal{L}) = 3$, $\omega_S \simeq \mathcal{L}^*$, and that $\mathcal{L}$ is effective were needed there.

Therefore, $\mathcal{L}$ is ample if and only if $(\mathcal{L}.L) > 0$ for every $\mathbb{P}^1 \simeq L \subset \mathrm{Bl}_{\{x_i\}}(\mathbb{P}^2)$ with $(L.L) = -1$. If L is one of the exceptional lines, then clearly, $(\mathcal{L}.L) = -(\mathcal{L}.E_i) = 1$. Otherwise, let $D := \tau(L)$ be its image, which is a member of a linear system $|\mathcal{O}_{\mathbb{P}^2}(d)|$ for some d. Denote by $m_i := \mathrm{mult}_{x_i}(D) = (E_i.L)$ the multiplicity of D at the point x_i. For example, $m_i = 0$ if $x_i \notin D$ and $m_i = 1$ if x_i is a smooth point of D. Moreover, L is the strict transform of D and $\tau^*D = L + \sum m_i E_i$. The latter shows

$$d^2 = (D.D) = (\tau^*D.\tau^*D) = -1 + 2\sum m_i - \sum m_i^2 = 5 - \sum(m_i - 1)^2$$

from which we deduce that $d = 1$ or $d = 2$, i.e. D is a line or a conic. Now, $(\mathcal{L}.L) \leq 0$ is equivalent to $3d \leq \sum m_i$, which reads $3 \leq \sum m_i$ for $d = 1$ and $6 \leq \sum m_i$ for $d = 2$. Hence, for $d = 1$, the line D passes through at least three of the points $x_1, \ldots, x_6$. If $d = 2$ and D is a smooth conic, then D contains all six points $x_1, \ldots, x_6$. If $d = 2$ and D is singular, i.e. D consists of two lines, then one of the two contains at least three of the points. However, for general points $x_1, \ldots, x_6$, these two situations are excluded. Hence, $\mathcal{L}$ is indeed ample.

In order to prove that $\mathcal{L}$ is very ample, consider a generic curve in the linear system $|\mathcal{I}_{\{x_i\}} \otimes \mathcal{O}_{\mathbb{P}^2}(3)|$, which, for simplicity, we will assume to be smooth.[1] In other words, we pick a smooth elliptic curve in $\mathbb{P}^2$ passing through $x_1, \ldots, x_6$. Let C be its strict transform, which is still a smooth elliptic curve. Next observe that the restriction map

$$H^0(\mathrm{Bl}_{\{x_i\}}(\mathbb{P}^2), \mathcal{L}) \twoheadrightarrow H^0(C, \mathcal{L}|_C)$$

is surjective, for $H^1(\mathrm{Bl}_{\{x_i\}}(\mathbb{P}^2), \mathcal{O}) = H^1(\mathbb{P}^2, \mathcal{O}) = 0$. Using that $\deg(\mathcal{L}|_C) = 3$ and that any line bundle of degree three on an elliptic curve is very ample, we know that $\mathcal{L}$ is base point free. Thus, since $h^0(\mathcal{L}) = 4$, the line bundle $\mathcal{L}$ defines a morphism $\phi_{\mathcal{L}} \colon \mathrm{Bl}_{\{x_i\}}(\mathbb{P}^2) \longrightarrow \mathbb{P}^3$ and as $(\mathcal{L}.\mathcal{L}) = 3$, it is either of degree one or three. However, the latter would imply that $S := \mathrm{Im}(\phi_{\mathcal{L}})$ is a plane contradicting $h^0(\mathcal{L}) = 4$. Hence, $\phi_{\mathcal{L}}$ is generically injective. The map $\phi_{\mathcal{L}}$ does not contract any curve, as $\mathcal{L}$ is ample, and is therefore the normalization of its image S, a possibly singular cubic surface. However, the natural injection $H^0(S, \mathcal{O}_S(m)) \hookrightarrow H^0(\mathrm{Bl}_{\{x_i\}}(\mathbb{P}^2), \mathcal{L}^m)$ is a bijection, as both spaces are of the same dimension. Using that $\mathcal{L}^m$ is very ample for $m \gg 0$, this suffices to conclude that indeed $\phi_{\mathcal{L}} \colon \mathrm{Bl}_{\{x_i\}}(\mathbb{P}^2) \xrightarrow{\sim} S$. □

For later use, let us record the following: if a cubic surface $S \subset \mathbb{P}^3$ with the restriction of the hyperplane line bundle $\mathcal{O}_S(1)$ is viewed as a blow-up $\tau \colon S = \mathrm{Bl}_{\{x_i\}}(\mathbb{P}^2) \longrightarrow \mathbb{P}^2$ with exceptional lines $E_1, \ldots, E_6$, then

$$\mathcal{O}_S(1) \simeq \tau^*\mathcal{O}(3) \otimes \mathcal{O}(-\textstyle\sum E_i). \tag{2.1}$$

Remark 2.8 The proof also reveals that whenever a smooth cubic surface S is viewed as a blow-up $S = \mathrm{Bl}_{\{x_i\}}(\mathbb{P}^2) \longrightarrow \mathbb{P}^2$, then the points $x_1, \ldots, x_6 \in \mathbb{P}^2$ have to be in general position. This allows one to produce smooth global deformations of cubic surfaces that themselves are not cubic surfaces any longer by letting six points in general position become special, cf. Corollary **1**.3.14.

[1] In fact, by Bertini's theorem with base points, cf. [223, III. Rem. 10.9.2] and [154, Thm. 2.1], this can indeed be achieved.

Dimension check. The choice of six generic points in $\mathbb{P}^2$ modulo the action of PGL(3) accounts for a parameter space of dimension $4 = \dim|\mathcal{O}_{\mathbb{P}^3}(3)| - \dim \mathrm{PGL}(4)$, the dimension of the moduli space of smooth cubic surfaces, cf. Section **1**.2.1.

A similar analysis can be done for blow-ups $\tau\colon S \longrightarrow \mathbb{P}^1 \times \mathbb{P}^1$ in five points. If the five points $x_1, \ldots, x_5 \in \mathbb{P}^1 \times \mathbb{P}^1$ are completely arbitrary, then $\mathcal{L} := \tau^*\mathcal{O}(2,2) \otimes \mathcal{O}(-\sum E_i)$ may not be ample. For example, if two points x_1, x_2 are contained in the same fibre $\bar{F}$ of one of the two projections, then $(\mathcal{L}.F) \leq 0$ for the strict transform F of that fibre. Similarly, not four of them can lie on the diagonal.

Exercise 2.9 Work out the exact conditions for the five points in $\mathbb{P}^1 \times \mathbb{P}^1$ that ensure that the blow-up is a cubic surface.

Dimension check. The choice of five general points on $\mathbb{P}^1 \times \mathbb{P}^1$ modulo the action of $\mathrm{Aut}(\mathbb{P}^1 \times \mathbb{P}^1)$ again accounts for a parameter space of the same dimension four of the moduli space of cubic surfaces.

2.3 Cubic Surfaces As Double Covers We now describe the projection from a point as discussed in general in Section **1**.5.2. So, fix a point $u \in S$ not contained in any line and consider the projection of S from u to a generic plane $\mathbb{P}^2 \subset \mathbb{P}^3$:

$$S \xleftarrow{\ \sigma\ } \tilde{S} := \mathrm{Bl}_u(S) \xrightarrow[2:1]{\ \phi\ } \mathbb{P}^2.$$

It corresponds to the linear system $|\mathcal{I}_u \otimes \mathcal{O}_S(1)|$ on S or, alternatively, to the complete, base point free linear system $|\sigma^*\mathcal{O}_S(1) \otimes \mathcal{O}(-E)|$ on $\tilde{S}$, where $E := \sigma^{-1}(u)$. The fibre $\phi^{-1}(y)$ over $y \in \mathbb{P}^2$ consists of the residual intersection $\{x_1, x_2\}$ of the line $\overline{uy}$ through u and y with S, i.e. $\overline{uy} \cap S = \{u, x_1, x_2\}$.

Thus, as we assumed that u is not contained in any line, $\phi\colon \tilde{S} \longrightarrow \mathbb{P}^2$ is a finite morphism of degree two ramified along the intersection of $S = V(F)$ with the *polar quadric*

$$P_uS := V(\textstyle\sum u_i\, \partial_i F).$$

To see the last assertion, choose coordinates such that $u = [1:0:0:0]$, $\mathbb{P}^2 = V(x_0)$, and $y = [0:1:0:0]$. Then $P_uS = V(\partial_0 F)$, while the intersection of S with the line $V(x_2, x_3)$ through u and y is singular at $z = [z_0 : z_1 : 0 : 0] \neq u$, i.e. z is a branch point of the projection ϕ, if and only if $\partial_0 F(z) = 0$.

Note that $C := S \cap P_uS$ is singular at u. Indeed, the tangent plane of P_uS at $u \in P_uS$ is given by $\sum_i x_i \sum_j u_j\, (\partial_i\partial_j F)(u) = \sum_i x_i \sum_j u_j\, (\partial_j\partial_i F)(u) = 2\sum_i x_i\, (\partial_i F)(u)$, which is also the equation for the tangent plane of S at u. Similarly, one checks that C is smooth at every other point. Thus, the strict transform $\tilde{C} \subset \tilde{S}$ of C is the branch curve of ϕ. Observe that this implies that $\tilde{C}$ is contained in the linear system of $\phi^*\omega_{\mathbb{P}^2}^* \otimes \omega_{\tilde{S}} \simeq \phi^*\mathcal{O}_{\mathbb{P}^2}(3) \otimes \sigma^*\mathcal{O}_S(-1) \otimes \mathcal{O}(E) \simeq \phi^*\mathcal{O}_{\mathbb{P}^2}(2) \simeq \sigma^*\mathcal{O}_S(2) \otimes \mathcal{O}(-2E)$. This confirms $C \in |\mathcal{O}_S(2)|$. Also note that the smoothness of S implies that $\tilde{C}$ is smooth, i.e. C is smooth away from u and has multiplicity two at u. Moreover, $D := \phi(\tilde{C}) \subset \mathbb{P}^2$ is a smooth quartic. The discussion can be seen as a special case of Proposition **1**.5.3.

Remark 2.10 (i) The covering involution of $\tilde{S} \longrightarrow \mathbb{P}^2$ corresponds to taking the inverse on an elliptic curve. Indeed, think of $u \in S$ as the zero point. Then the covering involution maps any other point $y \in S$ to the residual point $z \in S$ of the intersection $\{u, y\} \subset \overline{uy} \cap S$, which is exactly the description of $y \longmapsto -y$ on an elliptic curve. Note that the covering involution on $\tilde{S}$ does not descend to an involution on S, for any point in the intersection $\mathbb{T}_u S \cap S$ would map to u. So, on S we only have a birational involution, the *Geiser involution*.

(ii) There is another type of birational involution, named after Bertini and associated with pairs $(u, y \in \mathbb{T}_u S \cap S)$. The image of a point $z \in S$ is determined by the group structure on the elliptic curve $\overline{uyz} \cap S \subset \overline{uyz} \simeq \mathbb{P}^2$. It turns out that the group of all birational transformations $\mathrm{Bir}(S)$ is generated by Bertini and Geiser involutions and the group $\mathrm{Aut}(S)$ of regular automorphisms, see [335, Thm. 33.7]. The latter has been studied intensively, viewing it as a subgroup of the Weyl group $W(E_6)$. We refer to [158, Ch. 9.5] for details and references. See also Remark **1**.3.20.

Summarizing, the blow-up of a smooth cubic surface in a point not contained in any of the 27 lines is a double cover of $\mathbb{P}^2$ ramified over a smooth quartic curve. The converse of the construction also holds true, as shown by the following.

Proposition 2.11 *Assume $k = \bar{k}$ and let $\phi \colon \tilde{S} \longrightarrow \mathbb{P}^2$ be a double cover ramified along a smooth quartic curve $D \subset \mathbb{P}^2$. Then there exists a (-1)-curve in $\tilde{S}$ the contraction $\tilde{S} \longrightarrow S$ of which is isomorphic to a smooth cubic surface S.*

Proof First note that $\omega_{\tilde{S}} \simeq \phi^*(\omega_{\mathbb{P}^2} \otimes \mathcal{O}_{\mathbb{P}^2}(2)) \simeq \phi^*\mathcal{O}_{\mathbb{P}^2}(-1)$. Next, let $E \subset \tilde{S}$ be an irreducible component of the pre-image of one of the 28 bitangents ℓ of D, cf. Section 3.7 below. We show that E is a (-1)-curve and that its contraction leads to a smooth cubic surface.

Compute the normal bundle $\mathcal{N}_{E/\tilde{S}}$ as the kernel of $\phi^*\mathcal{N}_{\ell/\mathbb{P}^2} \longrightarrow \mathcal{O}_{(D\cap\ell)_{\mathrm{red}}}$ to see that indeed $(E.E) = -1$. Let $\sigma \colon \tilde{S} \longrightarrow S$ be the contraction of E. If $\mathcal{O}_S(1)$ denotes the dual of ω_S, then $\sigma^*\mathcal{O}_S(1) \otimes \mathcal{O}(-E) \simeq \phi^*\mathcal{O}_{\mathbb{P}^2}(1)$. To conclude, one argues as in the proof of Proposition 2.7. First, twisting the structure sequence for $E \subset \tilde{S}$ with $\phi^*\mathcal{O}_{\mathbb{P}^2}(1) \otimes \mathcal{O}(E)$ shows that $h^0(S, \mathcal{O}_S(1)) = 4$. Therefore, the associated linear system defines a map $S \longrightarrow \mathbb{P}^3$, which is readily seen to be regular. Using the ampleness of $\phi^*\mathcal{O}_{\mathbb{P}^2}(1)$, one shows that it is an embedding. Eventually, observe that $(\mathcal{O}_S(1).\mathcal{O}_S(1)) = (\phi^*\mathcal{O}_{\mathbb{P}^2}(1) \otimes \mathcal{O}(E).\phi^*\mathcal{O}_{\mathbb{P}^2}(1) \otimes \mathcal{O}(E)) = 2\,(\mathcal{O}_{\mathbb{P}^2}(1).\mathcal{O}_{\mathbb{P}^2}(1)) + 2\,(\phi^*\mathcal{O}_{\mathbb{P}^2}(1).E) + (E.E) = 3$. □

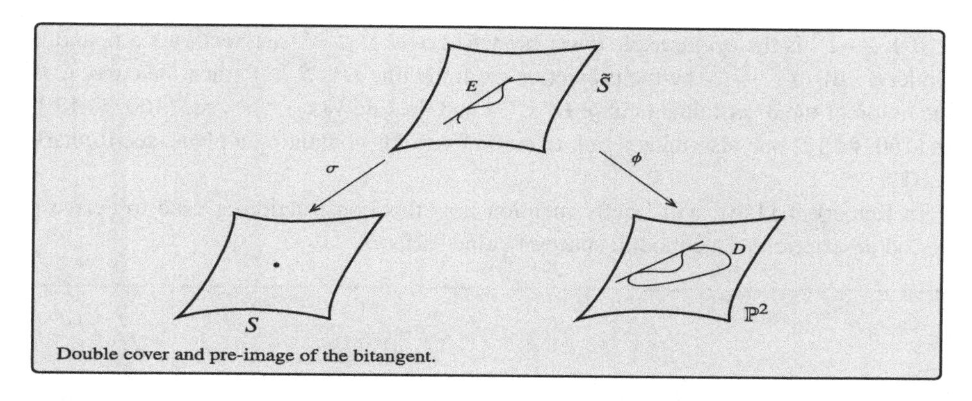

Double cover and pre-image of the bitangent.

Dimension check. The moduli space of smooth curves of genus $g = 3$ is of dimension $3g - 3 = 6$. Canonical embeddings of the non-hyperelliptic ones define smooth plane curves $D \subset \mathbb{P}^2$ of degree four. Cubic surfaces together with the choice of the additional point $u \in S$ needed for the passage to plane quartic curves also make up for a six-dimensional family.

2.4 Conic Fibrations of Cubic Surfaces We apply the general construction of Section **1**.5.1. So, pick a line $L \subset S$ in a smooth cubic surface and consider the linear projection

$$\phi\colon S \twoheadrightarrow \mathbb{P}^1$$

from L to a generic line $\mathbb{P}^1 \subset \mathbb{P}^3$. Usually, the linear projection is only a rational map, but, as L is of codimension one, it is regular in this case or, equivalently, $\mathrm{Bl}_L(S) \xrightarrow{\sim} S$ is an isomorphism. The fibres $\phi^{-1}(y)$ are the residual conics of the intersection $L \subset \overline{yL} \cap S$. In particular, $(\phi^{-1}(y).L) = 2$ and, therefore, $\phi\colon L \twoheadrightarrow \mathbb{P}^1$ is of degree two.

According to Proposition **1**.5.3 there are exactly five singular fibres. Furthermore, the singular fibres $\phi^{-1}(y_i)$, $i = 1, \ldots, 5$, consist of two distinct lines intersecting each other and both intersecting L, see Remark **1**.5.8, (ii), and Section 3.3 below.

Dimension check. The conic fibrations obtained in this way are given by a section of $S^2(\mathcal{F})$ with $\mathcal{F} \simeq \mathcal{O}_{\mathbb{P}^1}(1) \oplus \mathcal{O}_{\mathbb{P}^1}^{\oplus 2}$, see Section **1**.5.1. Now, $\dim \mathbb{P}(H^0(S^2(\mathcal{F}))) = 9$ on which $\mathrm{Aut}(\mathbb{P}^1)$ acts with one-dimensional orbits. The additional action of $\mathrm{Aut}(\mathcal{F})$ eventually cuts the space down to a four-dimensional space.

Remark 2.12 The above construction associates two divisors in $\mathbb{P}^1$ with a line in a cubic surface $L \subset S$: the discriminant divisor $D_L = \{y_1, \ldots, y_5\} \subset \mathbb{P}^1$, which is of degree five, and the branch divisor $R_L \subset \mathbb{P}^1$ of the projection $\phi\colon L \twoheadrightarrow \mathbb{P}^1$, which is of degree two. Dolgachev, van Geemen, and Kondô [160] write these divisors as $D_L = V(F_5(x_0, x_1))$ and $R_L = V(F_2(x_0, x_1))$ and consider the K3 surface obtained as the minimal resolution of the double cover of $\mathbb{P}^2$ branched over the sextic curve $C \subset \mathbb{P}^2$ defined by the reducible curve $x_2 \cdot (x_2^3 \cdot F_2(x_0, x_1) + F_5(x_0, x_1))$.

If $Y \longrightarrow \mathbb{P}^3$ is the cyclic triple cover branched over $S \subset \mathbb{P}^3$, see Section **1**.5.6, and if we let $\phi\colon \mathrm{Bl}_L(Y) \longrightarrow \mathbb{P}^2$ be the projection from the line $L \subset S \subset Y$, then the curve C is the union of the discriminant curve $D_L \subset \mathbb{P}^2$ and the line $V(x_2) \subset \mathbb{P}^2$, see [160, §4.13]. In [160, §4.12] one also finds a link to cubic fourfolds containing a plane, see Remark **6**.1.17.

In Remark 4.11 we will briefly mention how this construction is used to derive a period description of the moduli space of cubic surfaces.

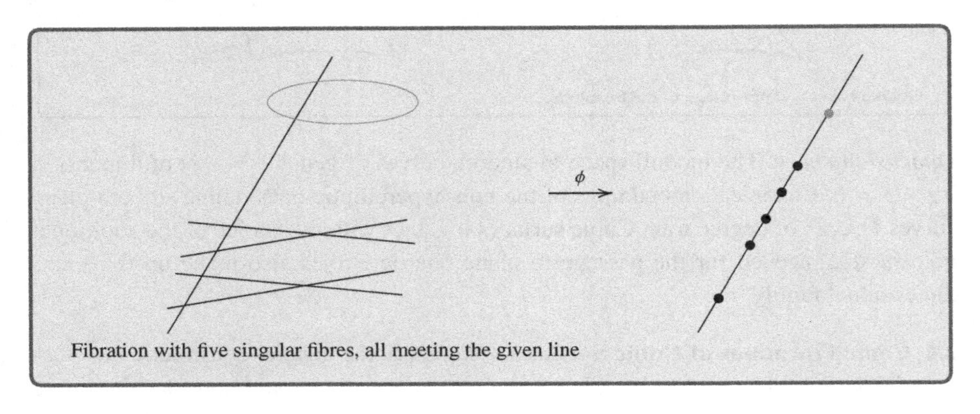

Fibration with five singular fibres, all meeting the given line

Remark 2.13 Lefschetz pencils provide another way of fibring a cubic surface S. As discussed in Remark **1**.5.20 in general, a Lefschetz pencil on S defines a fibration

$$\mathcal{S} \twoheadrightarrow \mathbb{P}^1$$

with 12 singular fibres, each with only one single ordinary double point. The smooth fibres $\mathcal{S}_t = S \cap H_t$, $t \in \mathbb{P}^1$, are smooth plane cubic curves and for a generic choice of the Lefschetz pencil the family is not isotrivial by Proposition **1**.5.18.

The other projection is the blow-up $\mathcal{S} \simeq \mathrm{Bl}_{x_1,x_2,x_3}(S) \twoheadrightarrow S$ in the three base points $x_1, x_2, x_3 \in S$ of the pencil. Alternatively, a Lefschetz pencil is determined by the choice of two generic points $x_1, x_2 \in S$ giving rise to a line $\mathbb{P}^1 \simeq |\mathcal{I}_{x_1,x_2} \otimes \mathcal{O}(1)| \subset |\mathcal{O}(1)|$.

Brown and Ryder [96] classify elliptic fibrations of smooth cubic surfaces up to birational correspondences.

2.5 Pfaffian Cubic Surfaces, Clebsch, and Cayley A cubic surface $S \subset \mathbb{P}^3$ is said to be *Pfaffian* if it is defined by a cubic equation F such that

$$F^2 = \det(A),$$

where A is an alternating matrix of size 6×6 with coefficients in $H^0(\mathbb{P}^3, \mathcal{O}_{\mathbb{P}^3}(1))$, i.e. up to sign the cubic polynomial $F(x_0, \dots, x_3)$ is the Pfaffian of A. Alternatively, the matrix A can be viewed as a map $\mathcal{O}_{\mathbb{P}^3}(-1)^{\oplus 6} \longrightarrow \mathcal{O}_{\mathbb{P}^3}^{\oplus 6}$ that induces a short exact sequence

$$0 \longrightarrow \mathcal{O}_{\mathbb{P}^3}(-1)^{\oplus 6} \xrightarrow{A} \mathcal{O}_{\mathbb{P}^3}^{\oplus 6} \longrightarrow i_*\mathcal{F} \longrightarrow 0.$$

Here, $i\colon S \hookrightarrow \mathbb{P}^3$ is the given closed embedding and $\mathcal{F}$ is a sheaf on S. In Section **6**.1.1 this is considered from a slightly different angle. There, the general alternating form is viewed as a map $W \otimes \mathcal{O}(-1) \longrightarrow W^* \otimes \mathcal{O}$ on the projective space $\mathbb{P}(\bigwedge^2 W^*)$ with $\dim(W) = 6$. Its degeneracy locus is a cubic hypersurface denoted by $\mathrm{Pf}(W^*) \subset \mathbb{P}(\bigwedge^2 W^*)$. Then a cubic surface $S \subset \mathbb{P}^3$ is Pfaffian if it can be written as a linear intersection $\mathrm{Pf}(W^*) \cap \mathbb{P}^3$, cf. Remark **6**.2.5.

A naive dimension count lets one expect every cubic to be Pfaffian, see the proof of [50, Prop. 7.6], and this is indeed the case as proved by Beauville. How to explicitly find the matrix A the Pfaffian of which is the defining equation $F(x_0, \ldots, x_3)$ has subsequently been addressed in [218, 379, 444]. Note that a cubic surface may be represented as a Pfaffian cubic in more than one way, see [50, Cor. 6.4] and [98]. The generic cubic surface can be represented in 72 different ways as Pfaffian and this is related to the choice of six pairwise disjoint lines (double sixes), of which there are 36, see Example 3.6.

Representing general and special cubic surfaces in various forms is a classical topic. The most famous representation of a generic cubic surface is provided by its *Sylvester (or pentahedral) form*:

$$\ell_0^3 + \ell_1^3 + \ell_2^3 + \ell_3^3 + \ell_4^3 = 0. \tag{2.2}$$

Here, $\ell_i = \ell_i(x_0, \ldots, x_3)$ are linear forms, pairwise independent, and uniquely determined up to scaling by cubic roots of unity. This is just a reformulation of Corollary **1**.5.19. The five planes $\mathbb{P}^2 \simeq V(\ell_i) \subset \mathbb{P}^3$ defined by the linear ℓ_i form the Sylvester pentahedron.

Exercise 2.14 Show that a smooth cubic surface described by (2.2), so the generic cubic surface is covered, is isomorphic to a cubic surface in $\mathbb{P}^4$ described by the equations

$$a_0 z_0^3 + \cdots + a_4 z_4^3 = z_0 + \cdots + z_4 = 0, \tag{2.3}$$

see [158, Cor. 9.4.2].

Example 2.15 The *Clebsch (diagonal) cubic surface* [117, §16] is the smooth cubic surface that is defined either by the equation

$$x_0^3 + x_1^3 + x_2^3 + x_3^3 - (x_0 + x_1 + x_2 + x_3)^3 = 0$$

in $\mathbb{P}^3$, i.e. $\ell_i = x_i$ for $i = 0, 1, 2, 3$ and $\ell_4 = -(x_0 + \cdots + x_3)$ in (2.2), or by the equations

$$y_0^3 + \cdots + y_4^3 = y_0 + \cdots + y_4 = 0$$

in $\mathbb{P}^4$. The surface is also called *Klein's icosahedral cubic surface*, since Klein explained how to obtain it as the blow-up of $\mathbb{P}^2$ in six points that correspond to opposite pairs of vertices of an icosahedron [276, §10].

The Clebsch cubic surface comes with a natural $\mathfrak{S}_5$-action. This allows one to realize the 27 lines as the union of two $\mathfrak{S}_5$-orbits. All lines are real and visible in the well-known model of the Clebsch surface. More precisely, starting with the line defined by

$$y_0 = y_1 + y_2 = y_3 + y_4 = 0,$$

permutation of coordinates produces 15 lines (all defined over $\mathbb{Q}$). In other words, 15 lines are described by $y_i = y_j + y_k = y_\ell + y_m$ with $\{i, j, k, \ell, m\} = \{0, 1, 2, 3, 4\}$. The remaining 12 lines are described by the equations

$$y_i + \varphi \cdot y_j + y_k = \varphi \cdot y_i + y_j + y_\ell = -\varphi \cdot (y_i + y_j) + y_4 = 0$$

with $0 \leq i < j \leq 3$ and $\{k, \ell\} \subset \{0, 1, 2, 3\} \setminus \{i, j\}$. Here, $\varphi = (1/2)(1 + \sqrt{5})$ is the golden ratio.

The Clebsch surface is also distinguished by its number of Eckardt points. It is the only smooth cubic surface with exactly 10 Eckardt points, see Section 3.8. The Eckardt points on the Clebsch surface are the points described by the 10 equations $y_i + y_j = y_k = y_\ell = y_m = 0$ for all $\{i, j, k, \ell, m\} = \{0, 1, 2, 3, 4\}$.

Remark 2.16 The modification

$$y_0^3 + y_1^3 + y_2^3 + y_3^3 + (1/4)\, y_4^3 = y_0 + \cdots + y_4 = 0$$

of the equations for the Clebsch surface describes the *Cayley cubic surface*. A more common description of it is given by the equation

$$\prod_{i=0}^{3} x_i \cdot \sum_{i=0}^{3} (1/x_i) = 0.$$

Up to coordinate change, the Cayley cubic is the only cubic surface with four nodes as singularities. Thus, it is the most singular nodal cubic surface, see Remark **1**.5.17.

The Cayley cubic is known to contain only nine lines and only eleven tritangent planes. It is often seen as a hyperplane section of the Segre cubic threefold, see Remark **1**.5.17. For more on the Cayley surface, see [243, Ch. 4], see also Remark 4.6.

Further ways of representing a cubic surface exist. For example, after coordinate change a smooth cubic surface $S \subset \mathbb{P}^3$ can be viewed as the zero set of a polynomial of the form $x_0 \cdot x_1 \cdot x_2 + x_3 \cdot (\sum x_i) \cdot \ell(x_0, \ldots, x_3)$ with ℓ linear [158, Cor. 9.3.3]. More generally, an equation of the type $\ell_1 \cdot \ell_2 \cdot \ell_3 = m_1 \cdot m_2 \cdot m_3$, with ℓ_i and m_j all linear, is called a Cayley–Salmon equation. The generic cubic surface can be written in 120 different ways in this form, see [216] for a recent account. Note that in this form certain lines contained in S can be spotted directly; the nine lines $V(\ell_i, m_j)$ are clearly contained in S. For more results on the representation of cubic surfaces, we refer to the classic [229] and [158, Ch. 9].

3 Lines on Cubic Surfaces

The 27 lines on a cubic surface are among the most studied geometric objects in mathematics. That smooth cubic surfaces contain at most finitely many lines was proved by Cayley in 1849 and then Salmon immediately observed that there are exactly 27 of them. Both papers, with identical titles, appeared in the same volume of the *Cambridge and Dublin Math. Journal* [111, 414]. We recommend the introduction to [229] and the essay [130] for more on the history and a discussion of the various notations.

Once the 27 lines have been found and described geometrically, one can study their configuration from various angles. We collect a few observations starting with the three types of lines (i)–(iii) as introduced in Remark 2.5. In the following, we let S be a smooth cubic surface with fixed six pairwise disjoint lines $E_1, \ldots, E_6$ viewed as the exceptional lines of a contraction $\tau\colon S \twoheadrightarrow \mathbb{P}^2$ over points $x_1, \ldots, x_6 \in \mathbb{P}^2$.

3.1 Lines Are Exceptional Any line $L \subset S$ can be realized as an exceptional line E'_1 of some blowdown $S \twoheadrightarrow \mathbb{P}^2$. In other words, for any line $L \subset S$ there exist five lines $E'_2, \ldots, E'_6$ such that $E'_1 := L, E'_2, \ldots, E'_6$ are pairwise disjoint lines.

This is clear if L is of type (i), i.e. if L is already one of the exceptional lines E_i. If L is of type (ii) or (iii) just observe that $L_{12}, L_{13}, L_{14}, L_{15}, E_6, L_6$ is a collection of pairwise disjoint lines involving at least one line of each type. That the first five are pairwise disjoint is easy and also that E_6 and L_6 are disjoint. To see that $L_6 \cap L_{1j} = \emptyset$ for $j \neq 6$, observe that the intersection of their images, a conic and a line in $\mathbb{P}^2$, satisfies $\bar{L}_6 \cap \bar{L}_{1j} = \{x_1, x_j\}$. Therefore, the intersection is transversal at both points and, hence, the intersection of the strict transforms L_6 and L_{1j} is empty.

Remark 3.1 Note that $L_i \cap L_{ij} \neq \emptyset \neq L_j \cap L_{ij}$. Indeed, either $\bar{L}_i \cap \bar{L}_{ij}$ consists of x_j and another point x distinct from $x_1, \ldots, x_6$ or of x_j with multiplicity two. In the first case, L_i and L_{ij} intersect in x (or rather in the unique point lying above x), while in the second case they meet in the point in E_j corresponding to the common tangent direction of $\bar{L}_i$ and $\bar{L}_{ij}$ at x_j.

3.2 Two Disjoint Lines I Any two disjoint lines are alike, i.e. any two disjoint lines L, L' can be completed to a collection of six pairwise disjoint lines $E'_1 := L, E'_2 := L', E'_3, \ldots, E'_6$, which then can be viewed as the exceptional lines of a blowdown of S to $\mathbb{P}^2$.

Indeed, by Remark 3.1, we only have to consider the following three cases: (i) $L = E_1$ and $L' = E_2$, (ii) $L = E_1$ and $L' = L_{23}$, and (iii) $L = E_1$ and $L' = L_1$. Of course, (i) can be completed by $E_3, \ldots, E_6$, and for (ii) and (iii), use a configuration of the type considered above already: $E_1, L_1, L_{23}, L_{24}, L_{25}, L_{26}$.

Remark 3.2 An elementary counting argument reveals that a smooth cubic surface contains exactly 216 unordered pairs of disjoint lines.

3.3 Ten Lines Intersecting a Given Line For every line $L \subset S$, there exist exactly ten further lines intersecting L. Moreover, these ten lines come in pairs $\{\ell_1, \ell'_1, \ldots, \ell_5, \ell'_5\}$ such that every two pairs, say $\{\ell_1, \ell'_1\}$ and $\{\ell_2, \ell'_2\}$, are disjoint, i.e. $(\ell_1 \cup \ell'_1) \cap (\ell_2 \cup \ell'_2) = \emptyset$. Furthermore, all triangles L, ℓ_i, ℓ'_i are coplanar, i.e. there exists a plane $\mathbb{P}^2 \subset \mathbb{P}^3$ with $S \cap \mathbb{P}^2 = L \cup \ell_i \cup \ell'_i$ and, in particular, $\ell_i \cap \ell'_i \neq \emptyset$.

According to Section 3.1, we may assume $L = E_6$. Going through the list, one finds that indeed E_6 intersects only $L_{16}, L_{26}, L_{36}, L_{46}, L_{56}$ and L_1, L_2, L_3, L_4, L_5. We let $\ell_i := L_{i6}$ and $\ell'_i := L_i$, $i = 1, \ldots, 5$.

Then check that for $i \neq j \in \{1, \ldots, 5\}$, for example $i = 1$, $j = 2$, one has $L_{i6} \cap L_{j6} = \emptyset$, $L_{i6} \cap L_j = \emptyset$ (see the arguments in Section 3.1), and $L_i \cap L_j = \emptyset$. For the last one use the fact that, for example, $\bar{L}_1 \cap \bar{L}_2$ consists of the four points $x_3, \ldots, x_6$. Hence, the intersection is transversal and, therefore, the intersection $L_1 \cap L_2$ of their strict transforms is empty. It remains to verify that L, ℓ_i, ℓ'_i are coplanar. For this, assume $i = 1$ and observe that

$$\begin{aligned} & \mathcal{O}(E_6) \otimes \mathcal{O}(L_{16}) \otimes \mathcal{O}(L_1) \\ \simeq \;& \mathcal{O}(E_6) \otimes (\tau^*\mathcal{O}(1) \otimes \mathcal{O}(-E_1 - E_6)) \otimes \left(\tau^*\mathcal{O}(2) \otimes \mathcal{O}(-\textstyle\sum_{i>1} E_i)\right) \\ \simeq \;& \tau^*\mathcal{O}(3) \otimes \mathcal{O}(-\textstyle\sum E_i) \simeq \mathcal{O}_{\mathbb{P}^3}(1)|_S, \end{aligned}$$

cf. the proof of Proposition 2.7. The implication $\ell_i \cap \ell'_i \neq \emptyset$ can be seen more directly and more geometrically. As an aside, note that the plane containing L, ℓ_i, ℓ'_i is tangent at the points of intersection of each pair of these three lines.

To prove the existence of the five pairs of lines intersecting $L \subset S$ one could alternatively use the linear projection $\phi\colon S \dashrightarrow \mathbb{P}^1$ from L, see Section 2.4. They occur as the five singular fibres $\phi^{-1}(y_i) = \ell_i \cup \ell'_i$.

Remark 3.3 (i) Each of the coplanar unions $L \cup \ell_i \cup \ell'_i$ is either a triangle, i.e. it has three singular points, or consists of three lines all going through one point. This corresponds to the two possibilities that the line $\bar{L}_{i6}$ and the conic $\bar{L}_i$ intersect transversally in x_6 or with multiplicity two, so that $\bar{L}_{i6}$ is tangent to $\bar{L}_i$ at x_6.

(ii) From the above count, we deduce that every smooth cubic surface admits exactly 45 *tritangent planes*, i.e. planes that intersect the cubic in the union of three pairwise distinct lines. The additional 15 constellations not of the above form are $L_{i_1i_2} \cup L_{i_3i_4} \cup L_{i_5i_6}$ with $\{i_1, \ldots, i_6\} = \{1, \ldots, 6\}$.

(iii) It is possible to show that there are 9 of the 45 tritangent planes that cut out all the 27 lines contained in S. In this sense the union of all lines on a cubic surface is described as the intersection with a (highly degenerate) surface of degree 9.

(iv) An elementary counting argument shows that a smooth cubic surface contains exactly 135 pairs of intersecting lines [229, Ch. 1.3]. Equivalently, on a smooth cubic surface without Eckardt points, see Section 3.8, there exist exactly 135 points that occur as the intersection of two lines contained in the surface.

3.4 Lines Generating the Picard Group There exist explicit bases of the lattices $\mathrm{Pic}(S) \simeq \mathrm{I}_{1,6}$ and $\mathcal{O}_S(1)^\perp \simeq \mathrm{E}_6(-1)$ that can be expressed as integral linear combinations of the 27 lines.

Let $f_0, \ldots, f_6$ denote the standard basis of $\mathrm{I}_{1,6}$, i.e. $f_i = -[E_i]$, $i = 1, \ldots, 6$, and f_0 corresponds to a line in $\mathbb{P}^2$. Then consider a tritangent plane, for example, $\overline{E_6 L_1 L_{16}}$. Its intersection with S gives the class $3f_0 + f_1 + \cdots + f_6$. So, the classes of $E_1, \ldots, E_6$ together with $[E_6] + [L_1] + [L_{16}]$ already generate a sublattice of $\mathrm{I}_{1,6}$ of index three. To generate all of $\mathrm{I}_{1,6}$ by lines, observe that $(L_1.E_i) = 1$, $i = 2, \ldots, 6$, $(L_1.E_1) = 0$, and, therefore, $L_1 = 2f_0 + 0f_1 + f_2 + \cdots + f_6$.

Spelling out the comments in the proof of Proposition **1**.1.21, a basis of the definite lattice $\mathcal{O}_S(1)^\perp \simeq E_6(-1)$ is then given by $e_1 = E_1 - E_2, e_2 = E_2 - E_3, e_3 = E_3 - E_4, e_5 = E_4 - E_5, e_6 = E_5 - E_6$, and $e_4 = E_1 - E_4 - E_5 + L_{16}$.

Remark 3.4 The lattice E_6 has 72 roots which can be written down explicitly in terms of the above bases, see [158, Sec. 8.2.3] or more explicitly [394, Remark 2.6].

3.5 Two Disjoint Lines II For any pair of disjoint lines L, L', there exist exactly five lines $\ell_1, \ldots, \ell_5$ meeting both. Moreover, those five lines are pairwise disjoint.

According to Section 3.2, we may assume $L = E_1$ and $L' = E_2$. The lines meeting E_1 are

$$L_{12}, L_{13}, L_{14}, L_{15}, L_{16}, L_2, L_3, L_4, L_5, L_6$$

and those meeting E_2 are

$$L_{12}, L_{23}, L_{24}, L_{25}, L_{26}, L_1, L_3, L_4, L_5, L_6.$$

Hence, the ones meeting both lines, E_1 and E_2, are precisely $L_{12}, L_3, L_4, L_5, L_6$, which we have seen to be pairwise disjoint already.

This collection of five pairwise disjoint lines is special and not at all like, for example, the lines $E_1, \ldots, E_5$. Namely, there is no further line disjoint to all of the lines $L_{12}, L_3, L_4, L_5, L_6$. Indeed, the lines E_i all intersect at least one of them. The lines L_{1j}, $j = 3, \ldots, 6$, intersect L_j, cf. Remark 3.1. The lines L_{ij}, $2 < i < j$, intersect L_{12}, and, again by Remark 3.1, L_1, L_2 also both intersect L_{12}. As a consequence of Lemma 2.1, we obtain the next result.

Corollary 3.5 *Any pair of disjoint lines $L, L' \subset S$ gives rise to a blowdown map*

$$S \twoheadrightarrow L \times L' \simeq \mathbb{P}^1 \times \mathbb{P}^1$$

contracting exactly the five lines intersecting both lines L and L'. □

There is a very geometric way of describing this blowdown, cf. [49]. Namely, for any point $x \in S \setminus (L \cup L')$, the plane $\overline{xL'}$ spanned by L' and x intersects L in exactly one point u_x. Similarly, $\overline{xL}$ intersects L' in a unique point u'_x. This defines a map

$$S \setminus (L \cup L') \longrightarrow L \times L', \ x \longmapsto (u_x, u'_x),$$

which can be extended to all of S by replacing $\overline{xL'}$ for $x \in L'$ by the tangent plane T_xS (which contains L'). Also, in this description one sees that exactly the lines $\ell_1, \ldots, \ell_5$ are contracted. Their images are the points (u_i, u'_i), where $\ell_i \cap L = \{u_i\}$ and $\ell_i \cap L' = \{u'_i\}$. The inverse of the birational map $L \times L' \dashrightarrow S$ studied in Example **1**.5.12.

3.6 Configuration of Lines Consider the *configuration*

$$\mathcal{L} := \mathcal{L}(S) := \{ \ell_1, \ldots, \ell_{27} \}$$

of all lines contained in a cubic surface S. By definition, it not only encodes the set of all lines, but also their intersection numbers (but not, for example, the intersection points and, in particular, not whether there are triple intersection points), and is independent of the actual surface S. Alternatively, view $\mathcal{L}$ as the graph with vertices corresponding to the 27 lines ℓ_i and with two vertices ℓ_i, ℓ_j connected if the two lines intersect. Its complement, i.e. the graph with the same set of vertices but with vertices connected by an edge if and only if they are not connected in $\mathcal{L}$, is the so-called *Schläfli graph*, checkout Wikipedia for graphical renderings of it.

Note that if the first six lines $\ell_1, \ldots, \ell_6$ are chosen to be the exceptional lines of a blow-up $S \longrightarrow \mathbb{P}^2$, i.e. $\ell_1 = E_1, \ldots, \ell_6 = E_6$, then all the remaining 21 lines are uniquely determined. For example, L_{12} is the unique line that intersects ℓ_1 and ℓ_2 but no $\ell_3, \ldots, \ell_6$ and L_1 is the unique line that intersects $\ell_2, \ldots, \ell_6$ but not ℓ_1. Moreover, according to Lemma 2.1, any subset $\{\ell_{i_1}, \ldots, \ell_{i_6}\}$ of six pairwise disjoint lines can be realized as the exceptional lines of a blow-up $S \longrightarrow \mathbb{P}^2$. In other words, for any two choices $\ell_1, \ldots, \ell_6$ and $\ell'_1, \ldots, \ell'_6$ of six pairwise disjoint lines, there exists a unique automorphism $g\colon \mathcal{L} \xrightarrow{\sim} \mathcal{L}$ of the configuration with $g(\ell_i) = \ell'_i$. Thus, choosing six pairwise disjoint lines $\ell_1, \ldots, \ell_6$ is equivalent to giving an element in $\mathrm{Aut}(\mathcal{L})$. This allows one to compute the order

$$|\mathrm{Aut}(\mathcal{L})| = 27 \cdot 16 \cdot 10 \cdot 6 \cdot 2 = 2^7 \cdot 3^4 \cdot 5 = 51\,840.$$

Indeed, there are 27 choices for E_1, then 16 choices for E_2, etc. Of course, it is no coincidence that this number is the order of the Weyl group $W(E_6)$, cf. Corollary 1.14 and Remark 3.8.

The configuration of lines presented by the entries of the matrix

$$\begin{pmatrix} E_1 & E_2 & E_3 & E_4 & E_5 & E_6 \\ L_1 & L_2 & L_3 & L_4 & L_5 & L_6 \end{pmatrix}$$

is what is called a *Schläfli double six*. It has the property that each of the 12 lines intersects exactly those lines in the matrix that are neither contained in the same row nor in the same column. As straightforward count reveals that there are exactly 30 points that occur as intersection points of two of the lines in a Schläfli double six.

Example 3.6 There exist 36 Schläfli double sixes in each smooth cubic surface, see [158, Ch. 9.1].

It is a classical fact that the choice of a double six of lines in $\mathbb{P}^3$ contained in a cubic surface S determines the surface uniquely. In fact, for any given five skew lines $\ell_1, \dots, \ell_5$ in $\mathbb{P}^3$ and one, say ℓ, that intersects them all (think of $E_1, \dots, E_5$ and L_6), there exists a unique cubic surface containing the six lines as part of a double six. Indeed, the curve $D := \ell \cup \bigcup \ell_i$ satisfies $h^0(\mathcal{O}_D(3)) = 19$, which together with $h^0(\mathbb{P}^3, \mathcal{O}(3)) = 20$ and using the short exact sequence $0 \longrightarrow \mathcal{I}_D(3) \longrightarrow \mathcal{O}_{\mathbb{P}^3}(3) \longrightarrow \mathcal{O}_D(3) \longrightarrow 0$ essentially implies the claim. See [231, Sec. 25] or [268] for further details.

In Section **1**.2.5 we have seen that the monodromy group of the family of all smooth cubics is the Weyl group $W(E_6)$, see also Corollary 1.14. As the discriminant divisor has degree 32, see Theorem **1**.2.2, the Weyl group is generated by 32 reflections. Coxeter [128] showed that $W(E_6)$ can also be generated by six reflections and one transformation that is given by interchanging the two rows of a double six.

3.7 Lines versus Bitangents for Double Covers Let us now make use of the description of a cubic surface S as a double cover of $\mathbb{P}^2$, cf. Section 2.3. We fix a point $u \in S$ not contained in any of the lines, consider the blow-up $\sigma\colon \tilde{S} = \mathrm{Bl}_u(S) \longrightarrow S$, and let

$$\phi\colon \tilde{S} = \mathrm{Bl}_u(S) \longrightarrow \mathbb{P}^2$$

be the projection onto a generic plane. We denote the exceptional line of the blow-up by E and the ramification curve by $D := \phi(\tilde{C}) \subset \mathbb{P}^2$, a smooth quartic curve. We will now establish the natural bijection between the 28 *bitangent lines* of D and the 27 lines in S together with E:

$$\{\, \ell \subset \mathbb{P}^2 \mid \text{bitangents to } D \,\} \longleftrightarrow \{\, \ell_1, \dots, \ell_{27} \subset S \mid \text{lines} \,\} \cup \{\, E \,\}. \tag{3.1}$$

First, observe that each line $\ell_i \subset S$, simultaneously considered as a curve in $\tilde{S}$, satisfies $1 = (\ell_i.\mathcal{O}_S(1)) = (\ell_i.\phi^*\mathcal{O}_{\mathbb{P}^2}(1))$. Hence, $\bar{\ell}_i := \phi(\ell_i) \subset \mathbb{P}^2$ is a line and $\phi\colon \ell_i \xrightarrow{\sim} \bar{\ell}_i$ is an isomorphism. Similarly, $(E.\phi^*\mathcal{O}_{\mathbb{P}^2}(1)) = 1$ and, therefore, $E \xrightarrow{\sim} \bar{E} := \phi(E) \subset \mathbb{P}^2$ is also a line. However, lines in $\mathbb{P}^2$ whose pre-images under ϕ split off a copy of the line cannot intersect D transversally at any point. Hence, all the lines $\bar{\ell}_i$ and $\bar{E}$ are bitangent to D.[2] Thus, (3.1) follows from the next result.[3]

Lemma 3.7 *Let $D \subset \mathbb{P}^2$ be a smooth quartic curve over an algebraically closed field k with* $\mathrm{char}(k) \neq 2$. *Then D admits exactly* 28 *bitangent lines.*

Proof As a first step, one observes that $\omega_D \simeq \mathcal{O}_{\mathbb{P}^2}(1)|_D$. Therefore, a bitangent through $x, y \in D$ (or a hyperflex through $x = y \in D$) corresponds to an invertible sheaf $N \in \mathrm{Pic}^2(D)$ with $H^0(D, N) \neq 0$ and $N^2 \simeq \omega_D$. Here, we tacitly use the fact that a line bundle N with $N^2 \simeq \omega_D$ satisfies $h^0(N) \leq 1$.

The number of square roots of ω_D, called theta characteristics, is of course $2^{2g(D)} = 64$. However, only 28 of them are effective. To deduce this from the general theory of theta characteristics one can use that on a smooth projective curve of genus g there are exactly $2^{g-1} \cdot (2^g + 1)$ even and $2^{g-1} \cdot (2^g - 1)$ odd theta characteristics, see [25, 358]. By definition, whether a theta characteristic N is even or odd is determined by the parity of $h^0(N)$. Therefore, in this case a theta characteristic is effective if and only if it is odd. Hence, there are exactly $2^2 \cdot (2^3 - 1) = 28$ of them.

Alternatively, one could use the Plücker formula for smooth curves $C \subset \mathbb{P}^2$ with only bitangents and simple flexes. It turns out that there exist 24 flexes and 28 bitangents, see [219, Sec. II]. □

As $\phi^{-1}(\bar{\ell}_i) \longrightarrow \bar{\ell}_i$ is of degree two, $\phi^{-1}(\bar{\ell}_i) = \ell_i \cup \ell_i'$ with $\phi\colon \ell_i' \xrightarrow{\sim} \bar{\ell}_i$. The two curves ℓ_i and ℓ_i' intersect in the pre-image of the points of contact $\bar{\ell}_i \cap D$. Note that ℓ_i' does not correspond to a line in S, as two lines in S intersect in at most one point and there transversally. Instead, $(\sigma(\ell_i').\mathcal{O}_S(1)) = 2$ and $(\ell_i'.E) = 1$, i.e. $u \in S$ is a smooth point of the curve $\sigma(\ell_i')$. Indeed, $\ell_i \cup \ell_i' = \phi^{-1}(\bar{\ell}_i)$ is a curve in the linear system of $\phi^*\mathcal{O}_{\mathbb{P}^2}(1) \simeq \sigma^*\mathcal{O}_S(1) \otimes \mathcal{O}(-E)$. Hence, $2 = (\ell_i \cup \ell_i'.\phi^*\mathcal{O}_{\mathbb{P}^2}(1)) = 1 + (\ell_i'.\sigma^*\mathcal{O}_S(1) \otimes \mathcal{O}(-E))$. As ℓ_i' is not a line and, hence, $(\ell_i'.\mathcal{O}_S(1)) > 1$, one has $(\ell_i'.E) \geq 1$ and in fact $(\ell_i'.E) = 1$, because the two lines $\phi(\ell_i') = \bar{\ell}_i$ and $\phi(E) = \bar{E}$ intersect in one point only and there transversally.

For example, for the line $\bar{L}_{12}$ there exists a unique conic Q through x_3, x_4, x_5, x_6 and $\tau(u)$ that intersects $\bar{L}_{12}$ in two points distinct from x_1, x_2. Here, as before, we view S as the blow-up $\tau\colon S \twoheadrightarrow \mathbb{P}^2$ in six points $x_1, \ldots, x_6 \in \mathbb{P}^2$. The strict transform $\tilde{Q} \subset \tilde{S}$ of Q is contained in the linear system of $\sigma^*(\tau^*\mathcal{O}(2) \otimes \mathcal{O}(-\sum_{i \neq 1,2} E_i)) \otimes \mathcal{O}(-E)$. This line

[2] By definition, a bitangent of D is a line in $\mathbb{P}^2$ that intersects D in two points x, y with multiplicity (at least) two. The case $x = y$ is allowed, in which the bitangent has multiplicity four at this point. This is sometimes also called a hyperflex. The locus of smooth quartic curves with a hyperflex is a divisor in the moduli space of curves of genus three, cf. [131, 239].

[3] The 27 lines contained on a cubic surfaces $S \subset \mathbb{P}^3$ and the 28 bitangent lines to plane quartic curve $D \subset \mathbb{P}^2$ are part of one of Arnold's trinity, joined by the 120 tritangent planes to a canonical curve of genus four $C \subset \mathbb{P}^3$. The latter will naturally come up again as curves of lines contained in a cubic fourfold passing through a fixed point. As in the proof here, they are accounted for by the odd theta characteristics on C.

bundle is indeed isomorphic to $\phi^*\mathcal{O}_{\mathbb{P}^2}(1) \otimes \mathcal{O}(-L_{12})$, cf. (2.1), and, hence, $L_{12} \cup \tilde{Q} = \phi^{-1}(\phi(L_{12}))$. Note that $\tilde{Q}$ is a (-1)-curve in $\tilde{S}$, but not its image in S.

Remark 3.8 For the universal family $\mathcal{D} \longrightarrow U := |\mathcal{O}_{\mathbb{P}^2}(4)|_{\rm sm}$ of smooth quartic curves in $\mathbb{P}^2$, the relative family of bitangents

$$B(\mathcal{D}/U) \longrightarrow U$$

is an étale map of degree 28. Its Galois group or, equivalently, its monodromy group $\mathrm{Im}(\pi_1(U) \longrightarrow \mathfrak{S}_{28})$ is isomorphic to $\mathrm{Sp}_6(\mathbb{Z}/2\mathbb{Z})$, which is of order $288 \cdot 7! = 2^9 \cdot 3^4 \cdot 5 \cdot 7$, see [219, Sec. II:4]. Compare this to Corollary 1.14 and the discussion in Section **1**.2.5. For example, to compute the monodromy group of the 27 lines on the universal family of smooth cubic surfaces one has to consider the stabilizer of one of the 28 bitangents, which amounts to fixing the exceptional curve blown down to the point $u \in S$. Hence, the order is $288 \cdot 7!/28 = 2^7 \cdot 3^4 \cdot 5$, which confirms Corollary 1.14.

For a short historic account of the interplay between lines on cubic surfaces and bitangents to quartic curves, we also recommend [462, Ch. 7].

3.8 Eckardt Points A point $x \in S$ in a smooth cubic surface S is called an *Eckardt point* if the tangent plane at $x \in S$ intersects S in three lines through x or, equivalently, if x is contained in three lines, cf. Section 3.3. How many Eckardt points can a smooth cubic surface have? Since there exist only 45 tritangent planes, see Remark 3.3, there are no more than 45 Eckardt points. In fact, by a result of Hirschfeld [233], in characteristic two this maximum is attained. However, in any other characteristic the maximum is 18, see below.

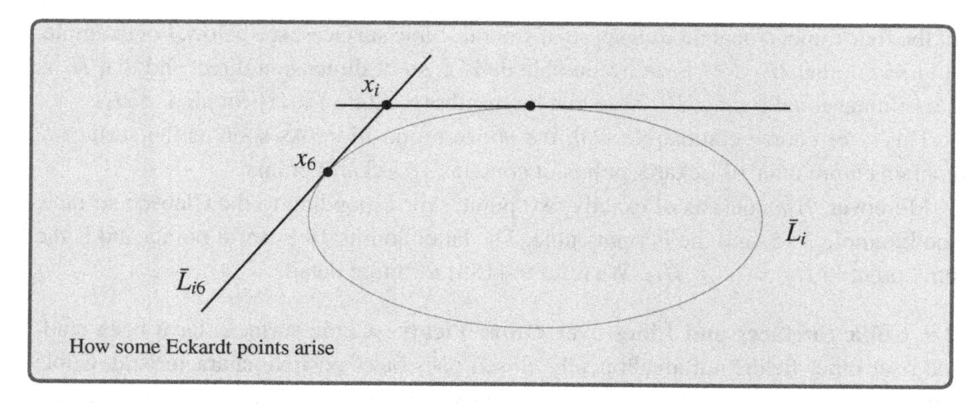

How some Eckardt points arise

In Remark 3.3 we have seen examples of Eckardt points, namely three lines consisting of an exceptional line E_6, the strict transform L_i, $i \neq 6$, of the conic $\bar{L}_i$, which contains x_6, and the strict transform L_{i6} of the line $\bar{L}_{i6}$ tangent to $\bar{L}_i$ (at x_6). As for each i there exist only two lines through x_i tangent to $\bar{L}_i$ at some point, each conic $\bar{L}_i$ will give rise to at most two Eckardt points. So altogether, there exist at most 12 Eckardt points of this type.

However, Eckardt points may also arise in a different way namely as the triple intersection $\bar{L}_{i_1 i_2} \cap \bar{L}_{i_3 i_4} \cap \bar{L}_{i_5 i_6}$ with $\{i_1, \dots, i_6\} = \{1, \dots, 6\}$. Generically, such triple intersection would be empty, but star-shaped configurations are of course possible.[4] Note that in this case the three lines $L_{i_1 i_2}$, $L_{i_3 i_4}$, and $L_{i_5 i_6}$ are not coplanar.

For generic choices of points $x_1, \dots, x_6 \in \mathbb{P}^2$, one does not expect any of these two possibilities to occur. Namely, neither will any of the conics $\bar{L}_i$ be tangent to any line $\bar{L}_{ij}$ nor will the line configuration show stars.

The following is a result of Eckardt [169]. Further numbers of Eckardt points can be realized over finite fields [67, 234].

Proposition 3.9 *The number of Eckardt points on a cubic surface S over an algebraically closed field of characteristic zero is* 0, 1, 2, 3, 4, 6, 9, 10, *or* 18.

Remark 3.10 First we want to mention the classical result that an Eckardt point of a cubic surface S induces a non-trivial automorphism of order two, see [158, Prop. 9.1.13].

The loci $H_k \subset |\mathcal{O}_{\mathbb{P}^3}(3)|_{\rm sm}$ of smooth cubic surfaces with at least k Eckardt points have been studied by Nguyen [455] (see also the author's thesis) and later in more detail by Keneshlou [271]. They are invariant under the action of PGL(4) and, thus, determine closed subschemes

$$\bar{H}_k := H_k/\mathrm{PGL}(4) \subset M = M_{3,2}$$

of the four-dimensional moduli space of smooth cubic surfaces, see below. For example, it turns out that $\bar{H}_1 \subset M$ is an irreducible divisor, so of dimension three, and that $\bar{H}_k$ is zero-dimensional for $k \geq 10$. Note that by the above, $\mathrm{Aut}(S) \neq \{1\}$ for all $S \in H_1$.

This is of course compatible with the above proposition. As soon as the surface S contains more than 10 Eckardt points, it contains 18 Eckardt points.

Moreover, $\bar{H}_{10}$ consists of exactly two points, corresponding to the Clebsch surface, see Example 2.15, and the Fermat cubic. The latter admits 18 Eckardt points and is the only point in $\bar{H}_{11} = \cdots = \bar{H}_{18}$. We refer to [158] for more details.

3.9 Cubic Surfaces and Lines over Other Fields Cubic surfaces have been studied over other fields, not algebraically closed ones or of positive characteristic. Cubic surfaces over the field of real numbers have received particular and sometimes artistic attention [263, 491].

[4] I would expect some combinatorial argument to show that at most six points can occur in this way. However, the next result may be valid without it. A priori it could happen that whenever there are more stars in the line configuration associated with the six points, then fewer conics $\bar{L}_i$ are tangent to those lines.

A classical result of Segre [420] states that the number N_S of lines contained in a smooth cubic surface $S \subset \mathbb{P}^3$ over an arbitrary field k is $N_S = 0, 1, 2, 3, 5, 7, 9, 15$, or 27 and all those numbers are realized by smooth cubic surfaces over $k = \mathbb{Q}$. For a modern account of the proof, see [350]. Kass and Wickelgren [269] developed a motivic count of lines over arbitrary fields.

Schläfli [417] had proved earlier that for $k = \mathbb{R}$ only $N_S = 3, 7, 15$, or 27 are possible. In the nineteenth century Clebsch [118], Klein [277], Schläfli [417], Cayley [112], and others were interested in actually constructing real cubic surfaces exhibiting all 27 lines in an instructive and appealing way, cf. [229, §20]. The plaster model of the Clebsch diagonal surface, see Example 2.15, can be found in many mathematics departments around the world.

4 Moduli Space

In higher dimensions, moduli spaces of cubic hypersurfaces are geometrically not completely understood. But the situation is much better for cubic surfaces. In this case, not only can the moduli space, as a quasi-projective GIT quotient, be described quite explicitly, but it also has been investigated as an arithmetic quotient of a period domain. We outline the main features but refer to the original literature for more details. It is fascinating that a classical object like the moduli space of cubic surfaces has been a topic of recent and quite beautiful research.

4.1 GIT Description Recall from Section **3**.1.3 that the moduli space of smooth cubic surfaces, say over an algebraically closed field of characteristic zero, is constructed as the quotient

$$M_2 := M_{3,2} := |\mathcal{O}_{\mathbb{P}^3}(3)|_{\mathrm{sm}} // \mathrm{SL}(4),$$

which is an open dense subset of the GIT quotient $|\mathcal{O}_{\mathbb{P}^3}(3)|^{\mathrm{ss}} // \mathrm{SL}(4)$ of the open subset of all semi-stable cubic surfaces. The latter has been introduced as the projective scheme $\mathrm{Proj}(k[a_I]^{\mathrm{SL}(4)})$. Here, the variables a_I, with $I = (i_0, \ldots, i_3)$ and $\sum i_j = 3$, are the universal coefficients of the universal cubic hypersurface, see Section **1**.2.1.

Note that, as $\dim |\mathcal{O}_{\mathbb{P}^3}(3)| = 19$ and $\dim \mathrm{SL}(4) = 15$, the moduli spaces M_2 is of dimension four.

The following result is a highlight of invariant theory in the nineteenth century. The art of computing rings of invariant polynomials has been almost forgotten and, indeed, it is typically quite difficult to perform in concrete situations. Salmon's original computation [416] of the ring of invariant quaternary cubic polynomials is nearly incomprehensible for the modern reader, but it was rewritten by Beklemishev [61] and given a detailed exposition with minor corrections by Reinecke [399].

Theorem 4.1 (Salmon) *The invariant ring* $S := k[a_I]^{\mathrm{SL}(4)}$ *is generated by polynomials* $I_8, I_{16}, I_{24}, I_{32}, I_{40}, I_{100} \in k[a_I]$ *of degree* $\deg(I_d) = d$. *Furthermore, the first five of these polynomials are algebraically independent, while for the last one we have* $I_{100}^2 \in k[I_8, I_{16}, I_{24}, I_{32}, I_{40}]$.

For cubic surfaces in Sylvester form (2.3), the polynomials I_d are explicitly given as $I_8 = \sigma_4^2 - 4\sigma_3\sigma_5$, $I_{16} = \sigma_1\sigma_5^3$, $I_{24} = \sigma_4\sigma_5^4$, $I_{32} = \sigma_2\sigma_5^6$, $I_{40} = \sigma_5^8$, and $I_{100} = \sigma_5^{18}v$. Here, σ_i are the elementary symmetric functions in the coefficients a_i of (2.3) and $v = \prod_{i<j}(a_j - a_i)$ is the Vandermonde determinant. Eventually, the ring of invariants $k[a_I]^{\mathrm{SL}(4)}$ is computed as a subring of the ring of invariants $k[a_0, \ldots, a_4]^H$ on the affine section of all cubics in Sylvester form under a finite group H.

Corollary 4.2 *The moduli space of semi-stable cubic surfaces is naturally isomorphic to the four-dimensional weighted projective space* $\mathbb{P}(1,2,3,4,5)$, *so*

$$|\mathcal{O}_{\mathbb{P}^3}(3)|^{ss}/\!/\mathrm{SL}(4) \simeq \mathbb{P}(1,2,3,4,5).$$

Proof Salmon's theorem shows that the invariant ring $S := k[a_I]^{\mathrm{SL}(4)}$ is generated by algebraically independent polynomials $I_8, I_{16}, I_{24}, I_{32}, I_{40}$ and a further polynomial I_{100} with $I_{100}^2 \in k[I_8, I_{16}, I_{24}, I_{32}, I_{40}]$. Since $8 \nmid 100$, one finds that the twisted ring $S^{(8)}$ is isomorphic to $k[I_8, I_{16}, I_{24}, I_{32}, I_{40}]$. This proves the assertion

$$\begin{aligned} |\mathcal{O}_{\mathbb{P}^3}(3)|^{ss}/\!/\mathrm{SL}(4) &\simeq \mathrm{Proj}(S) \simeq \mathrm{Proj}(S^{(8)}) \\ &\simeq \mathrm{Proj}(k[I_8, I_{16}, I_{24}, I_{32}, I_{40}]) \\ &\simeq \mathrm{Proj}(k[y_1 : y_2 : y_3 : y_4 : y_5]) \\ &\simeq \mathbb{P}(1,2,3,4,5), \end{aligned}$$

where the y_i are algebraically independent variables of degree i. The last isomorphism is simply the definition of the weighted projective space $\mathbb{P}(1,2,3,4,5)$. □

As seen by the above description as a weighted projective space, the GIT compactification $|\mathcal{O}_{\mathbb{P}^3}(3)|^{ss}/\!/\mathrm{SL}(4)$ of the moduli space $M_{3,2}$ of smooth cubic surfaces has only finite quotient singularities. In higher dimensions, this is no longer the case.

According to Section **1**.2.3, the discriminant $\Delta = \Delta_{3,2}$ is an invariant homogeneous polynomial of degree 32 in the coefficients a_I of a cubic surface $S = V(\sum a_I x^I)$. Thus, Δ can be written in terms of I_8, I_{16}, I_{24}, and I_{32}, which was already done by Salmon [415] with minor corrections by Edge [171], see also [132, Sec. 6.4].

For all practical purposes, one can in fact replace I_{32} as a coordinate by the discriminant such that $|\mathcal{O}(3)|_{\mathrm{sm}}/\!/\mathrm{SL}(4) \subset \mathbb{P}(1,2,3,4,5) = \mathrm{Proj}(k[y_1, y_2, y_3, y_4, y_5])$ is the open complement of the hyperplane $V(y_4) \simeq \mathbb{P}(1,2,3,5)$.

The hypersurface in $\mathbb{P}(1,2,3,4,5)$ defined by I_{100} parametrizes generically smooth cubic surfaces with exactly one Eckardt point, cf. Remark 3.10 and [158, Exa. 9.1.3 & Ch. 9.4.5].

Corollary 4.3 *The moduli space of smooth cubic surfaces $M_{3,2}$ is isomorphic to the quotient $\mathbb{A}^4/\mu_4$, where the cyclic group $\mu_4 = \langle t \mid t^4 = 1\rangle$ acts by $t \cdot (a_1, a_2, a_3, a_4) = (ta_1, t^2a_2, t^3a_3, ta_4)$.* □

From the above one deduces that $M_{3,2}$ is smooth outside the origin. The surface corresponding to the only singularity of $M_{3,2}$ was described by Naruki [370], see also [157, Ch. 10.7]. The other singular points of $\mathbb{P}(1, 2, 3, 4, 5)$ all correspond to singular cubic surfaces, see [132, Sec. 6.9 & 6.10]. Naruki's description [370] of the moduli space was used by Colombo and van Geemen [123] to describe the Chow group of a certain moduli space of marked cubic surfaces.

Warning: the moduli stack $\mathcal{M}_{3,2}$ of smooth cubic surfaces is not isomorphic to the natural Deligne–Mumford stack with $\mathbb{P}(1, 2, 3, 4, 5)$ as underlying coarse moduli space. Indeed, there is a divisor of smooth cubic surfaces with non-trivial automorphisms, see Remark 3.10, while the weighted projective space has non-trivial stabilizers only in the origin of the open subset $M_{3,2} \simeq \mathbb{A}^4/\mu_4 \subset\subset \mathbb{P}(1, 2, 3, 4, 5)$ and its complement.

Remark 4.4 Dardanelli and van Geemen [132] exploit a classical construction to describe the moduli space of cubic surfaces. If a cubic surface $S \subset \mathbb{P}^3$ is given by the cubic polynomial $F(x_0, x_1, x_2, x_3)$, then the determinant of the Hessian $(\partial_i\partial_j F)$ describes a quartic surface $T \subset \mathbb{P}^3$. For a generic cubic surface, the Hessian is nodal and, therefore, its minimal resolution is a K3 surface, the transcendental and Picard lattices of which depend on the geometry of the cubic surface S. The approach was pursued further by Koike [280].

4.2 Stable Cubic Surfaces After having described the GIT quotient $|\mathcal{O}(3)|^{ss}/\!/\mathrm{SL}(4)$, one wants to know, of course, what it parametrizes. We know that all smooth cubic surfaces define stable points in $|\mathcal{O}(3)|$, but can one also understand all semi-stable cubic surfaces in analogy to the case of plane cubic curves? See Section **3**.2.2 for the latter. This is a classical result already discussed by Hilbert [230], see also [157, 361] for the statement and [61, 356, 399] for complete proofs.

Theorem 4.5 (Hilbert) *Let $S \subset \mathbb{P}^3$ be a cubic surface.*

(i) *The surface S is stable if and only if S is nodal, i.e. all its singularities are ordinary double points.*
(ii) *The surface S is semi-stable if and only if every singularity of S is either an ordinary double point or an A_2-singularity.*[5]

[5] An A_2-singularity, also called a cusp, is a singularity that locally analytically is given by $x_0^2 + x_1^2 + x_2^3$.

We will not go into the details of the proof, but let us at least indicate how to use the Hilbert–Mumford criterion, see Theorem **3**.1.10, to prove that cubic surfaces $S \subset \mathbb{P}^3$ with at most ordinary double points as singularities are stable.

Assume that $S := V(F) \subset \mathbb{P}^3$ is integral and defines a point $x \in |\mathcal{O}(3)|$ that is not stable, i.e. such that there exists a non-trivial $\lambda\colon \mathbb{G}_m \longrightarrow \mathrm{SL}(4)$ with $\mu(x,\lambda) \leq 0$. After a linear coordinate change we may assume that the action is diagonal and the induced action on the linear coordinates is given by $\lambda(t)(x_j) = t^{r_j}x_j$ with $r_0 \leq \cdots \leq r_3$ and $\sum r_j = 0$. Then, on a cubic polynomial $F = \sum a_I x^I$ the action is given by $\lambda(t)(F) = \sum a_I t^{rI} x^I$, where $I = (i_0, \ldots, i_3)$, $\sum i_j = 3$, and $rI := \sum r_j i_j$. By an elementary computation, see e.g. [53, Prop. 6.5], one shows that $\mu(x,\lambda) \leq 0$ implies that $[1:0:0:0] \in S$ is either a non-ordinary double point or a cusp.

Remark 4.6 (i) In fact, it turns out that all semi-stable cubic surfaces that are not stable give the same point in $|\mathcal{O}_{\mathbb{P}^3}(3)|^{ss}/\!/\mathrm{SL}(4)$. More precisely, the only semi-stable non-stable cubic surface with a closed orbit is the surface $V(x_0^3 - x_1 \cdot x_1 \cdot x_2)$ up to coordinate change, see [356, Thm. 7.24].

(ii) Also recall from Remark 2.16 and Remark **1**.5.17, that the maximal number of ordinary double points of an otherwise smooth cubic surface is four. The maximal number four is only achieved by the Cayley cubic which with the above convention corresponds to a point of the form $[y_0 : y_1 : y_3 : 1 : y_5] \in \mathbb{P}(1,2,3,4,5)$. The explicit computation of the coordinates is easy, but not particularly enlightening.

(iii) The loci of semi-stable cubic surfaces with a given number of nodes and cusps has been investigated by Nguyen [456]. For example, the maximal number of cusps is three and surfaces with one node and two cusps can be realized as a specialization of a cubic with two cusps or alternatively of a cubic with two nodes and one cusp.

4.3 Period Description via Cubic Threefolds The Hodge structure of a smooth cubic surface S carries no information. Indeed, $H^2(S,\mathbb{Z}) \simeq \mathrm{NS}(S)$ is independent of the particular surface S. Nevertheless, a simple trick allows one to associate with S a Hodge structure of weight one which knows everything about S. This beautiful idea was first exploited by Allcock, Carlson, and Toledo [13] and later further explored by Dolgachev, van Geemen, and Kondô [160], Kudla and Rapoport [289], Achter [1], Zheng [499], and others. Eventually, it leads to a description of the moduli space of smooth cubic surfaces as an open subset of an arithmetic ball quotient. We will use results that will be explained only in Section **5**.4. So, the reader might want to skip this section and come back to it later. Similar ideas allowing the pass from cubic threefolds to cubic fourfolds will be discussed in Section **6**.6.1.

We follow the construction in Section **1**.5.6 and consider the cyclic triple cover

$$Y \longrightarrow \mathbb{P}^3 \supset S$$

branched over the smooth cubic surface $S \subset \mathbb{P}^3$. We view the Hodge structure $H^3(Y,\mathbb{Z})$ as naturally associated with S. It comes with the additional structure of a Hodge isometry $\iota\colon H^3(Y,\mathbb{Z}) \xrightarrow{\sim} H^3(Y,\mathbb{Z})$ of order three induced by the natural covering action of $\rho := e^{2\pi i/3}$ on Y. In other words, $H^3(Y,\mathbb{Z})$ is a free $\mathbb{Z}[\rho]$-module of rank five. Note that ι acts as a Hodge isometry.

Exercise 4.7 Assume $f\colon S \xrightarrow{\sim} S'$ is an isomorphism between two smooth cubic surfaces. Show that there exists an equivariant isomorphism $\tilde{f}\colon Y \xrightarrow{\sim} Y'$ between their associated cubic threefolds, and hence a Hodge isometry $H^3(Y,\mathbb{Z}) \xrightarrow{\sim} H^3(Y',\mathbb{Z})$ of $\mathbb{Z}[\rho]$-modules. The isomorphism $\tilde{f}$ is canonical up to the action of ρ and restricts to $\tilde{f}|_S = f$.

The following discussion is based on some basic linear algebra which we spell out for the reader's convenience. First, recall that the primitive third root of unity ρ satisfies $1+\rho+\rho^2 = 0$ and $\rho^2 = \rho^{-1} = \bar{\rho}$. Then the ring of Eisenstein integers $\mathbb{Z}[\rho] \subset \mathbb{C}$ contains $\theta := \rho - \rho^2 = \sqrt{-3} = i\sqrt{3}$, which can also be written as $\theta = 2\rho + 1$. Its complex conjugate satisfies $\bar{\theta} = -\theta$. Also recall that the units in $\mathbb{Z}[\rho]$ are ± 1, $\pm\rho$, and $\pm\rho^2$. In other words, $\mathbb{Z}[\rho]^* = \langle -\rho \rangle \simeq \mu_6$.

Note that for ι the analogous equation $\mathrm{id} + \iota + \iota^2 = 0$ holds. As the invariant part of its action on $H^3(Y,\mathbb{C})$ is $H^3(\mathbb{P}^3,\mathbb{C}) = 0$, its eigenspace decomposition has the form

$$H^3(Y,\mathbb{C}) = H_\rho \oplus H_{\rho^2} = (H_\rho^{2,1} \oplus H_\rho^{1,2}) \oplus (H_{\rho^2}^{2,1} \oplus H_{\rho^2}^{1,2}), \tag{4.1}$$

where the two eigenspaces are complex conjugates of each other and both are of dimension five.

Abusing the notation, we introduce another endomorphism of $H^3(Y,\mathbb{Z})$ as $\theta := \iota - \iota^2 = 2\iota + \mathrm{id}$. For its action we write $\theta(\alpha)$, which should not be confused with the simple multiplication $\theta \cdot \alpha$ for classes $\alpha \in H^3(Y,\mathbb{C})$. Observe that (4.1) is also the eigenspace decomposition for the action of θ, more precisely $\theta(\alpha) = \theta \cdot \alpha$ for $\alpha \in H_\rho$ and $\theta(\alpha) = -\theta \cdot \alpha$ for $\alpha \in H_{\rho^2}$. In particular, $\theta^2(\alpha) = \theta^2 \cdot \alpha = -3\alpha$ for all classes $\alpha \in H^3(Y,\mathbb{Z})$.

The endomorphism ι is a symplectic isometry, i.e. it satisfies $(\iota(\alpha).\iota(\beta)) = (\alpha.\beta)$ or, equivalently, $(\iota(\alpha).\beta) = (\alpha.\iota^2(\beta))$, which for θ becomes $(\theta(\alpha).\beta) = -(\alpha.\theta(\beta))$.

To start, we will only use the fact that $\Gamma := H^3(Y,\mathbb{Z})$ is a unimodular symplectic lattice with an action of $\mathbb{Z}[\rho]$ enjoying the above properties, the rank is of no importance for now. Then, following [13, §4], cf. [53, §3], one defines the pairing

$$h\colon \Gamma \times \Gamma \longrightarrow \mathbb{Z}[\rho], \quad h(\alpha,\beta) := \frac{(\theta(\alpha).\beta) + \theta \cdot (\alpha.\beta)}{2},$$

for which one proves a number of easy facts.

- The first thing to prove is that h really takes values in $\mathbb{Z}[\rho]$. This is left to the reader.
- Next, one checks that

$$h(\alpha,\beta) = \overline{h(\beta,\alpha)}.$$

Indeed, $(\theta(\alpha).\beta) + \theta\cdot(\alpha.\beta) = -(\beta.\theta(\alpha)) - \theta\cdot(\beta.\alpha) = (\theta(\beta).\alpha) + \bar{\theta}\cdot(\beta.\alpha) = \overline{(\theta(\beta).\alpha) + \theta \cdot (\beta.\alpha)}$.

• The form $h(\ ,\ )$ is $\mathbb{Z}[\rho]$-linear in the first component, i.e. $h(\iota(\alpha),\beta) = \rho \cdot h(\alpha,\beta)$ or, equivalently,

$$h(\theta(\alpha),\beta) = \theta \cdot h(\alpha,\beta).$$

Indeed, $(\theta^2(\alpha).\beta) + \theta \cdot (\theta(\alpha).\beta) = (\theta^2 \cdot \alpha.\beta) + \theta \cdot (\theta(\alpha).\beta) = \theta \cdot (\theta \cdot (\alpha.\beta) + (\theta(\alpha).\beta))$.

Altogether, we now have proved that h is a $\mathbb{Z}[\rho]$-hermitian form, i.e. its bilinear extension under $\mathbb{Z}[\rho] \subset \mathbb{C}$ is a hermitian form on the complex vector space $\Gamma \otimes_{\mathbb{Z}[\rho]} \mathbb{C}$ of dimension $(1/2) \cdot \mathrm{rk}_{\mathbb{Z}}(\Gamma)$.

• The eigenspaces H_ρ and H_{ρ^2} are isotropic with respect to the $\mathbb{C}$-linear extension of the symplectic pairing $(\ .\ )$ and, therefore, also with respect to h. For example, if $\alpha,\beta \in H_\rho$, then $\theta \cdot (\alpha.\beta) = (\theta \cdot \alpha.\beta) = (\theta(\alpha).\beta) = -(\alpha.\theta(\beta)) = -(\alpha.\theta \cdot \beta) = -\theta \cdot (\alpha.\beta)$ and hence $(\alpha.\beta) = 0$. For the second part, use the fact that the endomorphism θ preserves H_ρ.

• Consider the $\mathbb{Z}[\rho]$-linear composition $j\colon \Gamma \hookrightarrow \Gamma \otimes_{\mathbb{Z}} \mathbb{C} \twoheadrightarrow H_\rho$. If the complex vector space H_ρ is endowed with the hermitian form

$$h'(\gamma,\delta) := \theta \cdot (\gamma.\bar{\delta}) = i\sqrt{3} \cdot (\gamma.\bar{\delta}),$$

then j is isometric with respect to h on Γ and h' on H_ρ. Indeed, since the two eigenspaces H_ρ and H_{ρ^2} are h-isotropic and $\alpha = j(\alpha) + \bar{j}(\alpha)$, etc., we have $h(\alpha,\beta) = h(j(\alpha),\bar{j}(\beta)) + h(\bar{j}(\alpha), j(\beta)) = \theta \cdot (j(\alpha).\bar{j}(\beta))$, where we use that $\theta(j(\alpha)) = \theta \cdot j(\alpha)$ and $\theta(\bar{j}(\alpha)) = -\theta \cdot \bar{j}(\alpha)$.

For dimension reasons, the linear extension gives rise to a hermitian isomorphism $\Gamma \otimes_{\mathbb{Z}[\rho]} \mathbb{C} \simeq H_\rho$, which in our geometric setting reads

$$H^3(Y,\mathbb{Z}) \otimes_{\mathbb{Z}[\rho]} \mathbb{C} \xrightarrow{\sim} H^3(Y)_\rho.$$

The next step is not a purely linear algebra statement, it needs the cubic threefold in the background.

• The decomposition $H_\rho = H_\rho^{2,1} \oplus H_\rho^{1,2}$ is h'-orthogonal. Furthermore, $H_\rho^{2,1}$ is of dimension four and positive definite with respect to h', while $H_\rho^{1,2}$ is an h'-negative line. The signs are deduced from a local computation showing that, for example, for $\alpha = dz_1 \wedge dz_2 \wedge d\bar{z}_3$ one has $i \cdot \alpha \wedge \bar{\alpha} = i \cdot dz_1 \wedge d\bar{z}_1 \wedge dz_2 \wedge d\bar{z}_2 \wedge dz_3 \wedge d\bar{z}_3$ which is positive when integrated over any disc. The dimensions are easily computed by applying a result of Griffiths, see Lemma **1**.4.23 and Example **1**.4.24: the residue gives an isomorphism $H^{2,1}(Y) \simeq H^0(\mathbb{P}^4, \mathcal{O}(1))$ which is compatible with the action of ρ. Since the invariant part of the right-hand side is of dimension four, this shows $\dim H_\rho^{2,1} = 4$. As $\dim H_\rho = 5$, it also proves $\dim H_\rho^{1,2} = 1$.

Exercise 4.8 As a continuation of Exercise 4.7, show that any automorphism of a smooth cubic surface S for which the induced automorphism of $H^3(Y,\mathbb{Z})$ is given by multiplication by a unit in $\mathbb{Z}[\rho]$ is itself the identity. This corresponds to the statement that automorphisms of a framed cubic surface are trivial, see [53, Cor. 3.4].

- There exists an isometry

$$(H^3(Y,\mathbb{Z}),h) \simeq \mathbb{Z}[\rho]^{4,1} := (\mathbb{Z}[\rho]^{\oplus 5}, h^{4,1})$$

of $\mathbb{Z}[\rho]$-modules such that under its $\mathbb{C}$-linear extension $H^3(Y,\mathbb{Z}) \otimes_{\mathbb{Z}[\rho]} \mathbb{C} \simeq \mathbb{C}^{\oplus 5}$ the hermitian form h corresponds to the standard form $-x_0\bar{y}_0 + \sum_{i=1}^4 x_i\bar{y}_i$, i.e. to $\mathbb{C}^{4,1}$. This relies on the fact that h is a unimodular $\mathbb{Z}[\rho]$-lattice and some general classification results for those. We refer to [53, Prop. 2.6] for details and references. In fact, with minor modifications, everything that follows would also work if the isomorphism type of the $\mathbb{Z}[\rho]$-lattice were something else, as long as the signature is unchanged.

Recall from Remark **1**.5.22 that the generic cubic surface S is uniquely determined by its associated cubic threefold Y. Combined with the global Torelli theorem for cubic threefolds, see Section **5**.4.2, this immediately proves already the following global Torelli type theorem [13] for generic smooth cubic surfaces.

Theorem 4.9 (Allcock–Carlson–Toledo) *For two smooth cubic surfaces $S, S' \subset \mathbb{P}^3$ and their associated cubic threefolds $Y, Y' \subset \mathbb{P}^4$, the following conditions are equivalent:*

(i) *There exists an isomorphism $S \simeq S'$.*
(ii) *There exists a $\mathbb{Z}[\rho]$-equivariant Hodge isometry $H^3(Y,\mathbb{Z}) \simeq H^3(Y',\mathbb{Z})$.*
(iii) *There exists a $\mathbb{Z}[\rho]$-equivariant isomorphism $(J(Y),\Xi) \simeq (J(Y'),\Xi')$ of polarized abelian varieties.*

The proof of the full result uses moduli spaces, period domains, and period maps. We briefly outline the argument. If $M_2 := M_{3,2} = U//\mathrm{PGL}(4)$ with $U = |\mathcal{O}_{\mathbb{P}^3}(3)|_{\mathrm{sm}}$ denotes the moduli space of smooth cubic surfaces, then one defines the moduli space of framed smooth cubic surfaces $\widetilde{M}_2 := \widetilde{U}//\mathrm{PGL}(4)$ as the quotient of the space

$$\widetilde{U} := \{(S,\varphi) \mid S \in U, \varphi\colon H^3(Y,\mathbb{Z}) \xrightarrow{\sim} \mathbb{Z}[\rho]^{4,1}\}$$

by the natural action of PGL(4), which according to Exercise 4.8 is free. Here, Y is the cubic threefold associated with S and φ is an isometry of the two hermitian $\mathbb{Z}[\rho]$-lattices up to the action of $\mu_6 \simeq \mathbb{Z}[\rho]^*$.

The period map for smooth cubic surfaces is the holomorphic map

$$\widetilde{\mathcal{P}}\colon \widetilde{M}_2 \longrightarrow \mathbb{B}^4 \subset \mathbb{P}(\mathbb{C}^{4,1})^*, \quad (S,\varphi) \longmapsto \varphi(H^{2,1}(Y)_\rho) \tag{4.2}$$

that sends a framed smooth cubic surface to the hyperplane given by $\varphi(H^{2,1}(Y)_\rho) \subset \mathbb{Z}[\rho]^{4,1} \otimes_{\mathbb{Z}[\rho]} \mathbb{C} \simeq \mathbb{C}^{4,1}$. Here, the open subset $\mathbb{B}^4$ parametrizes all positive hyperplanes. It is naturally identified (anti-holomorphically) with the complex four-dimensional ball $\{z \mid \sum |z_i|^2 < 1\} \subset \mathbb{C}^4$. Indeed, a positive hyperplane $H \subset \mathbb{C}^{4,1}$ is determined by its hermitian orthogonal, which is a negative line spanned by a vector z (unique up to scaling) with $\sum_{i=1}^4 |z_i|^2 < |z_0|^2$. After normalizing $z_0 = 1$, the latter becomes $\sum |z_i|^2 < 1$.

The period map is compatible with the action of the discrete projective unitary group $\mathrm{PU} := \mathrm{U}(\mathbb{Z}[\rho]^{4,1})/\mu_6$ on both sides and thus descends to the period map

$$\mathcal{P}\colon M_2 \longrightarrow \mathrm{PU} \setminus \mathbb{B}^4. \tag{4.3}$$

The key observation is that with any point in $\mathbb{B}^4$ one can associate a point in the five-dimensional upper half plane $\mathbb{H}_5$. In terms of Hodge structures of weight one, it is described by sending a positive hyperplane $H \subset \mathbb{C}^{4,1}$ to $H \oplus \bar{H}^\perp \subset \mathbb{C}^{10}$, where the orthogonal complement is taken with respect to the hermitian structure. This causes the map $\mathbb{B}^4 \longrightarrow \mathbb{H}_5$ to depend holomorphically on H. Since, for $H = H^{2,1}(Y)_\rho$, one has $\bar{H}^\perp = (H^{1,2}_{\rho^2})^\perp = H^{2,1}(Y)_{\rho^2}$ and hence $H \oplus \bar{H}^\perp = H^{2,1}(Y)$, this gives rise to a commutative diagram

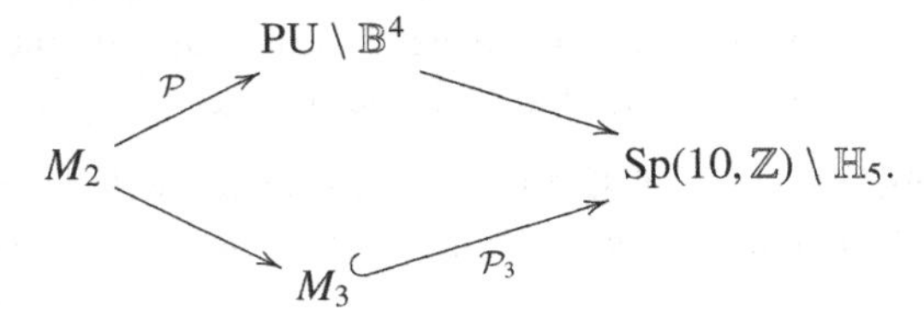

Here, $\mathcal{P}_3\colon M_3 \longrightarrow A_5 \simeq \mathrm{Sp}(10,\mathbb{Z}) \setminus \mathbb{H}_5$ is the period map for cubic threefolds that sends a smooth cubic threefold Y to its intermediate Jacobian $J(Y)$. The global Torelli theorem proves that this map is closed embedding, see Example **3**.3.4 and Section **5**.2.

Since $M_2 \longrightarrow M_3$, sending S to the associated triple cover $Y \longrightarrow \mathbb{P}^3$, is generically injective, cf. Remark **1**.5.22, the same holds true for the composition $M_2 \longrightarrow \mathrm{Sp}(10,\mathbb{Z}) \setminus \mathbb{H}_5$ and hence for (4.3). On the other hand, a version of the infinitesimal Torelli theorem implies that (4.2) is unramified, i.e. its tangent maps are injective. Since both $\widetilde{M}_2$ and $\mathbb{B}^4$ are smooth of dimension four, one finds that (4.2) is a local isomorphism. Altogether this proves that (4.3) is an open embedding of complex analytic spaces or, by applying Baily–Borel, of quasi-projective varieties.

In fact, instead of working with the coarse moduli scheme M_2 of smooth cubic surfaces, one can refine the above discussion to work on the level of stacks. Again the period map defines an open embedding of smooth analytic Deligne–Mumford stacks

$$\mathcal{M}_2 \hookrightarrow [\mathrm{PU} \setminus \mathbb{B}^4],$$

see Section **3**.1.6. The crucial observation here is a result by Zheng [499, Prop. 6.1] proving that the natural map that associates with an automorphism of a cubic surface S, an automorphism of the corresponding triple cover $Y \longrightarrow \mathbb{P}^3$ and hence an automorphism of its intermediate Jacobian, induces an isomorphism

$$\mathrm{Aut}(S) \xrightarrow{\sim} \mathrm{Aut}(J(Y), \iota)/\mu_6.$$

Here, $\mathrm{Aut}(J(Y), \iota)$ is the group of polarized automorphisms commuting with the action of the covering endomorphism ι.

Remark 4.10 The preceding discussion is complemented by the description of the image of (4.3). According to Allcock, Carlson, and Toledo [13], see also Beauville's exposition [53], one has

$$\mathcal{P}\colon M_2 \xrightarrow{\sim} \mathrm{PU} \setminus \left(\mathbb{B}^4 \setminus \mathcal{H}\right).$$

Here, $\mathcal{H} := \bigcup \delta^\perp$ is the set of all hyperplanes $H \subset \mathbb{C}^{4,1}$ containing some element $\delta \in \mathbb{Z}[\rho]^{4,1}$ with $h^{4,1}(\delta.\delta) = 1$. It turns out that all such classes δ are contained in the same orbit of the $\mathrm{U}(\mathbb{Z}[\rho]^{4,1})$-action, so that M_2 is isomorphic to the complement of an irreducible divisor $\mathrm{PU} \setminus \mathcal{H} \subset \mathrm{PU} \setminus \mathbb{B}^4$.

(i) To show that $\mathcal{P}(M_2)$ is contained in the complement of $\bigcup \delta^\perp$, observe first that for a class $\delta \in H^3(Y,\mathbb{Z})$ with $j(\delta) \in H^{1,2}(Y)$, the map $\mathbb{C} \longrightarrow H^{1,2}(Y)$, $1 \longmapsto j(\delta)$ defines a morphism

$$E_\rho := \mathbb{C}/\mathbb{Z}[\rho] \longrightarrow J(Y) = H^{1,2}(Y)/H^3(Y,\mathbb{Z}). \tag{4.4}$$

By Exercise 5.4.8, the intermediate Jacobian $J(Y)$ is an irreducible principally polarized abelian variety and, in particular, it cannot split off the elliptic curve E_ρ with its canonical principal polarization. Thus, (4.4) cannot respect the principal polarizations. Since the principal polarizations on the two sides are given by the intersection pairing on $H^1(E_\rho,\mathbb{Z})$, respectively $H^3(Y,\mathbb{Z})$, the morphism (4.4) is polarized if and only if the images $\delta, \iota(\delta) \in H^3(Y,\mathbb{Z})$ of $1, \rho \in \mathbb{Z}[\rho] = H^1(E_\rho,\mathbb{Z})$ satisfy $(\delta.\iota(\delta)) = -1$. The latter is equivalent to $-\theta \cdot (j(\delta).\bar{j}(\delta)) = (\rho^2 - \rho) \cdot (j(\delta).\bar{j}(\delta)) = (j(\delta).\bar{j}(\iota(\delta))) + (\bar{j}(\delta).j(\iota(\delta))) = -1$. A computation reveals that $h(\delta,\delta) = 1$ is equivalent to $\theta \cdot (j(\delta).\bar{j}(\delta)) = 1$, which shows that for a smooth cubic surface S the associated cubic threefold Y does not admit a class $\delta \in H^3(Y,\mathbb{Z})$ with $j(\alpha) \in H^{1,2}(Y)_\rho$ and $h(\delta,\delta) = 1$.

(ii) The other inclusion is more subtle, see [53, §7] for a detailed exposition or the original [13, §9]. For this, the period map has to be extended to an isomorphism of the stable locus

$$M_2 = |\mathcal{O}_{\mathbb{P}^3}(3)|_{\mathrm{sm}}/\mathrm{PGL}(4) \subset |\mathcal{O}_{\mathbb{P}^3}(3)|^{\mathrm{s}}/\!/\mathrm{PGL}(4) \xrightarrow{\sim} \mathrm{PU} \setminus \mathbb{B}^4$$

and in fact to an isomorphism

$$|\mathcal{O}_{\mathbb{P}^3}(3)|^{\mathrm{ss}}/\!/\mathrm{PGL}(4) \xrightarrow{\sim} \mathrm{PU} \setminus \widehat{\mathbb{B}}^4$$

of the projective moduli space of all semi-stable cubic surfaces with the Baily–Borel compactification. This makes use of the description of all stable and semi-stable cubic surfaces as outlined in Section 4.2.

For a comparison of various natural compactifications of the moduli space of smooth cubic surfaces, see Zhang's thesis [498] or the recent article [107].

Remark 4.11 Another approach to a period description of the moduli space of cubic surfaces was proposed by Dolgachev, van Geemen, and Kondō [160]. Instead of associating with a cubic surface $S \subset \mathbb{P}^3$ a cubic threefold $Y \twoheadrightarrow \mathbb{P}^3$, they consider a K3 surface $\tilde{T}$ obtained as the minimal resolution of the double plane $T \twoheadrightarrow \mathbb{P}^2$ branched over the curve defined by $x_2 \cdot (x_2^3 \cdot F_2(x_0, x_1) + F_5(x_0, x_1))$, see Remark 2.12. Here, F_5 defines the discriminant divisor of the quadric fibration $\phi \colon S \twoheadrightarrow \mathbb{P}^1$ from a chosen line $L \subset S$ and F_2 describes the branch locus of the degree two morphism $\phi \colon L \twoheadrightarrow \mathbb{P}^1$.

In terms of Hodge structures, the difference between the two approaches is that the one by Allcock, Carlson, and Toledo uses Hodge structures of weight one, while the one by Dolgachev, van Geemen, and Kondō works with Hodge structures of weight two. A detailed period description of the moduli space of all elliptic K3 surfaces with a section, can be found in [160, §6]. The final results describe the moduli space of smooth cubic surfaces again as an open subset of a ball quotient.

5

Cubic Threefolds

This chapter is devoted to cubic hypersurfaces $Y \subset \mathbb{P}(V) \simeq \mathbb{P}^4$ of dimension three. We will be mostly interested in smooth ones, but (mildly) singular ones will also make an appearance. Cubic threefolds and their Fano surfaces of lines have a long and distinguished history in algebraic geometry, going back to the Italian school, especially Gino Fano [177], and to the landmark article of Clemens and Griffiths [119]. The latter proves irrationality of all smooth cubic threefolds and introduces the intermediate Jacobian as an effective tool in complex algebraic geometry. Cubic threefolds have also served as a testing ground for the Weil conjectures already in [80] and their geometry has been investigated in detail in the series of papers of Beauville [41, 42], Tyurin [446–448], and Murre [364, 365].

Before getting started, we collect the basic facts on cubic threefolds that follow from the general theory as presented in Chapters 1 and 2. We will typically work over $\mathbb{C}$.

0.1 Invariants of Cubic Threefolds The canonical bundle of a smooth cubic threefold $Y \subset \mathbb{P}^4 = \mathbb{P}(V)$ is easily computed as $\omega_Y \simeq \mathcal{O}_Y(-2)$, which is the square of the dual of the ample generator of $\mathrm{Pic}(Y) \simeq \mathbb{Z} \cdot \mathcal{O}_Y(1)$, see Lemma **1**.1.6 and Corollary **1**.1.9. The non-trivial Betti numbers of Y are

$$b_0(Y) = b_2(Y) = b_4(Y) = b_6(Y) = 1 \quad \text{and} \quad b_3(Y) = 10$$

and, therefore, its Euler number is $e(Y) = -6$, see Section **1**.1.3. For the even Betti numbers, one can be more precise:

$$H^2(Y, \mathbb{Z}) \simeq \mathbb{Z} \cdot h \quad \text{and} \quad H^4(Y, \mathbb{Z}) = \mathbb{Z} \cdot (h^2/3),$$

where h is the restriction of the hyperplane class. Note that $h^2/3$ is an integral algebraic class, see Remark **1**.1.3. Furthermore, the middle degree Hodge numbers are

$$h^{3,0}(Y) = h^{0,3}(Y) = 0 \quad \text{and} \quad h^{2,1}(Y) = h^{1,2}(Y) = 5.$$

The linear system of all cubic threefolds is $|\mathcal{O}_{\mathbb{P}^4}(3)| \simeq \mathbb{P}^{34}$ and the moduli space of smooth cubic threefolds is of dimension 10, see Section **1**.1.2. Furthermore, the universal deformation space $\mathrm{Def}(Y)$ of a smooth cubic threefold is smooth of dimension 10. More precisely, $\dim H^1(Y, \mathcal{T}_Y) = 10$ and $H^2(Y, \mathcal{T}_Y) = 0$, see Remark **1**.3.13.

0.2 Invariants of Their Fano Variety As for cubic surfaces and maybe even more so, the geometry of lines on cubic threefolds is particularly rich and interesting. In dimension three, however, every point is contained in a line and a generic point is contained in exactly six lines, cf. Example **2**.5.12. As for a smooth cubic threefold $\mathrm{Pic}(Y) \simeq \mathbb{Z}\cdot\mathcal{O}_Y(1)$, there are no planes contained in Y, see also Remark **2**.1.7.

The general theory of Fano varieties of lines as outlined in the Chapter **2** provides us with detailed and useful information:

(i) The Fano variety of lines $F := F(Y)$ of a smooth cubic threefold Y is an irreducible, smooth, projective surface, the *Fano surface of* Y.

(ii) The canonical bundle ω_F of F is ample. It is isomorphic to the Plücker polarization induced by $F \hookrightarrow \mathbb{G}(1, \mathbb{P}^4) \hookrightarrow \mathbb{P}(\bigwedge^2 V)$, see Lemma **2**.3.1:

$$\omega_F \simeq \mathcal{O}_F(1). \tag{0.1}$$

(iii) The degree of the Fano variety F with respect to the natural Plücker polarization $g = \mathrm{c}_1(\mathcal{O}_F(1))$ is, cf. Section **2**.4.3:

$$\deg(F) = \int_F g^2 = 45. \tag{0.2}$$

(iv) The Euler number of the Fano surface is $e(F) = 27$, see Proposition **2**.4.6 and Section 2.1 below.

(v) The Hodge diamond of F up to the middle is, cf. Section **2**.4.6:

$b_0(F(Y)) = 1$			1		
$b_1(F(Y)) = 10$		5		5	
$b_2(F(Y)) = 45$	10		25		10.

Note that the Noether formula $\chi(\mathcal{O}_F) = \int_F (1/12)(\mathrm{c}_1^2(F) + \mathrm{c}_2(F))$ combined with the last two assertions provides another proof of the degree formula (0.2).

(vi) The universal family of lines on Y comes with two projections

$$F(Y) \xleftarrow{\ p\ } \mathbb{L} \xrightarrow{\ q\ } Y,$$

where the morphism $q \colon \mathbb{L} \twoheadrightarrow Y$ is generically finite of degree six, cf. Lemma **2**.5.11 and Example **2**.5.12. It can be shown that at least for the generic cubic threefold the Galois group of $q \colon \mathbb{L} \twoheadrightarrow Y$ is the symmetric group $\mathfrak{S}_6$.[1]

[1] I wish to thank F. Gounelas for providing an argument that will appear in a forthcoming paper.

(vii) The Fano correspondence $\varphi = p_* \circ q^* \colon H^3(Y, \mathbb{Q}) \xrightarrow{\sim} H^1(F, \mathbb{Q})(-1)$ has the following property, see Proposition **2**.5.5:

$$(\alpha.\beta) = -\frac{1}{6}\int_F \varphi(\alpha) \cdot \varphi(\beta) \cdot g.$$

(viii) There exist isomorphisms of Hodge structures, cf. Section **2**.4.6,

$$H^3(Y, \mathbb{Q})(1) \simeq H^1(F(Y), \mathbb{Q}) \tag{0.3}$$

$$\text{and} \quad \bigwedge^2 H^3(Y, \mathbb{Q})(2) \simeq \bigwedge^2 H^1(F(Y), \mathbb{Q}) \simeq H^2(F(Y), \mathbb{Q}). \tag{0.4}$$

The isomorphism in (0.3) can be obtained via the Fano correspondence, see Proposition **2**.5.5, or, alternatively, via the motivic approach in Section **2**.4.6. The first isomorphism in (0.4) is deduced by taking exterior products of (0.3) and for the second see Lemma 2.5 below. The isomorphisms (0.3) and (0.4) will be upgraded to isomorphisms of integral Hodge structures in the course of this chapter.

For a very general cubic threefold Y, the only rational Hodge classes in $\bigwedge^2 H^3(Y, \mathbb{Q})$ are multiples of the one given by the intersection product on Y, see Remark **1**.2.13. Hence, by virtue of (0.4), the Picard number of the Fano surface is $F(Y)$

$$\rho(F(Y)) = \operatorname{rk}\operatorname{NS}(F(Y)) = 1$$

in this case.

0.3 Chow Groups and Chow Motives The rational Chow motive of a smooth cubic threefold Y splits as, see Remark **1**.1.11:

$$\mathfrak{h}(Y) \simeq \bigoplus_{j=0}^{6} \mathfrak{h}^j(Y) \simeq \bigoplus_{i=0}^{3} \mathbb{Q}(-i) \oplus \mathfrak{h}(Y)_{\mathrm{pr}}.$$

For the Chow groups, we have

$$\mathrm{CH}^0(Y) \simeq \mathrm{CH}^1(Y) \simeq \mathrm{CH}^3(Y) \simeq \mathbb{Z}.$$

The only non-trivial Chow group is $\mathrm{CH}^2(Y)$ respectively $\mathrm{CH}^2(Y) \otimes \mathbb{Q} \simeq \mathrm{CH}(\mathfrak{h}(Y)_{\mathrm{pr}})$, which can be determined using arguments specific to cubic threefolds. However, its structure can also be deduced from general principles, see Corollary 3.16.

The Chow motive of the Fano surface $F(Y)$ was described in Section **2**.4.2 by

$$\mathfrak{h}(F(Y))(-2) \oplus \mathfrak{h}(Y) \oplus \mathfrak{h}(Y)(-3) \simeq S^2\mathfrak{h}(Y).$$

1 Lines on the Cubic and Curves on Its Fano Surface

The Fano surface $F(Y)$ of a cubic threefold Y parametrizes all lines $L \subset \mathbb{P}^4$ contained in the cubic Y. Such lines can be of the first or of the second type, see Section **2**.2, i.e.

$$\mathcal{N}_{L/Y} \simeq \mathcal{O}_L \oplus \mathcal{O}_L \quad \text{or} \quad \mathcal{N}_{L/Y} \simeq \mathcal{O}_L(1) \oplus \mathcal{O}_L(-1).$$

We study natural curves in $F(Y)$ of a smooth cubic threefold $Y \subset \mathbb{P}^4$ over an arbitrary algebraically closed field. Firstly, there is the curve of lines of the second type

$$R = F_2(Y) \subset F(Y),$$

cf. Section **2**.2.2. Its pre-image in $\mathbb{L}$ is the ramification divisor of the second projection $q\colon \mathbb{L} \longrightarrow Y$, see Corollary **2**.2.15 and Section 1.1 below. Secondly, for each line $L \subset Y$ one considers the closure $C_L \subset F(Y)$ of the curve of all lines $L \neq L' \subset Y$ intersecting L. It comes with a natural fixed point free involution, the quotient of which is the discriminant curve of the linear projection of Y from L.

1.1 Lines of the Second Type To understand the geometry of $F = F(Y)$, we need to study the surjective morphism $q\colon \mathbb{L} \twoheadrightarrow Y$. Note that both varieties are smooth projective and of dimension three. Therefore, the ramification locus $R(q) \subset \mathbb{L}$ of q, i.e. the closed set of points in which q fails to be smooth, is a surface (or, possibly, empty, which it is not).

According to Corollary **2**.2.15, we know already that $R(q)$ is the restriction $\mathbb{L}_2$ of the universal line $\mathbb{L}$ to the subvariety $R = F_2(Y) \subset F(Y)$ of lines of the second type, but it is instructive to see the traditional line of arguments in dimension three which provides us with more precise information.

Proposition 1.1 *The ramification divisor $R(q) \subset \mathbb{L}$ of the morphism $q\colon \mathbb{L} \twoheadrightarrow Y$ is an element in the linear system $|p^*\mathcal{O}_F(2)|$. It is the pre-image of a curve $R \subset F(Y)$ in the linear system $|\mathcal{O}_F(2)|$.*

Proof Note that $R(q)$, and hence R, cannot be empty. Indeed, otherwise $\mathbb{L} \longrightarrow Y$ would be an étale covering of degree six of the simply connected threefold Y. As $F(Y)$ and $\mathbb{L}$ are connected, see Proposition **2**.3.4, Exercise **2**.3.7, and Example **2**.4.21, this is absurd.

We consider the differential of q as a morphism of sheaves $dq\colon \mathcal{T}_{\mathbb{L}} \longrightarrow q^*\mathcal{T}_Y$. Then by definition, $R(q)$ is the zero locus of $\det(dq)\colon\ \det(\mathcal{T}_{\mathbb{L}}) \longrightarrow q^*\det(\mathcal{T}_Y)$, which we consider as a section of $\omega_{\mathbb{L}} \otimes q^*\omega_Y^*$.

Now, applying $q^*\omega_Y^* \simeq q^*\mathcal{O}_Y(2) \simeq \mathcal{O}_p(2)$, cf. the proof of Lemma **2**.5.1, and using (0.1) gives

$$\omega_{\mathbb{L}} \simeq \omega_p \otimes p^*\omega_F \simeq p^*\det(\mathcal{S}_F^*) \otimes \mathcal{O}_p(-2) \otimes p^*\mathcal{O}_F(1),$$

where we also make use of the Euler sequence $0 \longrightarrow \mathcal{O}_{\mathbb{L}} \longrightarrow p^*\mathcal{S}_F \otimes \mathcal{O}_p(1) \longrightarrow \mathcal{T}_p \longrightarrow 0$ for the projective bundle $p\colon \mathbb{L} \simeq \mathbb{P}(\mathcal{S}_F) \longrightarrow F$. Therefore, $\omega_{\mathbb{L}} \otimes q^*\omega_Y^* \simeq p^*\mathcal{O}_F(2)$ and hence $\det(dq) \in H^0(\mathbb{L}, p^*\mathcal{O}_F(2)) \simeq H^0(F, \mathcal{O}_F(2))$. □

Remark 1.2 Proposition 1.1 goes back to Fano. In [119, Sec. 10] the argument uses the observation that the pre-image $q^{-1}(S)$ of the generic hyperplane section $S := Y \cap \mathbb{P}^3$ is the blow-up $p\colon q^{-1}(S) \longrightarrow F(Y)$ in the 27 points $\ell_i \in F(Y)$ corresponding to the 27 lines $\ell_i \subset S$ contained in the cubic surface S:[2]

$$\mathrm{Bl}_{\{\ell_1,\ldots,\ell_{27}\}}(F(Y)) \xrightarrow{6:1} S = Y \cap \mathbb{P}^3.$$

The kernel of the tangent map $T_{(L,x)}\mathbb{L} \longrightarrow T_xY$ at a point $(L, x) \in \mathbb{L} \subset F \times Y$ is the space of first order deformations of $L \subset Y$ through $x \in L$. This space is naturally isomorphic to the subspace $H^0(L, \mathcal{N}_{L/Y} \otimes \mathcal{I}_x) \subset H^0(L, \mathcal{N}_{L/Y})$, cf. [281, Thm. II.1.7] or Remark **2**.1.11. As the ideal sheaf $\mathcal{I}_x$ of $x \in L \simeq \mathbb{P}^1$ is isomorphic to $\mathcal{O}_L(-1)$, this space is non-zero if and only if $L \subset Y$ is a line of the second type, i.e. $\mathcal{N}_{L/Y} \simeq \mathcal{O}_L(1) \oplus \mathcal{O}_L(-1)$, cf. Lemma **2**.1.13.

Note that the non-vanishing of $H^0(L, \mathcal{N}_{L/Y} \otimes \mathcal{I}_x)$ only depends on $L \subset Y$ and not on the point $x \in L$. In particular, $q\colon \mathbb{L} \longrightarrow Y$ is ramified along all of $L = p^{-1}[L] \subset \mathbb{L}$ or not at all. This confirms that $R(q) = p^{-1}(R)$ for the curve $R = F_2(Y) \subset F(Y)$. More precisely, one has the following characterization, cf. the more general Corollary **2**.2.15.

Corollary 1.3 *The morphism $q\colon \mathbb{L} \longrightarrow Y$ is smooth at $(L, x) \in \mathbb{L}$ if and only if $L \subset Y$ is a line of the first type, i.e. $R = F_2(Y) \subset F(Y)$ and $R(q) = p^{-1}(F_2(Y)) = \mathbb{L}_2 \subset \mathbb{L}$.* □

Exercise 1.4 Show that every point $x \in Y$ that is not contained in any line of the second type is contained in exactly six lines.

Exercise 1.5 Show that every *Eckardt point* $x \in Y$, i.e. a point contained in infinitely many lines, is contained in infinitely many lines of the second type. In fact, according to Murre [364, Lem. 1.18] every point contained in a line of the first type is only contained in finitely many lines and, therefore, all lines containing an Eckardt point are of the second type.

Exercise 1.6 It turns out that the ramification of $q\colon \mathbb{L} \longrightarrow Y$ along $R(q)$ is generically simple, see [119, Lem. 10.18]. Consider the projection $q\colon R(q) \longrightarrow \bar{R} := q(R(q)) \subset Y$ and denote its degree by d. Show that $\bar{R} \in |\mathcal{O}_Y(30/d)|$, which will be confirmed in Remark 1.18. In fact, it turns out that at least for generic cubic threefolds one has $d = 1$ and hence $\bar{R} \in |\mathcal{O}_Y(30)|$, see Remark 1.17. Show that R is a curve of arithmetic genus

$$p_a(R) = 136.$$

[2] The monodromy of this cover coincides with the monodromy of $\mathbb{L} \longrightarrow Y$ which is $\mathfrak{S}_6$, see page 194.

Remark 1.7 (i) According to Proposition **2**.2.13, for generic Y the curve $R = F_2(Y)$ is smooth. It seems that in [364, Cor. 1.9] a local computation is used to show that R is always smooth. However, special smooth cubic threefolds may contain Eckardt points and for those the morphism $q\colon \mathbb{L} \longrightarrow Y$ may contract curves $E_x = q^{-1}(x) \subset \mathbb{L}$. The image curves $p(E_x)$, which are smooth elliptic, are irreducible components of R. Therefore, a smooth cubic $Y \subset \mathbb{P}^4$ can admit at most finitely many Eckardt points. This is a result of Clemens and Griffiths [119, Lem. 8.1], see also Coskun–Starr [125, Cor. 2.2]. According to Roulleau [402], a smooth cubic threefold contains at most 30 Eckardt points. Furthermore, as $R \in |\mathcal{O}(2)|$, it is ample and hence connected. Thus, whenever R is reducible, it is singular.

Note that conversely the smoothness of $F_2(Y)$ for generic Y implies that the generic cubic does not contain any Eckardt points. For explicit computations in the case of the Fermat cubic threefold, see [73, 404], Bockondas and Boissière [73] also show that the singularities of $R = F_2(Y)$ are exactly the points corresponding to triple lines, i.e. lines $L \in F(Y)$ for which there exists a plane with $\mathbb{P}^2 \cap Y = 3L$. That there are at most finitely many triple lines was already observed by Clemens and Griffiths [119, Lem. 10.15].

(ii) The interpretation of R as $F_2(Y)$, which in turn by Remark **2**.2.11 can be thought of as the degeneracy locus $M_2(\psi)$ of the natural map $\psi\colon \mathcal{Q}_F \longrightarrow S^2(\mathcal{S}_F^*)$, allows one to deduce all at once that $R \neq \emptyset$, $\dim(R) = 1$, and $R = \{L \in F \mid \det(\psi_L) = 0\} \in |\mathcal{O}_F(2)|$, for $\det(\mathcal{Q})^* \otimes \det(S^2(\mathcal{S}_F^*)) \simeq \mathcal{O}_F(2)$. Note that the scheme structures of $R = F_2(Y) = M_2(\psi)$ all coincide.

Clearly, two distinct lines $L_1, L_2 \subset \mathbb{P}^4$, contained in the cubic Y or not, do not intersect at all or in exactly one point (and there transversally). In the second case, they are contained in a unique plane. For infinitesimal deformations also, one distinguishes between these two cases:

(i) Let $L \subset Y$ be a line of the first type. Then the lines $L_t \subset Y$ corresponding to $t \in F(Y)$ close to $L \in F(Y)$ are disjoint to L. Indeed, a first order deformation L_ε of L with non-trivial intersection with L would fix some point x and, therefore, define a non-trivial global section of $\mathcal{N}_{L/Y} \otimes \mathcal{I}_x \simeq \mathcal{O}_L(-1) \oplus \mathcal{O}_L(-1)$, which is absurd.

(ii) If $L \subset Y$ is of the second type, then for each point $x \in L$, the subspace

$$H^0(L, \mathcal{N}_{L/Y} \otimes \mathcal{I}_x) \subset H^0(L, \mathcal{N}_{L/Y}) \simeq T_{[L]}F(Y)$$

is one-dimensional and corresponds to a deformation $\operatorname{Spec} k[\varepsilon] \times \{x\} \subset L_\varepsilon \subset \operatorname{Spec} k[\varepsilon] \times Y$. The image of L_ε in $\mathbb{P}^4$ spans a plane $P_L \simeq \mathbb{P}^2$ and, therefore, is contained in $Y \cap P_L$. Then L_ε is the double line $2L \subset P_L$ and hence $2L \subset Y \cap P_L$. Note that a priori the plane P_L depends on the choice of the point $x \in L$, but from the fact that $2L \subset Y \cap P_L$ one deduces that it does not. So, we have reproved Corollary **2**.2.6 in this case, which we state again as follows.

Lemma 1.8 *Let $L \subset Y$ be a line in a smooth cubic threefold. Then L is of the second type if and only if there exists a (unique) plane $\mathbb{P}^2 \simeq P_L \subset \mathbb{P}^4$, that is tangent to Y at every point of L, i.e. $2L \subset P_L \cap Y$. For a triple line, one has $3L = P_L \cap Y$.* □

Corollary 1.9 *The generic line $L \subset Y$ is of the first type. In fact, for a dense open subset of lines $L \in F(Y)$ and any plane $\mathbb{P}^2 \subset \mathbb{P}^4$ containing L the intersection $Y \cap \mathbb{P}^2$ is reduced.*

Proof Lines of the second type are parametrized by the curve $R = F_2(Y)$. So, any line $L_1 \in F(Y) \setminus R$ is of the first type. Furthermore, for any line of the second type $L_2 \subset Y$, there exists a unique plane $\mathbb{P}^2 \simeq P_{L_2} \subset \mathbb{P}^4$ which intersects Y along L_2 with multiplicity at least two. The points $L \in F$ corresponding to the residual line $L \subset Y$ of $L_2 \subset P_{L_2} \cap Y$ for L_2 moving in R sweep out a curve $R' \subset F(Y)$. Then any line corresponding to a point in $F(Y) \setminus (R \cup R')$ has the required property. See Remark **6**.4.19 for a similar construction in dimension four and some information about R'. □

The curve R comes with a natural double cover $\widetilde{R} \longrightarrow R$. Indeed, for a line of the second type $L \subset Y$, so $L \in R$, the restriction of the Gauss map $\gamma\colon L \longrightarrow \gamma(L)$ is of degree two with two ramification points, see Exercise **2**.2.10. If we let $\widetilde{R}$ be the curve that parametrizes lines of the second type together with these ramification points, then the projection to R is étale of degree two. The situation will be studied in more detail for cubic fourfolds in Section **6**.4.4.

Remark 1.10 The curve $R' \subset F(Y)$ of lines that are residual to lines of the second type considered in the proof above has been studied in more detail by Lahoz, Naranjo, and Rojas. They show [303, Thm. C] that $[R'] = 8g \in H^2(F(Y), \mathbb{Q})$ and that for generic Y the curve R' is irreducible with exactly 1 485 nodes. It turns out that for the generic cubic threefold Y, the map $R \longrightarrow R'$, $L \longmapsto L'$, that sends $L \in R$ to the residual line L' of $2L \subset P_L \cap Y$ is of degree one, i.e. R is the normalization of R'. See Remark 1.25 for more details.

1.2 Lines Intersecting a Given Line We move on to the next class of curves in $F(Y)$. For a fixed line $L \in F$, we define the curve $C_L \subset F(Y)$ as the closure of the curve of all lines $L' \subset Y$ different from L but with non-empty intersection $\emptyset \neq L \cap L'$:

$$\{L' \neq L \mid L' \cap L \neq \emptyset\} \subset C_L \subset F(Y).$$

Note that taking the closure adds at most the point L. To define C_L rigorously with a natural scheme structure, use the formalism of Section **2**.5.5 and consider

$$\varphi = p_* \circ q^* \colon \mathrm{CH}^2(Y) \longrightarrow \mathrm{CH}^1(F(Y)) \simeq \mathrm{Pic}(F(Y)).$$

By definition, p_* is trivial on components of $q^{-1}(L)$ with positive fibre dimension over $F(Y)$, e.g. the class of the component $p^{-1}[L] \subset q^{-1}(L)$ is mapped to 0 under p_*. The image of the class of the line $L \subset Y$ under φ is a line bundle $\mathcal{O}(C_L)$ that comes with a natural section (up to scaling) vanishing along C_L.

The curve C_L is endowed with a natural involution

$$\iota\colon C_L \longrightarrow C_L, \quad L' \longmapsto L'' \tag{1.1}$$

determined by the condition $L \cup L' \cup L'' = \overline{LL'} \cap Y$. The quotient $\pi\colon C_L \longrightarrow C_L/\iota$ will play a key role in the rest of this chapter.

Note that at this point it is not clear whether C_L is smooth for generic choice of L. This will be deduced later as a consequence of Lemma 1.26.

The point $L \in F(Y)$ corresponding to a line $L \subset Y$ may or may not be contained in its associated curve C_L. This is the content of the next result, cf. [119, Lem. 10.7].

Lemma 1.11 *A line $L \subset Y$ is of the second type if and only if $L \in C_L$.*

Proof Assume $L \in C_L$. Then, as discussed above, a first order deformation of $L \subset Y$ given by deforming L along C_L defines a non-trivial class in $H^0(L, \mathcal{N}_{L/Y}(-1))$ and, therefore, L is of the second type. More geometrically, let $L' \in C_L$ specialize to $L \in C_L$. Then the plane $\overline{LL'}$ specializes to a plane P with $2L \subset Y \cap P$ and, therefore, L is of the second type.

For the converse, consider the curve R of all lines L of the second type. To prove that $L \in C_L$ holds for all $L \in R$, it suffices to prove it for the generic $L \in R$. In particular, we may assume that L is not one of the finitely many triple lines, see Remark 1.7. In other words, we can assume that the residual line L_0 of $2L \subset P_L \cap Y$ is not L, i.e. $3L \neq P_L \cap Y$. A small deformation $L_t \in C_L$ of $L_0 \in C_L$, leads to a deformation $P_t := \overline{LL_t}$ of the plane $P_L = \overline{LL_0}$. The residual line L_t' of $L \cup L_t \subset P_t \cap Y$ also defines a point in C_L and specializes to L. Hence, $L \in C_L$. □

Assume now that L is of the first type, i.e. $L \in F(Y) \setminus C_L$. In this case, $q^{-1}(L)$ is the disjoint union of $p^{-1}[L] \simeq L$ and a curve mapping isomorphically onto C_L:

$$q^{-1}(L) = p^{-1}[L] \sqcup C_L. \tag{1.2}$$

Indeed, for $L' \in C_L$, the line $L' = q(p^{-1}[L'])$ intersects L transversally in exactly one point. Hence, $p^{-1}[L']$ and $q^{-1}(L)$ intersect with multiplicity one.

Remark 1.12 Any curve in $F(Y)$ intersects the ample curve $R \subset F(Y)$ of all lines of the second type. Applied to C_L, this shows that any line $L \subset Y$ intersects some line $L' \subset Y$ of the second type. This can also be deduced from the fact that $q\colon \mathbb{L} \longrightarrow Y$ has to ramify along a divisor in Y which necessarily intersects every line.

Remark 1.13 It is not difficult to show that distinct lines L yield distinct curves C_L:

$$L \neq L' \Rightarrow C_L \neq C_{L'}.$$

For example, if L and L' intersect in a point $x \in Y$, in other words $\overline{LL'} \simeq \mathbb{P}^2$, then the intersection $C_L \cap C_{L'}$ parametrizes the residual line of $L \cup L' \subset \overline{LL'} \cap Y$ (which may coincide with L or L') and all lines passing through x. The latter set is finite unless x is an Eckardt point. Thus, if $C_L = C_{L'}$, then all lines intersecting L would go through one of the finitely many Eckardt points contained in L, which is absurd.

In the case of two disjoint lines $L \cap L' = \emptyset$, the intersection of $\overline{LL'} \simeq \mathbb{P}^3$ with Y defines a cubic surface $S \subset Y$. If S is smooth, then it contains only finitely many lines and hence the intersection $C_L \cap C_{L'}$ is finite. If S is normal and not a cone, then S has only rational double points as singularities and still contains only finitely many lines [157, Ch. 9.2.2]. If S is a cone over a cubic curve, then we may assume that L is a line through the vertex and L' is a component of the cubic curve. In particular, one finds a line intersecting L but not L'. Finally, if S is neither normal nor a cone, then S is reducible or projectively equivalent to one of two specific surfaces [157, Thm. 9.2.1]. One has to argue separately in the two cases.

Lemma 1.14 *For any two lines $L_1, L_2 \subset Y$, the curves $C_{L_1}, C_{L_2} \subset F(Y)$ are algebraically equivalent. Moreover, $\mathcal{O}(C_L)^{\otimes 3}$ is algebraically equivalent to $\mathcal{O}_F(1) \simeq \omega_F$. In particular, $3 \cdot [C_L] = g \in H^2(F(Y), \mathbb{Z})$ and $(C_L.C_L) = 5$.*

Proof By construction, $\mathcal{O}(C_L)$ is the image of $[L] \in \mathrm{CH}^2(Y)$ under

$$\varphi\colon \mathrm{CH}^2(Y) \longrightarrow \mathrm{CH}^1(F(Y)) \simeq \mathrm{Pic}(F(Y)).$$

Since all lines parametrized by the connected Fano surface $F(Y)$ are algebraically equivalent to each other, the same is true for the invertible sheaves $\mathcal{O}(C_L)$, i.e. the algebraic equivalence class of C_L is independent of L.

The second statement has been proved in general already in Remark **2**.5.2, where the curve C_L was denoted by F_L. Let us briefly sketch the argument again for threefolds. The class $h^2 \in \mathrm{CH}^2(Y)$ is represented by the intersection with an arbitrary plane $\mathbb{P}^2 \subset \mathbb{P}^4$. Choosing a plane that intersects Y in three lines, see Exercise **2**.1.20, this shows that $h^2 = [L_1] + [L_2] + [L_3]$. For example, for $\mathbb{P}^2 = P_L$ for L of the second type, $h^2 = 2[L] + [L']$ with L' the residual line of $2L \subset P_L \cap Y$.

Hence, by Lemma **2**.5.1, $g = \mathrm{c}_1(\mathcal{O}_F(1)) = \varphi(h^2) = [C_{L_1}] + [C_{L_2}] + [C_{L_3}]$, which is algebraically equivalent to $3 \cdot [C_{L_i}]$. □

Exercise 1.15 Let us make the last step of the above proof more explicit. We represent a plane $\mathbb{P}^2 \subset \mathbb{P}^4$ with $\mathbb{P}^2 \cap Y = L_1 \cup L_2 \cup L_3$ as the intersection $V(s_1) \cap V(s_2)$, where $s_1, s_2 \in V^* = H^0(\mathbb{P}^4, \mathcal{O}(1)) \simeq H^0(F, \mathcal{S}_F^*)$ are two linearly independent sections. Then the zero set of the image of $s_1 \wedge s_2$ under the natural map $\bigwedge^2 V^* \longrightarrow H^0(F(Y), \bigwedge^2 \mathcal{S}_F^*) \simeq H^0(F(Y), \mathcal{O}_F(1))$ is the set of all lines $L = \mathbb{P}(W)$ with $(s_1 \wedge s_2)|_W = 0$. The latter is equivalent to $L \cap \mathbb{P}^2 \neq \emptyset$ or, equivalently, to $L \cap (L_1 \cup L_2 \cup L_3) \neq \emptyset$, which in turn just says $L \in C_{L_1} \cup C_{L_2} \cup C_{L_3}$. Thus, once again, $\mathcal{O}(1) \simeq \mathcal{O}(C_{L_1} + C_{L_2} + C_{L_3})$.

Exercise 1.16 Show that the self-intersection and the (arithmetic) genus of $C_L \subset F(Y)$ are given by

$$(C_L.C_L) = 5 \quad \text{and} \quad g(C_L) = 11.$$

Relate this to the fact that for two disjoint lines on a smooth cubic surface there exist exactly five lines intersecting both, see Section **4**.3.5.

Remark 1.17 We know that for the generic cubic threefold Y the curve R is smooth and irreducible with $[R] = 2g$ in $H^2(F, \mathbb{Z})$. On the other hand, for any line L, Lemma 1.14 says $[C_L] = (1/3)g$. This proves that R cannot be contained in C_L and that, therefore, R and C_L intersect in only finitely many points. Applied to a line L of the second type, it shows that through the generic point of L there is no other line of the second type passing through it.[3] In Exercise 1.6, this fact was alluded to already and used to prove that for the generic cubic threefold $q\colon R(q) \twoheadrightarrow \bar{R}$ is generically injective.

Remark 1.18 Let $L_0 \subset Y$ be a fixed generic line and so, in particular, of the first type.

(i) As $q\colon \mathbb{L} \twoheadrightarrow Y$ is of degree six, sending $L \in C_{L_0} \subset F$ to the point of intersection of L_0 and L defines a morphism of degree five which by (1.2) is nothing but q:

$$q\colon C_{L_0} \xrightarrow{5:1} L_0, \;\; L \longmapsto L_0 \cap L.$$

The ramification points of $C_{L_0} \twoheadrightarrow L_0$, i.e. the points in the intersection $R(q) \cap C_{L_0}$, correspond to lines of the second type intersecting L_0. The Hurwitz formula applied to $q\colon C_{L_0} \twoheadrightarrow L_0$ shows $d \cdot (\bar{R}.L_0) = (R(q).C_{L_0}) = 30$, where d is the degree of $R(q) \twoheadrightarrow \bar{R} = q(R(q))$. This confirms $\bar{R} \in |\mathcal{O}_Y(30/d)|$, see Exercise 1.6, and as explained in Remark 1.17, $R(q) \twoheadrightarrow \bar{R}$ is generically injective, at least for the generic cubic threefold Y. Since the generic line L_0 avoids the locus where $R(q) \twoheadrightarrow \bar{R}$ is not injective, we can conclude that for every point $x \in L_0$ there exists at most one line of the second type passing through x or, equivalently, that the morphism $C_{L_0} \twoheadrightarrow L_0$ is not étale in at most one point in each fibre.

In fact, Clemens and Griffiths [119, Lem. 10.18] showed that the ramification of $\mathbb{L} \twoheadrightarrow Y$ is generically simple along $R(q)$ which implies that for the generic line L_0 the morphism $C_{L_0} \twoheadrightarrow L_0$ has simple ramification only. Together with the above, this shows that a fibre of $C_{L_0} \twoheadrightarrow L_0$ consists of either six lines of the first type or of four lines of the first type and one of the second.

It can be shown that the monodromy group of $C_{L_0} \twoheadrightarrow L_0$ is $\mathfrak{S}_5$.[4]

(ii) Let $L \in C_{L_0}$ and denote by L' the residual line of $L_0 \cup L \subset \overline{L_0 L} \cap Y$ which may coincide with L or L_0. Then

$$q^*\mathcal{O}_{L_0}(1) \simeq \mathcal{O}(C_L)|_{C_{L_0}} \otimes \mathcal{O}_{C_{L_0}}(L - L'). \tag{1.3}$$

[3] I wish to thank A. Rojas for the argument.
[4] I wish to thank F. Gounelas for the argument.

Indeed, if $\{x\} = L_0 \cap L$, then $q^{-1}(x)$ parametrizes the five lines (with multiplicities) $L_1 := L, L_2, \ldots, L_5$ distinct from L_0 containing x. Hence, $q^*\mathcal{O}_{L_0}(1) \simeq \mathcal{O}_{C_{L_0}}(\sum L_i)$. On the other hand, $C_{L_0} \cap C_L = \{L', L_2, \ldots, L_5\}$ and, therefore, $q^*\mathcal{O}_{L_0}(1) \simeq \mathcal{O}(C_L)|_{C_{L_0}} \otimes \mathcal{O}_{C_{L_0}}(L - L')$.

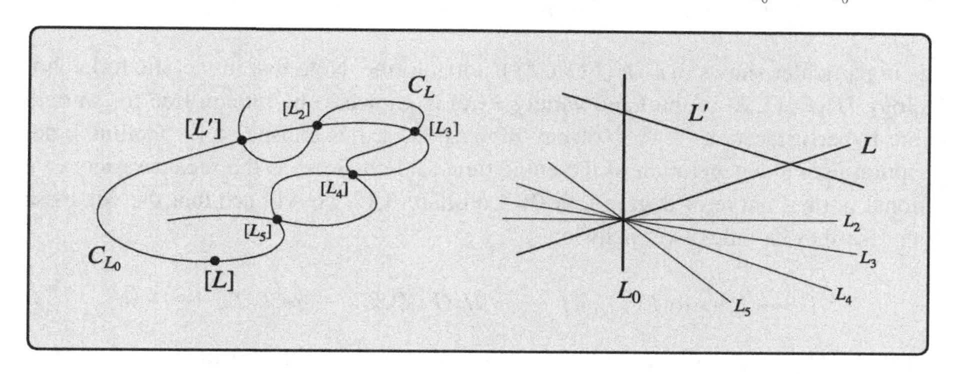

In other words, the two sets $q^{-1}(x) = \{L_0, L, L_2, \ldots, L_5\}$ and $(C_{L_0} \cap C_L) \cup \{L_0\} = \{L_0\} \cup \{L', L_2, \ldots, L_5\}$ differ only by L getting swapped for its residual line L'. In the proof of Corollary 1.30, we will see that $h^0(q^*\mathcal{O}_{L_0}(1)) = 2$.

(iii) As a consequence of (1.3), one obtains for any two points $L_1, L_2 \in C_{L_0}$ an isomorphism

$$\mathcal{O}(C_{L_2} - C_{L_1})|_{C_{L_0}} \simeq \mathcal{O}_{C_{L_0}}(L_1 - L_1' - (L_2 - L_2')), \tag{1.4}$$

which will be crucial in the proof of Corollary 3.12.

Exercise 1.19 Let L_0 be (as above) a generic line in a smooth cubic threefold. Apply the Hurwitz formula to the projection $q\colon C_{L_0} \longrightarrow L_0$ to prove

$$\omega_{C_{L_0}} \simeq q^*\mathcal{O}(-2) \otimes \mathcal{O}_F(2)|_{C_{L_0}}.$$

Combine this with the adjunction formula $\omega_{C_{L_0}} \simeq \mathcal{O}_F(1)|_{C_{L_0}} \otimes \mathcal{O}(C_{L_0})|_{C_{L_0}}$ to deduce

$$\mathcal{O}(C_{L_0})|_{C_{L_0}} \simeq q^*\mathcal{O}(-2) \otimes \mathcal{O}_F(1)|_{C_{L_0}}.$$

Corollary 1.20 *The Plücker class $g = c_1(\mathcal{O}_F(1)) \in H^2(F(Y), \mathbb{Z})$ is divisible by three and so is the Hodge–Riemann pairing $\int_F \gamma_1 \cdot \gamma_2 \cdot g$ on $H^1(F(Y), \mathbb{Z})$, cf. Proposition* 1.5.5. □

Remark 1.21 The fundamental group of $F(Y)$ has been computed by Collino [121]:[5] There exists a non-split short exact sequence

$$0 \longrightarrow \mathbb{Z}/2\mathbb{Z} \longrightarrow \pi_1(F(Y)) \longrightarrow \mathbb{Z}^{\oplus 10} \longrightarrow 0. \tag{1.5}$$

[5] Thanks to S. Stark for the reference.

In other words, $[\pi_1(F(Y)), \pi_1(F(Y))] \simeq \mathbb{Z}/2\mathbb{Z}$ and $H_1(F(Y),\mathbb{Z}) \simeq \mathbb{Z}^{\oplus 10}$. By the universal coefficient theorem, i.e. the short exact sequence, see [155, p. 186]:

$$0 \longrightarrow \mathrm{Ext}^1(H_1(F(Y),\mathbb{Z}),\mathbb{Z}) \longrightarrow H^2(F(Y),\mathbb{Z}) \longrightarrow \mathrm{Hom}(H_2(F(Y),\mathbb{Z}),\mathbb{Z}) \longrightarrow 0,$$

this in particular shows that $H^2(F(Y),\mathbb{Z})$ is torsion free. Note that in fact the full cohomology $H^*(F(X),\mathbb{Z})$ of the Fano variety $F(X)$ is known to be torsion free for smooth cubic hypersurfaces $X \subset \mathbb{P}^{n+1}$ of any dimension, see Exercise **2**.4.13. Collino's description uses a degeneration of the cubic threefold obtained as the secant variety of a rational normal curve of degree four. In Corollary 3.12 we will see that the Albanese map describes an index two inclusion

$$0 \longrightarrow H^2(\mathrm{Alb}(F(Y)),\mathbb{Z}) \xrightarrow{a^*} H^2(F(Y),\mathbb{Z}) \longrightarrow \mathbb{Z}/2\mathbb{Z} \longrightarrow 0$$

and in [121] it is the non-triviality of this cokernel $\mathbb{Z}/2\mathbb{Z}$ that ensures that a certain natural map $\mathbb{Z}/2\mathbb{Z} \twoheadrightarrow [\pi_1(F(Y)), \pi_1(F(Y))]$ is indeed non-zero. The surjection in (1.5) is the natural map $\pi_1(F(Y)) \longrightarrow \pi_1(\mathrm{Alb}(F(X)))$.

Remark 1.22 For a general cubic threefold $Y \subset \mathbb{P}^4$, the line bundle $\mathcal{O}(C_L)$ generates the Néron–Severi group

$$\mathrm{NS}(F(Y)) \simeq \mathbb{Z} \cdot \mathcal{O}(C_L).$$

Indeed, we have remarked that $\mathrm{NS}(F(Y))$ is of rank one for general Y and, since we have $(C_L.C_L) = 5$, the line bundle $\mathcal{O}(C_L)$ defines a primitive class in $\mathrm{NS}(F(Y))$.

1.3 Conic Fibration We study the linear projection from a line $L \subset Y$ as a special case of the construction in Section **1**.5.1.

Let $L = \mathbb{P}(W) \subset \mathbb{P}^4 = \mathbb{P}(V)$ be a line contained in the smooth cubic hypersurface $Y \subset \mathbb{P}^4$. Assume $\mathbb{P}^2 \subset \mathbb{P}^4$ is a plane disjoint to L, of which we think as $\mathbb{P}(V/W)$. The linear projection $Y \setminus L \dashrightarrow \mathbb{P}^2$ from L onto this plane is the rational map associated with the linear system $|\mathcal{O}_Y(1) \otimes \mathcal{I}_L| \subset |\mathcal{O}_Y(1)|$. It is resolved by a simple blow-up $\tau\colon \mathrm{Bl}_L(Y) \longrightarrow Y$ and the induced morphism $\phi\colon \mathrm{Bl}_L(Y) \longrightarrow \mathbb{P}^2$ is then associated with the complete linear system $|\tau^*\mathcal{O}_Y(1) \otimes \mathcal{O}(-E)|$, where E is the exceptional divisor.

The fibre over a point $y \in \mathbb{P}^2$ is the residual conic of $L \subset \overline{yL} \cap Y \subset \overline{yL} \simeq \mathbb{P}^2$. The conic is smooth or a union of two lines L_1, L_2, possibly non-reduced, i.e. $L_1 = L_2$, or with $L_i = L$. Note that in the case that $L_i = L$, the plane $\overline{yL} \simeq \mathbb{P}^2$ intersects Y with higher multiplicity along L and hence L is of the second type. Therefore, if L was chosen to be of the first type, then the fibres of $\phi\colon \mathrm{Bl}_L(Y) \longrightarrow \mathbb{P}^2$ are either smooth conics or possibly non-reduced unions of two lines, both different from L.

Corollary 1.23 *Let $L \subset Y$ be a line of the first type.*

(i) *Then the linear projection from L defines a morphism*

$$\phi\colon \mathrm{Bl}_L(Y) \twoheadrightarrow \mathbb{P}^2,$$

with a discriminant curve $D_L \subset \mathbb{P}^2$ of degree five and arithmetic genus six.

(ii) *The fibre over a point $y \in D_L$ is the possibly non-reduced union of two lines $\phi^{-1}(y) = L_1 \cup L_2$ with L, L_1, L_2 coplanar.*

(iii) *With the notation in* (ii), *$L_1 = L_2$ if and only if y is a singular point of D_L, which then is an ordinary double point.*

(iv) *For $L \subset Y$ generic in the sense of Corollary* 1.9, *i.e. $L \in F(Y) \setminus (R \cup R')$, then $\phi^{-1}(y) = L_1 \cup L_2$ with $L_1 \neq L_2$ and $L_i \neq L$ for all $y \in D_L$. In particular, D_L is smooth for generically chosen L.*

Proof Most of this has been verified already, see also Section **1**.5.2. For (iii), see [41, Prop. 1.2] or [80, Lem. 2]. The last assertion is a consequence of Proposition **1**.5.3. The fibre over $y \in D_L$ cannot be a double line $L_1 = L_2$, as then $\mathbb{P}^2 \cap Y = L \cup 2L_1$, and so L_1 would be of the second type, which is excluded for L generic. The smoothness of D_L for the generic choice of the pair $L \subset Y$ also follows from Remark **1**.5.8. □

Remark 1.24 The abstract approach matches nicely with the intuitive picture. Here are two comments in this direction.

(i) That D_L is of degree five, i.e. $D_L \in |\mathcal{O}(5)|$, can be linked to the fact that a line in a smooth cubic surface $S \subset \mathbb{P}^3$ is intersected by five pairwise disjoint pairs of lines, see Section **4**.3.3. Indeed, if $Y \subset \mathbb{P}^4$ is intersected with a generic hyperplane $\mathbb{P}^3 \subset \mathbb{P}^4$ containing L, then $D_L \subset \mathbb{P}^2$ is intersected with a generic line $\mathbb{P}^1 \subset \mathbb{P}^2$. The fibres over the intersection points $y \in D_L \cap \mathbb{P}^1$ are the pairs of lines in Y contained in the cubic surface $S := Y \cap \mathbb{P}^3$ intersecting L, of which there are exactly five.

(ii) For a line of the first type, the exceptional divisor

$$L \times \mathbb{P}^1 \simeq \mathbb{P}(\mathcal{N}_{L/Y} \simeq \mathcal{O}_L^{\oplus 2}) \simeq E \subset \mathrm{Bl}_L(Y)$$

has normal bundle $\mathcal{O}(0,-1)$. Hence, the restriction of $\phi^*\mathcal{O}(1) \simeq \tau^*\mathcal{O}(1) \otimes \mathcal{O}(-E)$ to $\mathbb{P}^1 \times \mathbb{P}^1$ is $\mathcal{O}(1,1)$. In particular, the composition $\mathbb{P}^1 \times \mathbb{P}^1 \simeq \mathbb{P}(\mathcal{N}_{L/Y}) \subset \mathrm{Bl}_L(Y) \longrightarrow \mathbb{P}^2$ is a morphism of degree two, which confirms the geometric description that $\tau(\phi^{-1}(y)) \cap L$ is the intersection of the residual conic of $L \subset \overline{yL} \cap Y$ with L.

Remark 1.25 A quick dimension count shows that (up to coordinate transformations) the generic quintic curve $D \subset \mathbb{P}^2$ is of the form D_L for some smooth cubic threefold Y with a line $L \subset Y$. Indeed, $\dim |\mathcal{O}_{\mathbb{P}^2}(5)| = 20$ and $\dim \mathrm{PGL}(3) = 8$, while the space of pairs $L \subset Y$ is of dimension $12 = 10 + 2$.

As a consequence, the generic singular $D_{L'}$ has just one node. Since a node of $D_{L'}$ corresponds to a line L of the second type with $2L \cup L' = \mathbb{P}^2 \cap Y$, in this case $L' \in R'$ and $R \twoheadrightarrow R'$ is of degree one, as mentioned already in Remark 1.10.[6]

We now consider the restriction $\mathrm{Bl}_L(Y)|_{D_L} = \phi^{-1}(D_L) \twoheadrightarrow D_L$ of the linear projection $\phi\colon \mathrm{Bl}_L(Y) \twoheadrightarrow \mathbb{P}^2$ to the discriminant curve $D_L \subset \mathbb{P}^2$. We assume that L is generic in the sense of Corollary 1.9, so that all fibres are reduced singular conics, i.e. unions of two distinct lines. Then the relative Fano scheme of lines

$$\pi\colon \widetilde{D}_L := F(\mathrm{Bl}_L(Y)|_{D_L}/D_L) \twoheadrightarrow D_L$$

parametrizing all lines in the fibres of ϕ is an étale cover. The morphism is indeed unramified which can be shown by abstract deformation theory or simply by arguing that a morphism from one curve onto a smooth curve with exactly two distinct points in each fibre is étale. As $\widetilde{D}_L$ parametrizes lines in Y, it comes with a classifying morphism $\widetilde{D}_L \hookrightarrow F(Y)$ which is easily seen to be a closed immersion.

Alternatively, $\widetilde{D}_L$ can be obtained as the Stein factorization of the composition of the normalization of $\phi^{-1}(D_L)$ with ϕ. The morphism to $F(Y)$ can then be viewed as follows: the natural rational map $\phi^{-1}(D_L) \dashrightarrow \mathbb{L}$ is regular on the complement of the section of $\phi^{-1}(D_L) \twoheadrightarrow D_L$ given by the intersection points of the two lines in each fibre. The image of the composition with $p\colon \mathbb{L} \twoheadrightarrow F(Y)$ is $\widetilde{D}_L \hookrightarrow F(Y)$.

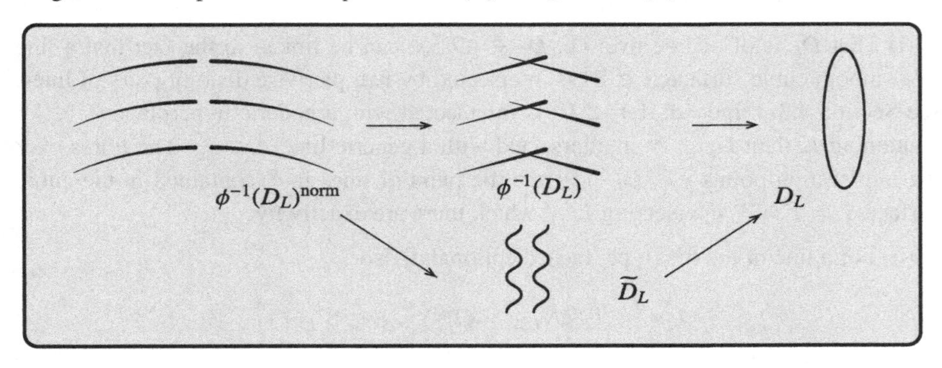

Lemma 1.26 *For a generic line $L \subset Y$, the two curves $\widetilde{D}_L \subset F(Y)$ and $C_L \subset F(Y)$ coincide. Furthermore, $\widetilde{D}_L = C_L$ is a smooth curve of genus* 11.

Proof Indeed, $\widetilde{D}_L$ and C_L both parametrize all lines $L \neq L' \subset Y$ intersecting L. We know that the (arithmetic) genus of C_L is $g(C_L) = 11$ and the same is true for $\widetilde{D}_L$ by Hurwitz's formula. This is enough to conclude equality. Smoothness of $\widetilde{D}_L$ follows from the smoothness of D_L. □

[6] Thanks to A. Rojas for the argument.

Corollary 1.27 *For a generic line $L \subset Y$, i.e. $L \in F(Y) \setminus (R \cup R')$, the curve $\widetilde{D}_L = C_L$ is connected or, equivalently, irreducible.*

Proof For a general cubic Y, one has $\rho(F(Y)) = 1$ and, in fact, $\mathrm{NS}(F(Y)) \simeq \mathbb{Z} \cdot \mathcal{O}(C_L)$, see Remark 1.22. Therefore, in this case C_L has to be irreducible. As under deformations of the pair $L \subset Y$ with $L \in F(Y) \setminus (R \cup R')$ the topology of the situation does not change, this proves the assertion in general.

Alternatively, one can use the smoothness of C_L and the general fact that effective ample divisors are connected, cf. [223, III. Cor. 7.9]. □

Exercise 1.28 Show that

$$\pi \colon C_L \simeq \widetilde{D}_L \longrightarrow D_L$$

is the quotient of the natural involution $L' \longmapsto L''$, see (1.1). Use Exercise 1.19 to deduce

$$\mathcal{O}_F(2)_{C_L} \simeq \pi^*\mathcal{O}(2) \otimes q^*\mathcal{O}(2).$$

In fact, using the explicit description of the morphisms q and π one finds

$$\mathcal{O}_F(1)|_{C_L} \simeq \pi^*\mathcal{O}(1) \otimes q^*\mathcal{O}(1).$$

Remark 1.29 See also [42, 202] for yet another proof that does not reduce to the case $\rho(F(Y)) = 1$ or uses the ampleness of C_L. The idea there is that sending a line $L' \in C_L$ to its intersection with L defines a map $C_L \longrightarrow L$ of degree five, cf. Remark 1.18. If C_L is not irreducible, then one of the irreducible components is rational or hyperelliptic. The first case would contradict the injectivity of the Albanese map in Corollary 2.8 and to exclude the second one uses that $F(Y)$ is not covered by hyperelliptic curves, cf. [202, Sec. 3].

If L is not generic but $L \in R \subset F(Y)$ is still a generic point of R, then the curve C_L is again irreducible, its genus is 11, and it comes with a morphism $C_L \twoheadrightarrow \mathbb{P}^1$ of degree 4, see [202, Lem. 3.3].

A Riemann–Roch computation reveals that $\chi(F(Y), \mathcal{O}(C_L)) = 1$. Although there is a priori no reason for the higher cohomology groups of $\mathcal{O}(C_L)$ to vanish, at least Kodaira vanishing does not imply anything in this direction, the following result was proved by Tyurin [447, Lem. 1.8].

Corollary 1.30 *For every line $L \subset Y$, the induced curve $C_L \subset F(Y)$ is unique in its linear system, i.e. $h^0(F(Y), \mathcal{O}(C_L)) = 1$. As a consequence, one obtains an injection*

$$F(Y) \hookrightarrow \mathrm{Pic}(F(Y)), \quad L \longmapsto \mathcal{O}(C_L). \tag{1.6}$$

Proof As a first step, we observe that for a generic $L_0 \in C_L$ the associated line bundle $\mathcal{O}(C_{L_0})$ is indeed not isomorphic to $\mathcal{O}(C_L)$. For this, it suffices to show that the morphism (1.6) does not contract the curve C_L. The latter follows from (1.6) being unramified which is a consequence of the discussion in Sections 2 and 3 below, see Corollary 2.9 and Lemma 3.1, which is independent of our discussion here.

Now pick a line $L_0 \in C_L$ with $\mathcal{O}(C_{L_0}) \not\simeq \mathcal{O}(C_L)$ and assume in addition that L_0 can be chosen generic such that in particular C_{L_0} is smooth. Then consider the exact sequence

$$0 \longrightarrow \mathcal{O}(C_L - C_{L_0}) \longrightarrow \mathcal{O}(C_L) \longrightarrow \mathcal{O}(C_L)|_{C_{L_0}} \longrightarrow 0$$

from which we deduce that it suffices to prove $h^0(C_{L_0}, \mathcal{O}(C_L)|_{C_{L_0}}) = 1$. Note that according to (1.3), this time applied to $q\colon C_{L_0} \longrightarrow L_0$, we have $\mathcal{O}(C_L)|_{C_{L_0}} \simeq q^*\mathcal{O}_{L_0}(1) \otimes \mathcal{O}_{C_{L_0}}(L' - L)$. A priori, for a line L of the second type it may happen that the residual line L' of $L_0 \cup L \subset \overline{L_0 L} \cap Y$ coincides with L and then one would have $h^0(C_{L_0}, \mathcal{O}(C_L)|_{C_{L_0}}) \geq 2$. However, by choosing $L_0 \in CL$ generically, the case $L = L'$ can be avoided.

Suppose now that $h^0(\mathcal{O}(C_L)|_{C_{L_0}}) \geq 2$. Then, consider the natural inclusion

$$\mathcal{O}(C_L)|_{C_{L_0}} \hookrightarrow \mathcal{L} := q^*\mathcal{O}_{L_0}(1) \otimes \mathcal{O}_{C_{L_0}}(L')$$

and observe that the only possible base point of $\mathcal{L}$ is L'.

If $L' \in C_{L_0}$ is indeed a base point of $\mathcal{L}$, then $q^*\mathcal{O}_{L_0}(1) \subset \mathcal{L}$ shows $h^0(q^*\mathcal{O}_{L_0}(1)) \geq 3$. For a generic two-dimensional linear system in $|q^*\mathcal{O}_{L_0}(1)|$, the image of the induced morphism $\zeta\colon C_{L_0} \longrightarrow \mathbb{P}^2$ cannot be a line and, therefore, $\deg(\zeta) = 1$. However, in this case $\operatorname{Im}(\zeta) \subset \mathbb{P}^2$ is a curve of degree five and, hence, of arithmetic genus six, which contradicts $g(C_{L_0}) = 11$. Thus, $\mathcal{L}$ is base point free.

Now choose a generic (hence base point free) two-dimensional linear subsystem of $|\mathcal{L}|$ and consider the induced morphism $\xi\colon C_{L_0} \longrightarrow \mathbb{P}^2$. Note that $\deg(\xi) = 1$, because otherwise $\xi^{-1}(\xi(L')) = \{L', L_1, \ldots\}$ (with multiplicities) and $q^*\mathcal{O}_{L_0}(1)$ would have L_1 as a base point, which is absurd. But $\deg(\xi) = 1$ implies that $\xi(C_{L_0}) \subset \mathbb{P}^2$ is of degree six and, therefore, of arithmetic genus 10. The latter again contradicts $g(C_{L_0}) = 11$.

If C_L does not contain a generic L_0, then the arguments have to be modified. For example, if L_0 is not generic but of the first type, then C_{L_0} is no longer smooth and in the discussion above it has to be replaced by its normalization. If all $L_0 \in C_L$ are of the second type, then one has to work with the morphism $q\colon C_{L_0} \longrightarrow L_0$ of strictly smaller degree than five and the description of $\mathcal{O}(C_L)|_{C_{L_0}}$ has to be adapted.

For the second assertion, use again Remark 1.13. □

2 Albanese of the Fano Surface

Fix a point $t_0 \in F = F(Y)$ corresponding to a line $L_0 \subset Y$ and consider the classical *Albanese morphism*

$$a\colon F \longrightarrow A := \operatorname{Alb}(F) = H^{1,0}(F)^*/H_1(F, \mathbb{Z}),\ t \longmapsto \left(\alpha \longmapsto \int_{t_0}^{t} \alpha\right).$$

According to the numerical results, see Section 0.2, A is an abelian variety of dimension five.

The goal of this section is to compare the following two pictures

$$\begin{array}{ccc} \mathbb{L} = \mathbb{P}(\mathcal{S}_F) & \xrightarrow{q} & Y \subset \mathbb{P}(V) \simeq \mathbb{P}^4 \\ {\scriptstyle p}\downarrow & & \\ F & & \end{array} \qquad \begin{array}{ccccc} \mathbb{P}(\mathcal{T}_F) & \dashrightarrow & \mathbb{P}(\mathcal{T}_A) & \longrightarrow & \mathbb{P}(T_0A) \simeq \mathbb{P}^4 \\ \downarrow & & \downarrow & & \\ F & \xrightarrow{a} & A. & & \end{array} \tag{2.1}$$

(with the curved arrow $\tilde{q}\colon \mathbb{P}(\mathcal{T}_F) \to \mathbb{P}(T_0A)$ on top.)

We will show that they describe the same geometric situation.

2.1 Tangent Bundle versus Universal Sub-bundle Let us begin with some preliminary comments:

(i) The natural inclusion $\mathcal{S}_F \subset V \otimes \mathcal{O}_F$ defines an embedding $\mathbb{L} \simeq \mathbb{P}(\mathcal{S}_F) \subset F \times \mathbb{P}(V)$, which is in fact nothing but the composition of the two inclusions $\mathbb{L} \subset F \times Y$ and $F \times Y \subset F \times \mathbb{P}(V)$. Thus, the relative tautological line bundle is described by $\mathcal{O}_p(1) \simeq q^*\mathcal{O}(1)$, cf. the proof of Lemma **2**.5.1, and the pullback describes a homomorphism

$$H^0(\mathbb{P}^4, \mathcal{O}(1)) \xrightarrow{\sim} H^0(Y, \mathcal{O}_Y(1)) \hookrightarrow H^0(\mathbb{L}, \mathcal{O}_p(1)) \simeq H^0(F, \mathcal{S}_F^*). \tag{2.2}$$

The injectivity holds, because $\mathbb{L} \twoheadrightarrow Y$ is surjective and $Y \subset \mathbb{P}(V)$ is not contained in any hyperplane. However, at this point it is not clear that the map is also surjective or, equivalently, that the morphism $q\colon \mathbb{L} \longrightarrow \mathbb{P}(V)$ is the morphism associated with the complete linear system $|\mathcal{O}_p(1)|$.

(ii) The differential of the Albanese morphism $a\colon F \longrightarrow A = \mathrm{Alb}(F)$ is a homomorphism $da\colon \mathcal{T}_F \longrightarrow a^*\mathcal{T}_A$ between the tangent sheaves. However, a priori, it may not induce a morphism $\mathbb{P}(\mathcal{T}_F) \longrightarrow \mathbb{P}(a^*\mathcal{T}_A) \longrightarrow \mathbb{P}(\mathcal{T}_A)$. For this we will have to argue that $da\colon T_tF \longrightarrow T_{a(t)}A$ is injective for all $t \in F$. Note that the tangent bundle $\mathcal{T}_A$ is trivial, which gives a natural projection $\mathbb{P}(\mathcal{T}_A) \simeq A \times \mathbb{P}(T_0A) \longrightarrow \mathbb{P}(T_0A)$.

(iii) Finally, note that there is indeed an isomorphism $V \simeq T_0A$. Namely, compose $T_0A \simeq H^{1,0}(F)^* \simeq H^{2,1}(Y)^*$ with the dual of $H^{2,1}(Y) \simeq R_1 \simeq V^*$ provided by Theorem **1**.4.21. Here, $R = \bigoplus R_i \simeq \mathbb{C}[V^*]/(\partial_i F)$ is the Jacobian ring of $Y = V(F)$, cf. [119, Sec. 12]. However, in the discussion in Section 2.3 below, the isomorphism between the two spaces will be obtained in a different manner.

Exercise 2.1 Show that the natural isomorphism $V^* \simeq H^{2,1}(Y)$ or, equivalently, $V \simeq H^{1,2}(Y)$, see Theorem **1**.4.21, is compatible with the action of $\mathrm{Aut}(Y)$. More precisely, if $g \in \mathrm{Aut}(Y)$ is the restriction of an automorphism $\tilde{g} \in \mathrm{PGL}(V)$ of the ambient projective space $\mathbb{P}(V)$, then the action g^* of g on $\mathbb{P}(H^{1,2}(Y))$ coincides indeed with $\tilde{g}$. See [499, Lem. 5.4] for details.

The first step is to show that $\mathcal{T}_F$ and $\mathcal{S}_F$ are naturally isomorphic. Evidence is provided by the following two numerical observations:

$$\det(\mathcal{T}_F) \simeq \omega_F^* \simeq \mathcal{O}_F(-1) \simeq \det(\mathcal{S}_F) \quad \text{and} \quad c_2(\mathcal{T}_F) = e(F) = 27 = c_2(\mathcal{S}_F).$$

The latter is shown by the following argument, which a posteriori explains geometrically the curious observation that $e(F) = 27$ is the number of lines on a cubic surface. Consider a generic hyperplane section $S := Y \cap V(s)$, $s \in H^0(\mathbb{P}^4, \mathcal{O}(1))$, which is a cubic surface $S \subset V(s) \simeq \mathbb{P}^3$. Let $\tilde{s}$ be the image of s under $H^0(\mathbb{P}(V), \mathcal{O}(1)) \twoheadrightarrow H^0(F, \mathcal{S}_F^*)$. Its zero set $V(\tilde{s}) \subset F$ is the set of lines $L \in F$ with $s|_L = 0$, i.e. the set of lines contained in the cubic surface S and, hence, $c_2(\mathcal{S}_F) = |V(\tilde{s})| = 27$, cf. Exercise 1.15.

Alternatively, one can use $c_2(F) = 6 \cdot g^2 - 9 \cdot c_2(\mathcal{S})$ (which actually holds in the Chow ring and not only in cohomology where for degree reasons it would simply be an equality of numbers), see Exercise **2**.3.2, and $\int_F g^2 = 45$.

Proposition 2.2 *Let $Y \subset \mathbb{P}^4$ be a smooth cubic threefold and $F = F(Y)$ its Fano variety of lines. Then there exists a natural isomorphism*

$$\mathcal{T}_F \simeq \mathcal{S}_F$$

between the tangent bundle $\mathcal{T}_F$ of F and the restriction $\mathcal{S}_F$ of the universal sub-bundle $\mathcal{S} \subset V \otimes \mathcal{O}_{\mathbb{G}}$ under the natural embedding $F \subset \mathbb{G}(1, \mathbb{P}^4)$.

Proof The result was originally proved by Clemens–Griffiths [119] and Tyurin [446] by very clever geometric arguments. We follow the more algebraic approach by Altman and Kleiman [15, Thm. 4.4]. As a first step, observe that the fibres of $\mathcal{T}_F$ and $\mathcal{S}_F$ at a point in $F(Y)$ corresponding to a line $L = \mathbb{P}(W) \subset Y$ are naturally isomorphic to $H^0(L, \mathcal{N}_{L/Y})$ and W. Thus, the following arguments can alternatively be seen as establishing an isomorphism $H^0(L, \mathcal{N}_{L/Y}) \simeq W$ that is natural and thus works in families. Start by observing that the first of the two spaces sits in the exact sequence

$$0 \longrightarrow H^0(L, \mathcal{N}_{L/Y}) \longrightarrow V/W \otimes W^* \longrightarrow S^3(W^*) \longrightarrow 0,$$

where the surjection is given by the partial derivatives $\partial_i F$ of the equation F defining Y, cf. Remarks **2**.2.2 and **2**.2.20.

Next, recall that $F \subset \mathbb{G} = \mathbb{G}(1, \mathbb{P}^4)$ is the zero set of the regular section $s_Y \in H^0(\mathbb{G}, S^3(\mathcal{S}^*))$. The latter is the image of the equation in $S^3(V^*)$ defining Y under the natural surjection $S^3(V^*) \twoheadrightarrow S^3(\mathcal{S}^*)$, see Section **2**.1.1. Hence, the normal bundle sequence for $F \subset \mathbb{G}$ has the form

$$0 \longrightarrow \mathcal{T}_F \longrightarrow \mathcal{T}_{\mathbb{G}}|_F \longrightarrow S^3(\mathcal{S}_F^*) \longrightarrow 0. \tag{2.3}$$

Deformation theory, see Section **2**.1.3, provides us with descriptions of the fibres of the two tangent bundles:

$$T_{[L]}F \simeq H^0(L, \mathcal{N}_{L/Y}) \quad \text{and} \quad T_{[L]}\mathbb{G} \simeq H^0(L, \mathcal{N}_{L/\mathbb{P}}).$$

Moreover, fibrewise (2.3) is described as the cohomology sequence of the exact sequence $0 \longrightarrow \mathcal{N}_{L/Y} \longrightarrow \mathcal{N}_{L/\mathbb{P}} \longrightarrow \mathcal{O}_L(3) \longrightarrow 0$ of normal bundles for the nested inclusion $L \subset Y \subset \mathbb{P}^4$. The global version of the latter is the exact sequence of normal bundles of the nested inclusion $\mathbb{L} \subset F \times Y \subset F \times \mathbb{P}$:

$$\begin{array}{ccccccc} 0 \longrightarrow \mathcal{N}_{\mathbb{L}/F\times Y} & \longrightarrow & \mathcal{N}_{\mathbb{L}/F\times\mathbb{P}} & \longrightarrow & \mathcal{N}_{F\times Y/F\times\mathbb{P}}|_{\mathbb{L}} & \longrightarrow & 0. \\ & & \simeq p^*\mathcal{Q}_F \otimes \mathcal{O}_p(1) & & \simeq q^*\mathcal{O}(3) \simeq \mathcal{O}_p(3) & & \end{array} \tag{2.4}$$

We use $\mathcal{N}_{\mathbb{L}_\mathbb{G}/\mathbb{G}\times\mathbb{P}} \simeq p^*\mathcal{Q} \otimes \mathcal{O}_p(1)$, which is the global version of the natural isomorphism $\mathcal{N}_{L/\mathbb{P}} \simeq V/W \otimes \mathcal{O}(1)$ for a line $L = \mathbb{P}(W) \subset \mathbb{P}(V)$, cf. the discussion in Section **2**.1.3. Here, $\mathcal{Q}$ is the universal quotient bundle on $\mathbb{G}$. Restricting to F, one obtains $\mathcal{N}_{\mathbb{L}/F\times\mathbb{P}} \simeq p^*\mathcal{Q}_F \otimes \mathcal{O}_p(1)$ and taking the direct image of (2.4) under $p\colon \mathbb{L} \longrightarrow F$, one recovers (2.3):

$$\begin{array}{ccccccc} 0 \longrightarrow p_*\mathcal{N}_{\mathbb{L}/F\times Y} & \longrightarrow & p_*\mathcal{N}_{\mathbb{L}/F\times\mathbb{P}} & \longrightarrow & p_*\mathcal{O}_p(3) & \longrightarrow & 0. \\ \simeq \mathcal{T}_F & & \simeq \mathcal{T}_\mathbb{G}|_F & & \simeq S^3(\mathcal{S}_F^*) & & \end{array} \tag{2.5}$$

See Exercise **2**.1.18 for the isomorphism $\mathcal{T}_F \simeq p_*\mathcal{N}$, where we use the shorthand $\mathcal{N} := \mathcal{N}_{\mathbb{L}/F\times Y}$. Taking determinants of (2.4) shows

$$\bigwedge^2 \mathcal{N} \simeq \det(\mathcal{N}) \simeq \det\left(p^*\mathcal{Q}_F \otimes \mathcal{O}_p(1)\right) \otimes \mathcal{O}_p(-3) \simeq p^*\det(\mathcal{Q}_F)$$

and applying $\bigwedge^2$ and $\otimes\,\mathcal{O}_p(-3)$ to (2.4), one obtains the exact sequence

$$0 \longrightarrow \bigwedge^2 \mathcal{N} \otimes \mathcal{O}_p(-3) \longrightarrow p^* \bigwedge^2 \mathcal{Q}_F \otimes \mathcal{O}_p(-1) \longrightarrow \mathcal{N} \longrightarrow 0.$$

As $p_*\mathcal{O}_p(-1) = 0 = R^1p_*\mathcal{O}_p(-1)$, taking direct images gives

$$\mathcal{T}_F \simeq p_*\mathcal{N} \simeq R^1p_*\left(\bigwedge^2\mathcal{N} \otimes \mathcal{O}_p(-3)\right) \simeq \det(\mathcal{Q}_F) \otimes R^1p_*\mathcal{O}_p(-3).$$

By relative Serre duality, cf. [223, III. Ex. 8.4], $R^1p_*\mathcal{O}_p(-3) \simeq p_*(\mathcal{O}_p(1))^* \otimes \det(\mathcal{S}_F)$ and, therefore, $\mathcal{T}_F \simeq p_*\mathcal{N} \simeq \det(\mathcal{Q}_F) \otimes \mathcal{S}_F \otimes \det(\mathcal{S}_F) \simeq \mathcal{S}_F$. □

Remark 2.3 In Remark **6**.4.6 we will give another, somewhat curious, argument to deduce an isomorphism $\mathcal{T}_F \simeq \mathcal{S}_F$ by viewing Y as a hyperplane section of a smooth cubic fourfold $X \subset \mathbb{P}^5$.

Note that $\mathcal{S}_F$ is naturally viewed as a sub-bundle $\mathcal{S}_F \subset V \otimes \mathcal{O}_F$ and, as we will see, $\mathcal{T}_F$ as a sub-bundle $\mathcal{T}_F \subset a^*\mathcal{T}_A \simeq T_0A \otimes \mathcal{O}_F$. However, the above result does not yet show the existence of an isomorphism $\mathcal{S}_F \simeq \mathcal{T}_F$ that would be compatible with these inclusions under some isomorphism $V \simeq T_0A$. This follows from the next result.

Corollary 2.4 *The natural map in* (2.2) *is an isomorphism*

$$V^* \xrightarrow{\sim} H^0(F, \mathcal{S}_F^*)$$

and $q\colon \mathbb{L} \twoheadrightarrow Y \subset \mathbb{P}(V)$ *is the morphism associated with the complete linear system* $|\mathcal{O}_p(1)|$.

Proof Use $\dim H^0(F, \mathcal{S}_F^*) = \dim H^0(F, \mathcal{T}_F^*) = \dim H^1(F, \mathcal{O}_F) = 5$ and the injectivity of the natural map $V^* \longrightarrow H^0(\mathbb{L}, q^*\mathcal{O}(1)) \simeq H^0(\mathbb{L}, \mathcal{O}_p(1)) \simeq H^0(F, \mathcal{S}_F^*)$, see (2.2).

One could also imagine a proof that uses spectral sequences as in Proposition **2**.3.12 and the isomorphism $V^* \simeq H^0(\mathbb{G}, \mathcal{S}^*)$. □

2.2 Cohomology Ring of the Fano Surface So far, we have shown that there exists an isomorphism $\mathbb{L} \simeq \mathbb{P}(\mathcal{S}_F) \simeq \mathbb{P}(\mathcal{T}_F)$, but not that the two morphisms in (2.1) are related. In fact, we have not yet even properly defined the morphism $\mathbb{P}(\mathcal{T}_F) \twoheadrightarrow \mathbb{P}(T_0A)$. This will be done next.

By virtue of Corollary **2**.5.15, there is an isomorphism $\bigwedge^2 H^1(F, \mathbb{Q}) \xrightarrow{\sim} H^2(F, \mathbb{Q})$ of Hodge structures. Recall that the discussion in Section **2**.4.4 only showed that there exist isomorphisms of Hodge structures

$$\textstyle\bigwedge^2 H^1(F, \mathbb{Q}) \simeq \bigwedge^2 H^3(Y, \mathbb{Q})(2) \simeq H^2(F, \mathbb{Q}),$$

but a priori not that the cup product induces such an isomorphism. We state the result again as the following lemma and present the traditional argument for it.

Lemma 2.5 *The exterior product defines isomorphisms*

$$\textstyle\bigwedge^2 H^{1,0}(F) \xrightarrow{\sim} H^{2,0}(F) \quad \text{and} \quad \bigwedge^2 H^1(F, \mathbb{Q}) \xrightarrow{\sim} H^2(F, \mathbb{Q}). \tag{2.6}$$

Proof We use the isomorphism $\mathcal{S}_F \simeq \mathcal{T}_F$, which turns the first assertion into the more geometric claim that the natural map $\bigwedge^2 H^0(F, \mathcal{S}_F^*) \longrightarrow H^0(F, \bigwedge^2 \mathcal{S}_F^*)$ is an isomorphism. For this, we use the commutative diagram

$$\begin{array}{ccc} \bigwedge^2 V^* & \xrightarrow{\sim} & \bigwedge^2 H^0(F, \mathcal{S}_F^*) \\ \wr\downarrow & & \downarrow \\ H^0(\mathbb{P}(\bigwedge^2 V), \mathcal{O}(1)) & \longrightarrow & H^0(F, \bigwedge^2 \mathcal{S}_F^*) \simeq H^0(F, \mathcal{O}_F(1)) \end{array}$$

and the fact that all spaces are of the same dimension 10. Thus, it suffices to show that the image of the Plücker embedding $F \subset \mathbb{P}(\bigwedge^2 V)$ is not contained in any hyperplane. This can either be argued geometrically [119, Lem. 10.2] or by using the Koszul complex as in the proofs of Propositions **2**.3.4 and **2**.3.12.

Let us turn to the second isomorphism in (2.6). The proof of Corollary **2**.5.15 in this particular case goes as follows: as the map $\bigwedge^2 H^1(F, \mathbb{Q}) \longrightarrow H^2(F, \mathbb{Q})$ is topologically defined, its injectivity is independent of the particular smooth cubic threefold. It is thus enough to check injectivity for one Fano surface $F = F(Y)$. However, for the very general cubic $\bigwedge^2 H^3(Y,\mathbb{Q})(2) \simeq \bigwedge^2 H^1(F,\mathbb{Q})$ is the direct sum $\mathbb{Q}(-1)\oplus H$ of two irreducible Hodge structures of weight two. The first summand is pure and spanned by the intersection product q_Y on $H^3(Y,\mathbb{Q})$, which by Proposition **2**.5.5 is mapped onto a non-trivial Hodge class. The irreducibility of $H = q_Y^\perp$ follows from the fact that $\mathrm{Sp}(H^3(Y))$ acts irreducibly on H, see Remark **1**.2.13.

The irreducibility of the Hodge structure H implies that the map $H \longrightarrow H^2(F,\mathbb{Q})$ is injective if and only if $\bigwedge^2 H^{1,0}(F) \longrightarrow H^{2,0}(F)$ is non-trivial, which we have shown above. Moreover, $H^{2,0}(F)$ is contained in $H^2(F,\mathbb{C})_{\mathrm{pr}}$ and, therefore, $H \hookrightarrow H^2(F,\mathbb{Q})_{\mathrm{pr}}$. Altogether, this proves the injectivity of $\bigwedge^2 H^1(F,\mathbb{Q}) \longrightarrow H^2(F,\mathbb{Q})$ and, for dimension reasons, its bijectivity. □

2.3 Albanese Morphism Geometrically, the first injectivity in (2.6) is equivalent to saying that the image of the Albanese morphism $a\colon F \longrightarrow A$ is a surface. Moreover, the pullback defines an isomorphism $a^*\colon H^2(A,\mathbb{Q}) \xrightarrow{\sim} H^2(F,\mathbb{Q})$, of which we will prove an integral version in Corollary 3.12 below.

Corollary 2.6 *The Albanese morphism $a\colon F \longrightarrow A$ is unramified, i.e. for all $t \in F$ the tangent map $da_t\colon T_tF \longrightarrow T_{a(t)}A$ is injective. In particular, the derivative of the Albanese map defines the morphism $\tilde{q}$ in* (2.1):

$$\tilde{q}\colon \mathbb{P}(\mathcal{T}_F) \longrightarrow \mathbb{P}(\mathcal{T}_A) \longrightarrow \mathbb{P}(T_0A).$$

Proof Assume da_t is not injective for some $t \in F$. Then the induced map

$$\bigwedge^2 T_tF \longrightarrow \bigwedge^2 T_{a(t)}A$$

is trivial. However, this map is the dual of the map

$$\bigwedge^2 T^*_{a(t)}A \simeq \bigwedge^2 H^{1,0}(A) \simeq \bigwedge^2 H^{1,0}(F) \xrightarrow{\sim} H^{2,0}(F) \simeq H^0(F,\omega_F) \longrightarrow \omega_F \otimes k(t),$$

which then is also trivial. As ω_F is very ample and, in particular, globally generated, this is absurd. □

Lemma 2.7 *The morphism $\tilde{q}\colon \mathbb{P}(\mathcal{T}_F) \longrightarrow \mathbb{P}(T_0A)$ is the morphism associated with the complete linear system $|\mathcal{O}_p(1)|$*

Proof First, $\tilde{q}^*\mathcal{O}(1) \simeq \mathcal{O}_p(1)$, as $\mathbb{P}(\mathcal{T}_F) \subset \mathbb{P}(a^*\mathcal{T}_0) \simeq \mathbb{P}(T_0A) \times F$ is induced by the inclusion $\mathcal{T}_F \hookrightarrow a^*\mathcal{T}_A \simeq T_0A \otimes \mathcal{O}_F$. It remains to show that the linear system is complete, i.e. that the pullback map $H^0(\mathbb{P}(T_0A),\mathcal{O}(1)) \longrightarrow H^0(\mathbb{P}(\mathcal{T}_F),\mathcal{O}_p(1))$ is a bijection. Both sides are of dimension five, so it suffices to prove the injectivity. If the map were not injective, then all tangent spaces $T_tF \hookrightarrow T_0A$ would be contained in a hyperplane. But this would contradict the bijectivity of the dual map $H^0(A,\Omega_A) \longrightarrow H^0(F,\Omega_F)$, which is the pullback of one-forms under the Albanese map $a\colon F \longrightarrow A$. □

This proves the main result of this section:

Proposition 2.8 *There exist isomorphisms $\mathcal{S}_F \simeq \mathcal{T}_F$ and $V \simeq T_0A$ inducing a commutative diagram*

$$\begin{array}{ccc} \mathbb{P}(\mathcal{S}_F) & \xrightarrow{q} & \mathbb{P}(V) \\ \wr\downarrow & & \downarrow\wr \\ \mathbb{P}(\mathcal{T}_F) & \xrightarrow[\tilde{q}]{} & \mathbb{P}(T_0A). \end{array}$$

□

Corollary 2.9 *The Albanese morphism $a\colon F \longrightarrow A$ is unramified and generically injective.*

Proof The first assertion is Corollary 2.6. To prove the injectivity generically, we choose the above isomorphism $V \simeq T_0A$ such that the two inclusions

$$\mathcal{T}_F \hookrightarrow a^*\mathcal{T}_A \simeq T_0A \otimes \mathcal{O}_F \quad \text{and} \quad \mathcal{S}_F \hookrightarrow V \otimes \mathcal{O}_F$$

coincide. Hence, the morphism $F \longrightarrow \mathbb{G}(1,\mathbb{P}(T_0A))$, $t \longmapsto [T_tF \subset T_0(A)]$ is identified with the Plücker embedding $F \hookrightarrow \mathbb{G}(1,\mathbb{P}(V))$. However, if for all points $s \in a(F)$ and distinct points $t_1 \neq t_2 \in a^{-1}(s)$ the tangent spaces $T_{t_1}F \subset T_0A$ and $T_{t_2}F \subset T_0A$ are different, then the generic fibre can only consist of just one point. □

In fact, Beauville [43, Thm. 4] has shown that $a\colon F \hookrightarrow A$ is injective and hence a closed immersion, see Corollary 3.5. We will see that in the end this assertion is equivalent to saying that the invertible sheaves $\mathcal{O}(C_{L_1})$ and $\mathcal{O}(C_{L_2})$ associated with two distinct lines $L_1 \neq L_2 \subset Y$ are never isomorphic, which we proved as Corollary 1.30 already.

As a side remark, we state the following observation by Voisin [479, Sec. 4]:

Corollary 2.10 *If Y does not admit any Eckardt point, then the cotangent bundle Ω_F of its Fano variety $F = F(Y)$ is ample.*

Proof By definition $\Omega_F \simeq \mathcal{T}_F^* \simeq \mathcal{S}_F^* \simeq p_*\mathcal{O}_p(1)$ is ample, if the relative tautological line bundle $\mathcal{O}_p(1)$ on $\mathbb{P}(\mathcal{T}_F) \longrightarrow F$ is ample. Now use the fact that $q\colon \mathbb{P}(\mathcal{T}_F) \simeq \mathbb{P}(\mathcal{S}_F) \simeq \mathbb{L} \longrightarrow Y$ is the morphism induced by the linear system $|\mathcal{O}_p(1)|$, which is finite unless Y contains Eckardt points. □

Remark 2.11 The ampleness of the cotangent bundle of an algebraic variety is a very strong condition and examples of such varieties are not easily produced. According to a result of Kobayashi, see e.g. [149, Prop. 3.1], ampleness of $\Omega_{F(Y)}$ implies that $F(Y)$ is hyperbolic, i.e. there are no non-constant holomorphic maps $\mathbb{C} \longrightarrow F(Y)$. In particular, $F(Y)$ does not contain any (singular) rational or elliptic curves.

Note that the ampleness of $\Omega_{F(Y)}$ is actually equivalent to the finiteness of $q\colon \mathbb{L} \longrightarrow Y$, i.e. to the non-existence of Eckardt points, and there certainly exist smooth cubic threefolds for which $\Omega_{F(Y)}$ is not ample, see [402] for a detailed discussion.

2.4 Geometric Global Torelli Theorem for Threefolds The following is a special case of the 'geometric global Torelli theorem', see Proposition **2**.3.12. The result in dimension three [119, 446] predates the general result and we present here its classical proof that relies on the preceding discussion.

Proposition 2.12 *Two smooth cubic threefolds $Y, Y' \subset \mathbb{P}^4$ are isomorphic if and only if their Fano surfaces $F(Y)$ and $F(Y')$ are isomorphic:*

$$Y \simeq Y' \quad \Leftrightarrow \quad F(Y) \simeq F(Y').$$

Proof For any smooth cubic threefold $Y \subset \mathbb{P}^4$, the Picard group $\mathrm{Pic}(Y)$ is generated by $\mathcal{O}_Y(1)$. Hence, any isomorphism $Y \simeq Y'$ is induced by an automorphism of the ambient $\mathbb{P}^4$ and, therefore, induces an isomorphism $F(Y) \simeq F(Y')$ between their Fano surfaces.

Conversely, any isomorphism $F(Y) \simeq F(Y')$ induces an isomorphism between the images Y and Y' of the natural morphisms

$$\mathbb{P}(\mathcal{T}_{F(Y)}) \longrightarrow \mathbb{P}(T_0\mathrm{Alb}(F(Y))) \quad \text{and} \quad \mathbb{P}(\mathcal{T}_{F(Y')}) \longrightarrow \mathbb{P}(T_0\mathrm{Alb}(F(Y'))),$$

given by the differentials of the Albanese maps. □

Exercise 2.13 Observe that the same techniques can be exploited to show that for any smooth cubic threefold $Y \subset \mathbb{P}^4$ there exists a natural isomorphism of finite groups, cf. Corollary **1**.3.9,

$$\mathrm{Aut}(Y) \xrightarrow{\sim} \mathrm{Aut}(F(Y)),$$

see (3.6) in Section **2**.3.3. Combined with Corollary **1**.3.18 and (0.4), this shows that the natural action $\mathrm{Aut}(F(Y)) \hookrightarrow \mathrm{Aut}(H^1(F(Y), \mathbb{Z}))$ is injective, see [383].

Quotients of $F(Y)$ by subgroups of $\mathrm{Aut}(F(Y))$ have been studied by Roulleau [406].

We complement Proposition 2.12 by the following infinitesimal statement.

Proposition 2.14 *Let $Y \subset \mathbb{P}^4$ be a smooth cubic threefold and let $F = F(Y)$ be its Fano surface. Then the natural map*

$$H^1(Y, \mathcal{T}_Y) \xrightarrow{\sim} H^1(F, \mathcal{T}_F)$$

is bijective.

Proof The injectivity of the map holds more generally for all smooth cubics of dimension at least three, see Corollary **2**.5.10. In order to prove bijectivity, it suffices to show that $h^1(F, \mathcal{T}_F) = 10$. Note that the Hirzebruch–Riemann–Roch formula shows $\chi(F, \mathcal{T}_F) = 30$. Hence, by applying the vanishing $H^0(F, \mathcal{T}_F) = 0$, see Corollary **2**.3.13, one obtains $h^1(\mathcal{T}_F) = h^2(\mathcal{T}_F) - 30$. Thus, it would be enough to show $h^2(\mathcal{T}_F) \leq 40$.

However, it is possible to compute $H^1(F, \mathcal{T}_F) \simeq H^1(F, \mathcal{S}_F)$ directly.[7] The Koszul resolution of $\mathcal{O}_F$, see (3.1) in the proof of Proposition **2**.3.4, tensored with $\mathcal{S}$ allows one to compute $H^1(F, \mathcal{S}_F)$ via the spectral sequence

$$E_1^{p,q} = H^q(\mathbb{G}, \bigwedge^{-p}(S^3(\mathcal{S})) \otimes \mathcal{S}) \Rightarrow H^{p+q}(F, \mathcal{S}_F).$$

After applying Bott–Borel–Weil theory to identify the various terms one eventually finds that $H^1(F, \mathcal{S}_F)$ is the kernel of the natural map $S^2(V) \otimes \det(V) \longrightarrow V^* \otimes \det(V)$ given by the equation of $Y \subset \mathbb{P}(V)$. For smooth Y, this map is surjective and, therefore, $h^1(F, \mathcal{S}_F) = \dim S^2(V) - \dim(V) = 15 - 5 = 10$. □

The result says that the Fano surface $F(Y)$ stays a Fano surface after small deformations, which is false, for example, for the Fano variety of lines on a smooth cubic fourfold, see Corollary **6**.3.12 and Remark **2**.3.14.

Remark 2.15 Note that the Picard number $\rho(F) := \mathrm{rk}(H^{1,1}(F, \mathbb{Z}))$ of the Fano variety satisfies $1 \leq \rho(F) \leq 25$. The general cubic threefold satisfies $\rho(F(Y)) = 1$, see the proof of Lemma 2.5, but in light of the moduli space of cubic threefolds being only 10-dimensional one is led to ask which other Picard numbers can be attained. Using (2.6) to see that the Albanese morphism induces an isomorphism of Hodge structures $H^2(A, \mathbb{Q}) \simeq H^2(F, \mathbb{Q})$, cf. Corollary 3.12 below for an integral version, the problem is linked to the analogous question for abelian varieties of dimension five.

According to Hulek and Laface [241] the Picard number of an abelian variety of dimension five satisfies $1 \leq \rho \leq 17$ or $\rho = 25$. Clearly, then the same holds true for the Fano variety of lines F of any smooth cubic threefold Y. For example, the upper bound $\rho(F) = 25$ is attained by the *Klein cubic*

$$Y = V(x_0^2\, x_1 + x_1^2\, x_2 + x_2^2\, x_3 + x_3^2\, x_4 + x_4^2\, x_0),$$

see [9, 55, 403]. In this case, $A \simeq E_1 \times \cdots \times E_5$ (as unpolarized abelian varieties), where the E_i are pairwise isogenous elliptic curves with CM, see [305, Exer. 5.6.10]. Roulleau [405] describes examples with $\rho(F) = 12, 13$ and observes that there exist infinitely many smooth cubic threefolds with $\rho(F) = 25$. However, which of the remaining values $1 \leq \rho(F) \leq 17$ are realized seems an open question. Note that the moduli space of abelian varieties of dimension 5 is of dimension 15. Over finite fields the problem has

[7] Thanks to A. Kuznetsov for providing the complete argument. His proof is too long and technical to be included here in full, so we only sketch the approach. As was pointed out to me by S. Stark, a complete proof can already be found in a paper by Wehler [489].

been studied by Debarre, Laface, and Roulleau [140]. This is then related to the Zeta function, cf. Remark **2**.4.22.

In the arithmetic setting, the link between the Fano surface and the intermediate Jacobian, cf. Corollary 3.3, has been used by Roulleau [407] to prove the Tate conjecture for the Fano surface over any finitely generated field not of characteristic two.

3 Albanese, Picard, and Prym

The general theory set up in Section **2**.5.5 provides us with a commutative diagram

$$\begin{array}{ccccc} \mathrm{CH}^2(F)_{\mathrm{alg}} & \longrightarrow & \mathrm{CH}^2(Y)_{\mathrm{alg}} & \longrightarrow & \mathrm{CH}^1(F)_{\mathrm{alg}} \\ \downarrow & & \downarrow & & \downarrow \wr \\ A(F) & \twoheadrightarrow & J(Y) & \twoheadrightarrow & \mathrm{Pic}^0(F). \end{array} \tag{3.1}$$

Note that $\mathrm{CH}^2(F)_{\mathrm{alg}}$ is known to be big (over $\mathbb{C}$), while $\mathrm{CH}^1(F) \simeq \mathrm{Pic}(F)$.

The intermediate Jacobian

$$J(Y) := J^3(Y) = \frac{H^{1,2}(Y)}{H^3(Y,\mathbb{Z})} \simeq \frac{H^{2,1}(Y)^*}{H_3(Y,\mathbb{Z})}$$

of Y is self-dual and the two maps on the bottom are dual to each other, cf. Section **2**.5.5. Indeed, they are induced by the Fano correspondence $\varphi\colon H^3(Y,\mathbb{Z}) \twoheadrightarrow H^1(F,\mathbb{Z})(-1)$ and its dual $\psi\colon H^3(F,\mathbb{Z}) \twoheadrightarrow H^3(Y,\mathbb{Z})$ (a priori up to torsion, but see Remark 1.21). Moreover, as the two maps induce isomorphisms between the cohomology groups with rational coefficients, they are isogenies of abelian varieties of dimension five.

The aim of this section is to show that all three abelian varieties, $A(F)$, $J(Y)$, and $\mathrm{Pic}^0(F)$, are isomorphic and, moreover, can be identified with the Prym variety of the morphism $C_L \simeq \widetilde{D}_L \twoheadrightarrow D_L$ in Section 1.3.

3.1 Albanese versus Intermediate Jacobian In order to understand the composition

$$A = A(F) \twoheadrightarrow J(Y) \twoheadrightarrow \mathrm{Pic}^0(F),$$

we first pre-compose it with the Albanese map $a\colon F \longrightarrow A$, which depends on the additional choice of a point $t_0 \in F$ corresponding to a line $L_0 \subset Y$. This map then factorizes through $F \longrightarrow \mathrm{CH}^2(F)_{\mathrm{alg}}$, $t \longmapsto [t] - [t_0]$ and the Abel–Jacobi map $\mathrm{CH}^2(F)_{\mathrm{alg}} \subset \mathrm{CH}^2(F)_{\mathrm{hom}} \longrightarrow A(F)$. According to Exercise **2**.4.13 and Remark 1.21, $H^2(F,\mathbb{Z})$ is torsion free. Hence, the notion of homological and algebraic equivalence for divisors on F coincide. A similar result holds for curves on Y, see the proof of Corollary 3.16 below. Thus,

$$\mathrm{CH}^2(Y)_{\mathrm{alg}} = \mathrm{CH}^2(Y)_{\mathrm{hom}} \quad \text{and} \quad \mathrm{Pic}^0(F) \simeq \mathrm{CH}^1(F)_{\mathrm{alg}} = \mathrm{CH}^1(F)_{\mathrm{hom}}.$$

Lemma 3.1 *The composition* $F \longrightarrow A(F) \twoheadrightarrow J(Y) \twoheadrightarrow \mathrm{Pic}^0(F)$ *sends a point* $L \in F$ *to the invertible sheaf* $\mathcal{O}(C_L - C_{L_0})$.

Proof Clearly, the class of the point $L \in F$ under $\psi\colon \mathrm{CH}^2(F) \longrightarrow \mathrm{CH}^2(Y)$ is mapped to the class $[L] \in \mathrm{CH}^2(Y)$ of the line $L \subset Y$. The image of the latter under the correspondence $\varphi\colon \mathrm{CH}^2(Y) \longrightarrow \mathrm{CH}^1(F)$ is by construction $\mathcal{O}(C_L)$. Then use the commutativity of the diagram (3.1). □

The result can be extended to a description of the composition

$$C \longrightarrow A(C) \longrightarrow A(F) \twoheadrightarrow J(Y) \twoheadrightarrow \mathrm{Pic}^0(F) \longrightarrow \mathrm{Pic}^0(C), \tag{3.2}$$

for an arbitrary (smooth) curve $C \subset F(Y)$ as $L \longmapsto \mathcal{O}(C_L - C_{L_0})|_C$, which combined with Remark 1.18 will come up again as the *Abel–Prym map* in Remark 3.9. The observation is particularly useful when C is ample, e.g. for $C = C_L$. In this case, one finds

$$A(C) \twoheadrightarrow A(F) \quad \text{and} \quad \mathrm{Pic}^0(F) \hookrightarrow \mathrm{Pic}^0(C).$$

The surjectivity follows from $H^1(F, \mathcal{O}_F(-C)) = 0$, a consequence of the Kodaira vanishing theorem, which proves the injectivity of the map $H^1(F, \mathcal{O}_F) \hookrightarrow H^1(C, \mathcal{O}_C)$ between their cotangent spaces.

For the injectivity, use the fact that for a line bundle $\mathcal{M}$ of degree 0 on F, the line bundle $\mathcal{O}(C) \otimes \mathcal{M}^*$ is still ample. Hence, again by Kodaira vanishing, $H^1(F, \mathcal{O}(-C) \otimes \mathcal{M}) = 0$ and, therefore, $H^0(F, \mathcal{M}) \twoheadrightarrow H^0(C, \mathcal{M}|_C)$ is surjective. The latter shows that with $\mathcal{M}|_C$ also $\mathcal{M}$ would be trivial.

Let $L \in F$ be generic and consider C_L as the étale cover

$$\pi\colon C_L \simeq \widetilde{D}_L \twoheadrightarrow D_L$$

of degree two, see Section 1.3.

Corollary 3.2 *Using the above notation, one has:*

(i) *The following composition is trivial:*

$$\mathrm{Pic}^0(D_L) \xrightarrow{\pi^*} \mathrm{Pic}^0(C_L) \simeq A(C_L) \twoheadrightarrow A(F) \twoheadrightarrow J(Y) \twoheadrightarrow \mathrm{Pic}^0(F). \tag{3.3}$$

(ii) *The image of the restriction map* $H^1(F, \mathbb{Z}) \longrightarrow H^1(C_L, \mathbb{Z})$ *is contained in the kernel of* $\pi_*\colon H^1(C_L, \mathbb{Z}) \longrightarrow H^1(D_L, \mathbb{Z})$.

Proof Observe that under the natural map

$$D_L \longrightarrow \mathrm{Pic}(D_L) \longrightarrow \mathrm{Pic}(C_L) \longrightarrow \mathrm{CH}^2(F) \longrightarrow \mathrm{CH}^2(Y)$$

a point $y \in D_L$ is mapped to $[L_1] + [L_2] \in \mathrm{CH}^2(Y)$, where L_1 and L_2 correspond to the two points of the fibre $\pi^{-1}(y)$.

Clearly, for the plane $\mathbb{P}^2 = \overline{yL}$, the following equality holds in $\mathrm{CH}^2(Y)$:

$$[L_1] + [L_2] = [L_1] + [L_2] + [L] - [L] = [\mathbb{P}^2]|_Y - [L].$$

As $[\overline{yL}] \in \mathrm{CH}^2(\mathbb{P}^4) \simeq \mathbb{Z}$ is independent of $y \in D_L$, also the class $[L_1] + [L_2] \in \mathrm{CH}^2(Y)$ is. Thus, $D_L \longrightarrow \mathrm{CH}^2(Y) \longrightarrow \mathrm{CH}^1(F)$ is constant and, therefore, the composition (3.3) is trivial.

The second assertion is equivalent to the vanishing of the composition

$$H^1(D_L, \mathbb{Z}) \xrightarrow{\pi^*} H^1(C_L, \mathbb{Z}) \xrightarrow{i_*} H^3(F, \mathbb{Z})(-1). \tag{3.4}$$

Composing (3.4) further with the injection $H^3(F,\mathbb{Z}) \hookrightarrow H^3(Y,\mathbb{Z}) \hookrightarrow H^1(F,\mathbb{Z})(-1)$ describes the map obtained by taking cohomology of (3.3). Hence, (i) implies (ii). □

Note that the purely topological assertion (ii) is deduced from a Chow theoretic argument. A more topological reasoning along the same lines is probably also possible but likely to be not quite as elegant.

By a purely topological description of étale coverings of degree two, one computes that the intersection pairing on $H^1(C_L, \mathbb{Z})$ restricted to the submodule

$$H^1(C_L, \mathbb{Z})^- := \mathrm{Ker}\left(\pi_* \colon H^1(C_L, \mathbb{Z}) \longrightarrow H^1(D_L, \mathbb{Z})\right)$$

is divisible by two, cf. Section 3.2. Hence, the Hodge–Riemann pairing $(\,.\,)_F$ on $H^1(F, \mathbb{Z})$ with respect to the Plücker polarization satisfies

$$(\gamma.\gamma')_F = \int_F \gamma \cdot \gamma' \cdot g = 3 \int_{C_L} \gamma|_{C_L} \cdot \gamma'|_{C_L} \in 6\,\mathbb{Z}, \tag{3.5}$$

where we use $3\,[C_L] = g \in H^2(F, \mathbb{Z})$, cf. Lemma 1.14. Then, according to Proposition **2**.5.5, the Fano correspondence provides an injection of integral(!) symplectic lattices

$$\varphi\colon (H^3(Y,\mathbb{Z}), (\,.\,)_Y) \hookrightarrow (H^1(F,\mathbb{Z}), (-1/6)(\,.\,)_F)$$

of finite index. As the left-hand side is unimodular, this map has to be an isomorphism. We thus have proved the following result, cf. Corollary **2**.5.8.

Corollary 3.3 *The Fano correspondence induces an isometry of Hodge structures*

$$\varphi\colon \left(H^3(Y,\mathbb{Z}), (\,.\,)_Y\right) \xrightarrow{\sim} \left(H^1(F,\mathbb{Z})(-1), (-1/6)(\,.\,)_F\right)$$

and, consequently, isomorphisms of the associated polarized abelian varieties

$$A(F) \xrightarrow{\sim} J(Y) \xrightarrow{\sim} \mathrm{Pic}^0(F).$$

□

Remark 3.4 By definition of the intermediate Jacobian of Y, there exists an isomorphism of Hodge structures

$$H^1(J(Y), \mathbb{Z})(-1) \simeq H^3(Y, \mathbb{Z}) \tag{3.6}$$

which gives rise to a Hodge class in $H^{2,2}(J(Y) \times Y, \mathbb{Z})$. It is an open question whether this class can be written as an integral(!) linear combination of algebraic classes. This question and its relation to the existence of a universal codimension two cycle and to the question whether cubic threefolds have universally trivial CH_0 has been studied by Voisin [481], cf. Section **7**.4.2.

Note that the Fano correspondence does give an algebraic correspondence

$$H^1(J(Y), \mathbb{Z}) \xrightarrow{\sim} H^1(A(F), \mathbb{Z}) \xrightarrow{\sim} H^1(F(Y), \mathbb{Z}) \xhookrightarrow{\cdot g} H^3(F(Y), \mathbb{Z}) \xrightarrow{\sim} H^3(Y, \mathbb{Z}),$$

which however is only an isomorphism of rational(!) Hodge structures, unlike (3.6). In Section **6**.3.8 a similar phenomenon will be discussed for cubic fourfolds.

The next result improves upon Corollary 2.9, cf. [43, Thm. 4].

Corollary 3.5 (Beauville) *The Albanese morphism*

$$a \colon F \hookrightarrow A(F) \simeq J(Y)$$

is a closed immersion. Equivalently, the morphism $F \hookrightarrow \mathrm{Pic}(F)$, $L \longmapsto \mathcal{O}(C_L)$ *is a closed immersion.*

Proof Indeed, by Corollary 2.9 the morphism $F \longrightarrow A(F)$ is unramified and its composition with $A(F) \longrightarrow J(Y) \longrightarrow \mathrm{Pic}(F)$, which is an isomorphism by Corollary 3.3, is the injective map $L \longmapsto \mathcal{O}(C_L)$, cf. Corollary 1.30. □

Remark 3.6 Some of the varieties we have considered, e.g. the Fano variety $F(Y)$ or its Albanese variety $A(F(Y)) \simeq J(Y)$, admit a modular description, i.e. are isomorphic to certain moduli spaces of sheaves on Y. Here are some facts and references:

(i) The blow-up $\tilde{J}(Y) \longrightarrow J(Y)$ in the codimension three subvariety $F(Y) \subset J(Y)$ admits a modular description. Druel [165] shows that it is isomorphic to the moduli space of semi-stable sheaves of rank two on Y with Chern classes $\mathrm{c}_1 = \mathrm{c}_3 = 0$ and $\mathrm{c}_2 = (2/3) \cdot h^2$. The complement of the exceptional divisor, a $\mathbb{P}^2$-bundle over $F(Y)$, is the open set parametrizing the locally free sheaves.

The original argument relies on work of Markushevich–Tikhomirov [341] and Iliev–Markushevich [258]. A simplified account was provided by Beauville [51] who also describes the proper transform of the theta divisor and a description in terms of matrix factorizations was provided by Böhning and von Bothmer [75].

(ii) The moduli space of semi-stable locally free sheaves of rank two on Y with Chern classes $\mathrm{c}_1 = h$ and $\mathrm{c}_2 = (2/3) \cdot h^2$ has been described as the Fano variety of lines $F(Y)$ by Beauville [57, Prop. 3] and by Biswas–Biswas–Ravindra [69, Thm. 1]. A similar result holds for the moduli space of semi-stable locally free sheaves of rank two with Chern classes $\mathrm{c}_1 = 0$ and $\mathrm{c}_2 = (1/3) \cdot h^2$.

(iii) Another moduli space appears naturally. As shown by Bayer et al. [39], the moduli space of semi-stable sheaves of rank three and Chern classes $c_1 = -h$, $c_2 = h^2$, and $c_3 = -h^3/3$ is the blow-up of the theta divisor $\Xi \subset J(Y)$ in its only singular point, see Section 4.1 and Remark **7**.2.7.

3.2 Reminder on Prym Varieties The next step is to relate the intermediate Jacobian $J(Y)$ to the Prym variety of the étale cover $C_L \twoheadrightarrow D_L$ for a generic line $L \subset Y$.

Let us begin with a reminder on Prym varieties. We recommend [41, 48, 305, 359] for a more detailed discussion and [179] for a historical account.

We consider an étale cover

$$\pi : C \twoheadrightarrow D$$

of degree two between smooth projective curves. Then C comes with a covering involution that we shall denote by ι. Note that according to the Hurwitz formula, we have $g(C) = 2g(D) - 1$. The double cover π corresponds to the two-torsion line bundle $\mathcal{L}_\pi \simeq \pi_*\mathcal{O}_C/\mathcal{O}_D \in \mathrm{Pic}^0(D)$ that satisfies $\pi^*\mathcal{L}_\pi \simeq \mathcal{O}_C$.

Lemma 3.7 *The pullback* $\pi^* : \mathrm{Pic}(D) \longrightarrow \mathrm{Pic}(C)$ *defines an isomorphism*

$$\mathrm{Pic}(D)/\langle \mathcal{L}_\pi \rangle \simeq \mathrm{Ker}\,(\mathrm{Pic}(C) \xrightarrow{1-\iota^*} \mathrm{Pic}(C)).$$

Proof The morphism $1 - \iota^*$ maps a line bundle $\mathcal{L}$ to $\mathcal{L} \otimes \iota^*\mathcal{L}^*$. Clearly, if $\mathcal{L} = \pi^*\mathcal{M}$, then $(1-\iota^*)(\mathcal{L}) \simeq \mathcal{O}_C$. For the other inclusion, use the fact that any ι^*-invariant invertible sheaf descends to an invertible sheaf on D.[8]

Following Beauville [41], the descent can be shown explicitly as follows: suppose $\mathcal{L} = \mathcal{O}(E)$ satisfies $\iota^*\mathcal{L} \simeq \mathcal{L}$. Write $E - i^*E$ as the principal divisor (f) for some $f \in K(C)$ and observe that then $f \cdot \iota^* f$ has neither 0s nor poles, so we may assume $f \cdot \iota^* f = 1$ (we need k to admit square roots for this). Pick an element $g \in K(C)$ with $\iota^* g = -g$ and set $f_0 := g \cdot (f - 1)$. Then, $f = f_0 \cdot (\iota^* f_0)^{-1}$ and, therefore, $E_0 := E - (f_0)$ is the pullback of a divisor on D. Alternatively, the existence of f_0 can be deduced from Hilbert's theorem 90. We leave it to the reader to verify that $\mathcal{O}_D$ and $\mathcal{L}_\pi$ are the only line bundles with trivial pullback to C. □

Definition 3.8 The *Prym variety* of an étale double over $\pi : C \twoheadrightarrow D$ is defined as

$$\mathrm{Prym}\,(C/D) := \mathrm{Im}\,(\mathrm{Pic}^0(C) \xrightarrow{1-\iota^*} \mathrm{Pic}(C)).$$

Hence, there exists a natural exact sequence, see [80, Thm. 2], [41, Sec. 2.6], or [365, Sec. 10.9]:

$$0 \longrightarrow \langle \mathcal{L}_\pi \rangle \longrightarrow \mathrm{Pic}^0(D) \longrightarrow \mathrm{Pic}^0(C) \longrightarrow \mathrm{Prym}(C/D) \longrightarrow 0. \tag{3.7}$$

[8] One could think that the absence of fixed points is important here. It is not, although for the descent of invariant invertible sheaves on surfaces, fixed points do cause problems.

In particular, the dimension of the Prym variety is

$$\dim \mathrm{Prym}(C/D) = g(D) - 1.$$

Next, we wish to show that the Prym variety can also be viewed as a connected component of the kernel of the norm map. Recall that the norm map $N\colon \mathrm{Pic}(C) \longrightarrow \mathrm{Pic}(D)$ is the push-forward map $\pi_*\colon \mathrm{CH}^1(C) \longrightarrow \mathrm{CH}^1(D)$, which, using $\mathrm{Pic}^0 \simeq \mathrm{Alb}$ for curves, on the identity component is described by the natural map $A(C) \longrightarrow A(D)$ between the Albanese varieties. Alternatively, $N(\mathcal{L}) \simeq \det \pi_*\mathcal{L} \otimes (\det \pi_*\mathcal{O}_C)^*$. For example, for the line bundle $\mathcal{L} = \mathcal{O}(x)$, $x \in C$, the exact sequence $0 \longrightarrow \mathcal{O}_C \longrightarrow \mathcal{L} \longrightarrow k(x) \longrightarrow 0$ indeed shows $\mathcal{O}(\pi(x)) \simeq \det \pi_* k(x) \simeq \det \pi_*\mathcal{L} \otimes (\det \pi_*\mathcal{O}_C)^*$. Clearly, N defined as π_* is a group homomorphism, which is not quite so apparent in the latter description. Note that $\mathrm{Ker}(N)$ has two connected components, non-canonically isomorphic to each other.

The claim now is that the connected component $\mathrm{Ker}(N)^{\mathrm{o}}$ of the kernel $\mathrm{Ker}(N)$ containing $\mathcal{O}_C$ is the Prym variety, i.e.

$$\mathrm{Prym}\,(C/D) = \mathrm{Im}\,(1 - \iota^*) \simeq \mathrm{Ker}\,(N)^{\mathrm{o}}.$$

For one inclusion, use $\pi_*\iota^* = \pi_*$ and compute $N(\mathcal{L} \otimes \iota^*\mathcal{L}^*) \simeq \det \pi_*(\mathcal{L}) \otimes (\det \pi_*\iota^*\mathcal{L})^* \simeq \mathcal{O}_D$. For the other inclusion, observe that $N\colon \mathrm{Pic}^0(C) \twoheadrightarrow \mathrm{Pic}^0(D)$ is surjective and hence $\mathrm{Prym}\,(C/D) \subset \mathrm{Ker}\,(N)^{\mathrm{o}}$ is an inclusion of abelian varieties of the same dimension $g(D) - 1$.

To summarize, in addition to the exact sequence (3.7) there is an exact sequence

$$\begin{array}{ccccccccc} 0 & \longrightarrow & \mathrm{Prym} \sqcup \mathrm{Prym}' & \longrightarrow & \mathrm{Pic}^0(C) & \xrightarrow{N} & \mathrm{Pic}^0(D) & \longrightarrow & 0. \\ & & & & \simeq A(C) & & \simeq A(D) & & \end{array} \tag{3.8}$$

It is useful to describe both points of view in terms of integral Hodge structures. For this, recall that

$$\mathrm{Pic}^0(C) \simeq \frac{H^1(C, \mathcal{O}_C)}{H^1(C, \mathbb{Z})} \simeq \frac{H^0(C, \omega_C)^*}{H_1(C, \mathbb{Z})} = \mathrm{Alb}(C)$$

and similarly for $\mathrm{Pic}^0(D)$. As explained in [305, Ch. 12.4], $H^1(C, \mathbb{Z})$ admits a symplectic basis of the form $\tilde{\lambda}_0, \tilde{\mu}_0, \lambda_i^{\pm}, \mu_i^{\pm}$, $i = 1, \ldots, g(D) - 1$ with $\tilde{\lambda}_0, \tilde{\mu}_0$ fixed by the action of ι^* on $H^1(C, \mathbb{Z})$ and $\iota^*(\lambda_i^{\pm}) = \lambda_i^{\mp}$, $\iota^*(\mu_i^{\pm}) = \mu_i^{\mp}$, see also the picture in [164]. In particular, the only non-zero intersection numbers are $(\tilde{\lambda}_0.\tilde{\mu}_0) = 1$ and $(\lambda_i^{\pm}.\mu_i^{\pm}) = 1$. This allows one to describe the eigenspaces $H^1(C, \mathbb{Z})^{\pm} \subset H^1(C, \mathbb{Z})$ of the involution ι^* as

$$H^1(C, \mathbb{Z})^+ = \langle \tilde{\lambda}_0, \tilde{\mu}_0, \lambda_i^+ + \lambda_i^-, \mu_i^+ + \mu_i^- \rangle \quad \text{and} \quad H^1(C, \mathbb{Z})^- = \langle \lambda_i^+ - \lambda_i^-, \mu_i^+ - \mu_i^- \rangle.$$

The latter implies the fact alluded to before that the intersection pairing on $H^1(C, \mathbb{Z})^-$ is divisible by two, which was used in the proof of Corollary 3.3. Moreover, the image of the pullback map

$$\pi^*\colon H^1(D, \mathbb{Z}) = \langle \lambda_i, \mu_i \rangle_{i=0,\ldots,g(D)-1} \hookrightarrow H^1(C, \mathbb{Z})^+,$$

which is given by $\lambda_0 \longmapsto \tilde{\lambda}_0$, $\mu_0 \longmapsto 2\tilde{\mu}_0$, $\lambda_i \longmapsto \lambda_i^+ + \lambda_i^-$, and $\mu_i \longmapsto \mu_i^+ + \mu_i^-$, for $i = 1, \ldots, g(D) - 1$, defines a sublattice of index two. With this notation, topologically the étale covering $C \longrightarrow D$ can be constructed by cutting D along the standard loop representing μ_0 and gluing two copies of D along μ_0. This explains why in particular the pullback of μ_0 gives $2\tilde{\mu}_0$.

Also observe that the image of

$$H^1(C, \mathbb{Z}) \longrightarrow H^1(C, \mathbb{Z})^+, \quad \alpha \longmapsto \alpha + \iota^*\alpha$$

is contained in $\pi^* H^1(D, \mathbb{Z}) \subset H^1(C, \mathbb{Z})^+$ with index two. On the other hand,

$$H^1(C, \mathbb{Z}) \twoheadrightarrow H^1(C, \mathbb{Z})^-, \quad \alpha \longmapsto \alpha - \iota^*\alpha$$

is surjective. As the sequence (3.7) is induced by the exact sequence

$$0 \longrightarrow H^1(C, \mathbb{Z})^+ \longrightarrow H^1(C, \mathbb{Z}) \xrightarrow{1-\iota^*} H^1(C, \mathbb{Z})^- \longrightarrow 0, \tag{3.9}$$

the Prym variety can be described as

$$\mathrm{Prym}(C/D) \simeq \frac{H^1(C, \mathcal{O}_C)^-}{H^1(C, \mathbb{Z})^-} \simeq \frac{H^0(C, \omega_C)^{-*}}{H_1(C, \mathbb{Z})^-}.$$

Remark 3.9 The Prym variety is commonly viewed as a principally polarized abelian variety: indeed, the last isomorphism together with the description of $H^1(C, \mathbb{Z})^-$ as $\langle \lambda_i^+ - \lambda_i^-, \mu_i^+ - \mu_i^- \rangle$ allows one to define a principal polarization

$$\Xi \in H^2(\mathrm{Prym}(C/D), \mathbb{Z}) \simeq \bigwedge^2 H^1(\mathrm{Prym}(C/D), \mathbb{Z}) \simeq \bigwedge^2 H^1(C, \mathbb{Z})^-$$

on $\mathrm{Prym}(C/D)$ explicitly given by the intersection pairing on $H^1(C, \mathbb{Z})^-$ scaled by the factor $(1/2)$.

For a fixed point $t_0 \in C$, one defines the *Abel–Prym map* as

$$\mathrm{AP}\colon C \longrightarrow \mathrm{Prym}(C/D), \quad t \longmapsto \mathcal{O}(t - t_0) \otimes \iota^*\mathcal{O}(t - t_0)^*.$$

It induces the canonical isomorphism $\mathrm{AP}^*\colon H^1(\mathrm{Prym}(C/D), \mathbb{Z}) \xrightarrow{\sim} H^1(C, \mathbb{Z})^-$. In particular, $\mathrm{AP}^*(\Xi) = (1/2) \sum (\lambda_i^+ \wedge \mu_i^+ + \lambda_i^- \wedge \mu_i^-) \in H^2(C, \mathbb{Z})$ and hence

$$\deg \mathrm{AP}^*(\Xi) = g(C) - 1 = 2g(D) - 2.$$

The kernel of $\mathrm{Pic}^0(C) \twoheadrightarrow \mathrm{Prym}(C/D)$ can be written as the degree two quotient

$$\mathrm{Pic}^0(D) \simeq \frac{H^1(C, \mathcal{O}_C)^+}{H^1(D, \mathbb{Z})} \twoheadrightarrow \frac{H^1(C, \mathcal{O}_C)^+}{H^1(C, \mathbb{Z})^+}.$$

As the last step of our general discussion, observe that (3.8) corresponds to

$$0 \longrightarrow H^1(C, \mathbb{Z})^- \longrightarrow H^1(C, \mathbb{Z}) \xrightarrow{1+\iota^*} (1 + \iota^*)H^1(C, \mathbb{Z}) \longrightarrow 0,$$

where we make use of the degree two quotient

$$\frac{H^1(C,\mathcal{O}_C)^+}{(1+\iota^*)H^1(C,\mathbb{Z})} \twoheadrightarrow \frac{H^1(C,\mathcal{O}_C)^+}{H^1(D,\mathbb{Z})} \simeq \mathrm{Pic}^0(D).$$

Let us now apply the general theory to the cover $C_L \twoheadrightarrow D_L$ associated with a generic line $L \subset Y$ in a smooth cubic threefold.

Proposition 3.10 (Mumford) *For a generic line $L \subset Y$, the curve $i\colon C_L \hookrightarrow F(Y)$ induces an isometry of Hodge structures*

$$\left(H^3(Y,\mathbb{Z}), -(\,.\,)_Y\right) \simeq \left(H^1(F,\mathbb{Z})(-1), (1/6)(\,.\,)_F\right) \xrightarrow[\iota^*]{\sim} \left(H^1(C_L,\mathbb{Z})^-, (1/2)(\,.\,)\right)$$

and an isomorphism of polarized abelian varieties

$$J(Y) \simeq A(F) \simeq \mathrm{Pic}^0(F) \xrightarrow[\iota^*]{\sim} \mathrm{Prym}(C_L/D_L).$$

Proof The first isomorphism is the content of Corollary 3.3. The second one follows from a comparison of (3.9) with the exact sequence

$$0 \longrightarrow \mathrm{Ker}(\xi) \longrightarrow H^1(C_L,\mathbb{Z}) \xrightarrow{\xi} H^3(Y,\mathbb{Z}) \longrightarrow 0.$$

Here, ξ is the composition of the push-forward map $i_*\colon H^1(C_L,\mathbb{Z}) \twoheadrightarrow H^3(F(Y),\mathbb{Z})$ induced by $i\colon C_L \hookrightarrow F(Y)$ and the dual $\psi\colon H^3(F(Y),\mathbb{Z}) \twoheadrightarrow H^3(Y,\mathbb{Z})$ of the Fano correspondence. The surjectivity of i_* is a consequence of the ampleness of C_L and the Lefschetz hyperplane theorem: $H^1(C_L,\mathbb{Z}) \simeq H_1(C_L,\mathbb{Z}) \twoheadrightarrow H_1(F(Y),\mathbb{Z}) \simeq H^3(F(Y),\mathbb{Z})$. As ψ is dual to $\varphi\colon H^3(Y,\mathbb{Z}) \xrightarrow{\sim} H^1(F(Y),\mathbb{Z})$, which is an isomorphism according to Corollary 3.3, ψ is surjective, too.

By virtue of Corollaries 3.2 and 3.3, we know that $H^1(D_L,\mathbb{Z}) \longrightarrow H^3(F(Y),\mathbb{Z})$ is trivial and, hence, $\pi^*H^1(D_L,\mathbb{Z}) \subset \mathrm{Ker}(\xi)$. Both are free $\mathbb{Z}$-modules of the same rank and $\mathrm{Ker}(\xi)$ is saturated, as its cokernel is the torsion free $H^3(Y,\mathbb{Z})$. However, $\pi^*H^1(D_L,\mathbb{Z})$ is also contained with finite index in $H^1(C_L,\mathbb{Z})^+$ and the latter is a saturated submodule of $H^1(C_L,\mathbb{Z})$, for its cokernel is the torsion free $H^1(C_L,\mathbb{Z})^-$. Hence, $\mathrm{Ker}(\xi)$ and $H^1(C_L,\mathbb{Z})^+$ both realize the saturation of $\pi^*H^1(D_L,\mathbb{Z}) \subset H^1(C_L,\mathbb{Z})$ and, therefore, coincide. So, altogether we obtain an isomorphism of Hodge structures

$$H^3(Y,\mathbb{Z})(1) \simeq H^1(C_L,\mathbb{Z})^-$$

and, hence, an isomorphism of abelian varieties $\mathrm{Prym}(C_L/D_L) \simeq J(Y)$. The compatibility with the various pairings and polarizations is easily checked. □

Remark 3.11 Not every smooth curve of genus 11 admits an étale double quotient. Furthermore, the space of étale double covers $C \longrightarrow D$ with $g(C) = 11$ arising as $C_L \longrightarrow D_L$ is of codimension three. Indeed, the quotient curve D_L is smooth of genus 6 and, therefore, its deformation space is of dimension 15. On the other hand, the space of all $(Y, L \subset Y)$ only accounts for 12 dimensions.[9]

3.3 Theta Divisor of the Albanese The following summarizes results in [119], [42, Prop. 4 & 7], and [43, Thm. 4]. Note that in [446, Sec. 2] it is wrongly claimed that the image of α is a divisor of degree two.

Corollary 3.12 (Beauville, Clemens–Griffiths) *Fix a point in $F = F(Y)$ and consider the Albanese embedding $a\colon F \hookrightarrow A = A(F) \simeq \mathrm{Prym}(C_L/D_L)$, see Corollary 3.5. Then*

(i) $[a(F)] = (1/3!) \cdot \Xi^3 \in H^6(A(F), \mathbb{Z}) \simeq H^6(\mathrm{Prym}(C_L/D_L), \mathbb{Z})$.

(ii) $a^*(\Xi) = (2/3) \cdot g \in H^2(F, \mathbb{Z})$ *and* $\deg_\Xi(a(F)) = \int_F \Xi^2|_F = 20$.

(iii) *The composition*

$$\alpha\colon F \times F \xrightarrow{a \times a} A \times A \xrightarrow{-} A$$

is generically finite of degree six and its image is the theta divisor $\Xi \subset A$.[10]

(iv) *In degree two, the pullback defines an index two inclusion*

$$0 \longrightarrow H^2(A, \mathbb{Z}) \xrightarrow{a^*} H^2(F, \mathbb{Z}) \longrightarrow \mathbb{Z}/2\mathbb{Z} \longrightarrow 0,$$

the cokernel of which is generated by the image of $(1/2) \cdot a^*(\Xi) = (1/3) \cdot g$.

Note that the Albanese map $F \longrightarrow A$ depends on the choice of a point in F, so a line in Y. However, the map $\alpha\colon F \times F \longrightarrow A$ does not. So, the intermediate Jacobian of Y or, equivalently, the Albanese of its Fano surface F comes with a distinguished divisor $\Xi \subset A$ representing the principal polarization.

Proof All assertions are invariant under deformations of Y, so we may choose Y general. Then, $H^{1,1}(F, \mathbb{Q}) = \mathbb{Q} \cdot g = \mathbb{Q} \cdot [C_L]$, see Remark 2.15, and $H^{3,3}(A, \mathbb{Q}) = \mathbb{Q} \cdot \Xi^3$. For the last equality, one has to use the description of the monodromy group and the isomorphism $H^6(A, \mathbb{Q}) \simeq \bigwedge^6 H^1(A, \mathbb{Q}) \simeq \bigwedge^6 H^3(Y, \mathbb{Q})(6)$ combined with the arguments in Remark 1.2.13. Therefore, (i) is equivalent to the second assertion in (ii), which in turn is is equivalent to the first one in (ii), for $\int_F g^2 = 45$.

In order to prove (ii), we use that the composition $C_L \hookrightarrow F \hookrightarrow A \simeq \mathrm{Prym}(C_L/D_L)$ is the Abel–Prym map, use (1.4) in Remark 1.18 and (3.2). This suffices to conclude, because $\deg \mathrm{AP}^*(\Xi) = 10$ by virtue of Remark 3.9.

For the verification of (iii), one first shows that α is of degree at least six. Pick a generic point $(L_1, L_2) \in F \times F$ and consider the points of intersections $C_{L_1} \cap C_{L_2} = \{M_1, \ldots, M_5\}$ and the residual lines M_{ki}, $k = 1, 2$, of $L_k \cup M_i \subset \overline{L_k M_i} \cap Y$. Then $\mathcal{O}(C_{L_1} +$

[9] This comment was prompted by a question of G. Oberdieck.
[10] So, both morphisms $q\colon \mathbb{L} \longrightarrow Y$ and $\alpha\colon F \times F \longrightarrow \Xi$ are of degree six, but it seems for different reasons.

$C_{M_i} + C_{M_{1i}}) \simeq \mathcal{O}(C_{L_2} + C_{M_i} + C_{M_{2i}})$, as both line bundles are given by the image of $\mathbb{P}^2 \cap Y$ under $\mathrm{CH}^2(Y) \twoheadrightarrow \mathrm{Pic}(F)$. Hence,

$$\mathcal{O}(C_{L_1} - C_{L_2}) \simeq \mathcal{O}(C_{M_{2i}} - C_{M_{1i}})$$

and, therefore, by Lemma 3.1 the points $(M_{2i}, M_{1i}) \in F \times F$, $i = 1, \ldots, 5$ are all contained in the fibre $\alpha^{-1}(\alpha(L_1, L_2))$. Hence, indeed $\deg(\alpha) \geq 6$.

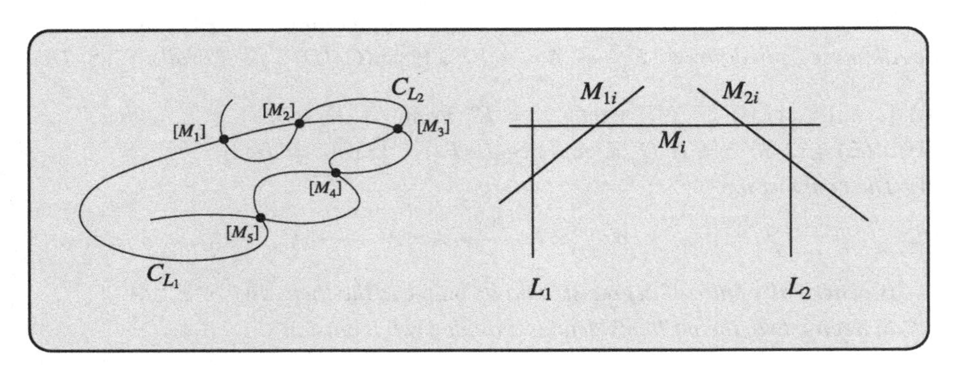

If we knew already that

$$\int_{F \times F} \alpha^*(\Xi^4) = 6 \cdot 5!, \tag{3.10}$$

then we could conclude that α is generically finite and, thus, its image is a divisor. As $H^{1,1}(A, \mathbb{Z}) = \mathbb{Z} \cdot \Xi$ for general Y and thus $[\alpha(F \times F)] = k \cdot \Xi$, $k \geq 1$, $\deg(\alpha) \geq 6$ and $\int_A \Xi^5 = 5!$ would then imply $\deg(\alpha) = 6$ and $[\alpha(F \times F)] = \Xi$, i.e. $k = 1$.

It remains to prove (3.10), for which one uses (i) and the fact that the Pontrjagin product $m_*\left(\frac{\Xi^3}{3!} \boxtimes \frac{\Xi^3}{3!}\right)$ equals $6 \cdot \Xi$. The latter is in [42] proved by evoking the geometric description of $(1/3!) \cdot \Theta^3$ for the theta divisor Θ on the Jacobian of a genus five curve. It is a special case of the general version of the Poincaré formula $m_*([W_n] \boxtimes [W_m]) = \binom{n+m}{m}[W_{n+m}]$, cf. [305, Ch. 16.5].

To prove (iv), we first claim that $a^* \colon H^2(A, \mathbb{Z}) \hookrightarrow H^2(F, \mathbb{Z})$ is a finite index inclusion. Indeed, a^* is the composition of $H^2(A, \mathbb{Z}) \simeq \bigwedge^2 H^1(A, \mathbb{Z}) \xrightarrow{\sim} \bigwedge^2 H^1(F, \mathbb{Z})$, cf. Proposition 3.10, and the cup product $\bigwedge^2 H^1(F, \mathbb{Z}) \hookrightarrow H^2(F, \mathbb{Z})$, which is injective by virtue of Lemma 2.5. Since both groups are of rank 45, this proves the claim. Furthermore, if the left-hand side is endowed with the quadratic form

$$(\alpha.\beta)_A := (1/3!) \int_A \alpha \cdot \beta \cdot \Xi^3,$$

which is integral as for the principal polarization Ξ the class $(1/3!) \cdot \Xi^3$ is integral, then a^* is an isometry by virtue of (i). Next, one shows that $\mathrm{disc}(\ .\ )_A = 4$. For this, consider the power E^5 of an elliptic curve with its natural principal polarization and write $H^2(E^5,\mathbb{Z})$ as a direct orthogonal sum of summands of the form $H^1(E,\mathbb{Z})\boxtimes H^1(E,\mathbb{Z})$ and $H^2(E,\mathbb{Z})\boxtimes H^0(E,\mathbb{Z})$. Then $(\ .\ )_{E^5}$ is unimodular on the first and of discriminant four on the second summand. Now, as the intersection pairing on $H^2(F,\mathbb{Z})$ is unimodular, the standard formula for the index of a finite index isometric embedding, see e.g. [249, 14.0.2], implies the result. The last statement follows from the fact that $\Xi \in H^2(A,\mathbb{Z})$ is not divisible. □

Remark 3.13 The two isomorphisms

$$H^1(A,\mathbb{Q}) \simeq H^1(F,\mathbb{Q}) \quad \text{and} \quad H^2(A,\mathbb{Q}) \simeq H^2(F,\mathbb{Q})$$

show that the natural map $H^*(A,\mathbb{Q}) \simeq \bigwedge^* H^1(A,\mathbb{Q}) \twoheadrightarrow H^*(F,\mathbb{Q})$ is surjective. This allows us to write the cohomology of the Fano variety as the quotient of $H^*(A,\mathbb{Q})$ by the ideal generated by the primitive classes in degree three:

$$H^*(A,\mathbb{Q})/(P_3) \xrightarrow{\sim} H^*(F,\mathbb{Q}).$$

The curve analogue is well known: the inclusion $C \hookrightarrow J(C)$ of a smooth curve into its Jacobian, after choosing a point $x_0 \in C$, induces a surjection $H^*(J(C),\mathbb{Q}) \twoheadrightarrow H^*(C,\mathbb{Q})$, the kernel of which is the ideal generated by primitive classes of degree two:

$$H^*(J(C),\mathbb{Q})/(P_2) \xrightarrow{\sim} H^*(C,\mathbb{Q}).$$

We will encounter a similar structure for cubic fourfolds, see Remark **6**.3.13 and Appendix 6.

Remark 3.14 For a principally polarized abelian variety (A,Ξ), the cohomology classes

$$(1/p!)\cdot \Xi^p \in H^{p,p}(A,\mathbb{Z}),$$

which are not divisible any further, are also called *minimal*.

For the Jacobian of a smooth projective curve, all minimal cohomology classes are effective. Furthermore, according to results of Matsusaka [345] and Ran [395], every principally polarized abelian variety that admits an effective minimal cohomology class of codimension $g-1$ is the Jacobian of a smooth projective curve of genus g.

In the above result, (i) says that for the intermediate Jacobian $J(Y)$ of a smooth cubic $Y \subset \mathbb{P}^4$, the minimal cohomology class of codimension three is effective and, according to a conjecture of Debarre [137], any irreducible principally polarized abelian variety A with an effective minimal cohomology class of codimension $1 < p < \dim(A)$ is either the Jacobian of a smooth projective curve or the intermediate Jacobian of a smooth cubic threefold. For recent progress on the conjecture in dimension five, see the work of Casalaina-Martin, Popa, and Schreieder [110]. In relation to this, Krämer [287] showed that the decomposition of Ξ as the sum of F and $-F$ is the only decomposition of the theta divisor. In Remark **7**.4.5 we will comment on $(1/4!)\cdot\Xi^4 \in H^{4,4}(J(Y),\mathbb{Z})$.

Remark 3.15 The class of $F \hookrightarrow A$ has another interesting feature. As shown by Collino, Naranjo, and Pirola [122], the difference of $[F]$ and $[-F]$ is homologically but not algebraically trivial. Here, by definition $-F$ is the image of the composition $(-1) \circ a$.

Although not stated as such explicitly, the result was independently and earlier proved by van der Geer and Kouvidakis [461] using degenerations to nodal cubic threefolds. They also observed that F and $(1/3!) \cdot \Xi^3$ are homologically but not algebraically equivalent.

3.4 Chow Groups As a further consequence, one obtains a description of the algebraically trivial part of the Chow group of one-dimensional cycles on Y.

Corollary 3.16 *Assume $L \subset Y$ is a line that is generic in the sense of Corollary* 1.9. *Then the Abel–Jacobi map gives an isomorphism of groups*

$$\mathrm{CH}^2(Y)_{\mathrm{alg}} \simeq \mathrm{CH}^2(Y)_{\mathrm{hom}} \xrightarrow{\sim} J(Y) \simeq \mathrm{Prym}(C_L/D_L).$$

Proof The result can be seen as an application of a result of Bloch and Srinivas [72, Thm. 1 (ii)]: if Y is a smooth complex projective variety with $\mathrm{CH}_0(Y) \simeq \mathbb{Z}$, then the Abel–Jacobi map induces isomorphisms of groups

$$\mathrm{CH}^2(Y)_{\mathrm{alg}} \simeq \mathrm{CH}^2(Y)_{\mathrm{hom}} \xrightarrow{\sim} J(Y). \tag{3.11}$$

Clearly, as on a cubic threefold any two points can be connected by a chain of lines, cubic threefolds satisfy the assumption.

However, in our case of a smooth cubic threefold more direct arguments for the isomorphism $\mathrm{CH}^2(Y)_{\mathrm{alg}} \simeq J(Y)$ exist, see [364–366] or [41, Thm. 3.1]. □

Remark 3.17 Just a few more comments on the motivic aspects of the above. For more information, see Section **7**.4.

(i) Note that the above result in particular shows the finite-dimensionality of the motive $\mathfrak{h}(Y)$ in the sense of Kimura and O'Sullivan [20, 272], cf. Section **7**.4.3. Since $\mathfrak{h}(F(Y))$ can be expressed in terms of $\mathfrak{h}(Y)$, see Section **2**.4.2, this implies that $\mathfrak{h}(F(Y))$ is also finite-dimensional. The result can also be deduced from the isomorphism of rational Chow motives

$$\mathfrak{h}^2(A) \simeq \mathfrak{h}^2(F(Y))$$

proved by Diaz [152, Thm. 1.2].

(ii) Bloch [71, Exa. 1.7] proves that the intersection product defines a surjection

$$\mathrm{Pic}^0(F(Y)) \otimes_{\mathbb{Z}} \mathrm{Pic}(F(Y)) \twoheadrightarrow \mathrm{CH}^2(F(Y))_0,$$

which also implies finite-dimensionality of the motive $\mathfrak{h}(F(Y))$.

(iii) The isomorphism of Chow groups

$$\mathrm{CH}^2(Y)_{\mathrm{alg}} \simeq \mathrm{Pic}^0(F(Y)) \simeq \mathrm{Prym}(C_L/D_L)$$

leads to an isomorphism of rational Chow motives

$$\mathfrak{h}^3(Y)(1) \simeq \mathfrak{h}^1(F(Y)) \simeq \mathfrak{h}^1(C_L)^-.$$

Here, the latter is the anti-invariant part of $\mathfrak{h}^1(C_L)$ under the action of the involution ι^*. In a slightly different form this result was first proved by Sermenev [486] and also a little later by Reid [398, App. 4.3], but see also Manin [334, Sec. 11] and the more general results of Nagel and Saito [368] on Chow motives of conic fibrations.

Remark 3.18 The arguments in the above discussion rely heavily on the ground field being $\mathbb{C}$. In fact, already the definition of the intermediate Jacobian $J(Y)$ over other fields is problematic. However, $\mathrm{CH}^2(Y)$ and $\mathrm{Prym}(C_L/D_L)$ make perfect sense over arbitrary fields and, indeed, Murre [364] describes an algebraic approach that defines the isomorphism $\mathrm{CH}^2(Y)_{\mathrm{alg}} \simeq \mathrm{Prym}(C_L/D_L)$ over arbitrary algebraically closed fields. In fact, the isomorphism was originally stated up to elements of order two, but the divisibility of $\mathrm{CH}^2(Y)_{\mathrm{alg}}$ (pointed out by Bloch, see the review of [364]), proves the full statement. Shen [427] discusses the isomorphism to the Prym for cubics over arbitrary fields.

Remark 3.19 Clemens and Griffiths [119, App. A] give a geometric argument that shows a weaker version of the first isomorphism in (3.11), namely that the difference between $\mathrm{CH}^2(Y)_{\mathrm{hom}}$ and $\mathrm{CH}^2(Y)_{\mathrm{alg}}$ is annihilated by six. More precisely, it is shown that $6 \cdot \mathrm{CH}^2(Y)_{\mathrm{hom}} = \mathrm{CH}^2(Y)_{\mathrm{alg}}$.

The idea is the following: let $C \subset Y$ be any curve. Then a surface $C \subset S \subset Y$ is constructed such that $6\,C$ on Y is rationally equivalent to the sum of $\mathbb{P}^2 \cap S$ and a sum of lines $\sum a_i L_i$. As $\mathrm{Pic}(Y) \simeq \mathbb{Z}$, this proves the assertion. The surface S is obtained as the image $q(\tilde{S})$ of the surface $\tilde{S} := p^{-1}(p(q^{-1}(C)))$ parametrizing pairs (L, x) consisting of a point x contained in the line L that intersects C. Clearly, $\tilde{S}$ is a $\mathbb{P}^1$-bundle over $p(q^{-1}(C))$, which comes with a natural multi-section $q^{-1}(C) \subset \tilde{S}$. Its image under q_* gives $6\,C$.

4 Global Torelli Theorem and Irrationality

In this section we survey the known arguments that, based on the results in the earlier parts of this chapter, prove two milestone results: the global Torelli theorem and the irrationality of all smooth cubic threefolds.

4.1 Torelli for Curves We begin by recalling the classical Torelli theorem for smooth projective curves over $\mathbb{C}$. The statement for cubic threefolds is literally the same and its original proof, by Clemens–Griffiths [119] and independently by Tyurin [446], mimics Andreotti's classical proof for curves [21]. However, other and easier proofs exist.

Theorem 4.1 (Torelli theorem) *For two smooth projective, irreducible curves C and C' over $\mathbb{C}$, the following assertions are equivalent:*

(i) *There exists an isomorphism $C \simeq C'$.*
(ii) *There exists a Hodge isometry $H^1(C,\mathbb{Z}) \simeq H^1(C',\mathbb{Z})$.*
(iii) *There exists an isomorphism $(J(C),\Theta) \simeq (J(C'),\Theta')$ of polarized varieties.*

Recall that a *Hodge isometry* is an isomorphism of Hodge structures that in addition is compatible with the intersection product $(\,.\,)$ naturally defined on the first cohomology $H^1(C,\mathbb{Z})$ of any smooth projective curve.

The theta divisor $\Theta \in H^2(J(C),\mathbb{Z})$ on $J(C) = \mathrm{Pic}^0(C)$ is given by the intersection form viewed as an element in

$$\bigwedge^2 H^1(C,\mathbb{Z}) \simeq \bigwedge^2 H^1(J(C),\mathbb{Z}) \simeq H^2(J(C),\mathbb{Z}).$$

The isomorphism in (iii) is an isomorphism of varieties $\varphi\colon J(C) \xrightarrow{\sim} J(C')$ such that the induced map $\varphi_*\colon H^2(J(C),\mathbb{Z}) \xrightarrow{\sim} H^2(J(C'),\mathbb{Z})$ satisfies $\varphi_*(\Theta) = \Theta'$.

As the intersection form on $H^1(C,\mathbb{Z})$ is unimodular, the theta divisor as a cohomology class on $J(C)$ satisfies $\int_{J(C)} \Theta^g = g!$ or, in other words, Θ is a principal polarization. For any line bundle with first Chern class Θ, we shall write $\mathcal{O}(\Theta)$. The Riemann–Roch formula shows $h^0(J(C),\mathcal{O}(\Theta)) = 1$, i.e. $\mathcal{O}(\Theta)$ is indeed the line bundle associated with a uniquely determined effective divisor which is also called Θ. As the line bundle $\mathcal{O}(\Theta)$ is only unique up to twisting by line bundles in $\mathrm{Pic}^0(J(C))$, the effective divisor Θ is only unique up to translation.

Geometrically, the (or, rather, a) theta divisor is described as the image

$$\Theta = W^0_{g-1}(C) \subset J(C)$$

of the following morphism which depends on the choice of a point $x \in C$

$$u\colon C^{g-1} \longrightarrow J(C) = \mathrm{Pic}^0(C),$$
$$(x_1,\ldots,x_{g-1}) \longmapsto \mathcal{O}\left(\textstyle\sum x_i - (g-1)x\right).$$

For any other choice of x, say x', the image of u is the translate by the line bundle $\mathcal{O}((g-1)(x-x')) \in J(C)$. Note that any isomorphism $\varphi\colon J(C) \xrightarrow{\sim} J(C')$ with $\varphi_*(\Theta) = \Theta' \in H^2(J(C'),\mathbb{Z})$ can be composed with a translation such that it in fact satisfies the equality $\varphi(\Theta) = \Theta'$ of effective divisors.

Remark 4.2 Note that it may very well happen that the Jacobians $J(C)$ and $J(C')$ of two curves are isomorphic as unpolarized (abelian) varieties without C and C' being isomorphic. It is unclear whether this can be reinterpreted purely in terms of C and C'. In any case, if $J(C) \simeq J(C')$ as unpolarized varieties, then the symmetric products of the two curves satisfy $[C^{(d)}] = [C'^{(d)}] \in K_0(\mathrm{Var}_{\mathbb{C}})$ for all $d \geq 2g - 2$.

4.2 Torelli for Cubic Threefolds The following is the analogue of the classical global Torelli theorem for curves.

Theorem 4.3 (Clemens–Griffiths, Tyurin) *For two smooth cubic hypersurfaces $Y, Y' \subset \mathbb{P}^4$ over $\mathbb{C}$, the following assertions are equivalent:*

(i) *There exists an isomorphism $Y \simeq Y'$.*
(ii) *There exists a Hodge isometry $H^3(Y,\mathbb{Z}) \simeq H^3(Y',\mathbb{Z})$.*
(iii) *There exists an isomorphism $(J(Y), \Xi) \simeq (J(Y'), \Xi')$ of polarized varieties.*

Remark 4.4 Unlike the Jacobian of a curve or of any variety, the intermediate Jacobian of a variety has usually no modular interpretation and, in fact, is not even necessarily an abelian variety.

For a cubic threefold $Y \subset \mathbb{P}^4$, the situation is better: $J(Y)$ is a principally polarized abelian variety and has a certain moduli interpretation provided by the isomorphism $J(Y) \simeq \mathrm{Pic}^0(F(Y))$ of polarized abelian varieties, see Corollary 3.3 and also Remark 3.6. However, the question whether the existence of an unpolarized isomorphism $J(Y) \simeq J(Y')$ reflects a geometric relation between Y and Y' remains. In fact, it is potentially even more interesting here than for curves.

Remark 4.5 According to Proposition 3.10, for a generic line $L \subset Y$ there exists an isomorphism of polarized abelian varieties $(\mathrm{Prym}(C_L/D_L), \Xi) \simeq (J(Y), \Xi)$. Of course, the polarized Prym variety is uniquely determined by the isomorphism class of the curve $C_L \subset F(Y)$ of all lines intersecting L and the involution ι. Thus, as an immediate consequence of the above global Torelli theorem, one can also state

$$(C_L, \iota) \simeq (C_{L'}, \iota') \;\Rightarrow\; Y \simeq Y',$$

where on the left-hand side one has an isomorphism of curves that commutes with the natural involutions on both sides and L and L' are generically chosen lines contained in Y and Y'. Note that the converse does not hold, i.e. $Y \simeq Y'$ is not expected to imply $C_L \simeq C_{L'}$, for the isomorphism type of $C_L \subset F(Y)$ varies with L for fixed Y.

Remark 4.6 Assume that Y is a very general cubic threefold. Then any other smooth cubic threefold Y' for which there exists an isomorphism of rational Hodge structures $H^3(Y,\mathbb{Q}) \simeq H^3(Y',\mathbb{Q})$ is actually isomorphic to Y. This can either be seen as a consequence of the variational Torelli theorem, see Corollary **3**.3.8, or by arguing that for the very general cubic threefold any such isomorphism would imply the existence of an integral Hodge isometry and then applying the above theorem.

Several proofs of Theorem 4.3 exist in the literature. The first one, see [42], seems the shortest and most instructive one, but it relies on the fact that the theta divisor $\Xi \subset J(Y)$ has only one singular point, which we will only prove in Corollary 4.7. The somewhat longer one to be outlined in Section 4.4 has the advantage of being closer to Andreotti's classical proof for curves.

Proof Clearly, it suffices to show that (iii) implies (i). First, note that a principal polarization of an abelian variety can be thought of as a cohomology class, but that the actual effective divisor is only determined up to translation. Nevertheless, from an isomorphism of principally polarized abelian varieties as in (iii) one obtains an isomorphism of the distinguished theta divisors $\Xi \simeq \Xi'$. Assuming that Ξ has only one singular point $0 \in \Xi$, cf. Corollary 4.7, this isomorphism sends $0 \in \Xi$ to $0' \in \Xi'$. Thus, it suffices to prove that the projective tangent cone of $0 \in \Xi$ is isomorphic to the cubic threefold:

$$\mathrm{TC}_0(\Xi) \simeq Y. \tag{4.1}$$

By virtue of the universal property of the blow-up [223, II. Cor. 7.15], the morphism $\alpha\colon F \times F \twoheadrightarrow A$ in Corollary 3.12 induces a diagram

$$\begin{array}{ccccccccc}
\mathbb{P}(\mathcal{T}_F) & \hookrightarrow & \mathrm{Bl}_\Delta(F \times F) & \twoheadrightarrow & \mathrm{Bl}_0(\Xi) & \hookrightarrow & \mathrm{Bl}_0(A) & \hookleftarrow & \mathbb{P}(T_0A) \\
\downarrow & & \downarrow & & \downarrow & & \downarrow & & \downarrow \\
\Delta & \hookrightarrow & F \times F & \overset{\alpha}{\twoheadrightarrow} & \Xi & \hookrightarrow & A & \hookleftarrow & \{0\}.
\end{array}$$

Here, we use the description of the exceptional divisor of $\mathrm{Bl}_\Delta(F \times F) \twoheadrightarrow F \times F$ as the projectivization $\mathbb{P}(\mathcal{N}_{\Delta/F\times F})$ of the normal bundle and the isomorphism $\mathcal{N}_{\Delta/F\times F} \simeq \mathcal{T}_F$.

The fibre of the blow-up $\mathrm{Bl}_0(\Xi) \twoheadrightarrow \Xi$ over the origin $0 \in \Xi$ is by definition the projective tangent cone of $0 \in \Xi$, which is regarded as a closed subscheme $\mathrm{TC}_0(\Xi) \subset \mathbb{P}(T_0A)$. As Ξ is irreducible, also the blow-up $\mathrm{Bl}_0(\Xi)$ is, see [223, II. Prop. 7.16]. Therefore, the induced morphism between the exceptional divisors $\mathbb{P}(\mathcal{T}_F) \twoheadrightarrow \mathrm{TC}_0(\Xi)$ is surjective. Composed with the isomorphism $\mathbb{L} \simeq \mathbb{P}(\mathcal{S}_F) \simeq \mathbb{P}(\mathcal{T}_F)$, see Proposition 2.2, and the inclusion $\mathrm{TC}_0(\Xi) \subset \mathbb{P}(T_0A)$, it gives a morphism

$$r\colon \mathbb{L} \simeq \mathbb{P}(\mathcal{S}_F) \simeq \mathbb{P}(\mathcal{T}_F) \twoheadrightarrow \mathrm{TC}_0(\Xi) \subset \mathbb{P}(T_0A) \simeq \mathbb{P}^4.$$

Up to a linear coordinate change, r is nothing but the projection $q\colon \mathbb{L} \twoheadrightarrow \mathbb{P}(V)$. Indeed, $r^*\mathcal{O}(1) \simeq \mathcal{O}_p(1)$, because the relative tautological line bundle of the blow-up $\mathrm{Bl}_0(A) \twoheadrightarrow A$ restricts to $\mathcal{O}(1)$ on the exceptional divisor $\mathbb{P}(T_0A)$ and pulls back to the relative tautological line bundle of the blow-up $\mathrm{Bl}_\Delta(F\times F) \twoheadrightarrow F\times F$. The latter restricts to $\mathcal{O}_p(1)$ on $\mathbb{L} \simeq \mathbb{P}(\mathcal{T}_F)$. Hence, r is indeed induced by a linear subsystem of $|\mathcal{O}_p(1)|$, but the only base-point free one is the complete linear system $|\mathcal{O}_p(1)|$.

In particular, we recover Y as the image of r. This concludes the proof of (4.1). □

It is tempting to try a shortcut here, involving the geometric global Torelli theorem, see Proposition **2**.3.12 or Proposition 2.12. If $(A, \Xi) \simeq (A', \Xi')$, can one use $\alpha\colon F \times F \twoheadrightarrow \Xi$ and $\alpha'\colon F' \times F' \twoheadrightarrow \Xi'$ to deduce directly $F \simeq F'$ and from there $Y \simeq Y'$? In the alternative proof below, the Fano surface will play a role, albeit in an indirect fashion.

4.3 Andreotti's Proof Next we give an outline of the main arguments of Andreotti's proof for non-hyperelliptic curves. We choose the notation to match the one we are using for cubic threefolds.

Let C be a smooth projective curve of genus g and let $V := H^{1,0}(C)^* \simeq H^0(C, \omega_C)^*$. The theta divisor

$$\Theta \subset J(C) \simeq H^{1,0}(C)^*/H_1(C,\mathbb{Z}) \simeq V/H_1(C,\mathbb{Z})$$

gives rise to the rational Gauss map $\gamma\colon \Theta \dashrightarrow \mathbb{P}(V^*)$. It is regular on the smooth locus $\Theta_{\rm sm} \subset \Theta$ and there given by $x \longmapsto \mathbb{P}(T_x\Theta)$. Here, the hyperplane $T_x\Theta \subset T_xJ(C) \simeq T_0J(C) \simeq V$ is considered as a point in $\mathbb{P}(V^*)$.

The Gauss map is studied via the canonical embedding of the non-hyperelliptic curve $i\colon C \hookrightarrow \mathbb{P}(V)$ given by the complete linear system $|\omega_C|$. The dual variety $C^* \subset \mathbb{P}(V^*)$ of this embedding is the hypersurface of all points $H \in \mathbb{P}(V^*)$ corresponding to hyperplanes $H \subset \mathbb{P}(V)$ tangent to C at at least one point, i.e. $C^* = \{H \mid |H \cap C| < 2g-2\}$. The key observation now is that the Gauss map $\gamma\colon \Theta_{\rm sm} \twoheadrightarrow \mathbb{P}(V^*)$ is a dominant map which is branched exactly over $C^* \subset \mathbb{P}(V^*)$:

$$\text{branch}(\gamma) = C^*.$$

As the Gauss map only depends on $\Theta \subset J(C)$ and C can be recovered from C^* as its dual variety [193], this immediately proves the global Torelli theorem.

To prove that C^* is indeed the branch divisor one studies the derivative of the morphism $u\colon C^{g-1} \twoheadrightarrow \Theta = W^0_{g-1} \subset J(C)$ at a point $\mathbf{x} = (x_1, \dots, x_{g-1}) \in C^{g-1}$. The image

$$du(T_{\mathbf{x}}C^{g-1}) \subset T_{u(\mathbf{x})}J(C) \simeq H^{1,0}(C)^* = \text{Hom}(H^0(C,\omega_C), \mathbb{C})$$

is nothing but the span of the linear maps $\alpha \longmapsto \alpha(x_i) \in \omega_C(x_i) \simeq \mathbb{C}$. Hence, $\mathbb{P}(T_{u(\mathbf{x})}\Theta)$ contains the span $\overline{i(x_1)\dots i(x_{g-1})}$. For a generic hyperplane $H \in \mathbb{P}(V^*)$, the intersection $H \cap i(C)$ consists of $2g-2$ distinct points $x_1, \dots, x_{2g-2}$ and there are exactly $\binom{2g-2}{g-1}$ choices of $(x_{i_1}, \dots, x_{i_{g-1}}) \in C^{g-1}$ that span H. In other words, the generic fibre of $C^{g-1} \overset{u}{\twoheadrightarrow} \Theta \overset{\gamma}{\dashrightarrow} \mathbb{P}(V^*)$ contains exactly $\binom{2g-2}{g-1}$ points.

Hence, the branch divisor of $\gamma \circ u$ is the locus with fewer than this number of points in the fibre. Away from the big diagonal in C^{g-1}, which maps into the singular locus $\Theta_{\rm sing}$, the cardinality of the fibre over $H \in \mathbb{P}(V^*)$ drops whenever H is tangent to C at one of the points, i.e. $H \in C^*$, or $g-1$ of the points, say $x_1, \dots x_{g-1}$ are linearly dependent. Note that the second case also leads to points in $\Theta_{\rm sing}$. Therefore, the branch divisor of γ is contained in C^*. As C^* is irreducible, this proves the claim and concludes this sketch of Andreotti's proof of the global Torelli theorem.

4.4 Singularity of the Theta Divisor The analogy between the original proof of the global Torelli theorem for cubic threefolds, which will be explained next, and Andreotti's for curves is visualized by the following picture:

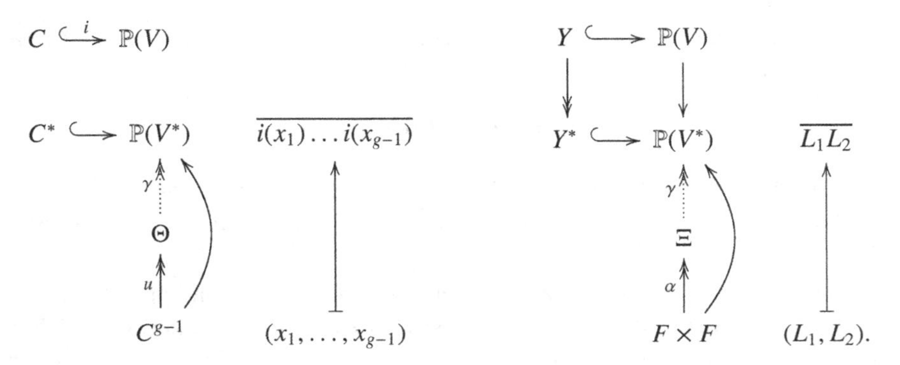

Here are the details, cf. [119, 446, 449]: let $\gamma\colon \Xi \dashrightarrow\!\!\!\!\to \mathbb{P}(V^*)$ be the Gauss map. It is regular on the smooth locus $\Xi_{\rm sm} \subset \Xi$ and there described by

$$x \longmapsto T_x\Xi \subset T_xJ(Y) \simeq T_0J(Y) \simeq T_0A(F) \simeq V,$$

using the identification $H^0(Y, \mathcal{O}(1))^* = V \simeq T_0A(F)$, cf. Corollary 2.9. The key step is to show that the branch divisor of the composition $F \times F \xrightarrow{\alpha} \Xi \overset{\gamma}{\dashrightarrow} \mathbb{P}(V^*)$ is (contained in) the dual variety $Y^* \subset \mathbb{P}(V^*)$. For this step, one uses the commutative diagram, see Corollary 2.9:

$$\begin{array}{ccc} \mathcal{S}_F & \overset{\delta_1}{\hookrightarrow} & V \otimes \mathcal{O}_F \\ \wr\| & & \wr\| \\ \mathcal{T}_F & \overset{\delta_2}{\hookrightarrow} & T_0A \otimes \mathcal{O}_F, \end{array}$$

which at a point $(L_1 = \mathbb{P}(W_1), L_2 = \mathbb{P}(W_2)) \in F \times F$ implies

$$\mathrm{Im}\,(d\alpha\colon T_{L_1,L_2}(F \times F) \longrightarrow T_{\alpha(L_1,L_2)}A) = \delta_2(T_{L_1}F) + \delta_2(T_{L_2}F) = W_1 + W_2.$$

Hence, for disjoint lines L_1 and L_2 or, equivalently, when $\overline{L_1L_2} \simeq \mathbb{P}^3$, one has

$$\gamma(\alpha(L_1, L_2)) = \overline{L_1L_2} \in \mathbb{P}(V^*).$$

The map $\gamma \circ \alpha$ can be extended to a morphism $F \times F \setminus \Delta \longrightarrow \mathbb{P}(V^*)$ by mapping a pair (L_1, L_2) of distinct lines with $L_1 \cap L_2 = \{x\}$ to the projective tangent space $\mathbb{T}_x Y \in \mathbb{P}(V^*)$. The generic fibre of $\gamma \circ \alpha$, say over $[H] \in \mathbb{P}(V^*) \setminus Y^*$, is the set of pairs $L_1 \neq L_2 \in F$ with $\overline{L_1 L_2} = H \subset \mathbb{P}(V)$ or, in other words, the set of pairs (L_1, L_2) of disjoint lines in the smooth cubic surface $S := Y \cap H$, of which there are exactly $27 \cdot 16 = 432$ independently of H, see Section **4**.3.2. Hence, the branch divisor of $\gamma \circ \alpha$ is contained in Y^*.

To conclude, observe

$$\text{branch}(\gamma) \subset \text{branch}(\gamma \circ \alpha) \subset Y^*.$$

Now, using $\deg(\alpha) = 6$, by virtue of Corollary 3.12, and $\deg(\gamma \circ \alpha) = 432$, one knows $\deg(\gamma \circ \alpha) > \deg(\alpha)$ and hence $\deg(\gamma) > 1$. Alternatively, one could use the fact that $\Xi \subset J(Y)$ is certainly not rational.

Then, as $\mathbb{P}(V^*)$ is simply connected, γ has a non-trivial branch divisor and, therefore,

$$\text{branch}(\gamma) = Y^*.$$

As the normalization of Y^* is the cubic threefold Y, this shows that Y is uniquely determined by $\Xi \subset J(Y)$, which concludes the second proof of Theorem 4.3. □

Extending the above considerations combined with an intersection theory computation one proves the following result.

Corollary 4.7 *The theta divisor $\Xi \subset J(Y)$ has only one singular point, namely $0 \in \Xi$ which has multiplicity three.*

Proof A naive idea to prove that $\Xi_{\text{sing}} = \{0\}$ is that the composition $\gamma \circ \alpha$ is regular in all points of the complement of $\Delta \subset F \times F$. The latter suggests that the Gauss map γ is regular in all points not contained in $\alpha(\Delta) = \{0\}$ and, hence, that 0 is the only singularity of Ξ. For a rigorous proof of $\Xi_{\text{sing}} = \{0\}$, we refer to [42, 439], see also Remark 4.14.

The assertion about the multiplicity follows essentially from (4.1). More precisely, a standard formula in intersection theory shows that $\deg(\alpha) \cdot \text{mult}_0(\Xi) = c_1^2(\Delta) - c_2(\Delta)$, where $\alpha^{-1}(0) = \Delta \simeq F$, see [188, Prop. 4.2(a)]. As the right-hand side equals 18 and $\deg(\alpha) = 6$, the result follows, cf. [188, Ex. 4.3.2]. □

Exercise 4.8 Observe the following immediate consequence of the above. The principally polarized abelian variety $J(Y)$ is irreducible, i.e. it cannot be written as a product $J(Y) \simeq A_1 \times A_2$ of two principally polarized varieties A_1 and A_2.

Casalaina-Martin and Friedman [106] prove a converse of the above result: a principally polarized abelian fivefold whose theta divisor has a unique singularity which in addition has multiplicity three is isomorphic to the intermediate Jacobian of a smooth cubic threefold.

Note that the two proofs of Theorem 4.3 sketched so far do not use the geometric Torelli theorem, see Proposition **2**.3.12.

Remark 4.9 (i) The strong form of the Torelli theorem for curves asserts that for two smooth projective, irreducible curves C and C' any isomorphism of principally polarized abelian varieties $(J(C), \Theta) \simeq (J(C'), \Theta')$ is up to a sign induced by a unique isomorphism $C \simeq C'$. This in particular applies to automorphisms and leads for a non-hyperelliptic curve C to the isomorphism

$$\mathrm{Aut}(C) \simeq \mathrm{Aut}(J(C), \Theta)/\{\pm \mathrm{id}\}.$$

If C is hyperelliptic, then $\mathrm{Aut}(C) \simeq \mathrm{Aut}(J(C), \Theta)$. This stronger form of the Torelli theorem is surprisingly poorly documented in the literature, but see Weil's Bourbaki talk [493], Milne's account of it [352, §13], or Serre's appendix to [310]. The result is needed to ensure that the map between the coarse moduli spaces of curves and principally polarized abelian varieties $M_g \hookrightarrow A_g$ is not only injective but in fact a locally closed embedding.

(ii) The analogous assertion for a smooth cubic threefold $Y \subset \mathbb{P}^4$ is the isomorphism

$$\mathrm{Aut}(Y) \times \{\pm \mathrm{id}\} \simeq \mathrm{Aut}(J(Y), \Xi),$$

which, of course, also shows $\mathrm{Aut}(Y) \simeq \mathrm{Aut}(J(Y), \Xi)/\{\pm \mathrm{id}\}$ as in the curve case.

Observe that, unlike the case of curves, the principally polarized abelian variety $J(Y)$ comes with a distinguished theta divisor, namely the one whose only singularity is at the origin. Hence, any polarized automorphism of $(J(Y), \Xi)$ restricts to an automorphism of Ξ, which in turn provides an automorphism of the projective tangent cone $Y \simeq \mathrm{TC}_0\Xi$ of its unique singular point. Note that $-\mathrm{id}$ acts as the identity on $\mathbb{P}(T_0J(Y))$ and hence on Y. More precisely, this map $\mathrm{Aut}(J(Y), \Xi) \longrightarrow \mathrm{Aut}(Y)$ has kernel $\{\pm \mathrm{id}\}$ and it provides an inverse of the natural injection $\mathrm{Aut}(Y) \hookrightarrow \mathrm{Aut}(J(Y), \Xi)$, see Exercise 2.1. This proves the assertion. In this form, the result was established by Beauville (private communication) and Zheng [499, Prop. 1.6].[11]

Remark 4.10 In [447] one finds another kind of Torelli theorem which instead of $(J(Y), \Xi)$ uses the algebraic equivalence class of $F \subset A(F) \simeq \mathrm{Pic}^0(F) \simeq J(Y)$. More precisely, Tyurin proves the following assertion: two smooth cubic threefolds $Y, Y' \subset \mathbb{P}^4$ are isomorphic if and only if there exists an isomorphism of varieties $A(F(Y)) \simeq A(F(Y'))$ such that under this isomorphism the two cycles $F(Y) \subset A(F(Y))$ and $F(Y') \subset A(F(Y'))$ are algebraically equivalent. With all the results proved in the previous sections, this consequence is not difficult to prove. We leave the details to the reader.

4.5 Cubic Threefolds Are Not Rational The description of the singular locus of $\Xi \subset J(Y)$ immediately leads to the following counterexample to the Lüroth problem for threefolds: smooth cubic threefolds are unirational but not rational. The result was

[11] Thanks to M. Rapoport for the references.

first proved in [119]. We recommend Beauville's survey in [56] for comments on the historical context and references to other examples.

Corollary 4.11 (Clemens–Griffiths) *A smooth cubic threefold $Y \subset \mathbb{P}^4$ is not rational.*

Proof First note that the blow-up of a smooth threefold Z in a point does not change its intermediate Jacobian $J(Z)$ while the blow-up along a smooth curve $C \subset Z$ changes the intermediate Jacobian. In the latter case, there is a natural isomorphism of principally polarized abelian varieties

$$J(\mathrm{Bl}_C(Z)) \simeq J(Z) \times J(C).$$

Hence, if a smooth cubic threefold $Y \subset \mathbb{P}^4$ is rational, then there exist smooth curves C_i and D_i and an isomorphism

$$J(Y) \times J(D_1) \times \cdots \times J(D_m) \simeq J(C_1) \times \cdots \times J(C_n)$$

of principally polarized abelian varieties. Using the principal polarization, this leads to an isomorphism of principally polarized abelian varieties

$$J(Y) \simeq J(C_1) \times \cdots \times J(C_k) \simeq J(C_1 \sqcup \ldots \sqcup C_k). \tag{4.2}$$

Classical Brill–Noether theory shows that the singular set Θ_{sing} of the theta divisor

$$\Theta = W^0_{g-1} = \{L \in \mathrm{Pic}^{g-1}(C) \mid h^0(L) > 0\} \subset \mathrm{Pic}^{g-1}(C) \simeq J(C)$$

of the Jacobian of a smooth curve C of genus g is the Brill–Noether locus

$$\Theta_{\mathrm{sing}} = W^1_{g-1} = \{L \in \mathrm{Pic}^{g-1}(C) \mid h^0(L) > 1\},$$

see [23, Ch. IV]. Furthermore, the determinantal description of W^1_{g-1} implies

$$\dim(W^1_{g-1}) \geq \rho(g-1, 1, g) = g - 4. \tag{4.3}$$

For $g = 5$, one finds $\dim(\Theta_{\mathrm{sing}}) \geq 1$. On the other hand, we know by Corollary 4.7 that $\dim(\Xi_{\mathrm{sing}}) = 0$ for the theta divisor $\Xi \subset J(Y)$. Hence, there is no polarized isomorphism $J(Y) \simeq J(C)$. Note that the fact that Ξ_{sing} is a point also shows that $J(Y)$ is irreducible and hence $k = 1$ in (4.2), see [56, Lem. 2] for some more details concerning this point. □

Exercise 4.12 Adapt the above arguments to show that two birational smooth cubic threefolds $Y, Y' \subset \mathbb{P}^4$ are always isomorphic.

Remark 4.13 The crucial observation in the above proof is that as a polarized abelian variety $J(Y)$ is not isomorphic to the Jacobian of a smooth curve $J(C)$. Independently of its application to the irrationality, this has other interesting consequences. For example, Schreieder [419, Cor. 25] shows that this implies that the surface $F(Y)$ is not dominated by a product of curves. In general, we are lacking the techniques to decide which surfaces can or cannot be dominated by product of curves, see Schoen's article [418] for some general results in this direction.

Remark 4.14 Since the original proof by Clemens and Griffiths [119], many other arguments have been found to prove the irrationality of smooth cubic threefolds.

(0) The irrationality of smooth cubic threefolds in arbitrary characteristic $\neq 2$ was proved by Murre in [365].

(i) In [119, App. C], Clemens and Griffiths explained how Mumford's theory of Prym varieties can alternatively be exploited to show that $J(Y) \simeq \mathrm{Prym}(C_L/D_L)$ cannot be the Jacobian of a curve.

Indeed, according to Mumford [359, Sec. 6], see also [42, Sec. 3] or [439, Sec. 5], with a short list of exceptions the theta divisor of $\mathrm{Prym}(C/L)$ has singularities in dimension $< g(D) - 4$, which would exclude it from being the Jacobian of a curve by (4.3). One exception is a quintic curve $D \subset \mathbb{P}^2$ for which $h^0(D, \mathcal{L}_\pi \otimes \mathcal{O}(1)|_D) \equiv 0\,(2)$, where $\mathcal{L}_\pi$ is the torsion line bundle defining the étale cover $C \longrightarrow D$. However, although in our case $D_L \subset \mathbb{P}^2$ is a quintic, one can show that $h^0(D_L, \mathcal{L}_\pi \otimes \mathcal{O}(1)|_{D_L}) = 1$. Indeed, $h^0(D_L, \mathcal{L}_\pi \otimes \mathcal{O}(1)|_{D_L}) = h^0(C_L, \pi^*\mathcal{O}(1)) - h^0(D_L, \mathcal{O}(1)|_{D_L}) = 4 - 3 = 1$. Here, we use the fact that $C_L \subset E \simeq \mathbb{P}(\mathcal{N}_{L/Y} \simeq \mathcal{O}_L^{\oplus 2}) \simeq L \times \mathbb{P}^1$, via the intersection point $L' \longmapsto L \cap L'$, and $\pi^*\mathcal{O}(1) \simeq \phi^*\mathcal{O}(1)|_{C_L} \simeq (\tau^*\mathcal{O}_Y(1) \otimes \mathcal{O}(-E))|_{C_L} \simeq (\mathcal{O}_{\mathbb{P}^1}(1) \boxtimes \mathcal{O}_{\mathbb{P}^1}(1))|_{C_L}$, cf. Remark 1.24. See also [43] for a detailed discussion.

(ii) Beauville [56, Thm. 3] considered the Klein cubic threefold $Y = V(F)$ defined by $F = x_0^2x_1 + x_1^2x_2 + x_2^2x_3 + x_3^2x_4 + x_4^2x_0$. Recall that its Fano variety of lines has the maximal Picard number $\rho(F(Y)) = 25$, see Remark 2.15. The Klein cubic comes with a group of automorphisms generated by two automorphisms of order 5 and 11. If Y were rational and hence $J(Y) \simeq J(C)$, these automorphisms would act on the curve C, which is shown to be geometrically impossible. No further information about the intermediate Jacobian $J(Y)$ and its theta divisor is needed in this example, which simplifies the argument considerably.

Similarly, Zarhin [497] used the natural automorphisms of a cubic threefold $Y \longrightarrow\!\!\!\!\!\rightarrow \mathbb{P}^3$ associated with a cubic surface $S \subset \mathbb{P}^3$, see Section **1**.5.6, to prove irrationality.

(iii) The argument by Markushevich and Roulleau [340] is more arithmetic. They exhibited a rather complicated cubic equation $F \in \mathbb{Z}[x_0, \ldots, x_4]$ with good reduction modulo three. The Weil conjectures can be used to prove that the reduction of the intermediate Jacobian cannot be the Jacobian of a curve over any finite extension of $\mathbb{F}_3$. This suffices to exclude the intermediate Jacobian $J(Y)$ of the complex cubic threefold Y from being the Jacobian of a curve.

(iv) The condition for cubic threefolds to be rational a priori defines a countable union of locally closed subsets in $|\mathcal{O}(3)|$. Thus, once an irrational smooth cubic threefold, like the Klein cubic threefold in [56], has been found, the very general cubic threefold will also be irrational. Note that according to a result of Kontsevich and Tschinkel [285], rationality is in fact closed under specialization.

(v) Specializations to mildly singular cubic threefolds have also been exploited to prove irrationality of very general cubic threefolds. Collino [120] and van der Geer and Kouvidakis [460] used degenerations to cubics with one ordinary double point and study the induced specialization of the intermediate Jacobian. Bardelli [35] relied on limiting mixed Hodge structures for a degeneration to a union of three hyperplanes, while Gwena [215] specialized in the unique cubic with 10 ordinary double points, see Remark **1**.5.17 and Remark **4**.2.16.

(vi) It is not known whether smooth cubic threefolds are stably rational. In fact, no example of a stably rational smooth cubic threefold has been found. Voisin [483] showed that if a smooth cubic threefold $Y \subset \mathbb{P}^4$ is stably rational, then the minimal curve class $(1/4!)\cdot\Xi^4$ on $J(Y)$ is effective. Note that for $\rho(F(Y)) = 1$, a minimal curve in $J(Y)$ cannot be contained in $F(Y)$ as it would then represent $(1/6)\cdot g$, which is not an integral class. As for rationality, it is known that stable rationality specializes [371]. We recommend [471, 482] for further information and references.

Remark 4.15 We briefly sketch an approach to the irrationality introduced by Galkin and Shinder [190]. Their approach is based on the motivic relation between a smooth cubic threefold $Y \subset \mathbb{P}^4$, its symmetric square $Y^{(2)}$, and its Fano surface $F(Y)$. Recall that according to Proposition **2**.4.2 one knows that

$$[Y^{(2)}] = (1 + \ell^3)\cdot[Y] + \ell^2\cdot[F(Y)] \tag{4.4}$$

in $K_0(\mathrm{Var}_k)$. On the other hand, if Y were rational, then $[Y] = [\mathbb{P}^3] + \ell\cdot\alpha$, where $\alpha = \sum a_i\,[C_i]$ is a linear combination of classes $[C_i]$ of smooth projective curves C_i, see [190, Cor. 2.2] for details. Taking symmetric products gives

$$[Y^{(2)}] = S^2[Y] = S^2[\mathbb{P}^3] + [\mathbb{P}^3]\cdot\ell\cdot\alpha + \ell^2\cdot S^2(\alpha). \tag{4.5}$$

Combining (4.4) and (4.5), one obtains

$$\begin{aligned}\ell^2\cdot[F(Y)] &= S^2[\mathbb{P}^3] + [\mathbb{P}^3]\cdot\ell\cdot\alpha + \ell^2\cdot S^2(\alpha) - (1+\ell^3)\cdot([\mathbb{P}^3] + \ell\cdot\alpha)\\ &= \ell^2\cdot[\mathbb{P}^2] - \ell^3 + \alpha\cdot(\ell^2 + \ell^3) + \ell^2\cdot S^2(\alpha).\end{aligned}$$

Assuming that the equation remains valid in $K_0(\mathrm{Var}_k)$ after dividing by ℓ^2, one would obtain

$$[F(Y)] = [\mathbb{P}^2] - \ell + \alpha\cdot(1+\ell) + S^2(\alpha) = S^2\left(\alpha + [\mathbb{P}^1]\right) - \ell. \tag{4.6}$$

Taking the image of this equation under

$$K_0(\mathrm{Var}_k) \twoheadrightarrow K_0(\mathrm{Var}_k)/(\ell) \simeq \mathbb{Z}[\mathrm{SB}_k],$$

see Remark 2.4.3, shows that the class of $[F(Y)]$ in $\mathbb{Z}[\mathrm{SB}_k]$ is a sum (with signs) of classes of the form $S^2[D] = [D^{(2)}]$ and $[D] \cdot [D'] = [D \times D']$, where D and D' are smooth projective curves.

However, one knows that whenever $[Z] = \sum \pm [Y_i]$ in $\mathbb{Z}[\mathrm{SB}_k]$ with irreducible varieties Z and Y_i, then Z is stably birational to one of the Y_i, see [306, Cor. 2.6]. Applied to our case, it shows that $F(Y)$ is stably birational to $D_1 \times D_2$ or to $D^{(2)}$, where D_1, D_2, and D are smooth projective curves. To conclude, one uses the fact that the Hodge numbers $h^{1,0}$ and $h^{2,0}$ are stable birational invariants. Hence,

$$1 + h^{1,0}(F(Y))\,t + h^{2,0}(F(Y))\,t^2 = \begin{cases} (1 + g(D_1)\,t) \cdot (1 + g(D_2)\,t) & \text{or} \\ 1 + h^{1,0}(D^{(2)})\,t + h^{2,0}(D^{(2)})\,t^2. \end{cases}$$

The left-hand side is $1 + 5\,t + 10\,t^2$, see (v) in Section 0.2. This immediately excludes the first case, as there is no solution to the two equations $g_1 + g_2 = 5$ and $g_1 \cdot g_2 = 10$. Numerically the second case is possible, as for $g(D) = 5$ one indeed has $h^{1,0}(D^{(2)}) = g(D) = 5$ and $h^{2,0}(D^{(2)}) = \binom{g(D)}{2} = 10$. However, if the two surfaces of general type $F(Y)$ and $D^{(2)}$ are stably birational, they are in fact birational. In fact, as their canonical bundles are both ample and minimal models are unique in dimension two, they are isomorphic $F(Y) \simeq D^{(2)}$. This leads to the contradiction $25 = h^{1,1}(F(Y)) = h^{1,1}(D^{(2)}) = 26$.

Of course, as it stands, this approach does not provide a complete proof, for it is not clear that (4.6) holds, because it was obtained by dividing by ℓ^2. Even worse, it is now known that ℓ is a zero divisor in $K_0(\mathrm{Var}_k)$, although there is no reason to expect that the difference of the two sides in (4.6) is really not trivial and annihilated by ℓ^2.

5 Nodal, Stable, and Other Special Cubic Threefolds

In this section we first discuss cubic threefolds with one ordinary double point as the only singularity. The irrationality as well as a global Torelli theorem are easy to prove for them and the Fano variety, albeit singular, has a simple geometric description.

This section also contains a collection of some famous special cubic threefolds, most of which have been mentioned before. The fact that cubic threefolds are Pfaffian will be mentioned in passing.

5.1 Torelli for Nodal Cubic Threefolds *One nodal cubic threefolds*, i.e. those with one ordinary double point as the only singularity, play a central role in the original article by Clemens and Griffiths [119]. The theory looks similar to the one for smooth cubic threefolds but simplifies at many points. The most notable difference between the two situations is that contrary to smooth cubic threefolds those with an ordinary double point are always rational, see Corollary **1**.5.16.

To fix the notation, let $Y \subset \mathbb{P}^4$ be a threefold with $y_0 \in Y$ an ordinary double point as its only singularity. Blowing up the singular point leads to a diagram, see Section **1**.5.4:

$$\begin{array}{ccccccc} E & \hookrightarrow & \mathrm{Bl}_{y_0}(Y) & \xrightarrow{\sim} & \mathrm{Bl}_C(\mathbb{P}^3) & \xleftarrow{j} & E' \\ \downarrow & & \downarrow{\scriptstyle \tau} & & {\scriptstyle \phi}\downarrow & & \downarrow{\scriptstyle p} \\ \{y_0\} & \hookrightarrow & Y & & \mathbb{P}^3 & \hookleftarrow & C. \end{array}$$

Here, $p\colon E' \twoheadrightarrow C$ is thought of as the family of all lines in Y passing through the node y_0. The curve $C \subset \mathbb{P}^3$ is a smooth complete intersection curve of type $(3,2)$ and genus $g(C) = 4$. If Y is obtained as the hyperplane section $\mathbb{T}_{y_0}X \cap X$ of a smooth cubic fourfold X, then C is simply the fibre of $q\colon \mathbb{L} \twoheadrightarrow X$ over $y_0 \in X$ parametrizing all lines in X through y_0, see Section **6**.0.2.

According to Exercise **1**.5.15, after a linear coordinate change, we may assume that $y_0 = [0:0:0:0:1] \in \mathbb{P}^4$ and Y is given by an equation of the form $F + x_4 \cdot G$, with $F \in H^0(\mathbb{P}^3, \mathcal{O}(3))$ and $G \in H^0(\mathbb{P}^3, \mathcal{O}(2))$ defining smooth hypersurfaces in $\mathbb{P}^3 \simeq V(x_4)$. Then $C = V(F) \cap V(G)$.

Alternatively, we can think of the exceptional divisor E as a quadric $E \subset \mathbb{P}(T_{y_0}\mathbb{P}^4) \simeq \mathbb{P}^3$ and then $C = Y \cap E$. Also, $\phi\colon E \cap E' \xrightarrow{\sim} C \subset \mathbb{P}^3$ and the quadric $E \subset \mathbb{P}^3$ can be reconstructed from $C \subset \mathbb{P}^3$ as the union of all lines $\ell \subset \mathbb{P}^3$ with $|C \cap \ell| = 3$ (with multiplicities), see [460] and also Section **6**.1.4 for more details in the situation of cubic fourfolds.

The Hodge structure of the smooth blow-up $\mathrm{Bl}_{y_0}(Y)$ can be described via the isomorphism $\mathrm{Bl}_{y_0}(Y) \simeq \mathrm{Bl}_C(\mathbb{P}^3)$ and $j_* \circ p^*$ as

$$H^3(\mathrm{Bl}_{y_0}(Y), \mathbb{Z}) \simeq H^1(C, \mathbb{Z})(-1),$$

which immediately defines an isomorphism

$$J(Y) := J(\mathrm{Bl}_{y_0}(Y)) \simeq J(C) = \mathrm{Pic}^0(C)$$

of principally polarized abelian varieties of dimension four. Note that in comparison the intermediate Jacobian of a smooth cubic threefold is of dimension five.

Remark 5.1 The analogue of Theorem 4.3 for nodal cubic threefolds is the following statement, the proof of which, however, is is much easier than in the smooth case: assume $Y, Y' \subset \mathbb{P}^4$ are two cubic threefolds each with an ordinary double point $y_0 \in Y$ and $y_0' \in Y'$ as their only singularities. Then the following conditions are equivalent:

(i) $Y \simeq Y'$ as (singular) complex varieties.
(ii) There exists a Hodge isometry $H^3(\mathrm{Bl}_{y_0}(Y), \mathbb{Z}) \simeq H^3(\mathrm{Bl}_{y_0'}(Y'), \mathbb{Z})$.

(iii) There exists an isomorphism $(J(Y), \Xi) \simeq (J(Y'), \Xi')$ of polarized abelian fourfolds.

To see that (iii) really implies (i), observe that a polarized isomorphism $J(Y) \simeq J(Y')$ induces a polarized isomorphism $J(C) \simeq J(C')$ between the Jacobians of the two associated genus four curves $C, C' \subset \mathbb{P}^3$ and, therefore, an isomorphism $C \simeq C'$. As the embeddings $C, C' \subset \mathbb{P}^3$, are canonical, any isomorphism $C \simeq C'$ extends to an isomorphism of the ambient projective spaces and thus leads to an isomorphism

$$\mathrm{Bl}_{y_0}(Y) \simeq \mathrm{Bl}_C(\mathbb{P}^3) \simeq \mathrm{Bl}_{C'}(\mathbb{P}^3) \simeq \mathrm{Bl}_{y_0'}(Y'). \tag{5.1}$$

The morphism (5.1) restricts to an isomorphism $E \simeq E'$ between the exceptional divisors of the blow-ups $\mathrm{Bl}_{y_0}(Y) \longrightarrow Y$ and $\mathrm{Bl}_{y_0'}(Y') \longrightarrow Y'$. Indeed, if their intersection is empty, then Y and Y' would have two nodes each. If their intersection is a curve, the blow-up τ would define a contraction of the quadric E. Thus, $Y \simeq Y'$.

The Fano variety $F(Y)$ of lines is not used in the proof of the global Torelli theorem for nodal cubic threefolds, but it is nevertheless interesting to understand its geometry.

Remark 5.2 Consider the Hilbert scheme of two points on C which is, of course, nothing but the symmetric product $C^{(2)}$. With each point in $C^{(2)}$ corresponding to a subscheme $Z \subset C$ of length two, one associates a unique line $\ell_Z \subset \mathbb{P}^3$ containing Z. Clearly, $Z \subset C \cap \ell_Z$ and if the inclusion is not an equality of schemes, then ℓ_Z is contained in the quadric E and $|C \cap \ell_Z| = 3$, cf. Section **6**.1.4 for more details. If ℓ_Z is not contained in E, then its strict transform in Y is a line L_Z and the induced map

$$C^{(2)} \setminus T \longrightarrow F(Y), \quad Z \longmapsto L_Z$$

is injective. Here, $T := \{Z \in C^{(2)} \mid |C \cap \ell_Z| = 3\}$ is the indeterminacy locus. However, mapping $Z \in T$ to the residual line E'_{s_3}, where $Z = \{s_1, s_2\} \subset C \cap \ell_Z = \{s_1, s_2, s_3\}$, extends the map to a surjective morphism $C^{(2)} \twoheadrightarrow F(Y)$. Observe that through every point of $C \subset E \simeq \mathbb{P}^1 \times \mathbb{P}^1 \subset \mathbb{P}^3$ there exist exactly two lines contained in E. This shows that T consists of two disjoint copies of C, which get identified under the map to $F(Y)$. In other words, $C^{(2)}$ is the normalization of $F(Y)$, which in this case can also be described as the blow-up

$$C^{(2)} \simeq \mathrm{Bl}_C(F(Y)) \longrightarrow F(Y) \quad \text{with} \quad C^{(2)} \setminus (C \sqcup C) \simeq F(Y) \setminus C$$

that glues two disjoint copies of C. In particular, this shows

$$[F(Y)] = [C^{(2)}] - [C] \quad \text{in } K_0(\mathrm{Var}),$$

see [190] and Corollary **6**.1.30 for the analogous result in dimension four.

The cohomological description of the blow-up $\mathrm{Bl}_{y_0}(Y) \simeq \mathrm{Bl}_C(\mathbb{P}^3)$ lifts to the level of Chow groups. Hence, the composition

$$\mathrm{Pic}^0(C) \longrightarrow \mathrm{CH}^2(\mathrm{Bl}_{y_0}(Y))_{\mathrm{alg}} \longrightarrow J(\mathrm{Bl}_{y_0}(Y))$$

of the pullback map and the Abel–Jacobi map induces $J(C) = \mathrm{Pic}^0(C) \simeq J(Y)$. Furthermore, the Albanese map

$$a\colon C^{(2)} \longrightarrow A(C^{(2)}) \simeq J(C) \simeq J(Y) \tag{5.2}$$

can be reinterpreted in terms of the Fano correspondence $C \xleftarrow{p} E' \xrightarrow{\tau} Y$ as the map $Z = \{s_1, s_2\} \longmapsto a([E_{s_1}] + [E_{s_2}])$. Over the open set $C^{(2)} \setminus (C \sqcup C) \simeq F(Y) \setminus C$, the map (5.2) coincides with $F(Y) \setminus C \longrightarrow J(Y)$, $L \longmapsto a(L)$.

The analogue of (i) in Corollary 3.12 is the classical fact

$$[a(C^{(2)})] = (1/2) \cdot \Xi^2 \in H^4(J(Y), \mathbb{Z}).$$

Note that in the above discussion we suppressed the choice of a reference point $s_0 \in C$ or a line $L_0 \subset Y$, needed to actually define the Albanese maps a for $C^{(2)}$ and $F(Y)$.

A very similar and more detailed discussion for nodal cubic fourfolds can be found in Section **6**.1.4.

5.2 Semi-stable Cubic Threefolds The moduli space of smooth cubic threefolds $M_3 = |\mathcal{O}_{\mathbb{P}^4}(3)|_{\mathrm{sm}} /\!/ \mathrm{SL}(5)$ is naturally compactified by the moduli space

$$M_3 \subset \bar{M}_3 = |\mathcal{O}_{\mathbb{P}^4}(3)|^{\mathrm{ss}} /\!/ \mathrm{SL}(5)$$

of semi-stable cubic threefolds. However, a clear geometric understanding of the singular cubics that need to be added is complicated in general. For cubic surfaces the situation was still fairly easy, see Section **4**.4.2, but for cubic threefolds it is already quite a bit more involved, before it gets really complicated for cubic fourfolds, see Section **6**.6.7, and essentially impossible in dimension five and beyond. In dimension three, we record what is known but refer to the original articles [12, 495] for details.

Theorem 5.3 (Allcock, Yokoyama) *Let $Y \subset \mathbb{P}^4$ be a cubic threefold. Then Y is stable if and only if all its singularities are of type A_n, i.e. locally analytically given by an equation $x_0^2 + x_1^2 + x_2^2 + x_3^{n+1}$, with $n \leq 4$.*

In [12] one also finds a complete description of all semi-stable cubic threefolds: Y is semi-stable if and only if all its singularities are of type (i) A_n with $n \leq 5$, (ii) of type D_4, i.e. locally described by $x_0^2 + x_1^2 + x_2^3 + x_3^3$, (iii) of type A_∞, i.e. locally described by $x_0^2 + x_1^2 + x_2^3$, or (iv) of type A_n with $n \geq 6$ but such that Y contains none of the planes containing the kernel of its quadratic part. The latter case is missing in [495].

The result is complemented by the observation that a semi-stable cubic threefold with an A_∞-singularity is isomorphic to the secant variety of a rational normal curve in $\mathbb{P}^4$. Furthermore, the locus of strictly semi-stable cubic threefolds in $\bar{M}_3$ consists of one component isomorphic to $\mathbb{P}^1$ and another one that consists of a single point represented by the cubic given by $x_0^3 + x_1^3 - x_2 \cdot x_3 \cdot x_4$ which has three singularities of type D_4. See also the discussion in Remark **4**.4.6.

In recent years, another notion of stability, especially for Fano varieties, has been studied. This notion of K-stability is linked to the existence of Kähler–Einstein metrics. Liu and Xu [324] show that for cubic threefolds both notions coincide. In particular, the GIT moduli space and the K-stability moduli space provide the same compactification of the moduli space of smooth cubic threefolds.

Other compactifications better suited for extending the link between smooth cubic threefolds $Y \subset \mathbb{P}^4$ to their intermediate Jacobians $J(Y)$ have been studied by Casalaina-Martin, Grushevsky, Hulek, and Laza [108, 109].

5.3 Moduli Space As hinted at already in Example 3.3.4, as a consequence of the global Torelli theorem one obtains a locally closed embedding

$$M_3 \hookrightarrow A_5$$

into the moduli space of principally polarized abelian varieties of dimension five. The description of the automorphism group of the intermediate Jacobian in Remark 4.9 allows one to state a similar result on the level of moduli stacks: mapping a smooth cubic threefold Y to its intermediate Jacobian defines a locally closed embedding of smooth Deligne–Mumford stacks

$$\mathcal{M}_3 \hookrightarrow \mathcal{A}_3 \simeq [\mathrm{Sp}(10, \mathbb{Z}) \setminus D].$$

Note that $\dim(\mathcal{M}) = 10$ while $\dim(\mathcal{A}_3) = 15$.

5.4 Pfaffian Cubics, Klein, and Segre We conclude with a review of some particular cubic threefolds. Since all smooth cubic threefolds are irrational, there has never been a good reason to study special cubic threefolds systematically. As we will see in Chapter 6, the situation is completely different in dimension four, where only special cubics are expected to be rational and the challenge to classify those has been a driving force in the theory. Of course, interesting special cubic threefolds exist and have been studied and we briefly mention some of them.

(i) The *Klein cubic threefold* is the smooth cubic threefold Y defined by the equation

$$x_0^2\, x_1 + x_1^2\, x_2 + x_2^2\, x_3 + x_3^2\, x_4 + x_4^2\, x_0 = 0.$$

Various aspects of it have been studied over the years [9, 55, 202, 208, 403]. It was mentioned already that its Fano surface attains the maximal Picard number $\rho(F(Y)) = 25$ and that its intermediate Jacobian $J(Y) \simeq A(F(Y))$ is of the form E^5 (unpolarized) with E a CM curve, cf. Remark 2.15. The automorphism group of Y has been determined by Adler [9] as $\mathrm{PGL}_2(\mathbb{F}_{11})$, which is of order 660, see also [56, Thm. 3] for explicit examples of automorphisms. Beauville [56] used the existence of these automorphisms to exclude Y from being rational directly, see Remark 4.14. Roulleau [403, Prop. 1] proved that the Klein cubic threefold is the only smooth cubic threefold with an automorphism of order 11.

Adler's result complements classical results of Klein who studied the Hessian of Y and observed that its singular set is the modular curve $X_0(11)$, see [10] for more information and references. Gross and Popescu [208] showed that the Klein cubic threefold is birational to the moduli space of $(1, 11)$ polarized abelian surfaces (with canonical level structure). In particular, this moduli space is unirational but not rational. An explicit resolution of the Hessian of the Klein cubic threefold was described by Gounelas and Kouvidakis [202, Sec. 6] who also proved that the intersection of Y with its Hessian is not uniruled.

(ii) It was mentioned before that every smooth cubic surface can be represented, in a non-unique way, as a Pfaffian, see Section **4**.2.5, and we will see that this is not true for cubic hypersurfaces of dimension four, see Remark **6**.2.6. For cubic threefolds, a naive dimension count, cf. the argument in Remark **6**.2.6, shows that in principle every smooth cubic threefold could be a Pfaffian. That this is indeed true has first been verified by Adler [10, App. V] for generic cubic threefolds, then extended by Beauville [50, Prop. 8.5] to all smooth cubic threefolds and by Comaschi [124] to singular ones. Beauville first proved a criterion that shows that a smooth cubic threefold is Pfaffian if it contains a normal elliptic quintic curve, for the analogue in dimension four, see Remark **6**.2.5.

(iii) The *Segre cubic threefold* (or *Segre cubic primal*) is given by

$$\sum_{i=0}^{5} x_i = \sum_{i=0}^{5} x_i^3 = 0.$$

Taking the hyperplane section $x_i = x_j$ of it, one obtains the Cayley cubic surface, see Remark **1**.5.17 and Remark **4**.2.16. The Segre cubic threefold has 10 singular points, all ordinary double points, and thus realizes the maximal number of ordinary double points. In fact, up to coordinate change, the Segre cubic is the only nodal cubic threefolds with 10 singular points. The singular points are explicitly given as the permutations of the point $[-1 : -1 : -1 : 1 : 1 : 1]$.

The Segre cubic threefold is isomorphic to the GIT quotient $(\mathbb{P}^1)^6/\!/\mathrm{SL}(2)$, see [158, Thm. 9.4.10], and can also be realized as a birational model of the moduli space $\mathcal{A}_2(2)$ of principally polarized abelian surfaces with a level two structure, see [242, Thm. IV.1.2]. The projective dual of the Segre cubic is the Igusa quartic, see [459] for a detailed study of the topology and [238] for a new duality perspective.

We recommend Dolgachev's survey [159] for further information on classical and modern aspects of the Segre cubic.

(iv) According to a result of Sylvester, see Corollary **1**.5.19 and Section **4**.2.5, the generic cubic surface $S \subset \mathbb{P}^3$ can be written in Sylvester form, i.e. $S = V(\ell_0^3 + \cdots + \ell_4^3)$, where ℓ_i are linear forms in the linear coordinates $x_0, \ldots, x_3$. This is no longer true for cubic threefolds. In fact, Edge showed [170] that the generic cubic threefold cannot even be written as the sum of seven(!) cubes of linear forms.

It is not difficult to see that the generic cubic polynomial cannot be written as the sum of six cubes of linear forms, but the naive dimension count does not include the possibility of seven cubes. Indeed, $h^0(\mathbb{P}^4, \mathcal{O}(3)) = 35$ and the choice of 7 linear forms also accounts for dimension 35.

6 Appendix: Comparison of Cubic Threefolds and Cubic Fourfolds

The goal of this appendix is twofold. We first briefly outline an approach to link cubic threefolds via the triple cover construction to cubic fourfolds and then to abelian varieties of dimension 11 and hyperkähler manifolds of dimension 4. Most parts of it are completely analogous to the lower-dimensional situation in Section **4**.4.3. It again leads to a description of the moduli space of cubic threefolds as an open set of a ball quotient. In the second part we will highlight similarities between the theory of cubic hypersurfaces of dimension three and four. This part can either be studied now, as a summary of this chapter and a preview of the next one, or later after having worked through the theory of cubic fourfolds in detail.

6.1 Passing from Threefolds to Fourfolds In Section **4**.4.3, we explained how to use the Hodge theory of the cubic threefold naturally associated with a smooth cubic surface via the triple cover construction to link the moduli space of smooth cubic surfaces, which a priori has no period description, with a certain moduli space of abelian varieties of dimension five. We will now explain a similar story for cubics of dimension three and four. This was again initiated by Allcock, Carlson, and Toledo [14] and investigated further by Looijenga and Swierstra [329], Kudla and Rapoport [289], and Boissière, Camere, and Sarti [76].

The discussion of the linear algebra is almost literally the same as the one in Section **4**.4.3, so we will state only the results and leave the adaptation of the arguments there to the case here as an exercise. The notation is chosen to match the two situations. Note however that the conventions in the references differ. We will mostly follow the original [14] (up to a scaling factor).

The geometric starting point is the triple cover

$$X \longrightarrow \mathbb{P}^4 \supset Y,$$

branched over a given smooth cubic threefold $Y \subset \mathbb{P}^4$, as described in Section **1**.5.6. The covering action induces a Hodge isometry $\iota\colon H^4(X,\mathbb{Z})_{\rm pr} \xrightarrow{\sim} H^4(X,\mathbb{Z})_{\rm pr}$. The main difference to the situation in Section **4**.4.3 is that now the intersection pairing is symmetric, but we still have the useful equalities $(\iota(\alpha).\beta) = (\alpha.\iota^2(\beta))$ and $(\theta(\alpha).\beta) = -(\alpha.\theta(\beta))$.

The eigenspace decomposition with respect to ι (or θ) now looks like

$$H^4(X,\mathbb{C})_{\rm pr} = H_\rho \oplus H_{\rho^2} = (H^{3,1}_\rho \oplus H^{2,2}_\rho) \oplus (H^{2,2}_{\rho^2} \oplus H^{1,3}_{\rho^2}). \tag{6.1}$$

In order to see that the one-dimensional $H^{3,1}$ is contained in H_ρ, one can use the residue description $H^{3,1} \simeq H^0(\mathbb{P}^5, \Omega^5_{\mathbb{P}^5}(2X))$, see Lemma **1**.4.23, and the fact that the covering action $[x_0 : \cdots : x_5] \longmapsto [x_0 : \cdots : \rho \cdot x_5]$ alters the differential form $\Omega = \sum_0^5 (-1)^i x_i \cdot dx_0 \wedge \cdots \wedge \widehat{dx_i} \wedge \cdots \wedge dx_5$ by a factor ρ. Since complex conjugation swaps H_ρ and H_{ρ^2}, this also shows that $H^{1,3}$ is contained in H_{ρ^2}. Using $b_4(X)_{\rm pr} = 22$ and again complex conjugation, we also conclude that $\dim(H^{2,2}_\rho) = \dim(H^{2,2}_{\rho^2}) = 10$.

Next, on the symmetric lattice $\Gamma := H^4(X,\mathbb{Z})_{\rm pr}$ one defines in complete analogy to the case of cubic threefolds the pairing[12]

$$h\colon \Gamma \times \Gamma \longrightarrow \mathbb{Z}[\rho], \quad h(\alpha,\beta) := \frac{(\theta(\alpha).\beta) + \theta \cdot (\alpha.\beta)}{2}.$$

- Using the fact that the lattice Γ is even, see Proposition **1**.1.21 or Section **6**.5.2, one checks that this pairing takes indeed values in the Eisenstein integers $\mathbb{Z}[\rho] = \{(1/2)(a + b\cdot\theta) \mid a,b \in \mathbb{Z},\ a \equiv b\,(2)\}$.
- The form $h(\ ,\ )$ is $\mathbb{Z}[\rho]$-linear in the first variable and satisfies $h(\alpha,\beta) = \overline{h(\beta,\alpha)}$.
- The two eigenspaces $H_\rho, H_{\rho^2} \subset \Gamma \otimes_{\mathbb{Z}} \mathbb{C} =$ are isotropic with respect to $(\ .\ )$ and h.
- Consider the $\mathbb{Z}[\rho]$-linear composition $j\colon \Gamma \hookrightarrow \Gamma \otimes_{\mathbb{Z}} \mathbb{C} \twoheadrightarrow H_\rho$. It is isometric with respect to h on Γ and $h'(\gamma,\delta) := i\sqrt{3}\cdot(\gamma.\delta)$ on H_ρ.
- The line $H^{3,1}_\rho \subset H_\rho$ is negative definite and the hyperplane $H^{2,2}_\rho \subset H_\rho$ is positive definite.

These facts are now applied to study the moduli space $M_3 = |\mathcal{O}_{\mathbb{P}^4}(3)|_{\rm sm}/\!/\mathrm{PGL}(5)$ of smooth cubic threefolds. First, one considers its natural cover $\widetilde{M}_3$ parametrizing pairs (Y,φ) consisting of a smooth cubic threefold Y and an isometry $\varphi\colon H^4(X,\mathbb{Z})_{\rm pr} \xrightarrow{\sim} \mathbb{Z}[\rho]^{10,1}$ up to the action of $\mu_6 \simeq \mathbb{Z}[\rho]^*$. Here, X is the cubic fourfold associated with Y and $\mathbb{Z}[\rho]^{10,1}$ simply denotes the $\mathbb{Z}[\rho]$-lattice that comes out of the above construction (its isomorphism type is independent of Y and X but its precise shape is of no importance). Altogether this leads to a holomorphic period map

$$\widetilde{\mathcal{P}}\colon \widetilde{M}_3 \longrightarrow \mathbb{B}^{10} \subset \mathbb{P}(\mathbb{C}^{10,1}), \quad (Y,\varphi) \longmapsto \varphi(H^{3,1}).$$

Here, $\mathbb{B}^{10}$ denotes the open set of negative lines which is biholomorphic to the 10-dimensional ball $\{z \mid \sum |z_i|^2 < 1\} \subset \mathbb{C}^{10}$. The infinitesimal Torelli theorem, see Corollary **1**.4.25, essentially shows that the map $\widetilde{\mathcal{P}}$ is immersive, i.e. its tangent map is injective at each point or, equivalently, it is a local isomorphism. Note that $\widetilde{M}_3$ is indeed smooth, see [14, Lem. 2.7].

[12] The pairing used in [14] is $\theta \cdot h$, which then takes values in $\theta \cdot \mathbb{Z}[\rho]$.

Taking quotients by $\mathrm{PU} := \mathrm{PU}(\mathbb{Z}[\rho]^{10,1})$, one obtains a holomorphic map from the coarse moduli space to a certain ball quotient:

$$\mathcal{P}\colon M_3 \longrightarrow \mathrm{PU} \setminus \mathbb{B}^{10}. \tag{6.2}$$

We phrase the analogue of Theorem **4**.4.9 as a statement about this period map.

Theorem 6.1 (Allcock–Carlson–Toledo, Looijenga–Swierstra) *The period map* (6.2) *is an open embedding of the coarse moduli space M_3 into the arithmetic quotient* $\mathrm{PU} \setminus \mathbb{B}^{10}$ *of the 10-dimensional ball* $\mathbb{B}^{10} \subset \mathbb{C}^{10}$.

Allcock, Carlson, and Toledo [14, Thm. 1.1] and independently Looijenga and Swierstra [329, Thm. 3.1] not only show that $\mathcal{P}$ is an open embedding, but also describe its complement.

Proof We restrict ourselves to prove the injectivity of (6.2) and the quickest way to do this uses the strong global Torelli theorem for cubic fourfolds, see Theorem **6**.3.17 and Remark **6**.3.18. The proof in [14] is more direct.

Assume that for the cubic fourfolds X and X' associated with two smooth cubic threefolds Y and Y' there exists an isomorphism $\xi\colon H^4(X,\mathbb{Z})_{\mathrm{pr}} \xrightarrow{\sim} H^4(X',\mathbb{Z})_{\mathrm{pr}}$ of $\mathbb{Z}[\rho]$-modules which is an isometry with respect to the form h on the two sides and satisfies $\xi(H^{3,1}(X)) = H^{3,1}(X')$. The real part of h and the compatibility with the action of θ gives back the intersection pairing. Hence, ξ is a standard Hodge isometry between the two cubic fourfolds and, therefore, ξ (up to sign) is induced by an isomorphism $X \simeq X'$ that is compatible with the covering action. This eventually proves $Y \simeq Y'$. □

The open immersion (6.2) of quasi-projective varieties can be upgraded to an open immersion of analytic Deligne–Mumford stacks (or orbifolds)

$$\mathcal{M}_3 \hookrightarrow [\mathrm{PU} \setminus \mathbb{B}^{10}].$$

This relies on a result of Zheng [499, Prop. 6.3] describing the automorphism group of a cubic threefold Y in terms of the associated triple cover X as

$$\mathrm{Aut}(Y) \simeq \mathrm{Aut}_{\mathbb{Z}[\rho]}(H^4(X,\mathbb{Z})_{\mathrm{pr}})/\mu_6,$$

where on the right-hand side, one has the group of all Hodge isometries of the primitive cohomology compatible with the $\mathbb{Z}[\rho]$-action. Implicitly, we have used this result already in the proof above.

The extension to the compactification of M_3 and the description of the boundary of the open inclusion $M_3 \hookrightarrow \mathrm{PU} \setminus \mathbb{B}^{10}$ are quite subtle. Roughly, there are two boundary components corresponding to nodal and cordal cubic threefolds. We refer to [14, 329] for details.

Kudla and Rapoport [289, §6] establish a link to abelian varieties. This is a special case of the general theory of half twists developed by van Geemen [463]. It comes down to the observation that the decomposition (6.1) can be rewritten as

$$H^4(X,\mathbb{Z})_{\rm pr}\otimes\mathbb{C}\simeq(H_\rho^{3,1}\oplus H_{\rho^2}^{2,2})\oplus(H_\rho^{2,2}\oplus H_{\rho^2}^{1,3})$$

and now the two summands are complex conjugate to each other. Hence, a Hodge structure of weight one on $H^4(X,\mathbb{Z})_{\rm pr}$ is defined by setting

$$H^{1,0}_{\mathbb{Z}[\rho]}(X) := H_\rho^{3,1}\oplus H_{\rho^2}^{2,2} \quad\text{and}\quad H^{0,1}_{\mathbb{Z}[\rho]}(X) := H_\rho^{2,2}\oplus H_{\rho^2}^{1,3}.$$

In particular, the quotient

$$A(Y) := H^4(X,\mathbb{Z})_{\rm pr}\setminus H^4(X,\mathbb{Z})_{\rm pr}/H^{1,0}_{\mathbb{Z}[\rho]}(X)$$

is a complex torus of dimension 11, and in fact an abelian variety, naturally associated with the cubic threefold Y. This eventually leads to a map from M_3 into a certain moduli space of abelian varieties of dimension 11. Note that in [289] the conventions are slightly different, which might lead to isogenous abelian varieties.

Finally we mention results by Boissière, Camere, and Sarti [76, Thm. 1.1]. Instead of working with the cubic fourfold X associated with the cubic threefold Y, they consider the hyperkähler fourfold provided by the Fano variety of lines $F(X)$. The covering automorphism of $X \twoheadrightarrow \mathbb{P}^4$ induces a non-symplectic automorphism of $F(X)$ of order three. This approach leads to the identification of the moduli space of hyperkähler fourfolds of $\mathrm{K3}^{[2]}$-type endowed with a non-symplectic automorphism of order three and the moduli space of smooth cubic threefolds. They also show [76, Prop. 5.1] that the image of the induced morphism $M_3 \to M_4$, $Y \mapsto X$, intersects the Hassett divisor $\mathcal{C}_{14}\cap M_4$, see Example **6**.5.8, in a generic Pfaffian cubic.

6.2 Summary The following table summarizes the central results concerning smooth cubic threefolds proved in this chapter. At the same time, it serves as a guide to the analogous results for smooth cubic fourfolds to be proved in Chapter 6.

$F(Y) \longleftarrow \mathbb{L} \xrightarrow{6:1} Y$ $\dim F(Y)=2$, surface of general type (Sect. **5**.0.1)	$F(X) \longleftarrow \mathbb{L} \xrightarrow{(2,3)} X$ $\dim F(X)=4$, hyperkähler fourfold (Thm. **6**.3.10)
$H^1(Y,\mathcal{T}_Y) \xrightarrow{\sim} H^1(F(Y),\mathcal{T}_{F(Y)})$ $\mathrm{Aut}(Y) \xrightarrow{\sim} \mathrm{Aut}(F(Y))$ (Prop. **5**.2.14 & Exer. **5**.2.13)	$H^1(X,\mathcal{T}_X) \hookrightarrow H^1(F(X),\mathcal{T}_{F(X)})$, corank = 1 $\mathrm{Aut}(X) \xrightarrow{\sim} \mathrm{Aut}(F(X),\mathcal{O}(1)) \hookrightarrow \mathrm{Aut}(F(X))$ (Cor. **6**.3.12 & (3.6) in Sec. **2**.3.3)
global Torelli: $H^3(Y,\mathbb{Z}) \longleftrightarrow Y \longleftrightarrow F(Y)$ (Prop. **5**.2.12 & Thm. **5**.4.3)	global Torelli: $H^4(X,\mathbb{Z})_{\rm pr} \longleftrightarrow X \longleftrightarrow F(X)$ (Prop. **2**.3.12 & Thm. **6**.3.17)

<table>
<tr>
<td>$\bigwedge^2 H^1(F(Y),\mathbb{Q}) \xrightarrow{\sim} H^2(F(Y),\mathbb{Q})$
$\bigwedge^2 H^1(F(Y),\mathbb{Z}) \hookrightarrow H^2(F(Y),\mathbb{Z}) \twoheadrightarrow \mathbb{Z}/2\mathbb{Z}$
$\bigwedge^* H^1(F(Y),\mathbb{Q})/(P_3) \xrightarrow{\sim} H^*(F(Y),\mathbb{Q})$
(Cor. **2**.5.15, Lem. **5**.2.5, & Rem. **5**.3.13)</td>
<td>$S^2H^2(F(X),\mathbb{Q}) \xrightarrow{\sim} H^4(F(X),\mathbb{Q})$
$S^2H^2(F,\mathbb{Z}) \hookrightarrow H^4(F,\mathbb{Z}) \twoheadrightarrow (\mathbb{Z}/2\mathbb{Z})^{23} \oplus \mathbb{Z}/5\mathbb{Z}$
$S^*H^2(F(X),\mathbb{Q})/(H_3) \xrightarrow{\sim} H^*(F(X),\mathbb{Q})$
(Exa. **6**.3.3, Cor. **6**.3.11, & Rem. **6**.3.13)</td>
</tr>
<tr>
<td>$R = F_2 \in |\mathcal{O}_F(2)|$
$(R.R) = 180,\ g(R) = 136$
(Prop. **5**.1.1 & Exer. **5**.1.6)</td>
<td>$[F_2] = 5 \cdot (g^2 - c_2(\mathcal{S}_F))$
$([F_2].[F_2]) = 1125,\ h^{2,0}(F_2) = 449$
(Prop. **6**.4.1 & Sec. **6**.4.4)</td>
</tr>
<tr>
<td>$H^3(Y,\mathbb{Z})(1) \simeq H^1(F(Y),\mathbb{Z}) \simeq H^1(C_L,\mathbb{Z})^-$
$J(Y) \simeq A(F(Y)) \simeq \mathrm{Prym}(C_L/D_L)$
$\mathrm{Pic}^0(D_L) \times \mathrm{Prym}(C_L/D_L) \xrightarrow{2:1} \mathrm{Pic}^0(C_L)$
$\mathrm{CH}^2(Y)_{\mathrm{alg}} \simeq \mathrm{Pic}^0(F(Y)) \simeq \mathrm{Prym}(C_L/D_L)$
$\mathfrak{h}^3(Y)(1) \simeq \mathfrak{h}^1(F(Y)) \simeq \mathfrak{h}^1(C_L)^-$
(Prop. **5**.3.10, Cor. **5**.3.16, & Rem. **5**.3.17)</td>
<td>$H^4(X,\mathbb{Z})_{\mathrm{pr}}(1) \simeq H^2(F(X),\mathbb{Z})_{\mathrm{pr}} \simeq H^2(F_L,\mathbb{Z})^-_{\mathrm{pr}}$

$\mathrm{CH}_0(F_L)^+_{\mathrm{hom}} \oplus \mathrm{CH}_0(F_L)^-_{\mathrm{hom}} = \mathrm{CH}_0(F_L)_{\mathrm{hom}}$
$\mathrm{CH}^3(X)_{\mathrm{alg}} \otimes \mathbb{Q} \simeq \mathrm{CH}^2(F(X))_{\mathrm{hom}} \otimes \mathbb{Q}$
$\mathfrak{h}(X)_{\mathrm{prim}}(1) \simeq \mathfrak{h}^2(F(X))_{\mathrm{tr}} \simeq \mathfrak{h}^2(F_L)^-_{\mathrm{pr}}$
(Cor. **6**.3.21, Rem. **6**.3.24, & Rem. **6**.4.14)</td>
</tr>
<tr>
<td>$\big(H^1(A,\mathbb{Z}), \Xi \in H^2(A,\mathbb{Z})\big)$
$a\colon F(Y) \hookrightarrow A,\ a^*(\Xi) = (-2/3) \cdot c_1(F(Y))$
(Lem. **5**.1.26)</td>
<td>$\big(H^2(F(X),\mathbb{Z}), \tilde{q}_F \in H^4(F(X),\mathbb{Q})\big)$
$\tilde{q}_F = (1/30) \cdot c_2(F(X))$
(Rem. **6**.4.2)</td>
</tr>
<tr>
<td>$\begin{array}{ccccc} & & \mathrm{Bl}_L(Y) & \longrightarrow & \mathbb{P}^2 \\ & & & & \cup \\ F(Y) & \supset & C_L & \xrightarrow[\text{étale}]{2:1} & D_L \\ & & {\scriptstyle 5:1}\downarrow & & \\ & & L & & \end{array}$
(Lem. **5**.1.26)</td>
<td>$\begin{array}{ccccc} & & \mathrm{Bl}_L(X) & \longrightarrow & \mathbb{P}^3 \\ & & & & \cup \\ F(X) & \supset & F_L & \xrightarrow[\text{16 fix. pts.}]{2:1} & D_L \\ & & {\scriptstyle (2,3)}\downarrow & & \\ & & L & & \end{array}$
(Prop. **6**.4.12)</td>
</tr>
<tr>
<td>$3\,[C_L] = g = \varphi(h^2),\ ([C_L].[C_L]) = 5$
$(\gamma)^2 = (1/2) \int_{C_L} \gamma^2$
(Lem. **5**.1.14, Exer. **5**.1.16, (3.5) in Sec. **5**.3.1)</td>
<td>$3\,[F_L] = \varphi(h^2),\ ([F_L].[F_L]) = 5$
$q_F(\alpha) = (1/2) \int_{F_L} \alpha^2$
(Prop. **6**.4.1 & Cor. **6**.4.3)</td>
</tr>
<tr>
<td>$H^1(C_L,\mathbb{Z})^+ \hookrightarrow H^1(C_L,\mathbb{Z}) \twoheadrightarrow H^1(C_L,\mathbb{Z})^-$
$\cup$ index=2
$H^1(D_L,\mathbb{Z})$ (Sect. **5**.3.2 (3.2))</td>
<td>$H^2(F_L,\mathbb{Z})^+ \hookrightarrow H^2(F_L,\mathbb{Z}) \twoheadrightarrow H^2(F_L,\mathbb{Z})^-$
$\cup$ index=?
$H^2(D_L,\mathbb{Z})$ (Rem. **6**.4.14)</td>
</tr>
<tr>
<td>nodal cubic $y_0 \in Y$, rational
$\mathrm{Bl}_{y_0}(Y) \simeq \mathrm{Bl}_C(\mathbb{P}^3)$, $C \subset \mathbb{P}^3$ compl. int. $(2,3)$
$C^{(2)} \simeq \mathrm{Bl}_C(F(Y)) \longrightarrow F(Y)$
$[C^{(2)}] = [F(Y)] + [C]$
(Cor. **1**.5.16, Rem. **5**.5.1, & Rem. **5**.5.2)</td>
<td>nodal cubic $x_0 \in X$, rational
$\mathrm{Bl}_{x_0}(X) \simeq \mathrm{Bl}_S(\mathbb{P}^4)$, $S \subset \mathbb{P}^4$ compl. int. $(2,3)$
$S^{[2]} \simeq \mathrm{Bl}_S(F(X)) \longrightarrow F(X)$
$[S^{(2)}] = [F(X)]$
(Cor. **1**.5.16, Prop. **6**.1.28, & Cor. **6**.1.30)</td>
</tr>
</table>

6

Cubic Fourfolds

In this chapter we turn to cubic hypersurfaces $X \subset \mathbb{P}(V) \simeq \mathbb{P}^5$ of dimension four. The key new feature is the unexpected appearance of K3 surfaces and hyperkähler fourfolds. K3 surfaces come in via their Hodge structures, which turn out to be very similar to Hodge structures of cubic fourfolds. The reason for the occurrence of hyperkähler fourfolds is more geometric: as shown by Beauville and Donagi [58], the Fano variety $F(X)$ of lines contained in a smooth cubic fourfold $X \subset \mathbb{P}^5$ is a hyperkähler manifold. More precisely, $F(X)$ is a hyperkähler manifold deformation equivalent to the four-dimensional Hilbert scheme $S^{[2]}$ of subschemes of length two of a K3 surface S.

We begin again by collecting immediate consequences of the results discussed in Chapters 1 and 2. As throughout these notes, most of the results hold for arbitrary (algebraically closed) ground fields. However, as Hodge theory will play an essential role in describing the geometry of the situation, we will work mostly over the complex numbers.

0.1 Invariants of Cubic Fourfolds The canonical bundle of a smooth cubic fourfold $X \subset \mathbb{P}^5$ is given by $\omega_X \simeq \mathcal{O}_X(-3)$. For the Picard group, one has $\mathrm{Pic}(X) \simeq \mathbb{Z} \cdot \mathcal{O}_X(1)$ and the non-trivial Betti numbers of X are given by

$$b_0(X) = b_2(X) = b_6(X) = b_8(X) = 1 \quad \text{and} \quad b_4(X) = 23,$$

see Section **1**.1. In particular, for the Euler number, one has $e(X) = 27$. If h denotes the restriction of the hyperplane class, then $H^2(X,\mathbb{Z}) = \mathbb{Z} \cdot h$ and $H^6(X,\mathbb{Z}) = \mathbb{Z} \cdot (h^3/3)$, see Exercise **1**.1.2. Any line $L \subset X$ satisfies $[L] = h^3/3$, so that the integral Hodge conjecture holds for X except possibly in degree four, but see Section 3.4.

The middle Hodge numbers are

$$h^{4,0}(X) = h^{0,4}(X) = 0, \quad h^{3,1}(X) = h^{1,3}(X) = 1, \quad \text{and} \quad h^{2,2}(X) = 21.$$

The lattice $H^4(X,\mathbb{Z})$ with its Hodge structure is of central importance and will be discussed in detail below.

The linear system of all cubic hypersurfaces in $\mathbb{P}^5$ is $|\mathcal{O}_{\mathbb{P}^5}(3)| \simeq \mathbb{P}^{55}$ and the moduli space of all smooth cubic fourfolds is of dimension 20, cf. Sections **1**.2 and **3**.1.5.

A smooth cubic fourfold $X \subset \mathbb{P}^5$ never contains a linear $\mathbb{P}^3 \subset \mathbb{P}^5$ and a generic one does not contain a linear $\mathbb{P}^2 \subset \mathbb{P}^5$, see Remark **2**.1.7 or (v) below for an argument using Hodge theory. However, some smooth cubic fourfolds do contain planes $\mathbb{P}^2 \subset \mathbb{P}^5$ and, as we will see, those are parametrized by a divisor in the moduli space of all smooth cubic fourfolds, see Remark 1.3.

0.2 Invariants of Their Fano Variety The Fano variety of lines $\mathbb{P}^1 \simeq L \subset X$ in a cubic fourfold X plays a central role in the theory. Recall that there are two types of lines which are distinguished by the splitting type of their normal bundles

$$\mathcal{N}_{L/X} \simeq \begin{cases} \mathcal{O}_L(1) \oplus \mathcal{O}_L \oplus \mathcal{O}_L & \text{if } L \text{ is of first type,} \\ \mathcal{O}_L(1) \oplus \mathcal{O}_L(1) \oplus \mathcal{O}_L(-1) & \text{if } L \text{ is of second type.} \end{cases} \tag{0.1}$$

Geometrically, the two types can be described in various ways, see Section **2**.2.2. For example, if L is of the first type, there exists a unique plane $\mathbb{P}^2 \subset \mathbb{P}^5$ such that $\mathbb{P}^2$ is tangent to X at every point of L, i.e. the intersection $\mathbb{P}^2 \cap X$ contains L with multiplicity at least two, i.e. $\mathbb{P}^2 \cap X = 2L \cup L'$ for some *residual line* $L' \subset X$ not necessarily distinct from L. For lines of the second type, there is a one-dimensional family of such planes all contained in a linear $\mathbb{P}^3 \subset \mathbb{P}^5$ that is tangent to X at every point of L, see Corollary **2**.2.6 and the discussion of the situation for cubic threefolds in Sections **5**.1.1. Lines of the second type are parametrized by a surface $F_2(X) \subset F(X)$, see Proposition **2**.2.13.

Here are some facts concerning the Fano variety $F(X)$ that can be deduced from the general discussion in Chapter **2**.

(i) The Fano variety $F(X)$ of lines contained in a smooth cubic fourfold $X \subset \mathbb{P}^5$ is an irreducible, smooth projective variety of dimension four with trivial canonical bundle $\omega_{F(X)} \simeq \mathcal{O}_{F(X)}$, see Proposition **2**.1.19 and Lemma **2**.3.1.

(ii) The degree of the Plücker embedding $F(X) \hookrightarrow \mathbb{G}(1, \mathbb{P}^5) \hookrightarrow \mathbb{P}(\bigwedge^2 V)$, i.e. the degree of $F(X)$ with respect to the Plücker polarization $g = c_1(\mathcal{O}_{F(X)}(1))$, is

$$\deg(F(X)) = \int_{F(X)} g^4 = 108,$$

cf. Section **2**.4.3. Later, we will show that $F(X)$ is a hyperkähler manifold, cf. Theorem 3.10, and its degree with respect to the Beauville–Bogomolov–Fujiki square will be computed as $q_{F(X)}(g) = 6$, see Lemma 2.12 and Remark 3.14.

(iii) The Euler number of $F(X)$ is $e(F(X)) = 324$ and its Hodge diamond up to the middle is given by, cf. Section **2**.4.6:

$$b_0(F(X)) = 1 \qquad\qquad 1$$

$$b_2(F(X)) = 23 \qquad\qquad 1 \quad 21 \quad 1$$

$$b_4(F(X)) = 276 \qquad\qquad 1 \quad 21 \quad 232 \quad 21 \quad 1.$$

The Betti numbers suggest that the map $S^2H^2(F(X), \mathbb{Q}) \xrightarrow{\sim} H^4(F(X), \mathbb{Q})$ may be an isomorphism. This is indeed true and can be proved by using the analogous result for the Hilbert scheme $S^{[2]}$ of a K3 surface S, cf. Corollary 2.10. It essentially also follows from Section **2**.5.3. The discussion there, not using any hyperkähler geometry, shows the injectivity of $S^2H^2(F(X), \mathbb{Q})_{\rm pr} \hookrightarrow H^4(F(X), \mathbb{Q})$ and can be extended to prove the full statement.

(iv) The projection $q\colon \mathbb{L} \longrightarrow X$ from the universal family $p\colon \mathbb{L} \longrightarrow F(X)$ of lines contained in X is surjective and its generic fibre $q^{-1}(x)$ is isomorphic to a smooth complete intersection curve of type $(2, 3)$ in $\mathbb{P}^3$ and, therefore, is of genus

$$g(q^{-1}(x)) = 4,$$

see Remark **2**.3.6 and Lemma **2**.5.11. There are at most finitely many fibres of dimension > 1, see e.g. [125, Cor. 2.2]. For the generic point $x \in X$, the cubic threefold $Y := \mathbb{T}_xX \cap X$ has a node at $x \in Y$ as its only singularity. Note that the fibre $q^{-1}(x)$ is nothing but the curve C in Section **5**.5.1 parametrizing all lines in Y passing through the node.

(v) The Fano correspondence induces an injective morphism of integral Hodge structures, see Proposition **2**.5.5:

$$\varphi = p_* \circ q^*\colon H^4(X, \mathbb{Z}) \hookrightarrow H^2(F(X), \mathbb{Z})(-1). \tag{0.2}$$

As both sides are of rank 23, the injection is of finite index. Moreover, φ maps $H^4(X, \mathbb{Z})_{\rm pr}$ to $H^2(F(X), \mathbb{Z})_{\rm pr}$, see Remark **2**.5.7, and for $\alpha, \beta \in H^4(X, \mathbb{Z})_{\rm pr}$, we have

$$(\alpha.\beta) = -\frac{1}{6}\int_{F(X)} \varphi(\alpha) \cdot \varphi(\beta) \cdot g^2.$$

By Deligne's invariant cycle theorem or its slightly stronger consequence Corollary **1**.2.12, we know that $H^{2,2}(X, \mathbb{Z})_{\rm pr} = 0$ for the very general smooth cubic fourfold $X \subset \mathbb{P}^5$. For its Fano variety of lines, this implies

$$\rho(F(X)) = \operatorname{rk} \operatorname{NS}(F(X)) = 1.$$

Another consequence of $H^{2,2}(X, \mathbb{Z})_{\rm pr} = 0$ for the very general smooth cubic fourfold, is the absence of any linear $\mathbb{P}^2 \subset X$. Indeed, a plane in the very general X would satisfy $[\mathbb{P}^2] = m \cdot h^2$. As $(h^2.h^2) = 3$ is square free, the class h^2 is not divisible any further and hence $m \in \mathbb{Z}$. This leads to the contradiction $1 = \int_{\mathbb{P}^2} h^2|_{\mathbb{P}^2} = ([\mathbb{P}^2].h^2) = (m \cdot h^2.h^2) = m \cdot \int_X h^4 = 3 \cdot m$.

(vi) The dual of the Fano correspondence (0.2) is given by

$$\psi := q_* \circ p^* \colon H^6(F(X), \mathbb{Z})(1) \twoheadrightarrow H^4(X, \mathbb{Z}),$$

see Section **2**.5.5. Tensoring with $\mathbb{Q}$ and using (0.2) gives an isomorphism of Hodge structures $H^6(F(X), \mathbb{Q})(1) \xrightarrow{\sim} H^4(X, \mathbb{Q})$ and, thus, a bijection between their spaces of Hodge classes $H^{3,3}(F(X), \mathbb{Q}) \twoheadrightarrow H^{2,2}(X, \mathbb{Q})$.

By the Lefschetz (1, 1)-theorem, all classes in $H^{1,1}(F(X), \mathbb{Q})$ are algebraic and hence, by applying the Lefschetz operator, also all classes in $H^{3,3}(F(X), \mathbb{Q})$ are algebraic. Since ψ maps algebraic classes to algebraic classes, this proves the Hodge conjecture for $H^{2,2}(X, \mathbb{Q})$. This was first established by Zucker [501] relying on ideas of Griffiths and drawing upon normal functions induced by hyperplane sections Y_t of X and the resulting family of Fano surfaces $F(Y_t)$. The use of the Fano correspondence simplifies the argument, but is unlikely to generalize to other classes of fourfolds. See also Corollary 3.29 for a refinement and further comments on the integral version.

0.3 Chow Groups and Chow Motives The rational Chow motive of a smooth cubic fourfold X splits as

$$\mathfrak{h}(X) \simeq \bigoplus_{j=0}^{8} \mathfrak{h}^j(X) \simeq \bigoplus_{i=0}^{4} \mathbb{Q}(-i) \oplus \mathfrak{h}(X)_{\mathrm{pr}},$$

see Remark **1**.1.11. For the Chow groups, we have

$$\mathrm{CH}^0(X) \simeq \mathrm{CH}^1(X) \simeq \mathrm{CH}^4(X) \simeq \mathbb{Z}.$$

From the Bloch–Srinivas principle [72, Thm. 1 (ii)] we deduce that the cycle class map $\mathrm{CH}^2(X) \hookrightarrow H^{2,2}(X, \mathbb{Z})$ is injective (and, as we will see, in fact bijective). So the only interesting Chow group is $\mathrm{CH}^3(X)$, parametrizing one-dimensional cycles up to rational equivalence, respectively $\mathrm{CH}(\mathfrak{h}(X)_{\mathrm{pr}}) \simeq H^{2,2}(X, \mathbb{Q})_{\mathrm{pr}} \oplus (\mathrm{CH}^3(X) \otimes \mathbb{Q})$. The Chow motive of the Fano variety $F(X)$ is described by

$$\mathfrak{h}(F(X))(-2) \oplus \mathfrak{h}(X) \oplus \mathfrak{h}(X)(-4) \simeq S^2\mathfrak{h}(X),$$

cf. Section **2**.4.2. See Remark 3.24 for some more information and Section **7**.4 for general comments on Chow groups and motives.

1 Geometry of Some Special Cubic Fourfolds

Special cubic fourfolds, e.g. those containing a plane or those that can be described in terms of Pfaffians, are not only geometrically rich and interesting, but they play a key role in the theory of general cubic fourfolds as well. In this first section, we discuss cubic fourfolds containing a special surface like a plane or a rational normal scroll and also comment on nodal cubic fourfolds. The case of Pfaffian cubic fourfolds will be dealt with in detail in Section 2. There is a well-defined meaning of the notion of special cubic fourfolds, to be discussed in detail in Section 5.

1.1 Cubic Fourfolds Containing a Plane: Lattice Theory Let us first consider smooth cubic fourfolds $X \subset \mathbb{P}^5$ containing a plane $\mathbb{P}^2 \simeq \mathrm{P} \subset X$. These cubics are central in the original proof of the global Torelli theorem for cubic fourfolds [472, 477] and have been used as a starting point for a number of considerations.

Lemma 1.1 *The sublattice*

$$K_8^- := \mathbb{Z} \cdot h^2 \oplus \mathbb{Z} \cdot [\mathrm{P}] \subset H^4(X, \mathbb{Z})$$

is saturated, i.e. its cokernel is torsion free, and its intersection matrix is

$$\begin{pmatrix} 3 & 1 \\ 1 & 3 \end{pmatrix}, \tag{1.1}$$

which is positive definite of discriminant eight.

Proof Clearly, $(h^2.h^2) = 3$ and $(h^2.[\mathrm{P}]) = 1$. To prove $([\mathrm{P}].[\mathrm{P}]) = \int_{\mathrm{P}} \mathrm{c}_2(\mathcal{N}_{\mathrm{P}/X}) = 3$, use either one of the two short exact sequences

$$0 \longrightarrow \mathcal{N}_{\mathrm{P}/X} \longrightarrow \mathcal{N}_{\mathrm{P}/\mathbb{P}} \longrightarrow \mathcal{O}_{\mathrm{P}}(3) \longrightarrow 0 \quad \text{or} \quad 0 \longrightarrow \mathcal{T}_{\mathrm{P}} \longrightarrow \mathcal{T}_X|_{\mathrm{P}} \longrightarrow \mathcal{N}_{\mathrm{P}/X} \longrightarrow 0,$$

from which one concludes $\mathrm{c}_1(\mathcal{N}_{\mathrm{P}/X}) = 0$ and $\mathrm{c}_2(\mathcal{N}_{\mathrm{P}/X}) = 3 \cdot h_{\mathrm{P}}^2$.

Assume $\alpha \in H^4(X, \mathbb{Z})$ is contained in the saturation of $\mathbb{Z} \cdot h^2 \oplus \mathbb{Z} \cdot [\mathrm{P}]$ and write

$$\alpha \overset{(*)}{=} s \cdot h^2 + t \cdot [\mathrm{P}] \quad \text{and} \quad \alpha \overset{(**)}{=} (s + (t/3)) \cdot h^2 + (t/3) \cdot v,$$

where $v := 3[\mathrm{P}] - h^2 \in H^4(X, \mathbb{Z})_{\mathrm{pr}}$. We observe that the class v is not divisible any further. To see this, we may assume that X is a cubic fourfold containing a second plane $\mathrm{P}' \subset X$ disjoint to P. Indeed, the family of smooth cubics fourfolds containing a fixed plane is connected and smooth cubic fourfolds containing two disjoint planes exist, see Example **1**.5.2 and Section **1**.5.3. Hence, $([\mathrm{P}'].v) = -1$ and thus v is not divisible.[1]

Then $(\alpha.h^2) \in \mathbb{Z}$ implies $3s + t \in \mathbb{Z}$. Using $(**)$ one finds $t \in \mathbb{Z}$ and then by $(*)$ also $s \in \mathbb{Z}$. □

The notation K_8^- shall be explained in Section 5.2, the intersection form there will be changed by a global sign which explains the minus sign in the notation here.

[1] I wish to thank X. Wei for a discussion of the argument.

Exercise 1.2 Let Q be the residual quadric of a generic linear intersection $\mathrm{P} \subset \mathbb{P}^3 \cap X$, so that cohomologically $[\mathrm{P}] + [Q] = h^2$. Imitate the above computation to show that the two classes $[\mathrm{P}], [Q] \in H^4(X, \mathbb{Z})$ describe a basis of the lattice K_8^- with the intersection matrix

$$\begin{pmatrix} 3 & -2 \\ -2 & 4 \end{pmatrix}. \tag{1.2}$$

Remark 1.3 A quick dimension count reveals that the set of smooth cubic fourfolds containing a plane forms a divisor in the moduli space of all smooth cubic fourfolds. Let us indicate two ways to verify this.

(i) Fix a plane $\mathbb{P}^2 \simeq \mathrm{P} \subset \mathbb{P} = \mathbb{P}^5$ and compute the linear space $|\mathcal{I}_{\mathrm{P}} \otimes \mathcal{O}_{\mathbb{P}}(3)|$ of all cubics passing through P. Its dimension is

$$h^0(\mathbb{P}, \mathcal{O}_{\mathbb{P}}(3)) - h^0(\mathrm{P}, \mathcal{O}_{\mathrm{P}}(3)) - 1 = 56 - 10 - 1 = 45.$$

The subgroup of PGL(6) preserving P as a subvariety (but not necessarily pointwise) is of dimension 26. This proves that within the 20-dimensional moduli space M_4 of all smooth cubic fourfolds, see Section **1**.2.1, the set of cubics containing a plane is an irreducible divisor, cf. Exercise **1**.5.2 and [472, §1 Lem. 1].

(ii) As explained above, h^2 and the class of a plane $[\mathrm{P}] \in H^4(X, \mathbb{Z})$ span a rank two sublattice. The first order deformations in $H^1(X, \mathcal{T}_X)$ preserving $[\mathrm{P}]$ as a $(2, 2)$-class, i.e.

$$\{v \in H^1(X, \mathcal{T}_X) \mid i_v[\mathrm{P}] = 0 \text{ in } H^{1,3}(X)\},$$

form a subspace of codimension one.

In (i) above and in Exercise **1**.5.2 we have seen that the set of smooth cubics X containing a plane inside the moduli space M_4 of all smooth cubic fourfolds forms an irreducible divisor. The Hodge theoretic condition on $H^{2,2}(X, \mathbb{Z})$ to contain a lattice isometric to K_8^- could a priori describe a union of several (Noether–Lefschetz) divisors in M_4. However, a result of Hassett [227, Prop. 3.2.4] says that this is not the case, cf. Proposition 5.6. Therefore, if X is very general among all smooth cubic fourfolds with $H^{2,2}(X, \mathbb{Z}) \simeq K_8^-$, then it contains a plane $\mathbb{P}^2 \subset X$. By specialization, this is then true for all X with $K_8^- \hookrightarrow H^{2,2}(X, \mathbb{Z})$ extending $h^2 \in H^{2,2}(X, \mathbb{Z})$.

Exercise 1.4 Adapt the techniques of (i) or (ii) above to show that the set of all smooth cubic fourfolds $X \in M_4$ that contain two disjoint planes forms an 18-dimensional subspace. This was first observed by Hassett [228, Sec. 1.2]. Recall from Corollary **1**.5.11 that every cubic fourfold containing two disjoint planes is rational.

Degtyarev, Itenberg, and Ottem [143] show that a smooth cubic fourfold can contain at most 405 planes. The maximum is attained only once, namely by the Fermat cubic.

1.2 Cubic Fourfolds Containing a Plane: Quadric Fibration Recall from Section 1.5 that the blow-up of X in a plane $\mathbb{P}^2 \simeq \mathrm{P} \subset X$ leads to a quadric surface fibration

$$\phi\colon \tilde{X} := \mathrm{Bl}_{\mathrm{P}}(X) \twoheadrightarrow \mathbb{P}^2$$

with the fibre over a point $y \in \mathbb{P}^2$ being the residual quadric surface Q_y of $\mathrm{P} \subset \overline{y\mathrm{P}} \cap X$. As X does not contain a linear $\mathbb{P}^3$, the morphism is flat. Furthermore, the discriminant divisor $D_{\mathrm{P}} \subset \mathbb{P}^2$ is a curve in the linear system $|\mathcal{O}_{\mathbb{P}^2}(6)|$.

Remark 1.5 (i) According to Exercise 1.5.2, the existence of an example with a smooth discriminant divisor D_{P} shows that for the generic choice of a pair $\mathbb{P}^2 \simeq \mathrm{P} \subset X$ the discriminant divisor is a smooth sextic curve. An explicit example of a Pfaffian cubic containing a plane can be found in [28, Thm. 9].[2]

Alternatively, but less explicitly, the existence can be shown as follows: as explained in Remark 1.5.8, it suffices to show that for a given plane $\mathbb{P}^2 \simeq \mathrm{P} \subset \mathbb{P}^5$ the sections in $H^0(\mathbb{P}^2, S^2(\mathcal{F}) \otimes \mathcal{O}(1))$ induced by the equations defining cubics $X \subset \mathbb{P}^5$ that contain the plane P are generic. This is clear, as $H^0(\mathbb{P}^2, S^2(\mathcal{F}) \otimes \mathcal{O}(1)) \simeq H^0(\mathbb{P}^5, \mathcal{O}(3) \otimes \mathcal{I}_{\mathrm{P}})$, see Section 1.5.2, and the existence of some smooth cubic containing P is known, see Exercise 1.8.

(ii) More precisely, one has the following criterion, cf. [472, §1 Lem. 2]: the discriminant curve $D_{\mathrm{P}} \subset \mathbb{P}^2$ of the projection from a plane $\mathrm{P} \subset X$ contained in a smooth cubic fourfold is smooth if and only if X does not contain a second plane with non-empty intersection with P. Indeed, $\overline{y\mathrm{P}} \cap X = \mathrm{P} \cup \phi^{-1}(y) \subset \overline{y\mathrm{P}} \simeq \mathbb{P}^3$ and, therefore, either $\phi^{-1}(y)$ is smooth, or it is a quadric cone with an isolated singularity, in which case D_{P} is smooth at y by Remark 1.5.8, or $\phi^{-1}(y)$ contains a plane. Note that there are indeed cases where D_{P} is singular and even reducible, see Exercise 1.5.7.

A plane $\mathrm{P} \subset X$ in a smooth cubic fourfold leads to two natural subvarieties of the Fano variety $F(X)$. First, there is the dual plane

$$\mathrm{P}^* := \{\, [L] \mid L \subset \mathrm{P} \,\} \subset F(X),$$

which is isomorphic to $\mathbb{P}^{2*} \simeq \mathbb{P}^2$. Second, there is the divisor F_{P} of all lines meeting P. As with C_L in Section 5.1.2, it has to be defined as the closure of

$$\{\, [L] \notin \mathrm{P}^* \mid L \cap \mathrm{P} \neq \emptyset \,\} \subset F_{\mathrm{P}} \subset F(X).$$

Alternatively, consider the Fano correspondence $F(X) \xleftarrow{\ p\ } \mathbb{L} \xrightarrow{\ q\ } X$ and the pre-image $q^{-1}(\mathrm{P}) \subset \mathbb{L}$ of dimension three. It breaks up into the $\mathbb{P}^1$-bundle $\mathbb{L}_{\mathrm{P}^*} \twoheadrightarrow \mathrm{P}^*$ and F'_{P}:

$$q^{-1}(\mathrm{P}) = \mathbb{L}_{\mathrm{P}^*} \cup F'_{\mathrm{P}},$$

[2] Thanks to M. Varesco for the reference.

where the latter maps under p generically injectively onto the hypersurface $F_{\mathrm{P}} \subset F(X)$. More precisely, $F'_{\mathrm{P}} \longrightarrow F_{\mathrm{P}}$ is an isomorphism over $F_{\mathrm{P}} \setminus \mathrm{P}^*$, because a line $L \subset X$ not contained in P intersects P transversally in one point or not at all.

Remark 1.6 (i) Note that the class $[F_{\mathrm{P}}] \in H^2(F(X), \mathbb{Z})$ is the image of $[\mathrm{P}] \in H^4(X, \mathbb{Z})$ under the Fano correspondence $\varphi\colon H^4(X, \mathbb{Z})(1) \longrightarrow H^2(F(X), \mathbb{Z})$, where we use that $p_*[\mathbb{L}_{\mathrm{P}^*}] = 0$. As φ is injective, this in particular proves that F_{P} is not empty, which of course also follows from the fact that $q^{-1}(x)$ is either a curve of genus four, of which there are none in $\mathrm{P}^* \simeq \mathbb{P}^2$, or of dimension two, see Remark **2**.2.16.

(ii) The restriction of the Plücker polarization $g \in H^2(F(X), \mathbb{Z})$ to $\mathbb{P}^2 \simeq \mathrm{P}^* \subset F(X)$ gives back the hyperplane class on $\mathbb{P}^2 \simeq \mathrm{P}^*$. Indeed, under $\mathrm{P}^* \simeq \mathbb{G}(1, \mathrm{P}) \subset \mathbb{G} = \mathbb{G}(1, \mathbb{P}^5)$ the Plücker polarization restricts to the Plücker polarization, i.e. $\mathcal{O}_{\mathbb{G}}|_{\mathrm{P}^*}(1) \simeq \mathcal{O}(1)$. As a consequence we find

$$\int_{\mathrm{P}^*} g^2 = 1.$$

In Remark 4.4 we will see that, $\mathcal{N}_{\mathrm{P}^*/F(X)} \simeq \Omega_{\mathrm{P}^*}$ and, therefore, $([\mathrm{P}^*].[\mathrm{P}^*]) = 3$. Thus, the two classes $[\mathrm{P}^*], g^2 \in H^4(F(X), \mathbb{Z})$ are not proportional and, in particular, $H^{2,2}(F(X), \mathbb{Z})$ is of rank at least two. In fact, using (4.2) in Section 4.1, one can prove that it is of rank at least four. Also note that a line $\mathbb{P}^1 \subset \mathrm{P}^*$ is also a line with respect to the Plücker embedding.

Exercise 1.7 Show that a line $L \subset \mathrm{P} \subset X$ is of the first type, i.e. $\mathcal{N}_{L/X} \simeq \mathcal{O}_L(1) \oplus \mathcal{O}_L^{\oplus 2}$, if and only if $\mathcal{N}_{\mathrm{P}/X}|_L \simeq \mathcal{O}_L^{\oplus 2}$.

Exercise 1.8 Consider the Fermat cubic $X = V\left(\sum x_i^3\right) \subset \mathbb{P}^5$ and the plane $\mathrm{P} = V(x_0 + x_1, x_2 + x_3, x_4 + x_5) \subset X$, see Exercise **1**.5.7. Use that the intersection $F_{\mathrm{P}} \cap \mathrm{P}^*$ consists of those lines in P that are contained in a residual quadric Q_y to show that $\mathrm{P}^* \cap F_{\mathrm{P}}$ consists of three lines.

Hence, $\mathcal{O}(F_{\mathrm{P}})|_{\mathrm{P}^*} \simeq \mathcal{O}(3)$ and we have $([F_{\mathrm{P}}]^2.[\mathrm{P}^*]) = 9$ for the two cohomology classes $[F_{\mathrm{P}}]^2, [\mathrm{P}^*] \in H^{2,2}(F(X), \mathbb{Z})$. Later we will see that $([F_{\mathrm{P}}]^2.[F_{\mathrm{P}}]^2) = \int_{F(X)} [F_{\mathrm{P}}]^4 = 12$, see Exercise 3.25.

Note that then $F_{\mathrm{P}} \cap \mathrm{P}^* \neq \emptyset$ for all smooth cubic fourfolds containing a plane. Anticipating the discussion in Section 4.1, express $[\mathrm{P}^*] \in H^{2,2}(F(X), \mathbb{Q})$ as a linear combination of $[F_{\mathrm{P}}]^2$, g^2, $[F_{\mathrm{P}}] \cdot g$, and $c_2(\mathcal{T}_F)$, which for the generic choice of the pair $\mathrm{P} \subset X$ generate $H^{2,2}(F(X), \mathbb{Q})$.

Next, we consider the relative Fano variety

$$\tilde{F}_{\mathrm{P}} := F(\tilde{X}/\mathbb{P}^2) \longrightarrow \mathbb{P}^2 \tag{1.3}$$

of lines contained in the fibres of $\phi\colon \tilde{X} \longrightarrow \mathbb{P}^2$. In other words, the fibre of (1.3) over a point $y \in \mathbb{P}^2$ is the Fano variety $F(Q_y) \subset F(X)$ of all lines contained in the residual quadric $Q_y \subset X$ of the intersection $\mathrm{P} \subset \overline{y\mathrm{P}} \cap X$. The natural morphism

$$\tilde{F}_{\mathrm{P}} \twoheadrightarrow F_{\mathrm{P}} \subset F(X) \tag{1.4}$$

maps onto F_{P}, see Proposition 1.10, and is injective over $F_{\mathrm{P}} \setminus \mathrm{P}^*$, cf. [472, §1]. However, it may fail to be injective over P^* for special smooth cubic fourfolds containing a plane, and indeed this happens in Exercise 1.8, but one certainly expect $\tilde{F}_{\mathrm{P}} \xrightarrow{\sim} F_{\mathrm{P}}$ for generic choices.

Let us look at the fibres of $\tilde{F}_{\mathrm{P}} \twoheadrightarrow \mathbb{P}^2$. For $y \in \mathbb{P}^2 \setminus D_{\mathrm{P}}$, the residual quadric Q_y is smooth, i.e. $Q_y \simeq \mathbb{P}^1 \times \mathbb{P}^1$, and the Fano variety $F(Q_y)$ consists of two connected components parametrizing the fibres of the two projections to $\mathbb{P}^1$:

$$F(Q_y) \simeq \mathbb{P}^1 \sqcup \mathbb{P}^1.$$

According to Remark 1.5, for the generic pair $\mathrm{P} \subset X$, the singular fibres Q_y are of the form $V(x_0^2 + x_1^2 + x_2^2) \subset \mathbb{P}^3$, i.e. they are isomorphic to a cone over a smooth quadric curve. Hence, in this case $F(Q_y) \simeq \mathbb{P}^1$, parametrizes the lines through the vertex of the cone.

Remark 1.9 The pullback of the Plücker polarization $g \in H^2(F(X), \mathbb{Z})$ to $\tilde{F}_{\mathrm{P}}$ defines a line bundle with fibre degree two, i.e. $\int_{\mathbb{P}^1} g = 2$ for $\mathbb{P}^1 \subset F(X)$ parametrizing fibres of one of the two rulings of a smooth quadric $Q_y \subset X$. This has nothing to do with the cubic X, but rather follows from an explicit computation of the Plücker embedding $\mathbb{P}^1 \subset F(Q_y) \subset F(\mathbb{P}^3) \subset F(\mathbb{P}^5) \subset \mathbb{P}(\bigwedge^2 V)$. Note that in particular $\mathbb{P}^1 \subset F(Q_y) \subset F(X)$ is cohomologically different from a line $\mathbb{P}^1 \subset \mathrm{P}^*$ in Remark 1.6, (ii).

Adding the universal line to (1.4) leads to a diagram

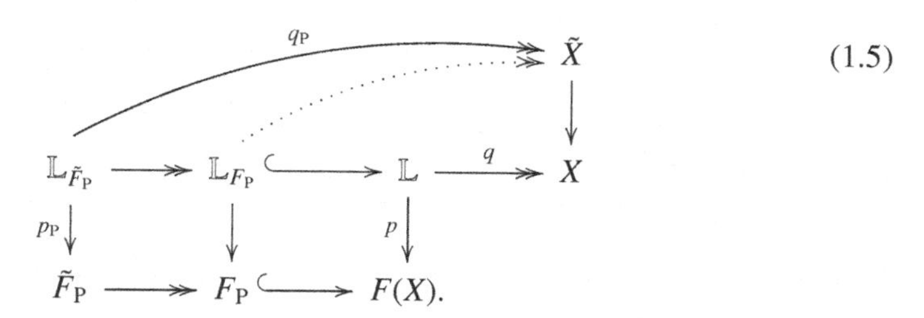

(1.5)

Recall that $q\colon \mathbb{L} \twoheadrightarrow X$ is surjective and its generic fibre is a curve of degree six with respect to the Plücker polarization on $F(X)$, see Lemma **2**.5.11 and Section 0.2, (iv). In contrast, the projection $q_{\mathrm{P}}\colon \mathbb{L}_{\tilde{F}_{\mathrm{P}}} \twoheadrightarrow \tilde{X} \twoheadrightarrow X$ is generically finite of degree two, as a generic point $x \in \tilde{X}$ is contained in exactly two lines in the fibre Q_y, $y = \phi(x)$, of the linear projection $\phi\colon \tilde{X} \twoheadrightarrow \mathbb{P}^2$.

Proposition 1.10 *Consider a smooth cubic fourfold $X \subset \mathbb{P}^5$ containing a plane* $\mathrm{P} \subset X$ *such that $D_{\mathrm{P}} \subset \mathbb{P}^2$ is smooth. Then:*

(i) *The relative Fano variety* $\tilde{F}_{\mathrm{P}}$ *is a* $\mathbb{P}^1$*-bundle over a smooth, polarized K3 surface* S_{P} *of degree two, obtained as the Stein factorization of* (1.3)*:*

$$\tilde{F}_{\mathrm{P}} \xrightarrow{\tilde{\pi}} S_{\mathrm{P}} \xrightarrow{\pi} \mathbb{P}^2.$$

Here, π *is a finite morphism of degree two ramified over the sextic curve* D_{P}.

(ii) *The relative Fano variety* $\tilde{F}_{\mathrm{P}}$ *is smooth and the morphism* (1.4) *is a birational map onto the uniruled divisor* $F_{\mathrm{P}} \subset F(X)$.

Proof If the discriminant curve $D_{\mathrm{P}} \subset \mathbb{P}^2$ is smooth, the double cover

$$\pi\colon S_{\mathrm{P}} \twoheadrightarrow \mathbb{P}^2 \tag{1.6}$$

ramified over D_{P} is a K3 surface naturally polarized by $\pi^*\mathcal{O}(1)$, which is of degree two, cf. [249, Exa. 1.1.3]. The smoothness of $\tilde{F}_{\mathrm{P}}$ follows from the vanishing of the obstruction space to deform a line in the fibres of $\phi\colon \tilde{X} \twoheadrightarrow \mathbb{P}^2$.

The morphism $\tilde{F}_{\mathrm{P}} \twoheadrightarrow F_{\mathrm{P}}$ is surjective. Indeed, any line $L \subset X$ intersecting P properly defines a linear space $\mathbb{P}^3 \simeq \overline{L\mathrm{P}}$, which can also be written as $\overline{y\mathrm{P}}$ for a unique $y \in \mathbb{P}^2$ and then L is contained in Q_y. As the map is generically injective and $\tilde{F}_{\mathrm{P}}$ is smooth (and irreducible), this proves (ii). □

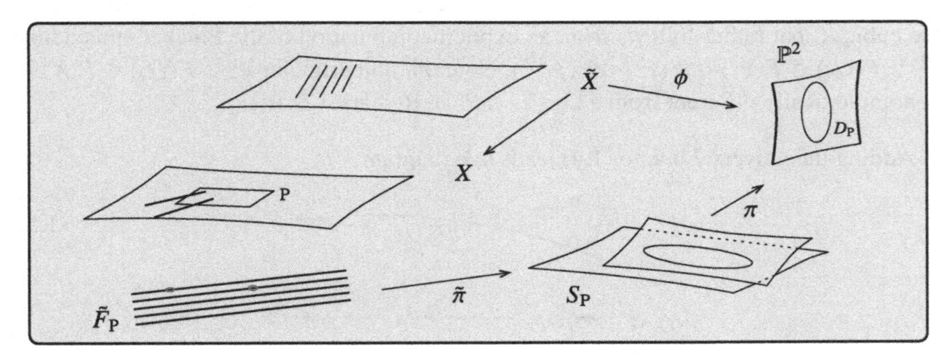

Remark 1.11 The double cover $S_{\mathrm{P}} \twoheadrightarrow \mathbb{P}^2$ can be understood more conceptually in terms of Clifford algebras, see [27, §1.5 & App. A] for more details.

Recall from Section **1**.5.1 that $\tilde{X} \twoheadrightarrow \mathbb{P}^2$ is described as the zero set in $\mathbb{P}(\mathcal{F}^*)$ of a quadratic form $q\colon \mathcal{F}^* \longrightarrow \mathcal{F} \otimes \mathcal{O}(1)$ associated with the defining equation. Here, $\mathcal{F} \simeq \mathcal{O}(1) \oplus \mathcal{O}^{\oplus 3}$. The sheaf of even *Clifford algebras* $\mathcal{C}_0$ associated with $(\mathcal{F}^*, q)$ is a locally free sheaf of algebras of rank four, explicitly $\mathcal{C}_0 \simeq \mathcal{O} \oplus \bigwedge^2 \mathcal{F}^*(-1) \oplus \bigwedge^4 \mathcal{F}^*(-2)$. Its centre $\mathcal{Z} \subset \mathcal{C}_0$ is a locally free sheaf of algebras of rank two. The relative affine spectrum gives back the covering K3 surface:

$$\pi\colon S_{\mathrm{P}} \simeq \mathrm{Spec}(\mathcal{Z}) \twoheadrightarrow \mathbb{P}^2.$$

Since $\mathcal{C}_0$ is a sheaf of algebras over $\mathcal{Z}$, it corresponds to a rank four sheaf $\mathcal{B}_0$ of $\mathcal{O}_{S_{\mathrm{P}}}$-algebras on S_{P}, i.e. $\pi_*\mathcal{B}_0 \simeq \mathcal{C}_0$. The two sheets of the covering parametrize the two half spinor representations and the fibres of $\tilde{\pi}\colon \tilde{F}_{\mathrm{P}} \longrightarrow S_{\mathrm{P}}$ are their projectivizations, cf. Example **2**.1.6.

Remark 1.12 There is a striking analogy between the curves $C_L \subset F(Y)$ in the Fano surface of lines in a cubic threefold $Y \subset \mathbb{P}^4$, see Section **5**.1.2, and the hypersurface $F_{\mathrm{P}} \subset F(X)$.

The similarities between the two situations can be pictured as follows:

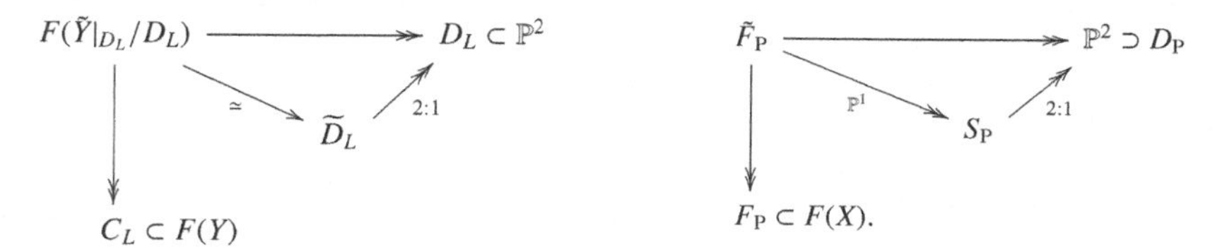

Here, the two vertical arrows are birational maps onto their images C_L and F_{P}. In fact, for the generic choice of a line $L \subset Y$ in a cubic threefold, one has $C_L \simeq \widetilde{D}_L$.

In Remark 4.11 we will discuss another fourfold analogue of the situation for cubic threefolds described by the left-hand diagram above, where instead of projecting the cubic fourfold from a plane one projects it from a line, as in dimension three.

All the fibres of the morphism $\tilde{\pi}\colon \tilde{F}_{\mathrm{P}} \longrightarrow S_{\mathrm{P}}$ are isomorphic to $\mathbb{P}^1$, but the fibration is in general not Zariski locally trivial. In other words, the Brauer–Severi variety $\tilde{F}_{\mathrm{P}} \longrightarrow S_{\mathrm{P}}$ is not trivial and, therefore, its Brauer class

$$\alpha_{\mathrm{P},X} \in \mathrm{Br}(S_{\mathrm{P}}),$$

which is always of order at most two, is in general not trivial.

Alternatively, $\alpha_{\mathrm{P},X}$ can be understood as the Brauer class of the natural sheaf of Azumaya algebras $\mathcal{B}_0$ on S_{P}, see Remark 1.11. Also, $\tilde{F}_{\mathrm{P}}$ can be written as the projectivization

$$\tilde{F}_{\mathrm{P}} \simeq \mathbb{P}(\mathcal{E})$$

of a locally free $\alpha_{\mathrm{P},X}$-twisted sheaf of rank two, the bundle of half spinor representations, and $\mathcal{B}_0 \simeq \mathcal{E}nd(\mathcal{E})$. See [249, Ch. 18] for general facts on Brauer groups of K3 surfaces.

Remark 1.13 The Brauer–Severi variety $\tilde{F}_{\mathrm{P}} \longrightarrow S_{\mathrm{P}}$ is typically not trivial, i.e. it is not the projectivization of an algebraic or holomorphic rank-two bundle, cf. Exercise 1.20 below. However, in the $\mathcal{C}^\infty$-setting it always is, i.e. $\tilde{F}_{\mathrm{P}} \simeq \mathbb{P}(E)$ for some $\mathcal{C}^\infty$-vector bundle of rank two $E \longrightarrow S_{\mathrm{P}}$.

A relative tautological class $g_0 \in H^2(\tilde{F}_{\mathrm{P}}, \mathbb{Z})$, fibrewise of degree one, is in this setting well defined up to translation by classes in $H^2(S_{\mathrm{P}}, \mathbb{Z})$ and by Leray–Hirsch

$$H^2(\tilde{F}_{\mathrm{P}}, \mathbb{Z}) \simeq H^2(S_{\mathrm{P}}, \mathbb{Z}) \oplus \mathbb{Z} \cdot g_0. \tag{1.7}$$

The pullback g_{P} of the Plücker polarization under $\tilde{F}_{\mathrm{P}} \longrightarrow F_{\mathrm{P}} \subset F(X)$ has fibre degree two, see Remark 1.9. Therefore, $g_0 = B \oplus (1/2) g_{\mathrm{P}}$ with $B \in (1/2) H^2(S_{\mathrm{P}}, \mathbb{Z}) \subset H^2(S_{\mathrm{P}}, \mathbb{Q})$ well defined up to translation by elements in $H^2(S_{\mathrm{P}}, \mathbb{Z})$. Warning: neither the class $g_0 \in H^2(\tilde{F}_{\mathrm{P}}, \mathbb{Z})$ nor the class $2B \in H^2(S_{\mathrm{P}}, \mathbb{Z})$ need to be of type $(1, 1)$.

The next result explains the role of the Brauer class $\alpha_{\mathrm{P},X}$ for the geometry of X, see Kuznetsov [298, Sec. 4.3].

Lemma 1.14 *The following conditions are equivalent:*

(i) *There exists a rational section of* $\phi\colon \tilde{X} \longrightarrow \mathbb{P}^2$.
(ii) *The Brauer class is trivial* $\alpha_{\mathrm{P},X} = 1 \in \mathrm{Br}(S_{\mathrm{P}})$.
(iii) *There exists a line bundle* L *on* $\tilde{F}_{\mathrm{P}}$ *of odd degree on all fibres of* $\tilde{\pi}$.

In this case, the cubic fourfold X *is rational, see Example* **1**.5.10.

Proof Let us first prove that $\alpha_{\mathrm{P},X} = 1$ if and only if $\tilde{\pi}\colon \tilde{F}_{\mathrm{P}} \longrightarrow S_{\mathrm{P}}$ has a rational section. Indeed, if $\alpha_{\mathrm{P},X} = 1$, then $\tilde{F}_{\mathrm{P}}$ is the projectivization of a locally free sheaf of rank two, which can be trivialized over Zariski open subsets. Each such trivialization leads to a rational section of $\tilde{\pi}\colon \tilde{F}_{\mathrm{P}} \longrightarrow S_{\mathrm{P}}$. As $\mathrm{Br}(S_{\mathrm{P}})$ is a subgroup of the Brauer group of the function field of S_{P}, so $\mathrm{Br}(S_{\mathrm{P}}) \subset \mathrm{Br}(K(S_{\mathrm{P}}))$, the converse also holds.

The generic fibre Q_y of $\phi\colon \tilde{X} \longrightarrow \mathbb{P}^2$ is a surface isomorphic to $\mathbb{P}^1 \times \mathbb{P}^1$. A rational section of ϕ picks out a point $x \in Q_y$ in the generic fibre, and, therefore, distinguishes the two lines given as the fibres of the two projections through x and hence a canonical point $[L_{z_i}] \in F_{\mathrm{P}}$ over each of the two points $z_i \in S_{\mathrm{P}}$ mapping to y under (1.6). This defines a rational section $z \longmapsto [L_z]$ of $\tilde{\pi}\colon \tilde{F}_{\mathrm{P}} \longrightarrow S_{\mathrm{P}}$. Conversely, if a rational section of $\tilde{\pi}$ is given, mapping $y \in \mathbb{P}^2$ to the point of intersection x of the two lines corresponding to the two points $z_1, z_2 \in S_{\mathrm{P}}$ over y, i.e. $L_{z_1} \cap L_{z_2} = \{x\}$, defines a rational section of $\tilde{X} \longrightarrow \mathbb{P}^2$. Thus, (i) and (ii) are equivalent.

As a Zariski locally trivial $\mathbb{P}^1$-bundle comes with a relative tautological line bundle, (i) and (ii) imply (iii). Conversely, a line bundle of odd fibre degree can be modified by powers of g_{P} to define a line bundle of fibre degree one. The dual of its direct image then gives a bundle of rank two, the projectivization of which describes $\tilde{\pi}$. □

Remark 1.15 Maybe a little surprisingly, a smooth cubic fourfold X containing a plane $\mathbb{P}^2 \simeq \mathrm{P} \subset X$ can be rational without the quadric fibration $\tilde{X} \longrightarrow \mathbb{P}^2$ having a rational section or, equivalently, without $\alpha_{\mathrm{P},X} \in \mathrm{Br}(S_{\mathrm{P}})$ being trivial, see [28, Thm. A].

Exercise 1.16 Show that the Brauer class $\alpha_{\mathrm{P},X}$ is trivial if there exists a class $\gamma \in H^{2,2}(X,\mathbb{Z})$ with $(\gamma.[Q]) = 1$, cf. [296, Prop. 4.7]. Here, $[Q]$ is the class of the residual quadric of a generic hyperplane section $\mathrm{P} \subset \mathbb{P}^3 \cap X$, see Exercise 1.2.

Remark 1.17 Dolgachev, van Geemen, and Kondō [160, §4.12] pointed out a curious link between cubic surfaces and cubic fourfolds containing a plane that is passed on to their associated K3 surfaces.

As explained in Remark **4**.2.12, with a line $L \subset S = V(F(x_0,\dots,x_3)) \subset \mathbb{P}^3$ in a smooth cubic surface, one can naturally associate a K3 surface as the minimal resolution $\tilde{T}$ of a double plane $T \longrightarrow \mathbb{P}^2$ branched over a certain sextic curve $C \subset \mathbb{P}^2$ derived from the discriminant divisor and the branch divisor of the projection $L \subset S \longrightarrow \mathbb{P}^1$. Then the cubic fourfold defined by $F(x_0,\dots,x_3) + x_4 \cdot x_5 \cdot (x_4 + x_5)$ contains the plane P spanned by $L \subset \mathbb{P}^3 \subset \mathbb{P}^5$ and the point $[0 : \dots : 0 : 1]$. Moreover, for the generic choice of S, the K3 surface $\tilde{T}$ is isomorphic to the minimal resolution of the double cover $S_{\mathrm{P}} \longrightarrow \mathbb{P}^2$ as considered in the discussion here.

1.3 Cubic Fourfolds Containing a Plane: Fano Correspondence The Fano correspondence composed with the restriction to $\tilde{F}_{\mathrm{P}} \twoheadrightarrow F_{\mathrm{P}} \subset F(X)$ defines a morphism of integral Hodge structures

$$\varphi_{\mathrm{P}} \colon H^4(X,\mathbb{Z}) \longrightarrow H^2(F(X),\mathbb{Z})(-1) \longrightarrow H^2(\tilde{F}_{\mathrm{P}},\mathbb{Z})(-1). \tag{1.8}$$

On the other hand, the pullback under $\tilde{\pi} \colon \tilde{F}_{\mathrm{P}} \longrightarrow S_{\mathrm{P}}$ defines an injective morphism of integral Hodge structures

$$\tilde{\pi}^* \colon H^2(S_{\mathrm{P}},\mathbb{Z}) \hookrightarrow H^2(\tilde{F}_{\mathrm{P}},\mathbb{Z}).$$

These two morphisms are linked to each other by the next result, cf. [472, §1, Prop. 1].

Proposition 1.18 (Voisin) *The Fano correspondence* (1.8) *is an injection of Hodge structures*

$$\varphi_{\mathrm{P}} \colon H^4(X,\mathbb{Z}) \hookrightarrow H^2(\tilde{F}_{\mathrm{P}},\mathbb{Z})(-1)$$

with finite cokernel. Furthermore, primitive classes $\alpha,\beta \in H^4(X,\mathbb{Z})_{\mathrm{pr}}$ *satisfy*

$$(\alpha.\beta) = -\frac{1}{2}\int_{\tilde{F}_{\mathrm{P}}} \varphi_{\mathrm{P}}(\alpha)\cdot\varphi_{\mathrm{P}}(\beta)\cdot g_{\mathrm{P}} \tag{1.9}$$

and, up to a global sign, the restriction to the sub-Hodge structure $K_8^{-\perp} \subset H^4(X,\mathbb{Z})$ *defines an index two, isometric embedding of integral Hodge structures*

$$K_8^{-\perp} \underset{-1}{\hookrightarrow} H^2(S_{\mathrm{P}},\mathbb{Z})_{\mathrm{pr}}(-1).$$

Proof First, observe that φ_{P} can alternatively be described via the correspondence $\tilde{F}_{\mathrm{P}} \xleftarrow{p_{\mathrm{P}}} \mathbb{L}_{\tilde{F}_{\mathrm{P}}} \xrightarrow{q_{\mathrm{P}}} \tilde{X} \longrightarrow X$ in (1.5). Arguments identical to the ones in the proof of Proposition **2**.5.5 ensure (1.9). The denominator six there, which is the degree of the generic fibre of $q\colon \mathbb{L} \twoheadrightarrow X$, is here replaced by the degree two of the morphism $q_{\mathrm{P}}\colon \mathbb{L}_{\tilde{F}_{\mathrm{P}}} \twoheadrightarrow X$. This clearly already proves the injectivity of φ_{P} on $H^4(X,\mathbb{Z})_{\mathrm{pr}}$.

For the very general pair $\mathrm{P} \subset X$, the space of Hodge classes $H^{2,2}(X,\mathbb{Z})$ is of rank two, see Remark 1.3, and coincides with K_8^-. As $K_8^{-\perp} \subset H^4(X,\mathbb{Z})$ is an irreducible Hodge structure, φ_{P} maps it into the transcendental part of $H^2(S_{\mathrm{P}},\mathbb{Z}) \subset H^2(\tilde{F}_{\mathrm{P}},\mathbb{Z})$, cf. (1.7), which for dimension reasons is $H^2(S_{\mathrm{P}},\mathbb{Z})_{\mathrm{pr}}$. As the Fano correspondence is invariant under deformations, the assertion then holds true for all $\mathrm{P} \subset X$ with smooth D_{P}.

Next, under the pullback $H^2(S_{\mathrm{P}},\mathbb{Z}) \longrightarrow H^2(\tilde{F}_{\mathrm{P}},\mathbb{Z})$ the intersection form on S_{P} corresponds to the intersection pairing $(1/2)\int_{\tilde{F}_{\mathrm{P}}} \gamma_1 \cdot \gamma_2 \cdot g_{\mathrm{P}}$, because g_{P} has fibre degree two, see Remark 1.9. Therefore, φ_{P} restricted to $K_8^{-\perp}$ is up to a global sign an isometric Hodge embedding into $H^2(S_{\mathrm{P}},\mathbb{Z})_{\mathrm{pr}}(-1)$. As $\mathrm{disc}(K_8^{-\perp}) = \mathrm{disc}(K_8^-) = 8$ and $\mathrm{disc}(H^2(S_{\mathrm{P}},\mathbb{Z})_{\mathrm{pr}}) = 2$, its index has to be two, cf. [249, Sec. 14.0.2]. □

In Example 3.25 similar arguments will be used to describe the Néron–Severi group of the Fano variety $F(X)$.

Remark 1.19 If $\alpha_{\mathrm{P},X} = 1$, then a rational section of $\tilde{F}_{\mathrm{P}} \twoheadrightarrow S_{\mathrm{P}}$ exists by Lemma 1.14. The image of such a section in $F_{\mathrm{P}} \subset F(X)$ defines a surface $\bar{S}_{\mathrm{P}}$ with the property that for all classes $\alpha \in H^2(F(X),\mathbb{Z})$ in the image of $K_8^{-\perp} \subset H^4(X,\mathbb{Z})$, so in particular for all classes in the transcendental part of $H^2(F(X),\mathbb{Z})$, one has

$$q_F(\alpha) = \int_{F(X)} [\bar{S}_{\mathrm{P}}] \cdot \alpha^2 = \int_{\bar{S}_{\mathrm{P}}} (\alpha|_{\bar{S}_{\mathrm{P}}})^2,$$

where q_F is the Beauville–Bogomolov–Fujiki pairing on $H^2(F(X),\mathbb{Z})$, see Section 3.1. Compare this result of Shen [428, Rem. 6.7] with (3.4) in Example 3.6.

Exercise 1.20 Show that $\varphi_{\mathrm{P}}(K_8^{-\perp}) \subset H^2(S_{\mathrm{P}},\mathbb{Z})_{\mathrm{pr}}$ is the kernel of the linear map

$$(2B.\ )\colon H^2(S_{\mathrm{P}},\mathbb{Z})_{\mathrm{pr}} \twoheadrightarrow \mathbb{Z}/2\mathbb{Z}, \tag{1.10}$$

where $2g_0 = 2B + g_{\mathrm{P}}$ as in Remark 1.13.

If S_{P} has Picard number one, then the kernel of (1.10) is the transcendental lattice $T(S_{\mathrm{P}},\alpha_{\mathrm{P},X})$ of the Hodge structure $\widetilde{H}(S_{\mathrm{P}},\alpha_{\mathrm{P},X},\mathbb{Z})$ of the twisted K3 surface $(S_{\mathrm{P}},\alpha_{\mathrm{P},X})$, cf. Remark 5.20 and see [249, Sec. 16.4.1] for the definition of the twisted Hodge structure. In particular, in this case $\alpha_{\mathrm{P},X} \neq 1$.

Exercise 1.21 Under the above assumptions, prove that the Fano correspondence φ_{P} induces an isomorphism of rational Chow motives

$$\mathfrak{h}^4(X)_{\mathrm{tr}} \simeq \mathfrak{h}^2(S_{\mathrm{P}})_{\mathrm{tr}}(1).$$

Remark 1.22 Let $\iota\colon S_P \xrightarrow{\sim} S_P$ denote the covering involution of the double cover $\pi\colon S_P \longrightarrow \mathbb{P}^2$. There then exists a fibre product diagram

$$\begin{array}{ccc} \mathbb{L}_{\tilde{F}_P} & \xrightarrow{q'_P} & \tilde{F}_P \\ {\scriptstyle q_P}\downarrow & & \downarrow{\scriptstyle \iota\circ\tilde{\pi}} \\ \tilde{F}_P & \xrightarrow{\tilde{\pi}} & S_P, \end{array}$$

where q'_P sends a point $x \in L$, $[L] \in \tilde{F}_P$, in the quadric $Q_y \simeq \mathbb{P}^1 \times \mathbb{P}^1$, $y = \phi(x)$, to the fibre through x of the other projection.

1.4 Nodal Cubic Fourfolds: Blow-up and Lattice Theory Another special class of cubic fourfolds is provided by *nodal cubic fourfolds*. They share many features of nodal cubic threefolds, see Section **5**.5.1, and were already discussed by Hassett [227]. Despite them being (mildly) singular, they show various features, in particular in their relation to K3 surfaces, that are similar to those observed for certain families of smooth cubic fourfolds. For some comments on the stability of nodal cubic fourfolds, see Section 6.7.

Assume $X \subset \mathbb{P}^5$ is a cubic hypersurface with an ordinary double point $x_0 \in X$ as its only singularity. As a consequence of the general discussion in Section **1**.5.4, we have the following picture

$$\begin{array}{ccccccc} E & \hookrightarrow & \mathrm{Bl}_{x_0}(X) & \xrightarrow{\sim} & \mathrm{Bl}_S(\mathbb{P}^4) & \xleftarrow{j} & E' \\ \downarrow & & \downarrow{\scriptstyle\tau} & & {\scriptstyle\phi}\downarrow & & \downarrow{\scriptstyle p} \\ \{x_0\} & \hookrightarrow & X & & \mathbb{P}^4 & \hookleftarrow & S. \end{array}$$

Here, $S \subset \mathbb{P}^4$ is the smooth complete intersection of a quadric and a cubic, so a K3 surface of degree six. Its normal bundle is $\mathcal{N}_{S/\mathbb{P}^4} \simeq \mathcal{O}_S(2) \oplus \mathcal{O}_S(3)$ and, therefore, the exceptional divisor of the blow-up ϕ is $E' \simeq \mathbb{P}(\mathcal{O}_S(2)\oplus\mathcal{O}_S(3))$. Furthermore, the surface S can be understood as the Fano variety of all lines in X passing through the ordinary double point $x_0 \in X$

$$S = \{\, \ell \mid x_0 \in \ell \,\} \subset F(X)$$

and $p\colon E' \longrightarrow S$ is the universal family over it. The image of $\tau\colon E' \longrightarrow X$ is the union of all lines in X that pass through x_0. It contracts the intersection $E' \cap E$ to x_0 and is an embedding elsewhere.

Under the identification $\mathbb{P}^4 \simeq \mathbb{P}(T_{x_0}\mathbb{P}^5)$, the morphism $\phi|_E\colon E \hookrightarrow \mathbb{P}^4$ is nothing but the natural embedding of the exceptional divisor $E = \tau^{-1}(x_0)$ as the quadric in $\mathbb{P}^4 \simeq \mathbb{P}(T_{x_0}\mathbb{P}^5)$. In fact, in the presentation of $S = V(t_1) \cap V(t_2)$ as the intersection of a cubic and a quadric, see Section **1**.5.4, $V(t_1)$ should be thought of as $X \cap \mathbb{P}^4$ and $V(t_2)$ as $E \subset \mathbb{P}^4 \simeq \mathbb{P}(T_{x_0}\mathbb{P}^5)$. Hence, $\phi\colon E \cap E' \xrightarrow{\sim} S$.

Example 1.23 The procedure can be reversed. Assume $S \subset \mathbb{P}^4$ is a smooth complete intersection of type $(3,2)$. So, we can write S as an intersection $S = Y \cap Q$ of a cubic Y and a quadric Q. However, the cubic $Y \subset \mathbb{P}^4$ is not unique. A computation reveals that $h^0(\mathbb{P}^4, \mathcal{I}_S(3)) = 6$ which yields a five-dimensional space $\mathbb{P}^5 \simeq |\mathcal{I}_S(3)|$ of cubics containing the sextic K3 surface S. Note that once Y is chosen, the quadric Q is unique. More explicitly, after choosing one representation $S = V(F) \cap V(G)$ with $\deg(F) = 3$ and $\deg(G) = 2$, the equations for all other cubics containing S are of the form $a\,F + \ell \cdot G$ with $\ell \in H^0(\mathbb{P}^4, \mathcal{O}(1))$.

In this situation a nodal cubic fourfold $X \subset \mathbb{P}^5$ occurs naturally, namely the one defined by the equation $F(x_0, \ldots, x_4) + x_5 \cdot G(x_0, \ldots, x_4)$, see Exercise **1.5.15**. It contains the various cubic threefolds Y containing S as the hyperplane sections $x_5 = \ell(x_0, \ldots, x_4)$.

The morphism ϕ is induced by the complete linear system $|\tau^* \mathcal{O}_X(1) \otimes \mathcal{O}(-E)|$, while τ is induced by the complete linear system $|\phi^* \mathcal{O}_{\mathbb{P}^4}(3) \otimes \mathcal{O}(-E')|$. Hence,

$$\phi^* \mathcal{O}_{\mathbb{P}^4}(1) \simeq \tau^* \mathcal{O}_X(1) \otimes \mathcal{O}(-E) \quad \text{and} \quad \tau^* \mathcal{O}_X(1) \simeq \phi^* \mathcal{O}_{\mathbb{P}^4}(3) \otimes \mathcal{O}(-E'), \tag{1.11}$$

which in turns gives $\mathcal{O}(E + E') \simeq \phi^* \mathcal{O}_{\mathbb{P}^4}(2)$ and $\mathcal{O}(3E + E') \simeq \tau^* \mathcal{O}_X(2)$.

The Hodge structure of the smooth blow-up $\mathrm{Bl}_{x_0}(X)$ can be described via the isomorphism $\mathrm{Bl}_{x_0}(X) \simeq \mathrm{Bl}_S(\mathbb{P}^4)$ as

$$H^4(\mathrm{Bl}_{x_0}(X), \mathbb{Z}) \simeq H^4(\mathbb{P}^4, \mathbb{Z}) \oplus H^2(S, \mathbb{Z})(-1), \tag{1.12}$$

where the second summand is embedded via $j_* \circ p^* \colon H^2(S, \mathbb{Z}) \hookrightarrow H^4(\mathrm{Bl}_S(\mathbb{P}^4), \mathbb{Z})$, cf. [474, Ch. 7]. Alternatively, the projection onto the second summand

$$\varphi_{x_0} \colon H^4(\mathrm{Bl}_{x_0}(X), \mathbb{Z}) \twoheadrightarrow H^2(S, \mathbb{Z})(-1)$$

can be viewed as being induced by the Fano correspondence $S \xleftarrow{p} E' \xrightarrow{\tau} X$ of all lines through $x_0 \in X$.

A standard computation reveals that the square of a class $\alpha \in H^4(\mathrm{Bl}_{x_0}(X), \mathbb{Z})$ written as $\alpha = \alpha' + j_*(p^*(\beta))$ with $\alpha' \in H^4(\mathbb{P}^4, \mathbb{Z})$ and $\beta \in H^2(S, \mathbb{Z})$ can be computed as

$$(\alpha.\alpha) = (\alpha'.\alpha')_{\mathbb{P}^4} - (\beta.\beta)_S,$$

where, of course, $H^4(\mathbb{P}^4, \mathbb{Z}) \simeq \mathbb{Z} \cdot h_{\mathbb{P}^4}^2$ with $(h_{\mathbb{P}^4}^2.h_{\mathbb{P}^4}^2) = 1$. The second isomorphism in (1.11) together with the equality

$$[E']^2 = -[S] + 5 \cdot h_S = -6 \cdot h_{\mathbb{P}^4}^2 + 5 \cdot h_S \in H^4(\mathbb{P}^4, \mathbb{Z}) \oplus H^2(S, \mathbb{Z})(-1),$$

which uses $-p_*(\mathrm{c}_1(\mathcal{O}_p(1))^2) = \mathrm{c}_1(\mathcal{N}_{S/\mathbb{P}^4}) = 5 \cdot h_S$, give $\tau^* h_X^2 = 9 \cdot h_{\mathbb{P}^4}^2 - 6 \cdot (h_{\mathbb{P}^4} \cdot [E']) + [E']^2 = 9 \cdot h_{\mathbb{P}^4}^2 - 6 \cdot j_*(p^*(h_S)) - 6 \cdot h_{\mathbb{P}^4}^2 + 5 \cdot j_*(p^* h_S) = 3 \cdot h_{\mathbb{P}^4}^2 - j_*(p^*(h_S))$, i.e.

$$\tau^* h_X^2 = 3 \cdot h_{\mathbb{P}^4}^2 - h_S \in H^4(\mathbb{P}^4, \mathbb{Z}) \oplus H^2(S, \mathbb{Z})(-1). \tag{1.13}$$

Note that the square of the class on the right-hand side is indeed $9 - (h_S.h_S) = 3 = (h_X^2.h_X^2)$.

Remark 1.24 Analogously to Remark 5.5.1, the description of the cohomology of the blow-up leads directly to the following global Torelli theorem for nodal cubic fourfolds. Assume $X, X' \subset \mathbb{P}^5$ are two cubic fourfolds each with an ordinary double point $x \in X$ and $x' \in X'$ as the only singularity. Then the following conditions are equivalent:

(i) $X \simeq X'$ as (singular) complex varieties.
(ii) There exists a Hodge isometry $\zeta\colon H^4(\mathrm{Bl}_x(X), \mathbb{Z}) \simeq H^4(\mathrm{Bl}_{x'}(X'), \mathbb{Z})$ with $\zeta(h_X^2) = h_{X'}^2$ and $\zeta(E_X^2) = E_{X'}^2$.

Indeed, any Hodge isometry as in (ii) defines a Hodge isometry $H^2(S, \mathbb{Z}) \simeq H^2(S', \mathbb{Z})$, $h_S \longmapsto h_{S'}$, which, by the global Torelli theorem for K3 surfaces, is then induced by a polarized isomorphism $(S, h_S) \simeq (S', h_{S'})$, cf. Section 6.3 and [249, Ch. 7.2]. As for both K3 surfaces, the inclusion $S, S' \subset \mathbb{P}^4$ is induced by the complete linear system defined by h_S and $h_{S'}$, the isomorphism extends to an isomorphism of the ambient projective spaces, of their blow-ups and, eventually, to $X \simeq X'$. See Theorem 3.17 for a global Torelli statement for smooth cubic fourfolds.

Note that in dimension four there is no analogue of (iii) in Remark 5.5.1, as there is no geometric object naturally associated with the Hodge structure of weight two that would replace the passage from Y to $J(Y)$ for cubics of dimension three.

Consider the sublattice of rank two of $H^{2,2}(\mathrm{Bl}_{x_0}(X), \mathbb{Z})$ spanned by the two Hodge classes $h_{\mathbb{P}^4}^2$ and h_S. The intersection form with respect to the two bases $h_{\mathbb{P}^4}^2, h_S$, respectively, $\tau^* h_X^2 = 3 \cdot h_{\mathbb{P}^4}^2 - h_S, 2 \cdot h_{\mathbb{P}^4}^2 - h_S$, see (1.13), is described by the matrices

$$\begin{pmatrix} 1 & 0 \\ 0 & -6 \end{pmatrix} \text{ and } \begin{pmatrix} 3 & 0 \\ 0 & -2 \end{pmatrix}. \tag{1.14}$$

We will denote this lattice by

$$K_6^- := \langle h_{\mathbb{P}^4}, h_S \rangle \subset H^4(\mathrm{Bl}_{x_0}(X), \mathbb{Z}),$$

in analogy to K_8^- for smooth cubic fourfolds containing a plane. Note, however, that K_8^- was positive definite while the matrices in (1.14) are indefinite. In fact, the real K_6^-, as in Section 5.2, would be positive definite with a diagonal intersection matrix diag(3, 2).[3]

The analogue of Proposition 1.18 is then the following result [227].

Proposition 1.25 (Hassett) *Let $X \subset \mathbb{P}^5$ be a cubic fourfold with an ordinary double point $x_0 \in X$ as its only singularity. Then, up to a global sign, the Fano correspondence of all lines through x_0 induces a Hodge isometry*

$$\varphi_{x_0} : K_6^{-\perp} \xrightarrow[-1]{\sim} H^2(S, \mathbb{Z})_{\mathrm{pr}}(-1).$$

[3] A similar sign issue occurs under the Fano correspondence $H^4(X, \mathbb{Z}) \longrightarrow H^2(F(X), \mathbb{Z})$ for smooth cubic fourfolds.

Proof As explained before, $\varphi_{x_0}\colon H^4(\mathrm{Bl}_{x_0}(X),\mathbb{Z})\longrightarrow H^2(S,\mathbb{Z})(-1)$ is nothing but the projection onto the second summand in (1.12), which is enough to conclude. □

In Remark 3.27 we will see that general K3 surfaces of degree $d=6$ can indeed only be associated with singular cubic fourfolds.

Continuing Example 1.23, we emphasize that for a generic polarized K3 surface (S,L) of degree six, so a $(2,3)$ complete intersection in $\mathbb{P}^4$, the incomplete linear system $|\mathcal{O}_{\mathbb{P}^4}(3)\otimes\mathcal{I}_S|$ defines a cubic fourfold with an ordinary double point $x_0\in X$ as its only singularity. For the very general such (S,L), the primitive cohomology $H^2(S,\mathbb{Z})_{\mathrm{pr}}(-1)$ then describes the transcendental part of $H^4(\mathrm{Bl}_{x_0}(X),\mathbb{Z})$.

Exercise 1.26 Prove that taking a generic hyperplane section $Y=H\cap X$ through the node $x_0\in X$ transforms $\mathrm{Bl}_{x_0}(X)\simeq\mathrm{Bl}_S(\mathbb{P}^4)$ to $\mathrm{Bl}_{x_0}(Y)\simeq\mathrm{Bl}_C(\mathbb{P}^3)$, where $C\subset\mathbb{P}^3$ is a complete intersection of type $(2,3)$, see Section **5**.5.1.

Remark 1.27 In Section 5.2 we will discuss the relation between certain families of smooth cubic fourfolds $X\subset\mathbb{P}^5$ and families of polarized K3 surfaces (S,L) of particular degrees d. We will see that for $d\equiv 0\,(6)$, there is an ambiguity in the choice of (S,L) for a given X. More precisely, generically there are two possibilities to associate a polarized K3 surface with a given X. However, for $d=6$, the situation is different, for the purely lattice theoretic reason explained in Remark 6.10. There is indeed a distinguished K3 surface associated with a nodal cubic fourfolds.[4]

1.5 Nodal Cubic Fourfolds: Fano Variety Assume that the nodal cubic $X\subset\mathbb{P}^5$ and its associated sextic K3 surface $S\subset\mathbb{P}^4$ are generic in the sense that X does not contain any planes $\mathbb{P}^2\subset\mathbb{P}^5$ and S does not contain any lines $\mathbb{P}^1\subset\mathbb{P}^4$. Then any subscheme of length two $Z=\{s_1,s_2\}\in S^{[2]}$ defines a line $\ell_Z\subset\mathbb{P}^4$ with $Z\subset S\cap\ell_Z$ that is not contained in S. See Example 3.3 for background on the Hilbert scheme $S^{[2]}$. Note that a non-reduced Z consists of a point $s\in S$ and a tangent direction $v\in T_sS$, which still defines a line ℓ_Z.

First we observe that either:

(i) Z is the scheme-theoretic intersection $S\cap\ell_Z$; or

(ii) $\ell_Z\subset E$ and $S\cap\ell_Z$ is a scheme of length three.

Indeed, if the length of the intersection is at least three, then ℓ_Z is contained in the quadric $E\subset\mathbb{P}(T_{x_0}\mathbb{P}^5)\simeq\mathbb{P}^4$ and $S\in|\mathcal{O}_E(3)|$ shows $(S.\ell_Z)=3$, where we use again that S does not contain any lines.

We denote by

$$T:=\{Z\in S^{[2]}\mid |S\cap\ell_Z|=3\}$$

the closed subset of subschemes Z of type (ii).

[4] I wish to thank E. Brakkee for a discussion related to this point.

Then for $Z = \{s_1, s_2\} \in S^{[2]} \setminus T$, the proper transform $\phi^{-1}(\ell_Z)$ is of the form

$$\phi^{-1}(\ell_Z) = E'_{s_1} \cup E'_{s_2} \cup \ell_Z,$$

where $E'_{s_i} := p^{-1}(s_i) \subset X$ are the lines through $x_0 \in X$ corresponding to $s_i \in S$, $i = 1, 2$, and, by abuse of notation, $\ell_Z \subset \mathrm{Bl}_{x_0}(X) \simeq \mathrm{Bl}_S(\mathbb{P}^4)$ also denotes the strict transform of $\ell_Z \subset \mathbb{P}^4$. Using $\tau^*\mathcal{O}_X(1) \simeq \phi^*\mathcal{O}_{\mathbb{P}^4}(3) \otimes \mathcal{O}(-E')$ in (1.11), the fact that $(E'.\phi^{-1}(\ell_Z)) = 0$ (as the proper transform of the generic line in $\mathbb{P}^4$ does not intersect E'), and $(\tau^*\mathcal{O}_X(1).E'_{s_i}) = 1$, one deduces $(\tau^*\mathcal{O}_X(1).\ell_Z) = 1$. Hence,

$$L_Z := \tau(\ell_Z) \subset X$$

is a line in $X \subset \mathbb{P}^5$. Alternatively, one may think of $L_Z \subset X$ as the residual line of $E'_{s_1} \cup E'_{s_2} \subset P_Z \cap X$, where $P_Z \subset \mathbb{P}^5$ is the plane spanned by the two intersecting lines $\tau(E'_{s_i}) \subset X$, $i = 1, 2$. We leave it to the reader to verify that all this makes sense also when Z consists of a point $s_1 = s_2$ together with a tangent direction. Also note that the strict transform of a line ℓ_Z contained in E would not give a line in X, because it gets contracted to $x_0 \in X$. We have seen a similar construction for nodal cubic threefolds already in Remark **5**.5.2.

Altogether, this defines a morphism $S^{[2]} \setminus T \longrightarrow F(X)$, $Z \longmapsto L_Z$. Conversely, if $L \subset X$ is a line not containing x_0, then L intersects $\tau(E') = \bigcup_{x_0 \in L'} L'$ in two points (counted with multiplicities) contained in two lines L_1 and L_2. Indeed, if L intersected three lines through x_0, then L and the three lines would all be contained in one plane necessarily contained in X, which is excluded by our genericity assumption on X. Therefore, the proper transform of L defines a line $\ell \subset \mathbb{P}^4$ intersecting S in the two points $s_1, s_2 \in S$ with $E'_{s_i} = L_i$. As a consequence, one finds an isomorphism

$$S^{[2]} \setminus T \simeq F(X) \setminus \{ L \mid x_0 \in L \}.$$

To push this a bit further, one observes that the rational map $Z \longmapsto \ell_Z$ extends to a morphism which on the exceptional set $T \subset S^{[2]}$ is given by $Z = \{s_1, s_2\} \longmapsto E'_{s_3}$, where $S \cap \ell_Z = \{s_1, s_2, s_3\}$. The resulting morphism $T \twoheadrightarrow S \hookrightarrow F(X)$ is a Zariski locally trivial $\mathbb{P}^1$-bundle over the image $S \subset F(X)$, because the lines through any fixed point of the quadric E form a $\mathbb{P}^1$ and $\mathbb{P}^1$-bundles over E are Zariski locally trivial.

The discussion essentially proves the following results, we refer to [227, Sec. 6.3] for the missing details. A less precise result but for cubic hypersurfaces with arbitrary isolated singularities has been proved by Lehn [318, Thm. 3.6].

Proposition 1.28 (Hassett) *Let $X \subset \mathbb{P}^5$ be a generic cubic fourfold with an ordinary double point $x_0 \in X$ as its only singularity. Then there exists an isomorphism*

$$S^{[2]} \simeq \mathrm{Bl}_S(F(X)).$$

Here, $S \subset \mathbb{P}^4$ is the associated K3 surface of degree six which viewed as $S \subset F(X)$ parametrizes all lines passing through x_0. □

Remark 1.29 (i) Consider a smooth hyperplane section $Y := X \cap \mathbb{P}^4$ and view S as the complete intersection $S = Y \cap Q$ of the cubic $Y \subset \mathbb{P}^4$ and a uniquely defined quadric $Q \subset \mathbb{P}^4$, see Example 1.23. Then the natural inclusion $F(Y) \hookrightarrow F(X)$, which by Lemma 4.5 describes a Lagrangian subvariety, can be reinterpreted as the inclusion

$$F(Y) \hookrightarrow S^{[2]}$$

that maps a line $\ell \subset Y \subset \mathbb{P}^4$ to the intersection $[\ell \cap Q] \in S^{[2]}$, see [57, Prop. 3].

Note that $F(Y) \subset S^{[2]} \simeq \mathrm{Bl}_S(F(X))$ avoids the exceptional locus of the contraction $S^{[2]} \longrightarrow F(X)$.

(ii) The moduli space $M(v)$ of semi-stable sheaves on S with Mukai vector $v = (2, 0, -2)$ is singular of dimension 10. Beauville [51, Prop. 8.4] showed that every bundle E in a dense open subset of $M(v)$ can be written as the cokernel of an alternating linear map $A\colon \mathcal{O}_Q(-2)^{\oplus 6} \longrightarrow \mathcal{O}_Q(-1)^{\oplus 6}$, cf. Section **4**.2.5. Associating with E the cubic $Y \subset \mathbb{P}^4$ determined by the Pfaffian of A, leads to a rational map

$$M(v) \dashrightarrow |\mathcal{I}_S(3)| \simeq \mathbb{P}^5, \tag{1.15}$$

which in fact is a rational Lagrangian fibration. The generic fibre over $Y \in |\mathcal{I}_S(3)|$ is the moduli space of sheaves of rank two with $c_1 = 0$ and $c_2 = (2/3) \cdot h^2$ on Y which is isomorphic to the blow-up $\tilde{J}(Y)$ of $J(Y)$ in $F(Y) \subset J(Y)$, see Remark **5**.3.6. The question raised by Beauville whether (1.15) can be compactified to a Lagrangian fibration of a hyperkähler manifold was answered affirmatively by Saccà [411, Rem. 1.18].

Note that the boundary divisor $\{Z \mid |\mathrm{supp}(Z)| = 1\} \subset S^{[2]}$, i.e. the exceptional divisor of the Hilbert–Chow morphism $S^{[2]} \longrightarrow S^{(2)}$ (also a $\mathbb{P}^1$-bundle over S), is not the $\mathbb{P}^1$-bundle contracted under $S^{[2]} \xrightarrow{\sim} \mathrm{Bl}_S(F(X)) \twoheadrightarrow F(X)$. Nevertheless, one has the following result which should be compared with the case of cubic threefolds, see Remark **5**.5.2.

Corollary 1.30 *Assume $X \subset \mathbb{P}^5$ is a cubic fourfold with an ordinary double point $x_0 \in X$ as its only singularity and let $S \subset \mathbb{P}^4$ be the associated K3 surface. Then the following equations hold in $K_0(\mathrm{Var}_k)$:*

$$[S^{[2]}] = [\mathrm{Bl}_S(F(X))] \quad \textit{and} \quad [S^{(2)}] = [F(X)].$$

Proof For generic X, the first equality is a consequence of the proposition. For the second, write

$$\begin{aligned}[S^{(2)}] - [S] &= [S^{[2]}] - [S] \cdot [\mathbb{P}^1] = [\mathrm{Bl}_S(F(X))] - [S] \cdot [\mathbb{P}^1] \\ &= [F(X)] + \ell \cdot [S] - [S] \cdot [\mathbb{P}^1] = [F(X)] - [S].\end{aligned}$$

At least morally, but not literally, the result for arbitrary nodal X can be obtained by specialization from the isomorphism for generic X and S in the sense of [371, 285]. Rigorously, specialization leads to a birational correspondence between the two hyperkähler manifolds $\mathrm{Bl}_S(F(X))$ and $S^{[2]}$ and those, in dimension four, are always described by Mukai flops which preserve the class in $K_0(\mathrm{Var}_k)$. Another argument was given by Galkin and Shinder [190]. □

Remark 1.31 Hassett's original paper [227] contains the discussion of another singular cubic fourfold, the *secant variety* $X_0 := \mathrm{Sec}(v_2(\mathbb{P}^2))$ of the Veronese surface $v_2(\mathbb{P}^2) \subset \mathbb{P}^5$, cf. [205, p. 178]. This is a singular cubic hypersurface, the equation F_0 of which is the determinant of

$$\begin{pmatrix} x_0 & x_1 & x_2 \\ x_1 & x_3 & x_4 \\ x_2 & x_4 & x_5 \end{pmatrix},$$

and the singular locus of X_0 is $v_2(\mathbb{P}^2)$. Note that, in this case, there is no K3 surface naturally attached to X_0.

However, K3 surfaces do play a role here as well, but only after the additional choice of a generic cubic equation F. The pre-image of the intersection $v_2(\mathbb{P}^2) \cap V(F)$ under the Veronese embedding $v_2 \colon \mathbb{P}^2 \hookrightarrow \mathbb{P}^5$ is a smooth sextic curve $C_F \subset \mathbb{P}^2$. The double cover $S_F \twoheadrightarrow \mathbb{P}^2$ branched over $C_F \subset \mathbb{P}^2$ is a K3 surface of degree two. It turns out that there is a close relationship between $H^2(S_F, \mathbb{Z})_{\mathrm{pr}}(-1)$ and the limiting mixed Hodge structure of the pencil of cubics fourfolds $V(F_0 + t \cdot F)$ at the fibre X_0, see [227, Thm. 4.4.1]. As an abstract lattice, $H^2(S_F, \mathbb{Z})_{\mathrm{pr}}(-1)$ is isomorphic to a lattice $K_2^\perp$ to be discussed in Section 5.2. In Remark 3.28 we will see that the lattice can indeed not be realized by any smooth cubic fourfold.

1.6 Normal Scrolls and Veronese Surfaces Without going into any details, we briefly mention smooth cubic fourfolds containing rational normal scrolls or the Veronese surface. Recall that a non-degenerate smooth projective surface of degree e in $\mathbb{P}^{e+1}$ is either a rational normal scroll or the image $v_2(\mathbb{P}^2)$ of the Veronese embedding $v_2 \colon \mathbb{P}^2 \hookrightarrow \mathbb{P}^5$, see also the discussion around Lemma 2.20. Cubic fourfolds containing one of those surfaces have been studied early on, see e.g. [227].

• The case $e = 2$ of a smooth quadric $Q \subset \mathbb{P}^3$ contained in a cubic fourfold corresponds to the special case of cubic fourfolds containing a plane $\mathrm{P} \simeq \mathbb{P}^2 \subset \mathbb{P}^3 \subset \mathbb{P}^5$, which was treated in Section 1.1. Indeed, any quadric $Q \subset \mathbb{P}^3$ contained in X determines a residual plane $\mathbb{P}^2 \simeq \mathrm{P} \subset X$, i.e. $\mathbb{P}^3 \cap X = Q \cup \mathrm{P}$, and vice versa. We have seen that the set of smooth cubic fourfolds $X \subset \mathbb{P}^5$ containing a plane, or equivalently a quadric, is an irreducible divisor in $|\mathcal{O}_{\mathbb{P}^5}(3)|_{\mathrm{sm}}$ and, similarly, in the moduli space M_4 of smooth cubic fourfolds, see Remark 1.3. The lattice spanned by h^2 and [P] in $H^{2,2}(X,\mathbb{Z})$ turned out to be the rank two lattice K_8^- given by the intersection matrix, see Lemma 1.1:

$$\begin{pmatrix} 3 & 1 \\ 1 & 3 \end{pmatrix}.$$

• The case $e = 4$ will be described in Section 2, where we explain that a generic Pfaffian cubic fourfold X_V contains a *quartic rational normal scroll* Σ_P; in fact, a family of such surfaces Σ_P depending on a point $P \in S_V$ in the associated K3 surface S_V. From a different angle, one could start with a quartic rational normal scroll $\Sigma \subset \mathbb{P}^5$ and consider all smooth cubic fourfolds $X \subset \mathbb{P}^5$ containing it. Once again, the condition describes an irreducible divisors in $|\mathcal{O}_{\mathbb{P}^5}(3)|_{\rm sm}$ and M_4. According to Corollary 2.21, h^2 and $[\Sigma_P]$ span a primitive sublattice K_{14}^- of $H^{2,2}(X,\mathbb{Z})$ given by the intersection matrix

$$\begin{pmatrix} 3 & 4 \\ 4 & 10 \end{pmatrix}.$$

Remark 1.32 For rational normal scrolls with $e = 2$ or $e = 4$, there are in each case two distinguished types of surfaces. For $e = 2$, one can either look at planes $\mathbb{P}^2 \simeq \mathrm{P} \subset X$ or at quadrics $Q \subset X$. They are residual in the sense that

$$\mathbb{P}^3 \cap X = Q \cup \mathrm{P}.$$

The residual quadrics for a given plane $\mathrm{P} \subset X$ are parametrized by $|\mathcal{I}_{\mathrm{P}}(1)| \simeq \mathbb{P}^2$, while the residual planes for a given quadric $Q \subset X$ are parametrized by $|\mathcal{I}_Q(1)| \simeq \mathbb{P}^1$.

Similarly, for $e = 4$, there are quartic rational normal scrolls $\Sigma_P \subset X$ and *quintic del Pezzo surfaces* $S \subset X$. They should be considered as residual, cf. Remark 2.22:

$$(\mathbb{P}^1 \times \mathbb{P}^2) \cap X = \Sigma_P \cup S.$$

The Veronese surface $\nu_2(\mathbb{P}^2) \subset \mathbb{P}^5$ also has degree $e = 4$ and, as shown by Hassett [227], the requirement on a cubic fourfold to contain the Veronese surface $\nu_2(\mathbb{P}^2)$ describes an irreducible divisor in the moduli space of smooth cubic fourfolds M_4. The intersection matrix of the lattice spanned by h^2 and $[\nu_2(\mathbb{P}^2)]$, denoted by K_{20}^-, is computed, similarly to the proof of Lemma 2.20, as

$$\begin{pmatrix} 3 & 4 \\ 4 & 12 \end{pmatrix}.$$

The case $e = 3$, also treated in [227], is the case of cubic fourfolds containing a cubic rational normal scroll. Once again, the condition describes irreducible divisors in $|\mathcal{O}_{\mathbb{P}^5}(3)|_{\rm sm}$ and M_4. The corresponding lattice in $H^{2,2}(X,\mathbb{Z})$ is denoted by K_{12}^- and described by the matrix

$$\begin{pmatrix} 3 & 3 \\ 3 & 7 \end{pmatrix}.$$

2 Pfaffian Cubic Fourfolds

Ever since the early work of Beauville and Donagi [58], Pfaffian cubic fourfolds have occupied a special place in the theory of cubic hypersurfaces. Pfaffian cubics are parametrized by a divisor in the moduli space of all cubic fourfolds and for them the link to K3 surfaces is particularly close and well understood.

2.1 Universal Pfaffian We start with a vector space W of dimension six and consider $\mathbb{P}(\bigwedge^2 W) \simeq \mathbb{P}^{14}$, which via the Plücker embedding contains the eight-dimensional Grassmann variety of planes in W:

$$\mathbb{G} := \mathbb{G}(1, \mathbb{P}(W)) \simeq G(2, W) \hookrightarrow \mathbb{P}(\textstyle\bigwedge^2 W).$$

We can also think of $\mathbb{G}$ as the subvariety of two-forms $\omega \in \bigwedge^2 W$, up to scaling, of rank two, i.e. such that the associated alternating linear map $\omega\colon W^* \longrightarrow W$ has an image of dimension two. Recall that $\dim(\mathbb{G}(1, \mathbb{P}(W))) = 8$.

But $\mathbb{P}(\bigwedge^2 W)$ contains another natural and bigger subvariety, the *Pfaffian hypersurface* $\mathrm{Pf}(W) \subset \mathbb{P}(\bigwedge^2 W)$ of all non-trivial two-forms ω that are degenerate, i.e. for which the associated linear map $\omega\colon W^* \longrightarrow W$ is not bijective or, equivalently, for which $\mathrm{Ker}(\omega)$ is either of dimension two or four. This hypersurface is of degree three, which can be seen either by describing it as the cubic

$$\mathrm{Pf}(W) = \left\{ \omega \in \mathbb{P}(\textstyle\bigwedge^2 W) \mid \omega \wedge \omega \wedge \omega = 0 \text{ in } \bigwedge^6 W \right\} \subset \mathbb{P}(\textstyle\bigwedge^2 W)$$

or, alternatively, by recalling that the determinant $\det(\omega)$ of an alternating matrix, a homogenous polynomial of degree six, has a canonical square root, the Pfaffian $\mathrm{Pf}(\omega)$, which then is a homogeneous polynomial of degree three. The two subvarieties are contained in each other:

$$\mathbb{G} = \mathbb{G}(1, \mathbb{P}(W)) \subset \mathrm{Pf}(W) \subset \mathbb{P}(\textstyle\bigwedge^2 W).$$

In fact, $\mathbb{G}$ is the singular locus of $\mathrm{Pf}(W)$ which can be proved by using the theory of degeneracy loci, see e.g. [23]. The smoothness of $\mathrm{Pf}(W) \setminus \mathbb{G}$, which is all we need, simply follows from the observation that it is homogenous under the action of $\mathrm{PGL}(W)$. Of course, the same picture exists for the dual space W^*:

$$\mathbb{G}^* := \mathbb{G}(1, \mathbb{P}(W^*)) \subset \mathrm{Pf}(W^*) \subset \mathbb{P}(\textstyle\bigwedge^2 W^*).$$

Remark 2.1 The varieties $\mathbb{G}$ and $\mathrm{Pf}(W^*)$ are dual to each other in the classical sense:

$$\mathbb{G}^\vee \simeq \mathrm{Pf}(W^*) \quad \text{and} \quad \mathbb{G}^* \simeq \mathrm{Pf}(W)^\vee.$$

To make this more precise, pick $\omega \in \mathbb{P}(\bigwedge^2 W^*)$ with kernel $\mathrm{Ker}(\omega) \subset W$ and the induced hyperplane $\omega^\perp \subset \mathbb{P}(\bigwedge^2 W)$. Now, consider a point in the hyperplane section $\mathbb{G} \cap \omega^\perp$

corresponding to a plane $P \subset W$. Then $\omega^\perp$ is tangent to $\mathbb{G}$ at $P \in \mathbb{G}$, i.e. $\mathbb{T}_P\mathbb{G} \subset \omega^\perp$, if and only if $P \subset \mathrm{Ker}(\omega)$. Hence, $\mathbb{G} \cap \omega^\perp$ is singular if and only if $\omega \in \mathrm{Pf}(W^*)$, i.e. $\mathrm{Pf}(W^*)$ is the dual variety of $\mathbb{G}$, see [88, Prop. 1.5] for the details of the elementary proof.[5]

Related to this, there is the natural correspondence $B \subset \mathbb{P}(W) \times \mathrm{Pf}(W^*)$:

$$\begin{array}{ccc} B := \{ (W_0, \omega) \mid W_0 \subset \mathrm{Ker}(\omega) \} & \xrightarrow{q} & \mathrm{Pf}(W^*) \\ {\scriptstyle p}\downarrow & & \\ \mathbb{P}(W). & & \end{array} \tag{2.1}$$

The projection p is a fibre bundle with fibre $\mathbb{P}(\bigwedge^2(W/W_0)^*)$ over the point in $\mathbb{P}(W)$ corresponding to a line $W_0 \subset W$. The projection q is a $\mathbb{P}^1$-bundle over the smooth locus $\mathrm{Pf}(W^*) \setminus \mathbb{G}^*$ of $\mathrm{Pf}(W^*)$ with fibre $\mathbb{P}(\mathrm{Ker}(\omega)) \simeq \mathbb{P}^1$ over $\omega \in \mathrm{Pf}(W^*) \setminus \mathbb{G}^*$.

Another natural correspondence is given by $\Sigma \subset \mathbb{G}(1, \mathbb{P}(W)) \times \mathrm{Pf}(W^*)$:

$$\begin{array}{ccc} \Sigma := \{(P, \omega) \mid P \cap \mathrm{Ker}(\omega) \neq 0\} & \longrightarrow & \mathrm{Pf}(W^*) \\ \downarrow & & \\ \mathbb{G}(1, \mathbb{P}(W)). & & \end{array} \tag{2.2}$$

Here, and in the following, we denote by $\omega|_P$ the restriction of a two-form $\omega \in \bigwedge^2 W^*$ on W to P, i.e. its image under the natural map $\bigwedge^2 W^* \longrightarrow \bigwedge^2 P^*$. Then for a plane $P \subset W$, the following two strict implications hold:

$$P \subset \mathrm{Ker}(\omega) \Rightarrow P \cap \mathrm{Ker}(\omega) \neq 0 \Rightarrow \omega|_P = 0.$$

Also note that the map $\omega \longmapsto \mathrm{Ker}(\omega)$ defines a section of the projection $\Sigma \longrightarrow \mathrm{Pf}(W^*)$ over the smooth locus $\mathrm{Pf}(W^*) \setminus \mathbb{G}^*$.

Remark 2.2 We collect some easy facts which will be used frequently below.

(i) Assume $P \neq Q \subset W$ are two distinct planes. Then the projective line

$$\overline{PQ} \subset \mathbb{P}(\textstyle\bigwedge^2 W)$$

through the two points $P, Q \in \mathbb{G} = \mathbb{G}(1, \mathbb{P}(W)) \subset \mathbb{P}(\bigwedge^2 W)$ is contained in $\mathbb{G}$ if and only if $\dim(P + Q) = 3$, which in turn is equivalent to $\dim(P \cap Q) = 1$ and, still equivalent, to $P \cap Q \neq 0$.

(ii) The Grassmann variety $F(\mathbb{G})$ of lines in $\mathbb{P}(\bigwedge^2 W)$ that are contained in $\mathbb{G}$ is related to the Grassmann variety $\mathbb{G}(2, \mathbb{P}(W))$ by the locally closed subset $\{(P, Q) \mid \dim(P+Q) = 3\}$ of $\mathbb{G} \times \mathbb{G}$ and the two projections $(P, Q) \longmapsto \overline{PQ} \in F(\mathbb{G})$ and $(P, Q) \longmapsto (P + Q) \in \mathbb{G}(2, \mathbb{P}(W))$. As the fibres of the two projections are of dimension two and three, one finds $\dim(F(\mathbb{G})) = 10$.

[5] Thanks to P. Belmans for the reference.

2.2 Generic Linear Sections of the Universal Pfaffian In order to make contact wiht cubic fourfolds, the Pfaffian $\mathrm{Pf}(W^*)$ has to be intersected with a five-dimensional linear projective space. At the same time, the Grassmann variety $\mathbb{G}$ intersected with a related linear space defines a K3 surface. The starting point is the following observation of Beauville and Donagi [58].

Lemma 2.3 *Consider generic linear subspaces*

$$\mathbb{P}^5 \simeq \mathbb{P}(V) \subset \mathbb{P}(\textstyle\bigwedge^2 W^*) \quad \textit{and} \quad \mathbb{P}^8 \simeq \mathbb{P}(U) \subset \mathbb{P}(\textstyle\bigwedge^2 W)$$

and let

$$X_V := \mathrm{Pf}(W^*) \cap \mathbb{P}(V) \quad \textit{and} \quad S_U := \mathbb{G}(1, \mathbb{P}(W)) \cap \mathbb{P}(U).$$

(i) *Then $X_V \subset \mathbb{P}(V)$ is a smooth cubic fourfold; and*
(ii) *S_U is a K3 surface, of degree* 14 *with respect to the polarization $\mathcal{O}(1)|_{S_U}$.*

Moreover, the fourfold X_V does not contain any plane $\mathbb{P}^2 \subset \mathbb{P}(V)$ and the surface S_U does not contain any line $\mathbb{P}^1 \subset \mathbb{P}(U)$.

Proof Assertion (i) follows from the classical Bertini theorem and the observation that the singular locus $\mathbb{G}(1, \mathbb{P}(W^*)) \subset \mathrm{Pf}(W^*)$ is of codimension five and, therefore, is not intersected by the generic linear subspace $\mathbb{P}(V)$ of codimension nine.

For (ii), use the fact that $\omega_{\mathbb{G}} \simeq \det(\mathcal{S} \otimes \mathcal{Q}^*) \simeq \mathcal{O}(-6)|_{\mathbb{G}}$, cf. Lemma **2**.3.1, and hence $\omega_{S_U} \simeq \omega_{\mathbb{G}}|_{S_U} \otimes \det(\mathcal{N}_{S_U/\mathbb{G}}) \simeq \mathcal{O}_{S_U}$, for $\mathcal{N}_{S_U/\mathbb{G}} \simeq \mathcal{O}(1)^{\oplus 6}$. The vanishing $H^1(S_U, \mathcal{O}_{S_U}) = 0$ follows from $H^1(\mathbb{G}, \mathcal{O}_{\mathbb{G}}) = 0$ and the standard Lefschetz theorem. Hence, S_U is indeed a K3 surface. To compute its degree with respect to the Plücker polarization observe that $\deg(S_U) = \deg(\mathbb{G})$ and use the classical formula $\deg(\mathbb{G}(1, \mathbb{P}^m)) = \frac{(2m-2)!}{m! \cdot (m-1)!}$, cf. [174, Prop. 4.12].

To show that S_U does not contain any lines, we consider the correspondence

$$T_1 := \{\, (L, \mathbb{P}(U)) \mid L \subset \mathbb{P}(U) \,\} \subset F(\mathbb{G}) \times \mathbb{G}(8, \mathbb{P}(\textstyle\bigwedge^2 W))$$

and recall that $\dim(F(\mathbb{G})) = 10$, see Remark 2.2. As the fibre of the first projection $T_1 \twoheadrightarrow F(\mathbb{G})$ over a line $L = \mathbb{P}(K) \in F(\mathbb{G})$ is the Grassmann variety $\mathbb{G}(6, \mathbb{P}((\bigwedge^2 W)/K))$, which is of dimension 42, one has $\dim(T_1) = 52$. Now use $\dim(\mathbb{G}(8, \mathbb{P}(\bigwedge^2 W))) = 54$ to conclude that the image of the second projection $T_1 \longrightarrow \mathbb{G}(8, \mathbb{P}(\bigwedge^2 W))$ cannot be surjective. In other words, for the generic $\mathbb{P}^8 \simeq \mathbb{P}(U) \subset \mathbb{P}(\bigwedge^2 W)$, there are no lines contained in $\mathbb{G} \cap \mathbb{P}(U) = S_U$.

For the proof of the remaining assertion, consider similarly the correspondence

$$T_2 := \{\, (\mathbb{P}^2, \mathbb{P}(V)) \mid \mathbb{P}^2 \subset \mathbb{P}(V) \,\} \subset F(\mathrm{Pf}(W^*)_{\mathrm{sm}}, 2) \times \mathbb{G}(5, \mathbb{P}(\textstyle\bigwedge^2 W^*))\,.$$

Here, $F := F(\mathrm{Pf}(W^*)_{\mathrm{sm}}, 2)$ is the Fano variety of planes $\mathbb{P}^2 \subset \mathbb{P}(\bigwedge^2 W^*)$ contained in the smooth locus $\mathrm{Pf}(W^*)_{\mathrm{sm}} = \mathrm{Pf}(W^*) \setminus \mathbb{G}(1, \mathbb{P}(W^*))$ of the cubic hypersurface $\mathrm{Pf}(W^*) \subset$

$\mathbb{P}(\bigwedge^2 W^*) \simeq \mathbb{P}^{14}$. According to Corollary **2**.1.5, the Fano variety of planes $\mathbb{P}^2 \subset \mathbb{P}^{14}$ contained in the generic cubic hypersurface in $\mathbb{P}^{14}$ is of dimension 26, but a priori there is no reason that the Pfaffian cubic has this property. However, it was shown by Manivel and Mezzetti [336, Cor. 5] that $F(\mathrm{Pf}(W^*)_{\mathrm{sm}}, 2)$ is indeed smooth of dimension 26 (with four irreducible components). The fibre of the projection $T_2 \twoheadrightarrow F$ over the point corresponding to a plane $\mathbb{P}^2 = \mathbb{P}(N) \subset \mathrm{Pf}(W^*)$ is the Grassmann variety $\mathbb{G}(2, \mathbb{P}((\bigwedge^2 W^*)/N))$, which has dimension 27. Hence, $\dim(T_2) = 53$ and, as $\dim(\mathbb{G}(5, \mathbb{P}(\bigwedge^2 W^*))) = 54$, one can again conclude that the second projection $T_2 \longrightarrow \mathbb{G}(5, \mathbb{P}(\bigwedge^2 W^*))$ is not surjective.

In Remark 2.24 we will give another argument for the last part relying on Hodge theory instead of the explicit description of $F(\mathrm{Pf}(W^*)_{\mathrm{sm}}, 2)$ provided by [336]. □

Definition 2.4 A smooth cubic fourfold $X \subset \mathbb{P}^5$ is a *Pfaffian cubic fourfold* if it is isomorphic to a cubic fourfold of the form X_V.

Remark 2.5 The universal map $\mathcal{O}(-1) \longrightarrow \bigwedge^2 W^* \otimes \mathcal{O}$ on $\mathbb{P}(\bigwedge^2 W^*)$ gives rise to an injective sheaf homomorphism $W \otimes \mathcal{O}(-1) \hookrightarrow W^* \otimes \mathcal{O}$. Restricted to a generic $\mathbb{P} := \mathbb{P}(V) \subset \mathbb{P}(\bigwedge^2 W^*)$ as in the above lemma, it leads to a short exact sequence on $\mathbb{P}$:

$$0 \longrightarrow W \otimes \mathcal{O}_{\mathbb{P}}(-1) \longrightarrow W^* \otimes \mathcal{O}_{\mathbb{P}} \longrightarrow i_*\mathcal{F} \longrightarrow 0. \tag{2.3}$$

Here, $\mathcal{F}$ is a locally free sheaf of rank two on the Pfaffian cubic $X := X_V$ and $i \colon X \hookrightarrow \mathbb{P}$ is the inclusion. By construction, $\mathcal{F}$ is globally generated. Hence, the zero set of a generic section $s \in H^0(X, \mathcal{F})$ is a smooth surface $S \subset X$. As shown by Beauville [50], S is a quintic del Pezzo surface.

To prove the last statement, one first computes the canonical bundle of the surface $S \subset X$ as $\omega_S \simeq (\omega_X \otimes \det(\mathcal{F}))|_S \simeq \mathcal{O}_S(-3+2) \simeq \mathcal{O}_S(-1)$, where the determinant $\det(\mathcal{F})$ on X is determined by dualizing (2.3) to give an alternating isomorphism $i_*\mathcal{F} \simeq \mathcal{E}xt^1(i_*\mathcal{F}, \mathcal{O}_{\mathbb{P}})(-1) \simeq i_*\mathcal{H}om(\mathcal{F}, \mathcal{O}_X) \otimes \mathcal{O}_{\mathbb{P}}(3-1)$ (by Grothendieck–Verdier duality, cf. [246, Thm. 3.34]), i.e. $\bigwedge^2 \mathcal{F} \simeq \det(\mathcal{F}) \simeq \mathcal{O}_X(2)$. To compute the degree of S, intersect it with a generic $\mathbb{P}^3 \subset \mathbb{P}$ and use the two short exact sequences

$$\mathcal{O}_{\mathbb{P}^3 \cap X_V} \xrightarrow{\ s\ } \mathcal{F}|_{\mathbb{P}^3 \cap X_V} \longrightarrow \mathcal{I}_{\mathbb{P}^3 \cap S}(2) \ \text{ and } \ \mathcal{I}_{\mathbb{P}^3 \cap S}(2) \longrightarrow \mathcal{O}_{\mathbb{P}^3}(2) \longrightarrow \mathcal{O}_{\mathbb{P}^3 \cap S}$$

to deduce

$$\begin{aligned} \deg(S) &= \chi(\mathcal{O}_{\mathbb{P}^3 \cap S}) = \chi(\mathcal{O}_{\mathbb{P}^3}(2)) - \chi(\mathcal{F}|_{\mathbb{P}^3 \cap X_V}) + \chi(\mathcal{O}_{\mathbb{P}^3 \cap X_V}) \\ &= \chi(\mathcal{O}_{\mathbb{P}^3}(2)) - \chi(W^* \otimes \mathcal{O}_{\mathbb{P}^3}) + \chi(\mathcal{O}_{\mathbb{P}^3 \cap X_V}) = 10 - 6 + 1 = 5. \end{aligned}$$

In fact, as shown by Beauville [50, Prop. 9.2], a smooth cubic fourfold $X \subset \mathbb{P}^5$ is Pfaffian if and only if it contains a quintic del Pezzo surface, see also [454].

Remark 2.6 Unlike the case of cubics of dimension two and three, see Sections **4**.2.5 and **5**.5.4, not every cubic fourfold is Pfaffian. In fact, a naive dimension count reveals that the space of Pfaffian cubics up to isomorphisms is of dimension

$$\dim(\mathbb{G}(5,\mathbb{P}(\textstyle\bigwedge^2 W^*))) - \dim(\mathrm{Aut}(\mathrm{Pf}(W^*))) = 54 - 35 = 19,$$

while the moduli space of all smooth cubic fourfolds is of dimension 20, see Section **1**.2.1.

A similar argument proves that the space of K3 surfaces constructed as S_U up to isomorphisms is of the maximal dimension

$$\dim(\mathbb{G}(8,\mathbb{P}(\textstyle\bigwedge^2 W))) - \dim(\mathrm{Aut}(\mathbb{G}(1,\mathbb{P}(W)))) = 54 - 35 = 19.$$

In both dimension counts one uses that automorphisms of $\mathrm{Pf}(W^*)$ and $\mathbb{G}(1,\mathbb{P}(W))$ extend uniquely to automorphisms of their ambient projective spaces $\mathbb{P}(\bigwedge^2 W^*)$ respectively $\mathbb{P}(\bigwedge^2 W)$ that are ultimately induced by automorphisms of W and W^*, cf. [220, Thm. 10.19].

2.3 Grassmannian Embeddings of X, $F(X)$, and $S^{[2]}$ In the following we will consider X_V and S_U for $U := V^\perp := \mathrm{Ker}(\bigwedge^2 W \twoheadrightarrow V^*)$. In this case, we shall also write S_V for the latter, which is then explicitly described as

$$S_V = \{P \in \mathbb{G} \mid \omega|_P = 0 \text{ for all } \omega \in V\}.$$

The next result provides closed embeddings into appropriate Grassmann varieties for all three four-dimensional varieties in the picture: the Pfaffian cubic X_V, the Hilbert scheme $S_V^{[2]}$ of subschemes of length two of the K3 surface S_V, and the Fano variety $F(X_V)$ of lines in X_V.

Corollary 2.7 *Let $\mathbb{P}^5 \simeq \mathbb{P}(V) \subset \mathbb{P}(\bigwedge^2 W^*)$ be a generic linear subspace. Then the maps*

$$\begin{aligned} i_{S^{[2]}}\colon\ & S_V^{[2]} \hookrightarrow \mathbb{G}(3,\mathbb{P}(W)),\ \{P,Q\} \longmapsto P+Q,\\ i_X\colon\ & X_V \hookrightarrow \mathbb{G}(1,\mathbb{P}(W)),\ \omega \longmapsto P_\omega := \mathrm{Ker}(\omega), \quad \text{and}\\ i_F\colon\ & F(X_V) \hookrightarrow \mathbb{G}(3,\mathbb{P}(W)),\ L \longmapsto W_L := \textstyle\sum_{\omega\in L} P_\omega \end{aligned}$$

define closed embeddings.[6]

Proof We shall explain that the natural maps are well defined and prove that they are injective. We will leave to the reader the verification that the maps are regular, and in fact closed immersions.

By virtue of Remark 2.2 and Lemma 2.3, we know that for any two $P \neq Q \in S_V$ the sum $P+Q$ is of dimension four. Hence, $i_{S^{[2]}}$ is certainly well defined on the open subset

[6] Thanks to Jia-Choon Lee for pointing out a mistake in the definition of W_L in an earlier version and for his help with fixing the proof.

in $S_V^{[2]}$ parametrizing reduced length-two subschemes of S_V. A non-reduced subscheme of length two in S_V is given by a point $P \in S_V$ and a tangent direction $v \in T_PS_V \subset T_P\mathbb{G}(1,\mathbb{P}(W)) \simeq \mathrm{Hom}(P, W/P)$. Sharpening the argument in Remark 2.2, one proves that the pre-image $\pi^{-1}(v(P))$ of $v(P) \subset W/P$ under the projection $\pi\colon W \twoheadrightarrow W/P$ is again a four-dimensional space. Setting $i_{S^{[2]}}(P,v) := \pi^{-1}(v(P)) \in \mathbb{G}(3,\mathbb{P}(W))$ extends $\{P,Q\} \longmapsto P+Q$ to a morphism $i_{S^{[2]}}\colon S_V^{[2]} \longrightarrow \mathbb{G}(3,\mathbb{P}(W))$. For the injectivity of $i_{S^{[2]}}$, use the arguments of Exercise 2.11 below.

As every $\omega \in X_V$ has rank four, $P_\omega \subset W$ is indeed a plane and, therefore, i_X is well defined. The fibre of i_X through ω is the linear space $\mathbb{P}(V) \cap \mathbb{P}(\bigwedge^2(W/P_\omega)^*)$. Hence, as $\mathrm{Pic}(X_V) \simeq \mathbb{Z}$, the map is either injective or constant. The latter can be excluded for generic V.

Since $L \subset X_V$, we have $\omega_1^2 \wedge \omega_2 = 0$ for all (linearly independent) $\omega_1, \omega_2 \in L$. As ω_1 is of rank four, we may pick a basis such that $\omega_1 = x^3 \wedge x^4 + x^5 \wedge x^6$. Then $P_{\omega_1} = \langle x_1, x_2\rangle$ and $\omega_2 = \sum a_{ij} x^i \wedge x^j$ with $a_{12} = 0$. Therefore, $\omega_2|_{P_{\omega_1}} = 0$ and then in fact $\omega_2|_{P_{\omega_1}+P_{\omega_2}} = 0$. Interchanging the role of ω_1 and ω_2, we find that the restriction of any $\omega \in L$, a linear combination of ω_1, ω_2, to the subspace $P_{\omega_1} + P_{\omega_2}$ is trivial. From the latter one concludes that $\omega|_{W_L} = 0$ for all $\omega \in L$, i.e. W_L is isotropic for all $\omega \in L$. As forms in L are of rank four, we have $\dim(W_L) \le 4$. Also, $\dim(W_L) > 2$ and in fact $\dim(P_{\omega_1} + P_{\omega_2}) > 2$, for otherwise $P_{\omega_1} = P_{\omega_2}$ and then $\omega_1 = \omega_2$, by the injectivity of i_X. Later it will become clear that $W_L = P_{\omega_1} + P_{\omega_2}$ for the generic line L, see Section 2.5.

The case $\dim(W_L) = 3$ can be excluded as follows. Assume L is generated by ω_1 and ω_2 and suppose $\dim(W_L) = 3$. Then $W_L = P_{\omega_1} + P_{\omega_2}$ and $P_\omega \subset P_{\omega_1} + P_{\omega_2}$ for all $\omega \in L$. We may choose a basis such that $\omega_1 = x^3 \wedge x^4 + x^5 \wedge x^6$, hence $P_{\omega_1} = \langle x_1, x_2\rangle$ and $P_{\omega_2} = \langle x_2, x_3\rangle$. Then clearly $x_2 \in P_\omega$ for all $\omega \in L$ and there exists a further non-trivial linear combination $x = \lambda_1 x_1 + \lambda_3 x_3 \in P_\omega$. Writing $\omega = \mu_1\omega_1 + \mu_2\omega_2$ and applying it to x then gives $\lambda_3\mu_1 x^4 + \lambda_1\mu_2\omega_2(x_1)$. Hence, up to scaling $\omega_2 = x^1 \wedge x^4 + \omega_2'$ with $\omega_2' \in \bigwedge^2\langle x^4, x^5, x^6\rangle$. Then, by subtracting an appropriate multiple of ω_1, we can modify ω such that it does not involve $x^5 \wedge x^6$ and, therefore, up to scaling $\omega = x^1 \wedge x^4 + x^4 \wedge f(x^3, x^5, x^6)$. But this implies that P_ω contains x_2 and the hyperplane defined by f in $\langle x_3, x_5, x_6\rangle$ which contradicts the assumption that all $\omega \in L \subset X_V$ are of rank four.

Hence, i_F is well defined. To prove the injectivity, observe that

$$V_L := V_{W_L} := \{\, \omega \in V \mid \omega|_{W_L} = 0 \,\} \tag{2.4}$$

defines a linear subspace $\mathbb{P}(V_L) \subset X_V$ containing L. Hence, as V was generic such that X_V does not contain any plane, in fact $L = \mathbb{P}(V_{W_L})$. Hence, the line L is uniquely determined by its image $W_L = i_F(L) \in \mathbb{G}(3,\mathbb{P}(W))$, i.e. i_F is injective. □

Remark 2.8 Here is a slightly less ad hoc way of introducing i_X. Restricting (2.3) to X_V and tensoring with $\mathcal{O}_{X_V}(1)$ leads to an exact sequence

$$0 \longrightarrow \mathcal{F}^* \longrightarrow W \otimes \mathcal{O}_{X_V} \longrightarrow W^* \otimes \mathcal{O}_{X_V}(1) \longrightarrow \mathcal{F} \otimes \mathcal{O}_{X_V}(1) \longrightarrow 0.$$

The fibre of this sequence at the point $\omega \in X_V$ is nothing but the short exact sequence

$$\mathrm{Ker}(\omega) \hookrightarrow W \xrightarrow{\omega} W^* \twoheadrightarrow \mathrm{Ker}(\omega)^*$$

and the morphism i_X is the classifying morphism for the inclusion $\mathcal{F}^* \hookrightarrow W \otimes \mathcal{O}_{X_V}$.

Exercise 2.9 Use Remark 2.5 to show that under $i_X \colon X_V \hookrightarrow \mathbb{G}(1, \mathbb{P}(W))$ the Plücker polarization pulls back to $\mathcal{O}(2)$, i.e.

$$i_X^* \mathcal{O}_{\mathbb{P}(\bigwedge^2 W)}(1) \simeq \mathcal{O}_{\mathbb{P}(V)}(2)|_{X_V}.$$

In particular, a line in $X_V \subset \mathbb{P}(V) \subset \mathbb{P}(\bigwedge^2 W^*)$ is of degree two in $X_V \subset \mathbb{G}(1, \mathbb{P}(W)) \subset \mathbb{P}(\bigwedge^2 W)$.

2.4 Fano Variety versus Hilbert Scheme Although, S_V and $X_V \simeq i_X(X_V)$ are both closed subvarieties of $\mathbb{G} = \mathbb{G}(1, \mathbb{P}(W))$, the link between the two as such becomes clear only after passing to the associated four-dimensional varieties $S_V^{[2]}$ and $F(X_V)$.

Corollary 2.10 (Beauville–Donagi) *Let $\mathbb{P}^5 \simeq \mathbb{P}(V) \subset \mathbb{P}(\bigwedge^2 W^*)$ be a generic linear subspace. Then the images of i_F and $i_{S^{[2]}}$ in $\mathbb{G}(3, \mathbb{P}(W))$ coincide and, therefore,*

$$F(X_V) \simeq S_V^{[2]}. \tag{2.5}$$

Proof As $S_V^{[2]}$ and $F(X_V)$ are both smooth subvarieties of $\mathbb{G}(3, \mathbb{P}(W))$ of dimension four, it suffices to show that for any (a generic one suffices) point $\{P \neq Q\} \in S_V^{[2]}$ the four-space $P + Q$ is of the form W_L for some line $L \subset X_V$. To this end, consider the linear subspace $\{\omega \in V \mid \omega|_{P+Q} = 0\} \subset V$ as in (2.4). By virtue of Lemma 2.3, it is of dimension at most two, as its projectivization defines a linear subspace of $\mathbb{P}(V)$ contained in X_V. To see that it is of dimension two and thus defines a line $L \subset X_V$, pick a basis such that $P = \langle x_1, x_2 \rangle$ and $Q = \langle x_3, x_4 \rangle$. As $P, Q \in S_V$, we know that $\omega|_P = \omega|_Q = 0$ for all $\omega \in V$. Hence, for $\omega = \sum a_{ij} x^i \wedge x^j$, the condition $\omega|_{P+Q} = 0$ translates into $a_{13} = a_{14} = a_{23} = a_{24} = 0$. They define a subspace of codimension at most four.

In order to show that $P + Q = W_L$, it suffices to prove that $P_\omega \subset P + Q$ for all $\omega \in L$. Since $\omega|_{P+Q} = 0$, the image of $P + Q \subset W$ under $\omega \colon W \longrightarrow W^*$ is contained in the two-dimensional kernel of $W^* \twoheadrightarrow (P + Q)^*$. For dimension reasons, $P + Q$ then contains the kernel P_ω.[7] □

Exercise 2.11 Show that an inverse map $F(X_V) \xrightarrow{\sim} S_V^{[2]}$ can be described as follows. For $L \in F(X_V)$, the intersection of the two linear spaces

$$\mathbb{P}^5 \simeq \mathbb{P}(\bigwedge^2 W_L) \quad \text{and} \quad \mathbb{P}^8 \simeq \mathbb{P}(V^\perp) \subset \mathbb{P}(\bigwedge^2 W) \simeq \mathbb{P}^{14}$$

[7] Thanks to Jia-Choon Lee for the last argument.

is a line $\mathcal{L} \subset \mathbb{P}(\bigwedge^2 W_L) \subset \mathbb{P}(\bigwedge^2 W)$. To see this use that the plane in V corresponding to the line $L \subset X_V \subset \mathbb{P}(V)$ is the kernel V_L of the projection $V \subset \bigwedge^2 W^* \twoheadrightarrow \bigwedge^2 W_L^*$, see the proof of Corollary 2.7.

The line $\mathcal{L}$ intersects the quadric $\mathbb{G}(1, \mathbb{P}(W_L)) \subset \mathbb{P}(\bigwedge^2 W_L)$ in two points corresponding to planes $P, Q \subset W_L$. They are both automatically contained in S_V and, therefore, define a point $\{P, Q\} \in S_V^{[2]}$. In other words, the line $L = \mathbb{P}(\langle \omega_1, \omega_2 \rangle) \in F_V$ corresponds to the point $\{P_1, P_2\} \in S_V^{[2]}$ if and only if

$$P_{\omega_1} + P_{\omega_2} = P_1 + P_2. \tag{2.6}$$

See [58] for more details.

The next lemma taken from [58] is a technical fact that will come in handy later. To state it, we use that $H^2(S^{[2]}, \mathbb{Z})$ of any K3 surface naturally decomposes as

$$H^2(S^{[2]}, \mathbb{Z}) \simeq H^2(S, \mathbb{Z}) \oplus \mathbb{Z} \cdot \delta,$$

where $2 \cdot \delta$ is the class of the exceptional divisor of the usual Hilbert–Chow morphism $S^{[2]} \longrightarrow S^{(2)}$, cf. Example 3.6. For $S = S_V$ and its Plücker polarization $g_S \in H^2(S, \mathbb{Z}) \subset H^2(S^{[2]}, \mathbb{Z})$, one proves:

$$\int_{S^{[2]}} g_S^4 = 3 \cdot 14^2, \quad \int_{S^{[2]}} g_S^3 \cdot \delta = \int_{S^{[2]}} g_S \cdot \delta^3 = 0,$$
$$\int_{S^{[2]}} g_S^2 \cdot \delta^2 = -28, \quad \text{and} \quad \int_{S^{[2]}} \delta^4 = 3 \cdot 4.$$

To perform the computation, one views $S^{[2]}$ as the quotient of $\mathrm{Bl}_\Delta(S \times S)$ by the natural $\mathbb{Z}/2\mathbb{Z}$-action and uses that the pullback of δ under $\mathrm{Bl}_\Delta(S \times S) \longrightarrow S^{[2]}$ is the class of the exceptional divisor $E = \mathbb{P}(\mathcal{T}_\Delta) \xrightarrow{\pi} \Delta$ and that the restriction $-\delta|_E$ is the tautological class u which in this case satisfies $u^2 = -\pi^* c_2(S)$. The above intersection numbers can equivalently be viewed and computed in terms of the Beauville–Bogomolov–Fujiki quadratic form, see Example 3.6.

Lemma 2.12 *The natural isomorphism of integral Hodge structures induced by* (2.5) *maps the Plücker polarization g on $F_V := F(X_V)$ to $2 \cdot g_S - 5 \cdot \delta$:*

$$H^2(F(X_V), \mathbb{Z}) \simeq H^2(S_V^{[2]}, \mathbb{Z}), \quad g \longmapsto 2 \cdot g_S - 5 \cdot \delta.$$

Here, g_S denotes the Plücker polarization on $S_V \subset \mathbb{G}(1, \mathbb{P}(W))$.

Proof It suffices to prove the claim for the very general Pfaffian cubic fourfold. Then, by Remark 2.24 below, $H^{2,2}(X, \mathbb{Z})$ is of rank two and so is $H^{1,1}(F_V, \mathbb{Z}) \simeq H^{1,1}(S_V^{[2]}, \mathbb{Z})$. Hence, the Plücker polarization g on F_V corresponds to some linear combination $a \cdot g_S + b \cdot \delta$ on $S_V^{[2]}$. Then from $\int_{S^{[2]}} (a \cdot g_S + b \cdot \delta)^4 = \int_F g^4 = 108$ one derives the quadratic equation $7 \cdot a^2 - b^2 = \pm 3$, which has infinitely many integral solutions. As the ampleness of g implies $a > 0 > b$, it suffices to determine one of the two coefficients a or b.

One possibility to go about this could be to show that the restriction of g to $\mathbb{P}^1 \simeq \mathbb{P}(T_P S_V) \subset S_V^{[2]} \simeq F_V$ is of degree five,[8] which then would show $b = -5$. Instead, we consider the natural embedding $S_V \setminus \{P\} \hookrightarrow S_V^{[2]} \simeq F_V$, cf. Example 3.6, and describe an isomorphism of the pullback of the Plücker polarization $\mathcal{O}_{F_V}(1)$ on F_V and the square of the Plücker polarization $\mathcal{O}_{S_V}(2)$ on $S_V \setminus \{P\}$, which then shows $a = 2$.

Here are the details for this approach. For $Q \in S_V \subset G(2, W)$ and $L = \mathbb{P}(K) \in F_V \subset \mathbb{G}(1, \mathbb{P}(\bigwedge^2 W^*))$, the fibres of the Plücker polarizations $\mathcal{O}_{S_V}(1)$ and $\mathcal{O}_{F_V}(1)$ at these points are naturally isomorphic to

$$\mathcal{O}_{S_V}(1)(Q) \simeq \det Q^* \quad \text{and} \quad \mathcal{O}_F(1)(L) \simeq \det K^*.$$

According to (2.6) in Exercise 2.11, the image $L = \mathbb{P}(K)$ of $Q \in S_V \setminus \{P\}$ in F_V is characterized by the property that $P + Q = P_{\omega_1} + P_{\omega_2} \subset W$, where $K = \langle \omega_1, \omega_2 \rangle$. Also, we know that $K = V_L = \mathrm{Ker}(V \subset \bigwedge^2 W \twoheadrightarrow \bigwedge^2 (P+Q)^*)$, see the arguments in the proof of Corollary 2.10. Furthermore, by definition of S_V, the natural maps

$$V \subset \bigwedge^2 W^* \twoheadrightarrow \bigwedge^2 P^* \quad \text{and} \quad V \subset \bigwedge^2 W^* \twoheadrightarrow \bigwedge^2 Q^*$$

are trivial. From the natural isomorphism $\bigwedge^2(P+Q) \simeq (P \otimes Q) \oplus \bigwedge^2 P \oplus \bigwedge^2 Q$ one obtains a natural short exact sequence $0 \longrightarrow K \longrightarrow V \longrightarrow P^* \otimes Q^* \longrightarrow 0$. Thus, for fixed P there indeed exists a natural isomorphism

$$\mathcal{O}_{F_V}(-1)(L) \simeq \det K \simeq (\det P)^2 \otimes (\det Q)^2 \simeq \mathcal{O}_S(-2)(Q).$$

We leave it to the reader to put these natural isomorphisms into a family to obtain an isomorphism $\mathcal{O}_{F_V}(1)|_{S_V \setminus \{P\}} \simeq \mathcal{O}_{S_V}(2)|_{S_V \setminus \{P\}}$, which then proves the assertion. □

Remark 2.13 If V is not chosen generically in the sense of Lemma 2.3, but X_V and S_V are nevertheless smooth, there is still a birational isomorphism

$$F(X_V) \sim S_V^{[2]}. \tag{2.7}$$

Indeed, the isomorphism in Corollary 2.10 specializes to a correspondence between $F(X_V)$ and $S_V^{[2]}$ and, using that both varieties have trivial canonical bundle, this proves that there is a unique irreducible component of the correspondence that gives (2.7).

An example, where the above birational correspondence between the two hyperkähler manifolds, $F(X_V)$ and $S_V^{[2]}$, does not extend to an isomorphism was described by Hassett [225, Sec. 6.1]. More concretely, one can take the Pfaffian cubic fourfold in Example 1.5.12, which is the closure of a rational parametrization $\mathbb{P}^2 \times \mathbb{P}^2 \dashrightarrow \mathbb{P}^5$. The K3 surface of degree 14 is a complete intersection of type $(2,1), (1,2)$ in $\mathbb{P}^2 \times \mathbb{P}^2$.

2.5 Correspondence: Pfaffian Cubics versus K3 Surfaces Let us now study the restriction of (2.2)

$$\Sigma_V := \Sigma \cap (S_V \times X_V) \subset S_V \times X_V,$$

[8] I have not actually done the computation.

which eventually links the cubic X_V and its associated K3 surface S_V more directly.

Lemma 2.14 *Let $\mathbb{P}^5 \simeq \mathbb{P}(V) \subset \mathbb{P}(\bigwedge^2 W^*)$ be a generic linear subspace.*

(i) *For $P \in S_V$ and a line $W_0 \subset P$, the set $\{\omega \in \mathbb{P}(V) \mid W_0 \subset \mathrm{Ker}(\omega)\}$ is a line in X_V.*
(ii) *If $(P, \omega) \in \Sigma_V$, then $P \cap \mathrm{Ker}(\omega) \subset P$ is a line.*

It is worth pointing out that the set defined in (i) only depends on W_0, only the proof uses the fact that there exists a $P \in S_V$ in the background. See (2.8) for a more geometric interpretation.

Proof The quickest way to prove (i) is by introducing a basis: $x_1, \ldots, x_6 \in W$ with $W_0 = \langle x_1 \rangle \subset P = \langle x_1, x_2 \rangle$. For $\omega = \sum_{i<j} a_{ij}\, x^i \wedge x^j \in \bigwedge^2 W^*$, the condition $W_0 \subset \mathrm{Ker}(\omega)$ is equivalent to imposing the five conditions $a_{12} = \cdots = a_{16} = 0$. As $P \in S_V$, the vanishing $a_{12} = 0$ is automatic for all $\omega \in V \subset \bigwedge^2 W^*$. Therefore, the remaining conditions $a_{13} = \cdots = a_{16} = 0$ define a subspace of dimension at least two and in fact of dimension exactly two, as by Lemma 2.3 we may assume that for generic V the cubic X_V does not contain any projective plane.

For (ii), it suffices to exclude that $P = \mathrm{Ker}(\omega)$ for any $(P, \omega) \in S_V \times X_V$ for generic choice of V. To this end, consider the correspondence

$$T \subset \mathbb{G}(1, \mathbb{P}(W)) \times \mathrm{Pf}(W^*) \times \mathbb{G}(5, \mathbb{P}(\textstyle\bigwedge^2 W^*))$$

of all (P, ω, V) with $P \subset \mathrm{Ker}(\omega)$, $\omega \in X_V$, and $P \in S_V$. Once P is fixed, ω varies in the six-dimensional subspace V_P of all forms with P contained in the kernel. If P and $\omega \in V_P$ are fixed, then V varies in a 40-dimensional Grassmann variety. Indeed, after picking appropriate coordinates, we may write $P = \langle x_1, x_2 \rangle$ and $\omega = x^3 \wedge x^4 + x^5 \wedge x^6$. Then $V/\langle \omega \rangle$ varies in the Grassmann variety of all five-dimensional subspaces of the space $\langle x^i \wedge x^j \mid (i, j) \neq (1, 2) \rangle / \langle \omega \rangle$. Counting dimensions shows $\dim(T) = 8 + 5 + 40 = 53 < 54 = \dim(\mathbb{G}(5, \mathbb{P}(\bigwedge^2 W^*)))$. Therefore, the projection $T \longrightarrow \mathbb{G}(5, \mathbb{P}(\bigwedge^2 W^*))$ is not surjective. □

With a fixed generic $\mathbb{P}(V) \subset \mathbb{P}(\bigwedge^2 W^*)$ as above, we associate the restriction of the correspondence (2.1), where we keep the notation for the projections:

$$\begin{array}{ccc} B_V := q^{-1}(X_V) & \xrightarrow{\;q\;} & X_V \\ {\scriptstyle p}\downarrow & & \\ \mathbb{P}(W). & & \end{array}$$

Note that $q\colon B_V \longrightarrow X_V$ is a $\mathbb{P}^1$-bundle, so that B_V is smooth and of dimension five. The fibre of p over the point in $\mathbb{P}(W)$ corresponding to a line $W_0 \subset W$ is the linear subspace of all $\omega \in \mathbb{P}(V)$ with $W_0 \subset \mathrm{Ker}(\omega)$. The situation is more precisely described as follows.

Lemma 2.15 *For a generic linear subspace $\mathbb{P}^5 \simeq \mathbb{P}(V) \subset \mathbb{P}(\bigwedge^2 W^*)$, the first projection $p\colon B_V \longrightarrow \mathbb{P}(W) \simeq \mathbb{P}^5$ is the blow-up in the smooth and irreducible subvariety*

$$Z_V := \{[W_0] \mid \dim(p^{-1}[W_0]) > 0\} \subset \mathbb{P}(W)$$

which satisfies $\dim(Z_V) = 3$ *and* $\deg(Z_V) = 9$.

Proof The morphism $p\colon B_V \twoheadrightarrow \mathbb{P}(W)$ is surjective and its fibres are linear subspaces of X_V of dimension 0 or 1. More precisely, over the point in $\mathbb{P}(W)$ corresponding to a line $W_0 \subset W$ the fibre is $\mathbb{P}((\bigwedge^2(W/W_0)^*) \cap V)$. In any case, together with the fact that B_V is smooth, irreducible, and of dimension five, this proves $\dim Z_V \leq 3$.

The degree of Z_V can be computed using Porteous formula, see [174, Ch. 12] or [188, Ch. 14]. Taking the second exterior power of the dual of the Euler sequence on $\mathbb{P} = \mathbb{P}(W)$ leads to the short exact sequence $0 \longrightarrow \Omega^2_{\mathbb{P}} \longrightarrow \bigwedge^2 W^* \otimes \mathcal{O}(-2) \longrightarrow \Omega_{\mathbb{P}} \longrightarrow 0$. Then the surjection composed with the inclusion $V \subset \bigwedge^2 W^*$ defines a generically surjective morphism $\eta\colon V \otimes \mathcal{O}(-2) \longrightarrow \Omega_{\mathbb{P}}$.

The fibre of η at the point $[W_0]$ has kernel $(\bigwedge^2(W/W_0)^*) \cap V$. Hence, Z is the degeneracy locus $M_4(\eta) = \{[W_0] \mid \operatorname{rk} \eta_{[W_0]} \leq 4\} \subset \mathbb{P}(W)$. Hence, $Z_V = M_4(\eta)$ has the expected codimension two and its class is given by $(\mathrm{c}_1^2 - \mathrm{c}_2)(\Omega_{\mathbb{P}}(2)) = 9 \cdot h^2$ and, therefore, $\deg(Z) = 9$. Note that the fact that the fibres of $p\colon B_V \longrightarrow \mathbb{P}(W)$ are of dimension at most one, translates into the fact that the rank of η is at least four at each point and, therefore, $M_4(\eta)$ is smooth.

Below we will describe Z_V explicitly as a $\mathbb{P}^1$-bundle over the K3 surface S_V, which in particular proves its irreducibility. Alternatively, one can use that $\Omega_{\mathbb{P}}(2)$ is ample and evoke the general connectivity criterion for degeneracy loci [189]. □

Lemma 2.14 can be rephrased by saying that the natural projection from Σ_V onto S_V can be written as the composition of two $\mathbb{P}^1$-bundles

$$\Sigma_V \longrightarrow \mathbb{P}(\mathcal{S}_{S_V}) \longrightarrow S_V,\ (P, \omega) \longmapsto (P, P \cap \operatorname{Ker}(\omega)) \longmapsto P. \tag{2.8}$$

Here, $\mathcal{S}_{S_V}$ is the restriction of the universal sub-bundle to $S_V \subset \mathbb{G}(1, \mathbb{P}(W))$. Furthermore, from the natural inclusion $\mathbb{P}(\mathcal{S}_{S_V}) \subset S_V \times \mathbb{P}(W)$, we obtain a projection

$$\iota_V\colon \mathbb{P}(\mathcal{S}_{S_V}) \longrightarrow \mathbb{P}(W). \tag{2.9}$$

Restricted to the fibre of $\mathbb{P}(\mathcal{S}_{S_V}) \longrightarrow S_V$ over the point in S_V corresponding to a plane $P \subset W$ it is the embedding of the line $\mathbb{P}^1 \simeq \mathbb{P}(P) \subset \mathbb{P}(W)$.

Corollary 2.16 *For a generic linear subspace $\mathbb{P}^5 \simeq \mathbb{P}(V) \subset \mathbb{P}(\bigwedge^2 W^*)$, the morphism* (2.9) *is a closed embedding with image Z_V:*

$$\iota_V\colon \mathbb{P}(\mathcal{S}_{S_V}) \xrightarrow{\sim} Z_V \subset \mathbb{P}(W).$$

Proof Combining Lemma 2.14, (i) and Lemma 2.15 we find that $[W_0] \in Z_V$ for any $W_0 \subset P \in S_V$. Similarly, one checks that for any $[W_0] \in Z_V$ there in fact exists a $P \in S_V$ with $W_0 \subset P$. This also follows from the fact that Z_V is irreducible and that ι_V is injective. To verify the latter, observe that for $P \neq Q \in S_V$ the two lines $\mathbb{P}(P), \mathbb{P}(Q) \subset \mathbb{P}(W)$ intersect if and only if the line $\overline{PQ} \subset \mathbb{P}(\bigwedge^2 W)$ is contained in $\mathbb{G}$ or, equivalently, in S_V, see Remark 2.2. However, by virtue of Lemma 2.3, for a generic choice of V the K3 surface S_V does not contain any lines. □

This leads to the following picture, where E denotes the exceptional divisor of p:

$$\begin{array}{ccccccc} \Sigma_V & \xrightarrow{\sim} & E & \hookrightarrow & B_V & \xrightarrow{\mathbb{P}^1} & X_V \\ \downarrow{\scriptstyle \mathbb{P}^1} & & \downarrow{\scriptstyle \mathbb{P}^1} & & \downarrow{\scriptstyle p} & & \\ \mathbb{P}(\mathcal{S}_{S_V}) & \xrightarrow[\iota_V]{\sim} & Z_V & \hookrightarrow & \mathbb{P}(W) & & \\ \downarrow{\scriptstyle \mathbb{P}^1} & & & & & & \\ S_V. & & & & & & \end{array} \tag{2.10}$$

Remark 2.17 In particular, to any $(W_0, \omega) \in E \subset B_V$ one naturally associates two planes $P_{W_0}, P_\omega \in \mathbb{G}(1, \mathbb{P}(W))$, cf. Corollary 2.7. Here, P_{W_0} is the unique plane defining a point in S_V with $W_0 \subset P_{W_0}$. Equivalently, there are two morphisms $E \longrightarrow \mathbb{G}(1, \mathbb{P}(W))$ corresponding to the compositions

$$E \subset B_V \twoheadrightarrow X_V \hookrightarrow \mathbb{G}(1, \mathbb{P}(W)) \quad \text{and} \quad E \twoheadrightarrow Z_V \twoheadrightarrow S_V \hookrightarrow \mathbb{G}(1, \mathbb{P}(W))$$

and they do not commute, cf. Corollary 2.7.

Note that if $P_{W_0} \neq P_\omega$ for $(W_0, \omega) \in E$, then $P_\omega \in X_V \setminus S_V \subset \mathbb{G}(1, \mathbb{P}(W))$. Indeed, otherwise, according to Remark 2.2, there would be a line contained in S_V, the existence of which is excluded for generic V by Lemma 2.3.

Remark 2.18 The $\mathbb{P}^1$-bundle $E \longrightarrow Z_V$ together with the projection $E \subset B_V \twoheadrightarrow X_V$ can be viewed as a family of lines in X_V parametrized by Z_V. The classifying morphism $Z_V \longrightarrow F(X_V)$ is a closed immersion and describes a uniruled divisor in the fourfold $F(X_V)$. For the injectivity of $Z_V \longrightarrow F(X_V)$. use the explicit description in the proof of Lemma 2.14, which shows that the linear subspace $\{\omega \in \mathbb{P}(V) \mid W_0 \subset \mathrm{Ker}(\omega)\}$ determines W_0. The base of the ruling is the K3 surface S_V.

As $Z_V \longrightarrow F(X_V)$ is injective, the family of lines is dominant, i.e. $\Sigma_V \simeq E \twoheadrightarrow X_V$ is surjective and, for dimension reasons, generically finite. The degree of the projection is four, see Lemma 2.20 below.

The commutative diagram (2.10) produces motivic relations between the cubic fourfold X_V and the K3 surface S_V, see Section **2**.4 for the notation.

Proposition 2.19 *Let* $\mathbb{P}^5 \simeq \mathbb{P}(V) \subset \mathbb{P}(\bigwedge^2 W^*)$ *be a generic linear subspace. Then*

$$[\mathbb{P}^1] \cdot [X_V] = [\mathbb{P}^5] + [S_V] \cdot \ell \cdot (\ell + 1) \tag{2.11}$$

in the Grothendieck ring of varieties $K_0(\mathrm{Var}_k)$ *and*

$$\mathfrak{h}(\mathbb{P}^1) \cdot \mathfrak{h}(X_V) \oplus \mathfrak{h}(Z_V)(-2) \simeq \mathfrak{h}(\mathbb{P}^5) + \mathfrak{h}(S_V)(-1) \cdot \mathfrak{h}(\mathbb{P}^1) \oplus \mathfrak{h}(Z_V)(-2) \tag{2.12}$$

in the category of rational Chow motives $\mathrm{Mot}(k)$. □

Presumably, the direct summand $\mathfrak{h}(Z_V)$ on both sides in (2.11) cancels out.

2.6 Family of Quartic Normal Scrolls The part $S_V \longleftarrow \Sigma_V \longrightarrow X_V$ of the above correspondence is of particular importance, for the geometry of the situation as well as for its derived aspects, see Section **7.3.2**. For each $P \in S_V$, one obtains a surface Σ_P which is described by the next result proved by Hassett [227].

Lemma 2.20 *Assume* $\mathbb{P}^5 \simeq \mathbb{P}(V) \subset \mathbb{P}(\bigwedge^2 W^*)$ *is a generic linear subspace.*

(i) *Then the fibre* Σ_P *of* $\Sigma_V \longrightarrow S_V$ *over a point* $P \in S_V$*, which is*

$$\Sigma_P = \{\omega \in X_V \subset \mathbb{P}(\textstyle\bigwedge^2 W^*) \mid P \cap \mathrm{Ker}(\omega) \neq 0\},$$

describes a quartic rational normal scroll contained in X_V.

(ii) *The self-intersection number of the surface* $\Sigma_P \subset X_V$ *is*

$$([\Sigma_P].[\Sigma_P]) = 10.$$

(iii) *The projection* $\Sigma_V \twoheadrightarrow X_V$ *is a generically finite morphism of degree four.*

Before proving the lemma, let us recall some basic facts concerning rational scrolls. Scrolls are classical objects in algebraic geometry. A scroll of dimension two is by definition a ruled surface $\pi\colon T := \mathbb{P}(\mathcal{E}) \longrightarrow C$ over some curve C together with an embedding $T \hookrightarrow \mathbb{P}^N$ such that all fibres of π are lines in $\mathbb{P}^N$:

$$\begin{array}{ccc} \mathbb{P}(\mathcal{E}) = T & \hookrightarrow & \mathbb{P}^N \\ \downarrow & & \\ C. & & \end{array}$$

The scroll is a rational scroll if the surface T is rational, i.e. birational to $\mathbb{P}^2$, or, equivalently, $C \simeq \mathbb{P}^1$. It is a rational normal scroll, if in addition the embedding is projectively normal, i.e. if the maps $H^0(\mathbb{P}^N, \mathcal{O}(n)) \longrightarrow H^0(T, \mathcal{O}_T(n))$ are surjective.

Alternatively, a two-dimensional rational normal scroll can be described as the union

$$T = \bigcup_{t \in \mathbb{P}^1} \overline{v_{a_1}(t) v_{a_2}(t)}$$

of all lines $\overline{v_{a_1}(t)v_{a_2}(t)}$. Here, $v_{a_i}\colon \mathbb{P}^1 \hookrightarrow \mathbb{P}^{a_i}$, $i = 1, 2$, are the Veronese embeddings into disjoint linear subspaces $\mathbb{P}^{a_i} \subset \mathbb{P}^N$ with $a_1 + a_2 = N - 1$, see [158, Ch. 8]. In this description, $\mathcal{E} \simeq \mathcal{O}(a_1) \oplus \mathcal{O}(a_2)$ and the restriction of $\mathcal{O}(1)$ to T is the relative $\mathcal{O}_\pi(1)$. In the context of cubic fourfolds, only the cases $N = 3, 4, 5$ are of interest. As we may assume $0 < a_1 \leq a_2$, the only possibilities are $(a_1, a_2) = (1, 1)$ for $N = 3$, in which case $T \simeq \mathbb{P}^1 \times \mathbb{P}^1$, $(a_1, a_2) = (1, 2)$ for $N = 4$ and then $T \simeq \mathbb{F}_1$, and the two cases $(a_1, a_2) = (2, 2)$ or $= (1, 3)$, for which $T \simeq \mathbb{P}^1 \times \mathbb{P}^1$ and $T \simeq \mathbb{F}_2$, for $N = 5$.

A classical result in algebraic geometry says that any smooth projective surface of degree e in $\mathbb{P}^{e+1}$ not contained in any hyperplane is either a rational normal scroll or the image of $\mathbb{P}^2$ under the Veronese embedding into $\mathbb{P}^5$. The case of a linear $\mathbb{P}^2$ in $\mathbb{P}^3$ can be considered as the case $e = 1$. Furthermore, for $e = 4$, which is the case of interest to us, the surface is either isomorphic to $\mathbb{P}(\mathcal{O}(2) \oplus \mathcal{O}(2)) \simeq \mathbb{P}(\mathcal{O} \oplus \mathcal{O}) \simeq \mathbb{P}^1 \times \mathbb{P}^1$ or to the Hirzebruch surface $\mathbb{P}(\mathcal{O}(1) \oplus \mathcal{O}(3)) \simeq \mathbb{P}(\mathcal{O}(-2) \oplus \mathcal{O}) \simeq \mathbb{F}_2$, the latter being obtained by specialization of the former. See [223, Ch. V.2] for more information and references.

Proof To compute the degree of Σ_P we intersect it with a generic $\mathbb{P}^3 \simeq \mathbb{P}(V_0) \subset \mathbb{P}(V)$. The quickest way to compute $\Sigma_P \cap \mathbb{P}(V_0)$ is by choosing coordinates. So, let us assume $P = \langle x_1, x_2 \rangle \subset W = \langle x_1, \ldots, x_6 \rangle$. For $\omega = \sum a_{ij}\, x^i \wedge x^j \in V_0$, we consider the composition $\omega'\colon P \subset W \xrightarrow{\ \omega\ } W^*$. Then $\Sigma_P \cap \mathbb{P}(V_0)$ is described by the additional linear condition $a_{12} = 0$, which holds automatically for $P \in S_V$, and $0 = \bigwedge^2 \omega'\colon \bigwedge^2 P \longrightarrow \bigwedge^2 W^*$, which amounts to the quadratic equations $a_{1k}\, a_{2\ell} = a_{1\ell}\, a_{2k}$. For dimension reasons, two of the quadratic equations are enough which readily proves $\deg(\Sigma_P) = |\Sigma_P \cap \mathbb{P}(V_0)| = 4$. This concludes the proof of (i).

To compute the self-intersection number, we assume for simplicity that $\Sigma_P \simeq \mathbb{P}^1 \times \mathbb{P}^1$, the case of the Hirzebruch surface $\mathbb{F}_2$ is similar. Here, the first factor corresponds to $\mathbb{P}(P)$ and $h|_{\Sigma_P} = 2\, h_1 + h_2$, where h_1 and h_2 are the hyperplane classes on the two factors. Therefore, $(\Sigma_P.\Sigma_P) = \mathrm{c}_2(\mathcal{N}_{\Sigma_P/X_V}) =: \mathrm{c}_2$ and $\mathrm{c}_1 := \mathrm{c}_1(\mathcal{N}_{\Sigma_P/X_V})$ can be computed by using the Euler sequence and the normal bundle sequences for $\Sigma_P \subset X_V$ and $X_V \subset \mathbb{P}$ by

$$
\begin{aligned}
(1 + 2h_1 + h_2)^6 &= \mathrm{c}(\mathcal{O}_{\Sigma_P}(1))^6 = \mathrm{c}(\mathcal{T}_{\mathbb{P}}|_{\Sigma_P}) \\
&= \mathrm{c}(\mathcal{T}_{X_V}|_{\Sigma_P}) \cdot \mathrm{c}(\mathcal{O}_{\Sigma_P}(3)) = \mathrm{c}(\mathcal{T}_{\Sigma_P}) \cdot \mathrm{c}(\mathcal{N}_{\Sigma_P/X_V}) \cdot \mathrm{c}(\mathcal{O}_{\Sigma_P}(3)) \\
&= (1 + 2\, h_1) \cdot (1 + 2\, h_2) \cdot (1 + \mathrm{c}_1 + \mathrm{c}_2) \cdot (1 + 3\, (2\, h_1 + h_2)).
\end{aligned}
$$

Hence,

$$
\mathrm{c}_1(\mathcal{N}_{\Sigma_P/X_V}) = 4\, h_1 + h_2 \quad \text{and} \quad \mathrm{c}_2(\mathcal{N}_{\Sigma_P/X_V}) = 10.
$$

It remains to compute the degree of $\Sigma_V \longrightarrow X$. The following argument for the computation of the degree was given by Addington and Lehn [7, Lem. 4]. As observed earlier, the projection $\Sigma_V \simeq E \twoheadrightarrow X_V$ is surjective and hence generically finite. The fibre over $\omega \in X_V$ is the intersection of the Schubert cycle $\{P \mid P \cap \mathrm{Ker}(\omega) \neq 0\} \subset \mathbb{G}(1, \mathbb{P}(W))$ with

S_V. In other words, the fibre is this Schubert cycle intersected with the linear subspace $\mathbb{P}(V^\perp) \simeq \mathbb{P}^5$ of codimension five. As the degree of the Schubert cycle is indeed four, this proves the last assertion. □

Corollary 2.21 *The two cycles h^2 and Σ_P on X_V, both of codimension two, span a rank two lattice with*

$$\begin{pmatrix} 3 & 4 \\ 4 & 10 \end{pmatrix}$$

as its intersection form. □

Remark 2.22 In Remark 2.5, we explained that a Pfaffian cubic $X \subset \mathbb{P}^5$ always contains a quintic del Pezzo surface $Y \subset X$. A straightforward numerical computation reveals that its class is $[Y] = 3 \cdot h^2 - [\Sigma_P]$, see [228]. A geometric interpretation of the relation between Y and Σ_P has been given by Tregub [454]. One should think of the two surfaces as residual to each other with respect to the intersection with the image of a Segre embedding $\mathbb{P}^1 \times \mathbb{P}^2 \hookrightarrow \mathbb{P}^5$, i.e.

$$(\mathbb{P}^1 \times \mathbb{P}^2) \cap X = Y \cup \Sigma_P.$$

Note that indeed $[\mathbb{P}^1 \times \mathbb{P}^2] = 3 \cdot h^2$. For another interpretation of Y, see Exercise 2.28.

Exercise 2.23 When working over $\mathbb{C}$, the lattice in Corollary 2.21 will be viewed as the sublattice

$$K_{14}^- := \mathbb{Z} \cdot h^2 \oplus \mathbb{Z} \cdot [\Sigma_P] \subset H^{2,2}(X, \mathbb{Z}) \subset H^4(X, \mathbb{Z}).$$

Imitate the arguments in the proof of Lemma 1.1 and show that $K_{14}^- \subset H^4(X, \mathbb{Z})$ is saturated.

Remark 2.24 For a very general Pfaffian cubic fourfold X, one has $K_{14}^- \simeq H^{2,2}(X, \mathbb{Z})$, cf. the arguments in Remark 1.3. Furthermore, as for purely numerical reasons K_{14}^- does not contain a class α with the properties $(h^2.\alpha) = 1$ and $(\alpha.\alpha) = 3$, the very general Pfaffian cubic fourfold X does not contain any plane $\mathbb{P}^2 \subset \mathbb{P}^5$, cf. Lemma 1.1.

Exercise 2.25 The morphism $\Sigma_V \longrightarrow X_V$ induces a rational map $X_V \dashrightarrow S^{[4]}$. Study this map and show that it defines a Lagrangian subvariety of the hyperkähler manifold $S^{[4]}$, cf. Section 3.1.

Exercise 2.26 Show that $\Sigma_P \subset X_V$ can also be described as the degeneracy locus $M_1(\eta) := \{\omega \in X_V \mid \operatorname{rk} \eta_\omega \leq 1\}$ of the composition

$$\eta \colon P \otimes \mathcal{O}_{X_V} \hookrightarrow W \otimes \mathcal{O}_{X_V} \longrightarrow W^* \otimes \mathcal{O}_{X_V}(1),$$

see Remark 2.8. Here, the last map is the universal alternating form restricted to X_V.

2.7 Pfaffian Cubic Fourfolds Are Rational The next step consists of cutting with a hyperplane. So let $H := \mathbb{P}(W') \subset \mathbb{P}(W)$ be a generic hyperplane section, $Z_{V,H} := Z_V \cap H$, and $B_{V,H} := B_V \cap (H \times X_V)$. Then, as the fibres of q are lines $\mathbb{P}(W)$, the diagram

$$\begin{array}{ccc} B_{V,H} & \xrightarrow{q_H} & X_V \\ {\scriptstyle p_H}\downarrow & & \\ H & & \end{array} \tag{2.13}$$

describes a birational correspondence between $H \simeq \mathbb{P}^4$ and X_V. More concretely, a rational parametrization of X_V is realized by the map

$$X_V \dashrightarrow \mathbb{P}^4 \simeq H \subset \mathbb{P}(W), \qquad \omega \longmapsto \mathrm{Ker}(\omega) \cap W'. \tag{2.14}$$

The result goes back to Beauville and Donagi [58] and in fact to Fano [177]. An independent and more direct argument was given by Bolognesi, Russo, and Staglianò [79]. The situation is described more precisely as follows.

Corollary 2.27 *Any Pfaffian cubic fourfold X_V is rational.*

(i) *The indeterminacy loci of the birational map $\mathbb{P}^4 \dashrightarrow X_V$ described above is a surface in $\mathbb{P}^4$ that is isomorphic to a blow-up of S_V in five points.*
(ii) *The indeterminacy locus of the inverse rational map $X_V \dashrightarrow \mathbb{P}^4$ is the surface $Y_H := X_V \cap \mathbb{G}(1, H)$, where $X_V \subset \mathbb{G}(1, \mathbb{P}(W))$ is as in Remark* 2.17.

Proof In the above arguments we have often chosen $\mathbb{P}(V)$ to be generic not only to ensure smoothness of X_V but also to exclude X_V from containing any planes. For the rational map (2.14) constructed above to be generically injective, planes in X_V or lines in S_V do not alter the argument.

The birational correspondence between X_V and $\mathbb{P}^4$ is provided by (2.13). Clearly, the map $B_{V,H} \longrightarrow H$ is the blow-up of H in the surface $Z_{V,H} \subset H$, which via $Z_{V,H} \subset Z_V \simeq \mathbb{P}(\mathcal{S}_{S_V}) \longrightarrow S_V$ projects onto S_V. More precisely, $Z_{V,H} \longrightarrow S_V$ is the blow-up of S_V in exactly those points that correspond to lines $\mathbb{P}(P) \subset \mathbb{P}(W)$ contained in H, i.e. in the intersection $S_V \cap \mathbb{G}(1, H)$. As $\deg(\mathbb{G}(1, H)) = 5$, this proves (i). Alternatively, the number of points that are blown up can be deduced from $\deg(S_V) - \deg(Z_{V,H}) = 5$.

To verify (ii), simply observe that the fibre of $B_{V,H} \longrightarrow X_V$ over ω is either a reduced point or the line $\mathbb{P}^1 \simeq \mathbb{P}(\mathrm{Ker}(\omega))$. The latter occurs exactly when $\mathbb{P}(\mathrm{Ker}(\omega)) \subset H$, i.e. when $\omega \in Y_H$. □

Exercise 2.28 (i) The hyperplane $H \subset \mathbb{P}(W)$ viewed as a linear form on W projects to a global section s_H of the rank two locally free sheaf $\mathcal{F}$ on X_V in Remark 2.5. Show that the zero set of this section is Y_H, i.e. $Z(s_H) = Y_H$. Thus, Y_H is a quintic del Pezzo surface and by Remark 2.22:

$$[Y_H] = 3 \cdot h^2 - [\Sigma_P].$$

(ii) Recall that the normal bundle of $\mathbb{G}(1, \mathbb{P}(W)) \subset \mathbb{P}(\bigwedge^2 W)$ is $\bigwedge^2 \mathcal{Q} \otimes \mathcal{O}(1)$. Then use Exercise 2.9 to show that the normal bundle of $Y_H \subset X_V$ has determinant $\mathcal{O}_{\mathbb{P}(V)}(2)|_{Y_H}$ and, therefore, $\omega_{Y_H} \simeq \mathcal{O}_{\mathbb{P}(V)}(-1)|_{Y_H}$ (confirming that Y_H is a del Pezzo surface).

For more on the surface Y_H, see [225, Sec. 1.2].

Remark 2.29 According to a result of Kontsevich and Tschinkel [285], which in turn was triggered by a result of Nicaise and Shinder [371], rationality is preserved under specialization. Thus, in the above proof one could have chosen $\mathbb{P}(V)$ as generic as needed and would still get rationality of all Pfaffian cubic fourfolds.

Remark 2.30 Recall that a Pfaffian cubic fourfold $X \subset \mathbb{P}^5$ contains quintic del Pezzo surfaces and quartic rational normal scrolls, see Remark 2.5 and Lemma 2.20. Thus, rationality of Pfaffian cubic fourfolds can also be seen as a consequence of either of the following two facts:

(i) Any cubic fourfold $X \subset \mathbb{P}^5$ containing a rational normal scroll of degree four is rational, see [453, §4] or [228, Prop. 4].
(ii) Similarly, any cubic fourfold $X \subset \mathbb{P}^5$ containing a quintic del Pezzo surface is rational, see [453, §3], [158, Thm. 8.5.5] or [228, Prop. 6].

Note that a smooth cubic fourfold containing a quintic del Pezzo surface is automatically Pfaffian, see Remark 2.5 and [50, Prop. 9.2]. Also, a smooth cubic fourfold containing a rational normal scroll is at least contained in the closure of the locus of all Pfaffian cubic fourfolds, see Proposition 5.6. For the cohomological relation between quartic rational normal scrolls and quintic del Pezzo surfaces, see Remark 2.22.

Exercise 2.31 Show that the above construction defines a rational and generically injective map $\mathbb{P}^5 \simeq \mathbb{P}(W^*) \hookrightarrow S_V^{[5]}$. Its closure is a rational and hence Lagrangian subvariety of the 10-dimensional hyperkähler manifold $S_V^{[5]}$, cf. Section 3.1.

Corollary 2.32 *Using the above notation one finds the following equation*

$$[X_V] = [\mathbb{P}^4] + [S_V] \cdot \ell + 5 \cdot \ell^2 - [Y_H] \cdot \ell \tag{2.15}$$

in the Grothendieck ring of varieties $K_0(\mathrm{Var}_k)$. □

Note that (2.15) describes the class of $[X_V]$, while (2.11) only encodes $[\mathbb{P}^1] \cdot [X_V]$.

Exercise 2.33 Assume we are in the situation described by Corollary 2.27.

(i) Apply the blow-up formula in Mot(k) to relate $\mathfrak{h}(X_V)$ and $\mathfrak{h}(S_V)$ similarly to (2.12).

(ii) Combine (2.15) and (2.11) to show that $[Y_H]\cdot\ell\cdot(1+\ell) = (1+5\cdot\ell+\ell^2)\cdot\ell\cdot(1+\ell)$ in the Grothendieck ring of varieties $K_0(\mathrm{Var}_k)$. Since we know already that Y_H is a quintic del Pezzo surface, the factor $\ell\cdot(1+\ell)$ on both sides cancels out.

2.8 Cohomology Let us from now on assume that the ground field is $\mathbb{C}$. Then the above constructions can be exploited to relate $H^4(X_V,\mathbb{Z})$ to $H^2(S_V,\mathbb{Z})$.

Let us start with (2.13), where $p_H\colon B := B_{V,H} \longrightarrow H$ is the blow-up of $H \simeq \mathbb{P}^4$ in the surface $\tilde{S} := Z_{V,H}$, which itself is the blow-up of the K3 surface S_V in five points, and $q_H\colon B \longrightarrow X_V$ is the blow-up in the surface $Y_H \subset X_V$. This leads to isomorphisms of Hodge structures

$$H^4(B,\mathbb{Z}) \simeq H^4(X_V,\mathbb{Z}) \oplus H^2(Y_H,\mathbb{Z})(-1) \text{ and} \tag{2.16}$$

$$\begin{aligned} H^4(B,\mathbb{Z}) &\simeq H^4(H,\mathbb{Z}) \oplus H^2(\tilde{S},\mathbb{Z})(-1) \\ &\simeq H^4(H,\mathbb{Z}) \oplus H^2(S_V,\mathbb{Z})(-1) \oplus \mathbb{Z}(-2)^{\oplus 5}. \end{aligned} \tag{2.17}$$

The equations show that $b_2(Y_H) = 5$, which confirms the description of $[Y_H]$ in Exercises 2.28 and 2.33. The inclusion $H^4(X_V,\mathbb{Z}) \hookrightarrow H^4(B,\mathbb{Z})$ in (2.16), which respects the intersection pairing, followed by the projection $H^4(B,\mathbb{Z}) \twoheadrightarrow H^2(S_V,\mathbb{Z})(-1)$ in (2.17) gives

$$\xi\colon H^4(X_V,\mathbb{Z}) \longrightarrow H^2(S_V,\mathbb{Z})(-1),$$

which coincides with the composition of $H^4(X_V,\mathbb{Z}) \longrightarrow H^4(B,\mathbb{Z}) \longrightarrow H^4(\Sigma_V,\mathbb{Z})$ followed by the push-forward $H^4(\Sigma_V,\mathbb{Z}) \longrightarrow H^2(Z_V,\mathbb{Z})(-1)$ and the map induced by restriction $H^2(Z_V,\mathbb{Z}) \longrightarrow H^2(\tilde{S},\mathbb{Z})$.

Alternatively, we can look directly at the correspondence, cf. (2.10):

$$\begin{array}{ccccc} & & \Sigma_V & \overset{q}{\twoheadrightarrow} & X_V \\ & \overset{\pi_1}{\swarrow} & \downarrow & & \\ Z_V & & \downarrow & & \\ & \underset{\pi_2}{\searrow} & S_V & & \end{array}$$

and describe ξ as the induced map (possibly up to sign)

$$H^4(X_V,\mathbb{Z}) \xrightarrow{q^*} H^4(\Sigma_V,\mathbb{Z}) \xrightarrow{\pi_{1*}} H^2(Z_V,\mathbb{Z})(-1) \xrightarrow{\cdot u} H^4(Z_V,\mathbb{Z}) \xrightarrow{\pi_{2*}} H^2(S_V,\mathbb{Z})(-1),$$

where $u := \mathrm{c}_1(\mathcal{O}_{\pi_2}(1))$.

Proposition 2.34 (Hassett) *The map ξ is a map of Hodge structures of weight four which restricts to a Hodge isometry*

$$\langle h^2, [\Sigma_P]\rangle^{\perp} \xrightarrow[-1]{\sim} H^2(S_V,\mathbb{Z})_{\mathrm{pr}}(-1) \subset H^2(S_V,\mathbb{Z})(-1).$$

On the right-hand side, the primitive cohomology is with respect to the Plücker polarization of degree 14*, see Lemma* 2.3*, and the intersection pairing is altered by a sign.*

Proof By construction, ξ is a morphism of Hodge structures. To prove the second assertion, we may pick V generic and even general, because all the correspondences constructed before come in families. As alluded to in Remark 2.24, for the general Pfaffian cubic fourfold $\mathrm{rk}(H^{2,2}(X_V,\mathbb{Z})) = 2$. Hence, $\langle h^2, [\Sigma_P]\rangle^\perp \subset H^4(X_V,\mathbb{Z})$ is an irreducible Hodge structure containing $H^{3,1}(X_V)$. In fact, using (2.16) it can be thought of as the smallest saturated sub-Hodge structure of $H^4(B,\mathbb{Z})$ containing $H^{3,1}(B)$. Using (2.17) it can also be seen as the Tate twist of the transcendental lattice of $H^2(S_V,\mathbb{Z})$. Hence, for the very general Pfaffian cubic fourfold X_V, the K3 surface S_V has Picard rank one and ξ induces an isomorphism of Hodge structures $\langle h^2, [\Sigma_P]\rangle^\perp \simeq H^2(S_V,\mathbb{Z})(-1)$.

To see that it is also an isometry (with the sign changed on the right-hand side), recall that $H^2(S_V,\mathbb{Z})(-1) \hookrightarrow H^4(B,\mathbb{Z})$ is induced by $\alpha \longmapsto i_*\pi^*\alpha$, where we denote by $\pi\colon D := \Sigma_{V,H} \twoheadrightarrow \tilde{S} = Z_{V,H}$ the projection from the exceptional divisor and $i\colon D \hookrightarrow B$ is the inclusion, cf. the discussion in Section **2**.4.5. Then for $\alpha,\beta \in H^2(S_V,\mathbb{Z})$,

$$\begin{aligned}\int_B i_*\pi^*\alpha \cdot i_*\pi^*\beta &= \int_D i^*i_*\pi^*\alpha \cdot \pi^*\beta = \int_D [D]|_D \cdot \pi^*(\alpha\cdot\beta)\\ &= \int_{\tilde{S}} \pi_*([D]|_D)\cdot\alpha\cdot\beta = -(\alpha.\beta)_S,\end{aligned}$$

as $\pi_*([D]|_D) = -[\tilde{S}]$. □

In Example 3.25, the picture will be complemented by a description of the Néron–Severi lattice of the Fano variety $F(X_V)$ of the very general Pfaffian cubic. The ideal sheaf of Σ can be used to establish an equivalence between the derived category $\mathrm{D^b}(S_V)$ of the K3 surface S_V and the Kuznetsov component $\mathcal{A}_{X_V}$, a distinguished full triangulated subcategory of the derived category $\mathrm{D^b}(X_V)$ of the Pfaffian cubic fourfold, see Section **7**.3.2.

Remark 2.35 Let us summarize the discussion. We have seen that for each Pfaffian cubic fourfold X_V there is naturally associated a K3 surface S_V. The geometry of X_V and S_V are related in various ways:

(i) The blow-up of S_V in five points is the locus of indeterminacies for a birational correspondence $\mathbb{P}^4 \dashrightarrow X_V$, see Corollary 2.27. Purely numerically, one could think of the five points as the difference between $e(X_V) = 27$ and $b_2(S_V) = 22$.
(ii) The K3 surface S_V is the base of a family of quartic rational normal scrolls $\Sigma_P \subset X_V$, see Lemma 2.20.
(iii) There exists a birational correspondence $F(X_V) \sim S_V^{[2]}$, see Corollary 2.10 and Remark 2.13.

Remark 2.36 In the course of the preceding discussion the linear subspace $\mathbb{P}(V) \subset \mathbb{P}(\bigwedge^2 W^*)$ had to be chosen generically. More precisely, this was necessary to ensure the following properties:

(i) X_V is a smooth cubic of dimension four not containing any plane, see Lemma 2.3.
(ii) S_V is a smooth surface not containing a line, see Lemma 2.3.
(iii) $P \cap \mathrm{Ker}(\omega) \neq P$ for all $(P, \omega) \in \Sigma_P$, see Lemma 2.14, (ii).
(iv) Z_V is a smooth threefold or, more technically, $V \otimes \mathcal{O} \subset \bigwedge^2 W^* \otimes \mathcal{O} \longrightarrow \Omega_{\mathbb{P}(W)} \otimes \mathcal{O}(2)$ is generically surjective, see the proof of Lemma 2.15.

Each of these conditions describes a Zariski dense open subset inside $\mathbb{G}(5, \mathbb{P}(\bigwedge^2 W^*))$. Therefore, they hold simultaneously for V in a dense open subset. Note that, for example, the Zariski open subset for which X_V is smooth is strictly larger than the set for which in addition there are no planes contained in X_V.

3 The Fano Variety as a Hyperkähler Fourfold

In this section we will briefly recall the main notions concerning hyperkähler manifolds and prove that the Fano variety of lines contained in a smooth cubic fourfold is a hyperkähler fourfold deformation equivalent to the Hilbert square of a K3 surface. Furthermore, following Charles [114], we will show how to use Verbitsky's global Torelli theorem for hyperkähler manifolds to deduce Voisin's global Torelli theorem for cubic fourfolds [472]. This section also contains a discussion of various special subvarieties of the Fano variety from the hyperkähler as well as from the Fano perspective.

3.1 Hyperkähler Fourfolds The Fano variety $F(X)$ of lines contained in a smooth cubic fourfold $X \subset \mathbb{P}^5$ is, as X itself, smooth, projective, and of dimension four. However, unlike X, its canonical bundle is trivial $\omega_{F(X)} \simeq \mathcal{O}_{F(X)}$ and as such $F(X)$ belongs to a distinguished class of varieties.

There are three types of smooth, projective varieties with trivial canonical bundle. They serve as the building blocks for all of them: abelian varieties, hyperkähler (or irreducible holomorphic symplectic) manifolds, and Calabi–Yau manifolds. For the precise meaning of this statement, we refer to [45]. As it turns out, the Fano variety $F(X)$ belongs to the class of hyperkähler manifolds and we will focus on those exclusively.

Definition 3.1 A smooth, complex projective variety Z is called *hyperkähler* or *irreducible holomorphic symplectic* if Z is simply connected and $H^0(Z, \Omega_Z^2)$ is spanned by an everywhere non-degenerate form.

Note that a hyperkähler manifold is always of even dimension $\dim(Z) = 2m$.

Remark 3.2 (i) If $0 \neq \sigma \in H^0(Z, \Omega_Z^2)$ for a hyperkähler manifold Z, then the naturally induced map $\mathcal{T}_Z \xrightarrow{\sim} \Omega_Z$ is an alternating isomorphism and its Pfaffian $\mathrm{Pf}(\omega)$ defines a trivialization of ω_Z and so $\omega_Z \simeq \mathcal{O}_Z$.

(ii) For a hyperkähler manifold Z of dimension $2m$, the space $H^0(Z, \Omega_Z^k)$ is 0 for k odd and of dimension 1 for $0 \leq k \leq 2m$ even, i.e. $H^*(Z, \mathcal{O}_Z) \simeq H^*(\mathbb{P}^m, \mathbb{C})$. Conversely, if a smooth, projective variety Z has this property and the generator of $H^0(Z, \Omega_Z^2)$ is symplectic, then Z is also simply connected and hence hyperkähler, see [254, Prop. A.1].[9]

Example 3.3 By definition, two-dimensional hyperkähler manifolds are nothing but K3 surfaces. In higher dimensions, examples are provided by Hilbert schemes of K3 surfaces. More precisely, if S is a projective K3 surface, then the Hilbert scheme $S^{[m]}$ of all subschemes of S of length m is a hyperkähler manifold of dimension $2m$. This result was first proved by Fujiki for $m = 2$ and then generalized by Beauville [45, Sec. 6], cf. [253, Thm. 6.2.4].

The four-dimensional Hilbert scheme $S^{[2]}$ can be described geometrically as the quotient of the blow-up $\mathrm{Bl}_\Delta(S \times S) \longrightarrow S \times S$ in the diagonal $\Delta \subset S \times S$ by the natural action of $\mathfrak{S}_2$. The rational cohomology can then easily be computed as

$$\begin{aligned} H^*(S^{[2]}, \mathbb{Q}) &\simeq H^*(\mathrm{Bl}_\Delta(S \times S), \mathbb{Q})^{\mathfrak{S}_2} \\ &\simeq (H^*(S \times S, \mathbb{Q}) \oplus H^*(\Delta, \mathbb{Q})(-1))^{\mathfrak{S}_2} \\ &\simeq S^2 H^*(S, \mathbb{Q}) \oplus H^*(S, \mathbb{Q})(-1). \end{aligned}$$

This describes the even part of the Hodge diamond of $S^{[2]}$ up to the middle as

$$\begin{array}{ccccc} & & 1 & & \\ & 1 & 21 & 1 & \\ 1 & 21 & 232 & 21 & 1. \end{array}$$

Note that it coincides with the one of $F(X)$, see Section 0.2, (iii), and thus confirms Corollary 2.10. The explicit description of the cohomology of $S^{[2]}$ also allows one to deduce that cup product defines an isomorphism

$$S^2 H^2(S^{[2]}, \mathbb{Q}) \xrightarrow{\sim} H^4(S^{[2]}, \mathbb{Q}).$$

The integral version is more complicated: the injection

$$S^2 H^2(S^{[2]}, \mathbb{Z}) \hookrightarrow H^4(S^{[2]}, \mathbb{Z})$$

[9] The first step in the proof of [254, Prop. A.1], which excludes the abelian factor in any finite étale cover $\pi\colon \tilde{Z} \longrightarrow Z$, is incorrect. Instead, use the fact, that, on the one hand, $\chi(\mathcal{O}_{\tilde{Z}}) = \deg(\pi) \cdot \chi(\mathcal{O}_Z) = \deg(\pi) \cdot (1 + \dim(Z)/2) \neq 0$ and, on the other hand, $\chi(\mathcal{O}_{\tilde{Z}}) = 0$ if $\tilde{Z} \simeq Y \times \mathbb{C}^n/\Gamma$.

has index $2^{23} \cdot 5$ and, more precisely, its quotient is $(\mathbb{Z}/2\mathbb{Z})^{\oplus 23} \oplus (\mathbb{Z}/5\mathbb{Z})$, see the original [77, Prop. 5.6] or [428, Lem. 2.12]. The cohomology $H^*(S^{[2]}, \mathbb{Z})$ is known to be torsion free, see [428, Cor. 2.9] or [452, Thm. 1.1].

Definition 3.4 Let Z be a hyperkähler manifold of complex dimension $2m$. Then the *Beauville–Bogomolov–Fujiki* form q_Z is an integral quadratic form on $H^2(Z, \mathbb{Z})$ for which there exists a positive constant c, the *Fujiki constant*, such that for all $\alpha \in H^2(Z, \mathbb{Z})$

$$c \cdot q_Z(\alpha)^m = \int_Z \alpha^{2m}.$$

The main properties of q are the following, see the original [45, 187] or [207]:

(i) The signature of q_Z is $(3, b_2 - 3)$.
(ii) If $b_2 \neq 6$, then there exists a unique primitive such q_Z.
(iii) Up to scaling,

$$q_Z(\alpha) = \lambda\bar{\lambda} + (n/2) \int_Z \beta^2 (\sigma\bar{\sigma})^{m-1} \tag{3.1}$$

for $\beta = \alpha - \lambda \cdot \sigma - \bar{\lambda} \cdot \bar{\sigma} \in H^{1,1}(Z)$ and assuming $\int_Z (\sigma\bar{\sigma})^m = 1$.
(iv) If $\gamma \in H^2(Z, \mathbb{Z})$ is primitive with respect to an ample class $\alpha \in H^2(Z, \mathbb{Z})$, then

$$\int_Z \alpha^{2m} \cdot q_Z(\gamma) = (2m - 1) \cdot q_Z(\alpha) \cdot \int_Z \gamma^2 \cdot \alpha^{2m-2}. \tag{3.2}$$

Remark 3.5 If $\alpha \in H^2(Z, \mathbb{Z})$ is an ample class, then $H^2(Z, \mathbb{Z})_{\rm pr}$ is a priori defined as the orthogonal complement of α with respect to the Hodge–Riemann pairing or, equivalently, as the kernel of $\alpha^{2m-1} \colon H^2(Z, \mathbb{Z}) \twoheadrightarrow H^{4m}(Z, \mathbb{Z})$. However, $H^2(Z, \mathbb{Z})_{\rm pr}$ can also be described as the kernel of the Beauville–Bogomolov–Fujiki pairing viewed as a linear map $q_Z(\alpha, \ ) \colon H^2(Z, \mathbb{Z}) \twoheadrightarrow \mathbb{Z}$, i.e.

$$H^2(Z, \mathbb{Z})_{\rm pr} = \alpha^{\perp_{\rm BBF}} \subset H^2(Z, \mathbb{Z}).$$

Indeed, both kernels define saturated sub-Hodge structures containing $H^{2,0}(Z)$. To conclude, use the fact that the two maps α^{2m-1} and $q_Z(\alpha, \ )$ are unchanged under deformations preserving α as a $(1, 1)$-class and that for a general such deformation both kernels are irreducible Hodge structures containing $H^{2,0}$ and, therefore, coincide. The formula in (iv) then shows that on $H^2(Z, \mathbb{Z})_{\rm pr}$ the Hodge–Riemann pairing and q_Z differ by a scalar factor only.

Example 3.6 The Beauville–Bogomolov–Fujiki form on the Hilbert scheme $S^{[2]}$ of a K3 surface S defines a lattice that is isometrically described as follows, cf. [45, Sec. 6]:

$$(H^2(S^{[2]}, \mathbb{Z}), q) \simeq H^2(S, \mathbb{Z}) \oplus \mathbb{Z} \cdot \delta. \tag{3.3}$$

Here, $H^2(S,\mathbb{Z})$ is endowed with the intersection form of the K3 surface S and the class δ is orthogonal to it with $q(\delta) = -2$. Geometrically, 2δ is the class of the exceptional divisor of the Hilbert–Chow morphism $S^{[2]} \longrightarrow S^{(2)}$.

Furthermore, the Fujiki constant in this case is $c = 3$, i.e.

$$3 \cdot q(\alpha)^2 = \int_{S^{[2]}} \alpha^4$$

for all $\alpha \in H^2(S^{[2]},\mathbb{Z})$. Indeed, if L is a line bundle on S and $L^{[2]}$ is the corresponding line bundle on $S^{[2]}$, then (3.3) says $q(\mathrm{c}_1(L^{[2]})) = (\mathrm{c}_1(L).\mathrm{c}_1(L))$. On the other hand, $L^{[2]}$ is the pullback of the line bundle $L^{(2)}$ on $S^{(2)}$, which by definition is the descent of $L \boxtimes L$ on $S \times S$ to $S^{(2)}$, and, therefore,

$$\begin{aligned}\int_{S^{[2]}} \mathrm{c}_1(L^{[2]})^4 &= \int_{S^{(2)}} \mathrm{c}_1(L^{(2)})^4 = (1/2) \cdot \int_{S\times S} \mathrm{c}_1(L \boxtimes L)^4 \\ &= (1/2) \cdot \binom{4}{2} \cdot \left(\int_S \mathrm{c}_1(L) \cdot \mathrm{c}_1(L) \right)^2 = 3 \cdot (\mathrm{c}_1(L).\mathrm{c}_1(L))^2 = 3 \cdot q(\mathrm{c}_1(L^{[2]}))^2.\end{aligned}$$

The blow-up $\tilde{S} := \mathrm{Bl}_x(S)$ of the surface S in an arbitrary point $x \in S$ is naturally contained in the Hilbert scheme $\tilde{S} \hookrightarrow S^{[2]}$ as the subvariety parametrizing all subschemes of length two containing x. The above computation shows that its cohomology class $[\tilde{S}]$ describes the Beauville–Bogomolov–Fujiki form q on $\delta^\perp$. More precisely, for all $\alpha \in H^2(S,\mathbb{Z}) = \delta^\perp \subset H^2(S^{[2]},\mathbb{Z})$, one has

$$q(\alpha) = \int_{\tilde{S}} \alpha|_{\tilde{S}} \cdot \alpha|_{\tilde{S}} = \int_{S^{[2]}} [\tilde{S}] \cdot \alpha \cdot \alpha. \tag{3.4}$$

For the class δ, one finds $q(\delta) = 2 \int_{S^{[2]}} [\tilde{S}] \cdot \delta \cdot \delta$, because the restriction $\delta|_{\tilde{S}}$ is the class of the exceptional divisor of the blow-up map $\tilde{S} \longrightarrow S$ and, hence, $\int_{\tilde{S}} \delta|_{\tilde{S}} \cdot \delta|_{\tilde{S}} = -1$.

Remark 3.7 Shen [428] defines a hyperkähler fourfold Z to be of *Jacobian type* if the form q_Z on transcendental classes is represented by an effective surface class as in (3.4). We have seen one other example already in Remark 1.19.

The Hodge structure of weight two provided by $H^2(Z,\mathbb{Z})$ together with the form q_Z determines much of the geometry of the hyperkähler manifold Z. The classical global Torelli theorem for K3 surfaces, a result by Pjateckiĭ-Šapiro and Šafarevič in the algebraic context and by Burns and Rapoport in the non-algebraic setting, is the most striking example, see Section 6.3 or [249] for the statement and references. A variant of the global Torelli theorem, proved by Verbitsky [467, 468], holds in higher dimensions also, see also the more recent work of Looijenga [328] and the Bourbaki survey [248]. The polarized version was established by Markman [337].

Theorem 3.8 (Verbitsky, Markman) *Assume* $\eta\colon (H^2(Z,\mathbb{Z}), q_Z) \xrightarrow{\sim} (H^2(Z',\mathbb{Z}), q_{Z'})$ *is a Hodge isometry between two hyperkähler manifolds* Z *and* Z' *that can be realized as* *parallel transport along a proper, smooth, connected family $\mathcal{Z} \longrightarrow C$ with fibres $Z = \mathcal{Z}_1$ and $Z' = \mathcal{Z}_2$. Then Z and Z' are birational.*

Moreover, if $\eta(\omega) = \omega'$ for some Kähler classes ω on Z and ω' on Z', then there exists an isomorphism $f \colon Z \xrightarrow{\sim} Z'$ such that $\eta = f_*$.

Remark 3.9 In general, the morphism f is not uniquely determined by its action $f_* = \eta$. However, according to a result of Beauville [44], it is or, equivalently, the representation $\mathrm{Aut}(Z) \longrightarrow \mathrm{Aut}(H^2(Z, \mathbb{Z}))$ is, faithful if Z is deformation equivalent to the Hilbert scheme $S^{[n]}$ of a K3 surface S.

3.2 Beauville–Donagi: Fano Variety versus Hilbert Scheme In Section 2 we have observed links between Pfaffian cubic fourfolds and K3 surfaces. A general link between cubic fourfolds and the hyperkähler world is established by the following result, which was originally proved by Beauville and Donagi [58] by means of the isomorphism $S^{[2]} \simeq F(X)$ for generic Pfaffian cubic fourfolds, see Corollary 2.10. The proof below is 'Pfaffian free'.

Theorem 3.10 (Beauville–Donagi) *The Fano variety $F = F(X)$ of lines contained in a smooth cubic fourfold $X \subset \mathbb{P}^5$ is a hyperkähler manifold of dimension four.*

In the sequel, q_F shall denote the Beauville–Bogomolov–Fujiki pairing on the hyperkähler fourfold $F = F(X)$.

Proof We know that $H^{2,0}(F)$ is one-dimensional, see Section **2**.4.6. Pick any $0 \neq \sigma \in H^{2,0}(F)$. We claim that $\sigma \wedge \sigma \in H^{4,0}(F) \simeq H^0(F, \Omega_F^4)$ is non-zero. Indeed, this follows immediately from Corollary **2**.5.15, see also the analogous statement for threefolds stated as Corollary **5**.2.5. Then, as $\Omega_F^4 \simeq \omega_F \simeq \mathcal{O}_F$, the form $\sigma \wedge \sigma$ defines a trivializing section of ω_F. Therefore, the induced map $\sigma \colon \mathcal{T}_F \longrightarrow \Omega_F$ is an isomorphism, i.e. σ is a holomorphic symplectic form.

In order to conclude that F is a hyperkähler manifold, it suffices to show that it is simply connected.[10] As we know the Hodge numbers of F, see Section **2**.4.6, this follows from Remark 3.2, (ii). □

The result combined with the description of $F(X_V)$ for the generic Pfaffian cubic fourfold, see Corollary 2.10, immediately leads to the following.

Corollary 3.11 *The Fano variety $F(X)$ of lines contained in a smooth cubic fourfold X is a hyperkähler manifold of dimension four. It is deformation equivalent (hence,*

[10] It is tempting to try to prove that $F(X)$ is simply connected by using the realization of $\mathbb{L} \subset \mathbb{P}(\mathcal{T}_X)$ as the zero set of a section of the bundle E given as the extension $\pi^*\mathcal{O}(3) \otimes \mathcal{O}_\pi(3) \longrightarrow E \longrightarrow \pi^*\mathcal{O}(3) \otimes \mathcal{O}_\pi(2)$, see Proposition **2**.3.10. However, neither one of the two involved invertible sheaves is ample. See also Remark **2**.4.19. However, arguments of Ottem in [244] may also work here.

diffeomorphic and homeomorphic) to the Hilbert scheme $S^{[2]}$ of a K3 surface and its Fujiki constant is $c = 3$. In particular, $F(X)$ is simply connected. □

Corollary 3.12 *For a smooth cubic fourfold, the natural map*

$$\mathbb{C}^{\oplus 20} \simeq H^1(X, \mathcal{T}_X) \hookrightarrow H^1(F(X), \mathcal{T}_{F(X)}) \simeq \mathbb{C}^{\oplus 21}$$

is injective of corank one.

Proof The injectivity is Corollary 2.5.10. For the dimensions, use $h^1(X, \mathcal{T}_X) = 20$ by Example 1.4.15 and $h^1(F(X), \mathcal{T}_{F(X)}) = h^1(F(X), \Omega_{F(X)}) = h^{1,1}(F(X)) = 21$. □

In particular, contrary to the case of cubic threefolds, see Proposition 5.2.14, there exists one additional direction of deformation in which $F(X)$ ceases to be the Fano variety of lines of a cubic fourfold.

Remark 3.13 For an arbitrary hyperkähler manifold Z of dimension $2m$, the subalgebra of $H^*(Z, \mathbb{C})$ generated by $H^2(Z, \mathbb{C})$, i.e. the image of the map $S^*H^2(Z, \mathbb{C}) \longrightarrow H^{2*}(Z, \mathbb{C})$, is isomorphic to the quotient $S^*H^2(Z, \mathbb{C})/(H_{m+1})$, cf. [74]. Here, $H_{m+1} \subset S^{m+1}H^2(Z, \mathbb{C})$ is the space of harmonic polynomials which is in fact spanned by all classes of the form α^{m+1} with $q(\alpha) = 0$ (at least over $\mathbb{C}$). For a hyperkähler manifold that is deformation equivalent to $S^{[2]}$, so for example $F(X)$, this becomes an isomorphism

$$S^*H^2(F(X), \mathbb{C})/(H_3) \simeq H^{2*}(F(X), \mathbb{C}).$$

Compare this to the situation for cubic threefolds explained in Remark 5.3.13 and Appendix 5.6. The space of harmonic forms $H_3 \subset S^3H^2(F(X), \mathbb{C})$ here corresponds there to the primitive forms $P_3 \subset \bigwedge^3 H^1(F(Y), \mathbb{C}) \simeq \bigwedge^3 H^1(J(Y), \mathbb{C}) \simeq H^3(J(Y), \mathbb{C})$.

Remark 3.14 (i) The Fano variety $F(X)$ comes with the natural polarization g provided by the Plücker embedding. As the purely topological quantity $q_F(g)$ stays constant under deformations, it can be computed on the Fano variety of an arbitrary smooth cubic fourfold X. Combining Lemma 2.12 and Example 3.6, we obtain an isometry

$$\left(H^2(F(X), \mathbb{Z}), q_F\right) \simeq \left(H^2(S^{[2]}, \mathbb{Z}), q\right) \simeq H^2(S, \mathbb{Z}) \oplus \mathbb{Z}(-2)$$

that maps g to $2 \cdot g_S - 5 \cdot \delta$ with $g_S \in H^2(S, \mathbb{Z})$ satisfying $(g_S.g_S) = 14$. This proves

$$q_F(g) = 6$$

and $2 \mid (g.\alpha)$ for all $\alpha \in H^2(F(X), \mathbb{Z})$, i.e. the divisibility of the Plücker polarization g is two.

According to a criterion of Eichler, any primitive class of the same square, the same divisibility, and the same class in the discriminant group is contained in the same orbit of the action of the orthogonal group of $H^2(S^{[2]}, \mathbb{Z})$.

(ii) Let us use the above to show that

$$H^2(F(X),\mathbb{Z})_{\rm pr} \oplus \mathbb{Z}\cdot g \subset H^2(F(X),\mathbb{Z}) \tag{3.5}$$

is of index three. Since the orthogonal group $\mathrm{O}(\Lambda)$ acts transitively on the set of all primitive elements of a fixed square in the unimodular lattice $\Lambda := H^2(S,\mathbb{Z}) \simeq U^{\oplus 3} \oplus E_8(-1)^{\oplus 2}$, see [249, Cor. 14.1.10], we may assume that $g = 2\cdot g_S - 5\cdot\delta$ corresponds to $2\cdot(e+7\cdot f) - 5\cdot\delta \in \Lambda\oplus\mathbb{Z}\cdot\delta$, where e, f is the standard basis of the first copy of U. Then

$$H^2(F(X),\mathbb{Z})_{\rm pr} \simeq g^\perp \simeq \langle e-7\cdot f, 5\cdot f-\delta\rangle \oplus U^{\oplus 2} \oplus E_8(-1)^{\oplus 2}.$$

The intersection matrix of the plane $\langle e-7\cdot f, 5\cdot f-\delta\rangle$ is

$$\begin{pmatrix} -14 & 5 \\ 5 & -2 \end{pmatrix}.$$

Hence, $\mathrm{disc}(H^2(F(X),\mathbb{Z})_{\rm pr}) = 3$ and, therefore, $\mathrm{disc}(H^2(F(X),\mathbb{Z})_{\rm pr}\oplus\mathbb{Z}\cdot g) = 18$. Since $\mathrm{disc}(H^2(F(X),\mathbb{Z})) = \mathrm{disc}(\Lambda\oplus\mathbb{Z}\cdot\delta) = -2$, we conclude that the index of the inclusion (3.5) is indeed three, cf. [249, Sec. 14.0.2].

Note that the above arguments in particular show that $g \in H^2(F(X),\mathbb{Z})$ is primitive. Otherwise the index in (ii) would be bigger or, more directly, because if $g = n\cdot g'$ for some integer $n \neq \pm 1$, then n^2 would divide $q_F(g) = 6$, which is absurd.

We will come back to the quadratic form q_F at the end of Section 3.4. For now we summarize the situation by the diagram

$$\begin{array}{ccccccccc} 0 & \longrightarrow & H^2(F(X),\mathbb{Z})_{\rm pr} & \longrightarrow & H^2(F(X),\mathbb{Z}) & \longrightarrow & \mathbb{Z} & \longrightarrow & 0 \\ & & & & \cup & & \cup & & \\ & & & & \mathbb{Z}\cdot g & \xrightarrow{\sim} & 3\mathbb{Z}. & & \end{array} \tag{3.6}$$

Remark 3.15 A smooth hyperplane section $Y := X\cap H$ of a smooth cubic fourfold $X \subset \mathbb{P}^5$ is a smooth cubic threefold $Y \subset H \simeq \mathbb{P}^4$ and its intermediate Jacobian $J(Y)$ is a principally polarized abelian variety of dimension five, see Section 5.2. Varying H, means $J(Y)$ varies also, which gives a smooth family $\pi\colon J(\mathcal{Y}/U) \longrightarrow U := |\mathcal{O}_X(1)|_{\rm sm}$.

This construction was first studied by Donagi and Markman [163] who in particular showed that $J(\mathcal{Y}/U)$ has a natural holomorphic symplectic structure for which π is a Lagrangian fibration. The natural reflex then was to try to compactify $J(\mathcal{Y}/U)$ to a hyperkähler manifold. This was eventually achieved by Laza, Saccà, and Voisin [314] for general X and for all smooth X by Saccà [411]. It turns out that the compactification is of OG10 type, i.e. deformation equivalent to the sporadic example of a hyperkähler manifold in dimension 10 constructed by O'Grady [374]. A twisted version was studied by Voisin [484].

Remark 3.16 Let us explain the modular description of a non-degenerate two-form $\sigma \in H^0(F(X), \Omega^2_{F(X)}) \simeq H^{2,0}(F(X))$ given by de Jong and Starr [135, Sec. 5]. A more explicit approach goes back to Iliev and Manivel [257, Sec. 2.1].

For a line $L \subset X$, the normal bundles of the nested inclusions $L \subset X \subset \mathbb{P} = \mathbb{P}^5$ sit in the natural short exact sequence

$$0 \longrightarrow \mathcal{N}_{L/X} \longrightarrow \mathcal{N}_{L/\mathbb{P}} \longrightarrow \mathcal{N}_{X/\mathbb{P}}|_L \longrightarrow 0 \tag{3.7}$$

with $\mathcal{N}_{X/\mathbb{P}} \simeq \mathcal{O}_X(3)$. Taking exterior products and twisting with $\mathcal{O}_L(-3)$ produces the short exact sequence

$$0 \longrightarrow \bigwedge^3 \mathcal{N}_{L/X} \otimes \mathcal{O}_L(-3) \longrightarrow \bigwedge^3 \mathcal{N}_{L/\mathbb{P}} \otimes \mathcal{O}_L(-3) \longrightarrow \bigwedge^2 \mathcal{N}_{L/X} \longrightarrow 0,$$

the boundary map δ of which is used to construct a map

$$\bigwedge^2 H^0(L, \mathcal{N}_{L/X}) \longrightarrow H^0(L, \bigwedge^2 \mathcal{N}_{L/X}) \xrightarrow{\delta} H^1(L, \bigwedge^3 \mathcal{N}_{L/X} \otimes \mathcal{O}_L(-3)) \simeq k. \tag{3.8}$$

For the last isomorphism, use the fact that the normal bundle is $\mathcal{N}_{L/X} \simeq \mathcal{O}_L(1) \oplus \mathcal{O}_L \oplus \mathcal{O}_L$ or $\mathcal{N}_{L/X} \simeq \mathcal{O}_L(1) \oplus \mathcal{O}_L(1) \oplus \mathcal{O}_L(-1)$, see Lemma **2**.1.13 or (0.1), and that in both cases $\bigwedge^3 \mathcal{N}_{L/X} \otimes \mathcal{O}(-3) \simeq \mathcal{O}_L(-2)$. Note that the boundary map δ is always surjective and that for L of the first type the natural map $\bigwedge^2 H^0(\mathcal{N}_{L/X}) \twoheadrightarrow H^0(\bigwedge^2 \mathcal{N}_{L/X})$ is surjective also. Composing (3.8) with the canonical isomorphism $T_{[L]}F(X) \simeq H^0(L, \mathcal{N}_{L/X})$, see Proposition **2**.1.10, we obtain a two-form at the point $[L] \in F = F(X)$:

$$\sigma_{[L]} \colon \bigwedge^2 T_{[L]}F \longrightarrow k.$$

Note that for lines of the first type $\sigma_{[L]}$ is by construction non-zero.

The construction globalizes as follows: first, replace (3.7) by

$$0 \longrightarrow \mathcal{N}_{\mathbb{L}/F\times X} \longrightarrow \mathcal{N}_{\mathbb{L}/F\times\mathbb{P}} \longrightarrow q^*\mathcal{O}_X(3) \longrightarrow 0. \tag{3.9}$$

Then use the global description of the tangent bundle $\mathcal{T}_F \simeq p_*\mathcal{N}_{\mathbb{L}/F\times X}$, cf. Exercise **2**.1.18, and apply Remark **2**.3.8 and Exercise **2**.3.9 to deduce

$$\begin{aligned}\bigwedge^3 \mathcal{N}_{\mathbb{L}/F\times X} \otimes q^*\mathcal{O}_X(-3) &\simeq \bigwedge^3 \varpi^*(\mathcal{T}_\pi \otimes \mathcal{O}_\pi(-1)) \otimes q^*\mathcal{O}_X(-3)\\ &\simeq \varpi^*(\pi^*\omega_X^* \otimes \mathcal{O}_\pi(1)) \otimes q^*\mathcal{O}_X(-3) \simeq \varpi^*\mathcal{O}_\pi(1)\\ &\simeq \mathcal{O}_p(-2) \otimes p^*\mathcal{O}_F(1).\end{aligned}$$

Finally, applying Grothendieck–Verdier duality to $p \colon \mathbb{L} \longrightarrow F = F(X)$, we obtain a two-form $\sigma \in H^0(F(X), \Omega^2_{F(X)})$ as the composition with the boundary map

$$\bigwedge^2 \mathcal{T}_F \simeq \bigwedge^2 p_*\mathcal{N}_{\mathbb{L}/F\times X} \longrightarrow p_* \bigwedge^2 \mathcal{N}_{\mathbb{L}/F\times X} \longrightarrow R^1p_*\mathcal{O}_p(-2) \otimes \mathcal{O}_F(1) \simeq \mathcal{O}_F$$

of $\bigwedge^3$ of (3.9). At each point $[L] \in F(X)$ the construction gives back $\sigma_{[L]}$ and, because $\sigma_{[L]} \neq 0$ for every line L of the first type and those do exist by Lemma **2**.2.12, σ is certainly not trivial.

As $F(X)$ is holomorphic symplectic and $H^0(F(X), \Omega^2_{F(X)})$ is one-dimensional, σ defines a holomorphic symplectic structure on $F(X)$. Alternatively, one can use the fact that ω_F is trivial and that σ is non-degenerate at the generic point. This in particular proves that $\sigma_{[L]}$ is also non-degenerate for lines L of the second type, which is not quite so obvious in the direct description above.

Beware that the above construction does not define a canonical symplectic form $\sigma \in H^0(F(X), \Omega^2_{F(X)})$ on $F(X)$. Indeed, implicitly in the above construction we have used an isomorphism $\det(\mathcal{T}_X) \otimes \mathcal{N}^*_{X/\mathbb{P}} \simeq \mathcal{O}_X$ or, equivalently, $\mathcal{N}^{\otimes 2}_{X/\mathbb{P}} \simeq \mathcal{O}_X(6)$, which does depend on the choice of an equation in $H^0(\mathbb{P}, \mathcal{O}(3))$ that defines X. Recall that from the Fano correspondence one obtains a canonical isomorphism $H^{3,1}(X) \simeq H^{2,0}(F(X))$, but that the isomorphism $H^{3,1}(X) \simeq R_{t(3)} \simeq \mathbb{C}$ in Theorem **1**.4.21 also depends on the choice of a defining equation for X which is only unique up to scaling.

Later we will describe a more categorical approach to the construction of the symplectic form on $F(X)$, see Section **7**.3.6.

3.3 Global Torelli Theorem for Cubic Fourfolds The global Torelli theorem is a cornerstone result in the theory of smooth cubic fourfolds. It was originally proved by Voisin in [472, 477], see also [255, 327, 499]. The proof we present here was given by Charles [114]. It makes use of the global Torelli theorem for hyperkähler manifolds, Theorem 3.8, and the geometric global Torelli theorem for cubic hypersurfaces, Proposition **2**.3.12.

Theorem 3.17 (Voisin) *Assume* $\zeta\colon H^4(X, \mathbb{Z}) \xrightarrow{\sim} H^4(X', \mathbb{Z})$ *is a Hodge isometry between two smooth cubic fourfolds with* $\zeta(h_X^2) = h_{X'}^2$*. Then there exists a unique isomorphism* $\phi\colon X \xrightarrow{\sim} X'$ *with* $\phi_* = \zeta$*.*

Proof From Section **1**.2.4 we know that any isometry $H^4(X, \mathbb{Z}) \simeq H^4(X', \mathbb{Z})$ mapping h_X^2 to $h_{X'}^2$ can be realized by parallel transport. In other words, there exists a smooth and projective family $\mathcal{X} \longrightarrow C$ of cubic fourfolds over a connected base C with fibres $X = \mathcal{X}_1$ and $X' = \mathcal{X}_2$ such that parallel transport from $\mathcal{X}_1$ to $\mathcal{X}_2$ describes ζ.

Parallel transport along the induced relative family of Fano varieties $F(\mathcal{X}/C) \longrightarrow C$ with special fibres $F(X) = F(\mathcal{X}/C)_1$ and $F(X') = F(\mathcal{X}/C)_2$ defines a bijective map $\eta\colon H^2(F(X), \mathbb{Z}) \xrightarrow{\sim} H^2(F(X'), \mathbb{Z})$, which is automatically an isometry with respect to the Beauville–Bogomolov–Fujiki pairings $q_{F(X)}$ and $q_{F(X')}$.

As the Fano correspondence is purely topological and exists in families, one has a commutative diagram

$$\begin{array}{ccc} H^4(X,\mathbb{Q}) & \xrightarrow[\zeta]{\sim} & H^4(X',\mathbb{Q}) \\ \varphi_X \downarrow \wr & & \wr \downarrow \varphi_{X'} \\ H^2(F(X),\mathbb{Q}) & \xrightarrow[\eta]{\sim} & H^2(F(X'),\mathbb{Q}). \end{array}$$

Moreover, since φ_X and $\varphi_{X'}$ by construction and ζ by assumption are isomorphisms of Hodge structures, then $\eta\colon H^2(F(X),\mathbb{Z}) \xrightarrow{\sim} H^2(F(X'),\mathbb{Z})$ is a Hodge isometry also.[11] Furthermore, the Plücker polarizations g_X and $g_{X'}$ on $F(X)$ and $F(X')$ satisfy $\varphi_X(h_X^2) = g_X$ and $\varphi_{X'}(h_{X'}^2) = g_{X'}$, see Lemma **2**.5.1.

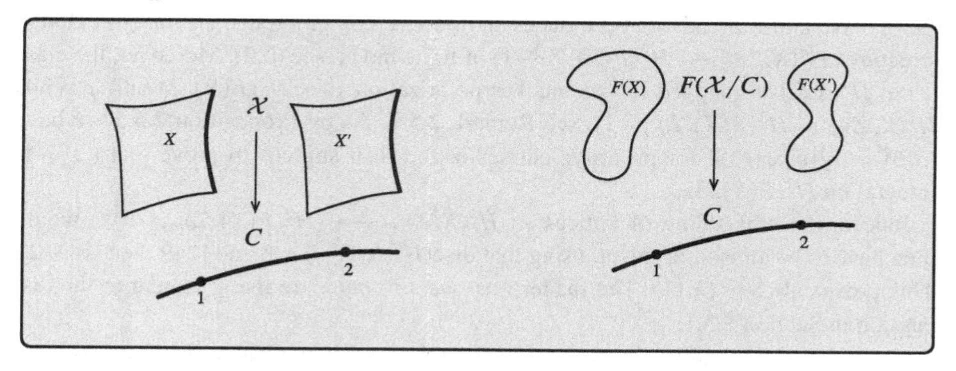

By assumption $\zeta(h_X^2) = h_{X'}^2$ and hence $\eta(g_X) = g_{X'}$. Hence, by Theorem 3.8, η is induced by an isomorphism $f\colon F(X) \xrightarrow{\sim} F(X')$ with $f_* = \eta$. The existence of a unique isomorphism $\phi\colon X \xrightarrow{\sim} X'$ inducing f, η, and the original ζ then follows from the geometric global Torelli theorem, Proposition **2**.3.12, see also the comments after its proof and Remark 3.9. □

Remark 3.18 Note that Voisin's global Torelli theorem can be rephrased by saying that two smooth cubic fourfolds X and X' are isomorphic if and only if there exists a Hodge isometry $H^4(X,\mathbb{Z})_{\rm pr} \simeq H^4(X',\mathbb{Z})_{\rm pr}$. Indeed, any such Hodge isometry, possibly after changing it by a sign, extends to a Hodge isometry $H^4(X,\mathbb{Z}) \simeq H^4(X',\mathbb{Z})$, as above. We will rephrase the result more in the spirit of Theorem **5**.4.3 and Remark **5**.4.5 in Proposition 4.15 and as the injectivity of the period map in Theorem 6.12.

3.4 Beauville–Donagi: Fano Correspondence The following result of Beauville–Donagi [58] can be proved via the isomorphism in Corollary 2.10 for generic Pfaffian cubic fourfolds. However, here we will keep the discussion closer to the one in Section

[11] It is worth pointing out that we are not using the fact that φ is an (integral Hodge) isometry, which it is only on the primitive part, see Proposition 3.19 below. The parallel transport η is automatically an isometry.

2.5 and the one for cubic threefolds in Section **5**.2, although Pfaffian cubics still play a role.

Proposition 3.19 (Beauville–Donagi) *The Fano correspondence, cf. Section* **2**.5*, induces an isomorphism of integral Hodge structures*

$$\varphi\colon H^4(X,\mathbb{Z}) \xrightarrow{\sim} H^2(F(X),\mathbb{Z})(-1), \tag{3.10}$$

which induces a Hodge isometry(!) between their primitive parts:

$$\varphi\colon H^4(X,\mathbb{Z})_{\rm pr} \xrightarrow{\sim} H^2(F(X),\mathbb{Z})_{\rm pr}(-1). \tag{3.11}$$

Here, $H^2(F(X),\mathbb{Z})_{\rm pr}$ is endowed with the quadratic form $(\gamma.\gamma')_F := (-1/6)\int_F \gamma\cdot\gamma'\cdot g^2$.

Proof We know already that φ induces an injective homomorphism of integral Hodge structures $H^4(X,\mathbb{Z}) \hookrightarrow H^2(F(X),\mathbb{Z})(-1)$ of finite index, see (0.2). Moreover, the class $h^2 \in H^4(X,\mathbb{Z})$ is mapped to the Plücker polarization $g \in H^2(F(X),\mathbb{Z})$ and φ sends $H^4(X,\mathbb{Z})_{\rm pr}$ to $H^2(F(X),\mathbb{Z})_{\rm pr}(-1)$, see Remark **2**.5.7. As by Proposition **2**.5.5 we have $(\alpha.\beta) = (\varphi(\alpha).\varphi(\beta))_F$ for primitive classes α and β, it suffices to prove that $(\ .\ )_F$ is integral on $H^2(F(X),\mathbb{Z})_{\rm pr}$.

Indeed, the embedding of lattices $\varphi\colon H^4(X,\mathbb{Z})_{\rm pr} \hookrightarrow (H^2(F(X),\mathbb{Z})_{\rm pr}, (\ .\ )_F)$ would then have to be an isomorphism, using that $\operatorname{disc}(H^4(X,\mathbb{Z})_{\rm pr}) = 3$ and [249, Sec. 14.0.2]. This proves already (3.11). The reader may want to compare the argument to the discussion in Section **5**.3.1.

To prove the integrality of $(\ .\ )_F$ on $H^2(F(X),\mathbb{Z})_{\rm pr}$, we may assume that X is a generic Pfaffian cubic X_V. Then, by virtue of (2.5) in Corollary 2.10 and the induced isomorphism of integral Hodge structures $H^2(F(X_V),\mathbb{Z}) \simeq H^2(S_V^{[2]},\mathbb{Z})$, which according to Lemma 2.12 maps g to $2\cdot g_S - 5\cdot\delta$, it suffices to show that for all $\gamma,\gamma' \in (2\cdot g_S - 5\cdot\delta)^{\perp_{\rm BBF}} \subset H^2(S_V^{[2]},\mathbb{Z})$ the integral $\int_{S_V^{[2]}} \gamma\cdot\gamma'\cdot(2\cdot g_S - 5\cdot\delta)^2$ is divisible by six. This follows from (3.2), see also the proof of Corollary 3.21 below. As an alternative to the argument used in the above proof, one can use (3.12) below and the explicit description of the lattice provided by the primitive cohomology $(H^2(F(X),\mathbb{Z})_{\rm pr}, q_F)$, see Remark 3.14.

To prove (3.10), consider the commutative diagram

$$\begin{array}{ccc} H^4(X,\mathbb{Z})_{\rm pr} \oplus \mathbb{Z}\cdot h^2 & \xrightarrow{\sim} & H^2(F(X),\mathbb{Z})_{\rm pr}(-1) \oplus \mathbb{Z}\cdot g \\ \cap & & \cap \\ H^4(X,\mathbb{Z}) & \hookrightarrow & H^2(F(X),\mathbb{Z})(-1). \end{array}$$

The upper horizontal map is an isomorphism of Hodge structures, which is an isometry only on the first summand. We know that the left vertical map is an inclusion of index three and the same is true for the right vertical map by Remark 3.14. Hence, the lower horizontal map is also an isomorphism of Hodge structures (but not an isometry). □

Remark 3.20 The use of Pfaffian cubic fourfolds and their relation to K3 surfaces seems unavoidable. Unlike the case of cubic threefolds, it seems difficult to prove integrality of $(\ .\ )_F$ on $H^2(F(X),\mathbb{Z})_{\rm pr}$ just by using the Fano description of $F(X)$. One could try to follow the idea of Corollary **5**.1.20, but g^2 is actually not even divisible by three. Indeed, as according to [77, Prop. 5.6] the quotient of the inclusion $S^2H^2(F(X),\mathbb{Z})\subset H^4(F(X),\mathbb{Z})$ is of the form $(\mathbb{Z}/2\mathbb{Z})^{\oplus 23}\oplus\mathbb{Z}/5\mathbb{Z}$, one would otherwise have that $g^2\in S^2H^2(F(X),\mathbb{Z})$ is divisible by three, which it clearly is not.

The difference between cubic threefolds and cubic fourfolds with respect to the divisibility will come up again in Lemma 4.5.

To emphasize the fact that $F(X)$ is a hyperkähler manifold, let us rephrase the above result in terms of the Beauville–Bogomolov–Fujiki form.

Corollary 3.21 *For any smooth cubic fourfold $X\subset\mathbb{P}^5$, the Fano correspondence induces a Hodge isometry*

$$\left(H^4(X,\mathbb{Z})_{\rm pr},(\ .\ )\right)\simeq\left(H^2(F(X),\mathbb{Z})_{\rm pr}(-1),-q_F\right).$$

In particular, it defines an isometry of lattices

$$\left(H^{2,2}(X,\mathbb{Z})_{\rm pr},(\ .\ )\right)\simeq\left(H^{1,1}(F(X),\mathbb{Z})_{\rm pr},-q_F\right)\simeq\left({\rm NS}(F(X))_{\rm pr},-q_F\right).$$

Furthermore, the inclusion

$$H^2(F(X),\mathbb{Z})_{\rm pr}\oplus\mathbb{Z}\cdot g\subset H^2(F(X),\mathbb{Z})$$

has index three and the two inclusions

$$H^{2,2}(X,\mathbb{Z})_{\rm pr}\oplus\mathbb{Z}\cdot h^2\subset H^{2,2}(X,\mathbb{Z})\quad\text{and}\quad H^{1,1}(F(X),\mathbb{Z})_{\rm pr}\oplus\mathbb{Z}\cdot g\subset H^{1,1}(F(X),\mathbb{Z})$$

are of the same index.

Recall from Remark 3.14 that there exists an isomorphism of lattices

$$(H^2(F(X),\mathbb{Z})_{\rm pr},q_F)\simeq\begin{pmatrix}-14&5\\5&-2\end{pmatrix}\oplus U^{\oplus 2}\oplus E_8(-1)^{\oplus 2}.$$

Proof It suffices to show that

$$q_F(\gamma)=\frac{1}{6}\int_F\gamma^2\cdot g^2\tag{3.12}$$

for all $\gamma\in H^2(F(X),\mathbb{Z})_{\rm pr}$.

There are various ways of confirming this. For example, one can use the general fact that q_Z and the Hodge–Riemann pairing on $H^2(Z,\mathbb{Z})_{\rm pr}$ of a hyperkähler manifold Z differ by a scalar, see Remark 3.5, and then compute this scalar by means of (3.2) combined with $\int g^4=108$ and $q_F(g)=6$. Alternatively, the scalar can be determined on a specific example. For instance, below in Example 3.25, (ii), the image $\gamma:=\varphi(\beta)$ of the class

$\beta := 4 \cdot h^2 - 3 \cdot [\Sigma_P]$ is the primitive generator $\pm(5 \cdot g_S - 14 \cdot \delta)$ of $(2 \cdot g_S - 5 \cdot \delta)^\perp \subset \mathbb{Z} \cdot g_S \oplus \mathbb{Z} \cdot \delta$ and hence $q_F(\gamma) = -42$ which equals $(1/6) \int \gamma^2 \cdot g^2 = -(\beta.\beta) = -42$.

The last assertion follows from $H^4(X,\mathbb{Z})_{\rm pr} \oplus \mathbb{Z} \cdot h^2 \subset H^4(X,\mathbb{Z})$ being of index three and (3.10). □

Warning: there is no Hodge isometry between $H^2(F(X),\mathbb{Z})$ and $H^4(X,\mathbb{Z})(1)$, but in Corollary 5.21 we will see that $H^2(F(X),\mathbb{Z})$ is Hodge isometric to a certain sub-Hodge structure of $\widetilde{H}(X,\mathbb{Z})$ that extends $H^4(X,\mathbb{Z})_{\rm pr}(1)$.

Remark 3.22 The *transcendental lattice* $T(X)$ of a smooth cubic fourfold is defined as the smallest saturated sub-Hodge structure $T(X) \subset H^4(X,\mathbb{Z})$ with $H^{3,1}(X) \subset T(X) \otimes \mathbb{C}$. Similarly, the transcendental lattice of the hyperkähler fourfold $F(X)$ is defined as the smallest saturated sub-Hodge structure $T(F(X)) \subset H^2(F(X),\mathbb{Z})$ with $H^{2,0}(F(X)) \subset T(F(X)) \otimes \mathbb{C}$.

Clearly, both transcendental lattices are contained in their respective primitive cohomologies and the Hodge isometry between them restricts to a Hodge isometry

$$(T(X), (\ .\ )) \simeq (T(F(X))(-1), -q_F).$$

3.5 Plücker Polarization We complement the above discussion by the following result which follows more or less directly from what has been explained earlier.

Proposition 3.23 *Let $X \subset \mathbb{P}^5$ be a smooth cubic fourfold. We view its Fano variety of lines $F = F(X)$ with the Plücker polarization as a polarized hyperkähler manifold with the Beauville–Bogomolov–Fujiki form q_F.*

(i) *Then $(g^3)^{\perp\perp} = H^6(F,\mathbb{Z})_{\rm pr}^\perp \subset H^6(F,\mathbb{Z})$ is generated by $G := (1/36)\, g^3 \in H^6(F,\mathbb{Z})$.*
(ii) *The class $g^2 \in H^4(F,\mathbb{Z})$ is indivisible.*
(iii) *Mapping $\gamma \in H^2(F,\mathbb{Z})$ to $q_F(\gamma,\ ) \in H^2(F,\mathbb{Z})^* \simeq H^6(F,\mathbb{Z})$ leads to a short exact sequence*

$$0 \longrightarrow H^2(F,\mathbb{Z}) \xrightarrow{\tilde{q}_F} H^6(F,\mathbb{Z}) \longrightarrow \mathbb{Z}/2\mathbb{Z} \longrightarrow 0.$$

(iv) *The map $\tilde{q}_F$ satisfies*

$$\tilde{q}_F(g) = (1/18)\, g^3 = 2\,G \quad \textit{and} \quad \tilde{q}_F(\gamma) = (1/6)\, g^2 \cdot \gamma$$

for all $\gamma \in H^2(F,\mathbb{Z})_{\rm pr}$.
(v) *The map $\tilde{q}_F$ or, equivalently, multiplication with $(1/6)\, g^2$ induces an isomorphism*

$$\tilde{q}_F = (1/6)\, g^2 \cdot\ : H^2(F,\mathbb{Z})_{\rm pr}(-1) \xrightarrow{\sim} H^6(F,\mathbb{Z})_{\rm pr}(1).$$

The map $\tilde{q}_F$ corresponds to multiplication with a class in $H^{2,2}(F(X),\mathbb{Z})$, also called $\tilde{q}_F$, that will be described in Remark 4.2. A slightly different proof of (i) is given at the end of Section 4.1 and concerning (ii) we here only prove that g^2 is at most divisible by two and use a result only proved later in Proposition 4.1 to exclude two also.

Proof As $\int_F g^4 = 108$, the class G satisfies $\int_F G \cdot g = 3$ and clearly $\int_F G \cdot \gamma = 0$ for all $\gamma \in H^2(F, \mathbb{Z})_{\mathrm{pr}}$. The description (3.6) then shows that G is integral and that there exists a class $\gamma \in H^2(F, \mathbb{Z})$ with $\int_F G \cdot \gamma = 1$. Hence, G is integral and indivisible. This proves (i).

To prove (ii), observe first that $\int_F g^4 = 108$ immediately shows that the only primes that could possibly divide g^2 are two and three. In Proposition 4.1 we will show that $\int_{F(Y)} g^2 = 45$, which excludes g^2 from being divisible by two. To conclude, we repeat the argument in Remark 3.20: the inclusion $S^2H^2(F(X), \mathbb{Z}) \subset H^4(F(X), \mathbb{Z})$ has index $2^{23} \cdot 5$, which follows from combining Example 3.3 with Corollary 2.10. Since g^2 is clearly indivisible as a class in $S^2H^2(F(X), \mathbb{Z})$, it cannot be divisible as a class in $H^4(F(X), \mathbb{Z})$ by any number $\neq 2, 5$. Hence, $g^2 \in H^4(F(X), \mathbb{Z})$ is indivisible.

The description of the abstract lattice $(H^2(F, \mathbb{Z}), q_F)$ as a direct sum of a unimodular lattice and $\mathbb{Z}(-2)$ as in Remark 3.14 immediately shows (iii).

Since g is orthogonal to all $\gamma \in H^2(F, \mathbb{Z})_{\mathrm{pr}}$ with respect to both quadratic forms, the Beauville–Bogomolov–Fujiki form q_F and the Hodge–Riemann pairing, we know that $\tilde{q}_F(g) \in \mathbb{Z} \cdot G$, i.e. $\tilde{q}_F(g) = \lambda \cdot g^3$ for some scalar λ. Then, $\int_F g^4 = 108$ and $q_F(g) = 6$ allow one to compute $\lambda = 1/18$. Now, combining $\int_F \tilde{q}_F(\gamma) \cdot \gamma = q_F(\gamma)$ and (3.2) together with $\int_F g^4 = 108$ and $q_F(g) = 6$, one finds $\tilde{q}_F(\gamma) = (1/6)\, g^2 \cdot \gamma$, which concludes the proof of (iv).

To prove the last assertion, we consider the commutative diagram

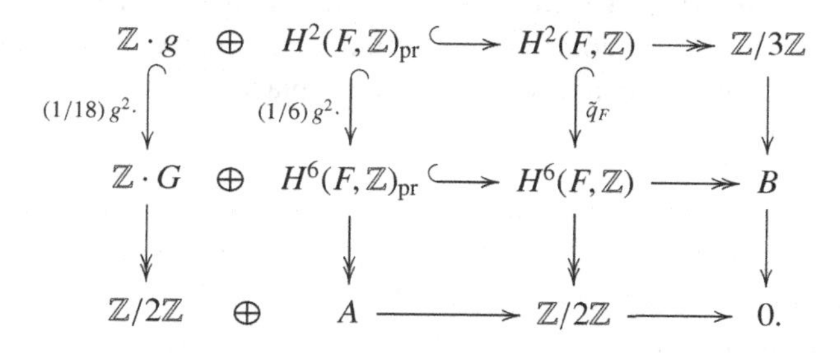

$$\begin{array}{ccccccc}
\mathbb{Z}\cdot g & \oplus & H^2(F,\mathbb{Z})_{\mathrm{pr}} & \hookrightarrow & H^2(F,\mathbb{Z}) & \twoheadrightarrow & \mathbb{Z}/3\mathbb{Z} \\
\downarrow{\scriptstyle \int (1/18)\,g^2\cdot} & & \downarrow{\scriptstyle \int (1/6)\,g^2\cdot} & & \downarrow{\scriptstyle \int \tilde{q}_F} & & \downarrow \\
\mathbb{Z}\cdot G & \oplus & H^6(F,\mathbb{Z})_{\mathrm{pr}} & \hookrightarrow & H^6(F,\mathbb{Z}) & \twoheadrightarrow & B \\
\downarrow & & \downarrow & & \downarrow & & \downarrow \\
\mathbb{Z}/2\mathbb{Z} & \oplus & A & \longrightarrow & \mathbb{Z}/2\mathbb{Z} & \longrightarrow & 0.
\end{array}$$

The first three columns and the first two rows are short exact sequences. Hence, $A = 0$ if and only if $B \simeq \mathbb{Z}/3\mathbb{Z}$. Otherwise, $A \simeq \mathbb{Z}/3\mathbb{Z}$ and $B = 0$. Thus, the assertion is equivalent to $B \neq 0$, and, since $\int_F g \cdot G = 3$, it therefore suffices to prove the existence of a class $\beta \in H^6(F, \mathbb{Z})$ with $\int_F g \cdot \beta = 1$. One example of such a class is provided by $\beta = [\mathbb{P}^1]$ for any line $\mathbb{P}^1 \subset \mathrm{P}^* \subset F$ contained in the dual of a plane $\mathbb{P}^2 \simeq \mathrm{P} \subset X$, see Remark 1.6, (ii). Since the assertion is topological, we may reduce to this case. □

Remark 3.24 We refrain from addressing here the topic of Chow groups and Chow motives of cubic fourfolds and their Fano varieties of lines. Here are just a few facts and pointers to the literature, see Section **7**.4 for more comments.

(i) As explained in Section **2**.5.5, the Fano correspondence also works on the level of Chow groups and even for motives. Bolognesi and Pedrini [78] showed that this leads to isomorphisms

$$\mathrm{CH}^3(X)_{\mathrm{alg}} \otimes \mathbb{Q} \simeq \mathrm{CH}^3(X)_{\mathrm{hom}} \otimes \mathbb{Q} \simeq \mathrm{CH}^2(F(X))_{\mathrm{hom}} \otimes \mathbb{Q},$$

which is the analogue of Corollary **5**.3.16. On the level of rational Chow motives this corresponds to

$$\mathfrak{h}(X)_{\mathrm{tr}} \simeq \mathfrak{h}^2(F(X))_{\mathrm{tr}}(1).$$

Here, $\mathfrak{h}(X)_{\mathrm{tr}}$ is the transcendental motive of X with $\mathfrak{h}(X)_{\mathrm{pr}} \simeq \mathfrak{h}(X)_{\mathrm{tr}} \oplus \mathbb{Q}(-2)^{\oplus\rho-1}$, where $\rho := \mathrm{rk}(H^{2,2}(X,\mathbb{Z}))$. Note that the isomorphism between the Chow motives in particular shows that $\mathfrak{h}(X)$ is finite-dimensional if and only if $\mathfrak{h}(F(X))$ is finite-dimensional. This is true in arbitrary dimension, see Section **2**.4.2.

(ii) The Chow ring and the Chow motive of the hyperkähler fourfold $F(X)$ have been studied intensively. Since the Fano varieties $F(X)$ are more accessible than general hyperkähler varieties, they have served as a good testing ground for general conjectures on the structure of the Chow ring. We refer to the work of Fu, Laterveer, Shen, and Vial [183, 429] for results in this direction and further references. The motive $\mathfrak{h}(F(X))$ for special cubic fourfolds has been studied Bülles [99] and by Laterveer [307, 308].

3.6 Néron–Severi for Picard Rank Two The reformulation of Proposition 3.19 as Corollary 3.21 allows one to compute the Néron–Severi lattice of the Fano variety of lines of the very general cubic fourfold containing a plane and of the very general Pfaffian cubic fourfold.

Example 3.25 (i) According to Lemma 1.1, the lattice $H^{2,2}(X,\mathbb{Z})$ of the very general cubic containing a plane $\mathbb{P}^2 \simeq \mathrm{P} \subset X \subset \mathbb{P}^5$ is K_8^-. The matrix (1.1) describes the intersection product with respect to the basis $h^2, [\mathrm{P}] \in H^{2,2}(X,\mathbb{Z})$.

The primitive part $H^{2,2}(X,\mathbb{Z})_{\mathrm{pr}}$ is spanned by the class $\beta := h^2 - 3\cdot[\mathrm{P}]$, which satisfies $(\beta.\beta) = 24$. For $\gamma := \varphi(\beta)$, which spans $H^{1,1}(F(X),\mathbb{Z})_{\mathrm{pr}}$, one then has $q_F(\gamma) = -24$. The inclusion $\mathbb{Z}\cdot g \oplus \mathbb{Z}\cdot\gamma \subset H^{1,1}(F(X),\mathbb{Z})$ is of index three.

Moreover, the intersection matrix of $(H^{1,1}(F(X),\mathbb{Z}), q_F)$ with respect to the basis g and $[F_{\mathrm{P}}] = \varphi([\mathrm{P}]) = (1/3)(g-\gamma)$ is

$$\left(H^{1,1}(F(X),\mathbb{Z}), q_F\right) \simeq \begin{pmatrix} 6 & 2 \\ 2 & -2 \end{pmatrix} \tag{3.13}$$

and hence disc $= -16$. Note that $q_F([F_{\rm P}]) = -2$ implies that $[F_{\rm P}]$ is not ample. As a consequence of the above, certain intersection numbers can be determined. For example, $([F_P]^2.[F_{\rm P}]^2) = \int_{F(X)}[F_{\rm P}]^4 = 3 \cdot q_F([F_{\rm P}])^2 = 12$. Also, $([F_P]^2.[{\rm P}^*]) = 9$, using $\mathcal{O}(F_{\rm P})|_{{\rm P}^*} \simeq \mathcal{O}(3)$, see Exercise 1.8.

(ii) The computation for the very general Pfaffian cubic fourfold is similar: by Corollary 2.21 and Remark 2.24, $H^{2,2}(X,\mathbb{Z})$ is in this case spanned by h^2 and $[\Sigma_P]$ with intersection form K_{14}^-. The primitive part $H^{2,2}(X,\mathbb{Z})_{\rm pr}$ is generated by $\beta := 4 \cdot h^2 - 3 \cdot [\Sigma_P]$ for which $(\beta.\beta) = 42$. Its image $\gamma := \varphi(\beta)$ then satisfies $q_F(\gamma) = -42$. Again, the inclusion $\mathbb{Z} \cdot g \oplus \mathbb{Z} \cdot \gamma \subset H^{1,1}(F(X),\mathbb{Z})$ is of index three and the intersection matrix of $(H^{1,1}(F(X),\mathbb{Z}), q_F)$ with respect to the basis g and $\varphi([\Sigma_P]) = (1/3)(4 \cdot g - \gamma)$ is

$$\left(H^{1,1}(F(X),\mathbb{Z}), q_F\right) \simeq \begin{pmatrix} 6 & 8 \\ 8 & 6 \end{pmatrix}, \tag{3.14}$$

with disc $= -28$. In this case, one could alternatively use $F(X) \simeq S^{[2]}$ for a K3 surface of degree 14 and Picard rank 1 and compute the pairing with respect to g_S and δ, cf. Example 3.6, as

$$\left(H^{1,1}(F(X),\mathbb{Z}), q_F\right) \simeq \begin{pmatrix} 14 & 0 \\ 0 & -2 \end{pmatrix},$$

which of course also has disc $= -28$. To see that these two descriptions are equivalent, recall that $g = 2 \cdot g_S - 5 \cdot \delta$, see Lemma 2.12.

Similarly to (i), it should be interesting to compare the square of the class $\varphi([\Sigma_P])$ in $H^{2,2}(F(X),\mathbb{Z})$ with the class of a natural surface in $F(X)$ associated with the situation.

The previous two examples are generalized by the following result the proof of which is more natural in the context of Section 5.2, cf. Remark 5.4.

Proposition 3.26 *Assume that the Fano variety $F(X)$ of a smooth cubic fourfold X has Picard rank two and let $\gamma \in H^{1,1}(F(X),\mathbb{Z})_{\rm pr}$ be a generator of its primitive part. Then the discriminant* disc $= -2d$ *of* NS$(F(X))$ *with respect to q_F is even and one of the two conditions holds true:*

(i) $\mathbb{Z} \cdot g \oplus \mathbb{Z} \cdot \gamma = H^{1,1}(F(X),\mathbb{Z})$, *$d \equiv 0\,(6)$, and $q_F(\gamma) = -d/3$, or*
(ii) $\mathbb{Z} \cdot g \oplus \mathbb{Z} \cdot \gamma \subset H^{1,1}(F(X),\mathbb{Z})$ *is of index three, $d \equiv 2\,(6)$, and $q_F(\gamma) = -3d$.*

Furthermore, the isomorphism type of the lattice NS$(F(X))$ *only depends on d.* □

The two cases in Example 3.25 correspond to $d = 8$ and $d = 14$, both covered by (ii).

3.7 Excluding $d = 2$ and $d = 6$ In the following two remarks we will indicate how the general theory of hyperkähler manifolds of K3$^{[2]}$-type allows one to exclude the cases $d = 6$ and $d = 2$ in Proposition 3.26.

The case $d = 6$ is rather easy to exclude, while the case $d = 2$ is surprisingly difficult to rule out. It was originally dealt with independently by Laza [312] and Looijenga

[327]. Their methods are of a different kind than those below which rely on the results of Markman [337] and Rieß [401] from the theory of hyperkähler manifolds. Although those were proved only later, it is nevertheless instructive to see how they can be effectively used in this context.

Remark 3.27 The main input to exclude $d = 6$ occurring in Proposition 3.26 is the following fact, see [338, Thm. 1.11]: if Z is a hyperkähler fourfold that is deformation equivalent to the Hilbert square $S^{[2]}$ of a K3 surface and $\alpha \in H^{1,1}(Z,\mathbb{Z})$ satisfies $q(\alpha) = -2$, then α or $-\alpha$ is effective.[12]

Taking this result for granted, the case $d = 6$, which would fall into (i) in Proposition 3.26, is excluded by observing that in this case γ would be such a (-2)-class and, therefore, γ or $-\gamma$ would be effective. However, the ample class g is orthogonal to γ, which leads to a contradiction. In other words, the primitive algebraic cohomology $H^{1,1}(F(X),\mathbb{Z})_{\rm pr}$ does not contain any (-2)-class, see also the proof of Corollary 4.16.

The reader may want to compare the situation to the discussion in Section 1.4, where it was explained how the case $d = 6$ is related to cubic fourfolds X with a single ordinary double point. In this case, the Fano variety $F(X)$ is singular along a K3 surface $S \subset F(X)$. For generic X, the blow-up ${\rm Bl}_S(F(X))$ is a hyperkähler fourfold, isomorphic to $S^{[2]}$ by Proposition 1.28, and the class $\pm\gamma$ above corresponds to the exceptional divisor of the blow-up ${\rm Bl}_S(F(X)) \twoheadrightarrow F(X)$, which is indeed orthogonal to the pullback of the Plücker polarization with respect to the pairing q on the blow-up.

Remark 3.28 Let us now show that there are no smooth cubic fourfolds $X \subset \mathbb{P}^5$ realizing $d = 2$ in Proposition 3.26.

Suppose $X \subset \mathbb{P}^5$ is a smooth cubic fourfold for which there exists a primitive class $\gamma \in H^{1,1}(F(X),\mathbb{Z})_{\rm pr}$ with $q_F(\gamma) = -6$. After passing to a generic deformation of X that leaves γ of type $(1,1)$, we may assume that there are no further primitive integral $(1,1)$-classes, i.e. $\langle g,\gamma\rangle^\perp \subset H^2(F(X),\mathbb{Z})$ is the transcendental lattice of $F(X)$, which up to Tate twist is Hodge isometric to the transcendental lattice of X.

From the lattice theory to be discussed in Section 5.2, see Proposition 5.10, one deduces that there exists a polarized K3 surface (S,L) of degree two and a Hodge isometry $H^2(S,\mathbb{Z})_{\rm pr} \simeq \langle g,\gamma\rangle^\perp \subset H^2(F(X),\mathbb{Z})$. Identifying $H^2(S,\mathbb{Z})_{\rm pr}$ with the transcendental lattice of the Hilbert scheme $S^{[2]}$ and using the existence of the hyperbolic plane $(H^0 \oplus H^4)(S,\mathbb{Z}) \subset \widetilde{H}(S,\mathbb{Z})$, the Hodge isometry for the primitive cohomology can be extended to a Hodge isometry $\widetilde{H}(S,\mathbb{Z}) \simeq \widetilde{H}(X,\mathbb{Z})$, cf. Corollary 5.18. Furthermore, as shown by Addington [3], see Corollary 5.21, it can be chosen to map $(1,0,-1)$ to $v(\lambda_1)$ and, therefore, induces a Hodge isometry

[12] Note that the Riemann–Roch formula (4.4) for the line bundle $\mathcal{L}$ with ${\rm c}_1(\mathcal{L}) = \delta$ shows $\chi(F(X),\mathcal{L}) = 1$, which suggests that $h^0(\mathcal{L}) \neq 0$ or $h^0(\mathcal{L}^*) = h^4(\mathcal{L}) \neq 0$ if only $h^2(\mathcal{L})$ could be shown to be 0. The actual proof first verifies the result for Hilbert schemes of K3 surfaces or, more generally, moduli spaces of sheaves on K3 surfaces and then uses the density of those in the space of deformations of the pair $(Z,\mathcal{L})$.

$$H^2(S^{[2]},\mathbb{Z}) \simeq (1,0,-1)^\perp \simeq v(\lambda_1)^\perp \simeq H^2(F(X),\mathbb{Z}).$$

By Markman's polarized global Torelli theorem [337, Thm. 9.8] this then is induced (up to a sign) by a birational map $F(X) \sim S^{[2]}$.

Recall that $\mathrm{NS}(S^{[2]}) \simeq \mathbb{Z}\cdot L \oplus \mathbb{Z}\cdot\delta$, where L denotes the polarization L on S as well as its symmetric square on $S^{[2]}$ and $2\cdot\delta$ is the class of the exceptional divisor of the Hilbert–Chow morphism $S^{[2]} \twoheadrightarrow S^{(2)}$, see Example 3.6. Viewing $g,\gamma \in \mathrm{NS}(F(X))$ as elements in $\mathrm{NS}(S^{[2]})$ and using $q(g.\delta) > 0$ and $q(g.L) > 0$, as L and δ are effective and g is ample on $F(X)$, a computation reveals that $g = 2\cdot L - \delta$ and, up to changing γ by a sign, $\gamma = L - 2\cdot\delta$. Note that this in particular confirms the assertion in Proposition 3.26 that $\mathbb{Z}\cdot g \oplus \mathbb{Z}\cdot\gamma \subset \mathrm{NS}(F(X))$ is of index three.

The closure of the ample cone of $S^{[2]}$ has been completely described by Rieß [401]: it is spanned by L, which is the pullback of an ample line bundle on $S^{(2)}$, and the class $3\cdot L - 2\cdot\delta$. In particular, g is contained in the ample cone of $S^{[2]}$, from which one deduces that the birational map is actually an isomorphism $F(X) \simeq S^{[2]}$. As an aside, the only other smooth birational model of $S^{[2]}$ is constructed by Mukai flopping the $\mathbb{P}^2 \hookrightarrow S^{[2]}$, obtained from the double cover $S \twoheadrightarrow \mathbb{P}^2$. Furthermore, the class $3\cdot L - 2\cdot\delta$ spans the wall between their ample cones, i.e. $3\cdot L - 2\cdot\delta$ is of degree 0 on the lines in $\mathbb{P}^2$. Moreover, according to [401, Thm. 6.1], the line bundle $g = 2\cdot L - \delta$ (which corresponds to $H + L$ in the notation there) contains $\mathbb{P}^2$ in its base locus. This contradicts the very ampleness of the Plücker polarization.

3.8 Integral Cohomology of a Cubic and Its Fano Variety Let us comment further on the (integral) Hodge conjecture for X and $F(X)$, cf. Section 0.1. According to [476, Thm. 18] or [480, Thm. 1.4], the integral Hodge conjecture holds for $H^4(X,\mathbb{Z})$. Voisin's proof is a refinement of [501]. Macrì and Stellari [333, Prop. 5.17] provide a proof relying on derived categories and Mongardi and Ottem [354] use hyperkähler geometry, see Remark 3.30. The following immediate consequence of Proposition 3.19 is a weaker version.

Corollary 3.29 *If $\alpha \in H^{2,2}(X,\mathbb{Z})_{\mathrm{pr}}$, then $6\cdot\alpha$ is algebraic, i.e. $6\cdot\alpha$ is an integral linear combination $\sum n_i[Z_i]$ of algebraic cycles $Z_i \subset X$. Furthermore, for any not necessarily primitive class $\alpha \in H^{2,2}(X,\mathbb{Z})$, the class $18\cdot\alpha$ is algebraic. In particular, the rational Hodge conjecture holds for any smooth cubic fourfold X.*

Proof Compare the following to the arguments in Section 0.2, (v).[13]

Let $\psi\colon H^6(F(X),\mathbb{Z})_{\mathrm{pr}}(1) \xrightarrow{\sim} H^4(X,\mathbb{Z})_{\mathrm{pr}}$ be the isomorphism of integral Hodge structures induced by the dual Fano correspondence. We want to compose ψ with the cor-

[13] Thanks to A. Auel for discussions concerning these arguments.

respondence $\tilde{q}_F = (1/6)g^2\cdot\colon H^2(F(X),\mathbb{Z})_{\rm pr}(-1) \xrightarrow{\sim} H^6(F(X),\mathbb{Z})_{\rm pr}(1)$, see Proposition 3.23. Note that the class class $(1/6)\,g^2$ itself is not integral, in fact, as we have seen before, g^2 is not divisible at all. We obtain an isomorphism of Hodge structures

$$H^2(F(X),\mathbb{Z})_{\rm pr}(-1) \xrightarrow[(1/6)\,g^2\cdot]{\sim} H^6(F(X),\mathbb{Z})_{\rm pr}(1) \xrightarrow[\psi]{\sim} H^4(X,\mathbb{Z})_{\rm pr}.$$

According to the Lefschetz $(1,1)$-theorem, any class $\beta \in H^{1,1}(F(X),\mathbb{Z})$ is algebraic. As g^2 and the dual Fano correspondence ψ are both algebraic, this shows that $6\cdot\alpha$ is indeed algebraic for any $\alpha \in H^{2,2}(X,\mathbb{Z})_{\rm pr}$.

To extend the result to non-primitive classes, observe that any $\alpha \in H^4(X,\mathbb{Z})$ can be written as $\alpha = \lambda\cdot\alpha' + \mu\cdot h^2$ with $3\,\lambda, 3\,\mu \in \mathbb{Z}$ and $\alpha' \in H^4(X,\mathbb{Z})_{\rm pr}$. □

Remark 3.30 Mongardi and Ottem [354] show that the integral Hodge conjecture holds for curve classes on hyperkähler manifolds of $\mathrm{K3}^{[n]}$-type. Thus, it holds, in particular, for $H^6(F(X),\mathbb{Z})$, i.e. all classes in $H^{3,3}(F(X),\mathbb{Z})$ are integral linear combination of curve classes.

The isomorphism of Hodge structures provided by the dual Fano correspondence $H^6(F(X),\mathbb{Z}) \xrightarrow{\sim} H^4(X,\mathbb{Z})$ then implies the integral Hodge conjecture for the middle cohomology $H^4(X,\mathbb{Z})$ of the cubic fourfold as in the above proof.

4 Geometry of the Very General Fano Variety

The Fano variety of lines $F(X)$ has a rich and interesting geometry. It has been studied widely, not only because its geometry sheds light on the cubic X, but also because the family of all $F(X)$ of smooth cubic fourfolds X is one of the very few families of polarized hyperkähler manifolds of maximal dimension that can be described and studied explicitly.

We will study divisors, surfaces, and curves in the Fano variety $F(X)$ of lines on the very general cubic X. More specific cases, e.g. for X containing a plane or for X being Pfaffian, have been discussed already in Sections 1 and 2 and in Exercise 3.25. Cubics on the other end of the spectrum, namely those for which $H^{2,2}(X,\mathbb{Z})$ is of maximal rank 21 or, equivalently, for which the Fano variety has maximal Picard number 21, have been addressed by Laza [313].

4.1 Algebraic Cohomology of the Very General Fano Variety We begin with a complete description of the algebraic part of $H^*(F(X),\mathbb{Z})$ for the very general cubic fourfold and we shall start with classes of degree two. Here,

$$H^{1,1}(F(X),\mathbb{Z}) \simeq \mathrm{NS}(F(X)) \simeq \mathbb{Z}\cdot g$$

is spanned by the Plücker polarization $g = -c_1(\mathcal{S}_F)$, see (v), Section 0.2. In addition, the ample generator g satisfies

$$\int_F g^4 = 108 \quad \text{and} \quad q_F(g) = 6.$$

We remarked already that the last equality can also be used to show that g is not divisible any further, see the comments after Remark 3.14.

It turns out that in degree four, the algebraic part $H^{2,2}(F(X), \mathbb{Z})$ is of rank two and there are quite a few classes that come to mind, for example,

$$g^2 = c_1(\mathcal{S}_F)^2,\ c_2(\mathcal{S}_F),\ c_2(\mathcal{T}_F),\ [F(Y)],\ [F_2(X)],\ \varphi(h^3), \quad \text{and} \quad [F_L]. \tag{4.1}$$

Here, $F(Y) \subset F(X)$ is the Fano surface of a generic hyperplane section $Y := X \cap \mathbb{P}^4$, see Section 4.3 for more on $F(Y)$, and

$$F_L := \{ [L'] \mid L \cap L' \neq \emptyset \} = p(q^{-1}(L))$$

is the surface of all lines intersecting a given (generic) line $L \subset X$. Note that unlike the case of cubic threefolds, the fibres $q^{-1}(x)$ are all connected and, therefore, F_L can indeed be defined directly in this way and not as the closure of the set of all lines $L' \neq L$ intersecting L, cf. Section **5**.1.2. For more on the geometry of F_L, see Section 4.5.

The next result is a combination of various results one finds in the literature notably in [428].

Proposition 4.1 *Let $X \subset \mathbb{P}^5$ be a very general cubic fourfold and $F = F(X)$ its Fano variety of lines.*

(i) *The cohomolovy $H^{2,2}(F(X), \mathbb{Z})$ is of rank two and the classes* (4.1) *satisfy*[14]

$$[F_2(X)] = 5 \cdot (g^2 - c_2(\mathcal{S}_F)),\ [F(Y)] = c_2(\mathcal{S}_F),$$

$$c_2(\mathcal{T}_F) = 5 \cdot g^2 - 8 \cdot c_2(\mathcal{S}_F),\ 3 \cdot [F_L] = \varphi(h^3),$$

$$\textit{and } \varphi(h^3) = g^2 - c_2(\mathcal{S}_F) = (1/8) \cdot (c_2(\mathcal{T}_F) + 3 \cdot g^2).$$

(ii) *The intersection numbers between (some of) these classes are*[15]

$$([F(Y)].[F(Y)]) = (c_2(\mathcal{S}_F).c_2(\mathcal{S}_F)) = 27,\ (g^2.g^2) = \deg(F(X)) = 108,$$

$$(c_2(\mathcal{S}_F).g^2) = ([F(Y)].g^2) = \deg(F(Y)) = 45,\ ([F_L].g^2) = \deg(F_L) = 21,$$

$$([F_2(X)].g^2) = 315,\ ([F_2(X)].[F_2(X)]) = 1\,125, \quad \textit{and} \quad (c_2(\mathcal{T}_F).c_2(\mathcal{T}_F)) = 828.$$

[14] All formulae not involving F_L actually hold in $\mathrm{CH}^2(F(X))$.

[15] There are various numerical coincidences to be observed, not all of which have a clear geometric interpretation. Firstly, why is $e(X) = 27$ the number of lines on a cubic surface? Further, $4 \cdot c_2(\mathcal{S})^2 - c_1(\mathcal{S})^4 = 0$ or, equivalently, $(2 \cdot c_2(\mathcal{S}) - c_1(\mathcal{S})^2) \cdot (2 \cdot c_2(\mathcal{S}) + c_1(\mathcal{S})^2) = 0$. Also, $e(F(X)) = 12 \cdot e(X) = 12 \cdot 27$ or, essentially equivalent, $e(\mathbb{L}) = 24 \cdot e(X) = e(\mathrm{K3}) \cdot e(X)$.

Furthermore, the two classes $[F_L]$ *and* $c_2(\mathcal{S}_F)$ *with the intersection matrix*

$$\begin{pmatrix} 5 & 6 \\ 6 & 27 \end{pmatrix}$$

provide an integral basis of $H^{2,2}(F(X), \mathbb{Z})$.

Proof To prove that $H^{2,2}(F(X), \mathbb{Q})$ is of dimension two, we use that cup product defines an isomorphism $S^2H^2(F(X), \mathbb{Q}) \xrightarrow{\sim} H^4(F(X), \mathbb{Q})$ of Hodge structures, which can either be verified by using (iii) in Section 0.2 or by exploiting the fact that $F(X)$ is homeomorphic to the Hilbert scheme $S^{[2]}$ of a K3 surface, cf. Corollary 3.11, for which the result is known classically, see Example 3.3. In any case, decomposing

$$H^2(F(X), \mathbb{Q}) \simeq \mathrm{NS}(F(X))_{\mathbb{Q}} \oplus T(F(X))_{\mathbb{Q}},$$

where the transcendental part $T(F(X))_{\mathbb{Q}}$ is the q_F-orthogonal complement of the algebraic part $\mathrm{NS}(F(X))_{\mathbb{Q}} \simeq H^{1,1}(F(X), \mathbb{Q})$, and taking the symmetric square defines a decomposition

$$H^4(F(X), \mathbb{Q}) \simeq S^2H^2(F(X), \mathbb{Q}) \simeq S^2\mathrm{NS}_{\mathbb{Q}} \oplus (\mathrm{NS}_{\mathbb{Q}} \otimes T_{\mathbb{Q}}) \oplus S^2T_{\mathbb{Q}}. \quad (4.2)$$

As $\rho(F(X)) = 1$ for the very general cubic X, the first summand, which is contained in $H^{2,2}(F(X), \mathbb{Q})$, is one-dimensional. The second summand, which is a Tate twist of $T_{\mathbb{Q}}$, does not contain any non-trivial $(2,2)$-class. Finally, $S^2(T_{\mathbb{Q}})$ is isomorphic to a Tate twist of $S^2H^4(X, \mathbb{Q})_{\mathrm{pr}}$ which contains only one algebraic class up to scaling, see Remark **1**.2.13.

The proof of the first equality computing the class of $F_2(X)$ follows arguments of Amerik [17, Sec. 3]: as in Remark **2**.2.11, we think of $F_2(X)$ as the locus $M_2(\psi)$ with $\psi\colon \mathcal{Q}_F \twoheadrightarrow S^2(\mathcal{S}_F^*)$. Then the classical Porteous formula, cf. [174, Thm. 12.4] or [188, Thm. 14.4], shows $[F_2(X)] = \gamma_1^2 - \gamma_2$ for $1 + \gamma_1 \cdot t + \gamma_2 \cdot t^2 = c_t(S^3(\mathcal{S}_F^*)) \cdot c_t(\mathcal{Q}_F)^{-1}$. A standard computation, using the splitting principle, then proves the assertion.

The argument to prove the second equality is similar to the one in Section **5**.2.1 showing $e(F(Y)) = \int c_2(F(Y)) = 27 = |F(Y \cap \mathbb{P}^3)|$. Indeed, if $Y \subset X \subset \mathbb{P}^5 = \mathbb{P}(V)$ is cut out by a section $s \in H^0(\mathbb{P}^5, \mathcal{O}(1)) \simeq V^*$, then the zero set of the image $\tilde{s}$ of s under the natural map $V^* \longrightarrow H^0(F(X), \mathcal{S}_F^*)$ is $F(Y) \subset F(X)$. To conclude, use $c_2(\mathcal{S}_F^*) = c_2(\mathcal{S}_F)$.

For the computation of $c_2(\mathcal{T}_F)$, see Exercise **2**.3.2. The argument to compute $[F_L]$ has been explained already in Remark **2**.5.2 and again in the proof of Lemma **5**.1.14. For fourfolds, this is a result of Voisin [472, Sec. 4]: first, observe that $[F_L]$ is independent of the choice of the (generic) line L. Then, by Proposition **1**.5.3, there exists a plane $\mathbb{P}^2 \subset \mathbb{P}^5$ with $X \cap \mathbb{P}^2 = L_1 \cup L_2 \cup L_2$. This implies (in cohomology!)

$$\varphi(h^3) = [p(q^{-1}(X \cap \mathbb{P}^2))] = [F_{L_1}] + [F_{L_2}] + [F_{L_3}] = 3 \cdot [F_L].$$

The last equality expressing $\varphi(h^3)$ is the content of Exercise **2**.5.3.

Of the claimed intersection numbers, only the first one needs a proof, all the others are then obtained by straightforward computations. We exploit the fact that the restriction of $\mathcal{S}$ to $F(Y) \subset F(X)$ is isomorphic to the tangent bundle, see Proposition 5.2.2. Hence,

$$([F(Y)].[F(Y)]) = ([F(Y)].c_2(\mathcal{S}_F)) = \int_{F(Y)} c_2(\mathcal{T}_{F(Y)}) = e(F(Y)) = 27.$$

Equivalently, one can compute $([F(Y)].[F(Y)])$ as the intersection number $|F(Y_1) \cap F(Y_2)|$ for two generic hyperplane sections $Y_i \subset X$, which is the number of lines in the cubic surface $Y_1 \cap Y_2$. Alternatively, one could start with proving the last equality first by using that $(c_2.c_2) = 828$ for $S^{[2]}$ and the fact that $F(X)$ is deformation equivalent to $S^{[2]}$.

For the last part, we observe that

$$([F_L].[F_L]) = 5 \quad \text{and} \quad (c_2(\mathcal{S}_F).[F_L]) = 6.$$

As the discriminant of the intersection matrix is 99, the lattice $\mathbb{Z} \cdot [F_L] \oplus \mathbb{Z} \cdot c_2(\mathcal{S}_F) \subset H^4(F(X), \mathbb{Z})$ either equals $H^{2,2}(F(X), \mathbb{Z})$ or is a sublattice of index three. In both cases, all classes $\alpha \in H^{2,2}(F(X), \mathbb{Z})$ are of the form $\alpha = (1/3)(a \cdot [F_L] + b \cdot c_2(\mathcal{S}_F)) \in H^{2,2}(F(X), \mathbb{Z})$ with $a, b \in \mathbb{Z}$. Intersecting with $[F_L]$ gives $(5/3)a + 2b = (\alpha.[F_L]) \in \mathbb{Z}$ and, therefore, $3 \mid a$ and then also $3 \mid b$, which excludes the index three option. □

The description of $H^{2,2}(F(X), \mathbb{Z})$ was given by Shen [428, Thm. 5.5], where it is also pointed out that it implies the integral Hodge conjecture for the Fano variety of lines of the very general cubic fourfold.

Remark 4.2 There is one other natural class in $H^{2,2}(F(X), \mathbb{Z})$, provided by the interpretation of $F(X)$ as a hyperkähler fourfold: the Beauville–Bogomolov–Fujiki form q_F determines a class $\tilde{q}_F \in H^{2,2}(F(X), \mathbb{Q})$ such that $q_F(\alpha) = \int_F \alpha^2 \cdot \tilde{q}_F$ for all $\alpha \in H^2(F(X), \mathbb{Z})$. Clearly, multiplication with this class is nothing but the map $\tilde{q}_F$ in Proposition 3.23. It turns out that $(\tilde{q}_F.\tilde{q}_F) = 23/25$ and more precisely

$$\tilde{q}_F = (1/30) \cdot c_2(\mathcal{T}_F),$$

which can be proved as follows: first, for the very general deformation F' of $F(X)$ as a hyperkähler manifold, $H^{2,2}(F', \mathbb{Q})$ is one-dimensional, which shows $\tilde{q}_{F'} = \lambda \cdot c_2(\mathcal{T}_{F'})$ for some scalar λ. Next consider a deformation of $F(X)$ of the form $S^{[2]}$ with (S, L) a general polarized K3 surface of degree two. Then $h^0(S, L) = 3$ and the associated line bundle $L^{[2]}$ on $S^{[2]}$ and its first Chern class ℓ satisfy $h^0(S^{[2]}, L^{[2]}) = 6$ and $q(\ell) = (L.L) = 2$. On the other hand, the Hirzebruch–Riemann–Roch formula gives

$$6 = h^0(S^{[2]}, L^{[2]}) = \chi(S^{[2]}, L^{[2]}) = 3 + (1/24) \int_{S^{[2]}} \ell^2 \cdot c_2(\mathcal{T}) + (1/24) \int_{S^{[2]}} \ell^4.$$

The vanishing of the higher cohomology is a consequence of Kawamata–Viehweg vanishing applied to the big and nef line bundle $L^{[2]}$. As the Fujiki constant is known for

$S^{[2]}$, see Remark 3.6, the last integral is $\int \ell^4 = 3 \cdot q(\ell)^2 = 12$. Altogether, this proves $\int \ell^2 \cdot c_2(\mathcal{T}_F) = 60 = 30 \cdot q(\ell) = 30 \cdot \int \ell^2 \cdot \tilde{q}$ and, therefore, $c_2(\mathcal{T}_F) = 30 \cdot \tilde{q}_F$.

For cubic threefolds, we know that $(1/6)\cdot(\gamma.\gamma')_{F(Y)} = (1/2) \int_{F(Y)} [C_L]\cdot\gamma\cdot\gamma'$, see Section 5.3.1. The following consequence should be seen as the analogue of this equation for cubic fourfolds.

Corollary 4.3 *For all primitive classes $\alpha \in H^2(F(X), \mathbb{Z})_{\rm pr}$, the Beauville–Bogomolov–Fujiki form q_F of the Fano variety of lines $F = F(X)$ satisfies*

$$q_F(\alpha) = (1/6) \int_F \alpha^2 \cdot g^2 = (1/2) \int_F [F_L] \cdot \alpha^2.$$

Proof From Proposition 4.1 and Remark 4.2 we deduce

$$\tilde{q}_F = (1/30) \cdot c_2(\mathcal{T}_F) = (1/2) \cdot [F_L] - (1/10) \cdot c_2(\mathcal{S}_F),$$

which proves $q_F(\alpha) = (1/2) \int_F [F_L] \cdot \alpha^2 - (1/10) \int_F c_2(\mathcal{S}_F) \cdot \alpha^2$ for all $\alpha \in H^2(F, \mathbb{Z})$. Then, restricting to primitive classes α and using $c_2(\mathcal{S}_F) = [F(Y)]$, Lemma 4.5 below and (3.12) in the proof of Corollary 3.21 allow us to conclude. □

It remains to describe algebraic classes in degree six, i.e. classes in $H^{3,3}(F(X), \mathbb{Z})$. Clearly, by Poincaré duality, the space is of rank one for general X and it contains g^3. However, g^3 is not the generator. Instead, we have

$$H^{3,3}(F(X), \mathbb{Z}) = \mathbb{Z} \cdot (1/36)\, g^3, \tag{4.3}$$

which we proved already in Proposition 3.23. Here is an alternative argument for this assertion. Differentiating the equation $\int_F (g + t \cdot \alpha)^4 = 3 \cdot q_F(g + t \cdot \alpha)^2$, see Corollary 3.11, at $t = 0$ shows $4 \cdot \int_F g^3 \cdot \alpha = 12 \cdot q_F(g) \cdot q_F(g, \alpha)$ and, therefore,

$$\int_F g^3 \cdot \alpha = 18 \cdot q_F(g, \alpha)$$

for all $\alpha \in H^2(F(X), \mathbb{Z})$. As q_F is an integral quadratic form on $H^2(F(X), \mathbb{Z})$, Poincaré duality implies that g^3 is divisible by 18. On the other hand, using the description of $H^2(S^{[2]}, \mathbb{Z})$ in Remark 3.6 and Lemma 2.12, we know that $q_F(g, \alpha)$ is even for all $\alpha \in H^2(F(X), \mathbb{Z})$ and that there exists a class α with $q_F(g, \alpha) = 2$. Thus, g^3 is actually divisible by 36 but not any further. The class $G = g^3/36$ is called the *minimal curve class*, see Section 4.6 for more on G and $2G$.

4.2 Some Explicit Realizations We want to look at some of the subvarieties realizing the classes above and start with divisors.

The Plücker polarization g is the first Chern class of the very ample line bundle $\mathcal{O}_F(1)$ induced by the Plücker embedding $F(X) \hookrightarrow \mathbb{P}(\bigwedge^2 V)$. According to Lemma 2.5.1, it can also be described as the image $\varphi(h^2)$ of the class $h^2 \in H^{2,2}(X, \mathbb{Z})$ under the Fano correspondence. Moreover, any actual linear intersection $X \cap \mathbb{P}^3$ induces an element in the linear system $|\mathcal{O}_F(1)|$. As $\dim(\mathbb{G}(3, \mathbb{P}(V))) = 8$, one describes in this way an eight-dimensional family of divisors in $|\mathcal{O}_F(1)|$.

Now, $F(X)$ as a hyperkähler fourfold that is deformation equivalent to a Hilbert scheme $S^{[2]}$ of a K3 surface S satisfies

$$\chi(F(X), \mathcal{L}) = \binom{q_F(\mathrm{c}_1(\mathcal{L}))/2 + 3}{2} \tag{4.4}$$

for any line bundle $\mathcal{L}$ on $F(X)$, cf. [207, Exa. 23.19].

Since by Kodaira vanishing the higher cohomology groups of the ample line bundle $\mathcal{O}_F(1)$ are trivial, i.e. $H^i(F(X), \mathcal{O}_F(1)) = 0$ for $i > 0$, and $q_F(g) = 6$ by Remark 3.14, (4.4) shows $h^0(F(X), \mathcal{O}_F(1)) = 15$. This gives a morphism

$$\mathbb{G}(3, \mathbb{P}(V)) \longleftrightarrow |\mathcal{O}_F(1)| \simeq \mathbb{P}(\textstyle\bigwedge^2 V^*) \simeq \mathbb{P}^{14},$$

which is nothing but the Plücker embedding $\mathbb{G}(3, \mathbb{P}(V)) \hookrightarrow \mathbb{P}(\bigwedge^4 V) \simeq \mathbb{P}(\bigwedge^2 V^*)$.

In Remark 4.19 one finds a discussion of a distinguished uniruled divisor in $F(X)$, which however only realizes a multiple of g.

Let us pass on to surfaces in $F(X)$. One distinguishes between surfaces that are Lagrangian and those that are not. A smooth surface $T \subset F(X)$ is called *Lagrangian* if the restriction map $H^{2,0}(F(X)) \longrightarrow H^{2,0}(T)$ is trivial. For singular surfaces, one uses the pullback to a desingularization. Clearly, any surface T with $H^{2,0}(T) = 0$, e.g. $T \simeq \mathbb{P}^2$, is Lagrangian, but there are other reasons that can cause T to be Lagrangian.

Remark 4.4 If $T \subset F(X)$ is a smooth Lagrangian surface, then $\mathcal{N}_{T/F(X)} \simeq \Omega_T$. This is a general fact for Lagrangians in hyperkähler manifolds and follows from the diagram

$$\begin{array}{ccccccccc} 0 & \longrightarrow & \mathcal{T}_T & \longrightarrow & \mathcal{T}_F|_T & \longrightarrow & \mathcal{N}_{T/F} & \longrightarrow & 0 \\ & & & \searrow & \downarrow \sigma \wr & & \vdots \wr & & \\ 0 & \longrightarrow & \mathcal{N}^*_{T/F} & \longrightarrow & \Omega_F|_T & \longrightarrow & \Omega_T & \longrightarrow & 0. \end{array}$$

The diagonal arrow is 0 if and only if $T \subset F$ is Lagrangian and in this case it induces the isomorphism on the right.

We are concerned with three types of surfaces, $F(Y)$, F_2, and F_L. Surfaces of the form $F(Y)$ are Lagrangian, while those of the form F_2 and F_L are not.[16]

[16] There is one other natural surface, the surface of all lines L for which there exists a plane with $\mathbb{P}^2 \cap X = 3L$. Until recently, almost nothing was known about it until F. Gounelas and A. Kouvidakis [200] computed its cohomology class.

4.3 Fano Variety of a Cubic Threefold Hyperplane Section We begin by coming back to the surface $F(Y) \subset F(X)$ of lines contained in a smooth hyperplane section $Y = X \cap \mathbb{P}^4 \subset \mathbb{P}^4$. As the hyperplane sections $Y_t \subset X$ vary in the five-dimensional family $|\mathcal{O}_X(1)|$, this provides a five-dimensional family of surfaces $F(Y_t)$ in the four-dimensional hyperkähler manifold $F(X)$. Since the generic line $L \subset X$ is contained in a smooth hyperplane section, the surfaces $F(Y_t) \subset F(X)$ cover a dense subset of $F(X)$.

The following was first observed by Voisin [473, Ex. 3.7], see [18, Sec. 1] for a stronger version.

Lemma 4.5 *The surface $F(Y) \subset F(X)$ is Lagrangian, i.e. the restriction*

$$\mathbb{C} \simeq H^{2,0}(F(X)) \longrightarrow H^{2,0}(F(Y)) \simeq \mathbb{C}^5$$

is trivial. In fact, all primitive classes restrict to 0 on $F(Y)$, i.e. the map

$$H^2(F(X), \mathbb{Z})_{\rm pr} \subset H^2(F(X), \mathbb{Z}) \longrightarrow H^2(F(Y), \mathbb{Z})$$

is trivial. Therefore,

$$0 = \alpha \cdot [F(Y)] = \alpha \cdot c_2(\mathcal{S}_F) \in H^6(F(X), \mathbb{Z})$$

for all $\alpha \in H^2(F(X), \mathbb{Z})_{\rm pr}$.

Proof The restriction maps induced by the inclusions $Y \subset X$ and $F(Y) \subset F(X)$ are compatible with the Fano correspondence and thus fit into a commutative diagram

$$\begin{array}{ccc} H^4(X, \mathbb{Q}) & \xrightarrow[\varphi_X]{\sim} & H^2(F(X), \mathbb{Q})(-1) \\ \downarrow & & \downarrow \\ H^4(Y, \mathbb{Q}) & \xrightarrow[\varphi_Y]{} & H^2(F(Y), \mathbb{Q})(-1). \end{array}$$

Since $H^4(Y, \mathbb{Q}) \simeq \mathbb{Q}(-2)$, the composition $H^{3,1}(X) \subset H^4(X, \mathbb{C}) \longrightarrow H^4(Y, \mathbb{C})$ is trivial. Hence, for $\tau \in H^{3,1}(X)$ with $\varphi_X(\tau) = \sigma$, one has $\sigma|_{F(Y)} = \varphi_X(\tau)|_{F(Y)} = \varphi_Y(\tau|_Y) = 0$. Using that $H^2(F(X), \mathbb{Z})_{\rm pr}$ is irreducible for very general X, see Corollary **1**.2.12, this also proves the second assertion. The last assertion is proved using the Gysin sequence: $\alpha \cdot [F(Y)] = i_*(\alpha|_{F(Y)})$ for the inclusion $i\colon F(Y) \hookrightarrow F(X)$ and $[F(Y)] = c_2(\mathcal{S}_F)$ by Proposition 4.1. □

Let us point out two consequences of the above. First,

$$(\sigma \wedge \bar{\sigma}.c_2(\mathcal{S}_F)) = (\sigma \wedge \bar{\sigma}.[F(Y)]) = \int_{F(Y)} \sigma \wedge \bar{\sigma} = 0.$$

Second, as the Plücker polarization g_X on $F(X)$ restricts to the Plücker polarization g_Y on $F(Y)$, there is a short exact sequence, cf. (3.6):

$$0 \longrightarrow H^2(F(X),\mathbb{Z})_{\mathrm{pr}} \longrightarrow H^2(F(X),\mathbb{Z}) \longrightarrow \mathbb{Z}\cdot(1/3)\,g_Y \longrightarrow 0,$$
$$\cup \qquad \mathbb{Z}\cdot g_X \hookrightarrow \mathbb{Z}\cdot(1/3)\,g_Y$$

where the diagonal arrow is an inclusion of index three. Here, we use the fact that the inclusion $H^2(F(X),\mathbb{Z})_{\mathrm{pr}} \oplus \mathbb{Z}\cdot g_X \subset H^2(F(X),\mathbb{Z})$ has to be proper, as $\mathrm{disc}(H^2(F(X),\mathbb{Z})) = -2$ and $q_F(g_X) = 6$, cf. Corollary 3.21. In particular, the image of the restriction map has to be bigger than just $\mathbb{Z}\cdot g_Y$, which leaves $\mathbb{Z}\cdot(1/3)\,g_Y$ as the only choice, see Corollary **5**.1.20. Note that in particular, for the very general cubic fourfold, $(1/3)\,g_Y$ cannot be lifted to an integral $(1,1)$-class on $F(X)$.

Remark 4.6 (i) For the generic hyperplane section $Y = X \cap \mathbb{P}^4$, the cotangent bundle $\Omega_{F(Y)}$ is ample, see Corollary **5**.2.10. As $F(Y) \subset F(X)$ is cut out by a section of $\mathcal{S}_F^*$, its normal bundle is $\mathcal{N}_{F(Y)/F(X)} \simeq \mathcal{S}_{F(Y)}^* \simeq \Omega_{F(Y)}$, see Proposition **5**.2.2 for the last isomorphism. Hence, the normal bundle sequence for the inclusion $F(Y) \subset F(X)$ is of the form:

$$0 \longrightarrow \mathcal{S}_{F(Y)} \longrightarrow \mathcal{T}_{F(X)}|_{F(Y)} \longrightarrow \mathcal{S}_{F(Y)}^* \longrightarrow 0.$$

Therefore, $F(Y) \subset F(X)$ is a surface with ample normal bundle and in many respects behaves like a complete intersection of two ample divisors. However, as shown by Voisin [479, Thm. 2.9], its class $[F(Y)] \in H^{2,2}(F(X),\mathbb{Z})$ is only at the boundary of the cone of effective classes.

In an attempt to describe the other extremal face of this cone, Ottem [381, Cor. 2.6] shows that for a very general cubic fourfold the class $c_2(\mathcal{T}_F) = 5\cdot g^2 - 8\cdot c_2(\mathcal{S}_F)$ is not big, i.e. it is not contained in the interior of the effective cone of codimension two cycles. Earlier attempts go back to Rempel [400].

(ii) Note that the above argument can be reversed: using the fact that $F(Y) \subset F(X)$ is a Lagrangian surface immediately provides us with an isomorphism $\mathcal{N}_{F(Y)/F(X)} \simeq \Omega_{F(Y)}$, see Remark 4.4. Combined with $\mathcal{N}_{F(Y)/F(X)} \simeq \mathcal{S}_{F(Y)}^*$, which was explained before, one finds

$$\mathcal{T}_{F(Y)} \simeq \mathcal{S}_{F(Y)}.$$

This results in an alternative proof of Proposition **5**.2.2.

(iii) By methods not discussed here, one can show that for no smooth hyperplane section Y, the Fano variety $F(Y) \subset F(X)$ can be contained in a smooth ample hypersurface [22, Sec. 8.2].

Remark 4.7 Iliev and Manivel [257, Sec. 2.2.5] investigated a different class of Lagrangian surfaces in $F(X)$ obtained by viewing X as a hyperplane section $X = Z \cap \mathbb{P}^5$ of a cubic fivefold $Z \subset \mathbb{P}^6$. For a generic cubic fivefold, the Fano variety $F(Z,2)$ of planes

$\mathbb{P}^2 \simeq P \subset Z$ is a smooth surface. Furthermore, intersecting a plane $P \subset \mathbb{P}^6$ with the fixed hyperplane $\mathbb{P}^5 \subset \mathbb{P}^6$ defines a map

$$i_Z \colon F(Z,2) \longrightarrow F(X), \ P \longmapsto P \cap \mathbb{P}^5,$$

which for generic choices is an immersion.[17] It turns out that as $F(Y) \subset F(X)$, $i_Z(F(Z,2)) \subset F(X)$ also describes a Lagrangian surface.

From Proposition 4.1 one deduces that the cohomology class of $i_Z(F(Z,2)) \subset F(X)$ is a multiple of $[F(Y)] = c_2(\mathcal{S}_F)$, i.e. $[i_Z(F(Z,2))] = \lambda \cdot [F(Y)]$. The coefficient λ can be computed from the degree $\int_{F(Z,2)} i_Z^* g^2$ of $F(Z,2)$ with respect to the Plücker embedding (the pullback of which to $F(Z,2)$ is again the Plücker polarization). Eventually one finds $[i_Z(F(Z,2))] = 63 \cdot [F(Y)]$, see [257, Lem. 6].

4.4 Lines of the Second Type Next, we comment on the surface $F_2 := F_2(X) \subset F := F(X)$ of lines of the second type. See Section **2**.2.2 for the general theory and recall in particular that the restriction $\mathbb{L}_2$ of the universal line $\mathbb{L} \longrightarrow F$ to F_2 is the non-smooth locus of the projection $q \colon \mathbb{L} \longrightarrow X$. Also, we know that for generic X, the surface F_2 is smooth and that $q \colon \mathbb{L}_2 \longrightarrow X$ is birational onto its image, see Remark **2**.2.16. From the computation of its cohomology class $[F_2] = 5 \cdot (g^2 - [F(Y)]) \in H^{2,2}(F, \mathbb{Z})$ in Proposition 4.1 we deduce

$$(\sigma \wedge \bar{\sigma}.[F_2]) = 5 \int_F (\sigma \wedge \bar{\sigma}) \cdot g^2 > 0.$$

Hence, $F_2 \subset F$ is not Lagrangian. The geometry of F_2 is not completely understood, but the following was essentially proved by Oberdieck, Shen, and Yin [373].

Proposition 4.8 *For generic X, the surface $F_2 \subset F$ is a smooth connected surface of general type. Its normal bundle $\mathcal{N}_{F_2/F}$ sits in an exact sequence*

$$0 \longrightarrow \mathcal{O}_{F_2}(-1)' \longrightarrow S^2(\mathcal{S}_{F_2}) \xrightarrow{\psi} \mathcal{Q}^*_{F_2} \longrightarrow \mathcal{N}_{F_2/F} \otimes \mathcal{O}_{F_2}(-1)' \longrightarrow 0,$$

where $\mathcal{O}_{F_2}(-1)'$ is a line bundle satisfying $\mathcal{O}_{F_2}(-1)' \otimes \mathcal{O}_{F_2}(-1)' \simeq \mathcal{O}_{F_2}(-2)$. Furthermore, $H^1(F_2, \mathcal{O}_{F_2}) = 0$ and the canonical bundle of F_2 satisfies $\omega_{F_2}^2 \simeq \mathcal{O}_{F_2}(6)$. In particular, F_2 is a minimal surface of general type.

By a recent work of Gounelas and Kouvidakis [201] all other Hodge numbers of F_2 are now also known:

$$h^{2,0}(F_2) = 449 \text{ and } h^{1,1}(F_2) = 1\,665,$$

but we refrain from reproducing this computation here.

[17] Beware that the preprint version of [257] claims i_Z to be a closed immersion. The published version explains that $i_Z(F(Z,2))$ has in fact 47 061 singular points.

Proof The proof in [373] uses the global version of F_2, while the proof of the connectedness below is based on a vanishing theorem.

We know already that F_2 is a smooth surface for the generic cubic fourfold, see Proposition **2**.2.13. Thus, to show that it is connected, it suffices to prove $h^0(F_2, \mathcal{O}_{F_2}) = 1$. For this, we use the description of F_2 as the zero locus $V(\tilde{\psi})$ of a certain global section $\tilde{\psi} \in H^0(\pi^*\mathcal{Q}_F^* \otimes \mathcal{O}_\pi(1))$ on $\pi\colon \mathbb{L}^{[2]} = \mathbb{P}(S^2(\mathcal{S}_F)) \longrightarrow F$, see Remark **2**.2.11. This allows one to exploit the Koszul resolution

$$0 \longrightarrow \bigwedge^4 \pi^*\mathcal{Q}_F \otimes \mathcal{O}_\pi(-4) \longrightarrow \cdots \longrightarrow \pi^*\mathcal{Q}_F \otimes \mathcal{O}_\pi(-1) \longrightarrow \mathcal{O} \longrightarrow \mathcal{O}_{F_2} \longrightarrow 0$$

and the associated spectral sequence

$$E_1^{p,q} = H^q(\mathbb{L}^{[2]}, \bigwedge^{-p} \pi^*\mathcal{Q}_F \otimes \mathcal{O}_\pi(p)) \Longrightarrow H^{p+q}(F_2, \mathcal{O}_{F_2}),$$

cf. the arguments in the proof of Proposition **2**.3.4. To conclude, it is enough to show that $H^q(\bigwedge^q \pi^*\mathcal{Q}_F \otimes \mathcal{O}_\pi(-q)) = 0$ for $q > 0$. For $q > 0$, the projection formula implies

$$H^q(\mathbb{L}^{[2]}, \bigwedge^q \pi^*\mathcal{Q}_F \otimes \mathcal{O}_\pi(-q)) \simeq H^{q-2}(F, \bigwedge^q \mathcal{Q}_F \otimes R^2\pi_*\mathcal{O}_\pi(-q))$$

and these cohomology groups are certainly trivial for $q = 1, 2$. For $q = 3$, the right-hand side becomes

$$H^1(F, \bigwedge^3 \mathcal{Q}_F \otimes \mathcal{O}(-3)) \simeq H^1(F, \mathcal{Q}_F^* \otimes \mathcal{O}(1) \otimes \mathcal{O}(-3)) \simeq H^3(F, \mathcal{Q}_F \otimes \mathcal{O}(2))^*,$$

where we use $R^2\pi_*\mathcal{O}_\pi(-3) \simeq \det(S^2(\mathcal{S}_F)) \simeq \mathcal{O}(-3)$, $\det(\mathcal{Q}_F) \simeq \mathcal{O}(1)$ and Serre duality. Now the Griffiths vanishing theorem applies, see [203, Thm. G] or [430, Thm. 5.52]. It asserts that $H^q(S^k(\mathcal{E}) \otimes \det(\mathcal{E}) \otimes \mathcal{L} \otimes \omega) = 0$ for $q > 0$ for every globally generated locally free sheaf $\mathcal{E}$ and any ample line bundle $\mathcal{L}$. This proves the desired vanishing for $q = 3$, because ω_F is trivial and as a quotient of $V \otimes \mathcal{O}_F$ the locally free sheaf $\mathcal{Q}_F$ is globally generated. For $q = 4$, one argues similarly:

$$\begin{aligned} H^2(F, \det(\mathcal{Q}_F) \otimes R^2\pi_*\mathcal{O}_\pi(-4)) &\simeq H^2(F, \mathcal{O}(1) \otimes S^2(\mathcal{S}_F) \otimes \mathcal{O}(-3)) \\ &\simeq H^2(F, S^2(\mathcal{S}_F^*) \otimes \mathcal{O}(1) \otimes \mathcal{O}(1))^* = 0 \end{aligned}$$

for $\det(\mathcal{S}_F^*) \simeq \mathcal{O}(1)$ and $\mathcal{S}_F^*$ is globally generated.

Similar computations prove the claimed vanishing $H^1(F_2, \mathcal{O}_{F_2}) = 0$.[18]

The normal bundle of $Z := V(\tilde{\psi}) \subset \mathbb{P}(S^2(\mathcal{S}_F))$ is $(\pi^*\mathcal{Q}_F^* \otimes \mathcal{O}_\pi(1))|_Z$ and, as the projection defines an isomorphism $\pi\colon Z \xrightarrow{\sim} F_2$, there is a commutative diagram of short exact rows and columns

[18] Thanks to F. Gounelas for pointing this out.

$$\begin{array}{ccccc} & & \mathcal{T}_\pi|_Z & \xrightarrow{\simeq} & \mathcal{T}_\pi|_Z \\ & & \downarrow & & \downarrow \\ \mathcal{T}_Z & \longrightarrow & \mathcal{T}_{\mathbb{L}^{[2]}}|_Z & \longrightarrow & (\pi^*\mathcal{Q}_F^* \otimes \mathcal{O}_\pi(1))|_Z \\ \downarrow\simeq & & \downarrow & & \downarrow \\ \mathcal{T}_{F_2} & \longrightarrow & \mathcal{T}_F|_{F_2} & \longrightarrow & \mathcal{N}_{F_2/F}. \end{array}$$

Now use the relative Euler sequence $0 \longrightarrow \mathcal{O} \longrightarrow \pi^*S^2(\mathcal{S}_F) \otimes \mathcal{O}_\pi(1) \longrightarrow \mathcal{T}_\pi \longrightarrow 0$ and observe that the composition

$$(\pi^*S^2(\mathcal{S}_F) \otimes \mathcal{O}_\pi(1))|_Z \twoheadrightarrow \mathcal{T}_\pi|_Z \hookrightarrow (\pi^*\mathcal{Q}_F^* \otimes \mathcal{O}_\pi(1))|_Z$$

is indeed the pullback of ψ tensored with $\mathcal{O}_\pi(1)$ to deduce the exact sequence

$$0 \longrightarrow \mathcal{O}_{F_2} \longrightarrow S^2(\mathcal{S}_{F_2}) \otimes \mathcal{O}_\pi(1)|_{F_2} \xrightarrow{\psi} \mathcal{Q}_{F_2}^* \otimes \mathcal{O}_\pi(1)|_{F_2} \longrightarrow \mathcal{N}_{F_2/F} \longrightarrow 0.$$

The last step consists of proving that $\mathcal{O}_\pi(2)|_Z$ is isomorphic to the restriction of the square of the Plücker polarization $\mathcal{O}_{F_2}(2)$ under $Z \simeq F_2$. The argument is similar to the proof of $\beta^*\mathcal{O}(2) \simeq \mathcal{O}_{F_2}(4)$ for the morphism $\beta\colon F_2 \longrightarrow \mathbb{G}(n-1, \mathbb{P})$, $L \longmapsto P_L$ in Remark **2**.2.20. At a point $L = \mathbb{P}(W) \in F_2 \simeq Z$ the fibre of $\mathcal{O}_\pi(-1)$ is naturally identified with the kernel of the map $S^2(W) \longrightarrow (V/W)^*$, the dual of which sits in the exact sequence

$$0 \longrightarrow H^0(\mathcal{N}_{L/X}(-1)) \longrightarrow V/W \longrightarrow S^2(W^*) \longrightarrow H^1(\mathcal{N}_{L/X}(-1)) \longrightarrow 0,$$

see Remark **2**.2.2. Hence, the fibre of $\mathcal{O}_\pi(1)$ at L is

$$\begin{aligned} H^1(\mathcal{N}_{L/X}(-1)) &\simeq \det(S^2(W^*)) \otimes \det(V/W)^* \otimes \det H^0(\mathcal{N}_{L/X}(-1)) \\ &\simeq \det(W^*)^3 \otimes \det(W) \otimes H^1(L, \mathcal{N}_{L/X}(-1))^* \\ &\simeq \det(W^*)^2 \otimes H^1(L, \mathcal{N}_{L/X}(-1))^*, \end{aligned}$$

where the second isomorphism uses (1.16) in Remark **2**.1.16. This leads to the desired natural isomorphism $H^1(\mathcal{N}_{L/X}(-1))^2 \simeq \det(W^*)^2$, which is straightforward to globalize to the desired isomorphism $\mathcal{O}_{F_2}(2) \simeq \mathcal{O}_\pi(2)|_Z$.

The last assertion concerning $\omega_{F_2}^2$ follows from the exact sequence describing $\mathcal{N}_{F_2/F}$ by taking determinants. □

Remark 4.9 It turns out that the natural guess that in fact $\mathcal{O}_\pi(1)|_Z \simeq \mathcal{O}_{F_2}(1)$ and, consequently, $\omega_{F_2} \simeq \mathcal{O}_{F_2}(3)$ is wrong.[19] In fact, in [201] it is shown that $H^2(F_2, \mathcal{O}_{F_2}(3)) = 0$, but, of course, $H^2(F_2, \omega_{F_2})$ is one-dimensional. In particular, the Picard group of F_2 contains elements of order two.

[19] I wish to thank for F. Gounelas for a fruitful discussion of this point.

The étale double cover that corresponds to the two-torsion line bundle $\mathcal{O}_{F_2}(1)' \otimes \mathcal{O}_{F_2}(-1)$ can geometrically be described as follows: the Gauss map restricted to a line L of the second type induces a cover of degree two $\gamma_X|_L : L \twoheadrightarrow \gamma(L)$ of a line, see Exercise **2**.2.10. Put in a family, one obtains an étale double cover

$$\tau : \widetilde{F}_2 \twoheadrightarrow F_2,$$

the fibre of which over the point $L \in F_2$ consists of the two ramification points of $\gamma_X|_L$. To see that this fits with the picture in the above proof, observe that the two ramification points, say $y_1, y_2 \in L$, correspond to two complementary lines $K_1, K_2 \subset W^*$, whose tensor product $K_1 \otimes K_2 \subset S^2(W^*)$ maps isomorphically onto $H^1(\mathcal{N}_{L/X}(-1))$. However,

$$K_1 \otimes K_2 \simeq K_1 \otimes (W^*/K_1) \simeq K_1 \otimes \det(W^*) \otimes K_1^* \simeq \det(W^*),$$

which gives an isomorphism $H^1(\mathcal{N}_{L/X}(-1)) \simeq \det(W^*)$. Changing the order of the two ramification points y_1, y_2 changes this isomorphism by a sign, which explains that in the above proof only the isomorphism $H^1(\mathcal{N}_{L/X}(-1))^2 \simeq \det(W^*)^2$ is natural. Note that

$$\omega_{\widetilde{F}_2} \simeq \tau^* \omega_{F_2} \simeq \tau^* \mathcal{O}_{F_2}(3).$$

Remark 4.10 (i) The surface $F_2 \subset F$ always comes equipped with a natural canonical divisor $C \in |\omega_{F_2}|$. Indeed, as $F_2 \subset F$ is not Lagrangian, the restriction of a generator $\sigma \in H^0(F, \Omega_F^2)$ to F_2 defines a non-trivial section $\sigma|_{F_2} \in H^0(F_2, \omega_{F_2})$ and its zero set is a curve C. Proposition 4.1 allows us to compute the cohomology class of this curve as $[C] = 315 \cdot (g^3/36) \in H^{3,3}(F(X), \mathbb{Z})$.

(ii) The surface $F_2 \subset F$ is not a constant cycle surface, i.e. general points $L \in F_2$ define distinct classes $[L] \in \mathrm{CH}_0(F)$. This will become clear in Section 4.6, where we will see that a certain $\mathbb{P}^1$-bundle over F_2 maps generically finite to F and the latter cannot contain constant cycle divisors. See [309] for some results on constant cycle subvarieties in $F(X)$.

4.5 Lines Intersecting a Given Line Last, but not least, we discuss the surfaces $F_L \subset F(X)$ of all lines intersecting a fixed line $L \subset X$.

As for $F_2 \subset F(X)$, the computation of its cohomology class (not in Chow) as

$$[F_L] = (1/3) \cdot (g^2 - c_2(\mathcal{S}_F)) = (1/3) \cdot (g^2 - [F(Y)])$$

in Proposition 4.1 shows that the surfaces $F_L \subset F(X)$ are not Lagrangian.

Assume $L' \in F_L$ is a line distinct from L. Mapping L' to its point of intersection with L defines a morphism $F_L \setminus \{L\} \twoheadrightarrow L$. This rational map can be extended to a morphism from the blow-up $\widetilde{F}_L := \mathrm{Bl}_L(F_L) \twoheadrightarrow L$, which is more directly realized as the restriction of the projection $q : \mathbb{L} \twoheadrightarrow X$ to the line L:

$$q : \widetilde{F}_L \simeq q^{-1}(L) \twoheadrightarrow L.$$

Alternatively, using Proposition **2**.3.10, $\widetilde{F}_L \subset \mathbb{P}(\mathcal{T}_X|_L)$ can be viewed as the zero set of a a section of a locally free sheaf of rank two E_L on $\mathbb{P}(\mathcal{T}_X|_L)$ which is an extension

$$0 \longrightarrow \pi^*\mathcal{O}(3) \otimes \mathcal{O}_\pi(3) \longrightarrow E_L \longrightarrow \pi^*\mathcal{O}(3) \otimes \mathcal{O}_\pi(2) \longrightarrow 0.$$

For the generic line L, the fibre $q^{-1}(x) \subset \widetilde{F}_L$ of the generic point $x \in L$ is a smooth complete intersection curve of degree $(2, 3)$, see Remark **2**.3.6 and Lemma **2**.5.11. Furthermore, all fibres project onto curves in F_L that pass through the point $L \in F_L$.

It is also interesting to study the plane $\mathbb{P}^2 \simeq \overline{L'L}$ associated with every $L \neq L' \in F_L$. It intersects X in the two lines L and L' and a residual line L'', which may coincide with L or L'. Alternatively, $\overline{L'L}$ intersects a linear subspace $\mathbb{P}^3 \subset \mathbb{P}^5$ disjoint from L in a unique point y, which is the image of the singular quadric $L' \cup L'' \subset X$ under the linear projection

$$\phi\colon \mathrm{Bl}_L(X) \longrightarrow \mathbb{P}^3$$

from L. As y is a point in the discriminant surface $D_L \subset \mathbb{P}^3$, see Section **1**.5.1, this defines a morphism $F_L \setminus \{L\} \longrightarrow D_L$. If there exists unique plane with $\mathbb{P}^2 \cap X = 2L \cup L'$, i.e. if L is of the first type, see Section 0.2 or Corollary **2**.2.6, it can be extended to a surjective morphism, confusingly also called

$$\pi\colon F_L \twoheadrightarrow D_L.$$

From its geometric description one finds that for L of the first type the morphism π is finite of degree two with the covering involution

$$\iota\colon F_L \longrightarrow F_L,\ L' \longmapsto L'',$$

mapping a line L' intersecting L to the residual line L'', i.e. $\overline{L'L} \cap X = L \cup L' \cup L''$. In particular, $L \in F_L$ is mapped to the unique line L'' for which $\overline{L''L} \cap X = 2L \cup L''$, cf. Section 4.6

Remark 4.11 Not surprisingly, certain features from cubic threefolds transfer to the situation here, see also Remark 1.12.

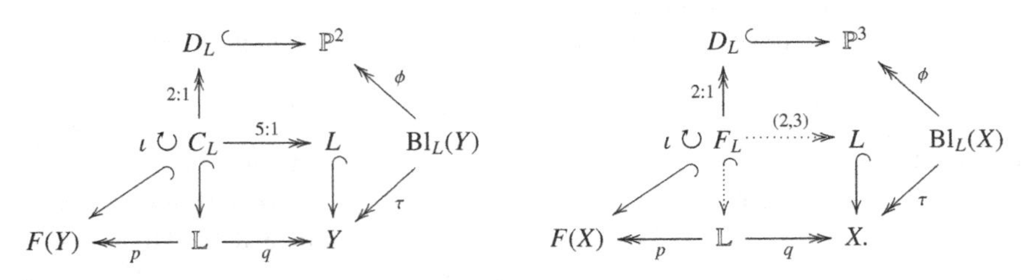

In both cases, the cubic is projected from a generic line L and the discriminant divisor $D_L \in |\mathcal{O}(5)|$ is a quintic curve, respectively, surface.

Recall that the smoothness of the discriminant quintic curve $D_L \subset \mathbb{P}^2$ for the generic line $L \subset Y$ in a smooth cubic threefold Y could not be deduced from general principles, see Remark **1**.5.8. One had to argue that a point in D_L is singular if the fibre over it has certain special geometric properties which are excluded for generic L.

The situation is similar for cubic fourfolds. According to Remark **1**.5.8, the discriminant quintic surface $D_L \subset \mathbb{P}^3$ is not expected to be smooth. However, if the defining symmetric equation could be chosen generic, all singularities of D_L would be just ordinary double points. There does not seem to be a way to argue directly that for generic choice of L also D_L is generic in this sense. Instead one has to evoke the conic fibration $\phi\colon \mathrm{Bl}_L(X) \twoheadrightarrow \mathbb{P}^3$ and the shape of the singular fibres over points in D_L. As a result, for the generic choice of a line $L \subset X$ in a cubic fourfold X the discriminant surface $D_L \subset \mathbb{P}^3$, which is singular, has only ordinary double points (nodes) as singularities. An explicit finite set of open conditions for the smoothness of F_L and the singularities of D_L being nodes was spelled out by Voisin [472, §3, Lem. 1].

The following facts are either classical [177, 450] or were proved by Voisin [472, §3, Lem. 3], see also the survey [244].

Lemma 4.12 *Let X be a smooth cubic fourfold. Then for a dense open subset of lines $L \in F(X)$ the following assertions hold true:*

(i) *The surface $F_L \subset F(X)$ is smooth, connected, and of general type.*
(ii) *The numerical invariants of F_L are as follows:*

$$\int_{F_L} c_1(\mathcal{T}_{F_L})^2 = 10, \quad \chi(\mathcal{O}_{F_L}) = 6, \ h^1(\mathcal{O}_{F_L}) = 0, \quad \text{and} \quad e(F_L) = 62.$$

(iii) *The covering involution ι of $\pi\colon F_L \twoheadrightarrow D_L$ has 16 fixed points, which via π are mapped to the 16 nodes of the quintic surface $D_L \subset \mathbb{P}^3$.*

Proof For a generic line L, the surface D_L has only ordinary double points and, as revealed by a local computation as in [36], those are exactly the points $y \in D_L$ at which the rank of q_y is one. The number of these points can be determined by a Grothendieck–Riemann–Roch computation as in [36] or by applying the symmetric version of Porteous formula developed in [221, Thm. 10]. The latter says that the number of points y where the rank of the quadratic equation q_y drops to one is $\int_{\mathbb{P}^3} 2^2(c_1c_2 - c_3)$. Here, the c_i are the Chern classes of the formal bundle $\mathcal{F} \otimes \mathcal{O}(1/2)$ on $\mathbb{P}^3$, i.e. $c_1 = (5/2)\,h$, $c_2 = (7/4)\,h^2$, and $c_3 = (3/8)\,h^3$. See Section **1**.5.1 for the notations, e.g. $q \in H^0(\mathbb{P}^3, S^2(\mathcal{F}) \otimes \mathcal{O}(1))$ describes $\mathrm{Bl}_L(X) \subset \mathrm{Bl}_L(\mathbb{P}^5)$, and for the fact that $\mathcal{F} \simeq \mathcal{O} \oplus \mathcal{O} \oplus \mathcal{O}(1)$. The computation then indeed shows that there are exactly 16 such points.

The rank of q_y, with $\{y\} = \overline{L'L} \cap \mathbb{P}^3$, is one if and only if L' coincides with the residual line L'' of $L' \cup L \subset \overline{L'L} \cap X$. The latter describes the fixed points of ι, i.e. $\pi\colon F_L \twoheadrightarrow D_L$ is unramified over the complement of the nodes $y_i \in D_L$, $i = 1, \ldots, 16$, with exactly one point over each of the y_i. Locally analytically, this looks like the quotient of $\mathbb{C}^2 \longrightarrow \mathbb{C}^2/\{\pm 1\}$ and, in particular, F_L is smooth and connected. Furthermore,

$$\omega_{F_L} \simeq \pi^*\omega_{D_L} \simeq \pi^*\mathcal{O}(1)$$

and, therefore, $\int_{F_L} c_1(\mathcal{T}_{F_L})^2 = 2 \cdot \deg(D_L) = 10$.

To compute the second Chern number of F_L, use the normal bundle sequence for $F_L \subset F = F(X)$ and the computation in Proposition 4.1 to obtain[20]

$$\begin{aligned} e(F_L) = c_2(F_L) &= c_2(\mathcal{T}_F)|_{F_L} - c_2(\mathcal{N}_{F_L/F}) - c_1(\mathcal{T}_{F_L}) \cdot c_1(\mathcal{N}_{F_L/F}) \\ &= (c_2(\mathcal{T}_F).[F_L]) - ([F_L].[F_L]) + 10 = 62. \end{aligned}$$

The Noether formula then shows $\chi(\mathcal{O}_{F_L}) = 6$. To prove that $H^1(F_L, \mathcal{O}_{F_L}) = 0$, one can use the following general fact:[21] a minimal surface S of general type satisfying $(\omega_S.\omega_S) = 2\chi(S, \mathcal{O}_S) - 2$ has a finite algebraic fundamental group of order ≤ 3. Indeed, if $S' \longrightarrow S$ is an étale cover of degree m, then Noether inequality $2\chi(S', \mathcal{O}_{S'}) - 6 \leq (\omega_{S'}.\omega_{S'})$ for S' and the two equalities $\chi(S', \mathcal{O}_{S'}) = m \cdot \chi(S, \mathcal{O}_S)$ and $(\omega_{S'}.\omega_{S'}) = m \cdot (\omega_S.\omega_S)$ give

$$2m \cdot \chi(S, \mathcal{O}_S) - 6 \leq m \cdot (\omega_S.\omega_S) = m \cdot (2\chi(S, \mathcal{O}_S) - 2)$$

and, therefore, $m \leq 3$. Hence, $|\pi_1^{\rm alg}(S)| \leq 3$ and, in particular, $H_1(S, \mathbb{Z})$ is torsion.

Voisin's original proof is more involved, but is a nice example of the interplay between cubic threefolds and cubic fourfolds, so we briefly sketch it too. Consider a Lefschetz pencil $\mathcal{Y} \longrightarrow \mathbb{P}^1$ of cubic threefolds $Y_t := X \cap H_t$ of $X \subset \mathbb{P}^5$. We we may choose L to be contained in its fixed locus. The hyperplanes H_t also define a pencil of hyperplane sections $D_L \cap H_t \subset H_t \cap \mathbb{P}^3$ of D_L and a moment's thought reveals that $\pi_t \colon C_L(Y_t) := \pi^{-1}(D_L \cap H_t) \longrightarrow D_L \cap H_t$ for the generic Y_t is exactly the situation described in Section **5**.3.2, i.e. $C_L(Y_t) \subset F(Y_t)$ is the Fano variety of lines in Y_t intersecting $L \subset Y_t$. By the Lefschetz hyperplane theorem, the restriction defines an injection

$$H^1(F_L, \mathbb{Q}) = H^1(F_L)^+ \oplus H^1(F_L)^- \longrightarrow H^1(C_L(Y_t), \mathbb{Q}) = H^1(C_L(Y_t))^+ \oplus H^1(C_L(Y_t))^-,$$

which is compatible with the action of the involution. Then we observe that $H^1(F_L)^+ \simeq H^1(D_L, \mathbb{Q}) = 0$ and that any class in the image of

$$H^1(F_L, \mathbb{Z})^- \longrightarrow H^1(C_L(Y_t), \mathbb{Z})^- \simeq H^3(Y_t, \mathbb{Z})$$

defines a monodromy invariant class on the fibre Y_t of $\mathcal{Y} \longrightarrow \mathbb{P}^1$ which by Deligne's invariant cycle theorem [474, V. Thm. 16.24] would come from a cohomology class on $\mathcal{Y}$. However, $H^3(\mathcal{Y}, \mathbb{Q}) = 0$, because $\mathcal{Y} \longrightarrow X$ is the blow-up in the cubic surface $S := X \cap H_0 \cap H_1$ and $H^1(S, \mathbb{Q}) = 0 = H^3(X, \mathbb{Q})$. □

[20] Voisin [472] computes the Euler number by a more topological argument for which she uses that $e(D) = 55$ for a smooth quintic $D \in |\mathcal{O}_{\mathbb{P}^3}(5)|$ and that each node reduces the Euler number by one.

[21] I wish to thank R. Pardini for the argument which goes back to Bombieri.

Remark 4.13 There exist three natural line bundles on F_L. First, the Plücker polarization $\mathcal{O}_{F_L}(1)$. Second, the line bundle associated with the genus four curve $C_x := q^{-1}(x) \subset F_L$ of all lines passing through a fixed point $x \in L$ and, third, $\pi^*\mathcal{O}(1) \simeq \omega_{F_L}$. They compare as follows [472, §3, Lem. 2]:

$$\pi^*\mathcal{O}(1) \simeq \mathcal{O}(C_x + \iota(C_x)) \quad \text{and} \quad \mathcal{O}_{F_L}(1) \simeq \mathcal{O}(2C_x + \iota(C_x)) \simeq \pi^*\mathcal{O}(1) \otimes \mathcal{O}(C_x). \tag{4.5}$$

This should be compared to Remark **5**.1.18 which describes the situation for threefolds. Note that the rational map $q \colon F_L \dashrightarrow L$ is associated with $\mathcal{O}_{F_L}(1) \otimes \pi^*\mathcal{O}(-1)$ and so

$$\mathcal{O}_{F_L}(1) \simeq \pi^*\mathcal{O}(1) \otimes q^*\mathcal{O}(1)$$

on the blow-up of F_L. Alternatively, this can be deduced from the last isomorphism in (4.5) and $\mathcal{O}(C_x) \simeq q^*\mathcal{O}(1)$. Compare this to the analogous formula for cubic threefolds in Exercise **5**.1.28. The intersection numbers are given by, cf. Proposition 4.1, Lemma **2**.5.11, and Lemma 4.12:

$$(\mathcal{O}_{F_L}(1).\mathcal{O}_{F_L}(1)) = \int_{F_L} g^2 = 21, \quad (\pi^*\mathcal{O}(1).\mathcal{O}(C_x)) = 5, \quad (\mathcal{O}_{F_L}(1).\mathcal{O}(C_x)) = 6,$$

$$(C_x.C_x) = 1, \quad (\mathcal{O}_{F_L}(1).\pi^*\mathcal{O}(1)) = 15, \quad \text{and} \quad (\pi^*\mathcal{O}(1).\pi^*\mathcal{O}(1)) = \int_{F_L} c_1(\mathcal{T}_{F_L})^2 = 10.$$

It would be interesting to know the Picard number of F_L for the very general line L.

Remark 4.14 In [244] one finds a survey of further observations concerning the geometry and topology of the surfaces F_L. For example, F_L is (algebraically) simply connected and $H^2(F_L, \mathbb{Z})$ is torsion free. Furthermore, the covering involution ι induces a short exact sequence, cf. (3.9) in Section **5**.3.2,

$$0 \longrightarrow H^2(F_L, \mathbb{Z})^+ \longrightarrow H^2(F_L, \mathbb{Z}) \longrightarrow H^2(F_L, \mathbb{Z})^- \longrightarrow 0,$$

where ι acts by ± 1 on $H^2(F_L, \mathbb{Z})^{\pm}$. At this point, the index of the natural inclusion $H^2(D_L, \mathbb{Z}) \hookrightarrow H^2(F_L, \mathbb{Z})^+$ is not known. Recall that in the analogous situation of cubic threefolds, $H^1(D_L, \mathbb{Z}) \hookrightarrow H^1(C_L, \mathbb{Z})^+$ has index two, see Section **5**.3.2.

The analogue of Proposition **5**.3.10 is the assertion that there exist Hodge isometries

$$\left(H^4(X, \mathbb{Z})_{\mathrm{pr}}(1), -(\ .\ )\right) \simeq \left(H^2(F(X), \mathbb{Z})_{\mathrm{pr}}, q_F\right) \simeq \left(H^2(F_L, \mathbb{Z})^-_{\mathrm{pr}}, (1/2)(\ .\ )\right). \tag{4.6}$$

Here, the primitive part $H^2(F_L, \mathbb{Z})^-_{\mathrm{pr}}$ is defined with respect to the Plücker polarization. On a motivic level, the composite Hodge isometry (4.6) is reflected by an isomorphism

$$\mathfrak{h}^4(X)_{\mathrm{pr}} \simeq \mathfrak{h}^2(F_L)^-_{\mathrm{pr}}$$

of rational Chow motives.

Analogously to the formulation of the global Torelli theorem for cubic threefolds in Theorem **5**.4.3 and Remark **5**.4.5, we combine the above results with Theorem 3.17 and Remark 3.18 to deduce the following version.

Proposition 4.15 *For two smooth cubic fourfolds $X, X' \subset \mathbb{P}^5$ over $\mathbb{C}$ the following conditions are equivalent:*

(i) *There exists an isomorphism $X \simeq X'$.*
(ii) *There exists a Hodge isometry $H^4(X,\mathbb{Z})_{\rm pr} \simeq H^4(X',\mathbb{Z})_{\rm pr}$.*
(iii) *There exists a Hodge isometry $H^2(F(X),\mathbb{Z})_{\rm pr} \simeq H^2(F(X'),\mathbb{Z})_{\rm pr}$.*
(iv) *There exists a Hodge isometry $H^2(F_L,\mathbb{Z})^-_{\rm pr} \simeq H^2(F_{L'},\mathbb{Z})^-_{\rm pr}$, where $L \subset X$ and $L' \subset X'$ are generic lines.*

Combining the results for the two types of surfaces $F(Y), F_L \subset F(X)$, Voisin proves in [472, Sec. 4] the following crucial technical result.

Corollary 4.16 (Voisin) *For a smooth cubic fourfold X, there is no primitive class $\delta \in H^{2,2}(X,\mathbb{Z})$ with $(\delta.\delta) = 2$.*

Proof From a hyperkähler point of view one could argue as follows. Under the Fano correspondence, the class δ would define a primitive $(1,1)$-class $\alpha = \varphi(\delta)$ on $F(X)$ which by Corollary 3.21 then satisfies $q_F(\alpha) = -(\delta.\delta) = -2$. However, general results for hyperkähler manifolds of $\mathrm{K3}^{[2]}$-type, see [338, Thm. 1.11] and Remark 3.27, show that the (-2)-class α is effective up to scaling, i.e. up to a non-zero factor it is of the form $[D]$ for some hypersurface $D \subset F(X)$. This leads to the contradiction $0 < \int_D g^3 = \int_F [D] \cdot g^3 = 0$, where the equality follows from α being primitive.

If one wants to avoid the general hyperkähler theory, which was not available at the time of [472], one could try to find a surface $S \subset F(X)$ such that $\alpha|_S$ is effective and primitive with respect to some polarization, which would result in a similar contradiction, as above. In [472] the surface F_L with the restriction of the Plücker polarization is used[22] and the rough idea goes as follows.

By virtue of Lemma 4.5, Proposition 4.1, and the primitivity of α, one has

$$\int_{F_L} g|_{F_L} \cdot \alpha|_{F_L} = \int_F [F_L] \cdot g \cdot \alpha = (1/3) \int_F (g^3 \cdot \alpha - g \cdot c_2(\mathcal{S}_F) \cdot \alpha) = 0,$$

i.e. $\alpha|_{F_L}$ is indeed primitive. If $\mathcal{L}$ is the line bundle corresponding to $\alpha|_{F_L}$, then the Riemann–Roch formula on F_L says $\chi(F_L, \mathcal{L}) = (1/2)\,c_1(\mathcal{L})(c_1(\mathcal{L}) - c_1(\pi^*\mathcal{O}_L(1))) + 6$. Ideally one would like to conclude from this that $\mathcal{L}$ or $\mathcal{L}^*$ has to be effective. However, the argument to conclude is more involved and uses the geometry of the surface F_L. We refer to [472] for the details. □

[22] There are only two other types of surfaces one could try, namely $F(Y)$ and F_2. The latter has the disadvantage of possibly being singular and for the former the fact that α is a (-2)-class is not reflected by any property of its restriction $\alpha|_{F(Y)}$.

4.6 Curves on the Fano Variety and Voisin's Endomorphism Next, let us look at natural curves contained in the Fano variety $F = F(X)$ of the very general cubic fourfold X. We will combine this with a discussion of a certain rational endomorphism of $F(X)$ introduced by Voisin.

Clearly, $H^{3,3}(F, \mathbb{Q}) \simeq \mathbb{Q} \cdot g^3$ for general X and, in fact, $H^{3,3}(F, \mathbb{Z}) \simeq \mathbb{Z} \cdot (g^3/36)$, see Proposition 3.23 or (4.3) in Section 4.1. To find out which multiples of the integral generator $(1/36) \cdot g^3$ are effective, one could first use Lemma **2**.5.11 to show that the fibres $q^{-1}(x) \subset F$ of $q \colon \mathbb{L} \twoheadrightarrow X$, which are curves of degree six in $F(X)$, satisfy

$$[q^{-1}(x)] = (1/18) \cdot g^3.$$

The latter shows again that at least $(1/18) \cdot g^3$ is indeed an integral class. See Remark 4.20 below for further comments. In particular, we will see that the integral generator $(1/36) \cdot g^3$ is also effective.

Assume that $X \subset \mathbb{P}^5$ is a generic smooth cubic fourfold, so that we may assume that $F_2 \subset F = F(X)$ is a smooth surface. In Remark **2**.2.19 we explained how to view the blow-up $\widetilde{F} := \widetilde{F}(X) \twoheadrightarrow F$ of F along $F_2 \subset F$ as the incidence variety of pairs $(L, \mathbb{P}^2)$ with L a line contained in X and $\mathbb{P}^2 \subset \mathbb{P}^5$ a plane tangent to X at every point of L, i.e. such that $\mathbb{P}^2 \cap X = 2L \cup L'$. The residual line $L' \subset X$ is generically distinct from L, but it may happen that $L' = L$. Mapping $(L, \mathbb{P}^2) \in \widetilde{F}$ to the residual line L' defines a morphism to F and thus a diagram

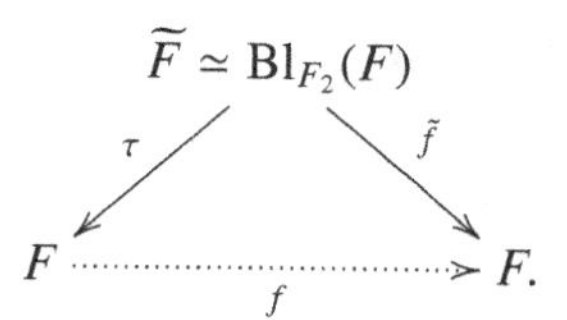

We will view f as a rational endomorphism of the Fano variety. It was introduced by Voisin [475] and further studied together with Amerik [17, 18].

Proposition 4.17 (Voisin, Amerik) *The rational endomorphism $f \colon F \dashrightarrow F$ of the Fano variety of the generic cubic fourfold X has the following properties:*

(i) *It is generically finite of degree* $\deg(f) = 16$.
(ii) *The pullback of the Plücker polarization is*

$$\tilde{f}^* \mathcal{O}(1) \simeq \tau^* \mathcal{O}(7) \otimes \mathcal{O}(-3E),$$

where E denotes the exceptional divisor of the blow-up $\tau \colon \widetilde{F} \twoheadrightarrow F$.

Proof The proof of the first assertion presented in [475] uses the interplay between Chow groups and cohomology. It can also be deduced as follows: fix a generic line $L' \in F$. Then lines $L \in F$ for which there exists a plane with $\mathbb{P}^2 \cap X = 2L \cup L'$, i.e. with $f(L) = L'$, correspond to points in $F_{L'}$ fixed by the covering involution of the projection $F_{L'} \twoheadrightarrow D_{L'}$. According to Lemma 4.12, there are exactly 16 such fixed points and, therefore, the generic fibre $f^{-1}(L')$ consists of 16 points, i.e. $\deg(f) = 16$. Note that τ identifies the fibre $\tilde{f}^{-1}(L')$ with $f^{-1}(L')$.

To confirm the second assertion, one first shows that the restriction of $\tilde{f}^*\mathcal{O}(1)$ to a fibre $\tau^{-1}(L)$ of the $\mathbb{P}^1$-bundle $E \simeq \mathbb{P}(\mathcal{N}_{F_2/F}) \twoheadrightarrow F_2$ is $\mathcal{O}(3)$. Recall that for L of the second type, i.e. $L \in F_2$, there exists a unique $P_L \simeq \mathbb{P}^3$ tangent to X at every point of L, see Corollary **2**.2.6. The fibre $\tau^{-1}(L)$ parametrizes all planes $L \subset \mathbb{P}^2 \subset P_L$ and can be identified with a complementary line $L_0 \subset P_L$. To compute the pullback of $\mathcal{O}(1)$ under $\tilde{f}\colon \mathbb{P}^1 \simeq \tau^{-1}(L) \simeq L_0 \twoheadrightarrow F$ it suffices to do this for one specific case. For example, one may consider $L = V(x_2, x_3) \subset P_L \simeq \mathbb{P}^3$ and the cubic equation $x_0x_2^2 + x_3^3$. In this case, we can identify $\tau^{-1}(L)$ with the line $L_0 = V(x_0, x_1)$. Then, for $\lambda := [0 : 0 : \lambda_2 : \lambda_3] \in L_0$ a straightforward computation shows that $\mathbb{P}^2 \simeq \overline{\lambda L}$ intersects X in $2L$ with the residual line spanned by e_1 and $-\lambda_3^3\, e_0 + \lambda_2^3\, e_2 + \lambda_3\lambda_2^2\, e_3$. Here, $e_0, \ldots, e_3$ is the standard basis. Its image under the Plücker embedding is a linear combination of $e_i \wedge e_j$ with coefficients that are of degree three in λ_2, λ_3. In other words, the composition of $\tilde{f}$ restricted to L_0 with the Plücker embedding $\tilde{f}\colon \tau^{-1}(L) \simeq L_0 \twoheadrightarrow F \hookrightarrow \mathbb{P}(\bigwedge^2 V)$ is of degree three.

Thus, as the Fano $F(X)$ of the very general cubic X has Picard rank one, see Section 4.1, the pullback $\tilde{f}^*\mathcal{O}(1)$ for any generic X is of the form $\tau^*\mathcal{O}(a) \otimes \mathcal{O}(-3E)$. In order to show $a = 7$, we use (i), which proves

$$\int_{\tilde{F}} (a \cdot \tau^* g - 3 \cdot [E])^4 = 16 \cdot \int_F g^4 = 16 \cdot 108.$$

To compute the left-hand side as a polynomial of degree four in a one observes

$$\begin{aligned}
\textstyle\int \tau^* g^3 \cdot [E] &= \textstyle\int_E \tau^* g^3 = 0,\\
\textstyle\int \tau^* g^2 \cdot [E]^2 &= \textstyle\int_E \tau^* g^2 \cdot [E]|_E = -\int_{F_2} g^2 = -([F_2].g^2) = -315,\\
\textstyle\int \tau^* g \cdot [E]^3 &= \textstyle\int_E \tau^* g \cdot [E]^2|_E = -\int_E \tau^* g \cdot u \cdot \tau^* \mathrm{c}_1 = -3\int_{F_2} g^2 = -3 \cdot 315, \text{ and}\\
\textstyle\int [E]^4 &= \textstyle\int_E [E]^3|_E = -\int_E u^3 = \int_E u \cdot \tau^*(\mathrm{c}_2 - \mathrm{c}_1^2)\\
&= \textstyle\int_{F_2} (\mathrm{c}_2 - \mathrm{c}_1^2) = 1\,125 - 9 \cdot 315 = -1\,710.
\end{aligned}$$

Here, u is the tautological class of the projective bundle $\mathbb{P}(\mathcal{N}_{F_2/F})$ and the c_i are the Chern classes of $\mathcal{N}_{F_2/F}$. Thus, $u^2 + u \cdot \tau^*\mathrm{c}_1 + \tau^*\mathrm{c}_2 = 0$ and $\mathrm{c}_1 = 3g|_{F_2}$, cf. Proposition 4.8. This eventually leads to the equation $108\,a^4 - 17\,010\,a^2 + 102\,060\,a - 140\,238 = 0$, the only integral solution of which is indeed $a = 7$. □

Remark 4.18 For the rational endomorphism f, one thus has $f^*g = 7 \cdot g$. Amerik [17] also computes the pullbacks under f of all algebraic cohomology classes on $F(X)$ of the very general cubic fourfold X and finds:

$$f^*g^2 = 4 \cdot g^2 + 45 \cdot c_2(\mathcal{S}_F), \quad f^*c_2(\mathcal{S}_F) = 31 \cdot c_2(\mathcal{S}_F), \quad \text{and} \quad f^*g^3 = 28 \cdot g^3.$$

More generally, according to Shen and Vial [429, Prop. 21.7], we have

$$(f^* + 2 \cdot \mathrm{id}) \cdot (f^* - 7 \cdot \mathrm{id}) = 0 \text{ on } H^2(F, \mathbb{Z}),$$

so $f^*|_{H^2}$ has only two eigenvalues, namely -2 and 7. Similarly,

$$(f^* - 28 \cdot \mathrm{id}) \cdot (f^* + 8 \cdot \mathrm{id}) = 0 \quad \text{on } H^6(F, \mathbb{Z}) \text{ and}$$
$$(f^* - 31 \cdot \mathrm{id}) \cdot (f^* + 14 \cdot \mathrm{id}) \cdot (f^* - 4 \cdot \mathrm{id}) = 0 \quad \text{on } H^4(F, \mathbb{Z}),$$

unless X contains a plane.

Remark 4.19 The rational endomorphism $f\colon F \dashrightarrow F$ produces a natural uniruled divisor $\tilde{f}(E) \subset F$. Here, as before, $E \simeq \mathbb{P}(\mathcal{N}_{F_2/F})$ is the exceptional divisor of the blow-up $\tau\colon \widetilde{F} \longrightarrow F$ in the surface of lines of the second type $F_2 \subset F$, which for the generic cubic X is smooth. According to a result of Oberdieck, Shen, and Yin [373, Sec. 3.1],

$$\tilde{f}_*[E] = 60 \cdot g \in H^{1,1}(F, \mathbb{Z}).$$

Indeed, the computations of $\int_{\widetilde{F}} \tau^*g^i \cdot [E]^j$ in the above proof also show

$$(\tilde{f}_*[E].g^3) = \int_E \tilde{f}^*g^3 = \int_E (7 \cdot \tau^*g - 3 \cdot [E])^3 = 60 \cdot 108 = (60 \cdot g.g^3),$$

which immediately proves the assertion.[23]

Remark 4.20 (i) As was argued in the proof above, for $L \in F_2$, the pullback of the Plücker polarization on F via the morphism $\tilde{f}\colon \tau^{-1}(L) \longrightarrow F$, which is easily seen to be injective, is of degree three. Hence, the curve class of its image is the generator

$$[\tilde{f}(\tau^{-1}(L))] = (1/36) \cdot g^3 \in H^{3,3}(F, \mathbb{Z}).$$

(ii) While the curves $q^{-1}(x)$, $x \in X$, representing twice the minimal curve class $(1/18)\cdot g^3$, cover the Fano variety $F(X)$, the curves $\tilde{f}(\tau^{-1}(L))$, $L \in F_2$, only cover the uniruled divisor $\tilde{f}(E) \subset F$. In [373, Thm. 0.2] it is shown that every rational curve $C \subset F(X)$ with $[C] = (1/36)\cdot g^3$ is of the form $\tilde{f}(\tau^{-1}(L))$ for some line $L \subset X$ of the second type. In fact, two proofs are given. One using Gromov–Witten theory and a second one studying the rational surface $S_C := q(p^{-1}(C)) \subset X$ for which $[S_C] = h^2 \in H^{2,2}(X, \mathbb{Z})$.

[23] Note however that there may be a difference between $\tilde{f}_*[E]$ and $[\tilde{f}(E)]$. In other words, the degree of $\tilde{f}\colon E \longrightarrow f(E)$ may a priori be >1. The analogous question for cubic threefolds has been answered: the map $R \longrightarrow F(Y)$ is generically injective, see Remark 5.1.10. Also, as M. Hartlieb informed me, $f^{-1}(f(E)) \longrightarrow f(E)$ might not be (generically) injective.

As mentioned earlier in Remark 3.30, Mongardi and Ottem [354] prove a less precise result but one which is valid for the Fano variety $F(X)$ of any smooth cubic fourfold X: every class $\alpha \in H^{3,3}(F(X),\mathbb{Z})$ can be written as an integral linear combination $\sum n_i [C_i]$ with curves $C_i \subset F(X)$ and $n_i \in \mathbb{Z}$ (possibly negative). In other words, the integral Hodge conjecture holds true for $F(X)$ in degree six.

5 Lattices and Hodge Theory of Cubics versus K3 Surfaces

In this chapter the main features of the curious relation between K3 surfaces and cubic fourfolds shall be outlined. For the details of most of the computations, we refer to the original paper by Hassett [227] or the lecture notes [252] from which certain arguments are taken verbatim.

5.1 Hodge and Lattice Theory of K3 Surfaces We begin by collecting some facts concerning K3 surfaces and their Hodge structures, see [249] for more information and references. Consider a K3 surface S and its cohomology $H^2(S,\mathbb{Z})$ with its intersection form and its natural Hodge structure. As an abstract lattice, it is independent of S, namely

$$H^2(S,\mathbb{Z}) \simeq \Lambda := E_8(-1)^{\oplus 2} \oplus U^{\oplus 3}.$$

We use the same notation as in Section **1**.1.5, see also [249, Ch. 14]. Note that Λ is the unique even, unimodular lattice of signature $(3,19)$. Adding another copy of the hyperbolic plane U to Λ, which geometrically corresponds to passing from $H^2(S,\mathbb{Z})$ to the full cohomology $H^*(S,\mathbb{Z})$, gives the unique even, unimodular lattice

$$\widetilde{\Lambda} := \Lambda \oplus U \simeq E_8(-1)^{\oplus 2} \oplus U^{\oplus 4}$$

of signature $(4,20)$. The hyperbolic plane U will usually be considered with its standard base consisting of two isotropic vectors e and f with $(e.f) = 1$. The corresponding bases in the first three copies of U inside $\widetilde{\Lambda}$ will be denoted e_i, f_i, $i = 1,2,3$, but for the last one it is more convenient to use a basis of isotropic vectors e_4, f_4 with $(e_4.f_4) = -1$. Indeed, one thinks of it as $(H^0 \oplus H^4)(S,\mathbb{Z})$ with the intersection form changed by a sign.

In fact, we will use $\widetilde{H}(S,\mathbb{Z})$ to denote the full cohomology $H^*(S,\mathbb{Z})$ with this altered intersection form, which is called the *Mukai pairing*, and a Hodge structure of weight two. The latter extends the natural one on $H^2(S,\mathbb{Z})$ by declaring $(H^0 \oplus H^4)(S)$ to be of type $(1,1)$.

Next, the plane A_2 with its standard base λ_1, λ_2 with $(\lambda_i.\lambda_i) = 2$ and $(\lambda_1.\lambda_2) = -1$ will be considered with the embedding $A_2 \hookrightarrow U_3 \oplus U_4$, $\lambda_1 \longmapsto e_4 - f_4$ and $\lambda_2 \longmapsto e_3 + f_3 + f_4$. Its orthogonal complement $\langle \lambda_1, \lambda_2 \rangle^\perp = A_2^\perp \subset \widetilde{\Lambda}$ is the lattice

$$A_2^\perp = E_8(-1)^{\oplus 2} \oplus U_1 \oplus U_2 \oplus A_2(-1) \tag{5.1}$$

of signature $(2,20)$. Here, $A_2(-1) \subset U_3 \oplus U_4$ is spanned by $\mu_1 := e_3 - f_3$ and $\mu_2 := -e_3 - e_4 - f_4$ satisfying $(\mu_i.\mu_i) = -2$ and $(\mu_1.\mu_2) = 1$. For later use, we observe that the natural inclusion

$$A_2 \oplus A_2^{\perp} \subset \widetilde{\Lambda} \tag{5.2}$$

has index three.

If our K3 surface S is polarized by a primitive ample line bundle L, then its degree $d := (L.L)$ is even. After changing the isometry $H^2(S,\mathbb{Z}) \simeq \Lambda$ by an element in the orthogonal group $\mathrm{O}(\Lambda)$ we can assume that the first Chern class $\mathrm{c}_1(L) \in H^2(S,\mathbb{Z})$ corresponds to the element $e_2 + (d/2)f_2 \in \Lambda$, cf. [249, Exa. 14.1.11]. For its orthogonal complement in Λ, we use the notation

$$\Lambda_d := (e_2 + (d/2)f_2)^{\perp} \subset \Lambda \subset \widetilde{\Lambda}.$$

As an abstract lattice, Λ_d is just $E_8(-1)^{\oplus 2} \oplus U^{\oplus 2} \oplus \mathbb{Z}(-d)$ and, in particular, has signature $(2,19)$. Note that for every positive even integer d there exists a polarized K3 surface (S,L) with $(L.L) = d$. The transcendental lattice $T(S)$ of the very general (S,L) is under the above lattice isomorphism $H^2(S,\mathbb{Z}) \simeq \Lambda$ identified with $T(S) \simeq H^{1,1}(S,\mathbb{Z})^{\perp} \simeq \Lambda_d$.

The Hodge decomposition for a K3 surface is of the form

$$H^2(S,\mathbb{Z}) \otimes \mathbb{C} \simeq H^2(S,\mathbb{C}) \simeq H^{2,0}(S) \oplus H^{1,1}(S) \oplus H^{0,2}(S), \tag{5.3}$$

where the outer summands are one-dimensional. Usually, a generator of $H^{2,0}(S)$ is denoted σ_S and thought of as a holomorphic or algebraic symplectic form. Note that (5.3) is determined by the line $H^{2,0}(S) \subset H^2(S,\mathbb{C})$ and the additional requirement that $H^{1,1}(S)$ and $H^{2,0}(S)$ be orthogonal.

5.2 Lattice Theory of Cubic Fourfolds The Hodge structure of a smooth cubic fourfold $X \subset \mathbb{P}^5$ looks surprisingly similar. Its middle cohomology $\widetilde{\Gamma} := H^4(X,\mathbb{Z})^-$ is the odd lattice $\mathrm{I}_{2,21}$, where the minus sign indicates that we changed the intersection form by a global sign. For example, $h^2 \in H^4(X,\mathbb{Z})^-$ has then square $(h^2.h^2) = -3$. The primitive part $H^4(X,\mathbb{Z})_{\mathrm{pr}}$ is an even lattice of rank 22 and signature $(2,20)$. More precisely,

$$H^4(X,\mathbb{Z})_{\mathrm{pr}}^- \simeq \Gamma := E_8(-1)^{\oplus 2} \oplus U^{\oplus 2} \oplus A_2(-1),$$

where again the intersection form on the primitive cohomology of X is changed by a sign, see Section **1**.1.5. Note that Γ is isometric to the lattice described by (5.1), i.e.

$$\widetilde{\Gamma} \supset \Gamma \simeq A_2^{\perp} \subset \widetilde{\Lambda}.$$

Exercise 5.1 Use Nikulin's classical results on embeddings of lattices, cf. [249, Sec. 14.1] and the (almost) surjectivity of the period map, see Theorem 6.19, to deduce results showing that many even definite lattices of small rank can be realized as $H^{2,2}(X,\mathbb{Z})$ of a smooth cubic fourfold. This has obvious consequences for the realization of lattices by the transcendental lattice of smooth cubic fourfolds. A systematic study in this direction has been conducted by Mayanskiy [348].

Remark 5.2 In a certain sense, the natural inclusion $A_2^\perp \simeq \Gamma \hookrightarrow H^4(X,\mathbb{Z})^-$ can be extended to an immersion of $\widetilde{\Lambda}$ into the rational(!) cohomology of X. More precisely, we let $A_2 \hookrightarrow H^*(X,\mathbb{Q})$ be given by sending the generators $\lambda_i \in A_2$ to $v(\lambda_i)$ with

$$v(\lambda_1) := 3 + \frac{5}{4}h - \frac{7}{32}h^2 - \frac{77}{384}h^3 + \frac{41}{2\,048}h^4 \text{ and}$$

$$v(\lambda_2) := -3 - \frac{1}{4}h + \frac{15}{32}h^2 + \frac{1}{384}h^3 - \frac{153}{2\,048}h^4.$$

This extends to an isometric embedding

$$A_2 \oplus A_2^\perp \subset \widetilde{\Lambda} \hookrightarrow H^*(X,\mathbb{Q}),$$

where the cohomology $H^*(X,\mathbb{Q})$ is endowed with the Mukai pairing

$$(\alpha.\alpha') := -\int_X \exp(3h/2)\cdot\alpha^*\cdot\alpha'. \tag{5.4}$$

Here, $\alpha^* := (-1)^i\alpha$ for $\alpha \in H^{2i}(X,\mathbb{Q})$. The Mukai pairing computes the Euler–Poincaré pairing, i.e. $\chi(E,F) = -(v(E).v(F))$, where the Mukai vector is given by

$$v\colon K_{\text{top}}(X) \longrightarrow H^*(X,\mathbb{Q}), \quad E \longmapsto \text{ch}(E)\sqrt{\text{td}(X)}.$$

Note that (5.4) on $H^*(X,\mathbb{Q})$ is not symmetric, but its restriction to the image of $\widetilde{\Lambda}$ is.

A topological interpretation of $\widetilde{\Lambda} \subset H^*(X,\mathbb{Q})$ was given by Addington and Thomas [8], cf. [252, Prop. 1.20]. It is the image of

$$K'_{\text{top}}(X) := \{[\mathcal{O}_X],[\mathcal{O}_X(1)],[\mathcal{O}_X(2)]\}^\perp \subset K_{\text{top}}(X)$$

under the Mukai vector, i.e.

$$v\colon K'_{\text{top}}(X) \xrightarrow{\sim} \widetilde{\Lambda} \subset H^*(X,\mathbb{Q}).$$

The orthogonal complement is here taken with respect to the Euler–Poincaré pairing and $K'_{\text{top}}(X)$ is endowed with $-\chi(E,F)$ (on algebraic classes).

To any primitive $v \in \Gamma \simeq A_2^\perp$ with $(v.v) < 0$ one associates two natural lattices, one on the cubic side and one on the K3 side:

$$K_v \subset \widetilde{\Gamma} \quad \text{and} \quad L_v \subset \widetilde{\Lambda}.$$

They are defined as the saturations of the two lattices

$$\mathbb{Z}\cdot h^2 \oplus \mathbb{Z}\cdot v \subset \widetilde{\Gamma} \quad \text{and} \quad A_2 \oplus \mathbb{Z}\cdot v \subset \widetilde{\Lambda},$$

which are of rank two and three, respectively.

Proposition 5.3 (Hassett) *For any primitive $v \in \Gamma \simeq A_2^\perp$ with $(v.v) < 0$, one of the following two assertions holds true:*

(i) *Either $\mathbb{Z} \cdot h^2 \oplus \mathbb{Z} \cdot v = K_v$, $A_2 \oplus \mathbb{Z} \cdot v = L_v$, and*

$$d := \operatorname{disc}(K_v) = \operatorname{disc}(L_v) = -3\,(v.v) \equiv 0\,(6),$$

(ii) *or the inclusions $\mathbb{Z} \cdot h^2 \oplus \mathbb{Z} \cdot v \subset K_v$, $A_2 \oplus \mathbb{Z} \cdot v \subset L_v$ are both of index three and*

$$d := \operatorname{disc}(K_v) = \operatorname{disc}(L_v) = -\frac{1}{3}(v.v) \equiv 2\,(6).$$

Furthermore, up to elements in the restricted orthogonal group $\tilde{\mathrm{O}}(\Gamma)$, see Section 1.2.4, *the isomorphism types of the lattices K_v and L_v only depend on $d = \operatorname{disc}(K_v) = \operatorname{disc}(L_v)$.*

Proof See [227, Prop. 3.2.2 & 3.2.4] or [252, Lem. 1.4 & Prop. 1.6]. □

In the following, we shall write K_d and L_d for the abstract isomorphism types of these lattices.

Remark 5.4 (i) In Lemma 1.1 and Exercise 2.23 we have seen already two of these lattices: K_8 and K_{14} (up to the sign change indicated by the minus sign in K_8^- and K_{14}^-). They were given by classes $v \in \Gamma \simeq A_2^\perp$ with $(v.v) = -24$ and $(v.v) = -42$, respectively. A variant of K_6 occurred in Section 1.4 and in Remark 1.31 the orthogonal $K_2^\perp$ was compared to the primitive cohomology of a polarized K3 surface of degree two.

(ii) Proposition 5.3 implies Proposition 3.26. To see this, use that $H^{2,2}(X,\mathbb{Z}) \simeq K_v$ is the saturation of $\mathbb{Z} \cdot h^2 \oplus \mathbb{Z} \cdot v$ and that the Fano correspondence defines an isomorphism (not an isometry!) $H^{2,2}(X,\mathbb{Z}) \simeq H^{1,1}(F(X),\mathbb{Z})$, which sends $v \in H^{2,2}(X,\mathbb{Z})_{\mathrm{pr}}$ to the generator γ of $H^{1,1}(F(X),\mathbb{Z})_{\mathrm{pr}}$.[24]

Remark 5.5 (i) The orthogonal complements of $K_v \subset \widetilde{\Gamma}$ and $L_v \subset \widetilde{\Lambda}$ are isomorphic, they are nothing but $v^\perp \subset \Gamma \simeq A_2^\perp$. As abstract lattices, they shall be denoted

$$\Gamma_d := K_d^\perp \simeq L_d^\perp.$$

Using the fact that K_d is negative definite or, alternatively, that L_d has signature $(2,1)$, one deduces that Γ_d is a lattice of signature $(2,19)$.

(ii) It is not difficult to write down explicit examples of primitive vectors $v_d \in \Gamma$ for any fixed discriminant $d > 0$, assuming d satisfies the numerical condition

$$d \equiv 0\,(6) \quad \text{or} \quad d \equiv 2\,(6). \tag{$*$}$$

[24] However, beware that the two rank two lattices K_d and $H^{1,1}(F(X),\mathbb{Z})$ are not the same, e.g. while K_d is negative definite $H^{1,1}(F(X),\mathbb{Z})$ is indefinite by the Hodge index theorem.

For example, one can use, see [252, Rem. 1.7]:

$$v_d := \begin{cases} e_1 - (d/6)f_1 \in U_1 \subset \Gamma & \text{if } d \equiv 0\,(6), \\ 3\,(e_1 - ((d-2)/6)\,f_1) + \mu_1 - \mu_2 \in U_1 \oplus A_2(-1) \subset \Gamma & \text{if } d \equiv 2\,(6). \end{cases} \tag{5.5}$$

(iii) The difference between the two cases $d \equiv 0\,(6)$ and $d \equiv 2\,(6)$ is also reflected by the discriminant groups of K_d and L_d. In the case $d \equiv 0\,(6)$, it is $\mathbb{Z}/3\mathbb{Z} \oplus \mathbb{Z}/(d/3)\mathbb{Z}$, while for $d \equiv 2\,(6)$, it is the cyclic group $\mathbb{Z}/d\mathbb{Z}$, cf. [252, Rem. 1.7].

Essentially, all values of d satisfying $(*)$ are realized geometrically. We state this as the following result for the moduli space $M = M_4$ of smooth cubic fourfolds, considered as a smooth Deligne–Mumford stack or as a quasi-projective variety with quotient singularities of dimension 19. The result was hinted at already in Remark 1.3 for $d = 8$ and a complete proof is further postponed to Proposition 6.15. The strange notation $C_d \cap M$ will become clear there also.

Proposition 5.6 (Hassett) *For a fixed discriminant $d > 6$ satisfying $(*)$, let $C_d \cap M \subset M$ be the set of cubics X for which $\mathbb{Z} \cdot h^2 \hookrightarrow H^{2,2}(X,\mathbb{Z})^-$ extends to a primitive isometric embedding $K_d \hookrightarrow H^{2,2}(X,\mathbb{Z})^-$. Then*

$$C_d \cap M \subset M$$

is an irreducible divisor.

More precisely, the result asserts that for all $d > 6$ satisfying $(*)$ there exist smooth cubic fourfolds such that $H^{2,2}(X,\mathbb{Z})^-$ is isomorphic to K_d and that the locus of those for which $H^{2,2}(X,\mathbb{Z})^-$ contains K_d is irreducible and of codimension one. In fact, once the period description of the moduli space $M \subset C$ is in place, see Section 6.4, the proof that $C_d \subset C$ and $C_d \cap M \subset M$ are of codimension one is standard.

A geometric reason for the irreducibility of Noether–Lefschetz divisors $C_d \cap M$ is only available for very special discriminants d, e.g. $d = 8$ and $d = 14$, corresponding to cubic fourfolds containing a plane or to Pfaffian cubics.

Definition 5.7 A smooth cubic fourfold $X \subset \mathbb{P}^5$ is called *special* (in the sense of Hassett) if for some d we have $X \in C_d \cap M$, i.e. if $h^2 \in H^{2,2}(X,\mathbb{Z})$ extends to a primitive isometric embedding $K_d \hookrightarrow H^{2,2}(X,\mathbb{Z})^-$.

In other words, a smooth cubic fourfold is special if it contains a surface that is not homologous to a multiple of h^2.

Example 5.8 We have seen examples of special cubic fourfolds before.

(i) A smooth cubic X is contained in $C_8 \cap M$ if and only if X contains a plane, see Lemma 1.1 and Remark 1.3.

(ii) A Pfaffian cubic X is contained in $C_{14} \cap M$, see Corollary 2.21 and Exercise 2.23.

Moreover, for dimension reasons and using the irreducibility of $\mathcal{C}_{14} \cap M$, the generic cubic in $\mathcal{C}_{14} \cap M$ is Pfaffian.

There are only few cases known in which special cubic fourfolds have been linked to the existence of special surfaces. For example, Hassett [225] shows that a smooth cubic fourfold X defines a point in $\mathcal{C}_{12}$ if and only if X contains a cubic scroll. Similarly, X defines a point in $\mathcal{C}_{20}$ if and only if X contains a Veronese surface, see the comments at the end of Section 1.6. Also, Farkas and Verra [180] show that the general cubic in $\mathcal{C}_{26} \cap M$ contains a two-dimensional family of 3-nodal septic scrolls.

Remark 5.9 There are two integers d that satisfy $(*)$ but are excluded in the above result by the additional requirement $d > 6$, namely $d = 2$ and $d = 6$. It turns out that those are indeed not realized by smooth cubics but admit a geometric realization in terms of singular cubic fourfolds. This will be made more precise using the period description of the moduli space, see Theorem 6.19.

(i) Assume $d = 2$. The intersection matrix describing K_2 is

$$\begin{pmatrix} -3 & 1 \\ 1 & -1 \end{pmatrix}.$$

There are various ways to exclude this case, see Section 3.7 and especially Remark 3.28. As explained in Remark 1.31, K3 surfaces with $H^2(S,\mathbb{Z})_{\rm pr} \simeq K_2^\perp$ nevertheless play a role for singular cubic fourfolds.

(ii) Assume $d = 6$. In this case, the intersection matrix describing K_6 is

$$\begin{pmatrix} -3 & 0 \\ 0 & -2 \end{pmatrix}.$$

In particular, if an isometric embedding $K_6 \hookrightarrow H^{2,2}(X,\mathbb{Z})^-$ for a smooth cubic fourfold X existed, then the image of the second basis vector would define a class $\delta \in H^{2,2}(X,\mathbb{Z})_{\rm pr}$ with $(\delta.\delta) = 2$, which is excluded by Corollary 4.16, see also Remark 3.27. The discussion in Section 1.4 can be rephrased by saying that for a cubic fourfold $X \subset \mathbb{P}^5$ with an ordinary double point $x_0 \in X$ as its only singularity, the blow-up $\mathrm{Bl}_{x_0}(X)$ admits a rank two lattice with the slightly modified intersection matrix

$$\begin{pmatrix} 3 & 0 \\ 0 & -2 \end{pmatrix}$$

in its algebraic part $H^{2,2}(\mathrm{Bl}_{x_0}(X),\mathbb{Z})$.

Proposition 5.10 (Hassett) *Assume d satisfies $(*)$. Then there exists an isometry*

$$\Gamma_d \simeq \Lambda_d \tag{5.6}$$

if and only if d satisfies the stronger numerical condition

$$d/2 = \prod p^{n_p} \quad \text{with } n_p = 0 \text{ for all } p \equiv 2\,(3) \quad \text{and} \quad n_3 \leq 1. \tag{$**$}$$

Proof See [227, Thm. 5.1.3] or [252, Lem. 1.10]. □

Remark 5.11 The numerical condition (∗∗) is equivalent to the condition that there exists a primitive vector $0 \neq w \in A_2$ with $(w.w) = d$ and also to the condition that $d = (2n^2 + 2n + 2)/a$ for certain integers a and n, see [252, Prop. 1.13].

Corollary 5.12 (Hassett) *Assume X is a smooth cubic fourfold and fix $d \in \mathbb{Z}$. Then the following two conditions are equivalent:*

(i) *The positive integer d satisfies (∗∗) and X is contained in $C_d \cap M \subset M$, i.e. the embedding $\mathbb{Z} \cdot h^2 \subset H^{2,2}(X, \mathbb{Z})^-$ can be extended to a primitive isometric embedding*

$$K_d \hookrightarrow H^{2,2}(X, \mathbb{Z})^-.$$

(ii) *There exists a polarized K3 surface (S, L) of degree d and a primitive Hodge isometric embedding*

$$H^2(S, \mathbb{Z})_{L\text{-pr}} \hookrightarrow H^4(X, \mathbb{Z})^-_{\mathrm{pr}}(1).$$

Proof Assume $K_d \hookrightarrow H^{2,2}(X, \mathbb{Z})^-$ is as in (i). Then its orthogonal complement in $H^4(X, \mathbb{Z})$ is isometric to Λ_d. By the surjectivity of the period map, see e.g. [249, Ch. 6] for proofs and references, there exists a K3 surface S with a primitive line bundle L such that $(L.L) = d$ and with $L^\perp \subset H^2(S, \mathbb{Z})$ Hodge isometric up to Tate twist to $K_d^\perp \subset H^4(X, \mathbb{Z})^-$. In order to ensure that we can choose L to be ample, it suffices to prove there is no (1, 1)-class $\delta \in L^\perp \subset H^2(S, \mathbb{Z})$ with $(\delta.\delta) = -2$ or, slightly stronger, that there is no $\delta \in H^{2,2}(X, \mathbb{Z})_{\mathrm{pr}}$ with $(\delta.\delta) = 2$. But this is the content of Corollary 4.16.

For the converse, observe that the orthogonal complement of an embedding as in (ii) gives an embedding $K_d \hookrightarrow H^{2,2}(X, \mathbb{Z})^-$. □

Here is a table of the first values of d satisfying (∗) and (∗∗):

(∗)	8	12	14	18	20	24	26	30	32	36	38	42
(∗∗)			14				26				38	42
(∗)	44	48	50	54	56	60	62	66	68	72	74	78
(∗∗)							62				74	

Remark 5.13 Assume X is a smooth cubic fourfold with $K_d \simeq H^{2,2}(X, \mathbb{Z})^-$. If d satisfies (∗∗) and $d \equiv 2\,(6)$, then there exists a unique polarized K3 surface (S, L) with $H^2(S, \mathbb{Z})_{\mathrm{pr}} \simeq K_d^\perp$. If d satisfies (∗∗) but $d \equiv 0\,(6)$, then there exist exactly two such polarized K3 surfaces. This phenomenon had been already observed by Hassett [227] and has been studied further by Brakkee [91]. See Corollary 5.22 for the global picture.

Definition 5.14 A polarized K3 surface (S, L) and a smooth cubic fourfold are *associated* if there exists an isometric embedding of Hodge structures

$$H^2(S, \mathbb{Z})_{L\text{-pr}} \hookrightarrow H^4(X, \mathbb{Z})^-_{\mathrm{pr}}(1).$$

We write $X \sim (S, L)$ in this case. For another point of view, see Corollary 5.18.

5.3 Hassett's Rationality Conjecture The link between cubic fourfolds and K3 surfaces, mainly via Hodge theory with the exception of a few special cases in Sections 1 and 2, is already mysterious, but it becomes even more interesting in the light of the following, see also Conjecture **7**.3.1 and the discussion following it.

Conjecture 5.15 (Hassett) *A smooth cubic fourfold $X \subset \mathbb{P}^5$ is rational if and only if X is contained in $C_d \cap M \subset M$ with d satisfying $(**)$, i.e. if and only if X is associated with a polarized K3 surface.*[25]

Remark 5.16 The conjecture is wide open. Although only cubic fourfolds in the countable union $\bigcup_{(**)} C_d$ of divisors are expected to be rational, so the very general cubic fourfold should be irrational, not a single cubic fourfold is known to be provably irrational. That smooth cubic fourfolds not contained in any Hassett divisor should be irrational was apparently conjectured already by Iskovskih, see [454, Sec. 1].

Here is a probably incomplete list of smooth cubic fourfolds that are known to be rational.

Codimension One:

(i) All Pfaffian cubic fourfolds are rational, see Corollary 2.27. Pfaffian cubics are parametrized by the Noether–Lefschetz divisor $C_{14} \cap M \subset M$. Pfaffian cubics always contain a quartic scroll, see Lemma 2.20, and a quintic del Pezzo surface, see Remark 2.5. (The existence of the latter is equivalent to the cubic being a Pfaffian.) It was known classically that the existence of either of the two types of surfaces is enough to conclude rationality, see [228, Prop. 4 & 6]. Note that originally only generic Pfaffian cubics were known to be rational but by the recent results on specializations [285, 371] they are all covered, see also [79] for a direct approach.

(ii) Russo and Staglianò [408, 409] prove rationality for smooth cubic fourfolds contained in C_{26}, C_{38} and C_{42}. The intersection of these divisors with any other C_d are studied by Yang and Yu [494]. The next divisor expected to parametrize rational cubics would be $C_{42} \cap M \subset M$, but it seems nothing is known in this case.

Codimension Two:

(iii) Cubic fourfolds containing a plane $\mathbb{P}^2 \simeq \mathrm{P} \subset X$ with trivial Brauer class $\alpha_{\mathrm{P},X}$ are rational, see Lemma 1.14. There is a countable union of codimension two subsets all contained in C_8 with this property, see [226]. The proof relies on Lemma 1.14 to show

[25] For comments on another and incompatible rationality criterion, see Remark **7**.3.2.

that under certain numerical conditions on d, e.g. d satisfying $(**)$, the intersection $\mathcal{C}_d \cap \mathcal{C}_8 \cap M$ has one irreducible component for which the Brauer class $\alpha_{P,X}$ is trivial. Note, however, that even for d satisfying $(**)$ not for every cubic in $\mathcal{C}_d \cap \mathcal{C}_8 \cap M$ this is the case. This has been studied in detail in [28].

(iv) Cubic fourfolds containing two disjoint planes are rational, see Corollary **1**.5.11 and Example **1**.5.12.

(v) There exists a countable union of subsets of codimension two in the moduli space M that are all contained in $\mathcal{C}_{18} \cap M$ parametrizing rational cubic fourfolds [6]. Their special geometric feature is the existence of a covering family of sextic del Pezzo surfaces.

Remark 5.17 We have seen already that every cubic hypersurfaces admits a unirational parametrization $\mathbb{P}^n \dashrightarrow X$ of degree two, see Corollary **2**.1.21. Those who admit a unirational parametrization of odd degree are good candidates for being rational.

However, it was shown by Hassett [228, Cor. 40] that generic cubic fourfolds $X \in \mathcal{C}_d \cap M$ with $d = 14, 18, 26, 30, 38$ admit an odd degree unirational parametrization. Later, Lai [304, Thm. 0.1] proved the result for $d = 42$. Note however that general smooth cubic fourfolds of 'degree' $d = 18$ and 30 are not expected to be rational.

5.4 Cubics versus K3 Surfaces via the Mukai Lattice There is a different way of looking at the curious relation between cubic fourfolds and K3 surfaces expressed by Corollary 5.12. At first, this is something that is just observed on the level of lattices and Hodge structures, but, as will be explained in Sections **7**.3.4 and **7**.3.5, it has a very clean interpretation in terms of derived categories. In comparison, the relation captured by Corollary 5.12, which at first looks more geometric, has not been given a satisfactory interpretation, for example, in terms of geometric correspondences.

As a first step, we introduce the Hodge structure $\widetilde{H}(X,\mathbb{Z})$ associated with any smooth cubic fourfold. As a lattice, we set

$$\widetilde{H}(X,\mathbb{Z}) := K'_{\rm top}(X) = \{\, [\mathcal{O}_X], [\mathcal{O}_X(1)], [\mathcal{O}_X(2)]\,\}^\perp \subset K_{\rm top}(X),$$

equipped with the Euler–Poincaré pairing changed by a sign, or, alternatively, as

$$\widetilde{H}(X,\mathbb{Z}) = v(K'_{\rm top}(X)) \subset H^*(X,\mathbb{Q})$$

together with the Mukai pairing. Recall that as abstract lattices $\widetilde{H}(X,\mathbb{Z}) \simeq \widetilde{\Lambda}$, see Remark 5.2. Next, $\widetilde{H}(X,\mathbb{Z})$ is endowed with a Hodge structure of weight two by setting

$$\widetilde{H}^{2,0}(X) := v^{-1}(H^{3,1}(X))$$

and requiring $\widetilde{H}^{1,1}(X)$ and $\widetilde{H}^{2,0}(X)$ to be orthogonal. With this definition, the two distinguished classes $v(\lambda_1), v(\lambda_2) \in \widetilde{H}(X,\mathbb{Z})$ span a sublattice

$$A_2 \subset \widetilde{H}^{1,1}(X,\mathbb{Z}).$$

Also, note that the transcendental part of $\widetilde{H}(X,\mathbb{Z})$, i.e. the orthogonal complement of $\widetilde{H}^{1,1}(X,\mathbb{Z}) := \widetilde{H}(X,\mathbb{Z}) \cap \widetilde{H}^{1,1}(X)$, is just the usual transcendental part up to Tate twist

$$\widetilde{H}^{1,1}(X,\mathbb{Z})^{\perp}(-1) \simeq T(X) = H^{2,2}(X,\mathbb{Z})^{\perp} \subset H^4(X,\mathbb{Z}).$$

The next result [8] should be viewed as the unpolarized analogue of Corollary 5.12.

Corollary 5.18 (Addington–Thomas) *Assume X is a smooth cubic fourfold. Then the following conditions are equivalent:*

(i) *There exists an isometric embedding*

$$U \hookrightarrow \widetilde{H}^{1,1}(X,\mathbb{Z}).$$

(ii) *There exists a projective K3 surface S and a Hodge isometry*

$$\widetilde{H}(X,\mathbb{Z}) \simeq \widetilde{H}(S,\mathbb{Z}).$$

(iii) *There exists an integer d satisfying* (∗∗) *such that* $X \in \mathcal{C}_d \cap M$ *or, equivalently according to Corollary* 5.12, *there exists a polarized K3 surface* (S',L') *and a Hodge isometric embedding*

$$H^2(S',\mathbb{Z})_{L'\text{-pr}} \hookrightarrow H^4(X,\mathbb{Z})^{-}_{\text{pr}}.$$

Proof For the proof, we refer to the original [8] or to [252], but see the proof of Corollary 6.23. In particular one uses that any Hodge isometry between the transcendental lattices $T(S) \simeq T(X)$ extends to a Hodge isometry $\widetilde{H}(S,\mathbb{Z}) \simeq \widetilde{H}(X,\mathbb{Z})$, cf. [249, Cor. 14.3.12]. □

Warning: the K3 surface in (ii) may not admit any polarization for which an isometric embedding as in (iii) exists.

Definition 5.19 We say that a smooth cubic fourfold X and an unpolarized(!) K3 surface S are *associated* if there exists a Hodge isometry $\widetilde{H}(S,\mathbb{Z}) \simeq \widetilde{H}(X,\mathbb{Z})$, as above.

Note that in this case, S itself may not admit a polarization L such that X and (S,L) are associated in the sense of Definition 5.14, but there always exists another polarized K3 surface (S',L') that is and for which $\widetilde{H}(S,\mathbb{Z}) \simeq \widetilde{H}(S',\mathbb{Z})$. See Section **7**.3.5 for an interpretation in terms of derived categories.

Remark 5.20 A twisted version of the above exists [252, 250] and it says that the following conditions are equivalent:

(i) There exists an isometric embedding $U(n) \hookrightarrow \widetilde{H}^{1,1}(X,\mathbb{Z})$.
(ii) There exists a twisted projective K3 surface $(S,\alpha \in \text{Br}(S))$ and a Hodge isometry $\widetilde{H}(X,\mathbb{Z}) \simeq \widetilde{H}(S,\alpha,\mathbb{Z})$.[26]

[26] For the definition of the lattice and the Hodge structure $\widetilde{H}(S,\alpha,\mathbb{Z})$, we refer to [249, Sec. 16.4.1].

(iii) There exists an integer d satisfying

$$d/2 = \prod p^{n_p} \quad \text{with } n_p \equiv 0 \text{ for all } p \equiv 2\,(3) \tag{$**'$}$$

such that $X \in \mathcal{C}_d$.

In this case, we say that X and the unpolarized but twisted (S, α) are associated. Again, an interpretation in terms of derived categories exists, see Section **7**.3.5.

Similar to Remark 5.11, one can show that the numerical condition $(**')$ is equivalent to the existence of a (not necessarily primitive) vector $0 \neq w \in A_2$ with $(w.w) = d$.

5.5 Fano Variety: Hilbert Schemes and Lagrangian Fibrations The advantage of working with the Hodge structure $\widetilde{H}(X, \mathbb{Z})$ rather than the middle cohomology of X is illustrated by the next result [3]. It explains that instead of extending the Hodge isometry $H^4(X, \mathbb{Z})_{\mathrm{pr}} \simeq H^2(F(X), \mathbb{Z})_{\mathrm{pr}}(-1)$, see Corollary 3.21, to an isomorphism of Hodge structures $H^4(X, \mathbb{Z}) \simeq H^2(F(X), \mathbb{Z})(-1)$ that is not an isometry anymore, one should realize $H^2(F(X), \mathbb{Z})$ isometrically inside $\widetilde{H}(X, \mathbb{Z})$ as the orthogonal complement of the distinguished vector $v(\lambda_1) \in H^{1,1}(X, \mathbb{Z})$.

Corollary 5.21 (Addington) *The Hodge isometry $H^4(X, \mathbb{Z})_{\mathrm{pr}}(1) \simeq H^2(F(X), \mathbb{Z})_{\mathrm{pr}}$ induced by the Fano correspondence extends naturally to a Hodge isometry*

$$\begin{array}{ccc} H^4(X, \mathbb{Z})_{\mathrm{pr}}(1) & \simeq & H^2(F(X), \mathbb{Z})_{\mathrm{pr}} \\ \cap & & \cap \\ v(\lambda_1)^{\perp} & \simeq & H^2(F(X), \mathbb{Z}) \\ \cap & & \\ \widetilde{H}(X, \mathbb{Z}). & & \end{array}$$

Here, $H^2(F(X), \mathbb{Z})$ is considered with the quadratic form q_F and $v(\lambda_1)^{\perp} \subset \widetilde{H}(X, \mathbb{Z})$ comes with the induced Mukai pairing.

Furthermore, under this Hodge isometry $v(\lambda_1 + 2\lambda_2) \in v(\lambda_1)^{\perp} \subset \widetilde{H}^{1,1}(X, \mathbb{Z})$ is mapped to the Plücker polarization $g \in H^{1,1}(F(X), \mathbb{Z})$.

Proof The Hodge isometry $v(\lambda_1)^{\perp} \simeq H^2(F(X), \mathbb{Z})$ is induced by the Fano correspondence but with a twist:

$$\alpha \longmapsto \mathrm{c}_1 \circ p_*(q^*(\alpha) \cdot \mathrm{td}(p)).$$

We refer to the original [3] for details, see also [252, Sec. 3.1]. □

The Fano version of Corollaries 5.12 and 5.18 is then the following result [3, 227].

Corollary 5.22 (Hassett, Addington) *Assume X is a smooth cubic fourfold with its Fano variety of lines $F(X)$. Then the following conditions are equivalent:*

(i) *There exists an isometric embedding*

$$U \hookrightarrow \widetilde{H}^{1,1}(X,\mathbb{Z}) \quad \text{with } v(\lambda_1) \in U.$$

(ii) *There exists a K3 surface S and a Hodge isometry*

$$H^2(S^{[2]},\mathbb{Z}) \simeq H^2(F(X),\mathbb{Z}). \tag{5.7}$$

In this case, the two four-dimensional hyperkähler manifolds $F(X)$ and $S^{[2]}$ in (ii) *are birational and S and X are associated in the sense of Corollary* 5.18.

Proof The key observation is that any Hodge isometry (5.7) extends to a Hodge isometry $\widetilde{H}(S,\mathbb{Z}) \simeq \widetilde{H}(X,\mathbb{Z})$, see [252, Lem. 3.3], which then proves the equivalence of (i) and (ii) and the assertion that S and X are associated.

Eventually, according to [337, Thm. 5.8], any Hodge isometry (5.7) is induced by a birational map between $S^{[2]}$ and $F(X)$. □

The existence of a hyperbolic plane $v(\lambda_1) \in U \subset \widetilde{H}^{1,1}(X,\mathbb{Z})$ can be rephrased in terms of the primitive sublattices $K_d \subset H^{2,2}(X,\mathbb{Z})$, see [3, Thm. 2]. It turns out that (i) is equivalent to the existence of K_d in $H^{2,2}(X,\mathbb{Z})$ with d satisfying the numerical condition

$$d = (2n^2 + 2n + 2)/a^2 \quad \text{for certain } a, n \in \mathbb{Z}. \tag{$***$}$$

For a comparison of the two conditions ($**$) and ($***$), we list the first few values.

($**$)	14	26	38	42	62	74	78	86	98	114	122	134
($***$)	14	26	38	42	62			86		114	122	134

In particular, the Fano variety $F(X)$ of a cubic fourfold can be birational, and even isomorphic, to the Hilbert scheme $S^{[2]}$ of a K3 surface without being associated directly with any K3 surface. In particular, Hassett's rationality conjecture, see Conjecture 5.15, and the one by Galkin and Shinder [190] are incompatible, see Remark **7**.3.2.

Remark 5.23 The Hilbert scheme $S^{[2]}$ of a K3 surface is a particular moduli space of stable sheaves on S, namely of the ideal sheaves I_Z of subschemes $Z \subset S$ of length two.

From this perspective, it is natural to wonder whether $F(X)$ is also birational to some moduli spaces of (twisted) sheaves on a K3 surface S and whether this is expressed as a numerical condition on d. For the untwisted case, this was worked out by Addington [3], for the twisted case, see [250]. It turns out that the resulting conditions are numerically expressed by ($**$) and ($**'$). The phenomenon is naturally explained by equivalences $\mathcal{A}_X \simeq \mathrm{D^b}(S)$ and $\mathcal{A}_X \simeq \mathrm{D^b}(S,\alpha)$, see Section **7**.3.4.

A precursor of this observation in the twisted case is the result of Macrì and Stellari [332, Thm. 1.3] that for a cubic fourfold containing a plane $\mathrm{P} \subset X$ the Fano variety $F(X)$ is a moduli space of twisted sheaves on S_{P}.

Also, certain geometric properties of the hyperkähler fourfold $F(X)$ are determined by numerical properties of the cubic X. Here is an example.

Corollary 5.24 *Assume X is a smooth cubic fourfold with its Fano variety of lines $F(X)$. Then the following two conditions are equivalent:*

(i) *$X \in \mathcal{C}_d \cap M$ with $d/2$ a perfect square.*
(ii) *The hyperkähler manifold $F(X)$ admits a rational Lagrangian fibration.*

Proof Since $F(X)$ is deformation equivalent to the Hilbert scheme of a K3 surfaces, the existence of a rational Lagrangian fibration is known to be equivalent to the existence of a non-trivial isotropic class $\ell \in H^{1,1}(F(X), \mathbb{Z})$, see [339, Thm. 1.3]. For an isotropic class ℓ, a primitive linear combination $v = a\,\ell + b\,g$ is a primitive cohomology class, i.e. $q(v, g) = 0$, if and only if $a = -6\,b/q(\ell, g)$. Then $q(v) = -6\,b^2$ and, in particular, $b \in \mathbb{Z}$. According to Proposition 5.3, this implies $X \in \mathcal{C}_d$ with $d = 18\,b^2$ or $d = 2\,b^2$. In both cases, $d/2$ is a perfect square. We leave the converse to the reader. □

In the spirit of condition (**), the condition for $F(X)$ to admit a Lagrangian fibration can be rephrased as

$$d/2 = \prod p^{n_p} \quad \text{with } n_p \equiv 0\,(2) \text{ for all } p \qquad (**'')$$

such that $X \in \mathcal{C}_d$.

Remark 5.25 We conclude this section by giving references to results concerning the *LLSvS eightfold* $Z(X)$, a hyperkähler manifold of dimension eight naturally associated with any smooth cubic fourfold $X \subset \mathbb{P}^5$ not containing a plane. The variety $Z(X)$ was constructed by Lehn, Lehn, Sorger, and van Straten [319] and further studied by Addington and Lehn [7], who in particular showed that $Z(X)$ is deformation equivalent to the Hilbert scheme $S^{[4]}$ of a K3 surface by establishing a birational correspondence between $Z(X_V)$ and $S_V^{[4]}$ for any Pfaffian cubic fourfold X_V not containing a plane and its associated K3 surface S_V, see Section 2.

The period computations for $F(X)$ were done by Addington and Giovenzana [5]. The numerical conditions for $Z(X)$ being birational to a moduli space of (twisted) sheaves on a K3 surface or to a Hilbert scheme $S^{[4]}$ are the same as those for $F(X)$, see Corollary 5.22 and Remark 5.23.

6 Period Domains and Moduli Spaces

We now compare the Hodge theory of K3 surfaces and cubic fourfolds in families. By means of period maps, this leads to an algebraic correspondence between the moduli space of polarized K3 surfaces of certain degrees and the moduli space of cubic fourfolds. The approach has been initiated by Hassett [227] and has turned out to be a beneficial point of view.

6.1 Baily–Borel Here is a very brief reminder of some results by Borel and by Baily–Borel on arithmetic quotients of orthogonal type. Let $(N, (\ .\ ))$ be a lattice of signature $(2, n_-)$ and set $V := N \otimes \mathbb{R}$. Then the *period domain* D_N associated with N is the real Grassmann variety of positive, oriented planes $W \subset V$, which also can be described as

$$\begin{aligned} D_N &\simeq \mathrm{O}(2, n_-)/(\mathrm{O}(2) \times \mathrm{O}(n_-)) \\ &\simeq \{\, x \mid (x.x) = 0,\ (x.\bar{x}) > 0 \,\} \subset \mathbb{P}(N \otimes \mathbb{C}). \end{aligned}$$

With the latter description, the period domain D_N associated with N has the structure of a complex manifold. This is turned into an algebraic statement by the following fundamental result [31]. It uses the fact that under the assumption on the signature of N the orthogonal group $\mathrm{O}(N)$ acts properly discontinuously on D_N.

Theorem 6.1 (Baily–Borel) *Assume $G \subset \mathrm{O}(N)$ is a torsion free subgroup of finite index. Then the quotient*

$$G \setminus D_N$$

has the structure of a smooth, quasi-projective complex variety.

As G also acts properly discontinuously, all the stabilizer subgroups are finite and hence, by the assumption on G, trivial. This already proves the smoothness of the quotient $G \setminus D_N$. The difficult part of the theorem is to find a Zariski open embedding into a complex projective variety.

Finite index subgroups $G \subset \mathrm{O}(N)$ with torsion are relevant, too. In this situation, one uses Minkowski's theorem stating that the map $\pi_p \colon \mathrm{GL}(n, \mathbb{Z}) \longrightarrow \mathrm{GL}(n, \mathbb{F}_p)$, $p \geq 3$, is injective on finite subgroups or, equivalently, that its kernel is torsion free. Hence, for every finite index subgroup $G \subset \mathrm{O}(N)$, there exists a normal and torsion free subgroup $G_0 := G \cap \mathrm{Ker}(\pi_p) \subset G$ of finite index.

Corollary 6.2 *Assume $G \subset \mathrm{O}(N)$ is a subgroup of finite index. Then the quotient*

$$G \setminus D_N$$

has the structure of a normal, quasi-projective complex variety with finite quotient singularities. □

Not only are these arithmetic quotients algebraic, also holomorphic maps into them are algebraic. This is the following remarkable GAGA style result, see [87].

Theorem 6.3 (Borel) *Assume $G \subset \mathrm{O}(N)$ is a torsion free subgroup of finite index. Then any holomorphic map $\varphi\colon Z \longrightarrow G \setminus D_N$ from a complex variety Z is regular.*

Remark 6.4 Often, the result is applied to holomorphic maps to singular quotients $G \setminus D_N$, i.e. in situations when G is not necessarily torsion free. This is covered by the above only when $Z \longrightarrow G \setminus D_N$ is induced by a holomorphic map $Z' \longrightarrow G_0 \setminus D_N$, where $Z' \longrightarrow Z$ is a finite quotient and $G_0 \subset G$ is a normal, torsion free subgroup of finite index.

6.2 Hassett Divisors of Special Fourfolds We shall be interested in (at least) three different types of period domains: for polarized K3 surfaces and for (special) smooth cubic fourfolds. These are the period domains associated with the three lattices Γ, Γ_d, and Λ_d, which all have signature $(2, n)$:

$$D \subset \mathbb{P}(\Gamma \otimes \mathbb{C}), \quad D_d \subset \mathbb{P}(\Gamma_d \otimes \mathbb{C}), \quad \text{and} \quad Q_d \subset \mathbb{P}(\Lambda_d \otimes \mathbb{C}).$$

These period domains are endowed with the natural action of the corresponding orthogonal groups $\mathrm{O}(\Gamma)$, $\mathrm{O}(\Gamma_d)$, and $\mathrm{O}(\Lambda_d)$ and we will be interested in the following quotients by distinguished finite index subgroups of those:[27]

$$\mathcal{C} := \tilde{\mathrm{O}}(\Gamma) \setminus D = \mathrm{O}(\Gamma) \setminus D, \quad \tilde{\mathcal{C}}_d := \tilde{\mathrm{O}}(\Gamma, K_d) \setminus D_d, \quad \tilde{\boldsymbol{\mathcal{C}}}_d := \tilde{\mathrm{O}}(\Gamma, v_d) \setminus D_d, \tag{6.1}$$

$$\text{and} \quad \mathcal{M}_d := \tilde{\mathrm{O}}(\Lambda_d) \setminus Q_d.$$

For the definition of $\tilde{\mathrm{O}}(\Gamma)$ and $\tilde{\mathrm{O}}(\Lambda_d)$, see Section **1**.2.4, while the groups

$$\tilde{\mathrm{O}}(\Gamma, v_d) \subset \tilde{\mathrm{O}}(\Gamma, K_d) \subset \tilde{\mathrm{O}}(\Gamma) \subset \mathrm{O}(\widetilde{\Gamma}) \tag{6.2}$$

are the subgroups of transformations g of $\widetilde{\Gamma}$ that in addition to h^2 also fix the vector v_d resp. the lattice K_d, see (5.5). The latter condition is equivalent to $g(v_d) = \pm v_d$. It is known that the inclusion $\tilde{\mathrm{O}}(\Gamma, v_d) \subset \tilde{\mathrm{O}}(\Gamma, K_d)$ is of index two for $d \equiv 0\,(6)$ and it is an equality for $d \equiv 2\,(6)$, cf. [252, Lem. 1.8].

Remark 6.5 For a lattice N if signature $(2, n_-)$, the period domain has two connected components. Yet, as all groups in (6.2) contain elements swapping the components, all quotients in (6.1) are irreducible quasi-projective varieties.

By virtue of Theorems 6.1 and 6.3, see also Remark 6.4, we know that the induced maps $\tilde{\boldsymbol{\mathcal{C}}}_d \twoheadrightarrow \tilde{\mathcal{C}}_d$ and $\tilde{\mathcal{C}}_d \longrightarrow \mathcal{C}$ are regular morphisms between normal quasi-projective varieties. The image in $\mathcal{C}$ shall be denoted by $\mathcal{C}_d$, so that

$$\tilde{\boldsymbol{\mathcal{C}}}_d \twoheadrightarrow \tilde{\mathcal{C}}_d \twoheadrightarrow \mathcal{C}_d \subset \mathcal{C}.$$

By construction, the $\mathcal{C}_d \subset \mathcal{C}$ are irreducible hypersurfaces, also called *Noether–Lefschetz divisors* or in this particular situation *Hassett divisors*. Their theory, in a more

[27] For the first equality, note that $\tilde{\mathrm{O}}(\Gamma) \subset \mathrm{O}(\Gamma)$ is of index two, but $-\mathrm{id} \in \mathrm{O}(\Gamma) \setminus \tilde{\mathrm{O}}(\Gamma)$ acts trivially on D.

general setting, i.e. using arbitrary lattices of signature $(2, n)$, has been studied intensively. We will concentrate here on those aspects that are relevant for our purposes.

Corollary 6.6 (Hassett) *Assume the positive integer d satisfies $(*)$. Then the naturally induced maps*

$$\tilde{\boldsymbol{\mathcal{C}}}_d \twoheadrightarrow \tilde{\mathcal{C}}_d \twoheadrightarrow \mathcal{C}_d$$

are surjective, finite, and algebraic.

Furthermore, $\tilde{\mathcal{C}}_d \twoheadrightarrow \mathcal{C}_d$ is the normalization of $\mathcal{C}_d$ and $\tilde{\boldsymbol{\mathcal{C}}}_d \to \tilde{\mathcal{C}}_d$ is a finite morphism between normal varieties, which is an isomorphism for $d \equiv 2\,(6)$ and of degree two for $d \equiv 0\,(6)$.

Proof If d satisfies $(*)$ and $d \equiv 2\,(6)$, then $\tilde{\mathrm{O}}(\Gamma, K_d) = \tilde{\mathrm{O}}(\Gamma, v_d)$ and, therefore, $\tilde{\boldsymbol{\mathcal{C}}}_d \simeq \tilde{\mathcal{C}}_d$. Otherwise, $\tilde{\boldsymbol{\mathcal{C}}}_d \twoheadrightarrow \tilde{\mathcal{C}}_d$ is the quotient by any $g \in \tilde{\mathrm{O}}(\Gamma, K_d) \setminus \tilde{\mathrm{O}}(\Gamma, v_d)$, which can in fact be chosen to be an involution, see the proof of [252, Lem. 1.8].

To prove that $\tilde{\mathcal{C}}_d \twoheadrightarrow \mathcal{C}_d$ is quasi-finite, use that $\tilde{\mathcal{C}}_d \to \mathcal{C}$ is algebraic with discrete and hence finite fibres. For a very general $x \in D_d$, i.e. such that there does not exist any proper primitive sublattice $N \subset \Gamma_d$ with $x \in N \otimes \mathbb{C}$, any $g \in \tilde{\mathrm{O}}(\Gamma)$ with $g(x) = x$ also satisfies $g(\Gamma_d) = \Gamma_d$ and, therefore, $g(K_d) = K_d$, i.e. $g \in \tilde{\mathrm{O}}(\Gamma, K_d)$. This proves that $\tilde{\mathcal{C}}_d \to \mathcal{C}$ is generically injective. Thus, once $\tilde{\mathcal{C}}_d \to \mathcal{C}$ is shown to be finite, and not only quasi-finite, it is the normalization of its image $\mathcal{C}_d$. We refer to [91, 227] for more details concerning the finiteness or, equivalently, the properness of the map. □

Remark 6.7 Note that while the fibre of $\tilde{\boldsymbol{\mathcal{C}}}_d \to \tilde{\mathcal{C}}_d$ consists of at most two points, the fibres of $\tilde{\mathcal{C}}_d \to \mathcal{C}_d$ may contain more points, depending on the singularities of $\mathcal{C}_d$. For fixed d, the cardinality of the fibres is of course bounded, although no explicit bound is known, but the maximal cardinality of the fibres grows with d.

The next result is the first step to link K3 surfaces and cubic fourfolds by means of their periods.

Corollary 6.8 *Assume $6 < d$ satisfies $(**)$ and choose an isomorphism $\varepsilon\colon \Gamma_d \xrightarrow{\sim} \Lambda_d$.*

(i) *If $d \equiv 0\,(6)$, then ε naturally induces an isomorphism $\mathcal{M}_d \simeq \tilde{\boldsymbol{\mathcal{C}}}_d$. Therefore, $\mathcal{M}_d$ comes with a finite morphism onto $\mathcal{C}_d$ which is generically of degree two:*

$$\Phi_\varepsilon\colon\ \mathcal{M}_d \simeq \tilde{\boldsymbol{\mathcal{C}}}_d \xrightarrow{2:1} \tilde{\mathcal{C}}_d \xrightarrow{\text{norm}} \mathcal{C}_d \subset \mathcal{C}.$$

(ii) *If $d \equiv 2\,(6)$, then ε naturally induces an isomorphism $\mathcal{M}_d \simeq \tilde{\boldsymbol{\mathcal{C}}}_d \simeq \tilde{\mathcal{C}}_d$. Therefore, $\mathcal{M}_d$ can be seen as the normalization of $\mathcal{C}_d \subset \mathcal{C}$:*

$$\Phi_\varepsilon\colon \mathcal{M}_d \simeq \tilde{\boldsymbol{\mathcal{C}}}_d \simeq \tilde{\mathcal{C}}_d \xrightarrow{\text{norm}} \mathcal{C}_d \subset \mathcal{C}.$$

Proof The key observation here is that the choice of ε naturally induces an isomorphism $\tilde{\mathrm{O}}(\Gamma, v_d) \simeq \tilde{\mathrm{O}}(\Lambda_d)$, see [252, Lem. 1.10]. □

Remark 6.9 As indicated by the notation, the morphism $\Phi_\varepsilon\colon \mathcal{M}_d \twoheadrightarrow \mathcal{C}_d \subset \mathcal{C}$, linking polarized K3 surfaces (S, L) of degree d with special cubic fourfolds X, depends on the choice of $\varepsilon\colon \Gamma_d \xrightarrow{\sim} \Lambda_d$. There is no distinguished choice for ε and, therefore, for general d one should not expect to find a distinguished morphism $\mathcal{M}_d \longrightarrow \mathcal{C}_d$ that can be described by a geometric procedure associating to a polarized K3 surface (S, L) a cubic fourfold X.[28]

Note however that for fixed d, the infinitely many choices of ε lead to only finitely many different maps Φ_ε.

Remark 6.10 For $d = 6$, the situation is a bit different. Unlike the case $6 < d \equiv 0\,(6)$, the map $\Phi_\varepsilon\colon \mathcal{M}_6 \simeq \tilde{\mathcal{C}}_6 \twoheadrightarrow \tilde{\mathcal{C}}_6$ is generically injective for $d = 6$, cf. Remark 1.27.

Geometrically this is reflected by the fact that with any nodal cubic fourfold there comes a distinguished K3 surface, see Sections 1.4 and 1.5. In terms of lattice theory, the difference between the two cases can be explained by the observation that the natural inclusion $\mathrm{Im}(\tilde{\mathrm{O}}(\Gamma, v_d) \longrightarrow \mathrm{O}(K_d^\perp)) \subset \mathrm{Im}(\tilde{\mathrm{O}}(\Gamma, K_d) \longrightarrow \mathrm{O}(K_d^\perp))$ of the two subgroups of $\mathrm{O}(K_d^\perp)$ is an equality for $d = 6$ and is of index two for $6 < d \equiv 0\,(6)$.

6.3 Reminder: K3 Surfaces We start by recalling the central theorem in the theory of K3 surfaces: the global Torelli theorem. In the situation at hand, it is a result of Pjateckiĭ-Šapiro and Šafarevič, see [249, Ch. 6] for details, generalizations, and references.

Consider the coarse moduli space M_d of polarized K3 surfaces (S, L) of degree $(L.L) = d$, which can be constructed as a quasi-projective variety either by (not quite) standard GIT methods, by using the theorem below, or as a Deligne–Mumford stack, cf. [249, Ch. 5].

The period map associates with any $(S, L) \in M_d$ a point in $\mathcal{M}_d$. For this, choose an isometry $H^2(S, \mathbb{Z}) \simeq \Lambda$, called a *marking*, that maps $\mathrm{c}_1(L)$ to $\ell_d := e_2 + (d/2) f_2$ and, therefore, induces an isometry $H^2(S, \mathbb{Z})_{L\text{-pr}} \simeq \Lambda_d$. Then the $(2, 0)$-part $H^{2,0}(S) \subset H^2(S, \mathbb{C}) \simeq \Lambda \otimes \mathbb{C}$ defines a point in the period domain Q_d. The image point in the quotient $\tilde{\mathrm{O}}(\Lambda_d) \setminus Q_d$ is then independent of the choice of any marking. This defines the period map $\mathcal{P}\colon M_d \longrightarrow \mathcal{M}_d$ which Hodge theory reveals to be holomorphic, see also the discussion in Section **3**.3. Note that both spaces, M_d and $\mathcal{M}_d$, are quasi-projective varieties with finite quotient singularities.

The global Torelli theorem for polarized K3 surfaces asserts that two polarized K3 surfaces (S, L) and (S', L') are isomorphic if and only if there exists a Hodge isometry $H^2(S, \mathbb{Z}) \simeq H^2(S', \mathbb{Z})$ that maps $\mathrm{c}_1(L)$ to $\mathrm{c}_1(L')$. The strong global Torelli theorem asserts the more precise statement that any such Hodge isometry lifts uniquely to a polarized isomorphism. In terms of moduli spaces, this leads to the injectivity part of the following celebrated result. The openness is a consequence of the local Torelli theorem, cf. [249, Cor. 6.4.3], and the algebraicity can be seen as a consequence of Theorem 6.3.

[28] I wish to thank E. Brakkee and P. Magni for discussions concerning this point.

Theorem 6.11 (Pjateckiĭ-Šapiro and Šafarevič) *The period map is an algebraic, open embedding*

$$\mathcal{P}\colon M_d \hookrightarrow \mathcal{M}_d = \tilde{\mathrm{O}}(\Lambda_d) \setminus Q_d. \tag{6.3}$$

It turns out that the complement $\mathcal{M}_d \setminus M_d$ consists of one or two irreducible divisors, see [252, Prop. 2.11]. Using the strong form of the global Torelli theorem, the open embedding (6.3) can be upgraded to an open embedding of the moduli space of polarized K3 surfaces viewed as a Deligne–Mumford stacks into $[\tilde{\mathrm{O}}(\Lambda_d) \setminus Q_d]$, cf. Remark **3**.3.3.

6.4 Period Map for Cubic Fourfolds We now switch to the cubic side. The moduli space M of smooth cubic fourfolds can be constructed by means of standard GIT methods, see Section **3**.1.5, as the quotient

$$M = |\mathcal{O}_{\mathbb{P}^5}(3)|_{\mathrm{sm}}//\mathrm{PGL}(6).$$

As in the case of K3 surfaces, mapping a smooth cubic fourfold X to its period, i.e. the line $H^{3,1}(X) \subset H^4(X,\mathbb{C})_{\mathrm{pr}}^- \simeq \Gamma \otimes \mathbb{C}$ considered as a point in the period domain $D \subset \mathbb{P}(\Gamma \otimes \mathbb{C})$, defines a holomorphic map $\mathcal{P}\colon M \longrightarrow C$, see Section **3**.3.

In analogy to the situation for K3 surfaces, the global Torelli theorem 3.17 originally proved by Voisin [472, 477], but with alternative proofs given later [114, 255, 327], leads to a realization of the moduli space as an open subset of the arithmetic quotient C.

Theorem 6.12 (Voisin, Looijenga, . . . , Charles, Huybrechts–Rennemo, . . .) *The period map is an algebraic, open embedding*

$$\mathcal{P}\colon M \hookrightarrow C = \mathrm{O}(\Gamma) \setminus D.$$

Remark 6.13 There is one potentially irritating point about stating the global Torelli theorem in this way. Recall that the monodromy group of cubic fourfolds is the finite index subgroup $\tilde{\mathrm{O}}^+(\Gamma) \subset \mathrm{O}(\Gamma)$, see Section **1**.2.4. So why are we allowed to divide out by the bigger group $\mathrm{O}(\Gamma)$ in the above statement? First, the period domain has two connected components $D = D^+ \sqcup D^-$, interchanged by complex conjugation, and $\mathrm{O}(\Gamma) \setminus D = \mathrm{O}^+(\Gamma) \setminus D^+$. Second, the cokernel of the index two inclusion $\tilde{\mathrm{O}}(\Gamma) \subset \mathrm{O}(\Gamma)$ is generated by $-\mathrm{id}$ which acts trivially on D. Hence,

$$\mathrm{O}(\Gamma) \setminus D = \mathrm{O}^+(\Gamma) \setminus D^+ = \tilde{\mathrm{O}}^+(\Gamma) \setminus D^+.$$

This is nothing but the global version of the argument in Remark 3.18.

In this context, we observe that the global Torelli theorem can also be phrased as an open immersion of smooth Deligne–Mumford stacks, cf. Section **3**.3:

$$\mathcal{M} \hookrightarrow [\tilde{\mathrm{O}}^+(\Gamma) \setminus D^+],$$

for which one needs to use the strong form of the global Torelli theorem 3.17.

Remark 6.14 Mapping a smooth cubic fourfold X to its Fano variety $F(X)$ together with the Plücker polarization works in families and thus leads to a natural morphism

$$M \longrightarrow M_{\mathrm{K3}^{[2]}}, \quad X \longmapsto (F(X), g).$$

Here, M is as before the moduli space of smooth cubic fourfolds and by $M_{\mathrm{K3}^{[2]}}$ we denote the moduli space of polarized hyperkähler fourfolds (Z, h) of $\mathrm{K3}^{[2]}$-type where the polarization h is equivalent to $2 \cdot g_S - 5 \cdot \delta$ for $Z = S^{[2]}$, see Remark 3.13.

Note that there exists a Hodge isometry $H^2(F(X), \mathbb{Z})_{\mathrm{pr}} \simeq H^4(X, \mathbb{Z})_{\mathrm{pr}}(1)$, see Corollary 5.21. In particular, the primitive lattice $H^2(Z, \mathbb{Z})_{\mathrm{pr}}$ of every $(Z, h) \in M_{\mathrm{K3}^{[2]}}$ is isometric to Γ and so the period of (Z, h) is also a point in $D \subset \mathbb{P}(\Gamma)$.

Verbitsky's global Torelli theorem [248, 328, 467, 468] combined with Markman's monodromy computations for hyperkähler manifolds of $\mathrm{K3}^{[2]}$-type [337] shows that the period map defines an open immersion. We recommend Debarre's survey [136] for further information and background.

As the period maps for X and $F(X)$ are compatible, we have a commutative diagram:

$$\begin{array}{ccccccc} X & & M & \hookrightarrow & M_{\mathrm{K3}^{[2]}} & & (Z,h) \\ \downarrow & & \downarrow & & \downarrow & & \downarrow \\ H^{3,1}(X) & & \mathrm{O}(\Gamma) \setminus D & = & \mathrm{O}(\Gamma) \setminus D & & H^{2,0}(Z). \end{array}$$

The theorem and the next result in particular explain the strange notation $\mathcal{C}_d \cap M$ used in previous sections, see e.g. Proposition 5.6.

Proposition 6.15 *The image of a smooth cubic fourfold $X \in M$ satisfies $\mathcal{P}(X) \in \mathcal{C}_d$ if and only if there exists a primitive isometric embedding $K_d \hookrightarrow H^{2,2}(X, \mathbb{Z})^-$ extending $h^2 \in H^{2,2}(X, \mathbb{Z})$. The locus $\mathcal{C}_d \cap M := \mathcal{P}^{-1}(\mathcal{C}_d)$ is an irreducible hypersurface in M.*

Proof Observe that $\mathcal{P}(X) \in \mathcal{C}_d$ if and only if $H^{3,1}(X) \subset \Gamma_d \otimes \mathbb{C}$ for some marking $H^4(X, \mathbb{Z})_{\mathrm{pr}} \simeq \Gamma$, which is equivalent to $K_d = \Gamma_d^{\perp} \subset \widetilde{\Gamma}$ being contained in $H^{2,2}(X, \mathbb{Z})$. This together with the uniqueness of $K_d \subset \widetilde{\Gamma}$ up to the action of $\mathrm{O}(\Gamma)$, see Proposition 5.3, proves the first assertion.

The second assertion follows from $\mathcal{C}_d$ being irreducible by construction and $M \subset \mathcal{C}$ being Zariski open. □

Remark 6.16 The variety $\mathcal{C}$, and hence the moduli space of smooth cubic fourfolds M, is unirational. This can be deduced from the description of M as a GIT quotient of $|\mathcal{O}_{\mathbb{P}^5}(3)|$. I am not aware of an argument that would not use the modular description. The situation is less clear for the Hassett divisors $\mathcal{C}_d \subset \mathcal{C}$.

(i) The unirationality of $\mathcal{C}_d$ or, equivalently, of $\mathcal{C}_d \cap M$ is often linked to explicit parametrizations of the cubic fourfolds in $\mathcal{C}_d \cap M$. Only for small values of d unirationality of $\mathcal{C}_d \cap M$ is expected. Nuer [372] completed classical results proving that

$C_d \cap M$ is unirational for all $d \leq 44$ satisfying $(*)$, but $d \neq 42$. Lai [304] shows that for $d = 42$ it is at least uniruled.

(ii) In the other direction, Tanimoto and Várilly-Alvarado [443, Thm. 1.1] show that $C_d \cap M$ is of general type for large d, with very explicit and rather small bounds.

Note that a priori one could try to approach the birational geometry of $C_d \cap M$, e.g. its Kodaira dimension, unirationality, etc., just by using the description of C as an orthogonal modular variety. There are indeed general results saying that those have negative Kodaira dimension for at most finitely many lattices. However, the results of Ma [330] concerning lattices of large rank a priori do not cover our cases here.

Remark 6.17 The behaviour of the intersection $C_d \cap C_{d'}$ of two Hassett divisors is not fully understood.

(i) For example, is the intersection $C_d \cap C_{d'}$ always irreducible? The answer is no in general, e.g. $C_8 \cap C_{14}$ has five irreducible components [28, Thm. A]. Moreover, for only three of these irreducible components, the Brauer class $\alpha_{P,X}$ is trivial, cf. Remark 5.16 and Section 1.1.

(ii) More geometrically, it is interesting to ask whether the intersection of $C_d \cap C_{d'}$ with the moduli space M is always non-empty. For example, for applications to derived categories it is important to ensure that $C_d \cap C_8 \cap M \neq \emptyset$, see Section **7**.3.14.[29] This is a result of Addington and Thomas [8, Thm. 4.1], which relies heavily on the (almost) surjectivity of the period map in Theorem 6.19.

Yang and Yu [494, Thm. 7] prove more generally that for any two d and d' satisfying the condition $(*)$, the intersection $C_d \cap C_{d'} \cap M$ is non-empty.

Remark 6.18 Hassett divisors $C_d \cap M \subset M$ are central to our understanding of cubic fourfolds. However, other natural divisors in M exist that are not of this type, i.e. that do not have a natural Hodge theoretic motivation or, more, concretely, for which the general member as $H^{2,2}(X, \mathbb{Z})_{\rm pr} = 0$. The first example was found by Ranestad and Voisin [396] and two other divisors suggested by them were later confirmed by Addington and Auel [4] to be indeed not one of the Hassett divisors. All three divisors are described by an apolarity condition with respect to a certain type of surface.

6.5 Laza and Looijenga Theorem 6.12 is complemented by a result of Laza [312] and Looijenga [327], which can be seen as an analogue of the description of $\mathcal{M}_d \setminus M_d$ for K3 surfaces. We know already that the period of a smooth cubic fourfold X cannot be contained in $C_2 \cup C_6$, see Section 3.7 and Remark 5.9. It turns out that any other period can indeed be realized, which answers affirmatively a question of Hassett [227,

[29] In fact, one needs a slightly more precise result, namely that there exists a smooth cubic fourfold $X \in C_d \cap C_8$ with trivial Brauer class $\alpha_{P,X}$. This is ensured by the existence of a class $\gamma \in H^{2,2}(X, \mathbb{Z})$ with $(\gamma.[Q]) = 1$ for the residual quadric, see Exercise 1.16.

Sec. 4.3]. The main input is the characterization of all semi-stable cubic hypersurfaces in $\mathbb{P}^5$ by Laza [311] and Yokoyama [496], see Section 6.7.

Theorem 6.19 (Laza, Looijenga) *The period map identifies the moduli space M of smooth cubic fourfolds with the complement of $C_2 \cup C_6 \subset C$, i.e.*

$$\mathcal{P}\colon M \xrightarrow{\sim} C \setminus (C_2 \cup C_6) \subset C.$$

It is natural to wonder what happens with the Fano variety $F(X)$, viewed as a polarized hyperkähler fourfold, when X approaches the boundary $C_2 \cup C_6$? For example, when X specializes to a generic nodal cubic fourfold X_0, then $F(X)$ specializes to either the singular $F(X_0)$ or to its blow-up $\mathrm{Bl}_S(F(X_0)) \simeq S_0^{[2]}$, see Proposition 1.28. Further results in this direction have been proved by van den Dries [458].

To complete the picture, we state the following result, but we refrain from giving a proof and refer to similar results in the theory of K3 surfaces, cf. [249, Prop. 6.2.9].

Proposition 6.20 *The countable union $\bigcup C_d \subset C$ of all C_d with d satisfying $(***)$ is analytically dense in C. Consequently, the union of all C_d for satisfying $(**')$* (or $(**)$ or $(*)$) is analytically dense.

Remark 6.21 On the level of moduli spaces, the theory of K3 surfaces is linked with the theory of cubic fourfolds in terms of the (non-canonical) morphism

$$\Phi_\varepsilon|_{M_d}\colon M_d \subset \mathcal{M}_d \xrightarrow{\Phi_\varepsilon} C_d \subset C,$$

cf. Corollary 6.8. Note that the image of a point $(S, L) \in M_d$ corresponding to a polarized K3 surface (S, L) can a priori be contained in the boundary $C \setminus M = C_2 \cup C_6$. However, unless $d = 2$ or $d = 6$, generically this is not the case and the map defines a rational map

$$\Phi_\varepsilon\colon M_d \dashrightarrow M,$$

which is of degree one or two.

6.6 Cubic Fourfolds and Polarized/Twisted K3 Surfaces In Section 5.2 we have linked the Hodge theory of K3 surfaces and the Hodge theory of cubic fourfolds. We will now cast this in the framework of period maps and moduli spaces, i.e. in terms of the maps $\Phi_\varepsilon|_{M_d}\colon M_d \dashrightarrow M$. For a categorical interpretation of the map Φ_ε, see the discussion in Section 7.3.4.

Proposition 6.22 *A smooth cubic fourfold X and a polarized K3 surface (S, L) are associated, $(S, L) \sim X$, in the sense of Definition 5.14 if and only if $\Phi_\varepsilon(S, L) = X$ for some choice of $\varepsilon\colon \Gamma_d \xrightarrow{\sim} \Lambda_d$,*

$$(S, L) \sim X \iff \exists \varepsilon\colon \Phi_\varepsilon(S, L) = X.$$

Proof Assume $\Phi_\varepsilon(S,L) = X$ and pick an arbitrary marking $H^2(S,\mathbb{Z}) \xrightarrow{\sim} \Lambda$ with the property that $c_1(L) \longmapsto \ell_d$. Composing the induced isometry $H^2(S,\mathbb{Z})_{L\text{-pr}} \xrightarrow{\sim} \Lambda_d$ with the inverse map $\varepsilon^{-1}\colon \Lambda_d \xrightarrow{\sim} \Gamma_d \subset \Gamma$ allows us to associate with (S,L) a point in $D_d \subset D$. There then exists a marking $H^4(X,\mathbb{Z})^-_{\text{pr}} \simeq \Gamma$ such that X defines the same period point in D, which thus leads to a Hodge isometric embedding $H^2(S,\mathbb{Z})_{L\text{-pr}} \hookrightarrow H^4(X,\mathbb{Z})^-_{\text{pr}}(1)$.

For the converse, observe that any such Hodge isometric embedding defines a sublattice of $\Gamma \simeq H^4(X,\mathbb{Z})^-_{\text{pr}}$ isomorphic to some $v^\perp$ which after applying some element in $\mathrm{O}(\Gamma)$ becomes Γ_d, see Proposition 5.3. Composing with a marking of (S,L) leads to the appropriate ε. □

The following was essentially already stated in Corollary 5.18 and Remark 5.20.

Corollary 6.23 *Let X be a smooth cubic fourfold.*

(i) *For fixed d, there exists a polarized K3 surface (S,L) of degree d with $X \sim (S,L)$ if and only if $X \in \mathcal{C}_d \cap M$ and d satisfies* $(**)$.
(ii) *There exists a twisted K3 surface (S,α) with $X \sim (S,\alpha)$ if and only if $X \in \mathcal{C}_d \cap M$ for some d satisfying* $(**')$.

Proof The 'only if' direction in (i) is clear. For the other, we first interpret $\mathcal{M}_d$ as the moduli space of quasi-polarized K3 surfaces (S,L), i.e. with L only big and nef but not necessarily ample. One then has to show that whenever there exists a Hodge isometric embedding $H^2(S,\mathbb{Z})_{L\text{-pr}} \hookrightarrow H^4(X,\mathbb{Z})^-_{\text{pr}}(1)$, the class of L is not orthogonal to any algebraic class $\delta_S \in H^2(S,\mathbb{Z})$ with $(\delta_S.\delta_S) = -2$. Indeed, then L would be automatically ample and, hence, $(S,L) \in M_d$. Now, if such a class δ_S existed, it would correspond to a class $\delta \in H^{2,2}(X,\mathbb{Z})_{\text{pr}}$ with $(\delta.\delta) = 2$, which contradicts $X \in M = \mathcal{C} \setminus (\mathcal{C}_2 \cup \mathcal{C}_6)$.

To prove (ii), observe that the period of X is contained in D_d if and only if one finds $L_d \hookrightarrow \widetilde{H}^{1,1}(X,\mathbb{Z})$. If d satisfies $(**')$, then there exists $U(n) \hookrightarrow L_d$ and we can conclude by Remark 5.20. Conversely, if $(S,\alpha) \sim X$ in the sense of Definition 5.19, one finds $U(n) \hookrightarrow \widetilde{H}^{1,1}(S,\alpha,\mathbb{Z}) \simeq \widetilde{H}^{1,1}(X,\mathbb{Z})$. As there also exists a positive plane $A_2 \hookrightarrow \widetilde{H}^{1,1}(X,\mathbb{Z})$, the lattice $U(n)$ is contained in a primitive sublattice of rank three in $\widetilde{H}^{1,1}(X,\mathbb{Z})$, which is then necessarily of the form L_d for some d satisfying $(**')$. □

Remark 6.24 Note that a given cubic fourfold X can be associated with more than one polarized K3 surface (S,L) and, in fact, sometimes even with infinitely many (S,L).

To start, there are the finitely many choices of $\bar{\varepsilon} \in \mathrm{O}(\Lambda_d)/\tilde{\mathrm{O}}(\Lambda_d)$. Then Φ_ε for d satisfying $(**)$, is only generically injective if $d \equiv 2\,(6)$ and, even worse, of degree two if $d \equiv 0\,(6)$. And finally, X could be contained in more than $\mathcal{C}_d$.[30] To be more precise, depending on the degree d, there may exist non-isomorphic K3 surfaces S and S' endowed

[30] In fact, it can happen that $X \in \mathcal{C}_d$ for infinitely many d satisfying $(**)$.

with polarizations L and L', respectively, both of degree d, such there nevertheless exists a Hodge isometry $H^2(S,\mathbb{Z})_{L\text{-pr}} \simeq H^2(S',\mathbb{Z})_{L'\text{-pr}}$. Indeed, the latter may not extend to a Hodge isometry $H^2(S,\mathbb{Z}) \simeq H^2(S',\mathbb{Z})$, see Section 6.3.

The situation is not quite as bad as it sounds. Although there may be infinitely many polarized K3 surfaces (S, L) associated with one X, only finitely many isomorphism types of unpolarized K3 surfaces S will be involved. Moreover, all of them will be derived equivalent, see Corollary **7**.3.19.

Remark 6.25 Brakkee [91] gives a geometric interpretation for the generic fibre of the map $\Phi_\varepsilon\colon M_d \dashrightarrow \mathcal{C}_d$ in the case $d \equiv 0\,(6)$. It turns out that $\Phi_\varepsilon(S,L) = \Phi_\varepsilon(S',L')$ implies that S' is isomorphic to $M(3, L, d/6)$, the moduli space of stable bundles on S with the indicated Mukai vector.

6.7 Semi-stable Cubic Fourfolds The moduli space of smooth cubic fourfolds M is a quasi-projective variety with finite quotient singularities. This can either be deduced from its GIT construction, cf. Corollary **3**.1.13, or from its identification via the period map with the open set $\mathcal{C}\setminus(\mathcal{C}_2 \cup \mathcal{C}_6)$ of the arithmetic quotient $\mathcal{C} = \tilde{\mathrm{O}}(\Gamma)\setminus D$, see Theorems 6.1 and 6.19.

Accordingly, M can be compactified to a projective variety by either adopting a modular approach, i.e. by defining an appropriate functor that is coarsely represented by a projective variety $\bar{M}$ containing M as an open subscheme, or by striving for a meaningful arithmetic compactification of the quotient $\mathcal{C} = \tilde{\mathrm{O}}(\Gamma) \setminus D$. Both approaches have a certain flexibility that makes it difficult to single out one compactification that should be preferred over any other.

However, the GIT construction of M suggests one quite natural solution. Clearly, the moduli space M viewed as the GIT quotient $|\mathcal{O}(3)|_{\mathrm{sm}}/\!/\mathrm{SL}(6)$ admits a natural compactification by the projective GIT quotient $|\mathcal{O}(3)|^{\mathrm{ss}}/\!/\mathrm{SL}(6)$, see the discussion in Section **3**.1.5. The only problem with this solution is that it is hard to describe the points in the boundary $\bar{M} \setminus M$ or, in other words, to decide which cubic hypersurfaces $X \subset \mathbb{P}^5$ are semi-stable. The analysis in dimension four has been carried out independently by Laza [311] and Yokoyama [496]. As a result, a complete characterization of all (semi-)stable cubic fourfolds is known. Since the general result is somewhat technical and nowhere used in these notes, we only state the following.

Theorem 6.26 (Laza) *Assume $X \subset \mathbb{P}^5$ is a cubic hypersurface with at worst isolated singularities. Then X is stable if and only if all its singularities are simple (of type ADE).*

So concretely, in dimension four the singularities of a stable hypersurface with only isotated singularities are locally analytically given by one of the following equations:

$$\begin{aligned}
\mathrm{A_n}:&\quad x_0^2+x_1^2+x_2^2+x_3^2+z_4^{n+1};\\
\mathrm{D_n}:&\quad x_0^2+x_1^2+x_2^2+x_3^2\cdot x_4+x_4^{n-1},\ n\geq 4;\\
\mathrm{E_6}:&\quad x_0^2+x_1^2+x_2^2+x_3^3+x_4^4;\\
\mathrm{E_7}:&\quad x_0^2+x_1^2+x_2^2+x_3^3+x_3\cdot x_4^3;\\
\mathrm{E_8}:&\quad x_0^2+x_1^2+x_2^2+x_3^3+x_4^5.
\end{aligned}$$

In recent years, the notion of K-stability for varieties has been studied intensively. Similar to the notion of slope-stability for vector bundles it is linked to the existence of special metrics. For cubic fourfolds, Liu [323] has shown that K-stability is equivalent to GIT stability.

The cohomology of the moduli space of stable cubic fourfolds has been investigated. For example, Si [438] computes its Poincaré polynomial.

7

Derived Categories of Cubic Hypersurfaces

In this final chapter, derived categories enter the scene. Investigating cubics from the point of view of their derived category of coherent sheaves is a rather recent development that opened new perspectives on hard classical questions. Many of the geometric phenomena studied in this book so far can be studied on various levels: cohomology, Chow groups, motives, etc. Lifting them to derived categories is the ultimate step and promises the deepest understanding.

It seems that for more than one reason a meaningful and interesting theory of derived categories of cubic hypersurfaces should only be expected in dimensions two, three, and four, and we will restrict to those.

In Section 1 we provide background on triangulated and derived categories. The main objective is to introduce the Kuznetsov component of a cubic hypersurface, which is also interpreted in terms of matrix factorizations. We also present a few general results that apply to cubics of all dimensions. After that we devote Sections 2 and 3 to cubics of dimension three and four, respectively. The role of the Kuznetsov category is similar in these two cases, but in dimension four it is more geometric. The attraction of derived categories in this dimension is the mysterious appearance of K3 surfaces which cannot be explained geometrically. The Hodge theory shadow was discussed in Section **6**.5 and Section **6**.6. Section 4 is a survey of results concerning Chow groups and Chow motives of cubic hypersurfaces.

Throughout, we work over an algebraically closed field and, for simplicity, often assume it to be $\mathbb{C}$, especially when Hodge theory is used.

0.1 Orlov's Formulae For a smooth projective variety X over a field k, the abelian category of coherent sheaves $\mathrm{Coh}(X)$ and its derived category $\mathrm{D^b}(X) := \mathrm{D^b}(\mathrm{Coh}(X))$, a triangulated category, will both be considered as k-linear categories. In the applications, one often encounters the more flexible notion of α-twisted coherent sheaves, with $\alpha \in \mathrm{Br}(X)$ a Brauer class, and then studies the abelian category $\mathrm{Coh}(X,\alpha)$ and its derived category $\mathrm{D^b}(X,\alpha) := \mathrm{D^b}(\mathrm{Coh}(X,\alpha))$. In fact, $\mathrm{Coh}(X,\alpha)$ does not only depend on α but on an additional choice like a cocycle $\{\alpha_{ijk}\}$ representing α or an Azumaya algebra $\mathcal{A}_\alpha$ whose class is α or the choice of a gerbe. For different choices, e.g. of $\{\alpha_{ijk}\}$, the abelian categories $\mathrm{Coh}(X, \{\alpha_{ijk}\})$ are equivalent but not canonically so. See the survey [247] for more details on this.

The behaviour of $\mathrm{Coh}(X)$ and $\mathrm{D^b}(X)$ under certain standard morphisms is well understood. Let us recall the following two classical facts first proved by Orlov [377]:

(i) Consider the projectivization $\pi\colon \mathbb{P}(\mathcal{E}) \twoheadrightarrow X$ of a locally free sheaf $\mathcal{E}$ of rank r. Then there exists a semi-orthogonal decomposition

$$\mathrm{D^b}(\mathbb{P}(\mathcal{E})) \simeq \langle \mathrm{D^b}(X), \mathrm{D^b}(X)(1), \ldots, \mathrm{D^b}(X)(r-1)\rangle, \tag{0.1}$$

see Section 1.3 below for a brief reminder on semi-orthogonal decompositions.

Here, $\mathrm{D^b}(X)$ is viewed as a full triangulated subcategory of $\mathrm{D^b}(\mathbb{P}(\mathcal{E}))$ via the fully faithful functor $\pi^*\colon \mathrm{D^b}(X) \hookrightarrow \mathrm{D^b}(\mathbb{P}(\mathcal{E}))$. Similarly, $\mathrm{D^b}(X)(j) \simeq \mathrm{D^b}(X)$ denotes the image of the fully faithful functor $\mathrm{D^b}(X) \hookrightarrow \mathrm{D^b}(\mathbb{P}(\mathcal{E}))$, $F \longmapsto \pi^*F \otimes \mathcal{O}_\pi(j)$, where $\mathcal{O}_\pi(1)$ is the relative tautological invertible sheaf, cf. [246, Cor. 8.36].

A natural generalization to the case of *Brauer–Severi varieties* $\mathbb{P}(\mathcal{E}) \twoheadrightarrow X$, i.e. where $\mathcal{E}$ is not a locally free coherent sheaf but an α-twisted locally free sheaf of rank r, has been established by Bernardara [64]:

$$\mathrm{D^b}(\mathbb{P}(\mathcal{E})) = \langle \mathrm{D^b}(X), \mathrm{D^b}(X,\alpha), \mathrm{D^b}(X,\alpha^2), \ldots, \mathrm{D^b}(X,\alpha^{r-1})\rangle. \tag{0.2}$$

The fully faithful embeddings $\mathrm{D^b}(X,\alpha^j) \hookrightarrow \mathrm{D^b}(\mathbb{P}(\mathcal{E}))$ are given by $F \longmapsto \pi^*F \otimes \mathcal{O}_\pi(j)$, where one uses the existence of the tautological $\pi^*\alpha^{-1}$-twisted line bundle $\mathcal{O}_\pi(1)$, see [247, Sec. 3] for details and references.

(ii) Let $\sigma\colon \mathrm{Bl}_Z(X) \twoheadrightarrow X$ be the blow-up of a smooth projective variety X in a smooth subvariety $Z \subset X$ of codimension c. Then there exists a natural semi-orthogonal decomposition

$$\mathrm{D^b}(\mathrm{Bl}_Z(X)) = \langle \mathrm{D^b}(Z)_{-c+1}, \ldots, \mathrm{D^b}(Z)_{-1}, \mathrm{D^b}(X)\rangle. \tag{0.3}$$

Here, $\mathrm{D^b}(Z)_{-j} \simeq \mathrm{D^b}(Z)$ is the image of the fully faithful functor $\mathrm{D^b}(Z) \hookrightarrow \mathrm{D^b}(\mathrm{Bl}_Z(X))$, $F \longmapsto i_*(\pi^*F \otimes \mathcal{O}_E(jE))$, where $\pi\colon E \simeq \mathbb{P}(\mathcal{N}_{Z/X}) \twoheadrightarrow Z$ is the exceptional divisor and $i\colon E \hookrightarrow \mathrm{Bl}_Z(X)$ its natural closed embedding.

0.2 Fourier–Mukai Functors Functors between bounded derived categories of coherent sheaves on smooth projective varieties are often described via the Fourier–Mukai formalism. With any $P \in \mathrm{D^b}(X \times Y)$ one associates the *Fourier–Mukai functor*, cf. [246, Ch. 5]:

$$\Phi_P\colon \mathrm{D^b}(X) \twoheadrightarrow \mathrm{D^b}(Y), \quad F \longmapsto q_*(p^*F \otimes P).$$

Here, p and q are the two projections and q_*, p^*, and $\otimes$ denote the derived functors. According to a result of Orlov [377] combined with results of Bondal and Van den

Bergh [83], any fully faithful functor $\Phi\colon \mathrm{D}^{\mathrm{b}}(X) \longrightarrow \mathrm{D}^{\mathrm{b}}(Y)$ is of Fourier–Mukai type, i.e. $\Phi \simeq \Phi_P$, and its Fourier–Mukai kernel $P \in \mathrm{D}^{\mathrm{b}}(X \times Y)$ is uniquely determined. The result was generalized to the twisted setting by Canonaco and Stellari [102].

1 Kuznetsov's Component

Some of the basic notions of the theory of derived and triangulated categories shall be reviewed, but for a thorough introduction into these topics we have to refer to the standard literature, e.g. [194, 246].

1.1 Bondal–Orlov The main result in [82] shows that the derived category $\mathrm{D}^{\mathrm{b}}(X)$ of the abelian category of coherent sheaves on a smooth projective variety X with ample canonical bundle ω_X or ample anti-canonical bundle ω_X^* determines the isomorphism type of X, cf. [246, Ch. 4]. For cubic hypersurfaces, this immediately implies the following theorem. Strictly speaking, the case of plane cubic curves is not covered by [82], but at least over $\mathbb{C}$ the result is rather easy in this case, cf. [246, Cor. 5.46].

Theorem 1.1 (Bondal–Orlov) *If two smooth projective cubic hypersurfaces $X, X' \subset \mathbb{P}^{n+1}$ are derived equivalent, i.e. there exists a linear exact equivalence $\mathrm{D}^{\mathrm{b}}(X) \simeq \mathrm{D}^{\mathrm{b}}(X')$, then X and X' are isomorphic.*

Furthermore, again by a result of Bondal and Orlov [82], the group of exact linear auto-equivalences of $\mathrm{D}^{\mathrm{b}}(X)$ of a smooth cubic hypersurface of dimension at least two is

$$\mathrm{Aut}(\mathrm{D}^{\mathrm{b}}(X)) \simeq \mathbb{Z} \times (\mathrm{Aut}(X) \ltimes \mathrm{Pic}(X)).$$

Here, the first factor acts by shifts and the last by tensor products.

From these two statements it seems that the derived category $\mathrm{D}^{\mathrm{b}}(X)$ of a smooth cubic is not a very interesting object to study, but nothing could be further from the truth. However, it is not $\mathrm{D}^{\mathrm{b}}(X)$ that encodes the relevant information about the hypersurface X itself, but a certain natural full triangulated subcategory $\mathcal{A}_X \subset \mathrm{D}^{\mathrm{b}}(X)$.

1.2 Admissible Subcategories Let $\mathcal{D}$ be a linear triangulated category over a field k. To simplify our discussion, we assume that $\mathcal{D}$ is of finite type or Hom-finite, i.e. for any two objects E and F in $\mathcal{D}$ the space $\bigoplus \mathrm{Hom}(E, F[j])$ has finite dimension.

If a full triangulated subcategory $\mathcal{D}_0 \subset \mathcal{D}$ is given, one often regards the inclusion as a fully faithful functor and, to appeal to the geometric intuition, denotes it by[1]

$$i_*\colon \mathcal{D}_0 \hookrightarrow \mathcal{D}.$$

[1] However, recall that for a closed immersion $i\colon Y \hookrightarrow X$, the derived direct image $i_*\colon \mathrm{D}^{\mathrm{b}}(Y) \longrightarrow \mathrm{D}^{\mathrm{b}}(X)$ is typically not fully faithful.

The subcategory $\mathcal{D}_0$ is called *admissible* if the functor i_* has a left and a right adjoint functor, which are automatically exact again, cf. [246, Prop. 1.41]:

$$i^* \dashv i_* \dashv i^!.$$

Note that an admissible subcategory is also closed under isomorphisms, so it is strictly full, and under direct summands, i.e. it is a thick subcategory.

Exercise 1.2 For a thick subcatgeory $\mathcal{D}_0 \subset \mathcal{D}$, the induced map

$$K(\mathcal{D}_0) \hookrightarrow K(\mathcal{D})$$

between the Grothendieck groups need not be injective. Show that for admissible subcategories it is.

For an admissible subcategory, the two functors $i^*, i^!: \mathcal{D} \longrightarrow \mathcal{D}_0$ come with natural adjunction maps

$$\mathrm{id}_{\mathcal{D}} \longrightarrow i_* \circ i^* \quad \text{and} \quad i_* \circ i^! \longrightarrow \mathrm{id}_{\mathcal{D}}.$$

Taking cones allows one to associate with any object $E \in \mathcal{D}$ two exact triangles

$$B_E \longrightarrow E \longrightarrow i_*i^*E \quad \text{and} \quad i_*i^!E \longrightarrow E \longrightarrow C_E. \tag{1.1}$$

If E is (isomorphic to) an object in $\mathcal{D}_0$, the adjunction maps lead to isomorphisms

$$E \xrightarrow{\sim} i_*i^*E \quad \text{and} \quad i_*i^!E \xrightarrow{\sim} E$$

because the natural embedding i_* is fully faithful, cf. [246, Cor. 1.22].

Lemma 1.3 *With the above notation, for every object E in $\mathcal{D}$ the naturally associated object B_E is contained in the full and triangulated subcategory*

$${}^{\perp}\mathcal{D}_0 := \{B \mid \forall F \in \mathcal{D}_0 : \mathrm{Hom}(B, i_*F) = 0\}.$$

Similarly, the object C_E is contained in the full and triangulated subcategory

$$\mathcal{D}_0^{\perp} := \{C \mid \forall F \in \mathcal{D}_0 : \mathrm{Hom}(i_*F, C) = 0\}.$$

The categories ${}^{\perp}\mathcal{D}_0$ and $\mathcal{D}_0^{\perp}$ are the *left and right orthogonal complements* of $\mathcal{D}_0$.

Proof Using the fact that $\mathcal{D}_0$ is a triangulated subcategory, one checks that ${}^{\perp}\mathcal{D}_0$ and $\mathcal{D}_0^{\perp}$ are also both full and triangulated subcategories of $\mathcal{D}$. To show that $\mathrm{Hom}(B_E, i_*F) = 0$ for all $F \in \mathcal{D}_0$ apply $\mathrm{Hom}(\ , i_*F)$ to the exact triangle defining B_E which gives the long exact sequence

$$\longrightarrow \mathrm{Hom}(B_E[1], i_*F) \longrightarrow \mathrm{Hom}(i_*i^*E, i_*F) \xrightarrow{\alpha} \mathrm{Hom}(E, i_*F) \longrightarrow \mathrm{Hom}(B_E, i_*F) \longrightarrow.$$

The map α in the middle can be written as the composition of two isomorphisms

$$\mathrm{Hom}(i_*i^*E, i_*F) \xrightarrow{\sim} \mathrm{Hom}(i^*E, F) \xrightarrow{\sim} \mathrm{Hom}(E, i_*F)$$

and, therefore, is itself an isomorphism. Here, the first isomorphism uses the fact that i_* is fully faithful and the second follows from the adjunction $i^* \dashv i_*$. This proves that the outer terms in the long exact sequence all vanish and hence B_E is contained in ${}^\perp\mathcal{D}_0$. The argument to prove that C_E is contained in $\mathcal{D}_0^\perp$ is similar and left to the reader. □

Example 1.4 Here is an example of a thick and full triangulated subcategory that is not admissible: consider the derived category $\mathcal{D} = \mathrm{D^b}(X)$ of coherent sheaves on a variety X and let $\mathcal{D}_0 \subset \mathcal{D}$ be the full subcategory of all objects with zero-dimensional support. Then clearly $\mathcal{D}_0$ is thick and triangulated but not admissible unless X itself is of dimension zero. For the non-admissibility use the fact that ${}^\perp\mathcal{D}_0 = 0$, cf. [246, Prop. 3.7].

Similarly, the subcategory $\mathrm{D}^{\mathrm{b}}_Z(X)$ of complexes with support contained in a proper closed subvariety $Z \subset X$ is thick and triangulated, but not admissible.

1.3 Semi-orthogonal Decompositions and Mutations Observe that for an admissible subcategory $\mathcal{D}_0 \subset \mathcal{D}$, the inclusions of its left and right orthogonal categories $k_* \colon {}^\perp\mathcal{D}_0 \hookrightarrow \mathcal{D}$ and $j_* \colon \mathcal{D}_0^\perp \hookrightarrow \mathcal{D}$ admit a right respectively a left adjoint functor. For example, $E \longmapsto B_E$ describes the right adjoint $k_* \dashv k^!$ and $E \longmapsto C_E$ is the left adjoint $j^* \dashv j_*$.[2] In particular, the exact triangles in (1.1) can be written as

$$k_*k^!E \longrightarrow E \longrightarrow i_*i^*E \quad \text{and} \quad i_*i^!E \longrightarrow E \longrightarrow j_*j^*E. \tag{1.2}$$

Applying (1.1) also shows $({}^\perp\mathcal{D}_0)^\perp = \mathcal{D}_0$ and ${}^\perp(\mathcal{D}_0^\perp) = \mathcal{D}_0$. A priori, a left adjoint $k^* \dashv k_*$ or a right adjoint $j_* \dashv j^!$ may not exist. However, in the situation below, where $\mathcal{D} = \mathrm{D^b}(X)$, the existence of both functors follows from the existence of a Serre functor, i.e. both categories ${}^\perp\mathcal{D}_0$ and $\mathcal{D}_0^\perp$ are admissible.

Definition 1.5 The *right and left mutations* of $i_* \colon \mathcal{D}_0 \hookrightarrow \mathcal{D}$ are the functors

$$\mathbb{R}_{\mathcal{D}_0} := k^! \colon \mathcal{D} \longrightarrow {}^\perp\mathcal{D}_0 \quad \text{and} \quad \mathbb{L}_{\mathcal{D}_0} := j^* \colon \mathcal{D} \longrightarrow \mathcal{D}_0^\perp.$$

Using this notation, the triangles in (1.1) and (1.2) are often written as

$$\mathbb{R}_{\mathcal{D}_0}E \longrightarrow E \longrightarrow i_*i^*E \quad \text{and} \quad i_*i^!E \longrightarrow E \longrightarrow \mathbb{L}_{\mathcal{D}_0}E, \tag{1.3}$$

where strictly speaking one should write $k_*\mathbb{R}_{\mathcal{D}_0}E$ and $j_*\mathbb{L}_{\mathcal{D}_0}E$.

[2] We leave it as an exercise for the reader to check that the usual problem with the non-uniqueness in the axiom TR3 of a triangulated category does not prevent $k^!$ and j^* from being actual functors.

Exercise 1.6 Show that the restrictions of $\mathbb{L}_{\mathcal{D}_0}$ and $\mathbb{R}_{\mathcal{D}_0}$ to $\mathcal{D}_0$ are both trivial. Furthermore, deduce a result of Bondal and Kapranov [81, Lem. 1.9] that says that the restrictions of $\mathbb{L}_{\mathcal{D}_0}$ to ${}^{\perp}\mathcal{D}_0$ and of $\mathbb{R}_{\mathcal{D}_0}$ to $\mathcal{D}_0^{\perp}$ are inverse to each other

$$\mathcal{D}_0^{\perp} \underset{\mathbb{L}}{\overset{\mathbb{R}}{\underset{\longleftarrow}{\overset{\sim}{\longrightarrow}}}} {}^{\perp}\mathcal{D}_0.$$

Eventually, show that

$$\Phi \circ \mathbb{L}_{\mathcal{D}_0} \simeq \mathbb{L}_{\Phi(\mathcal{D}_0)} \circ \Phi \quad \text{and} \quad \Phi \circ \mathbb{R}_{\mathcal{D}_0} \simeq \mathbb{R}_{\Phi(\mathcal{D}_0)} \circ \Phi$$

for any auto-equivalence $\Phi\colon \mathcal{D} \xrightarrow{\sim} \mathcal{D}$.

Example 1.7 An object E_0 in $\mathcal{D}$ is called *exceptional* if it satisfies

$$\mathrm{Hom}(E_0, E_0[m]) \simeq \begin{cases} k & \text{if } m = 0, \\ 0 & \text{if } m \neq 0. \end{cases}$$

In this case, $\mathcal{D}_0 := \langle E_0 \rangle \subset \mathcal{D}$ denotes the full triangulated subcategory of all objects that are isomorphic to finite direct sums $\bigoplus E_0[m_i]$. As an abstract k-linear triangulated category, $\mathcal{D}_0$ is the bounded derived category of finite-dimensional k-vector spaces:

$$\langle E_0 \rangle \simeq \mathrm{D^b}(\mathrm{Vec_{fd}}(k)) \simeq \bigoplus_m \mathrm{Vec_{fd}}(k)[m].$$

The category $\langle E_0 \rangle \subset \mathcal{D}$ is admissible and the adjoint functors i^* and $i^!$ are explicitly described as follows:

$$i^*E := \bigoplus_m \mathrm{Hom}(E, E_0[m])^* \otimes E_0[m] \quad \text{and} \quad i^!E := \bigoplus_m \mathrm{Hom}(E_0, E[m]) \otimes E_0[-m].$$

For example, with this definition we indeed have

$$\begin{aligned} \mathrm{Hom}(i^*E, E_0) &\simeq \mathrm{Hom}\Big(\bigoplus \mathrm{Hom}(E, E_0[m])^* \otimes E_0[m], E_0\Big) \\ &\simeq \mathrm{Hom}(\mathrm{Hom}(E, E_0)^* \otimes E_0, E_0) \\ &\simeq \mathrm{Hom}(E, E_0) \otimes \mathrm{End}(E_0) \simeq \mathrm{Hom}(E, i_*E_0). \end{aligned}$$

Furthermore, the adjunction maps $E \longrightarrow i_*i^*E$ and $i_*i^!E \longrightarrow E$ are given by the coevaluation map and the evaluation map

$$E \xrightarrow{\mathrm{coev}} \bigoplus \mathrm{Hom}(E, E_0[m])^* \otimes E_0[m] \quad \text{and} \quad \bigoplus \mathrm{Hom}(E_0, E[m]) \otimes E_0[-m] \xrightarrow{\mathrm{ev}} E.$$

If we write $\mathbb{R}_{E_0}\colon \mathcal{D} \longrightarrow {}^{\perp}\langle E_0 \rangle$ and $\mathbb{L}_{E_0}\colon \mathcal{D} \longrightarrow \langle E_0 \rangle^{\perp}$ for the two mutation functors associated with the admissible subcategory $\langle E_0 \rangle \subset \mathcal{D}$, then this becomes

$$\mathbb{R}_{E_0}(E) = \mathrm{cone}(\mathrm{coev})[-1] \quad \text{and} \quad \mathbb{L}_{E_0}(E) = \mathrm{cone}(\mathrm{ev}). \tag{1.4}$$

Definition 1.8 A *semi-orthogonal decomposition* of a k-linear triangulated category $\mathcal{D}$ consists of admissible full and triangulated subcategories $\mathcal{D}_1, \ldots, \mathcal{D}_m \subset \mathcal{D}$ satisfying $\mathrm{Hom}(\mathcal{D}_j, \mathcal{D}_i) = 0$ for $j > i$ and such that for each object $E \in \mathcal{D}$ there exists a sequence of exact triangles

$$\begin{array}{ccccccccc} 0 = E_m & \longrightarrow & E_{m-1} & \longrightarrow \cdots \longrightarrow & E_1 & \longrightarrow & E_0 = E \\ & \nwarrow \quad \swarrow & & & & \nwarrow \quad \swarrow & \\ & A_m & & & & A_1 & \end{array}$$

with $A_i \in \mathcal{D}_i$. In this case, we write

$$\mathcal{D} = \langle \mathcal{D}_1, \ldots, \mathcal{D}_m \rangle.$$

Example 1.9 (i) The standard example is obtained from taking the left or the right orthogonal complement ${}^{\perp}\mathcal{D}_0$ respectively $\mathcal{D}_0^{\perp}$ of an admissible subcategory $\mathcal{D}_0 \subset \mathcal{D}$:[3]

$$\mathcal{D} = \langle \mathcal{D}_0^{\perp}, \mathcal{D}_0 \rangle \quad \text{and} \quad \mathcal{D} = \langle \mathcal{D}_0, {}^{\perp}\mathcal{D}_0 \rangle.$$

In particular, any exceptional object $E_0 \in \mathcal{D}$ induces the two semi-orthogonal decompositions $\mathcal{D} = \langle \langle E_0 \rangle^{\perp}, \langle E_0 \rangle \rangle$ and $\mathcal{D} = \langle \langle E_0 \rangle, {}^{\perp}\langle E_0 \rangle \rangle$. To ease the notation, one also simply writes $\mathcal{D} = \langle E_0^{\perp}, E_0 \rangle$ and $\mathcal{D} = \langle E_0, {}^{\perp}E_0 \rangle$.

(ii) More generally, one can use exceptional collections to produce semi-orthogonal decompositions. By definition, an *exceptional collection* consists of exceptional objects $E_1, \ldots, E_m$ with $\mathrm{Hom}(E_j, E_i[*]) = 0$ for all $j > i$. Then the subcategories $\mathcal{D}_i = \langle E_i \rangle \subset \mathcal{D}$ are all admissible and provide semi-orthogonal decompositions[4]

$$\mathcal{D} = \langle \{E_1, \ldots, E_m\}^{\perp}, \langle E_1 \rangle, \ldots, \langle E_m \rangle \rangle \quad \text{and} \quad \mathcal{D} = \langle \langle E_1 \rangle, \ldots, \langle E_m \rangle, {}^{\perp}\{E_1, \ldots, E_m\} \rangle.$$

The exceptional collection is full, if $\mathcal{D} = \langle \langle E_1 \rangle, \ldots, \langle E_m \rangle \rangle = \langle E_1, \ldots, E_m \rangle$. For example, $\mathcal{O}, \ldots, \mathcal{O}(n)$ is a full exceptional collection on $\mathbb{P}^n$ as is $\mathcal{O}(-n), \ldots, \mathcal{O}$, cf. (0.1):

$$\mathrm{D^b}(\mathbb{P}^n) = \langle \mathcal{O}, \ldots, \mathcal{O}(n) \rangle = \langle \mathcal{O}(-n), \ldots, \mathcal{O} \rangle.$$

(iii) In the geometric context, $\mathrm{D^b}(X) = \mathrm{D^b}(\mathrm{Coh}(X))$ admits a natural semi-orthogonal decomposition for X a projective bundle or a blow-up, see (0.1) and (0.3).

Remark 1.10 Assume an admissible subcategory $\mathcal{D}_0 \subset \mathcal{D}$ itself comes with a semi-orthogonal decomposition $\mathcal{D}_0 = \langle \mathcal{D}_1, \mathcal{D}_2 \rangle$. In this case, $\mathcal{D} = \langle \mathcal{D}_0^{\perp}, \mathcal{D}_1, \mathcal{D}_2 \rangle$ and $\mathcal{D} =$

[3] We omit the subtlety that in general the complements may not be admissible. In our applications, this will be automatic, see Corollary 1.15.

[4] Again, strictly speaking, one would need to make sure that $\{E_1, \ldots, E_m\}^{\perp}$ is an admissible subcategory, which is automatic if $\mathcal{D}$ has a Serre functor, see Corollary 1.15.

$\langle\mathcal{D}_1, \mathcal{D}_2, {}^{\perp}\mathcal{D}_0\rangle$. Furthermore, the left mutation $\mathbb{L}_{\mathcal{D}_0}\colon \mathcal{D} \longrightarrow \mathcal{D}_0^{\perp}$ of $\mathcal{D}_0$ can be written as the composition

$$\mathcal{D} \xrightarrow{\mathbb{L}_{\mathcal{D}_2}} \mathcal{D}_2^{\perp} = \langle\mathcal{D}_0^{\perp}, \mathcal{D}_1\rangle \xrightarrow{\mathbb{L}_{\mathcal{D}_1}} \mathcal{D}_0^{\perp},$$

where we consider $\mathcal{D}_0^{\perp} \subset \mathcal{D}_2^{\perp} = \langle\mathcal{D}_0^{\perp}, \mathcal{D}_1\rangle$ as the right orthogonal of $\mathcal{D}_1 \subset \mathcal{D}_2^{\perp} = \langle\mathcal{D}_0^{\perp}, \mathcal{D}_1\rangle$. Similarly, the right mutation $\mathbb{R}_{\mathcal{D}_0}\colon \mathcal{D} \longrightarrow {}^{\perp}\mathcal{D}_0$ is written as the composition

$$\mathcal{D} \xrightarrow{\mathbb{R}_{\mathcal{D}_1}} {}^{\perp}\mathcal{D}_1 = \langle\mathcal{D}_2, {}^{\perp}\mathcal{D}_0\rangle \xrightarrow{\mathbb{R}_{\mathcal{D}_2}} {}^{\perp}\mathcal{D}_0.$$

More generally, for a semi-orthogonal decomposition $\mathcal{D}_0 = \langle\mathcal{D}_1, \dots, \mathcal{D}_m\rangle$ and the corresponding left and right mutations, one has

$$\mathbb{L}_{\mathcal{D}_0} \simeq \mathbb{L}_{\mathcal{D}_1} \circ \cdots \circ \mathbb{L}_{\mathcal{D}_m} \quad \text{and} \quad \mathbb{R}_{\mathcal{D}_0} \simeq \mathbb{R}_{\mathcal{D}_m} \circ \cdots \circ \mathbb{R}_{\mathcal{D}_1}.$$

Exercise 1.11 Prove that a semi-orthogonal decomposition $\mathcal{D} = \langle\mathcal{D}_1, \dots, \mathcal{D}_m\rangle$ leads to further semi-orthogonal decompositions

$$\mathcal{D} = \langle\mathcal{D}_1, \dots, \mathcal{D}_{i-1}, \mathbb{L}_{\mathcal{D}_i}(\mathcal{D}_{i+1}), \mathcal{D}_i, \mathcal{D}_{i+2}, \dots, \mathcal{D}_m\rangle$$

and

$$\mathcal{D} = \langle\mathcal{D}_1, \dots, \mathcal{D}_{i-1}, \mathcal{D}_{i+1}, \mathbb{R}_{\mathcal{D}_{i+1}}(\mathcal{D}_i), \mathcal{D}_{i+2}, \dots, \mathcal{D}_m\rangle.$$

For a full exceptional collection $E_1, \dots, E_m$ this is expressed by saying that

$$E_1, \dots, E_{i-1}, \mathbb{L}_{E_i}(E_{i+1}), E_i, E_{i+2}, \dots, E_m \quad \text{and} \quad E_1, \dots, E_{i-1}, E_{i+1}, \mathbb{R}_{E_{i+1}}(E_i), E_{i+2}, \dots, E_m$$

are again full exceptional collections.

Remark 1.12 Probably the best way to think of left and right mutations is via their action on the collection of all semi-orthogonal decompositions of fixed length m. By the results of Bondal and Kapranov [81], viewed as such they satisfy the *braid relations*

$$\mathbb{R}_i \circ \mathbb{R}_{i+1} \circ \mathbb{R}_i = \mathbb{R}_{i+1} \circ \mathbb{R}_i \circ \mathbb{R}_{i+1} \quad \text{and} \quad \mathbb{R}_i \circ \mathbb{R}_j = \mathbb{R}_j \circ \mathbb{R}_i \quad \text{for } |i-j| > 1$$

and similarly for $\mathbb{L}_i$. See also [292, Sec. 2.4].

1.4 Serre Functor Recall that a *Serre functor* for a Hom-finite k-linear triangulated category $\mathcal{D}$ is a k-linear functor $\mathcal{S}_{\mathcal{D}}\colon \mathcal{D} \longrightarrow \mathcal{D}$ with functorial isomorphisms (Serre duality)

$$\mathrm{Hom}(E, F) \simeq \mathrm{Hom}(F, \mathcal{S}_{\mathcal{D}}E)^*$$

for all objects E and F, cf. [246, Ch. 1]. A Serre functor is always an equivalence and automatically exact [81, Prop. 3.3], cf. [246, Prop. 1.46]. If it exists, a Serre functor is unique, cf. [246, Lem. 1.30].

For a smooth projective variety X of dimension n, a Serre functor for $\mathrm{D}^{\mathrm{b}}(X)$ is given by $E \longmapsto E \otimes \omega_X[n]$, cf. [246, Thm. 3.12].

Exercise 1.13 Show that for an admissible subcategory $\mathcal{D}_0 \subset \mathcal{D}$ of a linear, triangulated category $\mathcal{D}$ a Serre functor $\mathcal{S}_{\mathcal{D}}$ (if it exists) and its inverse induce equivalences

$$\mathcal{D}_0^{\perp} \underset{\mathcal{S}_{\mathcal{D}}}{\overset{\mathcal{S}_{\mathcal{D}}^{-1}}{\rightleftarrows}} {}^{\perp}\mathcal{D}_0.$$

In other words, with $\mathcal{D} = \langle \mathcal{D}_0, \mathcal{D}_1 \rangle$ also

$$\mathcal{D} = \langle \mathcal{S}_{\mathcal{D}}(\mathcal{D}_1), \mathcal{D}_0 \rangle \quad \text{and} \quad \mathcal{D} = \langle \mathcal{D}_1, \mathcal{S}_{\mathcal{D}}^{-1}(\mathcal{D}_0) \rangle$$

are semi-orthogonal decompositions. Compare this to Exercise 1.6 and observe that in the geometric setting $\mathcal{D} = \mathrm{D}^{\mathrm{b}}(X)$ the assertion can be rephrased as

$$\mathcal{D}_0^{\perp} = \mathbb{L}_{\mathcal{D}_0}({}^{\perp}\mathcal{D}_0) = \mathcal{S}_{\mathcal{D}}({}^{\perp}\mathcal{D}_0) = {}^{\perp}\mathcal{D}_0 \otimes \omega_X \text{ and } {}^{\perp}\mathcal{D}_0 = \mathbb{R}_{\mathcal{D}_0}(\mathcal{D}_0^{\perp}) = \mathcal{S}_{\mathcal{D}}^{-1}({}^{\perp}\mathcal{D}_0) = \mathcal{D}_0^{\perp} \otimes \omega_X^{-1}.$$

Lemma 1.14 *Assume $i_*\colon \mathcal{D}_0 \hookrightarrow \mathcal{D}$ is an admissible subcategory with left and right adjoint functors i^* and $i^!$. If $\mathcal{D}$ admits a Serre functor $\mathcal{S}_{\mathcal{D}}$, then $\mathcal{D}_0$ also admits a Serre functor which, moreover, satisfies*

$$\mathcal{S}_{\mathcal{D}_0} \simeq i^! \circ \mathcal{S}_{\mathcal{D}} \circ i_* \quad \textit{and} \quad i^! \simeq \mathcal{S}_{\mathcal{D}_0} \circ i^* \circ \mathcal{S}_{\mathcal{D}}^{-1}.$$

Furthermore, a Serre functor on $\mathcal{D}_0^{\perp}$ exists as well and satisfies

$$\mathcal{S}_{\mathcal{D}_0^{\perp}} \simeq \mathcal{S}_{\mathcal{D}} \circ \mathbb{R}_{\mathcal{D}_0} \quad \textit{and} \quad \mathcal{S}_{\mathcal{D}_0^{\perp}}^{-1} \simeq \mathbb{L}_{\mathcal{D}_0} \circ \mathcal{S}_{\mathcal{D}}^{-1}.$$

Proof The first isomorphism is verified by composing a number of functorial isomorphisms:

$$\mathrm{Hom}_{\mathcal{D}_0}(E, F) \simeq \mathrm{Hom}_{\mathcal{D}}(i_*E, i_*F) \simeq \mathrm{Hom}_{\mathcal{D}}(i_*F, \mathcal{S}_{\mathcal{D}} i_*E)^* \simeq \mathrm{Hom}_{\mathcal{D}_0}(F, i^! \mathcal{S}_{\mathcal{D}} i_*E)^*.$$

Here, the first isomorphism follows from i_* being fully faithful, the second from Serre duality for $\mathcal{D}$, and the last one uses the adjunction $i_* \dashv i^!$. Similarly, for the second isomorphism we use

$$\mathrm{Hom}_{\mathcal{D}}(i_*E, F) \simeq \mathrm{Hom}_{\mathcal{D}}(\mathcal{S}_{\mathcal{D}}^{-1} F, i_*E)^* \simeq \mathrm{Hom}_{\mathcal{D}_0}(i^* \mathcal{S}_{\mathcal{D}}^{-1} F, E)^* \simeq \mathrm{Hom}_{\mathcal{D}_0}(E, S_{\mathcal{D}_0} i^* S_{\mathcal{D}}^{-1} F).$$

To describe the Serre functor for $\mathcal{D}_0^{\perp}$, use that for objects E and F in $\mathcal{D}_0^{\perp}$ one has

$$\mathrm{Hom}(E, \mathcal{S}_{\mathcal{D}} \mathbb{R}_{\mathcal{D}_0} F) \simeq \mathrm{Hom}(\mathbb{R}_{\mathcal{D}_0} F, E)^* \simeq \mathrm{Hom}(\mathbb{L}_{\mathcal{D}_0} \mathbb{R}_{\mathcal{D}_0} F, E)^* \simeq \mathrm{Hom}(F, E)^*,$$

where we use $\mathrm{Hom}(i_* i^! \mathbb{R}_{\mathcal{D}_0} F, E) = 0$ for $E \in \mathcal{D}_0^{\perp}$ and Exercise 1.6. We leave the verification of the last isomorphism to the reader. □

Corollary 1.15 *Assume $\mathcal{D}_0 \subset \mathcal{D}$ is an admissible subcategory of a triangulated category $\mathcal{D}$ with a Serre functor $\mathcal{S}_{\mathcal{D}}$. Then $\mathcal{D}_0^{\perp}$ and ${}^{\perp}\mathcal{D}_0$ are also admissible.*

Proof We have seen that the inclusion $j_*\colon \mathcal{D}_0^\perp \hookrightarrow \mathcal{D}$ always admits a left adjoint $j^* \dashv j_*$ and for the existence of the right adjoint $j_* \dashv j^!$ use the above lemma twice: first, to see that $\mathcal{D}_0^\perp$ admits a Serre functor, namely $\mathcal{S}_{\mathcal{D}_0^\perp} = \mathcal{S}_{\mathcal{D}} \circ \mathbb{R}_{\mathcal{D}_0}$, and then to express $j^!$ as $\mathcal{S}_{\mathcal{D}_0^\perp} \circ j^* \circ \mathcal{S}_{\mathcal{D}}^{-1}$. The argument to prove that ${}^\perp\mathcal{D}_0$ is admissible is similar. Alternatively, one may use Exercise 1.13. □

Definition 1.16 A linear triangulated category is a *Calabi–Yau n-category* if it admits a Serre functor $\mathcal{S}$ which is isomorphic to the translation $E \longmapsto E[n]$. It is a *fractional Calabi–Yau* (p/q)*-category* if it admits a Serre functor whose qth power $\mathcal{S}^q$ is isomorphic to the translation $E \longmapsto E[p]$.

Note that a Calabi–Yau (np/nq)-category need not be a Calabi–Yau (p/q)-category. Typically, when one speaks of a Calabi–Yau (p/q)-category, the positive number q is chosen minimal.

Standard examples of Calabi–Yau categories are provided by $\mathrm{D^b}(X)$ of a smooth projective variety X with trivial canonical bundle $\omega_X \simeq \mathcal{O}_X$, as in this case, the Serre functor is simply the translation $E \longmapsto E[\dim(X)]$. For example, $\mathrm{D^b}(S)$ of a K3 surface is a Calabi–Yau 2-category. The derived category $\mathrm{D^b}(S)$ of an Enriques surface is a fractional $(4/2)$-category, but not a Calabi–Yau 2-category.

Exercise 1.17 (i) Assume $\mathcal{D} = \langle \mathcal{D}_1, \mathcal{D}_2 \rangle$ is a semi-orthogonal decomposition of a Calabi–Yau n-category $\mathcal{D}$. Show that then $\mathcal{D}_1 = \mathcal{D}_2^\perp = {}^\perp\mathcal{D}_2$, or, in other words, $\mathcal{D} = \mathcal{D}_1 \oplus \mathcal{D}_2$.

(ii) Assume that $\mathcal{D}$ is a Calabi–Yau (p/q)-category with $q \nmid p$, $q > 1$. Show that then $\mathcal{D}$ is not geometric, i.e. not equivalent to $\mathrm{D^b}(X)$ for any smooth projective variety X.

1.5 Derived Categories of Hypersurfaces The triangulated categories that are relevant for us are the bounded derived categories of coherent sheaves on smooth hypersurfaces and their admissible subcategories.

Concretely, consider a smooth hypersurface $X \subset \mathbb{P} = \mathbb{P}^{n+1}$ of degree d and its bounded derived category $\mathrm{D^b}(X) = \mathrm{D^b}(\mathrm{Coh}(X))$ of coherent sheaves on X. This is a k-linear triangulated category with a Serre functor $\mathcal{S}_X$ that is explicitly given by

$$\mathcal{S}_X\colon E \longmapsto E \otimes \mathcal{O}_X(d-(n+2))[n],$$

use the adjunction formula $\omega_X \simeq \mathcal{O}_X(d-(n+2))$, see Lemma **1**.1.6. The line bundle $\mathcal{O}_X(i) := \mathcal{O}(i)|_X$ as an object in the bounded derived category $\mathrm{D^b}(X)$ is exceptional if and only if $H^m(X, \mathcal{O}_X) = 0$ for all $m > 0$. By Bott vanishing, this is the case exactly when $d \le n+1$, cf. Section **1**.1.2, which we will assume from now on.

Next, one would like to know when the line bundle $\mathcal{O}_X(i)$ is contained in $\langle \mathcal{O}_X(j) \rangle^\perp$. As $\mathrm{Hom}(\mathcal{O}_X(j), \mathcal{O}_X(i)[*]) \simeq \mathrm{Ext}^*(\mathcal{O}_X(j), \mathcal{O}_X(i)) \simeq H^*(X, \mathcal{O}_X(i-j))$, Bott vanishing provides the answer again:

$$\mathcal{O}_X(i) \in \langle \mathcal{O}_X(j) \rangle^\perp \text{ if and only if } j - (n+2-d) < i < j.$$

So, the longest sequence of line bundles with $\mathrm{Hom}(\mathcal{O}_X(j), \mathcal{O}_X(i)[*]) = 0$ for all $j > i$ is $\mathcal{O}_X, \mathcal{O}_X(1), \ldots, \mathcal{O}_X(n+1-d)$ and any line bundle twist of it. This immediately provides us with a semi-orthogonal decomposition

$$\mathrm{D}^\mathrm{b}(X) = \langle \mathcal{A}_X, \mathcal{O}_X, \ldots, \mathcal{O}(n+1-d) \rangle, \tag{1.5}$$

where $\mathcal{A}_X$ is by definition the right orthogonal

$$\mathcal{A}_X := \langle \mathcal{O}_X, \ldots, \mathcal{O}_X(n+1-d) \rangle^\perp \subset \mathrm{D}^\mathrm{b}(X)$$

of the full subcategory $\mathcal{D}_0 := \langle \mathcal{O}_X, \ldots, \mathcal{O}_X(n+1-d) \rangle$ or, simply, the right orthogonal of the set $\{\mathcal{O}_X, \ldots, \mathcal{O}_X(n+1-d)\}$.

Note that $\mathcal{A}_X$ really is an admissible subcategory, i.e. $j_* \dashv j^!$ exists, see Corollary 1.15. The left adjoint of the inclusion $j_* \colon \mathcal{A}_X \hookrightarrow \mathrm{D}^\mathrm{b}(X)$ is the left mutation

$$\mathbb{L}_{\langle \mathcal{O}_X, \ldots, \mathcal{O}_X(n+1-d) \rangle} = j^* \colon \mathrm{D}^\mathrm{b}(X) \longrightarrow \mathcal{A}_X.$$

Exercise 1.18 According to Remark 1.10,

$$\mathbb{L}_{\langle \mathcal{O}_X, \ldots, \mathcal{O}_X(n+1-d) \rangle} \simeq \mathbb{L}_{\mathcal{O}_X} \circ \cdots \circ \mathbb{L}_{\mathcal{O}_X(n+1-d)}.$$

Show that each of the left mutations $\mathbb{L}_{\mathcal{O}_X(i)}$ is a Fourier–Mukai functor

$$\mathbb{L}_{\mathcal{O}_X(i)} \simeq \Phi_{C(\mathrm{tr}_i)},$$

the kernel of which is the cone of the trace map $\mathrm{tr}_i \colon \mathcal{O}_X(-i) \boxtimes \mathcal{O}_X(i) \longrightarrow \mathcal{O}_\Delta$.

Definition 1.19 The full triangulated subcategory $\mathcal{A}_X \subset \mathrm{D}^\mathrm{b}(X)$ in (1.5) is called the *residual* or *Kuznetsov component* of $\mathrm{D}^\mathrm{b}(X)$.

We think of $\mathcal{A}_X$ as the non-trivial part of $\mathrm{D}^\mathrm{b}(X)$.

Example 1.20 The easiest non-trivial case is that of a smooth quadric $X = Q \subset \mathbb{P}^{n+1}$. In this case, the semi-orthogonal decomposition takes the form

$$\mathrm{D}^\mathrm{b}(Q) = \langle \mathcal{A}_Q, \mathcal{O}_Q, \ldots, \mathcal{O}_Q(n-1) \rangle$$

and the Kuznetsov component $\mathcal{A}_Q$ can be described explicitly. The answer depends on the parity of n. One has $\mathcal{A}_Q = \langle \Sigma \rangle$ for $n = 2m+1$ and $\mathcal{A}_Q = \langle \Sigma^+, \Sigma^- \rangle = \langle \Sigma^-, \Sigma^+ \rangle$ for $n = 2m$, which leads to a result by Kapranov [266]:

$$\mathcal{A}_Q \simeq \begin{cases} \mathrm{D}^\mathrm{b}(\mathrm{Vec}_{\mathrm{fd}}(k)) & \text{if } n = 2m+1, \\ \mathrm{D}^\mathrm{b}(\mathrm{Vec}_{\mathrm{fd}}(k)) \oplus \mathrm{D}^\mathrm{b}(\mathrm{Vec}_{\mathrm{fd}}(k)) & \text{if } n = 2m. \end{cases}$$

Here, Σ and $\Sigma^\pm$ are the spinor bundles on the quadric Q, for which various descriptions are available, see [380] or [301, Sec. 2.2]. For example, for $n = 1$, i.e. for a conic

$\mathbb{P}^1 \simeq Q \subset \mathbb{P}^2$, one finds $\Sigma \simeq \mathcal{O}(-1)$ and for $n = 2$, i.e. for a quadric $\mathbb{P}^1 \times \mathbb{P}^1 \simeq Q \subset \mathbb{P}^3$, one has $\Sigma^+ \simeq \mathcal{O}(-1,0)$ and $\Sigma^- \simeq \mathcal{O}(0,-1)$.

In general, Σ is the restriction of the universal sub-bundle with respect to a certain natural embedding of $Q \subset \mathbb{P}^{2m+2}$ into $\mathrm{Gr}(2^m - 1, 2^{m+1} - 1)$. Alternatively, the dual Σ^* is isomorphic to the image $q_* p^* \mathcal{O}(1/2)$ under the Fano correspondence for $F(Q, m)$, see Exercise **2.1.6**. It is also known that $\Sigma|_{Q'} \simeq \Sigma^+ \oplus \Sigma^-$ under the embedding of an even dimensional quadric $Q' := Q \cap \mathbb{P}^{2m+1} \subset Q \subset \mathbb{P}^{2m+2}$ into an odd-dimensional one. For $Q' := Q \cap \mathbb{P}^{2m} \subset Q \subset \mathbb{P}^{2m+1}$, one knows $\Sigma^{\pm}|_{Q'} \simeq \Sigma$.

Exercise 1.21 Let $X \subset \mathbb{P}^{n+1}$ be a smooth cubic hypersurface of dimension at least three. Then $\mathrm{D^b}(X) = \langle \mathcal{A}_X, \mathcal{O}_X, \dots, \mathcal{O}_X(n-2) \rangle$. Show that the ideal sheaf $\mathcal{I}_L$ of a line $L \subset X$ is right orthogonal to $\mathcal{O}_X$ and $\mathcal{O}_X(1)$ but not to $\mathcal{O}_X(k)$, $k > 1$. In particular, for cubic threefolds $Y \subset \mathbb{P}^4$, one finds

$$\mathcal{I}_L \in \mathcal{A}_Y,$$

but for cubic fourfolds $X \subset \mathbb{P}^5$ we only have $\mathcal{I}_L \in \langle \mathcal{O}_X, \mathcal{O}_X(1) \rangle^\perp = \langle \mathcal{A}_X(-1), \mathcal{O}_X(-1) \rangle$. The latter is more conveniently rephrased as

$$\mathcal{I}_L(1) \in \langle \mathcal{O}_X(1), \mathcal{O}_X(2) \rangle^\perp.$$

Furthermore, its projection $\mathbb{L}_{\mathcal{O}_X}(\mathcal{I}_L(1)) = j^*(\mathcal{I}_L(1)) \in \mathcal{A}_X$ is $\mathcal{F}_L[1]$, where $\mathcal{F}_L$ is the kernel of the surjective evaluation map

$$0 \longrightarrow \mathcal{F}_L \longrightarrow H^0(X, \mathcal{I}_L(1)) \otimes \mathcal{O}_X \overset{\mathrm{ev}}{\twoheadrightarrow} \mathcal{I}_L(1) \longrightarrow 0,$$

see (1.4). These results were proved by Kuznetsov and Markushevich [300]. Note that instead of the ideal sheaf $\mathcal{I}_L(1)$ of a line, one could also consider its structure sheaf $\mathcal{O}_L(1)$, which, however, is only right orthogonal to $\mathcal{O}_X(2)$. The first step of projecting it into $\mathcal{A}_X$ consists of making it orthogonal to $\mathcal{O}_X(1)$, which means passing to $\mathcal{I}_L(1)$.

Note that for $n + 1 - d < d$, the category $\langle \mathcal{O}_X, \dots, \mathcal{O}_X(n+1-d) \rangle$ is equivalent to $\langle \mathcal{O}_{\mathbb{P}}, \dots, \mathcal{O}_{\mathbb{P}}(n+1-d) \rangle$ and, in particular, independent of any specific X, see [246, Thm. 8.34] for an explicit description of this category and for references. The numerical condition holds for all cubic hypersurfaces of dimension $n < 5$.

Exercise 1.22 The left orthogonal ${}^\perp\mathcal{A}_X$ of $\mathcal{A}_X$ is the category $\langle \mathcal{O}_X, \dots, \mathcal{O}_X(n+1-d) \rangle$. Show that its right orthogonal $\mathcal{A}_X^\perp$ is generated by $\mathcal{O}_X(d-n-2), \dots, \mathcal{O}_X(-1)$, cf. Exercise 1.13. Hence, there are two natural semi-orthogonal decompositions

$$\mathrm{D^b}(X) = \langle \mathcal{A}_X, \mathcal{O}_X, \dots, \mathcal{O}(n+1-d) \rangle = \langle \mathcal{O}_X(d-n-2), \dots, \mathcal{O}_X(-1), \mathcal{A}_X \rangle.$$

Deduce from Exercise 1.2 that the Grothendieck groups of X or, equivalently, of $\mathrm{D^b}(X)$ and of $\mathcal{A}_X$ satisfy

$$K(X) \simeq K(\mathrm{D^b}(X)) \simeq K(\mathcal{A}_X) \oplus \mathbb{Z}^{n+2-d}.$$

1.6 Serre Functor of Kuznetsov's Component As a consequence of Lemma 1.14, one finds the following description of the (inverse) Serre functor of $\mathcal{A}_X$ [299, Thm. 3.5].

Corollary 1.23 (Kuznetsov) *Under the above assumption, one has*

$$\mathcal{S}_{\mathcal{A}_X}^{-1} \simeq \mathrm{O}^{n+2-d} \circ [-n] \simeq j^* \circ (\mathcal{O}_X(n+2-d) \otimes \ ) \circ [-n]$$

where O *is the composition* $\mathbb{L}_{\mathcal{O}_X} \circ (\mathcal{O}_X(1) \otimes \ ) \colon \mathrm{D^b}(X) \longrightarrow \langle \mathcal{O}_X \rangle^\perp$.

Proof On the one hand, by virtue of Lemma 1.14, $\mathcal{S}_{\mathcal{A}_X}^{-1} \simeq \mathbb{L}_{\langle \mathcal{O}_X, \ldots, \mathcal{O}_X(n+1-d) \rangle} \circ \mathcal{S}_{\mathrm{D^b}(X)}^{-1}$ or, equivalently, the Serre functor is given by

$$\mathcal{S}_{\mathcal{A}_X} \colon E \longmapsto j^! \,(E \otimes \mathcal{O}_X(d-(n+2)))[n]. \tag{1.6}$$

On the other hand, according to Remark 1.10, we have $\mathbb{L}_{\langle \mathcal{O}_X, \ldots, \mathcal{O}_X(n+1-d) \rangle} \simeq \mathbb{L}_{\mathcal{O}_X} \circ \cdots \circ \mathbb{L}_{\mathcal{O}_X(n+1-d)}$. Combined with Exercise 1.6, this gives

$$\begin{aligned} \mathrm{O}^{n+2-d} &\simeq (\mathbb{L}_{\mathcal{O}_X} \circ (\mathcal{O}_X(1) \otimes \ )) \circ \cdots \circ (\mathbb{L}_{\mathcal{O}_X} \circ (\mathcal{O}_X(1) \otimes \ )) \\ &\simeq \mathbb{L}_{\langle \mathcal{O}_X, \ldots, \mathcal{O}_X(n+1-d) \rangle} \circ (\mathcal{O}_X(n+2-d) \otimes \ ) \\ &\simeq j^* \circ (\mathcal{O}_X(n+2-d) \otimes \ ), \end{aligned}$$

which proves the assertion. □

Remark 1.24 Motivated by the description of $\mathcal{A}_X$ as a category of matrix factorizations, see Section 1.7, the restriction T of O to $\mathcal{A}_X$ is also called the *degree shift functor*

$$T := \mathrm{O}|_{\mathcal{A}_X} \colon \mathcal{A}_X \xrightarrow{\sim} \mathcal{A}_X, \quad E \longmapsto j^*(j_* E \otimes \mathcal{O}(1)).$$

The corollary and its proof show that T is indeed an auto-equivalence of $\mathcal{A}_X$ and that we can write

$$\mathcal{S}_{\mathcal{A}_X} \simeq T^{d-n-2} \circ [n]. \tag{1.7}$$

Here is an alternative way of proving this isomorphism: use the adjunction $j^* \dashv j_* \dashv j^!$ and the description of the Serre functor (1.6), to deduce functorial isomorphisms

$$\begin{aligned} \mathrm{Hom}(\mathcal{S}_{\mathcal{A}_X}^{-1} E, F) &\simeq \mathrm{Hom}(E, \mathcal{S}_{\mathcal{A}_X} F) \simeq \mathrm{Hom}(E, j^!(j_* F \otimes \mathcal{O}_X(d-(n+2)))[n]) \\ &\simeq \mathrm{Hom}(j_* E \otimes \mathcal{O}_X(n+2-d)[-n], j_* F) \\ &\simeq \mathrm{Hom}(j^*(j_* E \otimes \mathcal{O}_X(n+2-d))[-n], F) \\ &\simeq \mathrm{Hom}(T^{n+2-d} E[-n], F), \end{aligned}$$

which by applying the Yoneda lemma proves $T^{n+2-d} \simeq \mathcal{S}_{\mathcal{A}_X^{-1}} \circ [n]$. Note that this second argument is not quite complete, as it does not explain why T^{n+2-d} is really given by $E \longmapsto j^*(j_* E \otimes \mathcal{O}(n+2-d))$.

Example 1.25 For cubic hypersurfaces $X \subset \mathbb{P}^{n+1}$ the condition $d \leq n+1$ reads $2 \leq n$ and $n+1-d < d$ becomes $n \leq 4$. In this light, everything that will follow works well for cubic surfaces S, cubic threefolds Y, and cubic fourfolds X. In these three cases, one has semi-orthogonal decompositions

$$\mathrm{D^b}(S) = \langle \mathcal{A}_S, \mathcal{O}_S \rangle, \ \ \mathrm{D^b}(Y) = \langle \mathcal{A}_Y, \mathcal{O}_Y, \mathcal{O}_Y(1) \rangle, \ \text{ and } \ \mathrm{D^b}(X) = \langle \mathcal{A}_X, \mathcal{O}_X, \mathcal{O}_X(1), \mathcal{O}_X(2) \rangle.$$

The proof of the next result [297] will be given at the end of this subsection.

Proposition 1.26 (Kuznetsov) *Assume $X \subset \mathbb{P}^{n+1}$ is a smooth hypersurface of degree d with $n+1 < 2d$ and let $c := \gcd(d, n+2)$. Then $\mathcal{A}_X$ is a fractional Calabi–Yau category. More precisely,*

$$\mathcal{S}_{\mathcal{A}_X}^{d/c} \colon E \longmapsto E[(n+2)(d-2)/c].$$

Thus, if $d \mid (n+2)$, then $\mathcal{A}_X$ is a Calabi–Yau N-category with $N = (n+2)(d-2)/d \in \mathbb{Z}$.

Example 1.27 To be explicit, for a cubic surface S, a cubic threefold Y, and a cubic fourfold X the result means

$$\mathcal{S}_{\mathcal{A}_S}^3 \simeq [4], \quad \mathcal{S}_{\mathcal{A}_Y}^3 \simeq [5], \quad \text{and} \quad \mathcal{S}_{\mathcal{A}_X} \simeq [2],$$

i.e. $\mathcal{A}_S$ is a Calabi–Yau $(4/3)$-category, $\mathcal{A}_Y$ is a Calabi–Yau $(5/3)$-category, and $\mathcal{A}_X$ is a Calabi–Yau 2-category.

Note that in dimension two and three, the fractional Calabi–Yau property excludes the Kuznetsov component from being geometric, see Exercise 1.17, i.e. for any smooth projective variety Z

$$\mathcal{A}_S \not\simeq \mathrm{D^b}(Z) \quad \text{and} \quad \mathcal{A}_Y \not\simeq \mathrm{D^b}(Z).$$

Remark 1.28 Observe that the number $\sigma := (n+2)(d-2)$ played a central role in the study of the Jacobian ring of a hypersurface, see Section **1**.4.1. This is not a coincidence and we will come back to this point, see (1.10) in Section 1.7.

We will need some preparations before entering the proof of Proposition 1.26. First observe that the functor T is isomorphic to the restriction of a Fourier–Mukai functor

$$T \simeq \Phi_{\mathcal{Q}_1}|_{\mathcal{A}_X} \ \text{ with kernel } \ \mathcal{Q}_1 := [\mathcal{O}_X(1) \boxtimes \mathcal{O}_X \longrightarrow \mathcal{O}_{\Delta_X}(1)]. \tag{1.8}$$

Indeed, since $T \simeq \mathbb{L}_{\mathcal{O}_X} \circ (\mathcal{O}_X(1) \otimes \ )$, $(\mathcal{O}_X(1) \otimes \ ) \simeq \Phi_{\mathcal{O}_{\Delta_X}(1)}$, and $\mathbb{L}_{\mathcal{O}_X} = \Phi_C$ with $C := C(\mathcal{O}_X \boxtimes \mathcal{O}_X \longrightarrow \mathcal{O}_{\Delta_X})$, see Exercise 1.18, standard rules for the convolution of Fourier–Mukai kernels give the result, cf. [246, Prop. 5.10].

Similarly, using the Euler sequence on $\mathbb{P}^{n+1}$, one proves that T^i, $i = 1, \ldots, n+1$, is isomorphic to the restriction of a Fourier–Mukai functor

$$T^i \simeq \Phi_{\mathcal{Q}_i}|_{\mathcal{A}_X} \ \text{ with kernel}$$

$$\mathcal{Q}_i := [\mathcal{O}_X(1) \boxtimes \Omega_{\mathbb{P}}^{i-1}(i-1)|_X \longrightarrow \cdots \longrightarrow \mathcal{O}_X(i-1) \boxtimes \Omega_{\mathbb{P}}(1)|_X \longrightarrow \mathcal{O}_X(i) \boxtimes \mathcal{O}_X \longrightarrow \mathcal{O}_{\Delta_X}(i)].$$

This will enter the proof of the following crucial step in the proof of Proposition 1.26.

Lemma 1.29 (Kuznetsov) *The degree shift functor $T\colon \mathcal{A}_X \longrightarrow \mathcal{A}_X$ satisfies the following:*

(i) *There exists an isomorphism $T^d \simeq [2]$.*
(ii) *The functor T is an equivalence.*

Proof The second assertion follows from the first one, but it was in fact already proved in Remark 1.24. The first assertion comes down to an isomorphism between the projections of the Fourier–Mukai kernels $\mathcal{Q}_d$ and $\mathcal{O}_{\Delta_X}[2]$ of the two functors.

First, one proves that the pullback $\varphi^*\mathcal{O}_{\Delta_{\mathbb{P}}}$ under $\varphi\colon X\times X \hookrightarrow \mathbb{P}\times\mathbb{P}$ is concentrated in degree 0 and -1 with cohomology sheaves $\mathcal{H}^0 \simeq \mathcal{O}_{\Delta_X}$ and $\mathcal{H}^{-1} \simeq \mathcal{O}_{\Delta_X}(-d)$. For this, view $\mathcal{H}^i$ as $\mathcal{H}^i(\mathcal{O}_{X\times X}\otimes_{\mathbb{P}\times\mathbb{P}}\mathcal{O}_{\Delta_{\mathbb{P}}})$ which is then computed by means of the Koszul resolution

$$[\mathcal{O}(-d,-d)\longrightarrow \mathcal{O}(-d,0)\oplus\mathcal{O}(0,-d)\longrightarrow \mathcal{O}_{\mathbb{P}\times\mathbb{P}}]\xrightarrow{\sim}\mathcal{O}_{X\times X}$$

as $\mathcal{H}^i \simeq \mathcal{H}^i[\mathcal{O}_{\Delta_{\mathbb{P}}}(-2d)\longrightarrow\mathcal{O}_{\Delta_{\mathbb{P}}}(-d)^{\oplus 2}\longrightarrow\mathcal{O}_{\Delta_{\mathbb{P}}}]$. In particular, $\mathcal{H}^i = 0$ for $i \neq -2,-1,0$. Also, since $\mathcal{O}_{\Delta_{\mathbb{P}}}(-2d)\longrightarrow\mathcal{O}_{\Delta_{\mathbb{P}}}(-2)^{\oplus 2}$ is clearly injective, one finds $\mathcal{H}^{-2} = 0$. Furthermore, $\mathcal{H}^{-1} = \mathrm{Coker}(\mathcal{O}_{\Delta_{\mathbb{P}}}(-2d)\longrightarrow\mathcal{O}_{\Delta_{\mathbb{P}}}(-d)) \simeq \mathcal{O}_{\Delta_X}(-d)$.

Second, writing out the above provides us with an exact triangle

$$\mathcal{H}^{-1}[1]\simeq\mathcal{O}_{\Delta_X}(-d)[1]\longrightarrow\varphi^*\mathcal{O}_{\Delta_{\mathbb{P}}}\longrightarrow\mathcal{O}_{\Delta_X}\simeq\mathcal{H}^0,$$

which after tensoring with $\mathcal{O}(d,0)$ and rotating becomes

$$\varphi^*\mathcal{O}_{\Delta_{\mathbb{P}}}(d)\longrightarrow\mathcal{O}_{\Delta_X}(d)\longrightarrow\mathcal{O}_{\Delta_X}[2].$$

This exact triangle is now compared to the natural exact triangle

$$\mathcal{Q}'_d\longrightarrow\mathcal{O}_{\Delta_X}(d)\longrightarrow\mathcal{Q}_d,$$

where by definition

$$\mathcal{Q}'_d := [\mathcal{O}_X(1)\boxtimes\Omega_{\mathbb{P}}^{d-1}(d-1)|_X\longrightarrow\cdots\longrightarrow\mathcal{O}_X(d-1)\boxtimes\Omega_{\mathbb{P}}(1)|_X\longrightarrow\mathcal{O}_X(d)\boxtimes\mathcal{O}_X].$$

To conclude the proof, it suffices to show that for $E\in\mathcal{A}_X$ one has $\Phi_{\mathcal{Q}'_d}(E)\simeq\Phi_{\varphi^*\mathcal{O}_{\Delta_{\mathbb{P}}}(d)}(E)$ as part of the commutative diagram of exact triangles

$$\begin{array}{ccccc}
\Phi_{\mathcal{Q}'_d}(E) & \longrightarrow & \Phi_{\mathcal{O}_{\Delta_X}(d)}(E) & \longrightarrow & \Phi_{\mathcal{Q}_d}(E)\simeq T^d(E)\\
\downarrow{\simeq} & & \downarrow{=} & & \vdots{\simeq}\\
\Phi_{\varphi^*\mathcal{O}_{\Delta_{\mathbb{P}}}(d)}(E) & \longrightarrow & \Phi_{\mathcal{O}_{\Delta_X}(d)}(E) & \longrightarrow & \Phi_{\mathcal{O}_{\Delta_X}[2]}(E)\simeq E[2].
\end{array}$$

This follows from the standard Koszul resolution of $\mathcal{O}_{\Delta_{\mathbb{P}}}$ tensored with $\mathcal{O}(d)\boxtimes\mathcal{O}$ and the observation that the Fourier–Mukai functor with kernel $\mathcal{O}_X(d-i)\boxtimes\Omega^i_{\mathbb{P}}(i)|_X$ applied to any $E\in\mathcal{A}_X$ is trivial for $i=d,\dots,n+1$, because then $H^*(X,E\otimes\mathcal{O}(d-i))=0$. □

Proof of Proposition 1.26. One combines the above results: raising the isomorphisms (1.7) in Remark 1.24 and Lemma 1.29 to the power d/c and $-(n+2-d)/c$, we obtain

$$S_{\mathcal{A}_X}^{d/c} \simeq T^{-(n+2-d)d/c} \circ [nd/c] \simeq [-2(n+2-d)/c] \circ [nd/c] \simeq [(n+2)(d-2)/c],$$

which is what we had to prove. □

Remark 1.30 Let us also explain how to view the projection $j^*\colon \mathrm{D^b}(X) \longrightarrow \mathcal{A}_X$ as a Fourier–Mukai functor. Indeed, there exists an object $P_0 \in \mathrm{D^b}(X \times X)$ such that the associated Fourier–Mukai functor

$$\Phi_{P_0}\colon \mathrm{D^b}(X) \longrightarrow \mathrm{D^b}(X), \quad E \longmapsto E \otimes p_{2*}(p_1^* E \otimes P_0)$$

is $\Phi_{P_0} \simeq j_* \circ j^*$. The Fourier–Mukai kernel P_0 is described as the projection of $\mathcal{O}_\Delta$, which is the Fourier–Mukai kernel of the identity functor, to the admissible subcategory

$$J_*\colon \mathcal{C} := \mathcal{A}_X(-(n+1-d)) \boxtimes \mathcal{A}_X \hookrightarrow \mathrm{D^b}(X \times X). \tag{1.9}$$

By definition, $\mathcal{C}$ is the right orthogonal category of the collection of all objects

$$\mathcal{O}_X(k) \boxtimes F,\ d-n-1 \le k \le 0, \quad \text{and} \quad F \boxtimes \mathcal{O}_X(\ell),\ 0 \le \ell \le n+1-d.$$

Alternatively, $\mathcal{C}_X$ is the intersection of $\mathcal{A}_X(-(n+1-d)) \boxtimes \mathrm{D^b}(X)$ and $\mathrm{D^b}(X) \boxtimes \mathcal{A}_X$.

Here, for any triangulated subcategory $\mathcal{B} \subset \mathrm{D^b}(X)$, one defines $\mathcal{B} \boxtimes \mathrm{D^b}(X) \subset \mathrm{D^b}(X \times X)$ as the full triangulated subcategory of all objects $E \in \mathrm{D^b}(X \times X)$ such that $p_{1*}(E \otimes p_2^* F) \in \mathcal{B}$ for all $F \in \mathrm{D^b}(X)$. The category $\mathrm{D^b}(X) \boxtimes \mathcal{B}$ is defined analogously. This theory has been developed by Kuznetsov in much broader generality [297].

For an object $P \in \mathrm{D^b}(X \times X)$ and its associated functor $\Phi_P\colon \mathrm{D^b}(X) \longrightarrow \mathrm{D^b}(X)$, cf. [292] or [255, Lem. 1.5], one then has:

(i) The object P is contained in $\mathrm{D^b}(X) \boxtimes \mathcal{A}_X$ if and only if the essential image of Φ_P is contained in $\mathcal{A}_X$.
(ii) The object P is contained in $\mathcal{A}_X(-(n+1-d)) \boxtimes \mathrm{D^b}(X)$ if and only if the functor Φ_P is the composition of $j^*\colon \mathrm{D^b}(X) \longrightarrow \mathcal{A}_X$ and a functor $\mathcal{A}_X \longrightarrow \mathrm{D^b}(X)$.

This leads to the Fourier–Mukai description of the left adjoint $j^*\colon \mathrm{D^b}(X) \longrightarrow \mathcal{A}_X$ as

$$j^* \simeq \Phi_{J^* \mathcal{O}_\Delta},$$

where $J^*\colon \mathrm{D^b}(X \times X) \longrightarrow \mathcal{C}$ is the left adjoint of (1.9).

1.7 Matrix Factorizations The residual category $\mathcal{A}_X = \langle \mathcal{O}_X, \dots, \mathcal{O}_X(n+1-d) \rangle^\perp \subset \mathrm{D^b}(X)$ admits an interpretation as a category of matrix factorizations. We briefly outline this side of the theory and provide references for further reading.

Example 1.31 Let us recall a few classical examples of matrix factorizations.

(i) The adjoint matrix of an invertible matrix $A \in \mathrm{GL}(n,k)$ is the matrix $A^{\mathrm{adj}} := (\det(A_{ji}))$ where A_{ij} is the ijth minor of the matrix A. The importance of this notion stems from the fact that

$$A \cdot A^{\mathrm{adj}} = \det(A) \cdot \mathrm{I}_n = A^{\mathrm{adj}} \cdot A.$$

In other words, the differentials ∂ in the two-periodic 'complex'

$$\cdots \longrightarrow k^n \xrightarrow{A^{\mathrm{adj}}} k^n \xrightarrow{A} k^n \xrightarrow{A^{\mathrm{adj}}} k^n \xrightarrow{A} \cdots$$

satisfy $\partial^2 = \det(A) \cdot \mathrm{id}$.

(ii) Let $F \in k[x_0,\dots,x_{n+1}]_d$ be a homogenous polynomial, defining a hypersurface $X \subset \mathbb{P}^{n+1}$ of degree d, and let K and L denote two copies of $S := k[x_0,\dots,x_{n+1}]$ as a module over itself. Then the differentials in the two-periodic sequence of S-modules

$$\cdots \longrightarrow L \xrightarrow{F} K \xrightarrow{\mathrm{id}} L \xrightarrow{F} K \xrightarrow{\mathrm{id}} \cdots$$

satisfy $\partial^2 = F \cdot \mathrm{id}$. If the grading is taken into account, this is written as

$$\cdots \longrightarrow L(-d) \xrightarrow{F} K \xrightarrow{\mathrm{id}} L \xrightarrow{F} K(d) \xrightarrow{\mathrm{id}} \cdots,$$

where for a graded S-module $M = \bigoplus M_i$ the graded S-module $M(m) := \bigoplus M(m)_i$ is defined by $M(m)_i := M_{m+i}$.

(iii) Let now $K := S$ and $L := K(d-1)^{\oplus n+2}$. Then the Euler equation $\sum x_i \cdot \partial_i F = d \cdot F$ for $F \in k[x_0,\dots,x_{n+1}]_d$ as before leads to another two-periodic sequence

$$\cdots \longrightarrow L(-d) \xrightarrow{(x_i)^t} K \xrightarrow{\frac{1}{d}(\partial_i F)} L \xrightarrow{(x_i)^t} K(d) \xrightarrow{\frac{1}{d}(\partial_i F)} \cdots.$$

Again, the differentials satisfy $\partial^2 = F \cdot \mathrm{id}$.

(iv) Let as before $F \in k[x_0,\dots,x_{n+1}]_d$ be a homogeneous polynomial and consider the graded quotient ring $R := S/(F)$. A finite R-module M is called *maximal Cohen–Macaulay* if $\mathrm{depth}_R(M) = \dim(R)$.

Although for us the hypersurface $X = V(F) \subset \mathbb{P}^{n+1}$ is typically smooth, the quotient ring R is not regular, as the affine hypersurface has a singularity at the origin. In particular, the Auslander–Buchsbaum formula does not necessarily hold and, thus, a maximal Cohen–Macaulay module may not be projective. However, the Auslander–Buchsbaum formula holds for M as an S-module and $\mathrm{depth}_S(M) = \mathrm{depth}_R(M)$, which altogether shows that a maximal Cohen–Macaulay module M over R admits a resolution

$$0 \longrightarrow K \xrightarrow{\alpha} L \longrightarrow M \longrightarrow 0$$

by finite free S-modules K and L. Since M is a module over $R = S/(F)$, multiplication with F defines an endomorphism of the complex $(K \longrightarrow L)$ which is homotopic to 0, i.e. there exists a morphism $\beta\colon L \longrightarrow K$ with $\beta \circ \alpha = F \cdot \mathrm{id} = \alpha \circ \beta$.

Thus, one associates with a maximal Cohen–Macaulay module over R a two-periodic complex

$$\cdots \longrightarrow L \xrightarrow{\beta} K \xrightarrow{\alpha} L \xrightarrow{\beta} K \xrightarrow{\alpha} \cdots$$

with $\partial^2 = F \cdot \mathrm{id}$. The stable category of maximal Cohen–Macaulay modules $\mathrm{MCM}(R)$ has as objects maximal Cohen–Macaulay modules M and as morphisms elements in $\mathrm{Hom}_R(M_1, M_2)/\sim$, where $g \sim g'\colon M_1 \longrightarrow M_2$ if $g - g'$ factors through a finite free R-module. The category $\mathrm{MCM}(R)$ was shown to be naturally triangulated by Buchweitz [97, Sec. 4.7].

Fix a homogenous polynomial $F \in S = k[x_0, \ldots, x_{n+1}]$ of degree d. Then the associated category of graded matrix factorizations $\mathrm{MF}(F, \mathbb{Z})$ is defined as follows: its objects are of the form

$$(K \xrightarrow{\alpha} L \xrightarrow{\beta} K(d)).$$

Here, K and L are finitely generated, graded, free S-modules, so isomorphic to a finite direct sum $\bigoplus S(n_i)$, and α and β are homomorphisms of graded S-modules such that the compositions satisfy

$$\beta \circ \alpha = F \cdot \mathrm{id} \quad \text{and} \quad \alpha(d) \circ \beta = F \cdot \mathrm{id}.$$

Morphisms in $\mathrm{MF}(F, \mathbb{Z})$ are homotopy classes of commutative diagrams

$$\begin{array}{ccccc} K & \xrightarrow{\alpha} & L & \xrightarrow{\beta} & K(d) \\ \downarrow{\scriptstyle\varphi} & & \downarrow{\scriptstyle\psi} & & \downarrow{\scriptstyle\varphi(d)} \\ K' & \xrightarrow[\alpha']{} & L' & \xrightarrow[\beta']{} & K'(d), \end{array}$$

where φ and ψ are homomorphisms of graded S-modules and $(\varphi, \psi) \sim 0$ if there exist graded S-module homomorphisms $h_K\colon K \longrightarrow L'(-d)$ and $h_L\colon L \longrightarrow K'$ such that

$$\varphi = h_L \circ \alpha + \beta'(-d) \circ h_K \quad \text{and} \quad \psi = h_K(d) \circ \beta + \alpha' \circ h_L.$$

For example, $(S \xrightarrow{\mathrm{id}} S \xrightarrow{F} S(d))$ is an object in $\mathrm{MF}(F, \mathbb{Z})$. However, since $(\mathrm{id}, \mathrm{id}) \sim 0$ via the homotopy $h_K = 0$ and $h_L = \mathrm{id}$, it is isomorphic to the zero object.

The category $\mathrm{MF}(F, \mathbb{Z})$ comes with a natural auto-equivalence, the shift functor,

$$[1]\colon \mathrm{MF}(F, \mathbb{Z}) \xrightarrow{\sim} \mathrm{MF}(F, \mathbb{Z})$$

that sends an object $(K \xrightarrow{\alpha} L \xrightarrow{\beta} K(d))$ to

$$(K \xrightarrow{\alpha} L \xrightarrow{\beta} K(d))[1] := (L \xrightarrow{-\beta} K(d) \xrightarrow{-\alpha(d)} L(d)).$$

The construction is similar to the one of the homotopy category of complexes and, indeed, with this structure the category $\mathrm{MF}(F, \mathbb{Z})$ becomes a triangulated k-linear category, see [376, Sec. 3.1] or [378, Sec. 3.1]. The exact triangles are those that are isomorphic to the standard triangles constructed by taking cones.

The fact that we work over a graded ring and with graded modules has not played a role so far. This comes in now. The category $\mathrm{MF}(F, \mathbb{Z})$ is endowed with an additional exact auto-equivalence, the *degree shift functor*

$$(1)\colon \mathrm{MF}(F, \mathbb{Z}) \xrightarrow{\sim} \mathrm{MF}(F, \mathbb{Z}),$$

that sends an object $(K \xrightarrow{\alpha} L \xrightarrow{\beta} K(d))$ to

$$(K \xrightarrow{\alpha} L \xrightarrow{\beta} K(d))(1) := (K(1) \xrightarrow{\alpha(1)} L(1) \xrightarrow{\beta(1)} K(d+1)).$$

If the auto-equivalences $[m]$ and (m) are defined as the m-fold compositions $[1] \circ \cdots \circ [1]$ and $(1) \circ \cdots \circ (1)$, then clearly

$$(d) \simeq [2],$$

which is reminiscent of Lemma 1.29, (i).

Remark 1.32 The category of matrix factorizations was first studied in the ungraded case. There also it leads to a triangulated category $\mathrm{MF}(F)$, which is naturally $\mathbb{Z}/2\mathbb{Z}$-graded. Observe that $(1) \simeq \mathrm{id} \simeq [2]$. Also, $(K \xrightarrow{\alpha} L \xrightarrow{\beta} K) \longmapsto \mathrm{Coker}(\alpha)$ describes an equivalence

$$\mathrm{MF}(F) \simeq \mathrm{MCM}(S/(F)).$$

Furthermore, Auslander [29] shows that $\mathrm{MF}(S/(F))$ is a Calabi–Yau category of dimension $\dim(X) = \dim(R) - 1$. An explicit description of the Calabi–Yau pairing was provided in physics terms by Kapustin and Li [267], see also [167, 363, 391].

For any object E in $\mathrm{MF}(F, \mathbb{Z})$, multiplication with $s \in k[x_0, \dots, x_{n+1}]_m$ defines a morphism $E \longrightarrow E(m)$. In other words, there exists a natural map

$$k[x_0, \dots, x_{n+1}]_m \longrightarrow \mathrm{Hom}(\mathrm{id}, (m)).$$

The crucial observation is that the induced map $k[x_0, \dots, x_{n+1}] \longrightarrow \mathrm{Hom}(\mathrm{id}, \bigoplus (m))$ factorizes via the surjection $k[x_0, \dots, x_{n+1}] \twoheadrightarrow R(F)$ onto the Jacobian ring, see Section 1.4.1. This is the next result.

Lemma 1.33 *For any object E in* $\mathrm{MF}(F,\mathbb{Z})$*, multiplication with $\partial_i F \in k[x_0,\ldots,x_{n+1}]$ defines the trivial morphism $E \longrightarrow E(d-1)$ in* $\mathrm{MF}(F,\mathbb{Z})$*.*

Proof If $E = (K \xrightarrow{\alpha} L \xrightarrow{\beta} K(d))$, then $(\varphi = \partial_i F, \psi = \partial_i F) \sim 0$ via the homotopy given by $h_L := \partial_i\beta\colon L \longrightarrow K(d-1)$ and $h_K := \partial_i\alpha\colon K(d) \longrightarrow L(d-1)$. Here, we think of α and β as matrices of polynomials to which partial derivatives can be applied. The verification of this observation is based on the product formula for derivatives, e.g. $F \cdot \mathrm{id} = \beta \circ \alpha$ implies $\partial_i F \cdot \mathrm{id} = \partial_i\beta \circ \alpha + \beta \circ \partial_i\alpha$. □

This eventually leads to a natural injection, see [255, Prop. 2.5]:

$$R(F) = k[x_0,\ldots,x_{n+1}]/(\partial_i F) \hookrightarrow \mathrm{Hom}(\mathrm{id}, \textstyle\bigoplus(m)). \tag{1.10}$$

The category of matrix factorizations $\mathrm{MF}(F,\mathbb{Z})$ seems a more concrete and manageable category associated with a homogenous polynomial F. However, by a result of Orlov [378, Thm. 2.13], this category is in fact equivalent to the Kuznetsov component.

Theorem 1.34 (Orlov) *Let $X \subset \mathbb{P}^{n+1}$ be a smooth hypersurface defined by a polynomial F of degree $d \leq n+2$. Then there exists an exact linear equivalence*

$$\mathrm{MF}(F,\mathbb{Z}) \simeq \mathcal{A}_X$$

between the category of matrix factorizations $\mathrm{MF}(F,\mathbb{Z})$ *and the Kuznetsov component* $\mathcal{A}_X = \langle \mathcal{O}_X,\ldots,\mathcal{O}_X(n+1-d)\rangle^\perp \subset \mathrm{D^b}(X)$.

Furthermore, the equivalence can be chosen such that the degree shift functor (1) *on* $\mathrm{MF}(F,\mathbb{Z})$ *corresponds to the functor T on $\mathcal{A}_X$, see Remark* 1.24 *and Lemma* 1.29.

For $d = n+2$, i.e. the case that ω_X is trivial, the assertion should be interpreted as saying $\mathrm{MF}(F,\mathbb{Z}) \simeq \mathrm{D^b}(X)$. There is also a corresponding statement for $d > n+2$. In this case, the roles are reversed and $\mathrm{D^b}(X)$ becomes an admissible subcategory of $\mathrm{MF}(F,\mathbb{Z})$.

The proof of this result is difficult and quite involved. It uses yet another natural triangulated category, namely the singularity category $\mathrm{D^b_{sing}}(\mathrm{Spec}(S/(F)))$ of the isolated singularity of the affine variety defined by F. The transition from one point of view to the other is not easy. For example, it is usually difficult to describe the matrix factorization that corresponds to an object in $\mathrm{D^b}(X)$, even to a very easy one like the structure sheaf $\mathcal{O}_X$ or a skyscraper sheaf $k(x)$. As we will not use the result in any essential way, we refrain from going into the details.

A similar equivalence in the ungraded case was established earlier by Buchweitz [97]:

$$\mathrm{MF}(F) \simeq \mathrm{MCM}(S/(F)) \simeq \mathrm{D^b_{sing}}(\mathrm{Spec}(S/(F))).$$

The crucial point here is a result of Eisenbud [173] that any finite $S/(F)$-module admits a free resolution that eventually becomes two-periodic.

The Calabi–Yau property for the $\mathbb{Z}/2\mathbb{Z}$-category $\mathrm{MF}(F)$ of ungraded matrix factorizations was established early on by Auslander [29], see also [167, 363, 391]. However, the corresponding result in the graded case, namely that $\mathrm{MF}(F,\mathbb{Z}) \simeq \mathcal{A}_X$ is a fractional Calabi–Yau category, see Proposition 1.26, has been proved relatively late and not directly for $\mathrm{MF}(F,\mathbb{Z})$ but for its geometric incarnation $\mathcal{A}_X$.

1.8 Derived Category of the Fano Variety Following the general philosophy of Section **2**.5, one should try to link $\mathrm{D^b}(X)$ and $\mathcal{A}_X$ via the Fano correspondence to the derived category of the Fano variety of lines $F(X)$.

The theory of Fourier–Mukai transformations suggests what to do: use the ideal sheaf $\mathcal{I}_{\mathbb{L}}$ of the universal line regarded as a subvariety $\mathbb{L} \subset F(X) \times X$ and define the Fourier–Mukai functor

$$\Phi := \Phi_{\mathcal{I}_{\mathbb{L}}} : \mathrm{D^b}(F(X)) \longrightarrow \mathrm{D^b}(X), \quad E \longmapsto q_*(p^*E \otimes \mathcal{I}_{\mathbb{L}}).$$

It maps $k([L])$ to the ideal sheaf $\mathcal{I}_L \in \mathrm{D^b}(X)$ and its left adjoint is the functor

$$\Psi \colon \mathrm{D^b}(X) \longrightarrow \mathrm{D^b}(F(X)), \quad G \longmapsto p_*(q^*G \otimes \mathcal{I}_{\mathbb{L}}^{\vee} \otimes \omega_X[3]).$$

However, these two functors do not behave particularly well. For example, Φ is neither full nor faithful. To see this, compare the dimensions of the two extension spaces $\mathrm{Ext}^i_{F(X)}(k([L]), k([L']))$ and $\mathrm{Ext}^i_X(\mathcal{I}_L, \mathcal{I}_{L'})$. Replacing $\mathcal{I}_{\mathbb{L}}$ by any of its twists $\mathcal{I}_{\mathbb{L}}(i)$ does not improve the situation, but see Section 3.6 for the case of cubic fourfolds.

The better strategy is to relate the two derived categories $\mathrm{D^b}(F(X))$ and $\mathrm{D^b}(X^{[2]})$ via the diagram (4.4) in Remark **2**.4.4

Applying Orlov's blow-up formula (0.3) to $B := \mathrm{Bl}_{\mathbb{L}}(\mathbb{L}_{\mathbb{G}}|_X) \simeq \mathrm{Bl}_{\mathbb{L}^{[2]}}(X^{[2]})$ leads to two semi-orthogonal decompositions

$$\langle \mathrm{D^b}(\mathbb{L})_{-2}, \mathrm{D^b}(\mathbb{L})_{-1}, \mathrm{D^b}(\mathbb{L}_{\mathbb{G}}|_X) \rangle = \mathrm{D^b}(B) = \langle \mathrm{D^b}(\mathbb{L}^{[2]})_{-1}, \mathrm{D^b}(X^{[2]}) \rangle. \tag{1.11}$$

Here, using the notation from Section **2**.4.5, we denote by $\mathrm{D^b}(\mathbb{L})_{-2}$ the image of the fully faithful functor $\mathrm{D^b}(\mathbb{L}) \hookrightarrow \mathrm{D^b}(B)$, $F \longmapsto j_*(\tau_1^*F \otimes \mathcal{O}(2E)|_E)$ and by $\mathrm{D^b}(\mathbb{L}^{[2]})_{-1}$ the image of the fully faithful functor $\mathrm{D^b}(\mathbb{L}^{[2]}) \hookrightarrow \mathrm{D^b}(B)$, $F \longmapsto j_*(\tau_2^*F \otimes \mathcal{O}(E)|_E)$, etc.

Furthermore, applying Orlov's formula (0.1) to the projective bundles

$$\mathbb{L}_{\mathbb{G}}|_X \longrightarrow X, \quad p \colon \mathbb{L} \longrightarrow F(X), \quad \text{and} \quad p^{[2]} \colon \mathbb{L}^{[2]} \longrightarrow F(X)$$

with fibres $\mathbb{P}^n$, $\mathbb{P}^1$, and $\mathbb{P}^2$, gives rise to three semi-orthogonal decompositions:

$$\mathrm{D^b}(\mathbb{L}_{\mathbb{G}}|_X) = \langle \mathrm{D^b}(X), \mathrm{D^b}(X)(1), \ldots, \mathrm{D^b}(X)(n) \rangle, \quad \mathrm{D^b}(\mathbb{L}) = \langle \mathrm{D}(F(X)), \mathrm{D^b}(F(X))(1) \rangle,$$

$$\text{and } \mathrm{D^b}(\mathbb{L}^{[2]}) = \langle \mathrm{D^b}(F(X)), \mathrm{D^b}(F(X))(1), \mathrm{D^b}(F(X))(2) \rangle.$$

Inserted into (1.11) one obtains semi-orthogonal decompositions

$$\langle \mathcal{D}_1, \ldots, \mathcal{D}_4, \mathcal{E}_0, \ldots, \mathcal{E}_n \rangle = \mathrm{D^b}(B) = \langle \mathcal{D}_5, \mathcal{D}_6, \mathcal{D}_7, \mathrm{D^b}(X^{[2]}) \rangle$$

with $\mathcal{D}_i \simeq \mathrm{D}^{\mathrm{b}}(F(X))$ and $\mathcal{E}_i \simeq \mathrm{D}^{\mathrm{b}}(X)$. This implies equality in the appropriate Grothendieck group as we explain next.

Bondal, Larsen, and Lunts [84] introduced the Grothendieck ring $K_0(\text{dg-cat}_k)$ of (dg-enhanced) triangulated categories over a field k. This ring is generated by classes $[\mathcal{D}]$ for each such category $\mathcal{D}$ with one relation $[\mathcal{D}] \sim [\mathcal{D}_1] + [\mathcal{D}_2]$ for each semi-orthogonal decomposition $\mathcal{D} = \langle \mathcal{D}_1, \mathcal{D}_2 \rangle$. The next result, cf. [394], is an immediate consequence and should be viewed as an analogue of the following three facts:

(i) $\ell^2 \cdot [F(X)] + [\mathbb{P}^n] \cdot [X] = [X^{[2]}]$ in $K_0(\mathrm{Var}_k)$, see Proposition **2**.4.2,

(ii) $\mathfrak{h}(F(X))(-2) \oplus \bigoplus_{i=0}^n \mathfrak{h}(X)(-i) \simeq \mathfrak{h}(X^{[2]})$ in $\mathrm{Mot}(k)$, see (4.8) in Section **2**.4.2, and

(iii) $H^*(F(X), \mathbb{Q}) \oplus (H^*(\mathbb{P}^n, \mathbb{Q}) \otimes H^*(X, \mathbb{Q}))(2) \simeq H^*(X^{[2]}, \mathbb{Q})(2)$ in $\mathrm{HS}_{\mathbb{Q}}$, see (4.19) in Section **2**.4.4.

Corollary 1.35 *The following equation*

$$[\mathrm{D}^{\mathrm{b}}(F(X))] + (n+1) \cdot [\mathrm{D}^{\mathrm{b}}(X)] = [\mathrm{D}^{\mathrm{b}}(X^{[2]})] \tag{1.12}$$

holds in the Grothendieck ring $K_0(\text{dg-cat}_k)$ *of triangulated categories.* □

Remark 1.36 The equality (1.12) can alternatively be deduced directly from the equation in (i), which is an equation in $K_0(\mathrm{Var}_k)$, by applying a result of Bondal–Lunts–Larson asserting the existence of a ring homomorphism $K_0(\mathrm{Var}_k) \longrightarrow K_0(\text{dg-cat}_k)$ that sends a smooth projective variety Z to the class of its derived category $[\mathrm{D}^{\mathrm{b}}(Z)]$, see [84, Prop. 7.5].

The corollary has recently been upgraded to an actual semi-orthogonal decomposition.

Proposition 1.37 (Belmans–Fu–Raedschelders) *For a smooth cubic hypersurface* $X \subset \mathbb{P}^{n+1}$, $n \geq 0$, *there exists a semi-orthogonal decomposition*

$$\mathrm{D}^{\mathrm{b}}(X^{[2]}) = \langle \mathrm{D}^{\mathrm{b}}(F(X)), \mathcal{E}_0, \ldots, \mathcal{E}_n \rangle \tag{1.13}$$

with $\mathcal{E}_i \simeq \mathrm{D}^{\mathrm{b}}(X)$.

Proof We follow the proof in [62, App. A]. The idea is to realize the two categories $\mathrm{D}^{\mathrm{b}}(F)$ and $\mathrm{D}^{\mathrm{b}}(X^{[2]})$ as subcategories $\mathrm{D}^{\mathrm{b}}(F) \subset \mathrm{D}^{\mathrm{b}}(B)$ and $\mathrm{D}^{\mathrm{b}}(X^{[2]}) \subset \mathrm{D}^{\mathrm{b}}(B)$ via

$$B \xleftarrow{\;j\;} E \xrightarrow{\;\tau_1\;} \mathbb{L} \xrightarrow{\;p\;} F(X) \;\text{ and }\; B \longrightarrow X^{[2]}$$

and then apply the formal mutation formalism.

To ease the notation, let $\mathcal{D} := \mathrm{D}^{\mathrm{b}}(F(X))$ and denote by $\mathcal{D}(a,b) \subset \mathrm{D}^{\mathrm{b}}(B)$ the full subcategory given by the image of

$$\mathrm{D}^{\mathrm{b}}(F(X)) \longrightarrow \mathrm{D}^{\mathrm{b}}(B),\; F \longmapsto j_*(\tau_1^* p^*(F) \otimes \mathcal{O}(a,b))$$

or, equivalently, as the image of

$$\mathrm{D^b}(F(X)) \longrightarrow \mathrm{D^b}(B), \quad F \longmapsto j_*(\tau_2^* p^{[2]*}(F) \otimes \mathcal{O}(a,b)),$$

where $\mathcal{O}(a,b) := \tau_1^*\mathcal{O}_p(a) \otimes \tau_2^*\mathcal{O}_{p^{[2]}}(b) \in \mathrm{Pic}(E)$.

It is not difficult to check that both functors define equivalences $\mathcal{D} \xrightarrow{\sim} \mathcal{D}(a,b)$. Also, for later use, observe that $\mathcal{O}(E)|_E \simeq \mathcal{O}(-1,-1)$ up to a line bundle on $F(X)$, which is of no importance in what follows.

Now, the right-hand side of (1.11) is refined to

$$\mathrm{D^b}(B) = \langle \mathrm{D^b}(\mathbb{L}^{[2]})_{-1}, \mathrm{D^b}(X^{[2]})\rangle = \langle \mathcal{D}(-1,-2), \mathcal{D}(-1,-1), \mathcal{D}(-1,0), \mathrm{D^b}(X^{[2]})\rangle,$$

where we use (0.1) to write $\mathrm{D^b}(\mathbb{L}^{[2]})_{-1} = \langle \mathcal{D}(0,-1), \mathcal{D}(0,0), \mathcal{D}(0,1)\rangle \otimes \mathcal{O}(E)$.

Next, one refines the left-hand side of (1.11), using $\mathrm{D^b}(\mathbb{L})_{-2} = \langle \mathcal{D}(0,0), \mathcal{D}(1,0)\rangle \otimes \mathcal{O}(2E)$ and $\mathrm{D^b}(\mathbb{L})_{-1} = \langle \mathcal{D}, \boldsymbol{\mathcal{D}}(1,0)\rangle \otimes \mathcal{O}(E)$,[5] and starts mutating while applying the rules in Exercise 1.11:

$$\begin{aligned}
\mathrm{D^b}(B) &= \langle \mathrm{D^b}(\mathbb{L})_{-2}, \mathrm{D^b}(\mathbb{L})_{-1}, \mathrm{D^b}(\mathbb{L}_{\mathbb{G}}|_X)\rangle \\
&= \langle \mathcal{D}(-2,-2), \underbrace{\mathcal{D}(-1,-2), \mathcal{D}(-1,-1), \boldsymbol{\mathcal{D}}(0,-1), \mathrm{D^b}(\mathbb{L}_{\mathbb{G}}|_X)}_{=:\mathcal{D}'}\rangle \\
&= \langle \mathcal{D}', \mathbb{R}_{\mathcal{D}'}\mathcal{D}(-2,-2)\rangle \\
&= \langle \mathcal{D}(-1,-2), \mathcal{D}(-1,-1), \boldsymbol{\mathcal{D}}(0,-1), \mathrm{D^b}(\mathbb{L}_{\mathbb{G}}|_X), \mathcal{D}(-1,0)\rangle \\
&= \langle \mathcal{D}(-1,-2), \mathcal{D}(-1,-1), \boldsymbol{\mathcal{D}}(0,-1), \mathcal{D}(-1,0), \mathbb{R}_{\mathcal{D}(-1,0)}\mathrm{D^b}(\mathbb{L}_{\mathbb{G}}|_X)\rangle \\
&= \langle \mathcal{D}(-1,-2), \mathcal{D}(-1,-1), \mathcal{D}(-1,0), \mathbb{R}_{\mathcal{D}(-1,0)}\boldsymbol{\mathcal{D}}(0,-1), \mathbb{R}_{\mathcal{D}(-1,0)}\mathrm{D^b}(\mathbb{L}_{\mathbb{G}}|_X)\rangle.
\end{aligned}$$

Here, we used $\mathbb{R}_{\mathcal{D}'}\mathcal{D}(-2,-2) = \mathcal{S}_B^{-1}\mathcal{D}(-2,-2) = \mathcal{D}(-1,0)$, which is a consequence of $\omega_B|_E \simeq \mathcal{O}(-1,-2)$ up to line bundles on $F(X)$ and Lemma 1.14. Altogether, one obtains

$$\mathrm{D^b}(X^{[2]}) = {}^{\perp}\langle \mathcal{D}(-1,-2), \mathcal{D}(-1,-1), \mathcal{D}(-1,0)\rangle = \langle \mathbb{R}_{\mathcal{D}(-1,0)}\boldsymbol{\mathcal{D}}(0,-1), \mathbb{R}_{\mathcal{D}(-1,0)}\mathrm{D^b}(\mathbb{L}_{\mathbb{G}}|_X)\rangle.$$

We conclude by using the two equivalences $\mathbb{R}_{\mathcal{D}(-1,0)}\mathrm{D^b}(\mathbb{L}_{\mathbb{G}}|_X) \simeq \langle \mathrm{D^b}(X), \ldots, \mathrm{D^b}(X)(n)\rangle$ and $\mathbb{R}_{\mathcal{D}(-1,0)}\boldsymbol{\mathcal{D}}(0,-1) \simeq \mathcal{D}(0,-1) \simeq \mathrm{D^b}(F(X))$.[6] □

Remark 1.38 In [62] it is also observed that for any smooth cubic hypersurface $X \subset \mathbb{P}^{n+1}$ of dimension at least three, the Hilbert scheme $X^{[2]}$ is a Fano variety, i.e. its canonical bundle is anti-ample.

[5] From now on, the component that will eventually provide the copy of $\mathrm{D^b}(F(X))$ in $\mathrm{D^b}(X^{[2]})$ appears in bold face.

[6] This composition of isomorphism describes the embedding $\mathrm{D^b}(F(X)) \hookrightarrow \mathrm{D^b}(X^{[2]})$ in the assertion of the proposition. Due to the right mutation, it is not obvious how to make this more explicit, e.g. for objects of the form $k([L])$.

This allows one to use (1.13) to say that $F(X)$ is a Fano visitor of the Fano host $X^{[2]}$. Of course, this is only an interesting statement for $n = 3$ and $n = 4$, as for $n \geq 5$ the variety $F(X)$ itself is a Fano variety, see Section **2**.3.1.

Note that every smooth projective variety is conjectured to be a Fano visitor, but this has been proved only in few cases, e.g. curves, complete intersections, cf. Remark 3.15.

See Remark 3.28 for a link between $\mathrm{D^b}(F(X))$ and the categorical symmetric square $\mathcal{A}_X^{[2]}$ of $\mathcal{A}_X$ for cubic fourfolds.

2 Cubic Surfaces and Cubic Threefolds

From dimension three on, the Kuznetsov component sheds new light on certain geometric phenomena encountered before. The role of the Kuznetsov component is clarified by the derived global Torelli theorem of Bernardara, Macrì, Mehrotra, and Stellari [65] asserting that the Kuznetsov component $\mathcal{A}_Y \subset \mathrm{D^b}(Y)$ determines the cubic threefold Y, see Proposition 2.4. This is not totally surprising in view of the Bondal–Orlov theorem that $\mathrm{D^b}(Y)$ determines Y, cf. Theorem 1.1. However, for cubic fourfolds the situation is more subtle which in turn makes the derived global Torelli for cubic threefolds quite interesting. We will also sketch a Hodge theoretic approach to the result inspired by techniques of Addington and Thomas [8] for cubic fourfolds, see Remark 2.6.

From the perspective of the global Torelli theorem, $\mathcal{A}_Y$ should be seen as a categorical version of the intermediate Jacobian $J(Y)$ or, equivalently, of the Prym variety $\mathrm{Prym}(C_L/D_L)$, see Section **5**.3.2. The Hodge theoretic reflection of this point of view is an isomorphism of Hodge structures of weight one:

$$\widetilde{H}^1(\mathcal{A}_Y, \mathbb{Z}) \simeq H^3(Y, \mathbb{Z})(1) \simeq H^1(\mathrm{Prym}(C_L/D_L), \mathbb{Z}) \simeq H^1(C_L, \mathbb{Z})^-.$$

In light of the discussion in Section **5**.5.1, it is natural to wonder whether nodal cubic threefolds allow for a similar, and perhaps easier, treatment. However, Kalck, Pavic, and Shinder [264] show that in this case $\mathrm{D^b}(Y)$ does not admit a decomposition in which the category $\mathcal{A}_Y$ would be contained in the category of perfect complexes.

2.1 Cubic Surfaces The theory is most interesting and most developed for cubic threefolds and cubic fourfolds. For cubic surfaces it is much easier but still quite interesting. Here are a few comments.

Viewing S as a blow-up $\tau\colon S \simeq \mathrm{Bl}_{\{x_i\}}(\mathbb{P}^2) \longrightarrow \mathbb{P}^2$ in six points $x_1, \ldots, x_6 \in \mathbb{P}^2$, the semi-orthogonal decomposition $\mathrm{D^b}(\mathbb{P}^2) = \langle \mathcal{O}(-2), \mathcal{O}(-1), \mathcal{O} \rangle$, see Example 1.9, together with Orlov's blow-up formula (0.3) immediately provides the semi-orthogonal decomposition

$$\mathrm{D^b}(S) = \left\langle \bigoplus_{i=1}^{6} \mathcal{O}_{E_i}(-1), \mathcal{O}_S(-2), \mathcal{O}_S(-1), \mathcal{O}_S \right\rangle \tag{2.1}$$

and hence a semi-orthogonal decomposition of the Kuznetsov component

$$\mathcal{A}_S = \left\langle \bigoplus_{i=1}^{6} \mathcal{O}_{E_i}(-1), \mathcal{O}_S(-2), \mathcal{O}_S(-1) \right\rangle.$$

Here, $\mathcal{O}_S(i) = \tau^*\mathcal{O}(i)$ and $E_i \simeq \mathbb{P}^1$ are the six exceptional lines.

2.2 Kuznetsov Component As a Category of Clifford Sheaves We begin by a rather technical result that allows one to view the Kuznetsov component $\mathcal{A}_Y \subset \mathrm{D^b}(Y)$, a fractional Calabi–Yau (5/3)-category, as a part of another semi-orthogonal decomposition.

Let $L \subset Y$ be a line and $\tau\colon \tilde{Y} := \mathrm{Bl}_L(Y) \longrightarrow Y$ its blow-up. By $\pi\colon E \longrightarrow L$ we denote the exceptional divisor and its projection to L. As in Section **1**.5.2, we consider the projection from L as a conic fibration $\phi\colon \tilde{Y} \longrightarrow \mathbb{P}^2$.

By applying a right mutation to Orlov's blow-up formula (0.3), one obtains the semi-orthogonal decomposition $\mathrm{D^b}(\tilde{Y}) = \langle \mathrm{D^b}(Y), \mathrm{D^b}(L)_{-1} \otimes \omega_{\tilde{Y}}^{-1} \rangle$, see Exercise 1.13. Since $\omega_{\tilde{Y}} \simeq \tau^*\omega_Y \otimes \mathcal{O}(E)$, this becomes $\mathrm{D^b}(\tilde{Y}) = \langle \mathrm{D^b}(Y), j_*\pi^*\mathrm{D^b}(L) \rangle$. Combined with $\mathrm{D^b}(L) = \langle \mathcal{O}_L, \mathcal{O}_L(1) \rangle$ and $\mathrm{D^b}(Y) = \langle \mathcal{A}_Y, \mathcal{O}_Y, \mathcal{O}_Y(1) \rangle$, this leads to

$$\mathrm{D^b}(\tilde{Y}) = \langle \mathcal{A}_Y, \mathcal{O}_Y, \mathcal{O}_Y(1), \mathcal{O}_L, \mathcal{O}_L(1) \rangle. \tag{2.2}$$

In the notation we omit the pullback and simply write $\mathcal{O}_Y(1)$ for $\tau^*\mathcal{O}_Y(1)$, $\mathcal{O}_{\mathbb{P}^2}(1)$ for $\phi^*\mathcal{O}_{\mathbb{P}^2}(1)$, $\mathcal{O}_L(1)$ for $j_*\pi^*\mathcal{O}_L(1)$, etc.

The next step consists of applying a number of left mutations, we follow [65, Sec. 2.1]. Exploiting the description (1.4) of the right mutation functor, the isomorphism $\mathcal{O}_{\mathbb{P}^2}(1) \simeq \mathcal{O}_Y(1)(-E)$, and the rules in Exercise 1.11, we find

$$\begin{aligned} \mathrm{D^b}(\tilde{Y}) &= \langle \mathcal{A}_Y, \mathcal{O}_Y, \mathcal{O}_Y(1), \mathcal{O}_L, \mathcal{O}_L(1) \rangle & (2.3)\\ &= \langle \mathcal{A}_Y, \mathcal{O}_Y, \mathbb{L}_{\mathcal{O}_Y(1)}\mathcal{O}_L \simeq \mathcal{O}_L, \mathcal{O}_Y(1), \mathcal{O}_L(1) \rangle \\ &= \langle \mathcal{A}_Y, \mathbb{L}_{\mathcal{O}_Y}\mathcal{O}_L \simeq \mathcal{O}(-E)[1], \mathcal{O}_Y, \mathcal{O}_Y(1), \mathcal{O}_L(1) \rangle \\ &= \langle \mathcal{A}_Y, \mathcal{O}(-E), \mathcal{O}_Y, \mathbb{L}_{\mathcal{O}_Y(1)}\mathcal{O}_L(1) \simeq \mathcal{O}_{\mathbb{P}^2}(1)[1], \mathcal{O}_Y(1) \rangle \\ &= \langle \mathcal{A}_Y, \mathcal{O}(-E), \mathcal{O}_Y, \mathcal{O}_{\mathbb{P}^2}(1), \mathcal{O}_Y(1) \rangle. & (2.4) \end{aligned}$$

This semi-orthogonal decomposition will be compared with the one that can be produced starting with the following result. For the definition of $\mathcal{C}_0$, see the proof of the next proposition.

Proposition 2.1 (Kuznetsov) *For the sheaf of even Clifford algebras $\mathcal{C}_0$ on $\mathbb{P}^2$ determined by $Y \subset \mathbb{P}^4$ there exists a semi-orthogonal decomposition*

$$\mathrm{D^b}(\tilde{Y}) = \langle \mathrm{D^b}(\mathbb{P}^2, \mathcal{C}_0), \mathrm{D^b}(\mathbb{P}^2) \rangle = \langle \mathrm{D^b}(\mathbb{P}^2, \mathcal{C}_0), \mathcal{O}_{\mathbb{P}^2}(-1), \mathcal{O}_{\mathbb{P}^2}, \mathcal{O}_{\mathbb{P}^2}(1) \rangle. \tag{2.5}$$

Proof The assertion in [294, Thm. 4.2] is much more general and its proof quite technical. We can only give a rough idea.

As a first approximation, the result should be seen as a combination of

$$\mathrm{D^b}(Q) = \langle \mathcal{A}_Q, \mathcal{O}_Q \rangle \quad \text{and} \quad \mathrm{D^b}(\mathbb{P}) = \langle \mathrm{D^b}(\mathbb{P}^2, \alpha), \mathrm{D^b}(X) \rangle,$$

where the first one is the semi-orthogonal decomposition of a conic $Q \subset \mathbb{P}^2$, see Example 1.20, and the second one is the decomposition for a Brauer–Severi variety $\mathbb{P} \longrightarrow \mathbb{P}^2$ with a Brauer class α of order two, cf. (0.2). The singular fibres of $\tilde{Y} \longrightarrow \mathbb{P}^2$ make the situation more complicated. So, the component $\mathrm{D^b}(\mathbb{P}^2, \mathcal{C}_0)$ should be seen as the relative version of $\mathcal{A}_Q$ and, at the same time, as a generalization of $\mathrm{D^b}(\mathbb{P}^2, \alpha)$. Its left orthogonal $\mathrm{D^b}(\mathbb{P}^2)$ is the relative version of $\mathcal{O}_Q$.

For some more details, recall first that $\tilde{Y} \subset \mathbb{P}(\mathcal{F}^*)$ is cut out by a symmetric equation $q\colon \mathcal{F}^* \longrightarrow \mathcal{F} \otimes \mathcal{O}(1)$ on $\mathbb{P}^2$ with $\mathcal{F} \simeq \mathcal{O}(1) \oplus \mathcal{O}^{\oplus 2}$, see Section **1**.5.2. The associated sheaf of Clifford algebras splits in an even and an odd part $\mathcal{C} = \mathcal{C}_0 \oplus \mathcal{C}_1$. Explicitly,

$$\mathcal{C}_0 \simeq \mathcal{O} \oplus (\textstyle\bigwedge^2 \mathcal{F}^*)(-1) \simeq \mathcal{O} \oplus \mathcal{O}(-1) \oplus \mathcal{O}(-2)^{\oplus 2} \quad \text{and}$$

$$\mathcal{C}_1 \simeq \mathcal{F}^* \oplus (\textstyle\bigwedge^3 \mathcal{F}^*)(-1) \simeq \mathcal{O}^{\oplus 2} \oplus \mathcal{O}(-1) \oplus \mathcal{O}(-2).$$

Clifford multiplication leads to a natural map $\varphi^*\mathcal{C}_0 \otimes \mathcal{O}_\varphi(-2) \longrightarrow \varphi^*\mathcal{C}_1 \otimes \mathcal{O}_\varphi(-1)$ on $\mathbb{P}(\mathcal{F}^*)$, where $\varphi\colon \mathbb{P}(\mathcal{F}^*) \longrightarrow \mathbb{P}^2$ is the projection. It is straightforward to check that this map drops rank only along $\tilde{Y} \subset \mathbb{P}(\mathcal{F}^*)$ and there exactly by two, cf. [294, Lem. 2.5]. This provides us with a short exact sequence

$$0 \longrightarrow \varphi^*\mathcal{C}_0 \otimes \mathcal{O}_\varphi(-2) \longrightarrow \varphi^*\mathcal{C}_1 \otimes \mathcal{O}_\varphi(-1) \longrightarrow \mathcal{E} \longrightarrow 0,$$

where $\mathcal{E}$ is a locally free sheaf of rank two on $\tilde{Y}$ and by construction a module over $\varphi^*\mathcal{C}_0$. Restricted to a fibre $P_y := \mathbb{P}(\mathcal{F}^*(y)) \simeq \mathbb{P}^2$ the short exact sequence is of the form $0 \longrightarrow \mathcal{O}_{P_y}(-2)^{\oplus 4} \longrightarrow \mathcal{O}_{P_y}(-1)^{\oplus 4} \longrightarrow \mathcal{E}_y \longrightarrow 0$ with $\mathcal{E}_y$ a locally free sheaf of rank two on the conic $Q_y \subset P_y \simeq \mathbb{P}^2$. Note that in particular $H^*(Q_y, \mathcal{E}_y) = 0$ and hence $\phi_*\mathcal{E} = 0$.

The functor $\mathrm{D^b}(\mathbb{P}^2) \longrightarrow \mathrm{D^b}(\tilde{Y})$, $F \longmapsto \phi^*F$ is fully faithful, since $\phi_*\mathcal{O}_{\tilde{Y}} \simeq \mathcal{O}_{\mathbb{P}^2}$. This describes the second component of (2.5). Similarly, one defines

$$\mathrm{D^b}(\mathbb{P}^2, \mathcal{C}_0) \longrightarrow \mathrm{D^b}(\tilde{Y}), \quad F \longmapsto \phi^*F \otimes_{\phi^*\mathcal{C}_0} \mathcal{E}$$

and shows that it is fully faithful. It is also not difficult to prove that with these definitions $\mathrm{D^b}(\mathbb{P}^2, \mathcal{C}_0) \subset \mathrm{D^b}(\mathbb{P}^2)^\perp$, which essentially follows from adjunction and $\phi_*\mathcal{E} = 0$.

The proof is then concluded by showing that the two subcategories together span $\mathrm{D^b}(\tilde{Y})$, which is a non-trivial statement. □

We now modify (2.5) by a number of right mutations, following again [65, Sec. 2.1]:

$$\begin{aligned}\mathrm{D^b}(\tilde{Y}) &= \langle \mathrm{D^b}(\mathbb{P}^2, \mathcal{C}_0), \mathcal{O}_{\mathbb{P}^2}(-1), \mathcal{O}_{\mathbb{P}^2}, \mathcal{O}_{\mathbb{P}^2}(1)\rangle \\ &= \langle \mathcal{O}_{\mathbb{P}^2}(-1), \underbrace{\mathbb{R}_{\mathcal{O}_{\mathbb{P}^2}(-1)}\mathrm{D^b}(\mathbb{P}^2, \mathcal{C}_0), \mathcal{O}_{\mathbb{P}^2}, \mathcal{O}_{\mathbb{P}^2}(1)}_{=:\mathcal{D}_1}\rangle \\ &= \langle \mathcal{D}_1, \mathcal{O}_{\mathbb{P}^2}(-1) \otimes \omega^*_{\tilde{Y}}\rangle \\ &= \langle \mathbb{R}_{\mathcal{O}_{\mathbb{P}^2}(-1)}\mathrm{D^b}(\mathbb{P}^2, \mathcal{C}_0), \mathcal{O}_{\mathbb{P}^2}, \mathcal{O}_{\mathbb{P}^2}(1), \mathcal{O}_Y(1)\rangle. \end{aligned} \tag{2.6}$$

Since by our convention $\mathcal{O}_Y \simeq \mathcal{O}_{\mathbb{P}^2}$ and clearly $\mathbb{R}_{\mathcal{O}_{\mathbb{P}^2}(-1)}\mathrm{D^b}(\mathbb{P}^2, \mathcal{C}_0) \simeq \mathrm{D^b}(\mathbb{P}^2, \mathcal{C}_0)$, comparing the results of the two mutations procedures (2.4) and (2.6) implies the following.

Corollary 2.2 *There exists a natural semi-orthogonal decomposition*

$$\mathrm{D^b}(\mathbb{P}^2, \mathcal{C}_0) = \langle \mathcal{A}_Y, \mathcal{O}(-E)\rangle.$$ □

This new presentation of $\mathcal{A}_Y$ as an admissible category has several advantages. For example, it allows one to prove the existence of stability conditions and, in particular, of bounded t-structures. This is the content of the next result [65], which will play a key technical role in the proof of the derived global Torelli theorem.

Corollary 2.3 (Bernardara–Macrì–Mehrotra–Stellari) *The Kuznetsov component $\mathcal{A}_Y$ of a smooth cubic threefold $Y \subset \mathbb{P}^4$ admits a bounded t-structure with heart $\mathcal{C}_Y$. Furthermore, $\mathcal{C}_Y$ can be chosen such that $\mathcal{I}_L \in \mathcal{C}_Y$ for every line $L \subset Y$ and* $\mathrm{Ext}^j(E_1, E_2) = 0$ *for all $E_1, E_2 \in \mathcal{C}_Y$ and $j > 2$.*

Proof The proof is quite ingenious, but too technical to be presented here in full without entering into the theory of stability conditions.

The main idea of [65, Sec. 3.2] is the following: as a derived category, $\mathrm{D^b}(\mathbb{P}^2, \mathcal{C}_0)$ has a natural bounded t-structure with heart $\mathrm{Coh}(\mathbb{P}^2, \mathcal{C}_0)$. This t-structure can be tilted with respect to a well-chosen torsion theory defined in terms of slope stability for sheaves on $\mathbb{P}^2$. The result is a bounded t-structure with a heart $\mathrm{Coh}^\sharp \subset \mathrm{D^b}(\mathbb{P}^2, \mathcal{C}_0)$ such that all objects in $\mathrm{Coh}^\sharp$ are concentrated in degree -1 and 0. A computation reveals that the image of $\mathcal{I}_L \in \mathcal{A}_Y$ under the inclusion $\mathcal{A}_Y \subset \mathrm{D^b}(\mathbb{P}^2, \mathcal{C}_0)$ is a sheaf contained in the torsion part of the torsion theory defining $\mathrm{Coh}^\sharp$, i.e. $\mathcal{I}_L \in \mathrm{Coh}^\sharp \cap \mathrm{Coh}(\mathbb{P}^2, \mathcal{C}_0)$. Since $\mathbb{P}^2$ is a surface, Ext-groups between objects in $\mathrm{Coh}(\mathbb{P}^2, \mathcal{C}_0)$ and also in $\mathrm{Coh}^\sharp$ are supported in degree 0, 1, and 2. The proof is concluded by defining $\mathcal{C}_Y$ as the subcategory $\mathcal{A}_Y \cap \mathrm{Coh}^\sharp$ and showing that it is the heart of a bounded t-structure on $\mathcal{A}_Y$. □

2.3 Categorical Global Torelli Theorem According to the result of Bondal and Orlov, see Theorem 1.1, the bounded derived category $\mathrm{D^b}(Y)$ of a smooth cubic threefold determines the cubic Y uniquely. From this perspective, the next result, the main result of [65], which replaces $\mathrm{D^b}(Y)$ by $\mathcal{A}_Y$, is maybe not so surprising. However, in dimension four the Kuznetsov component stops determining the cubic uniquely, which draws a clear line between cubics of dimension three and those of dimension four.

Proposition 2.4 (Bernardara–Macrì–Mehrotra–Stellari) *Assume $Y, Y' \subset \mathbb{P}^4$ are two smooth cubic threefolds. Then there exists a linear exact equivalence $\mathcal{A}_Y \simeq \mathcal{A}_{Y'}$ between their Kuznetsov components if and only if $Y \simeq Y'$.*

Proof We sketch the original proof [65]. The Grothendieck group $K(Y) = K(\mathrm{D^b}(Y))$ is naturally endowed with the non-symmetric quadratic form $\chi(\ ,\ )$. Its quotient by the radical $\{\alpha \mid \chi(\ ,\alpha) \equiv 0\}$ is by definition the numerical Grothendieck group $N(\mathrm{D^b}(Y))$, which is of rank four, since the algebraic cohomology of Y is $H^{\mathrm{ev}}(Y, \mathbb{Q}) = (H^0 \oplus H^2 \oplus H^4 \oplus H^6)(Y, \mathbb{Q}) \simeq \mathbb{Q}^{\oplus 4}$.

An explicit identification is described by the Mukai vector

$$N(\mathrm{D^b}(Y)) \otimes \mathbb{Q} \xrightarrow{\sim} H^{\mathrm{ev}}(Y, \mathbb{Q}), \qquad [E] \longmapsto v(E) := \mathrm{ch}(E) \cdot \sqrt{\mathrm{td}(Y)}.$$

Similarly, the numerical Grothendieck group of $\mathcal{A}_Y$ can be defined as the quotient of $K(\mathcal{A}_Y)$ by its radical or, alternatively, as the subgroup of $N(\mathrm{D^b}(Y))$

$$N(\mathcal{A}_Y) := \{\, [\mathcal{O}_Y], [\mathcal{O}_Y(1)] \,\}^{\perp} := \{\, [E] \mid \chi(\mathcal{O}_Y, E) = \chi(\mathcal{O}_Y(1), E) = 0 \,\},$$

which is of rank two. To compute the pairing on $N(\mathcal{A}_Y)$, recall from Exercise 1.21 that for a line $L \subset Y$ we have $\mathcal{I}_L \in \mathcal{A}_Y$ and consider the two classes $[\mathcal{I}_L]$ and $[\mathcal{S}_{\mathcal{A}_Y}(\mathcal{I}_L)]$. For example,

$$\begin{aligned}
\chi(\mathcal{I}_L, \mathcal{I}_L) &= \chi([\mathcal{O}_Y] - [\mathcal{O}_L], [\mathcal{O}_Y] - [\mathcal{O}_L]) \\
&= \chi(\mathcal{O}_Y, \mathcal{O}_Y) - \chi(\mathcal{O}_Y, \mathcal{O}_L) - \chi(\mathcal{O}_L, \mathcal{O}_Y) + \chi(\mathcal{O}_L, \mathcal{O}_L) \\
&= \chi(\mathcal{O}_Y) - \chi(\mathcal{O}_L) + \chi(\mathcal{O}_L(-2)) = -1,
\end{aligned}$$

where we used Serre duality and $\chi(\mathcal{O}_L, \mathcal{O}_L) = 0$, a consequence of $v(\mathcal{O}_L) \in H^4 \oplus H^6$.

By the properties of the Serre functor $\mathcal{S}_{\mathcal{A}_Y}$, this also gives $\chi(\mathcal{S}_{\mathcal{A}_Y}(\mathcal{I}_L), \mathcal{S}_{\mathcal{A}_Y}(\mathcal{I}_L)) = \chi(\mathcal{I}_L, \mathcal{I}_L) = -1$, as well as $\chi(\mathcal{I}_L, \mathcal{S}_{\mathcal{A}_Y}(\mathcal{I}_L)) = \chi(\mathcal{I}_L, \mathcal{I}_L) = -1$. Eventually, one computes

$$\chi(\mathcal{S}_{\mathcal{A}_Y}(\mathcal{I}_L), \mathcal{I}_L) = \chi(\mathcal{I}_L, \mathcal{S}^2_{\mathcal{A}_Y}(\mathcal{I}_L)) = \chi(\mathcal{I}_L, \mathcal{S}^{-1}_{\mathcal{A}_Y}(\mathcal{I}_L)[5]) = -\chi(\mathcal{S}_{\mathcal{A}_Y}(\mathcal{I}_L), \mathcal{I}_L),$$

where we used $\mathcal{S}^3_{\mathcal{A}_Y} \simeq [5]$, see Example 1.27, and then concludes from it the vanishing $\chi(\mathcal{S}_{\mathcal{A}_Y}(\mathcal{I}_L), \mathcal{I}_L) = 0$.

This immediately implies that $N(\mathcal{A}_Y)$ is generated by $[\mathcal{I}_L]$ and $[\mathcal{S}_{\mathcal{A}_Y}(\mathcal{I}_L)]$ and that with respect to this basis the (non-symmetric!) intersection matrix is

$$\begin{pmatrix} -1 & -1 \\ 0 & -1 \end{pmatrix}.$$

Observe that only the classes $v = \pm[\mathcal{I}_L], \pm[\mathcal{S}_{\mathcal{A}_Y}(\mathcal{I}_L)], \pm[\mathcal{S}^2_{\mathcal{A}_Y}(\mathcal{I}_L)]$ satisfy $\chi(v,v) = -1$. Here, we use the intersection numbers computed above to conclude that $[\mathcal{S}^2_{\mathcal{A}_Y}(\mathcal{I}_L)] = [\mathcal{S}_{\mathcal{A}_Y}(\mathcal{I}_L)] - [\mathcal{I}_L]$.

Now, any exact equivalence $\Phi\colon \mathcal{A}_Y \xrightarrow{\sim} \mathcal{A}_{Y'}$ induces naturally an isometry

$$\Phi^N\colon N(\mathcal{A}_Y) \xrightarrow{\sim} N(\mathcal{A}_{Y'})$$

and we may assume that $\Phi^N[\mathcal{I}_L] = [\mathcal{I}_{L'}]$ after composing Φ with $\mathcal{S}_{\mathcal{A}_{Y'}}$, $\mathcal{S}^2_{\mathcal{A}_{Y'}}$, and [1] if necessary. In fact, using Lemma 2.5 below, we conclude that $\Phi(\mathcal{I}_L) \in \mathcal{C}_{Y'}$ for all lines $L \subset Y$. It is not difficult to see that the shift appearing there is independent of the line L.

The next step consists of showing that actually

$$\Phi\colon \{\, \mathcal{I}_L \,\} \xrightarrow{\sim} \{\, \Phi(\mathcal{I}_L) \,\} = \{\, \mathcal{I}_{L'} \,\}. \tag{2.7}$$

This is the technical heart of the proof which makes use of Corollary 2.3. We have to refer to [65] for the complete argument. Geometrically, one has to prove that the moduli space of (stable) objects in $\mathcal{C}_{Y'}$ with numerical class $[\mathcal{I}_{L'}]$ is connected. This is reminiscent of a similar fact for moduli spaces of stable sheaves on K3 surfaces, see [253, Thm. 6.1.8]. However, the situation here is more complicated because $\mathcal{C}_Y$ is not a Calabi–Yau 2-category.

We conclude the proof under the simplifying assumption that Φ is a Fourier–Mukai functor so that the bijection (2.7) immediately defines an isomorphism $F(Y) \simeq F(Y')$, which by the geometric global Torelli theorem, see Proposition **2**.3.12, is enough to conclude $Y \simeq Y'$. □

Lemma 2.5 *Assume $E \in \mathcal{A}_Y$ satisfies* $\dim \operatorname{Ext}^1(E,E) = 2$. *Then some shift $E[n]$ of E is contained in the heart $\mathcal{C}_Y$.*

Proof The proof uses the standard spectral sequence, cf. [194, IV.2.2]:

$$E_2^{p,q} = \bigoplus \operatorname{Ext}^p(\mathcal{H}^i(E), \mathcal{H}^{i+q}(E)) \Rightarrow \operatorname{Ext}^{p+q}(E,E),$$

where the $\mathcal{H}^j := \mathcal{H}^j(E) \in \mathcal{C}_Y$ are the cohomology objects with respect to the bounded t-structure with heart $\mathcal{C}_Y$. Since $\operatorname{Ext}^j(E_1,E_2) = 0$ for all $E_1, E_2 \in \mathcal{C}_Y$ and $j > 2$ or $j < 0$, the $E_2^{1,0}$ survives and thus injects into $\operatorname{Ext}^2(E,E)$. Therefore, $\sum \dim \operatorname{Ext}^1(\mathcal{H}^i,\mathcal{H}^i) \leq \dim \operatorname{Ext}^1(E,E) = 2$.

Let us assume $\mathcal{H}^i \neq 0$ and then write $[\mathcal{H}^i] = a\,[\mathcal{I}_L] + b\,[\mathcal{S}_{\mathcal{A}_Y}(\mathcal{I}_L)]$. Then, using $\dim \operatorname{Ext}^0(\mathcal{H}^i,\mathcal{H}^i) \geq 1$ and $\chi(\mathcal{H}^i,\mathcal{H}^i) = -a^2 - ab - b^2 < 0$, we find $\dim \operatorname{Ext}^1(\mathcal{H}^i,\mathcal{H}^i) \geq 2$. Hence, at most one of the cohomology objects $\mathcal{H}^i$ can be non-zero, i.e. $E[n] \in \mathcal{C}_Y$ for some n. □

Remark 2.6 Here is an outline of an approach to the above result that is closer in spirit to the arguments of Addington and Thomas [8] for cubic fourfolds, see Section **6**.5.4 and Section 3.5. Instead of passing through an isomorphism $F(Y) \simeq F(Y')$ between their

Fano varieties one uses Hodge theory to see that an equivalence $\mathcal{A}_Y \simeq \mathcal{A}_{Y'}$ implies a polarized isomorphism $J(Y) \simeq J(Y')$ between the intermediate Jacobians and then uses the global Torelli theorem **5**.4.3.[7] The argument proceeds in several steps as follows.

(i) We aim at defining a Hodge structure $\widetilde{H}^1(\mathcal{A}_Y, \mathbb{Z})$ of weight one. For this, recall that the graded topological K-theory $K^*_{\text{top}}(Y) = K_{\text{top}}(Y) \oplus K^1_{\text{top}}(Y)$ of Y is torsion free and that the Chern character defines a graded isomorphism, see [26, §2.5]:

$$\text{ch}\colon K^*_{\text{top}}(Y) \otimes \mathbb{Q} \xrightarrow{\sim} H^{\text{ev}}(Y, \mathbb{Q}) \oplus H^{\text{odd}}(Y, \mathbb{Q}).$$

(ii) Consider the orthogonal complement $K'_{\text{top}}(Y)$ of $[\mathcal{O}_Y], [\mathcal{O}_Y(1)]$ in $K_{\text{top}}(Y)$ and in $K^*_{\text{top}}(Y)$. The latter simply is $K'_{\text{top}}(Y) \oplus K^1_{\text{top}}(Y)$. Then define $\widetilde{H}(\mathcal{A}_Y, \mathbb{Z})$ as $K'_{\text{top}}(Y) \oplus K^1_{\text{top}}(Y)$ endowed with the Euler–Poincaré pairing, cf. Remark **6**.5.2. It contains a natural Hodge structure $\widetilde{H}^1(\mathcal{A}_Y, \mathbb{Z}) \subset \widetilde{H}(\mathcal{A}_Y, \mathbb{Z})$ of weight one by declaring $\widetilde{H}^{1,0}(\mathcal{A}_Y) := v^{-1}(H^{2,1}(Y))$, cf. Section **6**.5.4. Then one needs to show that $\widetilde{H}^1(\mathcal{A}_Y, \mathbb{Z}) \simeq H^3(Y, \mathbb{Z})(1)$ as polarized Hodge structures of weight one. There are technical details to be checked, but similar techniques have recently been exploited by Perry [385, §5.3].

(iii) Any equivalence $\mathcal{A}_Y \xrightarrow{\sim} \mathcal{A}_{Y'}$ of Fourier–Mukai type induces a graded isomorphism of lattices $K'_{\text{top}}(Y) \oplus K^1_{\text{top}}(Y) \xrightarrow{\sim} K'_{\text{top}}(Y') \oplus K^1_{\text{top}}(Y')$, see [232, Rem. 3.4]. Furthermore, since it is compatible with the given Hodge structure, it provides an isometry $\widetilde{H}(\mathcal{A}_Y, \mathbb{Z}) \simeq \widetilde{H}(\mathcal{A}_{Y'}, \mathbb{Z})$ and a Hodge isometry

$$\widetilde{H}^1(\mathcal{A}_Y, \mathbb{Z}) \xrightarrow{\sim} \widetilde{H}^1(\mathcal{A}_{Y'}, \mathbb{Z}).$$

The latter then gives rise to an isomorphism of polarized abelian varieties

$$J(Y) \xrightarrow{\sim} J(Y').$$

A more algebraic argument to deduce a polarized isomorphism $J(Y) \simeq J(Y')$ from any equivalence $\mathcal{A}_Y \simeq \mathcal{A}_{Y'}$ was worked out by Bernardara and Tabuada [66].

Remark 2.7 Instead of studying $F(Y)$ from the perspective of $\mathcal{A}_Y$ via the ideal sheaves $\mathcal{I}_L \in \mathcal{A}_Y$ one could look at Y itself, realized as the collection of point sheaves $\{k(x) \mid x \in Y\} \subset \mathrm{D^b}(Y)$. This has been recently undertaken by Bayer et al. [39].

(i) First, observe that the objects $k(x)$ are not contained in $\mathcal{A}_Y$ but one can always project them via $j^*\colon \mathrm{D^b}(Y) \longrightarrow \mathcal{A}_Y$. As it turns out, for every $x \in Y$, the natural map

$$k^{\oplus 3} \simeq \mathrm{Ext}^1_Y(k(x), k(x)) \longrightarrow \mathrm{Ext}^1_{\mathcal{A}_Y}(j^*k(x), j^*k(x)) \simeq k^{\oplus 4}$$

is still injective, but the target space is four-dimensional. Also, in $N(\mathcal{A}_Y)$ we have

$$[j^*k(x)] = [\mathcal{I}_L] + [\mathcal{S}_{\mathcal{A}_Y}(\mathcal{I}_L)].$$

[7] In fact, in the original proof [65], the argument was concluded by passing from $F(Y) \simeq F(Y')$ to $J(Y) \simeq J(Y')$, for the result of Charles [114], cf. Proposition **2**.3.12, was not yet available.

(ii) Thus, the moduli space of (stable) objects in $\mathcal{A}_Y$ with numerical class $[\mathcal{I}_L] + [\mathcal{S}_{\mathcal{A}_Y}(\mathcal{I}_L)]$ admits a four-dimensional component M containing all objects of the form $j^*k(x)$. This can be viewed as a closed immersion $Y \hookrightarrow M$.

(iii) The main result in [39] now shows that $\mathcal{M}$ is isomorphic to the blow-up $\mathrm{Bl}_0(\Xi)$ of the theta divisor $\Xi \subset J(Y)$ in the intermediate Jacobian with $Y \subset \mathcal{M}$ corresponding to the exceptional divisor of $\mathrm{Bl}_0(\Xi) \longrightarrow \Xi$, see Section **5**.4.2 and Corollary **5**.4.7.

See Remark 3.24 for the analogous phenomenon in dimension four.

3 Cubic Fourfolds

For a smooth cubic fourfold $X \subset \mathbb{P}^5$, the standard semi-orthogonal decomposition of its derived category is of the form

$$\mathrm{D^b}(X) = \langle \mathcal{A}_X, \mathcal{O}_X, \mathcal{O}_X(1), \mathcal{O}_X(2) \rangle,$$

for which the category $\mathcal{A}_X$ is a Calabi–Yau 2-category, i.e. the double shift $E \longmapsto E[2]$ defines a Serre functor. Note that according to Exercise 1.17 any further semi-orthogonal decomposition $\mathcal{A}_X = \langle \mathcal{D}_1, \mathcal{D}_2 \rangle$ is actually orthogonal, i.e. $\mathcal{A}_X \simeq \mathcal{D}_1 \oplus \mathcal{D}_2$. However, it can be shown that $\mathcal{A}_X$ cannot be a non-trivial direct sum[8] and so it does not admit any non-trivial semi-orthogonal decomposition.

The following conjecture [296, Conj. 1.1], see also [298, Conj. 4.2], has triggered a lot of interest in derived categories among classical algebraic geometers.

Conjecture 3.1 (Kuznetsov) *A smooth cubic fourfold $X \subset \mathbb{P}^5$ is rational if and only if there exists a K3 surface S and an exact linear equivalence $\mathcal{A}_X \simeq \mathrm{D^b}(S)$.*

The conjecture is wide open. As mentioned before, at this moment not a single smooth cubic fourfold is known to be irrational.

Kuznetsov's rationality conjecture coexisted with Hassett's rationality conjecture, see Conjecture **6**.5.15, until the two conjectures were eventually shown to be equivalent, see Section 3.4.

Remark 3.2 There is a rivaling criterion for rationality of cubic fourfolds put forward by Galkin and Shinder [190]. It predicts that X is rational if and only if $F(X)$ is birational to the Hilbert scheme $S^{[2]}$ of a K3 surface. The argument that lends credibility to this belief is similar to the one for cubic threefolds explained in Remark **5**.4.15. This prediction is not compatible with the Hassett–Kuznetsov criterion as was shown by Addington [3]. The numerical condition is $(***)$ in Corollary **6**.5.22.

[8] Geometrically this is very intuitive, for in many cases $\mathcal{A}_X$ is equivalent to the derived category of a (twisted) K3 surface, which is indecomposable. In general, one can use Hochschild cohomology and more precisely the fact that $HH^0(\mathcal{A}_X)$ is one-dimensional, cf. [292]. Possibly, the description of $\mathcal{A}_X$ as a category of matrix factorization could be used alternatively.

For three types of cubic fourfolds, Kuznetsov [296] established a direct link between the category $\mathcal{A}_X$ and the derived category of a (twisted) K3 surface: for cubics containing a plane, see Section **6**.1.2, for nodal cubics, see Section **6**.1.4, and for Pfaffian cubics, see Section **6**.2. For other cubic fourfolds, essentially for all smooth cubics satisfying the Hassett condition (∗∗) or (∗∗′) in Section **6**.5, a link between $\mathcal{A}_X$ and the derived category of a (twisted) K3 surface has been obtained via deformation [8, 250].

For cubics containing a plane and for Pfaffian cubics, the basic construction is similar. The ideal sheaf of the two natural correspondences, see (1.5) in Section **6**.1.1 and (2.10) in Section **6**.2.5,

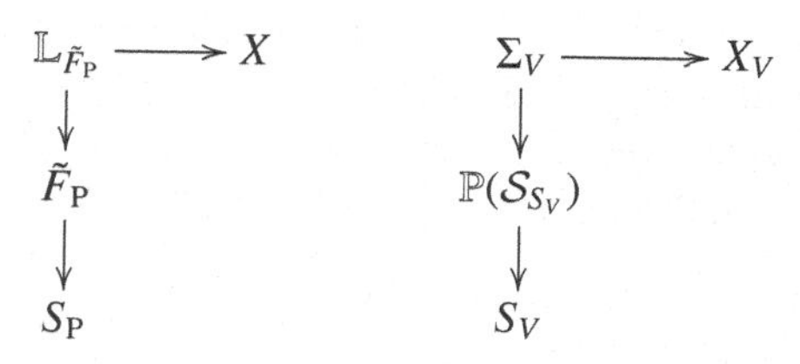

are used as Fourier–Mukai kernels to link $\mathrm{D^b}(X)$ with $\mathrm{D^b}(S)$. For nodal cubics we follow Kuznetsov's original argument based on a sequence of mutations.

3.1 Fourfolds Containing a Plane Consider a smooth cubic fourfold $X \subset \mathbb{P}^5$ containing a plane $\mathrm{P} \subset X$. Projecting from P defines a quadric surface fibration $\phi\colon \tilde{X} := \mathrm{Bl}_{\mathrm{P}}(X) \longrightarrow \mathbb{P}^2$, which has been discussed at length in Section **6**.1.2.

In particular, we saw that its relative Fano variety $\tilde{F} := \tilde{F}_{\mathrm{P}} \longrightarrow \mathbb{P}^2$ factors through a K3 surface which is a double cover $\pi\colon S_{\mathrm{P}} \longrightarrow \mathbb{P}^2$ ramified over the discriminant sextic curve $D_{\mathrm{P}} \subset \mathbb{P}^2$ of ϕ. The projection $\tilde{\pi}\colon \tilde{F} \longrightarrow S_{\mathrm{P}}$ is a Brauer–Severi variety and its Brauer class was denoted by $\alpha_{\mathrm{P},X} \in \mathrm{Br}(S_{\mathrm{P}})$.

The goal of this subsection is to establish a categorical link between cubic fourfolds containing a plane and K3 surfaces expressed by the following result.

Proposition 3.3 (Kuznetsov) *For the generic cubic fourfold $X \subset \mathbb{P}^5$ containing a plane $\mathbb{P}^2 \simeq \mathrm{P} \subset X$ there exists an exact linear equivalence*

$$\mathrm{D^b}(S_{\mathrm{P}}, \alpha_{\mathrm{P},X}) \simeq \mathcal{A}_X$$

between the bounded derived category of $\alpha_{\mathrm{P},X}$-twisted coherent sheaves on S_{P} and the Kuznetsov component $\mathcal{A}_X \subset \mathrm{D^b}(X)$.

Remark 3.4 The assumption that X is generic is simply the condition that the discriminant curve $D_{\mathrm{P}} \subset \mathbb{P}^2$ is smooth. The case when D_{P} is nodal has been investigated by Moschetti [355]. The double cover $S_{\mathrm{P}} \longrightarrow \mathbb{P}^2$ ramified over D_{P} is singular in this case, but replacing S_{P} by its minimal resolution, a smooth K3 surface, the above equivalence still holds.

Since by Lemma **6**.1.14 the Brauer class $\alpha_{\mathrm{P},X}$ is trivial if and only if the quadric fibration has a rational section, one finds that in this case the Kuznetsov component is equivalent to the bounded derived category of a K3 surface.

Corollary 3.5 *If the quadric fibration $\phi\colon \tilde{X} \longrightarrow \mathbb{P}^2$ has a rational section, then there exists an exact linear equivalence $\mathrm{D^b}(S_{\mathrm{P}}) \simeq \mathcal{A}_X$.* □

For Kuznetsov's original proof we refer to [296, Prop. 4.3] or the outline in [298, Prop. 4.10]. We shall describe a geometrically more intuitive argument, inspired by the treatment of Pfaffian cubics by Addington and Lehn [7], cf. Section 3.2 below.

Let $p\colon \mathbb{L}_{\tilde{F}} \longrightarrow \tilde{F}$ be the universal line over the relative Fano variety, which is the pullback of the universal line $\mathbb{L} \longrightarrow F(X)$. We think of $\mathbb{L}_{\tilde{F}}$ as a closed subscheme of $\tilde{F} \times X$ and denote its ideal sheaf by $\mathcal{I}$. Then we use $\mathcal{I}(1) := \mathcal{I} \otimes (\mathcal{O}_{\tilde{F}} \boxtimes \mathcal{O}_X(1))$ as a Fourier–Mukai kernel to define the exact linear functor

$$\Phi := \Phi_{\mathcal{I}(1)}\colon \mathrm{D^b}(X) \longrightarrow \mathrm{D^b}(\tilde{F}).$$

As recalled in Section 0.1, the bounded derived category $\mathrm{D^b}(\tilde{F})$ of the Brauer–Severi variety $\tilde{\pi}\colon \tilde{F} \longrightarrow S_{\mathrm{P}}$ admits a semi-orthogonal decomposition

$$\mathrm{D^b}(\tilde{F}) = \langle \mathrm{D^b}(S_{\mathrm{P}}, \alpha_{\mathrm{P},X}), \mathrm{D^b}(S_{\mathrm{P}}) \rangle.$$

Here, $\mathrm{D^b}(S_{\mathrm{P}})$ is embedded via the pullback, so $\mathrm{D^b}(S_{\mathrm{P}}) \simeq \tilde{\pi}^*\mathrm{D^b}(S_{\mathrm{P}}) \subset \mathrm{D^b}(\tilde{F})$, and its right orthogonal is the image of $\mathrm{D^b}(S_{\mathrm{P}}, \alpha_{\mathrm{P},X}) \simeq \mathrm{D^b}(S_{\mathrm{P}}, \alpha_{\mathrm{P},X}^{-1})$ under the fully faithful embedding $E \longmapsto \tilde{\pi}^*E \otimes \mathcal{O}_{\tilde{\pi}}(-1)$ obtained by composing the pullback and the tensor product with the twisted relative $\mathcal{O}_{\tilde{\pi}}(-1)$.[9] For example, the structure sheaf $k(x) \in \mathrm{D^b}(S_{\mathrm{P}}, \alpha_{\mathrm{P},X})$ of $x \in S_{\mathrm{P}}$ viewed as a twisted sheaf corresponds to $\mathcal{O}_{F_x}(-1) \in \mathrm{D^b}(\tilde{F})$, where $F_x := \tilde{\pi}^{-1}(x) \simeq \mathbb{P}^1$.

Lemma 3.6 *Composing the left adjoint functor $\Psi\colon \mathrm{D^b}(\tilde{F}) \longrightarrow \mathrm{D^b}(X)$ of Φ with the inclusion $\mathrm{D^b}(S_{\mathrm{P}}, \alpha_{\mathrm{P},X}) \subset \mathrm{D^b}(\tilde{F})$ defines an exact linear functor*

$$\Psi'\colon \mathrm{D^b}(S_{\mathrm{P}}, \alpha_{\mathrm{P},X}) \longrightarrow \mathrm{D^b}(X).$$

Its image is contained in the Kuznetsov component $\mathcal{A}_X = \{\mathcal{O}_X, \mathcal{O}_X(1), \mathcal{O}_X(2)\}^{\perp}$.

Proof Since by Serre duality $\mathcal{A}_X = {}^{\perp}\{\mathcal{O}_X(-3), \mathcal{O}_X(-2), \mathcal{O}_X(-1)\}$, the assertion is equivalent to $\mathrm{Hom}(\Psi(E), \mathcal{O}_X(j)) = 0$ for $-3 \leq j \leq -1$ and all $E \in \mathrm{D^b}(S_{\mathrm{P}}, \alpha_{\mathrm{P},X}) \subset \mathrm{D^b}(\tilde{F})$. By adjunction, this is equivalent to $\Phi(\mathcal{O}_X(j)) \in \mathrm{D^b}(S_{\mathrm{P}}, \alpha_{\mathrm{P},X})^{\perp} \subset \mathrm{D^b}(\tilde{F})$ for $-3 \leq j \leq -1$.

To prove this, we use the short exact sequence

$$0 \longrightarrow \mathcal{I} \longrightarrow \mathcal{O}_{\tilde{F}\times X} \longrightarrow \mathcal{O}_{\mathbb{L}_{\tilde{F}}} \longrightarrow 0$$

[9] Note that compared to (0.2) we have passed from the semi-orthogonal decomposition $\langle \mathrm{D^b}(S_{\mathrm{P}}), \mathrm{D^b}(S_{\mathrm{P}}, \alpha_{\mathrm{P},X}) \rangle$ to $\langle \mathrm{D^b}(S_{\mathrm{P}}, \alpha_{\mathrm{P},X}^{-1}), \mathrm{D^b}(S_{\mathrm{P}}) \rangle$ using the Serre functor of $\mathrm{D^b}(\tilde{F})$, cf. Exercise 1.13.

and Bott vanishing on the fibres $\mathbb{P}^1 \simeq p^{-1}(t)$ of $p\colon \mathbb{L}_{\tilde{F}} \longrightarrow \tilde{F}$. This immediately gives $\Phi(\mathcal{O}_X(-2)) = 0$, since $\Phi_{\mathcal{O}_{\tilde{F}\times X}}(\mathcal{O}_X(-1)) = 0 = \Phi_{\mathcal{O}_{\mathbb{L}_{\tilde{F}}}}(\mathcal{O}_X(-1))$, and, for the same reason, $\Phi(\mathcal{O}_X(-1)) = 0$, since $\Phi_{\mathcal{O}_{\tilde{F}\times X}}(\mathcal{O}_X) \simeq \mathcal{O}_{\tilde{F}} \simeq \Phi_{\mathcal{O}_{\mathbb{L}_{\tilde{F}}}}(\mathcal{O}_X)$. Finally,

$$\Phi(\mathcal{O}_X(-3)) \simeq \Phi_{\mathcal{O}_{\mathbb{L}_{\tilde{F}}}}(\mathcal{O}_X(-2))[-1] \simeq p_*\mathcal{O}_p(-2)[-1] \simeq \det(\mathcal{S}_{\tilde{F}})[-2] \simeq \mathcal{O}_{\tilde{F}}(-1)[-2].$$

By Serre duality, the desired vanishing $\mathrm{Hom}(E, \Phi(\mathcal{O}_X(-3))) = 0$ for all $E \in \mathrm{D^b}(S_{\mathrm{P}}, \alpha_{\mathrm{P},X})$ is equivalent to $\mathrm{Hom}(\mathcal{O}_{\tilde{F}}(-1), E \otimes \omega_{\tilde{F}}) = 0$, which one only needs to prove for the spanning class of point sheaves $E = k(x) \in \mathrm{D^b}(S_{\mathrm{P}}, \alpha_{\mathrm{P},X})$ and their shifts. The latter then follows from $\mathrm{Hom}^*(\mathcal{O}_{\tilde{F}}(-1), \mathcal{O}_{F_x}(-1)\otimes\omega_{\tilde{F}}) \simeq H^*(F_x, \mathcal{O}_{F_x}(-1)\otimes\mathcal{O}_{F_x}(-2)\otimes\mathcal{O}_{\tilde{F}}(1)|_{F_x}) \simeq H^*(F_x, \mathcal{O}_{F_x}(-1)) = 0$, where we use that the Plücker polarization on $F(X)$ restricts to $\mathcal{O}_{F_x}(2)$ on $\mathbb{P}^1 \simeq F_x$, see Remark 6.1.9. □

To prove the faithfulness of Ψ', we need to compute the image $\Psi'(k(x))$ of the structure sheaf of a point $x \in S_{\mathrm{P}}$. The Fourier–Mukai kernel of the left adjoint functor Ψ is $\mathcal{I}(1)^\vee \otimes (\omega_{\tilde{F}} \boxtimes \mathcal{O}_X)[3]$, cf. [246, Prop. 5.9]. Hence, the image $\Psi'(k(x))$ is the direct image of $\mathcal{I}_x(1)^\vee \otimes ((\omega_{\tilde{F}} \otimes \mathcal{O}_{F_x}(-1)) \boxtimes \mathcal{O}_X)[3]$ under the projection $r\colon F_x \times X \longrightarrow X$. Here, $\mathcal{I}_x$ is the ideal sheaf of the surface $\mathbb{L}_{F_x} \subset F_x \times X$.

Using again Remark 6.1.9 and applying Grothendieck–Verdier duality, one finds

$$\begin{aligned}\Psi'(k(x)) &\simeq r_*\mathcal{H}om(\mathcal{I}_x \otimes (\mathcal{O}_{F_x} \boxtimes \mathcal{O}_X(1)), \omega_r \otimes (\mathcal{O}_{F_x}(-1) \boxtimes \mathcal{O}_X))[3]\\ &\simeq r_*\mathcal{H}om(\mathcal{I}_x \otimes (\mathcal{O}_{F_x}(1) \boxtimes \mathcal{O}_X(1)), \omega_r)[3]\\ &\simeq (r_*(\mathcal{I}_x \otimes (\mathcal{O}_{F_x}(1) \boxtimes \mathcal{O}_X)))^\vee \otimes \mathcal{O}_X(-1)[2].\end{aligned}$$

The short exact sequence $0 \longrightarrow \mathcal{I}_x \longrightarrow \mathcal{O}_{F_x\times X} \longrightarrow \mathcal{O}_{\mathbb{L}_{F_x}} \longrightarrow 0$ allows one to compute $r_*(\mathcal{I}_x \otimes (\mathcal{O}_{F_x}(1) \boxtimes \mathcal{O}_X))$ as

$$\Psi'(k(x))^\vee \otimes \mathcal{O}_X(-1)[2] \simeq \mathcal{K}_x := \mathrm{Ker}\left(H^0(F_x, \mathcal{O}_{F_x}(1)) \otimes \mathcal{O}_X \twoheadrightarrow \mathcal{O}_{Q_x}(1)\right).$$

Here, $Q_x \subset X$ is the image of $\mathbb{L}_{F_x}$ which is nothing but the residual quadric surface Q_y of $\mathrm{P} \subset \overline{y\mathrm{P}} \cap X$ with $y := \pi(x) \in \mathbb{P}^2$, and by definition the line bundle $\mathcal{O}_{Q_x}(1)$ is the pullback of $\mathcal{O}_{F_x}(1)$ under $Q_x \longrightarrow F_x$. Recall that Q_y is a quadric cone for y contained in the discriminant curve $D_{\mathrm{P}} \subset \mathbb{P}^2$ and $Q_y \simeq \mathbb{L}_{F_x} \simeq \mathbb{P}(\mathcal{S}_{F_x}) \simeq \mathbb{P}^1 \times \mathbb{P}^1$ otherwise.

Lemma 3.7 *The functor* $\Psi'\colon \mathrm{D^b}(S_{\mathrm{P}}, \alpha_{\mathrm{P},X}) \longrightarrow \mathcal{A}_X$ *is fully faithful.*

Proof Adapting the criterion of Bondal and Orlov, cf. [246, Prop. 7.1], to the twisted case, it suffices to show that for $x \in S_{\mathrm{P}}$ the image object $\Psi'(k(x))$ is simple and that

$$\mathrm{Ext}^i(\Psi'(k(x_1)), \Psi'(k(x_2))) = 0$$

for $i < 0$, $i > 2$, and for any i when $x_1 \neq x_2 \in S_{\mathrm{P}}$.

By the comments above, we have natural isomorphisms $\mathrm{Ext}^i(\Psi'(k(x_1)), \Psi'(k(x_2))) \simeq \mathrm{Ext}^i(\mathcal{K}_{x_2}, \mathcal{K}_{x_1})$. Thus, $\Psi'(k(x))$ is simple if $\mathcal{K}_x$ is simple, which in turn follows from the observation that the reflexive hull of $\mathcal{K}_x$ is $\mathcal{O}_X^{\oplus 2}$. The vanishing for $i < 0$ is clear, since the $\mathcal{K}_{x_i}$ are sheaves, and the vanishing for $i > 2$ then follows from Serre duality in $\mathcal{A}_X$, see Example 1.27. To prove the vanishing for $x_1 \neq x_2$, one has to deal with two cases: first, $\pi(x_1) = \pi(x_2)$, in which case $Q_{x_1} = Q_{x_2}$, but the two line bundles $\mathcal{O}_{Q_{x_1}}(-1), \mathcal{O}_{Q_{x_2}}(-1)$ are different, since they are pulled back by the two projections of $Q_{x_i} \simeq \mathbb{L}_{F_{x_i}} \simeq F_{x_1} \times F_{x_2}$. Second, $\pi(x_1) \neq \pi(x_2)$, in which case the two quadrics are distinct.

In both cases, the vanishing for $i = 0$ is immediate and for $i = 2$ it follows again from Serre duality in $\mathcal{A}_X$. The remaining vanishing $\mathrm{Ext}^1(\mathcal{K}_{x_2}, \mathcal{K}_{x_1}) = 0$ is proved by a Riemann–Roch computation using $[\mathcal{K}_x] = 2[\mathcal{O}_X] - [\mathcal{O}_{Q_x}(1)]$ in the Grothendieck ring:

$$\begin{aligned}\chi(\mathcal{K}_{x_2}, \mathcal{K}_{x_1}) &= 4\chi(\mathcal{O}_X, \mathcal{O}_X) - 2\chi(\mathcal{O}_X, \mathcal{O}_{Q_x}(1)) - 2\chi(\mathcal{O}_{Q_x}(1), \mathcal{O}_X) + \chi(\mathcal{O}_{Q_x}, \mathcal{O}_{Q_x}) \\ &= 4 - 2 \cdot 2 - 2\chi(\omega_X \otimes \mathcal{O}_{Q_x}(1)) + 4 = 0,\end{aligned}$$

where we use $\omega_X|_{Q_x} \simeq \mathcal{O}(-3,-3)$ on the quadric $Q_x \simeq \mathbb{P}^1 \times \mathbb{P}^1$ and $([Q_x].[Q_x]) = 4$, see Exercise **6**.1.2. □

The last step in the proof of Proposition 3.3 consists of arguing that the fully faithful Fourier–Mukai functor $\mathrm{D^b}(S_{\mathrm{P}}, \alpha_{\mathrm{P},X}) \hookrightarrow \mathcal{A}_X$ between the two Calabi–Yau 2-categories is automatically an equivalence. This follows from $\mathcal{A}_X$ being indecomposable, as mentioned at the beginning of this section, and the fact that the image of a fully faithful Fourier–Mukai functor is always admissible. □

Remark 3.8 The twisted derived category $\mathrm{D^b}(S_{\mathrm{P}}, \alpha_{\mathrm{P},X})$ can also be realized as the derived category $\mathrm{D^b}(S_{\mathrm{P}}, \mathcal{B}_0)$ of coherent sheaves over a certain sheaf of Azumaya algebras $\mathcal{B}_0$ of rank four, see Remark **6**.1.11. The latter can in turn be interpreted as the derived category $\mathrm{D^b}(\mathbb{P}^2, \mathcal{C}_0)$ of coherent sheaves over the direct image $\mathcal{C}_0 \simeq \pi_*\mathcal{B}_0$ under the projection $\pi\colon S_{\mathrm{P}} \longrightarrow \mathbb{P}^2$. Thus,

$$\mathcal{A}_X \simeq \mathrm{D^b}(S_{\mathrm{P}}, \alpha_{\mathrm{P},X}) \simeq \mathrm{D^b}(S_{\mathrm{P}}, \mathcal{B}_0) \simeq \mathrm{D^b}(\mathbb{P}^2, \mathcal{C}_0).$$

The last equivalence is given by the direct image functor $E \longmapsto \pi_*E$ and its inverse by $G \longmapsto \pi^*G \otimes_{\pi^*\mathcal{C}_0} \mathcal{B}_0$.

Recall from Remarks **6**.1.11 and **6**.1.12 that the sheaf of Azumaya algebras $\mathcal{C}_0$ on $\mathbb{P}^2$ is the sheaf of even Clifford algebras associated with the quadratic form $q\colon \mathcal{F}^* \longrightarrow \mathcal{F} \otimes \mathcal{O}(1)$ on $\mathbb{P}^2$. The Brauer class of $\mathcal{B}_0$ is $\alpha_{\mathrm{P},X}$.

The general theory of derived categories in quadric fibrations has been developed by Kuznetsov [294].

3.2 Pfaffian Cubic Fourfolds We now come back to Pfaffian cubic fourfolds $X_V \subset \mathbb{P}(V)$ and their associated K3 surfaces $S_V \subset \mathbb{P}(V^\perp)$ discussed in detail in Section **6**.2. According to Hassett's Hodge theoretic considerations, see Proposition **6**.2.34, there is a close link between the cubic fourfold X_V and the K3 surface S_V. This is lifted to the level of derived categories by the following result.

Proposition 3.9 (Kuznetsov) *There exists a linear exact equivalence*

$$\mathrm{D}^{\mathrm{b}}(S_V) \simeq \mathcal{A}_{X_V}$$

between the bounded derived category of the K3 surface S_V and the Kuznetsov component $\mathcal{A}_{X_V} \subset \mathrm{D}^{\mathrm{b}}(X_V)$.

Kuznetsov's original proof [293] uses the machinery of homological projective duality. We shall outline the more direct and geometrically intuitive argument, valid at least for generic Pfaffian cubic fourfolds, that was given by Addington and Lehn [7].

We will stick to the notation introduced in Section **6**.2 and in particular denote by $\Sigma_V \subset S_V \times X_V$ the incidence correspondence of all (P, ω) with $P \cap \mathrm{Ker}(\omega) \neq 0$. We denote the two projections from $S_V \times X_V$ by p_S and p_X. According to Lemma **6**.2.14 and Lemma **6**.2.20, the fibres Σ_P of the first projection $p_S : \Sigma_V \longrightarrow S_V$ are quartic rational normal scrolls in X_V, so all isomorphic either to $\mathbb{P}^1 \times \mathbb{P}^1$ or to the Hirzebruch surface $\mathbb{F}_2$ and the second projection $p_X : \Sigma_V \longrightarrow X_V$ is generically finite of degree four. In the following, we often simplify the notation and just write $S = S_V$, $X = X_V$, and $\Sigma = \Sigma_V$.

The twist of the ideal sheaf

$$\mathcal{I}(-1) := \mathcal{I}_\Sigma \otimes p_X^* \mathcal{O}_X(-1)$$

of the inclusion $\Sigma \subset S \times X$ induces a Fourier–Mukai functor

$$\Phi := \Phi_{\mathcal{I}(-1)} : \mathrm{D}^{\mathrm{b}}(X) \longrightarrow \mathrm{D}^{\mathrm{b}}(S).$$

Its right adjoint $\Psi : \mathrm{D}^{\mathrm{b}}(S) \longrightarrow \mathrm{D}^{\mathrm{b}}(X)$ is the Fourier–Mukai functor with kernel $\mathcal{I}(-1)^\vee \otimes p_X^* \mathcal{O}_X(-3)[4]$, cf. [246, Prop. 5.9].

Lemma 3.10 *The image of the right adjoint functor $\Psi : \mathrm{D}^{\mathrm{b}}(S_V) \longrightarrow \mathrm{D}^{\mathrm{b}}(X_V)$ is contained in the Kuznetsov component $\mathcal{A}_{X_V} = \{\mathcal{O}_X, \mathcal{O}_X(1), \mathcal{O}_X(2)\}^\perp$.*

Proof We have to show that $\mathrm{Hom}(\mathcal{O}_X(j), \Psi(E)) = 0$ for $j = 0, 1, 2$ and all $E \in \mathrm{D}^{\mathrm{b}}(S)$. By adjunction, this is equivalent to proving $\Phi(\mathcal{O}_X(j)) = 0$ for $j = 0, 1, 2$, which follows from the short exact sequence $0 \longrightarrow \mathcal{I}_\Sigma \longrightarrow \mathcal{O}_{S \times X} \longrightarrow \mathcal{O}_\Sigma \longrightarrow 0$.

Indeed, the vanishing $\Phi(\mathcal{O}_X) = 0$ is deduced from $H^*(X, \mathcal{O}_X(-1)) = 0$ and the fibrewise vanishing $H^*(\Sigma_P, \mathcal{O}(-2, -1)) = 0$. Here, restricting to the case $\Sigma_P \simeq \mathbb{P}^1 \times \mathbb{P}^1$ for simplicity, we use the fact that $\mathcal{O}_X(1)|_{\Sigma_V} \simeq \mathcal{O}(2, 1)$, see the proof of Lemma **6**.2.20. Similarly, the vanishing $\Phi(\mathcal{O}_X(1)) = 0$ follows from $H^{*>0}(\Sigma_P, \mathcal{O}_{\Sigma_P}) = 0$ and the isomorphism $H^0(X, \mathcal{O}_X) \xrightarrow{\sim} H^0(\Sigma_P, \mathcal{O}_{\Sigma_P})$. Eventually, the proof of the vanishing $\Phi(\mathcal{O}_X(2)) = 0$ uses $H^{*>0}(\Sigma_P, \mathcal{O}_{\Sigma_P}(2, 1)) = 0$ and the projective normality of $\Sigma_P \subset X$ which ensures that $H^0(X, \mathcal{O}_X(1)) \xrightarrow{\sim} H^0(\Sigma_P, \mathcal{O}_{\Sigma_P}(2, 1))$ is an isomorphism. □

Lemma 3.11 *The functor $\Psi : \mathrm{D}^{\mathrm{b}}(S_V) \longrightarrow \mathcal{A}_{X_V} \subset \mathrm{D}^{\mathrm{b}}(X_V)$ is fully faithful.*

Proof According to the criterion of Bondal and Orlov, cf. [246, Prop. 7.1], it suffices to show that for any point $P \in S$ the image object $\Psi(k(P)) \simeq \mathcal{I}_{\Sigma_P}(-1)^\vee(-3)[4]$ is simple, which follows from $\mathcal{I}_{\Sigma_P}$ being simple, and that

$$\mathrm{Ext}^i(\Psi(k(P_1)), \Psi(k(P_2))) = 0$$

when $i < 0$, $i > 2$, or $P_1 \neq P_2 \in S$. To prove this vanishing, first observe that

$$\mathrm{Ext}^i(\Psi(k(P_1)), \Psi(k(P_2))) \simeq \mathrm{Ext}^i(\mathcal{I}_{\Sigma_{P_1}}(-1)^\vee(-3)[4], \mathcal{I}_{\Sigma_{P_2}}(-1)^\vee(-3)[4])$$
$$\simeq \mathrm{Ext}^i(\mathcal{I}_{\Sigma_{P_2}}, \mathcal{I}_{\Sigma_{P_1}}).$$

Now, since there are no negative non-trivial extensions between sheaves, we certainly have $\mathrm{Ext}^{<0}(\mathcal{I}_{\Sigma_{P_2}}, \mathcal{I}_{\Sigma_{P_1}}) = 0$, which by Serre duality in $\mathcal{A}_X$, see Example 1.27, also implies the vanishing for $i > 2$ Now assume $P_1 \neq P_2$. Then $\mathrm{Hom}(\mathcal{I}_{\Sigma_{P_2}}, \mathcal{I}_{\Sigma_{P_1}}) = 0$, since $\Sigma_{P_1}, \Sigma_{P_2} \subset X$ are two distinct irreducible surfaces, which by Serre duality in $\mathcal{A}_X$ also implies the vanishing for $i = 2$. It remains to prove $\mathrm{Ext}^1(\mathcal{I}_{\Sigma_{P_2}}, \mathcal{I}_{\Sigma_{P_1}}) = 0$ which follows from a Riemann–Roch computation

$$\chi(\mathcal{I}_{\Sigma_{P_1}}, \mathcal{I}_{\Sigma_{P_2}}) = \chi(\mathcal{I}_{\Sigma_P}, \mathcal{I}_{\Sigma_P}) = \chi(\mathcal{O}_X, \mathcal{O}_X) - \chi(\mathcal{O}_X, \mathcal{O}_{\Sigma_P}) - \chi(\mathcal{O}_{\Sigma_P}, \mathcal{O}_X) + \chi(\mathcal{O}_{\Sigma_P}, \mathcal{O}_{\Sigma_P})$$
$$= 1 - 1 - \chi(\Sigma_P, \omega_X|_{\Sigma_P}) + \chi(\mathcal{O}_{\Sigma_P}, \mathcal{O}_{\Sigma_P}) = -\chi(\mathbb{P}^1, \mathcal{O}(-6)) \cdot \chi(\mathbb{P}^1, \mathcal{O}(-3)) + 10 = 0,$$

where we use $([\Sigma_P].[\Sigma_P]) = 10$ to determine $\chi(\mathcal{O}_{\Sigma_P}, \mathcal{O}_{\Sigma_P})$, see Lemma **6**.2.20. □

The last step in the proof of Proposition 3.9 consists of arguing that the fully faithful Fourier–Mukai functor $\mathrm{D^b}(S) \hookrightarrow \mathcal{A}_X$ between the two Calabi–Yau 2-categories is automatically an equivalence, which follows from $\mathcal{A}_X$ being indecomposable as in the proof of Proposition 3.3. □

3.3 Nodal Cubic Fourfolds The third class of cubics for which $\mathcal{A}_X$ is directly accessible is provided by nodal cubic fourfolds $X \subset \mathbb{P}^5$, i.e. cubics with one ordinary double point $x_0 \in X$ as the only singularity. We will use the notation introduced in Section **6**.1.4.

The derived category $\mathrm{D^b}(\tilde{X})$ of the blow-up $\tau\colon \tilde{X} = \mathrm{Bl}_{x_0}(X) \longrightarrow X$ admits two natural semi-orthogonal decompositions. First, (0.3) applied to the blow-up $\phi\colon \tilde{X} \longrightarrow \mathbb{P}^4$ gives

$$\mathrm{D^b}(\tilde{X}) = \langle \mathrm{D^b}(S)_{-1}, \phi^*\mathrm{D^b}(\mathbb{P}^4)\rangle$$
$$= \langle \mathrm{D^b}(S)_{-1}, \phi^*\mathcal{O}(-3), \phi^*\mathcal{O}(-2), \phi^*\mathcal{O}(-1), \phi^*\mathcal{O}, \phi^*\mathcal{O}(1)\rangle. \quad (3.1)$$

Next, from (0.3) applied to $\tau\colon \tilde{X} \longrightarrow X$, one obtains

$$\mathrm{D^b}(\tilde{X}) = \langle i_*\mathcal{O}_E(-2), i_*\mathcal{O}_E(-1), \mathcal{A}_X, \tau^*\mathcal{O}_X, \tau^*\mathcal{O}_X(1), \tau^*\mathcal{O}_X(2)\rangle. \quad (3.2)$$

Here, $i\colon E \hookrightarrow \tilde{X}$ is the inclusion of the exceptional divisor of τ and the category $\mathcal{A}_X$ is in this case defined as the intersection

$$\mathcal{A}_X := {}^\perp\langle i_*\mathcal{O}_E(-2), i_*\mathcal{O}_E(-1)\rangle \cap \langle \tau^*\mathcal{O}_X, \tau^*\mathcal{O}_X(1), \tau^*\mathcal{O}_X(2)\rangle^\perp,$$

which mimics the situation of the blow-up of a smooth point.[10] Implicitly we are using the fact that $i_*\mathcal{O}_E(-2), i_*\mathcal{O}_E(-1) \in \langle \tau^*\mathcal{O}_X, \tau^*\mathcal{O}_X(1), \tau^*\mathcal{O}_X(2)\rangle^\perp$, which is easy to check.

Proposition 3.12 (Kuznetsov) *There exists an exact linear equivalence* $\mathrm{D}^\mathrm{b}(S) \simeq \mathcal{A}_X$.

Proof The idea of the proof is to relate the two semi-orthogonal decompositions (3.1) and (3.2) in order to identify $\mathrm{D}^\mathrm{b}(S)_{-1}$ and $\mathcal{A}_X$. Instead of projection $\mathrm{D}^\mathrm{b}(S)_{-1}$ to $\mathcal{A}_X$ directly via one of the projections $j^!$ or j^*, one uses a more systematic way by applying mutations and in this way pass from one semi-orthogonal decomposition to the other. We sketch Kuznetsov's original proof [296, Cor. 5.7].

To simplify the notation, let $\mathcal{A}_S := \mathrm{D}^\mathrm{b}(S)_{-1} \subset \mathrm{D}^\mathrm{b}(\tilde{X})$ and $\mathcal{O}(k) := \phi^*\mathcal{O}(k)$. Then, using the standard formulae for mutations, see Exercise 1.11, one computes:

$$\mathrm{D}^\mathrm{b}(\tilde{X}) = \langle \mathcal{A}_S, \mathcal{O}(-3), \mathcal{O}(-2), \mathcal{O}(-1), \mathcal{O}, \mathcal{O}(1)\rangle$$
$$= \langle \mathbb{L}_{\mathcal{A}_S}\mathcal{O}(-3), \mathcal{A}_S, \mathcal{O}(-2), \mathcal{O}(-1), \mathcal{O}, \mathcal{O}(1)\rangle$$
$$= \langle \mathbb{L}_{\mathcal{A}_S}\mathcal{O}(-3), \mathbb{L}_{\mathcal{A}_S}\mathcal{O}(-2), \mathcal{A}_S, \mathcal{O}(-1), \mathcal{O}, \mathcal{O}(1)\rangle$$
$$= \langle \mathbb{L}_{\mathcal{A}_S}\mathcal{O}(-3), \mathbb{L}_{\mathcal{A}_S}\mathcal{O}(-2), \mathbb{L}_{\mathcal{A}_S}\mathcal{O}(-1), \mathcal{A}_S, \mathcal{O}, \mathcal{O}(1)\rangle$$
$$= \langle \mathbb{L}_{\mathcal{A}_S}\mathcal{O}(-3), \mathbb{L}_{\mathcal{A}_S}\mathcal{O}(-2), \mathbb{L}_{\mathcal{A}_S}\mathcal{O}(-1), \mathcal{O}, \mathbb{R}_{\mathcal{O}}(\mathcal{A}_S), \mathcal{O}(1)\rangle$$
$$= \langle \mathbb{L}_{\mathcal{A}_S}\mathcal{O}(-3), \mathbb{L}_{\mathcal{A}_S}\mathcal{O}(-2), \mathbb{L}_{\mathcal{A}_S}\mathcal{O}(-1), \mathcal{O}, \mathcal{O}(1), \mathbb{R}_{\mathcal{O}(1)}\mathbb{R}_{\mathcal{O}}(\mathcal{A}_S)\rangle$$
$$= \langle \mathbb{L}_{\mathcal{A}_S}\mathcal{O}(-3), \mathbb{L}_{\mathcal{A}_S}\mathcal{O}(-2), \mathcal{O}, \mathbb{L}_{\mathcal{A}_S}\mathcal{O}(-1), \mathcal{O}(1), \mathbb{R}_{\mathcal{O}(1)}\mathbb{R}_{\mathcal{O}}(\mathcal{A}_S)\rangle$$
$$= \langle \mathbb{L}_{\mathcal{A}_S}\mathcal{O}(-3), \mathcal{O}, \mathbb{R}_{\mathcal{O}}\mathbb{L}_{\mathcal{A}_S}\mathcal{O}(-2), \mathbb{L}_{\mathcal{A}_S}\mathcal{O}(-1), \mathcal{O}(1), \mathbb{R}_{\mathcal{O}(1)}\mathbb{R}_{\mathcal{O}}(\mathcal{A}_S)\rangle$$
$$= \langle \mathbb{L}_{\mathcal{A}_S}\mathcal{O}(-3), \mathcal{O}, \mathbb{R}_{\mathcal{O}}\mathbb{L}_{\mathcal{A}_S}\mathcal{O}(-2), \mathcal{O}(1), \mathbb{R}_{\mathcal{O}(1)}\mathbb{L}_{\mathcal{A}_S}\mathcal{O}(-1), \mathbb{R}_{\mathcal{O}(1)}\mathbb{R}_{\mathcal{O}}(\mathcal{A}_S)\rangle$$
$$= \langle \underbrace{\mathbb{L}_{\mathcal{A}_S}\mathcal{O}(-3), \mathcal{O}, \mathbb{L}\,\mathcal{O}(1)}_{=:\mathcal{D}_1}, \underbrace{\mathbb{R}_{\mathcal{O}}\mathbb{L}_{\mathcal{A}_S}\mathcal{O}(-2), \mathbb{R}_{\mathcal{O}(1)}\mathbb{L}_{\mathcal{A}_S}\mathcal{O}(-1), \mathbb{R}_{\mathcal{O}(1)}\mathbb{R}_{\mathcal{O}}(\mathcal{A}_S)}_{=:\mathcal{D}_2}\rangle.$$

Here, we used the shorthand $\mathbb{L} := \mathbb{L}_{\mathbb{R}_{\mathcal{O}}\mathbb{L}_{\mathcal{A}_S}\mathcal{O}(-2)}$ in the definition of $\mathcal{D}_2$ and in line seven that $\mathcal{O}$ and $\mathbb{L}_{\mathcal{A}_S}\mathcal{O}(-1)$ are orthogonal, so that their order can be changed. For the latter, we claim that $\mathbb{L}_{\mathcal{A}_S}\phi^*F \simeq \phi^*F \otimes \mathcal{O}(E')$, where $E' \subset \tilde{X}$ is the exceptional divisor of $\phi\colon \tilde{X} \longrightarrow \mathbb{P}^4$. This is proved as follows: the right adjoint $\Phi_{\mathcal{P}_R}$ of the Fourier–Mukai functor

$$\Phi_{\mathcal{P}=\mathcal{O}_{E'}(-1)}\colon \mathrm{D}^\mathrm{b}(S) \xrightarrow{\sim} \mathrm{D}^\mathrm{b}(S)_{-1} \hookrightarrow \mathrm{D}^\mathrm{b}(\tilde{X}),\ \mathcal{F} \longmapsto j_*p^*\mathcal{F} \otimes \mathcal{O}(E')$$

is the Fourier–Mukai functor $[2] \circ \Phi_{\mathcal{P}^\vee}\colon \mathrm{D}^\mathrm{b}(\tilde{X}) \longrightarrow \mathrm{D}^\mathrm{b}(S)$, cf. [246, Rem. 5.8]. The dual $\mathcal{P}^\vee$ of $\mathcal{P} = \mathcal{O}_{E'}(E')$ is computed as $\mathcal{O}_{E'}[-3]$ using [246, Cor. 3.40]. Hence, $[2] \circ \Phi_{\mathcal{P}^*}$ is

[10] Note that for the blow-up of a smooth point one, would also expect $i_*\mathcal{O}_E(-3)$ to be part of the semi-orthogonal decomposition. However, for the blow-up of a node, $i_*\mathcal{O}_E(-3)$ is not orthogonal to $\mathcal{O}_{\tilde{X}}$.

$G \longmapsto p_*(G|_{E'})[-1]$. Thus, on the one hand, $\Phi_{\mathcal{P}} \circ \Phi_{\mathcal{P}_R}$, which we identify with $i_* \circ i^!$ for the inclusion $i_* \colon \mathcal{A}_S \hookrightarrow \mathrm{D}^{\mathrm{b}}(\tilde{X})$, is $G \longmapsto (p^* p_* G|_{E'}) \otimes \mathcal{O}(E')[-1]$.

On the other hand, the adjunction map $i_* \circ i^! \longrightarrow \mathrm{id}$ is induced by the first map in the exact triangle $\mathcal{O}_{E'}(E')[-1] \longrightarrow \mathcal{O}_{\tilde{X}} \longrightarrow \mathcal{O}_{\tilde{X}}(E')$, which in turn is obtained by rotating the natural short exact sequence $0 \longrightarrow \mathcal{O}_{\tilde{X}} \longrightarrow \mathcal{O}_{\tilde{X}}(E') \longrightarrow \mathcal{O}_{E'}(E') \longrightarrow 0$. Hence, for $G \simeq \phi^* F$ one obtains $C(i_* i^! \phi^* F \longrightarrow \phi^* F) \simeq \phi^* F \otimes \mathcal{O}(E')$ as claimed.

Thus, we found the following semi-orthogonal decomposition

$$\mathrm{D}^{\mathrm{b}}(\tilde{X}) = \langle \mathcal{D}_1, \mathcal{D}_2 \rangle = \langle \mathbb{L}_{\mathcal{D}_1}(\mathcal{D}_2), \mathcal{D}_1 \rangle = \langle \mathcal{D}_2 \otimes \omega_{\tilde{X}}, \mathcal{D}_1 \rangle$$
$$= \langle \mathcal{D}_2 \otimes \omega_{\tilde{X}} \otimes \tau^* \mathcal{O}(1), \mathcal{D}_1 \otimes \tau^* \mathcal{O}(1) \rangle.$$

The final step in the proof is the verification of the following isomorphisms (up to shift) for which we refer to [296]:

$$\mathbb{R}_{\mathcal{O}} \mathbb{L}_{\mathcal{A}_S} \mathcal{O}(-2) \otimes \omega_{\tilde{X}} \otimes \tau^* \mathcal{O}(1) \simeq i_* \mathcal{O}_E(-2),$$

$$\mathbb{R}_{\mathcal{O}(1)} \mathbb{L}_{\mathcal{A}_S} \mathcal{O}(-1) \otimes \omega_{\tilde{X}} \otimes \tau^* \mathcal{O}(1) \simeq i_* \mathcal{O}_E(-1)$$

and, using $\mathbb{L}_{\mathcal{A}_S} \phi^* F \simeq \phi^* F \otimes \mathcal{O}(E')$ as above and the isomorphisms (1.11) in Section **6**.1.4,

$$\mathbb{L}_{\mathcal{A}_S} \mathcal{O}(-3) \otimes \tau^* \mathcal{O}(1) \simeq \mathcal{O} \quad \text{and} \quad \mathbb{L}\mathcal{O}(1) \otimes \tau^* \mathcal{O}(1) \simeq \tau^* \mathcal{O}(2).$$

Together with $\mathbb{R}_{\mathcal{O}(1)} \mathbb{R}_{\mathcal{O}}(\mathcal{A}_S) \otimes \omega_{\tilde{X}} \otimes \tau^* \mathcal{O}(1) \simeq \mathcal{A}_S$ this proves $\mathcal{A}_S \simeq \mathcal{A}_X$. □

The equivalence $\mathrm{D}^{\mathrm{b}}(S) \xrightarrow{\sim} \mathcal{A}_X$ in the proof above is the functor

$$F \longmapsto \mathbb{R}_{\mathcal{O}(1)} \mathbb{R}_{\mathcal{O}}(j_* p^* F \otimes \mathcal{O}(E')) \otimes \omega_{\tilde{X}} \otimes \tau^* \mathcal{O}(1).$$

Since the right mutations $\mathbb{R}_{\mathcal{O}(1)}$ and $\mathbb{R}_{\mathcal{O}}$ are cones of coevaluation maps, see Example 1.7, this can, at least in principle, be explicitly described as a Fourier–Mukai functor. However, unlike the case of cubics containing a plane and Pfaffian cubics the Fourier–Mukai kernel does not seem to be simply the twist of the ideal sheaf of some natural correspondence.

Remark 3.13 The subcategory $\langle \mathcal{A}_X, \tau^* \mathcal{O}_X, \tau^* \mathcal{O}_X(1), \tau^* \mathcal{O}_X(2) \rangle \subset \mathrm{D}^{\mathrm{b}}(\tilde{X})$ is a crepant categorical resolution of the singular variety as introduced by Kuznetsov [295]. Similarly, $\mathcal{A}_X \simeq \mathrm{D}^{\mathrm{b}}(S)$ can be viewed as a crepant categorical resolution of the perfect right orthogonal $\langle \mathcal{O}_X, \mathcal{O}_X(1), \mathcal{O}_X(2) \rangle^{\perp} \subset \mathrm{Perf}(X)$, see [296, Thm. 5.2].

3.4 Addington–Thomas The rationality conjectures of Hassett and Kuznetsov, see Conjecture **6**.5.15 and Conjecture 3.1, are equivalent as shown by the next result. The result was first proved by Addington and Thomas [8] for generic cubic fourfolds in each Noether–Lefschetz divisor and later extended to all in [38].

Theorem 3.14 (Addington–Thomas) *Let $X \subset \mathbb{P}^5$ be a smooth cubic fourfold. There exists a K3 surface S and an exact linear equivalence $\mathcal{A}_X \simeq \mathrm{D}^{\mathrm{b}}(S)$ if and only if X is contained in a Hassett divisor $\mathcal{C}_d$ with d satisfying condition (∗∗) in Proposition* **6**.5.10.

The idea to prove the result is to start with a Hassett divisor $\mathcal{C}_d$ and a cubic $X \in \mathcal{C}_d \cap \mathcal{C}_8$. In particular, X contains a plane $\mathbb{P}^2 \simeq \mathrm{P} \subset X$ which by Proposition 3.3 shows $\mathcal{A}_X \simeq \mathrm{D}^{\mathrm{b}}(S_{\mathrm{P}}, \alpha_{\mathrm{P},X})$ provided the discriminant curve $D_{\mathrm{P}} \subset \mathbb{P}^2$ is smooth. The latter can be assured by choosing a general $X \in \mathcal{C}_d \cap \mathcal{C}_8$, i.e. such that $H^{2,2}(X, \mathbb{Z})$ is of rank two, and using Remark **6**.1.5, (ii). Moreover, one can pick $X \in \mathcal{C}_d \cap \mathcal{C}_8$ such that $\alpha_{\mathrm{P},X}$ is trivial, see Remark **6**.6.17.

After possibly modifying the given equivalence, using the hypothesis (∗∗), deformation theory of complexes applied to the Fourier–Mukai kernel of the equivalence shows that the kernel deforms sideways with X within the Hassett divisor $\mathcal{C}_d$. This is enough to prove the assertion for the generic cubic in $\mathcal{C}_d$ and stability conditions are used to specialize to every other cubic in $\mathcal{C}_d$.

Theorem 3.14 can be rephrased in terms of the rational map $\Phi_\varepsilon \colon M_d \dashrightarrow M$, see Remark **6**.6.21. If Φ_ε, which depends on the additional choice of an isometry $\varepsilon \colon \Gamma_d \xrightarrow{\sim} \Lambda_d$, is regular at a point $(S, L) \in M_d$ with $X = \Phi_\varepsilon(S, L)$, then $\mathcal{A}_X \simeq \mathrm{D}^{\mathrm{b}}(S)$.

Remark 3.15 The K3 surfaces S of degree d satisfying (∗∗) and occurring in the above theorem, so $\mathrm{D}^{\mathrm{b}}(S) \simeq \mathcal{A}_X$ for some cubic fourfold X, form a Zariski dense open subset of the moduli space of all polarized K3 surfaces of degree d, see Remark **6**.6.21.

Thus, via $\mathrm{D}^{\mathrm{b}}(S) \simeq \mathcal{A}_X \subset \mathrm{D}^{\mathrm{b}}(X)$ the generic polarized K3 surface of degree d satisfying (∗∗), is realized as a *Fano visitor* of the Fano host X. This is not known for (generic) K3 surfaces of arbitrary degree. See also Remark 1.38

Remark 3.16 The following twisted version of the theorem was established in [250] for the generic cubic in each Noether–Lefschetz divisor $\mathcal{C}_d$:

If for a smooth cubic fourfold $X \subset \mathbb{P}^5$ there exists a K3 surfaces S with a Brauer class $\alpha \in \mathrm{Br}(S)$ and an exact linear equivalence

$$\mathcal{A}_X \simeq \mathrm{D}^{\mathrm{b}}(S, \alpha),$$

then $X \in \mathcal{C}_d$ for some d satisfying (∗∗′), see Remark **6**.5.20. Conversely, for each d satisfying (∗∗′) there exists a Zariski dense open subset $U_d \subset \mathcal{C}_d$ such that for all $X \in U_d$ one finds a twisted K3 surface (S, α) with $\mathcal{A}_X \simeq \mathrm{D}^{\mathrm{b}}(S, \alpha)$.

3.5 Fourfolds with Equivalent Kuznetsov Components Recall that for a projective K3 surface S, the Hodge structure $\widetilde{H}(S, \mathbb{Z})$ determines the derived category $\mathrm{D}^{\mathrm{b}}(S)$ and a twisted version of the result exists. The general situation is summarized as follows, see [249, Prop. 16.3.5 & 16.4.2] for references.

Theorem 3.17 (Mukai, Orlov, Huybrechts–Stellari) *Two twisted K3 surfaces (S,α) and (S',α') are derived equivalent, i.e. there exists an exact linear equivalence $\mathrm{D}^{\mathrm{b}}(S,\alpha) \simeq \mathrm{D}^{\mathrm{b}}(S',\alpha')$, if and only if there exists a Hodge isometry $\widetilde{H}(S,\alpha,\mathbb{Z}) \simeq \widetilde{H}(S',\alpha',\mathbb{Z})$ that respects the natural orientation of the four positive directions.*

The result has the consequence that the set of Fourier–Mukai partner is always finite, i.e. for a fixed K3 surface S there exist at most finitely many isomorphism classes of K3 surfaces S' with $\mathrm{D}^{\mathrm{b}}(S) \simeq \mathrm{D}^{\mathrm{b}}(S')$, cf. [249, Prop. 16.3.10]. A similar result holds for twisted K3 surfaces, see [256, Cor. 0.5].

It is natural to wonder whether any of these results well known for K3 surfaces also hold for the Calabi–Yau 2-categories $\mathcal{A}_X$ naturally associated with smooth cubic fourfolds $X \subset \mathbb{P}^5$. The result has not been established in full generality but almost, see [250, Thm. 1.4]. In order to formulate the result we write

$$\widetilde{H}(\mathcal{A}_X,\mathbb{Z}) := \widetilde{H}(X,\mathbb{Z}),$$

as introduced in Section **6**.5.4. The next result can be seen as a categorical global Torelli theorem for general and Hassett general cubic fourfolds.

Theorem 3.18 (Huybrechts) *Let X and X' be two smooth cubic fourfolds.*

(i) *If X is not contained in any Noether–Lefschetz divisor $\mathcal{C}_d$, then $\mathcal{A}_X \simeq \mathcal{A}_{X'}$ if and only if $X \simeq X'$.*
(ii) *If $X \in \mathcal{C}_d$ is general, then $\mathcal{A}_X \simeq \mathcal{A}_{X'}$ if and only if there exists a Hodge isometry $\widetilde{H}(\mathcal{A}_X,\mathbb{Z}) \simeq \widetilde{H}(\mathcal{A}_{X'},\mathbb{Z})$.*[11]

The key observation to prove these results is that any Fourier–Mukai equivalence $\mathcal{A}_X \simeq \mathcal{A}_{X'}$ induces a Hodge isometry $\widetilde{H}(\mathcal{A}_X,\mathbb{Z}) \simeq \widetilde{H}(\mathcal{A}_{X'},\mathbb{Z})$. Since the transcendental part of $\widetilde{H}(\mathcal{A}_X,\mathbb{Z})$ up to Tate twist is Hodge isometric to the transcendental part of $H^4(X,\mathbb{Z})$, the assertion in (i) follows from the global Torelli theorem for cubic fourfolds, see Theorem **6**.3.17. The proof of (ii) is more involved.

Note that unlike the case of K3 surfaces, any Hodge isometry $\widetilde{H}(\mathcal{A}_X,\mathbb{Z}) \simeq \widetilde{H}(\mathcal{A}_{X'},\mathbb{Z})$ can be modified to be orientation preserving, so this condition does not show up in (ii).

As a consequence of the discussion in this and in the last section, let us also mention the following categorical interpretation of the results in Section **6**.5.4, see Remark **6**.5.20.

Corollary 3.19 *Assume X is a smooth cubic fourfold and (S,α) is a twisted K3 surface. Then there exists an equivalence $\mathcal{A}_X \simeq \mathrm{D}^{\mathrm{b}}(S,\alpha)$ if and only if there is a Hodge isometry $\widetilde{H}(\mathcal{A}_X,\mathbb{Z}) \simeq \widetilde{H}(S,\alpha,\mathbb{Z})$.* □

[11] For technical reasons, all exact equivalences $\mathcal{A}_X \simeq \mathcal{A}_{X'}$ are assumed to be of Fourier–Mukai type, i.e. induced by an object in $\mathrm{D}^{\mathrm{b}}(X \times X')$. This is automatic, as has been shown by Li, Pertusi, and Zhao [320].

Remark 3.20 (i) The analogue of the finiteness result mentioned above also holds true [250, Thm. 1.1]: for a fixed smooth cubic fourfold $X \subset \mathbb{P}^5$, there exist at most finitely many isomorphism classes of smooth cubic fourfolds $X' \subset \mathbb{P}^5$ with $\mathcal{A}_X \simeq \mathcal{A}_{X'}$. Using the above theorem, this is reduced to a finiteness argument for Hodge structures. In the same spirit, Pertusi [386] uses the result to count cubic fourfolds with equivalent Kuznetsov components.

(ii) Unlike the case of projective K3 surfaces, for which the group of exact autoequivalences $\mathrm{Aut}(\mathrm{D}^{\mathrm{b}}(S))$ is always highly non-trivial, cf. [249, Conj. 16.3.14], the very general cubic fourfold is much easier in this respect [250, Thm. 1.2]: for the very general smooth cubic fourfold X, the group of all exact auto-equivalences of $\mathcal{A}_X$ that act trivially on $\widetilde{H}^{2,0}(\mathcal{A}_X)$ is an infinite cyclic group containing the subgroup of even shifts $[2n]$ as a subgroup of index three:

$$\mathrm{Aut}_s(\mathcal{A}_X)/\mathbb{Z} \cdot [2] \simeq \mathbb{Z}/3\mathbb{Z}.$$

In contrast to the case of cubic threefolds, see Proposition 2.4, there do exist non-isomorphic smooth cubic fourfolds $X, X' \subset \mathbb{P}^5$ with equivalent Kuznetsov components $\mathcal{A}_X \simeq \mathcal{A}_{X'}$.[12] At this point it is not clear what this means geometrically, but in view of Conjecture 3.1 one might venture a guess and propose the following.

Conjecture 3.21 *Assume $\mathcal{A}_X \simeq \mathcal{A}_{X'}$. Then X and X' are birational.*

Note that the converse does not hold, as, for example, all Pfaffian cubics are rational but according to Theorem 3.18 the Kuznetsov components $\mathcal{A}_X$ and $\mathcal{A}_{X'}$ of two generic Pfaffian cubics are certainly not equivalent. Some evidence for this conjecture has been provided by Fan and Lai [176].

Remark 3.22 Recall that for cubic threefolds, the Kuznetsov component $\mathcal{A}_Y$ determines the cubic threefold Y. So, why exactly do the arguments to prove Proposition 2.4 not work for cubic fourfolds?

There are two main reasons: first, the numerical Grothendieck group $N(\mathcal{A}_X)$ is of rank $\dim H^{2,2}(X, \mathbb{Q})_{\mathrm{pr}} + 2$, which varies between 2 and 22. This, in general, makes it impossible to modify a given equivalence $\Phi\colon \mathcal{A}_X \xrightarrow{\sim} \mathcal{A}_{X'}$ to one that respects the class of $[\mathcal{F}_L] = -[j^*\mathcal{I}_L] \in N(\mathcal{A}_X)$. In other words, Φ may not lead to an isomorphism between the Fano varieties $F(X) \xrightarrow{\sim} F(X')$. But even if it does, in order to apply the geometric global Torelli theorem, see Proposition **2**.3.12, one would need it to be polarized which might not be possible.

[12] For example, if $X \in \mathcal{C}_d$ is general with d satisfying (∗∗) in Proposition **6**.5.10, $d \equiv 2\,(6)$ and $d/2$ not prime, then $\mathcal{A}_X \simeq \mathrm{D}^{\mathrm{b}}(S)$ for a K3 surface S with non-isomorphic Fourier–Mukai partner S' which is associated with a non-isomorphic cubic $X' \in \mathcal{C}_d$. So, $\mathcal{A}_X \simeq \mathrm{D}^{\mathrm{b}}(S) \simeq \mathrm{D}^{\mathrm{b}}(S') \simeq \mathcal{A}_{X'}$.

3.6 Fano Variety As a Moduli Space in $\mathcal{A}_X$ As seen in Exercise 1.21, the projection $j^*(\mathcal{I}_L(1)) \in \mathcal{A}_X$ of the twist of the ideal sheaf of a line $L \subset X$ is, up to shift, the torsion free sheaf of rank three $\mathcal{F}_L$ which by definition, is the kernel of the evaluation map $H^0(X, \mathcal{I}_L(1)) \otimes \mathcal{O}_X \twoheadrightarrow \mathcal{I}_L(1)$. This leads to a map

$$F(X) \longrightarrow \mathrm{Ob}(\mathcal{A}_X), \quad [L] \longmapsto \mathcal{F}_L \simeq j^*(\mathcal{I}_L(1))[-1], \tag{3.3}$$

which lets us recover the symplectic structure on the Fano variety, cf. Remark **6**.3.16.

Proposition 3.23 (Kuznetsov–Markushevich) *The map* (3.3) *is injective and induces an isomorphism*

$$\mathrm{Ext}^1(\mathcal{I}_L, \mathcal{I}_L) \xrightarrow{\sim} \mathrm{Ext}^1(\mathcal{F}_L, \mathcal{F}_L) \tag{3.4}$$

on the level of first order deformations.

Proof The map (3.3) is injective, as $\mathcal{F}_L$ is locally free at a point $x \in X$ if and only if $x \in X \setminus L$. Applying $\mathrm{Hom}(\mathcal{I}_L(1), \ )$ and $\mathrm{Hom}(\ , \mathcal{F}_L)$ to the short exact sequence

$$0 \longrightarrow \mathcal{F}_L \longrightarrow \mathcal{O}_X^{\oplus 4} \longrightarrow \mathcal{I}_L(1) \longrightarrow 0, \tag{3.5}$$

one obtains horizontal and vertical exact sequences

$$\begin{array}{ccccccc} & & & \mathrm{Ext}^2(\mathcal{O}_X, \mathcal{F}_L)^{\oplus 4} & & \\ & & & \uparrow & & \\ \mathrm{Ext}^1(\mathcal{I}_L(1), \mathcal{O}_X)^{\oplus 4} & \longrightarrow & \mathrm{Ext}^1(\mathcal{I}_L, \mathcal{I}_L) & \stackrel{\gamma}{\hookrightarrow} & \mathrm{Ext}^2(\mathcal{I}_L(1), \mathcal{F}_L) & \longrightarrow & \mathrm{Ext}^2(\mathcal{I}_L(1), \mathcal{O}_X)^{\oplus 4} \\ & & & & \wr \uparrow \beta & & \\ & & & & \mathrm{Ext}^1(\mathcal{F}_L, \mathcal{F}_L) & & \\ & & & & \uparrow & & \\ & & & & \mathrm{Ext}^1(\mathcal{O}_X, \mathcal{F}_L)^{\oplus 4}. & & \end{array}$$

In the vertical direction, use $H^1(X, \mathcal{F}_L) = 0 = H^2(X, \mathcal{F}_L)$, which also follows from (3.5), to check that β is bijective. For the horizontal exact sequence, use Serre duality $\mathrm{Ext}^1(\mathcal{I}_L(1), \mathcal{O}_X) \simeq H^3(X, \mathcal{I}_L(-2))^* = 0$ to see that γ is injective.

Again by virtue of the short exact sequence (3.5), the sheaves $\mathcal{F}_L$ are all stable, and hence $\dim \mathrm{Hom}(\mathcal{F}_L, \mathcal{F}_L) = 1$. By Serre duality in $\mathcal{A}_X$, we then also have $\dim \mathrm{Ext}^2(\mathcal{F}_L, \mathcal{F}_L) = 1$ and $\mathrm{Ext}^{>2}(\mathcal{F}_L, \mathcal{F}_L) = 0$. A simple Riemann–Roch computation for the sheaf $\mathcal{F}_L$ finally reveals that $\chi(\mathcal{F}_L, \mathcal{F}_L) = -2$ and, therefore, $\dim \mathrm{Ext}^1(\mathcal{F}_L, \mathcal{F}_L) = 4$. Hence, (3.4) is an isomorphism. □

Remark 3.24 There is an analogous but different story for the LLSvS eightfold $Z(X)$, see Remark **6**.5.25. As proved in [302], its blow-up in the naturally embedded $X \subset Z(X)$ can be realized as a moduli space of objects in $\mathcal{A}_X$. Roughly, the projection of the objects

$k(x) \in \mathrm{D^b}(X)$ into the Kuznetsov component $\mathcal{A}_X$ deform in a larger eight-dimensional family, which is shown to be birational to $Z(X)$. So, the analogue of Proposition 3.23 for $\mathcal{I}_L(1)$ replaced by $k(x)$ does not hold.

The embedding $X \subset Z(X)$ for cubic fourfolds not containing a plane was studied by Addington and Lehn [7] and for those containing a plane by Ouchi [382].

Observe the similarity with the phenomenon in dimension three in Remark 2.7.

Combined with the fact that $\mathcal{A}_X$ is a Calabi–Yau 2-category, which can be interpreted as saying that for all $\mathcal{F} \in \mathcal{A}_X$ the space of first order deformations $\mathrm{Ext}^1(\mathcal{F}, \mathcal{F})$ is endowed with a natural symplectic structure, the result provides a new perspective on the symplectic nature of $F(X)$, see Remark **6**.3.16.

Corollary 3.25 *The natural pairing*

$$\mathrm{Ext}^1(\mathcal{I}_L, \mathcal{I}_L) \times \mathrm{Ext}^1(\mathcal{I}_L, \mathcal{I}_L) \longrightarrow \mathrm{Ext}^1(\mathcal{F}_L, \mathcal{F}_L) \times \mathrm{Ext}^1(\mathcal{F}_L, \mathcal{F}_L) \longrightarrow k$$

is non-degenerate. In particular, $F(X)$ is naturally endowed with a closed symplectic form $\sigma \in H^0(F(X), \Omega^2_{F(X)})$.

Proof The pairing on $\mathrm{Ext}^1(\mathcal{F}_L, \mathcal{F}_L)$ is non-degenerate by Serre duality in $\mathcal{A}_X$. Using

$$T_{[L]}F(X) \simeq \mathrm{Hom}(\mathcal{I}_L, \mathcal{O}_L) \simeq \mathrm{Ext}^1(\mathcal{I}_L, \mathcal{I}_L) \simeq \mathrm{Ext}^1(\mathcal{I}_L(1), \mathcal{I}_L(1)),$$

see Section **2**.1.3, and the isomorphism (3.4), one obtains a non-degenerate alternating pairing on $T_{[L]}F(X)$, i.e. a symplectic form σ.

Since $F(X)$ is a smooth projective variety, any differential form is closed. Kuznetsov and Markushevich [300] show this fact by a local argument similar to the standard one for moduli spaces of sheaves on K3 surfaces, cf. [253, Ch. 10]. □

Remark 3.26 The equivalence $\mathcal{A}_X \simeq \mathrm{D^b}(S)$ or $\mathcal{A}_X \simeq \mathrm{D^b}(S, \alpha)$ also explains why the numerical conditions (∗∗) and (∗∗′) in Section **6**.1.1 also decide whether $F(X)$ is birational to a moduli space of (twisted) sheaves on S, cf. Remark **6**.5.23.

3.7 Spherical Functor and the Hilbert Square $\mathcal{A}_X^{[2]}$ We add two further comments that shed some light on the derived category of the Fano variety.

Remark 3.27 Although the Fano correspondence does not provide any useful link between $\mathrm{D^b}(X)$ and $\mathrm{D^b}(F(X))$ directly, it can be refined for cubic fourfolds to provide a spherical functor.

This has been worked out by Addington [2, §4]: consider the structure sheaf $\mathcal{O}_{\mathbb{L}}$ of the Fano correspondence $\mathbb{L} \subset F(X) \times X$ and the associated Fourier–Mukai functor $\Psi := p_* \circ q^* \colon \mathrm{D^b}(X) \longrightarrow \mathrm{D^b}(F(X))$. Restricted to the Kuznetsov component it gives

$$\Psi \colon \mathcal{A}_X \longrightarrow \mathrm{D^b}(F(X)).$$

For the right adjoint $\Phi\colon \mathrm{D}^{\mathrm{b}}(F(X)) \longrightarrow \mathcal{A}_X$, the adjunction morphism $\mathrm{id} \longrightarrow \Phi \circ \Psi$ splits and its cone is the shift functor $[-2]$. In other words, Ψ is a spherical functor, i.e.

$$\Phi \circ \Psi \simeq \mathrm{id} \oplus [-2],$$

see [2, §4.2]. In particular, Ψ is faithful. It cannot also be full for more than one reason, e.g. because the derived category of the Calabi–Yau variety $F(X)$ does not admit non-trivial semi-orthogonal decompositions, cf. Exercise 1.17.

In this situation, the cone $T := C(\Psi \circ \Phi \longrightarrow \mathrm{id})$ of the other adjunction morphism is an auto-equivalence $T\colon \mathrm{D}^{\mathrm{b}}(F(X)) \xrightarrow{\sim} \mathrm{D}^{\mathrm{b}}(F(X))$, the associated spherical twist, cf. [351] for a streamlined argument.

Remark 3.28 Ganter and Kapranov [192] introduced the notion of the symmetric square $\mathcal{D}^{[2]}$ of a triangulated category $\mathcal{D}$.

(i) If $\mathcal{D}$ is the derived category $\mathrm{D}^{\mathrm{b}}(X)$, then the symmetric square $\mathrm{D}^{\mathrm{b}}(X)^{[2]}$ is the equivariant category $\mathrm{D}^{\mathrm{b}}_{\mathfrak{S}_2}(X \times X)$ of all linearized objects $(\mathcal{F}, \varphi\colon \mathcal{F} \xrightarrow{\sim} \iota^*\mathcal{F})$, where $\mathcal{F} \in \mathrm{D}^{\mathrm{b}}(X \times X)$ and ι swaps the two factors of $X \times X$.

The construction has the property that for a smooth projective surface S there exists an exact equivalence $\mathrm{D}^{\mathrm{b}}(S^{[2]}) \simeq \mathrm{D}^{\mathrm{b}}(S)^{[2]}$.

More generally, by a result of Krug, Ploog, and Sosna [288, Thm. 4.1], for a smooth projective variety X of dimension $n \geq 2$, there exists a semi-orthogonal decomposition

$$\mathrm{D}^{\mathrm{b}}(X^{[2]}) = \langle \mathrm{D}^{\mathrm{b}}(X)^{[2]}, \mathcal{D}_1, \ldots, \mathcal{D}_{n-2} \rangle$$

with $\mathcal{D}_i \simeq \mathrm{D}^{\mathrm{b}}(X)$. This decomposition can also be used to define $\mathrm{D}^{\mathrm{b}}(X)^{[2]}$ alternatively as the right orthogonal of the images of the naturally defined Fourier–Mukai functors $\mathrm{D}^{\mathrm{b}}(X) \xrightarrow{\sim} \mathcal{D}_i \subset \mathrm{D}^{\mathrm{b}}(X^{[2]})$. In any case, in the Grothendieck ring of (dg-enhanced) triangulated categories, see Section 1.8, one has

$$[\mathrm{D}^{\mathrm{b}}(X^{[2]})] = [\mathrm{D}^{\mathrm{b}}(X)^{[2]}] + (n-2) \cdot [\mathrm{D}^{\mathrm{b}}(X)] \in K_0(\text{dg-cat}_k).$$

(ii) The construction of the symmetric square also applies to the Kuznetsov component $\mathcal{A}_X$ and provides us with a triangulated category $\mathcal{A}_X^{[2]}$. Once the square $\mathcal{A}_X \boxtimes \mathcal{A}_X$ is defined, the symmetric square $\mathcal{A}_X^{[2]}$ is simply the equivariant category $(\mathcal{A}_X \boxtimes \mathcal{A}_X)_{\mathfrak{S}_2}$ with respect to the $\mathfrak{S}_2$-action on $\mathcal{A}_X \boxtimes \mathcal{A}_X$. Viewing $\mathcal{A}_X$ as an admissible subcategory of $\mathrm{D}^{\mathrm{b}}(X)$, the square $\mathcal{A}_X \boxtimes \mathcal{A}_X \subset \mathrm{D}^{\mathrm{b}}(X \times X)$ can be explicitly realized as the right orthogonal complement of all objects of the form $\mathcal{O}_X(i) \boxtimes \mathcal{F}, \mathcal{F} \boxtimes \mathcal{O}_X(i) \in \mathrm{D}^{\mathrm{b}}(X \times X)$ with $i = 0, \ldots, n-2$. In particular, $\mathcal{A}_X^{[2]} \subset \mathrm{D}^{\mathrm{b}}(X)^{[2]}$ and one can show that in $K_0(\text{dg-cat}_k)$ we have $[\mathrm{D}^{\mathrm{b}}(X)^{[2]}] = [\mathcal{A}_X^{[2]}] + (n-1) \cdot [\mathrm{D}^{\mathrm{b}}(X)]$.

(iii) Comparing the result with Corollary 1.35, we deduce that for $n = 4$, one has

$$[\mathcal{A}_X^{[2]}] = [\mathrm{D}^{\mathrm{b}}(F(X))] \in K_0(\text{dg-cat}_k). \tag{3.6}$$

Note that for a cubic threefold $Y \subset \mathbb{P}^4$ one finds $[\mathcal{A}_Y^{[2]}] = [\mathrm{D}^{\mathrm{b}}(F(Y))] + [\mathrm{D}^{\mathrm{b}}(Y)]$ and in dimension $n \geq 4$ one has $[\mathcal{A}_X^{[2]}] + (n-4) \cdot [\mathrm{D}^{\mathrm{b}}(X)] = [\mathrm{D}^{\mathrm{b}}(F(X))]$. To arrive at (3.6), one can alternatively use the comparison of symmetric products $Z^{(2)}$, or rather their images under the map $K_0(\mathrm{Var}_k) \longrightarrow K_0(\text{dg-cat}_k)$ mentioned in Remark 1.36, and $[\mathrm{D^b}(Z)^{[2]}]$ as worked out by Galkin and Shinder [191], see [394, Sec. 4].

(iii) It has been conjectured that the equality (3.6) can be lifted to an exact equivalence $\mathrm{D^b}(F(X)) \simeq \mathcal{A}_X^{[2]}$ and a possible approach has been put forward by Galkin. The question remains open in general, but evidence for it is provided by the special case of cubic fourfolds X for which $F(X)$ is birational to a Hilbert scheme $S^{[2]}$, e.g. for all Pfaffian cubic fourfolds. In this case, there exists an equivalence

$$\mathrm{D^b}(F(X)) \simeq \mathrm{D^b}(S^{[2]}) \simeq \mathrm{D^b}(S)^{[2]} \simeq \mathcal{A}_X^{[2]},$$

as the birational correspondence between $F(X)$ and $S^{[2]}$, see Corollary **6**.5.22, is automatically a composition of Mukai flops [246, Prop. 11.28].

4 Chow Groups and Chow Motives

In this very last section we summarize the most important results on Chow groups and Chow motives of cubic hypersurfaces. Some of the results have been mentioned and partially proved before and for others we refer to the literature.

4.1 Chow Groups Recall that the integral Chow ring of a smooth projective variety X of dimension n is naturally graded $\mathrm{CH}^*(X) \simeq \bigoplus_{i=0}^{n} \mathrm{CH}^i(X)$. Neither the kernel nor the image of the cycle class map (for complex varieties) $\mathrm{CH}^i(X) \longrightarrow H^{2i}(X, \mathbb{Z})(i)$ are well understood in general. The kernel, i.e. the homological trivial part, comes with the Abel–Jacobi map $\mathrm{CH}^i(X)_{\mathrm{hom}} \longrightarrow J^{2i-1}(X)$, cf. [474, §19.2], but its kernel, if not 0, is hard to analyze.

For smooth cubic hypersurfaces of dimension two, three, and four, the Chow ring decomposes as follows

$$\mathrm{CH}^*(S) = \bigoplus_{i=0}^{2} \mathrm{CH}^i(S) \simeq \mathbb{Z} \oplus \mathbb{Z}^{\oplus 7} \oplus \mathbb{Z},$$

$$\mathrm{CH}^*(Y) = \bigoplus_{i=0}^{3} \mathrm{CH}^i(Y) \simeq \mathbb{Z} \oplus \mathbb{Z} \oplus (J(Y) \oplus \mathbb{Z}) \oplus \mathbb{Z}, \quad \text{and}$$

$$\mathrm{CH}^*(X) = \bigoplus_{i=0}^{4} \mathrm{CH}^i(X) \simeq \mathbb{Z} \oplus \mathbb{Z} \oplus \mathbb{Z}^{\oplus \rho(X)} \oplus (\mathrm{CH}^3(X)_{\mathrm{hom}} \oplus \mathbb{Z}) \oplus \mathbb{Z}.$$

The assertion for cubic surfaces is clear and for cubic threefolds, the identification $\mathrm{CH}^2(Y)_{\mathrm{hom}} \simeq J(Y)$ is the content of Corollary **5**.3.16.

For cubic fourfolds, $\rho(X)$ denotes the rank of $H^{2,2}(X,\mathbb{Z})$. Note that by [476, Thm. 18] or [480, Thm. 1.4] the cycle class map $\mathrm{CH}^2(X) \twoheadrightarrow H^{2,2}(X,\mathbb{Z})$ is indeed surjective, i.e. the integral Hodge conjecture holds, cf. Section **6**.3.4. In fact, the cycle class map is an isomorphism, i.e. $\mathrm{CH}^2(X)_{\mathrm{hom}} = 0$, an assertion that holds for all smooth cubic hypersurfaces of dimension $n \geq 4$.

Indeed, by a result of Bloch and Srinivas [72, Thm. 1 (ii)] ones knows that for a smooth complex projective variety X with $\mathrm{CH}_0(X) \simeq \mathbb{Z}$ the Abel–Jacobi map induces isomorphisms of groups $\mathrm{CH}^2(X)_{\mathrm{alg}} \simeq \mathrm{CH}^2(X)_{\mathrm{hom}} \simeq J^3(X)$, cf. (3.11) in the proof of Corollary **5**.3.16. This indeed proves

$$\mathrm{CH}^2(X)_{\mathrm{hom}} = 0 \quad \text{if } \dim(X) \geq 4,$$

as $H^3(X,\mathbb{Q}) = 0$ for cubics of dimension at least four.

For cubic hypersurfaces of dimension $n \geq 5$, Paranjape [384] shows that $\mathrm{CH}^{n-1}(X) \otimes \mathbb{Q} \simeq \mathbb{Q}$, see also [427] and the discussion after (4.1) below. Furthermore, one expects

$$\mathrm{CH}_i(X) \otimes \mathbb{Q} \simeq \mathbb{Q} \text{ for } 3(i+1) \leq n+1,$$

see [384, Conj. 1.8].

A quadratic bound is implied by a result of Esnault, Levine, and Viehweg [175]: for a smooth cubic hypersurface $\mathrm{CH}_i(X) \otimes \mathbb{Q} \simeq \mathbb{Q}$ holds for all i with $\binom{i+3}{2} \leq n+1$. For cubic hypersurfaces of dimension five, only $\mathrm{CH}^3(\mathrm{X})$ is non-trivial. Its structure has been determined by Fu and Tian [185].

The ring structure of $\mathrm{CH}^*(X)$ is another interesting object of study. Not much is known but recently Diaz [153] showed that the multiplication map

$$(\mathrm{CH}^i(X) \otimes \mathbb{Q}) \otimes (\mathrm{CH}^j(X) \otimes \mathbb{Q}) \longrightarrow \mathrm{CH}^{i+j}(X) \otimes \mathbb{Q}$$

is of rank at most one.

Following the general philosophy of these notes that cubic hypersurfaces should be studied by means of their Fano varieties of lines, one would like to describe the Chow group $\mathrm{CH}_1(X)$ of one-dimensional cycles and the subgroup generated by lines. As first observed by Paranjape [384, Sec. 4] for cubic fivefolds and later reproved and extended by Shen [427, Thm. 1.1], lines do generate the full $\mathrm{CH}_1(X)$ or, equivalently, for a smooth cubic hypersurface $X \subset \mathbb{P}^{n+1}$ of dimension $n > 1$ the Fano correspondence induces a surjection

$$\mathrm{CH}_0(F(X)) \twoheadrightarrow \mathrm{CH}_1(X). \tag{4.1}$$

Paranjape's original idea uses quadric fibrations of cubics as discussed in Section **1**.5.2 and the classical fact that Chow groups of one-cycles on quadric hypersurfaces are generated by lines.

By Section **2**.3.1, for smooth cubics of dimension at least five the Fano variety $F(X)$ is Fano, i.e. the dual of its canonical bundle is ample, and hence $F(X)$ is rationally connected. Thus, $\mathrm{CH}_0(F(X)) \simeq \mathbb{Z}$ and, therefore, (4.1) implies $\mathrm{CH}_1(X) \simeq \mathbb{Z}$. For smooth cubic hypersurfaces of dimension $n \geq 5$, the group of two-dimensional cycles $\mathrm{CH}_2(X)$ has been investigated by Mboro [349] showing that it is generated by rational surfaces and in fact by planes contained in X for $n \geq 7$.

Remark 4.1 For cubics of dimension three and four, (4.1) is more interesting. In particular, the map is not injective.

(i) For example, for smooth cubic threefolds, we have the decomposition $\mathrm{CH}_1(Y) = \mathrm{CH}^2(Y) \simeq \mathbb{Z} \oplus \mathrm{CH}^2(Y)_{\mathrm{hom}} \simeq \mathbb{Z} \oplus J(Y)$, while the Chow group $\mathrm{CH}_0(F(Y))$ of the Fano surface $F(Y)$ is non-representable, i.e. the kernel of (4.1) cannot be parametrized by a scheme of finite type.

(ii) For a smooth cubic fourfold $X \subset \mathbb{P}^5$ the kernel of (4.1) is divisible. This was observed by J. Shen and Yin [425, Lem. 1.2] relying on a result by M. Shen and Vial [429, Sec. 20]. Since $\mathrm{CH}_0(F(X))$ is torsion free by Roitman's theorem, this proves that $\mathrm{CH}_1(X)$ is torsion free also. The kernel of (4.1) was described more explicitly by M. Shen and Vial as the homologically trivial part of the subgroup generated by $[L_1] + [L_2] + [L_3]$ for triangles $\mathbb{P}^2 \cap X = L_1 \cup L_2 \cup L_3$. As was previously mentioned in Remark **6**.3.24, $\mathrm{CH}^3(X)_{\mathrm{alg}} \otimes \mathbb{Q} \simeq \mathrm{CH}^3(X)_{\mathrm{hom}} \otimes \mathbb{Q} \simeq \mathrm{CH}^2(F(X))_{\mathrm{hom}} \otimes \mathbb{Q}$.

The Chow ring of the Fano variety $F(X)$ of a smooth cubic fourfold has been investigated from the hyperkähler perspective in great detail by M. Shen and Vial [429]. In particular, they define a natural filtration of $\mathrm{CH}^i(F(X))$, $i = 0, \ldots, 4$ and introduce a certain splitting. The filtration should be seen as an instance of the Bloch–Beilinson filtration conjectured to exist, with natural functoriality properties, for arbitrary smooth projective varieties. The splitting of the filtration is, however, a special feature of hyperkähler manifolds. This was initiated by Beauville [52] and further studied by Voisin [478]. In particular, the cycle class map injects the subring of $\mathrm{CH}^*(F(X))$ generated by $\mathrm{CH}^1(F(X))$ (and all Chern classes of $F(X)$) into cohomology. This 'weak splitting property' was first observed for K3 surfaces by Beauville and Voisin [59], but it remains an open question for general hyperkähler manifolds.

4.2 Decomposition of the Diagonal A smooth projective variety X of dimension n is said to have an integral Chow theoretic *decomposition of the diagonal* if

$$[\Delta] = [X \times \{x\}] + [Z] \quad \text{in } \mathrm{CH}^n(X \times X) \tag{4.2}$$

for some point $x \in X$ and such that the first projection $Z \twoheadrightarrow X$ is not surjective. If (4.2) only holds with rational coefficients, so

$$N \cdot [\Delta] = N \cdot [X \times \{x\}] + [Z] \quad \text{in } \mathrm{CH}^n(X \times X) \tag{4.3}$$

for some $N > 0$ and with Z as before, one says that X has a rational Chow theoretic decomposition of the diagonal. Further weaker versions can be introduced by requiring (4.2) to hold only up to algebraic equivalence or in integral or rational cohomology.

This notion has its origin in the paper by Bloch and Srinivias [72] and has more recently been explored in a series of papers by Voisin [480, 481, 483]. We summarize some of the results that concern cubic hypersurfaces.

• First, Bloch and Srinivas [72] prove that any smooth complex projective variety of dimension n with $\mathrm{CH}_0(X) = \mathrm{CH}^n(X) \simeq \mathbb{Z}$, so for example, a rationally connected variety, admits a rational Chow theoretic decomposition of the diagonal. This certainly applies to unirational varieties and so to all cubic hypersurfaces of dimension at least two. A rational or stably rational variety even admits an integral Chow theoretic decomposition, which is not true for general rationally connected varieties.

• Second, a smooth projective variety X admits an integral Chow theoretic decomposition of the diagonal if and only if its CH_0 is universally trivial, i.e. $\mathrm{CH}_0(X_K) = \mathbb{Z}$ for all field extensions K/k.

• Third, if there exists a dominant rational map $\mathbb{P}^n \dashrightarrow X$ of degree N then (4.3) exists with this N. In particular, using Corollary **2**.1.21, for any smooth cubic hypersurface $X \subset \mathbb{P}^{n+1}$ of dimension $n \geq 2$ there exists a decomposition of the form

$$2 \cdot [\Delta] = 2 \cdot [X \times \{x\}] + [Z] \quad \text{in } \mathrm{CH}^n(X \times X)$$

with Z, as before.

Example 4.2 If a cubic hypersurface also admits a parametrization $\mathbb{P}^n \dashrightarrow X$ of odd degree N, then it admits an integral Chow theoretic decomposition of the diagonal.

Thus, by the results of Hassett [228, Cor. 40] and Lai [304, Thm. 0.1] the generic cubic fourfold $X \in \mathcal{C}_d$ with $d = 14, 18, 26, 30, 38, 42$ admits an integral Chow theoretic decomposition of the diagonal, cf. Remark **6**.5.17.

Note that rationality of X implies the existence of (4.3), but at least conjecturally the generic $X \in \mathcal{C}_d$ for $d = 18$ or $d = 30$ provides a counterexample to the converse.

The first result of Voisin [483, Thm. 1.1] to be mentioned here states that the possible variants of integral decompositions of the diagonal essentially all coincide for cubic hypersurfaces (with certain restrictions on the dimensions). This makes use of the fact that the residual map $X^{[2]} \dashrightarrow \mathbb{L}_{\mathbb{G}}|_X \longrightarrow X$ is rationally a $\mathbb{P}^n$-bundle, see Section **2**.4.1.

Theorem 4.3 (Voisin) *Let $X \subset \mathbb{P}^{n+1}$ be a smooth cubic fourfold or a cubic hypersurface of odd dimension n. Then X admits an integral Chow theoretic decomposition of the diagonal if and only if it admits an integral cohomological decomposition of the diagonal.*

Another concept of interest is the essential CH_0-dimension, defined as the minimal m such that there exists a closed subset $Z \subset X$ of dimension m for which the push-forward $\mathrm{CH}_0(Z_K) \twoheadrightarrow \mathrm{CH}_0(X_K)$ surjective for all field extensions K/k. Clearly, if X admits an integral Chow theoretic decomposition of the diagonal, then its essential CH_0-dimension is 0. According to [483, Thm. 1.3], only one other essential CH_0-dimension is possible for very general cubic hypersurfaces.

Theorem 4.4 (Voisin) *If $X \subset \mathbb{P}^{n+1}$ is a very general cubic hypersurface of dimension n, then its essential CH_0-dimension is 0 or n.*

Remark 4.5 An integral (Chow theoretic or, equivalently, cohomological) decomposition of the diagonal is known to exist for certain cubic hypersurfaces of dimension three and four:

(i) According to [483, Thm. 5.6], a smooth cubic threefold $Y \subset \mathbb{P}^4$ admits a decomposition of the diagonal if and only if the class $(1/4!) \cdot [\Xi]^4 \in H^8(J(Y), \mathbb{Z})$ is algebraic, cf. Section **5**.3.3 and Remark **5**.4.14. Moreover, Voisin shows [483, Thm. 4.5] that there exists a countable union of closed subsets of codimension at most three in $|\mathcal{O}_{\mathbb{P}^4}(3)|_{\mathrm{sm}}$ for which this condition is satisfied and which, therefore, parametrize cubics that admit an integral Chow theoretic decomposition of the diagonal. In these cases, $J(Y)$ is shown to be isogenous to a Jacobian of a curve via an isogeny of odd degree.

The cohomological condition on the intermediate Jacobian $J(Y)$ is satisfied for stably rational cubic threefolds, but at this point no smooth stably rational cubic threefold is known.

(iii) Generalizing the examples mentioned in Example 4.2, Voisin proves [483, Thm. 5.6] that every cubic fourfold $X \in \mathcal{C}_d$ with $d \equiv 2\,(4)$ admits an integral decomposition of the diagonal. Note that this numerical condition is implied by the stronger condition $(**)$ in Section **6**.5.2, which by Conjecture **6**.5.15 is expected to be equivalent to the rationality of X. As a rational cubic fourfold certainly admits an integral decomposition of the diagonal, the picture is consistent. However, it also shows again that rationality should be strictly stronger than the existence of an integral decomposition of the diagonal.

4.3 Finite-dimensional Motives Recall from Remark **1**.1.11 that the motive $\mathfrak{h}(X)$ of a smooth hypersurface $X \subset \mathbb{P}^{n+1}$ of degree d in the category of rational Chow motives $\mathrm{Mot}(k)$ admits a (multiplicative) decomposition, cf. [153, 184, 387]:

$$\mathfrak{h}(X) \simeq \mathfrak{h}(X)_{\mathrm{pr}} \oplus \bigoplus_{i=0}^{n} \mathbb{Q}(-i),$$

for which the primitive part $\mathfrak{h}(X)_{\rm pr}$ has cohomology concentrated in degree n and its Chow group $\mathrm{CH}^*(\mathfrak{h}(X)_{\rm pr})$ contains the homological trivial part of $\mathrm{CH}^*(X)$. Guletskiĭ [214] uses a constant cycle surface in $F(X)$ constructed by Voisin [478] to prove that for cubic fourfolds the decomposition is in fact defined integrally.

In earlier parts of these notes we have studied links between the motive of the cubic $X \subset \mathbb{P}^{n+1}$ and the motive of its Fano variety $F(X)$. For example, in Section **2**.4.2 we expressed $\mathfrak{h}(F(X))$ as a direct summand of the symmetric product $S^2\mathfrak{h}(X)$. More specifically, in Remark **5**.3.17 we observed an isomorphism of rational Chow motives

$$\mathfrak{h}^3(Y) \simeq \mathfrak{h}^1(F(Y))$$

for a smooth cubic threefold $Y \subset \mathbb{P}^4$ and in Remark **6**.3.24 the isomorphism

$$\mathfrak{h}(X)_{\rm tr} \simeq \mathfrak{h}^2(F(X))_{\rm tr}(1)$$

for smooth cubic fourfolds $X \subset \mathbb{P}^5$ was mentioned.

The rational Chow motive $\mathfrak{h}(X)$ of any smooth projective variety X is conjectured to be finite-dimensional in the sense of Kimura and O'Sullivan [20, 272]. But even in small dimensions, this is still an open problem.

For example, only very few K3 surfaces are known to have motives of finite dimension. For complex projective varieties the following criterion, observed by Vial [469, Thm. 4], is sometimes useful.

Proposition 4.6 *Assume that X is a smooth complex projective variety such that all Abel–Jacobi maps $\mathrm{CH}^i(X)_{\rm hom} \otimes \mathbb{Q} \hookrightarrow J^{2i-1}(X) \otimes \mathbb{Q}$ are injective. Then the rational Chow motive $\mathfrak{h}(X)$ of X is finite-dimensional.*

Remark 4.7 (i) Recall from the discussion in Section **2**.4.2 that with $\mathfrak{h}(X)$ for a cubic hypersurface X, the motive $\mathfrak{h}(F(X))$ of its Fano variety of lines is finite-dimensional also [307].

(ii) For smooth cubic hypersurfaces, finite-dimensionality is known in dimension $n \leq 3$. For cubic curves and cubic surfaces this is obvious and for cubic threefolds this was discussed already in Remark **5**.3.17, tacitly relying on Proposition 4.6.

For cubic threefolds, one can alternatively follow arguments by Diaz [152] and first use the Albanese map $a\colon F = F(Y) \hookrightarrow A(F)$ in conjunction with Corollary **5**.3.16 to produce a split surjection $\mathfrak{h}(A(F)) \twoheadrightarrow \mathfrak{h}(F)$. Since the motive of an abelian variety is finite-dimensional, one obtains the finite-dimensionality of $\mathfrak{h}(F)$ and, consequently, also of $\mathfrak{h}(Y)$.

(iii) For cubic fourfolds the question is wide open. In particular, it is known that the Abel–Jacobi map $\mathrm{CH}^3(X)_{\rm hom} \longrightarrow J^5(X) = 0$ is not injective and so Proposition 4.6 does not apply.

Laterveer [307, 308] proves finite-dimensionality in a series of interesting examples. However, maybe not surprisingly, since the Kimura–O'Sullivan conjecture is open for general polarized K3 surfaces of any degree, nothing is known for the general cubic in any of the Hassett divisors $M \cap \mathcal{C}_d \subset M$.

Just to mention one example, a cubic fourfold that can be described as a triple cover $X \twoheadrightarrow \mathbb{P}^4$ branched over a cubic threefold $Y \subset \mathbb{P}^4$, see Section **1.5.6**, has a finite-dimensional Chow motive. Indeed, Laterveer [308] deduces this from the surjective morphism $\mathrm{Bl}_{S\times X_0}(Y\times E) \twoheadrightarrow X$ constructed in Section **1.5.6**. Here, E is an elliptic curve and X_0 consists of three points, which both have finite-dimensional motives. Cubic fourfolds of this type account for a 10-dimensional family in the moduli space of all cubic fourfolds and intersecting with the family with Hassett divisors leads to 9-dimensional families of K3 surfaces of unbounded degree with finite-dimensional motives.

Similarly, the motive of a smooth cubic fourfold given by an equation of the form $F = f(x_0, x_1, x_2, x_3) + x_4^3 + x_5^3$ is finite-dimensionals [307, Cor. 17].

(iii) Somewhat surprisingly, smooth cubic hypersurfaces $X \subset \mathbb{P}^6$ of dimension five are also known to have finite-dimensional motives. This follows again from Proposition 4.6 and $\mathrm{CH}^i(X) \otimes \mathbb{Q} \simeq \mathbb{Q}$, $i = 0, \ldots, 5$, mentioned before.[13]

(iv) Not much is known in this respect for cubics of dimension at least six. Of course, by induction and using the triple cover construction, one can construct smooth cubic hypersurfaces with finite-dimensional motives in any dimension.

4.4 Kuznetsov Component versus Motives The Kuznetsov component $\mathcal{A}_X$ of a cubic hypersurface $X \subset \mathbb{P}^{n+1}$ is the most interesting part of the semi-orthogonal decomposition $\mathrm{D^b}(X) = \langle \mathcal{A}_X, \mathcal{O}_X, \ldots, \mathcal{O}_X(n-2)\rangle$. For example, one always has

$$K(X) \simeq K(\mathrm{D^b}(X)) \simeq K(\mathcal{A}_X) \oplus \mathbb{Z}^{\oplus n-1},$$

as was noted in Exercise 1.22 already.

Example 4.8 Let us first examine the case of cubic surfaces $S \subset \mathbb{P}^3$ and cubic threefolds $Y \subset \mathbb{P}^4$.

(i) For a smooth cubic surface S one finds

$$K(\mathcal{A}_S) \simeq \mathbb{Z}^{\oplus 8}.$$

Indeed, this follows from the semi-orthogonal decomposition (2.1) in Section 2.1 obtained from viewing S as the blow-up of $\mathbb{P}^2$ and the resulting semi-orthogonal decomposition of the Kuznetsov componen $\mathcal{A}_S = \langle \bigoplus_{i=1}^6 \mathcal{O}_{E_i}(-1), \mathcal{O}_S(-2), \mathcal{O}_S(-1)\rangle$.

(ii) The situation is more interesting for cubic threefolds. Here, one has

$$K(\mathcal{A}_Y) \simeq J(Y) \oplus \mathbb{Z},$$

[13] Thanks to R. Laterveer for the explanation.

which follows from $\mathrm{CH}^2(Y)_{\mathrm{hom}} \simeq J(Y)$. As $K(Y)$ and $\mathrm{CH}^*(Y)$ are typically compared via the Chern character, the result is a priori only valid after tensoring with $\mathbb{Q}$.

Example 4.9 Assume $X \subset \mathbb{P}^5$ is a smooth cubic fourfold with $\mathcal{A}_X \simeq \mathrm{D}^{\mathrm{b}}(S)$ for some K3 surface S. As observed by Bülles [99, Thm. 0.3], in this case, a more geometric analogue of the isomorphism of Grothendieck groups is the isomorphims

$$\mathfrak{h}_{\mathrm{tr}}(S) \simeq \mathfrak{h}(X)_{\mathrm{tr}}(1) \tag{4.4}$$

in the category of rational Chow motives $\mathrm{Mot}(k)$. Here,

$$\mathfrak{h}(S) \simeq \mathfrak{h}_{\mathrm{alg}}(S) \oplus \mathfrak{h}_{\mathrm{tr}}(S) \quad \text{and} \quad \mathfrak{h}(X) = \mathfrak{h}_{\mathrm{alg}}(X) \oplus \mathfrak{h}_{\mathrm{tr}}(X)$$

with $\mathfrak{h}_{\mathrm{alg}}(S) \simeq \mathbb{Q} \oplus \mathbb{Q}(-1)^{\oplus \rho(S)} \oplus \mathbb{Q}(-2)$ and $\mathfrak{h}_{\mathrm{alg}}(X) = \bigoplus_{i=0}^4 \mathbb{Q}(-i) \oplus \mathbb{Q}(-2)^{\oplus \rho(X)-1}$, where as before $\rho(X) = \dim H^{2,2}(X, \mathbb{Q})$.

Taking Chow groups of (4.4) leads to an isomorphism

$$\mathrm{CH}_0(S)_{\mathrm{hom}} \otimes \mathbb{Q} \simeq \mathrm{CH}_1(X)_{\mathrm{hom}} \otimes \mathbb{Q}.$$

This should really be an isomorphism between the integral Chow groups. Indeed, both groups $\mathrm{CH}_0(S)$ and $\mathrm{CH}_1(X)$ are divisible and the former is also torsion free (and probably also the latter).

Remark 4.10 There are well-known and conjectural links between the various notions of motives summarized by the diagram

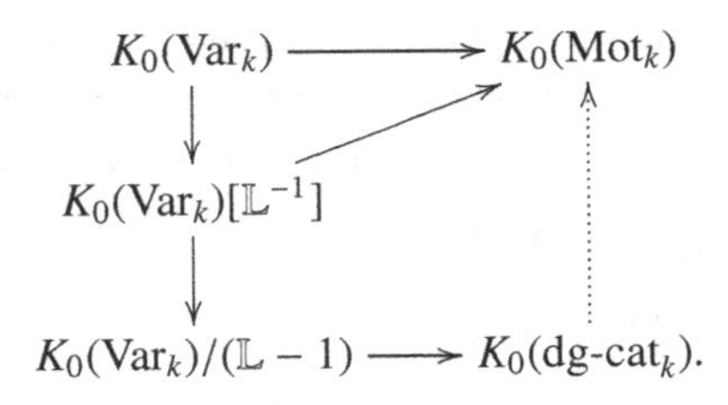

The existence of the dotted arrow on the right is unclear and should, in any case, only exist on the image of the horizontal arrow on the bottom.

For two smooth cubic hypersurfaces $X, X' \subset \mathbb{P}^{n+1}$ with equivalent Kuznetsov components $\mathcal{A}_X \simeq \mathcal{A}_{X'}$, the diagram reflects the expectation that the classes of their motives coincide, i.e. $[\mathfrak{h}(X)] = [\mathfrak{h}(X')] \in K_0(\text{dg-cat}_k)$, or even that their motives are isomorphic, i.e. $\mathfrak{h}(X) \simeq \mathfrak{h}(X')$. Note that the general conjecture that $\mathfrak{h}(Z) \simeq \mathfrak{h}(Z')$ for any two smooth projective varieties with equivalent derived categories $\mathrm{D}^{\mathrm{b}}(Z) \simeq \mathrm{D}^{\mathrm{b}}(Z')$ is still very much open. But it has been proved for K3 surfaces [251] and the combination of this result with Example 4.9 lends further evidence to the general conjectural picture. In fact, Fu and Vial [186] prove that for smooth cubic fourfolds, $\mathcal{A}_X \simeq \mathcal{A}_{X'}$ implies $\mathfrak{h}(X) \simeq \mathfrak{h}(X')$ as Frobenius algebra objects.

References

[1] Jeffrey Achter. On the abelian fivefolds attached to cubic surfaces. *Math. Res. Lett.*, 20(5):805–824, 2013. (Cited on page 186.)

[2] Nicolas Addington. New derived symmetries of some hyperkähler varieties. *Algebr. Geom.*, 3(2):223–260, 2016. (Cited on pages 397 and 398.)

[3] Nicolas Addington. On two rationality conjectures for cubic fourfolds. *Math. Res. Lett.*, 23(1):1–13, 2016. (Cited on pages 308, 340, 341, and 384.)

[4] Nicolas Addington and Asher Auel. Some non-special cubic fourfolds. *Doc. Math.*, 23:637–651, 2018. (Cited on page 349.)

[5] Nicolas Addington and Francesco Giovenzana. On the period of Lehn, Lehn, Sorger, and van Straten's symplectic eightfold. *arXiv:2003.10984*. (Cited on page 342.)

[6] Nicolas Addington, Brendan Hassett, Yuri Tschinkel, and Anthony Várilly-Alvarado. Cubic fourfolds fibered in sextic del Pezzo surfaces. *Amer. J. Math.*, 141(6):1479–1500, 2019. (Cited on page 338.)

[7] Nicolas Addington and Manfred Lehn. On the symplectic eightfold associated to a Pfaffian cubic fourfold. *J. Reine Angew. Math.*, 731:129–137, 2017. (Cited on pages 286, 342, 386, 389, and 397.)

[8] Nicolas Addington and Richard Thomas. Hodge theory and derived categories of cubic fourfolds. *Duke Math. J.*, 163(10):1885–1927, 2014. (Cited on pages 332, 339, 349, 377, 382, 385, and 392.)

[9] Allan Adler. Some integral representations of $\mathrm{PSL}_2(\mathbb{F}_p)$ and their applications. *J. Algebra*, 72(1):115–145, 1981. (Cited on pages 216 and 244.)

[10] Allan Adler and S. Ramanan. *Moduli of abelian varieties*, Vol. 1644 of *Lecture Notes in Math.* Berlin: Springer-Verlag, 1996. (Cited on page 245.)

[11] Daniele Agostini, Ignacio Barros, and Kuan-Wen Lai. On the irrationality of moduli spaces of K3 surfaces. *arXiv:2011.11025*. (Cited on page 148.)

[12] Daniel Allcock. The moduli space of cubic threefolds. *J. Algebraic Geom.*, 12(2):201–223, 2003. (Cited on page 243.)

[13] Daniel Allcock, James Carlson, and Domingo Toledo. The complex hyperbolic geometry of the moduli space of cubic surfaces. *J. Algebraic Geom.*, 11(4):659–724, 2002. (Cited on pages 74, 186, 187, 189, and 191.)

[14] Daniel Allcock, James A. Carlson, and Domingo Toledo. The moduli space of cubic threefolds as a ball quotient. *Mem. Amer. Math. Soc.*, 209(985):xii+70, 2011. (Cited on pages 74, 246, 247, and 248.)

[15] Allen Altman and Steven Kleiman. Foundations of the theory of Fano schemes. *Compositio Math.*, 34(1):3–47, 1977. (Cited on pages 75, 101, 103, 108, 113, 114, and 210.)

[16] Paolo Aluffi and Carel Faber. Linear orbits of arbitrary plane curves. *Michigan Math. J.*, 48(1):1–37, 2000. Dedicated to William Fulton on the occasion of his 60th birthday. (Cited on page 147.)

[17] Ekaterina Amerik. A computation of invariants of a rational self-map. *Ann. Fac. Sci. Toulouse Math. (6)*, 18(3):445–457, 2009. (Cited on pages 95, 312, 327, and 328.)

[18] Ekaterina Amerik and Claire Voisin. Potential density of rational points on the variety of lines of a cubic fourfold. *Duke Math. J.*, 145(2):379–408, 2008. (Cited on pages 316 and 327.)

[19] Yves André. Mumford–Tate groups of mixed Hodge structures and the theorem of the fixed part. *Compositio Math.*, 82(1):1–24, 1992. (Cited on page 26.)

[20] Yves André. *Une introduction aux motifs (motifs purs, motifs mixtes, périodes)*, Vol. 17 of *Panoramas et Synthèses*. Paris: Société Mathématique de France, 2004. (Cited on pages 4, 110, 112, 228, and 404.)

[21] Aldo Andreotti. On a theorem of Torelli. *Amer. J. Math.*, 80:801–828, 1958. (Cited on page 230.)

[22] Fabrizio Anella and Daniel Huybrechts. Characteristic foliations. *arXiv:2201.07624*. (Cited on page 317.)

[23] Enrico Arbarello, Maurizio Cornalba, Phillip Griffiths, and Joe Harris. *Geometry of algebraic curves. Vol. I*, Vol. 267 of *Grundlehren der Mathematischen Wissenschaften*. New York: Springer-Verlag, 1985. (Cited on pages 94, 96, 237, and 273.)

[24] Michael Artin and David Mumford. Some elementary examples of unirational varieties which are not rational. *Proc. London Math. Soc. (3)*, 25:75–95, 1972. (Cited on page 88.)

[25] Michael Atiyah. Riemann surfaces and spin structures. *Ann. Sci. École Norm. Sup. (4)*, 4(1):47–62, 1971. (Cited on page 180.)

[26] Michael Atiyah and Friedrich Hirzebruch. Vector bundles and homogeneous spaces. In *Proc. Sympos. Pure Math., Vol. III*, pp. 7–38. Providence, RI: American Mathematical Society, 1961. (Cited on page 383.)

[27] Asher Auel, Marcello Bernardara, and Michele Bolognesi. Fibrations in complete intersections of quadrics, Clifford algebras, derived categories, and rationality problems. *J. Math. Pures Appl. (9)*, 102(1):249–291, 2014. (Cited on pages 63 and 260.)

[28] Asher Auel, Marcello Bernardara, Michele Bolognesi, and Anthony Várilly-Alvarado. Cubic fourfolds containing a plane and a quintic del Pezzo surface. *Algebr. Geom.*, 1(2):181–193, 2014. (Cited on pages 257, 262, 338, and 349.)

[29] Maurice Auslander. Functors and morphisms determined by objects. In *Representation theory of algebras (Proc. Conf., Temple Univ., Philadelphia, Pa., 1976)*, Vol. 37, pp. 1–244. Lecture Notes in Pure Appl. Math., 1978. (Cited on pages 372 and 374.)

[30] Lucian Badescu. *Algebraic surfaces*. New York: Universitext, Springer-Verlag, 2001. Translated from the 1981 Romanian original by Vladimir Maşek and revised by the author. (Cited on pages 161 and 164.)

[31] Walter Baily and Armand Borel. Compactification of arithmetic quotients of bounded symmetric domains. *Ann. of Math. (2)*, 84:442–528, 1966. (Cited on page 343.)

[32] Benjamin Bakker, Yohan Brunebarbe, and Jacob Tsimerman. Quasiprojectivity of images of mixed period maps. *arXiv:2006.13709*. (Cited on page 152.)

[33] Edoardo Ballico, Fabrizio Catanese, and Ciro Ciliberto (eds.). *Classification of irregular varieties*, Vol. 1515 of *Lecture Notes in Mathematics*. Berlin: Springer-Verlag, 1992. Minimal models and abelian varieties. (Cited on page 2.)

[34] Barinder Banwait, Francesc Fité, and Daniel Loughran. Del Pezzo surfaces over finite fields and their Frobenius traces. *Math. Proc. Cambridge Philos. Soc.*, 167(1):35–60, 2019. (Cited on pages 158, 159, and 160.)

[35] Fabio Bardelli. Polarized mixed Hodge structures: on irrationality of threefolds via degeneration. *Ann. Mat. Pura Appl. (4)*, 137:287–369, 1984. (Cited on page 239.)

[36] Wolf Barth. Counting singularities of quadratic forms on vector bundles. In *Vector bundles and differential equations (Proc. Conf., Nice, 1979)*, Vol. 7 of *Progr. Math.*, pp. 1–19. Boston, MA: Birkhäuser, 1980. (Cited on pages 62, 63, and 323.)

[37] Wolf Barth and Antonius Van de Ven. Fano varieties of lines on hypersurfaces. *Arch. Math. (Basel)*, 31(1):96–104, 1978/79. (Cited on pages 75, 103, and 130.)

[38] Arend Bayer, Martí Lahoz, Emanuele Macrì, et al. Stability conditions in families. *Publ. Math. Inst. Hautes Études Sci.*, 133:157–325, 2021. (Cited on page 392.)

[39] Arend Bayer, Beentjes Sjoerd, Soheyla Feyzbakhsh, et al. The desingularization of the theta divisor of a cubic threefold as a moduli space. *arXiv:2011.12240*. (Cited on pages 221, 383, and 384.)

[40] Arnaud Beauville. Sur les hypersurfaces dont les sections hyperplanes sont à module constant. In *The Grothendieck Festschrift, I*, Vol. 86 of *Progr. Math.*, pp. 121–133. (Cited on pages 68 and 69.)

[41] Arnaud Beauville. Variétés de Prym et jacobiennes intermédiaires. *Ann. Sci. École Norm. Sup. (4)*, 10(3):309–391, 1977. (Cited on pages 61, 63, 193, 205, 221, and 228.)

[42] Arnaud Beauville. Les singularités du diviseur Θ de la jacobienne intermédiaire de l'hypersurface cubique dans $\mathbb{P}^4$. In *Algebraic threefolds (Varenna, 1981)*, Vol. 947 of *Lecture Notes in Math.*, pp. 190–208. Berlin, New York: Springer, 1982. (Cited on pages 193, 207, 225, 226, 232, 235, and 238.)

[43] Arnaud Beauville. Sous-variétés spéciales des variétés de Prym. *Compositio Math.*, 45(3):357–383, 1982. (Cited on pages 214, 220, 225, and 238.)

[44] Arnaud Beauville. Some remarks on Kähler manifolds with $c_1 = 0$. In *Classification of algebraic and analytic manifolds (Katata, 1982)*, Vol. 39 of *Progr. Math.*, pp. 1–26. Boston, MA: Birkhäuser 1983. (Cited on page 296.)

[45] Arnaud Beauville. Variétés Kähleriennes dont la première classe de Chern est nulle. *J. Differential Geom.*, 18(4):755–782 (1984), 1983. (Cited on pages 292, 293, and 294.)

[46] Arnaud Beauville. Le groupe de monodromie des familles universelles d'hypersurfaces et d'intersections complètes. In *Complex analysis and algebraic geometry (Göttingen, 1985)*, Vol. 1194 of *Lecture Notes in Math.*, pp. 8–18. Berlin: Springer, 1986. (Cited on pages 23, 28, 29, and 30.)

[47] Arnaud Beauville. Le problème de Torelli. Number 145–146, pages 3, 7–20. 1987. Séminaire Bourbaki, Vol. 1985/86. (Cited on pages 148 and 154.)

[48] Arnaud Beauville. Prym varieties: a survey. In *Theta functions–Bowdoin 1987, Part 1 (Brunswick, ME, 1987)*, Vol. 49 of *Proc. Sympos. Pure Math.*, pp. 607–620. Providence, RI: Amer. Math. Soc., 1989. (Cited on page 221.)

[49] Arnaud Beauville. *Complex algebraic surfaces*, Vol. 34 of *London Mathematical Society Student Texts*. Cambridge University Press, Cambridge, 2nd ed., 1996. Translated from the 1978 French original by R. Barlow, with assistance from N. I. Shepherd-Barron and M. Reid. (Cited on pages 156, 164, 167, and 178.)

[50] Arnaud Beauville. Determinantal hypersurfaces. *Michigan Math. J.*, 48:39–64, 2000. Dedicated to William Fulton on the occasion of his 60th birthday. (Cited on pages 173, 245, 276, and 289.)

[51] Arnaud Beauville. Vector bundles on the cubic threefold. In *Symposium in Honor of C. H. Clemens (Salt Lake City, UT, 2000)*, Vol. 312 of *Contemp. Math.*, pp. 71–86. Providence, RI: Amer. Math. Soc., 2002. (Cited on pages 220 and 270.)

[52] Arnaud Beauville. On the splitting of the Bloch–Beilinson filtration. In *Algebraic cycles and motives. Vol. 2*, Vol. 344 of *London Math. Soc. Lecture Note Ser.*, pp. 38–53. Cambridge: Cambridge University Press, 2007. (Cited on page 401.)

[53] Arnaud Beauville. Moduli of cubic surfaces and Hodge theory (after Allcock, Carlson, Toledo). In *Géométries à courbure négative ou nulle, groupes discrets et rigidités*, Vol. 18 of *Sémin. Congr.*, pp. 445–466. Paris: Soc. Math. France, 2009. (Cited on pages 186, 187, 188, 189, and 191.)

[54] Arnaud Beauville. The primitive cohomology lattice of a complete intersection. *C. R. Math. Acad. Sci. Paris*, 347(23–24):1399–1402, 2009. (Cited on page 13.)

[55] Arnaud Beauville. Some surfaces with maximal Picard number. *J. Éc. polytech. Math.*, 1:101–116, 2014. (Cited on pages 216 and 244.)

[56] Arnaud Beauville. The Lüroth problem. In *Rationality problems in algebraic geometry*, Vol. 2172 of *Lecture Notes in Math.*, pp. 1–27. Cham: Springer, 2016. (Cited on pages 237, 238, 239, and 244.)

[57] Arnaud Beauville. Vector bundles on Fano threefolds and K3 surfaces. *Boll. Unione Mat. Ital.*, 15(1–2):43–55, 2022. (Cited on pages 220 and 270.)

[58] Arnaud Beauville and Ron Donagi. La variété des droites d'une hypersurface cubique de dimension 4. *C. R. Acad. Sci. Paris Sér. I Math.*, 301(14):703–706, 1985. (Cited on pages 122, 127, 251, 273, 275, 280, 288, 296, and 301.)

[59] Arnaud Beauville and Claire Voisin. On the Chow ring of a $K3$ surface. *J. Algebraic Geom.*, 13(3):417–426, 2004. (Cited on page 401.)

[60] Roya Beheshti. Lines on projective hypersurfaces. *J. Reine Angew. Math.*, 592:1–21, 2006. (Cited on page 89.)

[61] Nikolay Beklemishev. Invariants of cubic forms of four variables. *Vestnik Moskov. Univ. Ser. I Mat. Mekh.*, (2):42–49, 116, 1982. (Cited on pages 183 and 185.)

[62] Pieter Belmans, Lie Fu, and Theo Raedschelders. Derived categories of flips and cubic hypersurfaces. *arXiv:2002.04940*. (Cited on pages 110, 375, and 376.)

[63] Olivier Benoist. Séparation et propriété de Deligne–Mumford des champs de modules d'intersections complètes lisses. *J. Lond. Math. Soc. (2)*, 87(1):138–156, 2013. (Cited on page 38.)

[64] Marcello Bernardara. A semiorthogonal decomposition for Brauer–Severi schemes. *Math. Nachr.*, 282(10):1406–1413, 2009. (Cited on page 355.)

[65] Marcello Bernardara, Emanuele Macrì, Sukhendu Mehrotra, and Paolo Stellari. A categorical invariant for cubic threefolds. *Adv. Math.*, 229(2):770–803, 2012. (Cited on pages 377, 378, 380, 381, 382, and 383.)

[66] Marcello Bernardara and Gonçalo Tabuada. From semi-orthogonal decompositions to polarized intermediate Jacobians via Jacobians of noncommutative motives. *Mosc. Math. J.*, 16(2):205–235, 2016. (Cited on page 383.)

[67] Anton Betten and Fatmas Karaoglu. The Eckardt point configuration of cubic surfaces revisited. *Des. Codes Cryptogr.*, 90:2159–2180, 2022. (Cited on page 182.)

[68] Gilberto Bini and Alice Garbagnati. Quotients of the Dwork pencil. *J. Geom. Phys.*, 75:173–198, 2014. (Cited on page 22.)

[69] Indranil Biswas, Jishnu Biswas, and G. Ṽ. Ravindra. On some moduli spaces of stable vector bundles on cubic and quartic threefolds. *J. Pure Appl. Algebra*, 212(10):2298–2306, 2008. (Cited on page 220.)

[70] Franziska Bittner. The universal Euler characteristic for varieties of characteristic zero. *Compositio Math.*, 140(4):1011–1032, 2004. (Cited on pages 113, 114, and 116.)

[71] Spencer Bloch. *Lectures on algebraic cycles.* Duke University Mathematics Series, IV. Durham, NC: Duke University Mathematics Department, 1980. (Cited on page 228.)

[72] Spencer Bloch and Vasudevan Srinivas. Remarks on correspondences and algebraic cycles. *Amer. J. Math.*, 105(5):1235–1253, 1983. (Cited on pages 228, 254, 400, and 402.)

[73] Gloire Grace Bockondas and Samuel Boissière. Triple lines on a cubic threefold. *arXiv:2201.08884.* (Cited on page 198.)

[74] Fjodor Bogomolov. On the cohomology ring of a simple hyper-Kähler manifold (on the results of Verbitsky). *Geom. Funct. Anal.*, 6(4):612–618, 1996. (Cited on page 297.)

[75] Christian Böhning and Hans-Christian Graf von Bothmer. Matrix factorizations and intermediate Jacobians of cubic threefolds. *arXiv:2112.10554.* (Cited on page 220.)

[76] Samuel Boissière, Chiara Camere, and Alessandra Sarti. Cubic threefolds and hyperkähler manifolds uniformized by the 10-dimensional complex ball. *Math. Ann.*, 373(3–4):1429–1455, 2019. (Cited on pages 246 and 249.)

[77] Samuel Boissière, Marc Nieper-Wißkirchen, and Alessandra Sarti. Smith theory and irreducible holomorphic symplectic manifolds. *J. Topol.*, 6(2):361–390, 2013. (Cited on pages 294 and 303.)

[78] Michele Bolognesi and Claudio Pedrini. The transcendental motive of a cubic fourfold. *J. Pure Appl. Algebra*, 224(8):106333, 16, 2020. (Cited on page 306.)

[79] Michele Bolognesi, Francesco Russo, and Giovanni Staglianò. Some loci of rational cubic fourfolds. *Math. Ann.*, 373(1–2):165–190, 2019. (Cited on pages 288 and 337.)

[80] Enrico Bombieri and Peter Swinnerton-Dyer. On the local zeta function of a cubic threefold. *Ann. Scuola Norm. Sup. Pisa (3)*, 21:1–29, 1967. (Cited on pages 16, 61, 113, 193, 205, and 221.)

[81] Alexei Bondal and Mikhai Kapranov. Representable functors, Serre functors, and reconstructions. *Izv. Akad. Nauk SSSR Ser. Mat.*, 53(6):1183–1205, 1337, 1989. (Cited on pages 359 and 361.)

[82] Alexei Bondal and Dmitri Orlov. Reconstruction of a variety from the derived category and groups of autoequivalences. *Compositio Math.*, 125(3):327–344, 2001. (Cited on page 356.)

[83] Alexei Bondal and Michel Van den Bergh. Generators and representability of functors in commutative and noncommutative geometry. *Mosc. Math. J.*, 3(1):1–36, 258, 2003. (Cited on page 356.)

[84] Alexey Bondal, Michael Larsen, and Valery Lunts. Grothendieck ring of pretriangulated categories. *Int. Math. Res. Not.*, (29):1461–1495, 2004. (Cited on page 375.)

[85] Ciprian Borcea. Deforming varieties of k-planes of projective complete intersections. *Pacific J. Math.*, 143(1):25–36, 1990. (Cited on pages 103 and 109.)

[86] Armand Borel. Sur l'homologie et la cohomologie des groupes de Lie compacts connexes. *Amer. J. Math.*, 76:273–342, 1954. (Cited on page 144.)

[87] Armand Borel. Some metric properties of arithmetic quotients of symmetric spaces and an extension theorem. *J. Differential Geometry*, 6:543–560, 1972. Collection of articles dedicated to S. S. Chern and D. C. Spencer on their 60th birthdays. (Cited on page 343.)

[88] Lev Borisov and Andrei Căldăraru. The Pfaffian–Grassmannian derived equivalence. *J. Algebraic Geom.*, 18(2):201–222, 2009. (Cited on page 274.)

[89] Lev Borisov and Anatoly Libgober. Elliptic genera of singular varieties, orbifold elliptic genus and chiral de Rham complex. In *Mirror symmetry, IV (Montreal, QC, 2000)*, Vol. 33 of *AMS/IP Stud. Adv. Math.*, pp. 325–342. Providence, RI: Amer. Math. Soc., 2002. (Cited on page 121.)

[90] Nicolas Bourbaki. *Algebra. II. Chapters 4–7.* Elements of Mathematics. Berlin: Springer-Verlag, 1990. Translated from the French by P. M. Cohn and J. Howie. (Cited on page 18.)

[91] Emma Brakkee. Two polarised K3 surfaces associated to the same cubic fourfold. *Math. Proc. Cambridge Philos. Soc.*, 171(1):51–64, 2021. (Cited on pages 336, 345, and 352.)

[92] Davide Bricalli, Filippo Favale, and Gian Pietro Pirola. A theorem of Gordan and Noether via Gorenstein rings. *arXiv:2201.07550*. (Cited on page 46.)

[93] Michel Brion. Equivariant cohomology and equivariant intersection theory. In *Representation theories and algebraic geometry (Montreal, PQ, 1997)*, volume 514 of *NATO Adv. Sci. Inst. Ser. C: Math. Phys. Sci.*, pp. 1–37. Dordrecht Kluwer Acad. Publ., 1998. Notes by Alvaro Rittatore. (Cited on page 145.)

[94] William Browder. *Surgery on simply-connected manifolds.* New York, Heidelberg: Springer-Verlag, 1972. Ergebnisse der Mathematik und ihrer Grenzgebiete, Band 65. (Cited on page 24.)

[95] William Browder. Complete intersections and the Kervaire invariant. In *Algebraic topology, Aarhus 1978 (Proc. Sympos., Univ. Aarhus, Aarhus, 1978)*, Vol. 763 of *Lecture Notes in Math.*, pp. 88–108. Berlin: Springer, 1979. (Cited on page 24.)

[96] Gavin Brown and Daniel Ryder. Elliptic fibrations on cubic surfaces. *J. Pure Appl. Algebra*, 214(4):410–421, 2010. (Cited on page 172.)

[97] Ragnar-Olaf Buchweitz. Maximal Cohen–Macaulay modules and Tate cohomology over Gorenstein rings. `https://tspace.library.utoronto.ca/handle/1807/16682`, 1986. (Cited on pages 371 and 373.)

[98] Anita Buckley and Tomaž Košir. Determinantal representations of smooth cubic surfaces. *Geom. Dedicata*, 125:115–140, 2007. (Cited on page 173.)

[99] Tim-Henrik Bülles. Motives of moduli spaces on K3 surfaces and of special cubic fourfolds. *Manuscripta Math.*, 161(1–2):109–124, 2020. (Cited on pages 306 and 406.)

[100] Dominic Bunnett and Hanieh Keneshlou. Determinantal representations of the cubic discriminant. *Matematiche (Catania)*, 75(2):489–505, 2020. (Cited on page 21.)

[101] Laurent Busé and Jean-Pierre Jouanolou. On the discriminant scheme of homogeneous polynomials. *Math. Comput. Sci.*, 8(2):175–234, 2014. (Cited on page 20.)

[102] Alberto Canonaco and Paolo Stellari. Twisted Fourier–Mukai functors. *Adv. Math.*, 212(2):484–503, 2007. (Cited on page 356.)

[103] James Carlson and Phillip Griffiths. Infinitesimal variations of Hodge structure and the global Torelli problem. In *Journées de Géometrie Algébrique d'Angers, Juillet 1979/Algebraic Geometry, Angers, 1979*, pp. 51–76. Alphen aan den Rijn: Sijthoff & Noordhoff, 1980. (Cited on pages 52 and 154.)

[104] James Carlson, Stefan Müller-Stach, and Chris Peters. *Period mappings and period domains*, Vol. 85 of *Cambridge Studies in Advanced Mathematics*. Cambridge: Cambridge University Press, 2nd ed., 2017. (Cited on pages 52, 57, 149, and 154.)

[105] James A. Carlson and Domingo Toledo. Compact quotients of non-classical domains are not Kähler. In *Hodge theory, complex geometry, and representation theory*, Vol. 608 of *Contemp. Math.*, pp. 51–57. Providence, RI: Amer. Math. Soc., 2014. (Cited on page 151.)

[106] Sebastian Casalaina-Martin and Robert Friedman. Cubic threefolds and abelian varieties of dimension five. *J. Algebraic Geom.*, 14(2):295–326, 2005. (Cited on page 235.)

[107] Sebastian Casalaina-Martin, Samuel Grushevsky, Klaus Hulek, and Radu Laza. Non-isomorphic smooth compactifications of the moduli space of cubic surfaces. *arXiv:2207.03533.* (Cited on page 191.)

[108] Sebastian Casalaina-Martin, Samuel Grushevsky, Klaus Hulek, and Radu Laza. Complete moduli of cubic threefolds and their intermediate Jacobians. *Proc. Lond. Math. Soc. (3)*, 122(2):259–316, 2021. (Cited on page 244.)

[109] Sebastian Casalaina-Martin and Radu Laza. The moduli space of cubic threefolds via degenerations of the intermediate Jacobian. *J. Reine Angew. Math.*, 633:29–65, 2009. (Cited on page 244.)

[110] Sebastian Casalaina-Martin, Mihnea Popa, and Stefan Schreieder. Generic vanishing and minimal cohomology classes on abelian fivefolds. *J. Algebraic Geom.*, 27(3):553–581, 2018. (Cited on page 227.)

[111] Arthur Cayley. On the triple tangent planes of surfaces of the third order. *Cambridge and Dublin Math. Journal*, IV:118–132, 1849. (Cited on page 175.)

[112] Arthur Cayley. A memoir on cubic surfaces. *Philos. Trans. R. Soc. Lond.*, 159:231–326, 1869. (Cited on page 183.)

[113] Antoine Chambert-Loir, Johannes Nicaise, and Julien Sebag. *Motivic integration*, Vol. 325 of *Progress in Mathematics*. New York: Birkhäuser/Springer, 2018. (Cited on pages 110 and 111.)

[114] François Charles. A remark on the Torelli theorem for cubic fourfolds. *arXiv:1209.4509.* (Cited on pages 107, 292, 300, 347, and 383.)

[115] Xi Chen, Xuanyu Pan, and Dingxin Zhang. Automorphism and cohomology II: complete intersections. *arXiv:1511.07906.* (Cited on pages 38 and 39.)

[116] Wei-Liang Chow. On the geometry of algebraic homogeneous spaces. *Ann. of Math. (2)*, 50:32–67, 1949. (Cited on page 108.)

[117] Alfred Clebsch. Ueber die Anwendung der quadratischen Substitution auf die Gleichungen 5 *ten* Grades und die geometrische Theorie des ebenen Fünfseits. *Math. Ann.*, 4(2):284–345, 1871. (Cited on page 173.)

[118] Alfred Clebsch. On Weiler's models. *Gött. Nachr.*, 1872:402–403, 1872. (Cited on page 183.)

[119] Herbert Clemens and Phillip Griffiths. The intermediate Jacobian of the cubic threefold. *Ann. of Math. (2)*, 95:281–356, 1972. (Cited on pages 94, 104, 127, 193, 197, 198, 200, 202, 209, 210, 212, 215, 225, 229, 230, 234, 237, 238, and 240.)

[120] Alberto Collino. A cheap proof of the irrationality of most cubic threefolds. *Boll. Un. Mat. Ital. B (5)*, 16(2):451–465, 1979. (Cited on page 239.)

[121] Alberto Collino. The fundamental group of the Fano surface. I, II. In *Algebraic threefolds (Varenna, 1981)*, Vol. 947 of *Lecture Notes in Math.*, pp. 209–218, 219–220. Berlin, New York: Springer, 1982. (Cited on pages 203 and 204.)

[122] Alberto Collino, Juan Carlos Naranjo, and Gian Pietro Pirola. The Fano normal function. *J. Math. Pures Appl. (9)*, 98(3):346–366, 2012. (Cited on page 228.)

[123] Elisabetta Colombo and Bert van Geemen. The Chow group of the moduli space of marked cubic surfaces. *Ann. Mat. Pura Appl. (4)*, 183(3):291–316, 2004. (Cited on page 185.)

[124] Gaia Comaschi. Pfaffian representations of cubic threefolds. *arXiv:2005.06593*. (Cited on page 245.)

[125] Izzet Coskun and Jason Starr. Rational curves on smooth cubic hypersurfaces. *Int. Math. Res. Not. IMRN*, (24):4626–4641, 2009. (Cited on pages 104, 198, and 253.)

[126] David Cox. Generic Torelli and infinitesimal variation of Hodge structure. In *Algebraic geometry, Bowdoin, 1985 (Brunswick, Maine, 1985)*, Vol. 46 of *Proc. Sympos. Pure Math.*, pp. 235–246. Providence, RI: Amer. Math. Soc., 1987. (Cited on page 49.)

[127] David Cox, John Little, and Donal O'Shea. *Using algebraic geometry*, Vol. 185 of *Graduate Texts in Mathematics*. New York: Springer, 2005. (Cited on pages 20 and 21.)

[128] Harold Coxeter. The polytopes with regular-prismatic vertex figures. *Proc. London Math. Soc. (2)*, 34(2):126–189, 1932. (Cited on page 179.)

[129] Harold Coxeter. Extreme forms. *Canadian J. Math.*, 3:391–441, 1951. (Cited on page 163.)

[130] Harold Coxeter. The twenty-seven lines on the cubic surface. In *Convexity and its applications*, pp. 111–119. Basel: Birkhäuser, 1983. (Cited on page 175.)

[131] Fernando Cukierman. Families of Weierstrass points. *Duke Math. J.*, 58(2):317–346, 1989. (Cited on page 180.)

[132] Elisa Dardanelli and Bert van Geemen. Hessians and the moduli space of cubic surfaces. In *Algebraic geometry*, Vol. 422 of *Contemp. Math.*, pp. 17–36. Providence, RI: Amer. Math. Soc., 2007. (Cited on pages 184 and 185.)

[133] Ronno Das. Cohomology of the universal smooth cubic surface. *Q. J. Math.*, 72(3):795–815, 2021. (Cited on pages 146 and 159.)

[134] Ronno Das. The space of cubic surfaces equipped with a line. *Math. Z.*, 298(1–2):653–670, 2021. (Cited on page 146.)

[135] Johan de Jong and Jason Starr. Cubic fourfolds and spaces of rational curves. *Illinois J. Math.*, 48(2):415–450, 2004. (Cited on page 299.)

[136] Olivier Debarre. Hyperkähler manifolds. *arXiv:1810.02087*. (Cited on page 348.)

[137] Olivier Debarre. Minimal cohomology classes and Jacobians. *J. Algebraic Geom.*, 4(2):321–335, 1995. (Cited on page 227.)

[138] Olivier Debarre. *Higher-dimensional algebraic geometry*. New York: Universitext, Springer-Verlag, 2001. (Cited on page 122.)

[139] Olivier Debarre. Variétés rationnellement connexes (d'après T. Graber, J. Harris, J. Starr, et A. J. de Jong). *Astérisque*, (290):Exp. No. 905, ix, 243–266, 2003. Séminaire Bourbaki. Vol. 2001/2002. (Cited on page 122.)

[140] Olivier Debarre, Antonio Laface, and Xavier Roulleau. Lines on cubic hypersurfaces over finite fields. In *Geometry over nonclosed fields*, Simons Symp., pp. 19–51. Cham: Springer, 2017. (Cited on pages 124 and 217.)

[141] Olivier Debarre and Laurent Manivel. Sur la variété des espaces linéaires contenus dans une intersection complète. *Math. Ann.*, 312(3):549–574, 1998. (Cited on pages 107 and 121.)

[142] Alex Degtyarev. Smooth models of singular K3-surfaces. *Rev. Mat. Iberoam.*, 35(1):125–172, 2019. (Cited on page 47.)

[143] Alex Degtyarev, Ilia Itenberg, and John Ottem. Planes in cubic fourfolds. *arXiv:2105.13951*. (Cited on page 256.)

[144] Pierre Deligne. Travaux de Griffiths. In *Séminaire Bourbaki, 22ème année (1969/70), Exp. No. 376*, Vol. 180, pp. 213–237. *Lecture Notes in Math.* Berlin: Springer, 1971. (Cited on page 148.)

[145] Pierre Deligne. La conjecture de Weil pour les surfaces K3. *Invent. Math.*, 15:206–226, 1972. (Cited on page 26.)

[146] Pierre Deligne. La conjecture de Weil. II. *Inst. Hautes Études Sci. Publ. Math.*, (52):137–252, 1980. (Cited on pages 1, 25, and 28.)

[147] Pierre Deligne and Luc Illusie. Relèvements modulo p^2 et décomposition du complexe de de Rham. *Invent. Math.*, 89(2):247–270, 1987. (Cited on pages 15, 53, 54, and 103.)

[148] Pierre Deligne and Nicholas Katz. Groupes de monodromie en géométrie algébrique (Séminaire de Géométrie Algébrique du Bois-Marie 1967–1969 (SGA 7 II)), Vol. 340 of Lecture Notes in Math. Berlin, New York: Springer, 1973. (Cited on pages 1, 9, 19, 21, and 71.)

[149] Jean-Pierre Demailly. Algebraic criteria for Kobayashi hyperbolic projective varieties and jet differentials. In *Algebraic geometry–Santa Cruz 1995*, Vol. 62 of *Proc. Sympos. Pure Math.*, pp. 285–360. Providence, RI: Amer. Math. Soc., 1997. (Cited on page 214.)

[150] Michel Demazure. Résultant, discriminant. *Enseign. Math. (2)*, 58(3–4):333–373, 2012. (Cited on page 20.)

[151] Benjamin Diamond. Smooth surfaces in smooth fourfolds with vanishing first Chern class. *J. Pure Appl. Algebra*, 222(5):1164–1188, 2018. (Cited on page 102.)

[152] Humberto Anthony Diaz. The motive of the Fano surface of lines. *C. R. Math. Acad. Sci. Paris*, 354(9):925–930, 2016. (Cited on pages 228 and 404.)

[153] Humberto Anthony Diaz. The Chow ring of a cubic hypersurface. *Int. Math. Res. Not. IMRN*, (22):17071–17090, 2021. (Cited on pages 5, 113, 400, and 403.)

[154] Steven Diaz and David Harbater. Strong Bertini theorems. *Trans. Amer. Math. Soc.*, 324(1):73–86, 1991. (Cited on page 168.)

[155] Jean Dieudonné. *Éléments d'analyse. Tome IX. Chapitre XXIV.* Cahiers Scientifiques, XL11. Paris: Gauthier-Villars, 1982. (Cited on pages 2 and 204.)

[156] Alexandru Dimca, Rodrigo Gondim, and Giovanna Ilardi. Higher order Jacobians, Hessians and Milnor algebras. *Collect. Math.*, 71(3):407–425, 2020. (Cited on page 46.)

[157] Igor Dolgachev. *Lectures on invariant theory*, Vol. 296 of *London Mathematical Society Lecture Note Series*. Cambridge: Cambridge University Press, 2003. (Cited on pages 135, 147, 185, and 201.)

[158] Igor Dolgachev. *Classical algebraic geometry*. Cambridge: Cambridge University Press, 2012. A modern view. (Cited on pages 40, 68, 70, 156, 170, 173, 174, 175, 177, 179, 182, 184, 245, 286, and 289.)

[159] Igor Dolgachev. Corrado Segre and nodal cubic threefolds. In *From classical to modern algebraic geometry*, Trends Hist. Sci., pp. 429–450. Cham: Birkhäuser/Springer, 2016. (Cited on page 245.)

[160] Igor Dolgachev, Bert van Geemen, and Shigeyuki Kondō. A complex ball uniformization of the moduli space of cubic surfaces via periods of K3 surfaces. *J. Reine Angew. Math.*, 588:99–148, 2005. (Cited on pages 171, 172, 186, 192, and 263.)

[161] Ron Donagi. Generic Torelli for projective hypersurfaces. *Compositio Math.*, 50(2–3):325–353, 1983. (Cited on pages 46, 47, 49, 50, and 154.)

[162] Ron Donagi and Mark Green. A new proof of the symmetrizer lemma and a stronger weak Torelli theorem for projective hypersurfaces. *J. Differential Geom.*, 20(2):459–461, 1984. (Cited on pages 49 and 50.)

[163] Ron Donagi and Eyal Markman. Spectral covers, algebraically completely integrable, Hamiltonian systems, and moduli of bundles. In *Integrable systems and quantum groups*

(Montecatini Terme, 1993), Vol. 1620 of *Lecture Notes in Math.*, pp. 1–119. Berlin: Springer, 1996. (Cited on page 298.)

[164] Ron Donagi and Roy Campbell Smith. The structure of the Prym map. *Acta Math.*, 146(1–2):25–102, 1981. (Cited on page 222.)

[165] Stéphane Druel. Espace des modules des faisceaux de rang 2 semi-stables de classes de Chern $c_1 = 0$, $c_2 = 2$ et $c_3 = 0$ sur la cubique de $\mathbb{P}^4$. *Internat. Math. Res. Notices*, (19):985–1004, 2000. (Cited on page 220.)

[166] Bernard Dwork. On the zeta function of a hypersurface. *Inst. Hautes Études Sci. Publ. Math.*, (12):5–68, 1962. (Cited on page 16.)

[167] Tobias Dyckerhoff and Daniel Murfet. The Kapustin–Li formula revisited. *Adv. Math.*, 231(3-4):1858–1885, 2012. (Cited on pages 372 and 374.)

[168] Wolfgang Ebeling. An arithmetic characterisation of the symmetric monodromy groups of singularities. *Invent. Math.*, 77(1):85–99, 1984. (Cited on page 29.)

[169] Friedrich Eckardt. Ueber diejenigen Flächen dritten Grades, auf denen sich drei gerade Linien in einem Punkte schneiden. *Math. Ann.*, 10(2):227–272, 1876. (Cited on page 182.)

[170] William Edge. Cubic primals in [4] with polar heptahedra. *Proc. Roy. Soc. Edinburgh Sect. A*, 77(1–2):151–162, 1977. (Cited on page 246.)

[171] William Edge. The discriminant of a cubic surface. *Proc. Roy. Irish Acad. Sect. A*, 80(1):75–78, 1980. (Cited on pages 21 and 184.)

[172] Alexander Efimov. Some remarks on L-equivalence of algebraic varieties. *Selecta Math. (N.S.)*, 24(4):3753–3762, 2018. (Cited on page 116.)

[173] David Eisenbud. Homological algebra on a complete intersection, with an application to group representations. *Trans. Amer. Math. Soc.*, 260(1):35–64, 1980. (Cited on page 373.)

[174] David Eisenbud and Joe Harris. *3264 and all that–a second course in algebraic geometry*. Cambridge: Cambridge University Press, 2016. (Cited on pages 70, 71, 75, 89, 93, 94, 96, 275, 283, and 312.)

[175] Hélène Esnault, Marc Levine, and Eckart Viehweg. Chow groups of projective varieties of very small degree. *Duke Math. J.*, 87(1):29–58, 1997. (Cited on page 400.)

[176] Yu-Wei Fan and Kuan-Wen Lai. New rational cubic fourfolds arising from Cremona transformations. *arXiv:2003.00366*. (Cited on page 395.)

[177] Gino Fano. Sul sistema ∞^2 di rette contenuto in una varietà cubica generale dello spazio a quattro dimensioni. *Math. Ann.*, 39(1–2):778–792, 1904. (Cited on pages 193, 288, and 323.)

[178] Barbara Fantechi, Lothar Göttsche, Luc Illusie, et al. *Fundamental algebraic geometry*, Vol. 123 of *Mathematical Surveys and Monographs*. Providence, RI: Amer. Math. Soc., 2005. Grothendieck's FGA explained. (Cited on pages 33, 34, 35, 36, 76, 81, and 82.)

[179] Gavril Farkas. Prym varieties and their moduli. In *Contributions to algebraic geometry*, EMS Ser. Congr. Rep., pp. 215–255. Zürich: Eur. Math. Soc., 2012. (Cited on page 221.)

[180] Gavril Farkas and Alessandro Verra. The universal K3 surface of genus 14 via cubic fourfolds. *J. Math. Pures Appl. (9)*, 111:1–20, 2018. (Cited on page 335.)

[181] Maksym Fedorchuk. GIT semistability of Hilbert points of Milnor algebras. *Math. Ann.*, 367(1–2):441–460, 2017. (Cited on page 139.)

[182] Lie Fu. Classification of polarized symplectic automorphisms of Fano varieties of cubic fourfolds. *Glasg. Math. J.*, 58(1):17–37, 2016. (Cited on page 40.)

[183] Lie Fu, Robert Laterveer, and Charles Vial. The generalized Franchetta conjecture for some hyper-Kähler varieties, II. *J. Éc. polytech. Math.*, 8:1065–1097, 2021. (Cited on pages 113 and 306.)

[184] Lie Fu, Robert Laterveer, and Charles Vial. Multiplicative Chow-Künneth decompositions and varieties of cohomological K3 type. *Ann. Mat. Pura Appl. (4)*, 200(5):2085–2126, 2021. (Cited on pages 5 and 403.)

[185] Lie Fu and Zhiyn Tian. 2-cycles sur les hypersurfaces cubiques de dimension 5. *Math. Z.*, 293(1–2):661–676, 2019. (Cited on page 400.)

[186] Lie Fu and Charles Vial. Cubic fourfolds, Kuznetsov components and Chow motives. *arXiv:2009.13173*. (Cited on page 406.)

[187] Akira Fujiki. On the de Rham cohomology group of a compact Kähler symplectic manifold. In *Algebraic geometry, Sendai, 1985*, Vol. 10 of *Adv. Stud. Pure Math.*, pp. 105–165. Amsterdam: North-Holland, 1987. (Cited on page 294.)

[188] William Fulton. *Intersection theory*, Vol. 2 of *Ergebnisse der Mathematik und ihrer Grenzgebiete. 3. Folge. A Series of Modern Surveys in Mathematics*. Berlin: Springer-Verlag, 2nd ed., 1998. (Cited on pages 235, 283, and 312.)

[189] William Fulton and Robert Lazarsfeld. On the connectedness of degeneracy loci and special divisors. *Acta Math.*, 146(3–4):271–283, 1981. (Cited on pages 94 and 283.)

[190] Sergey Galkin and Evgeny Shinder. The Fano variety of lines and rationality problem for a cubic hypersurface. *arXiv:1405.5154*. (Cited on pages 110, 111, 113, 118, 123, 124, 239, 242, 271, 341, and 384.)

[191] Sergey Galkin and Evgeny Shinder. On a zeta-function of a dg-category. *arXiv:1506.05831*. (Cited on page 399.)

[192] Nora Ganter and Mikhail Kapranov. Symmetric and exterior powers of categories. *Transform. Groups*, 19(1):57–103, 2014. (Cited on page 398.)

[193] Israel M. Gelfand, Mikhail Kapranov, and Andrei Zelevinsky. *Discriminants, resultants and multidimensional determinants*. Modern Birkhäuser Classics. Boston, MA: Birkhäuser Boston, Inc., 2008. Reprint of the 1994 ed. (Cited on pages 20 and 233.)

[194] Sergei Gelfand and Yuri Manin. *Methods of homological algebra*. Springer Monographs in Mathematics. Berllin: Springer-Verlag, 2nd ed. 2003. (Cited on pages 356 and 382.)

[195] Víctor González-Aguilera and Alvaro Liendo. Automorphisms of prime order of smooth cubic n-folds. *Arch. Math. (Basel)*, 97(1):25–37, 2011. (Cited on page 40.)

[196] Víctor González-Aguilera and Alvaro Liendo. On the order of an automorphism of a smooth hypersurface. *Israel J. Math.*, 197(1):29–49, 2013. (Cited on pages 32 and 40.)

[197] Roe Goodman and Nolan Wallach. *Symmetry, representations, and invariants*, Vol. 255 of *Graduate Texts in Mathematics*. Dordrecht: Springer, 2009. (Cited on page 26.)

[198] Ulrich Görtz and Torsten Wedhorn. *Algebraic geometry I*. Advanced Lectures in Mathematics. Vieweg + Teubner, Wiesbaden, 2010. Schemes with examples and exercises. (Cited on page 108.)

[199] Victor Goryunov. Symmetric quartics with many nodes. In *Singularities and bifurcations*, Vol. 21 of *Adv. Soviet Math.*, pp. 147–161. Providence, RI: Amer. Math. Soc., 1994. (Cited on page 68.)

[200] Frank Gounelas and Alexis Kouvidakis. Geometry of lines on a cubic fourfold. *arXiv:2109.08493*. (Cited on pages 97 and 315.)

[201] Frank Gounelas and Alexis Kouvidakis. On some invariants of cubic fourfolds. *arXiv:2008.05162*. (Cited on pages 318 and 320.)

[202] Frank Gounelas and Alexis Kouvidakis. Measures of irrationality of the Fano surface of a cubic threefold. *Trans. Amer. Math. Soc.*, 371(10):7111–7133, 2019. (Cited on pages 207, 244, and 245.)

[203] Phillip Griffiths. Hermitian differential geometry, Chern classes, and positive vector bundles. In *Global Analysis (Papers in Honor of K. Kodaira)*, pp. 185–251. Tokyo: University of Tokyo Press, 1969. (Cited on page 319.)

[204] Phillip Griffiths, editor. *Topics in transcendental algebraic geometry*, Vol. 106 of *Annals of Mathematics Studies*, Princeton, NJ: Princeton University Press, 1984. (Cited on pages 148 and 149.)

[205] Phillip Griffiths and Joseph Harris. *Principles of algebraic geometry*. New York: Wiley-Interscience [John Wiley & Sons], 1978. Pure and Applied Mathematics. (Cited on pages 45, 80, and 271.)

[206] Phillip Griffiths, Colleen Robles, and Domingo Toledo. Quotients of non-classical flag domains are not algebraic. *Algebr. Geom.*, 1(1):1–13, 2014. (Cited on page 151.)

[207] Mark Gross, Daniel Huybrechts, and Dominic Joyce. *Calabi–Yau manifolds and related geometries*. Berlin: Universitext, Springer-Verlag, 2003. Lectures from the Summer School held in Nordfjordeid, June 2001. (Cited on pages 294 and 315.)

[208] Mark Gross and Sorin Popescu. The moduli space of $(1, 11)$-polarized abelian surfaces is unirational. *Compositio Math.*, 126(1):1–23, 2001. (Cited on pages 244 and 245.)

[209] Isabell Grosse-Brauckmann. The Fano variety of lines. `www.math.uni-bonn.de/people/huybrech/Grosse-BrauckmannBach.pdf`. Bachelor thesis, Univ. Bonn. 2014. (Cited on page 104.)

[210] Alexander Grothendieck. *Cohomologie locale des faisceaux cohérents et théorèmes de Lefschetz locaux et globaux*. (Séminaire de Géométrie Algébrique du Bois-Marie 1962), Vol. 2 of *Advanced Studies in Pure Math.* Amsterdam: North-Holland, 1973. (Cited on pages 1 and 4.)

[211] Alexander Grothendieck. Sur quelques points d'algèbre homologique. *Tôhoku Math. J. (2)*, 9:119–221, 1957. (Cited on page 114.)

[212] Alexander Grothendieck. *Fondements de la géométrie algébrique. [Extraits du Séminaire Bourbaki]*. Paris: Secrétariat mathématique, 1962. (Cited on pages 76 and 82.)

[213] Alexander Grothendieck. On the de Rham cohomology of algebraic varieties. *Inst. Hautes Études Sci. Publ. Math.*, (29):95–103, 1966. (Cited on page 55.)

[214] Vladimir Guletskiĭ. Motivic obstruction to rationality of a very general cubic hypersurface in $\mathbb{P}^5$. *arXiv:1605.09434*. (Cited on page 404.)

[215] Tawanda Gwena. Degenerations of cubic threefolds and matroids. *Proc. Amer. Math. Soc.*, 133(5):1317–1323, 2005. (Cited on page 239.)

[216] Marvin Hahn, Sara Lamboglia, and Alejandro Vargas. A short note on Cayley–Salmon equations. *Matematiche (Catania)*, 75(2):559–574, 2020. (Cited on page 174.)

[217] Helmut Hamm and Lê Dũng Tráng. Un théorème de Zariski du type de Lefschetz. *Ann. Sci. École Norm. Sup. (4)*, 6:317–355, 1973. (Cited on page 27.)

[218] Frédéric Han. Pfaffian bundles on cubic surfaces and configurations of planes. *Math. Z.*, 278(1–2):363–383, 2014. (Cited on page 173.)

[219] Joe Harris. Galois groups of enumerative problems. *Duke Math. J.*, 46(4):685–724, 1979. (Cited on pages 29, 163, 164, 180, and 181.)

[220] Joe Harris. *Algebraic geometry*, Vol. 133 of *Graduate Texts in Mathematics*. New York: Springer-Verlag, 1995. A first course, Corrected reprint of the 1992 original. (Cited on pages 108 and 277.)

[221] Joe Harris and Loring W. Tu. On symmetric and skew-symmetric determinantal varieties. *Topology*, 23(1):71–84, 1984. (Cited on page 323.)

[222] Robin Hartshorne. *Residues and duality*. Lecture notes of a seminar on the work of A. Grothendieck, given at Harvard 1963/64. With an appendix by P. Deligne. *Lecture Notes in Math.*, No. 20. Berlin, New York: Springer-Verlag, 1966. (Cited on page 45.)
[223] Robin Hartshorne. *Algebraic geometry*, Vol. 52 of *Graduate Texts in Mathematics*. New York, Heidelberg: Springer-Verlag, 1977. (Cited on pages 19, 21, 35, 47, 156, 163, 164, 167, 168, 207, 211, 232, and 286.)
[224] Robin Hartshorne. *Deformation theory*, Vol. 257 of *Graduate Texts in Mathematics*. New York: Springer, 2010. (Cited on page 81.)
[225] Brendan Hassett. *Special cubic hypersurfaces of dimension four*. ProQuest LLC, Ann Arbor, MI, 1996. PhD thesis, Harvard University. (Cited on pages 281, 289, and 335.)
[226] Brendan Hassett. Some rational cubic fourfolds. *J. Algebraic Geom.*, 8(1):103–114, 1999. (Cited on page 337.)
[227] Brendan Hassett. Special cubic fourfolds. *Compositio Math.*, 120(1):1–23, 2000. (Cited on pages 13, 256, 265, 267, 269, 271, 272, 285, 330, 333, 336, 340, 343, 345, and 349.)
[228] Brendan Hassett. Cubic fourfolds, K3 surfaces, and rationality questions. In *Rationality problems in algebraic geometry*, Vol. 2172 of *Lecture Notes in Math.*, pp. 29–66. Cham: Springer, 2016. (Cited on pages 65, 256, 287, 289, 337, 338, and 402.)
[229] Archibald Henderson. *The twenty-seven lines upon the cubic surface*. Reprinting of Cambridge Tracts in Mathematics and Mathematical Physics, No. 13. New York: Hafner Publishing Co., 1960. (Cited on pages 156, 175, 177, and 183.)
[230] David Hilbert. Ueber die vollen Invariantensysteme. *Math. Ann.*, 42(3):313–373, 1893. (Cited on page 185.)
[231] David Hilbert and Stefan Cohn-Vossen. *Anschauliche Geometrie*. Wissenschaftliche Buchgesellschaft, Darmstadt, 1973. Mit einem Anhang: "Einfachste Grundbegriffe der Topologie" von Paul Alexandroff, Reprint der 1932 Ausgabe. (Cited on page 179.)
[232] Lutz Hille and Michel Van den Bergh. Fourier–Mukai transforms. In *Handbook of tilting theory*, Vol. 332 of *London Math. Soc. Lecture Note Ser.*, pp. 147–177. Cambridge: Cambridge University Press, 2007. (Cited on page 383.)
[233] James Hirschfeld. The double-six of lines over $PG(3, 4)$. *J. Austral. Math. Soc.*, 4:83–89, 1964. (Cited on page 181.)
[234] James Hirschfeld. *Finite projective spaces of three dimensions*. Oxford Mathematical Monographs. New York: Oxford University Press, 1985. Oxford Science Publications. (Cited on page 182.)
[235] Friedrich Hirzebruch. *Topological methods in algebraic geometry*. Classics in Mathematics. Berlin: Springer-Verlag, 1995. (Cited on pages 8 and 10.)
[236] Toshio Hosoh. Automorphism groups of cubic surfaces. *J. Algebra*, 192(2):651–677, 1997. (Cited on page 40.)
[237] Alan Howard and Andrew Sommese. On the orders of the automorphism groups of certain projective manifolds. In *Manifolds and Lie groups (Notre Dame, Ind., 1980)*, Vol. 14 of *Progr. Math.*, pp. 145–158. Boston, MA: Birkhäuser, 1981. (Cited on page 40.)
[238] Benjamin Howard, John Millson, Andrew Snowden, and Ravi Vakil. The geometry of eight points in projective space: representation theory, Lie theory and dualities. *Proc. Lond. Math. Soc. (3)*, 105(6):1215–1244, 2012. (Cited on page 245.)
[239] Xuntao Hu. The locus of plane quartics with a hyperflex. *Proc. Amer. Math. Soc.*, 145(4):1399–1413, 2017. (Cited on page 180.)
[240] Klaus Hulek and Remke Kloosterman. The L-series of a cubic fourfold. *Manuscripta Math.*, 124(3):391–407, 2007. (Cited on page 65.)

[241] Klaus Hulek and Roberto Laface. On the Picard numbers of Abelian varieties. *Ann. Sc. Norm. Super. Pisa Cl. Sci. (5)*, 19(3):1199–1224, 2019. (Cited on page 216.)

[242] Klaus Hulek and Gregory Sankaran. The geometry of Siegel modular varieties. In *Higher dimensional birational geometry (Kyoto, 1997)*, Vol. 35 of *Adv. Stud. Pure Math.*, pp. 89–156. Tokyo: Math. Soc. Japan, 2002. (Cited on page 245.)

[243] Bruce Hunt. *The geometry of some special arithmetic quotients*, Vol. 1637 of *Lecture Notes in Mathe*. Berlin: Springer-Verlag, 1996. (Cited on page 174.)

[244] Daniel Huybrechts. Nodal quintic surfaces and lines on cubic fourfolds. *arXiv:2108.10532*. (Cited on pages 106, 296, 323, and 325.)

[245] Daniel Huybrechts. *Complex geometry*. Berlin: Universitext, Springer-Verlag, 2005. An introduction. (Cited on pages 8 and 10.)

[246] Daniel Huybrechts. *Fourier–Mukai transforms in algebraic geometry*. Oxford Mathematical Monographs. Oxford: Oxford University Press, 2006. (Cited on pages 83, 276, 355, 356, 357, 358, 361, 365, 367, 387, 389, 390, 391, and 399.)

[247] Daniel Huybrechts. The global Torelli theorem: classical, derived, twisted. In *Algebraic geometry—Seattle 2005. 1*, Vol. 80 of *Proc. Sympos. Pure Math.*, pp. 235–258. Providence, RI: Amer. Math. Soc., 2009. (Cited on page 355.)

[248] Daniel Huybrechts. A global Torelli theorem for hyperkähler manifolds [after M. Verbitsky], Séminaire Bourbaki, Exposé 1040, 2010/2011. *Astérisque*, 348:375–403, 2012. (Cited on pages 295 and 348.)

[249] Daniel Huybrechts. *Lectures on K3 surfaces*, Vol. 158 of *Cambridge Studies in Advanced Mathematics*. Cambridge: Cambridge University Press, 2016. (Cited on pages 13, 23, 34, 140, 141, 142, 143, 148, 158, 162, 227, 260, 261, 264, 267, 295, 298, 302, 330, 331, 336, 339, 346, 350, 393, 394, and 395.)

[250] Daniel Huybrechts. The K3 category of a cubic fourfold. *Compositio Math.*, 153(3):586–620, 2017. (Cited on pages 339, 341, 385, 393, 394, and 395.)

[251] Daniel Huybrechts. Motives of derived equivalent K3 surfaces. *Abh. Math. Semin. Univ. Hambg.*, 88(1):201–207, 2018. (Cited on page 406.)

[252] Daniel Huybrechts. Hodge theory of cubic fourfolds, their Fano varieties, and associated K3 categories. In Andreas Hochenegger, Manfred Lehn, and Paolo Stellari (eds.), *Birational geometry of hypersurfaces*, Vol. 26 of *Lect. Notes Unione Mat. Ital.*, pp. 165–198. Cham: Springer, 2019. (Cited on pages 330, 332, 333, 334, 336, 339, 340, 341, 344, 345, and 347.)

[253] Daniel Huybrechts and Manfred Lehn. *The geometry of moduli spaces of sheaves*. Cambridge Mathematical Library. Cambridge: Cambridge University Press, 2nd ed., 2010. (Cited on pages 33, 34, 76, 81, 82, 293, 382, and 397.)

[254] Daniel Huybrechts and Marc Nieper-Wisskirchen. Remarks on derived equivalences of Ricci-flat manifolds. *Math. Z.*, 267(3–4):939–963, 2011. (Cited on page 293.)

[255] Daniel Huybrechts and Jørgen Rennemo. Hochschild cohomology versus the Jacobian ring and the Torelli theorem for cubic fourfolds. *Algebr. Geom.*, 6(1):76–99, 2019. (Cited on pages 300, 347, 369, and 373.)

[256] Daniel Huybrechts and Paolo Stellari. Equivalences of twisted K3 surfaces. *Math. Ann.*, 332(4):901–936, 2005. (Cited on page 394.)

[257] Atanas Iliev and Laurent Manivel. Cubic hypersurfaces and integrable systems. *Amer. J. Math.*, 130(6):1445–1475, 2008. (Cited on pages 299, 317, and 318.)

[258] Atanas Iliev and Dimitri Markushevich. The Abel–Jacobi map for a cubic threefold and periods of Fano threefolds of degree 14. *Doc. Math.*, 5:23–47, 2000. (Cited on page 220.)

[259] Vasiliĭ Iskovskikh and Yuri Prokhorov. Fano varieties. In *Algebraic geometry, V*, Vol. 47 of *Encyclopaedia Math. Sci.*, pp. 1–247. Berlin: Springer, Berlin, 1999. (Cited on page 37.)

[260] Elham Izadi. A Prym construction for the cohomology of a cubic hypersurface. *Proc. London Math. Soc. (3)*, 79(3):535–568, 1999. (Cited on page 128.)

[261] Wilhelmus Janssen. Skew-symmetric vanishing lattices and their monodromy groups. *Math. Ann.*, 266(1):115–133, 1983. (Cited on page 29.)

[262] Ariyan Javanpeykar and Daniel Loughran. The moduli of smooth hypersurfaces with level structure. *Manuscripta Math.*, 154(1–2):13–22, 2017. (Cited on page 39.)

[263] Rainer Kaenders. Die Diagonalfläche aus Keramik. *Mitteilungen der Deutschen Mathematiker-Vereinigung*, 4:16–21, 1999. (Cited on page 182.)

[264] Martin Kalck, Nebojsa Pavic, and Evgeny Shinder. Obstructions to semiorthogonal decompositions for singular threefolds I: K-theory. *Mosc. Math. J.*, 21(3):567–592, 2021. (Cited on page 377.)

[265] Antonius Kalker. Cubic fourfolds with fifteen ordinary double points. PhD thesis, Leiden University, 1986. (Cited on page 68.)

[266] Mikhai Kapranov. On the derived categories of coherent sheaves on some homogeneous spaces. *Invent. Math.*, 92(3):479–508, 1988. (Cited on page 364.)

[267] Anton Kapustin and Yi Li. D-branes in topological minimal models: the Landau–Ginzburg approach. *J. High Energy Phys.*, (7):045, 26, 2004. (Cited on page 372.)

[268] Edward Kasner. The double-six configuration connected with the cubic surface, and a related group of Cremona transformations. *Amer. J. Math.*, 25(2):107–122, 1903. (Cited on page 179.)

[269] Jesse Kass and Kirsten Wickelgren. An arithmetic count of the lines on a smooth cubic surface. *Compositio Math.*, 157(4):677–709, 2021. (Cited on pages 156 and 183.)

[270] Nicholas Katz and Peter Sarnak. *Random matrices, Frobenius eigenvalues, and monodromy*, Vol. 45 of *American Mathematical Society Colloquium Publications*. Providence, RI: Amer. Math. Soc., 1999. (Cited on pages 18, 32, 34, 137, and 138.)

[271] Hanieh Keneshlou. Cubic surfaces on the singular locus of the Eckardt hypersurface. *Matematiche (Catania)*, 75(2):507–516, 2020. (Cited on page 182.)

[272] Shun-Ichi Kimura. Chow groups are finite dimensional, in some sense. *Math. Ann.*, 331(1):173–201, 2005. (Cited on pages 228 and 404.)

[273] Frances Kirwan. *Cohomology of quotients in symplectic and algebraic geometry*, Vol. 31 of *Mathematical Notes*. Princeton, NJ: Princeton University Press, 1984. (Cited on page 143.)

[274] Frances Kirwan. Partial desingularisations of quotients of nonsingular varieties and their Betti numbers. *Ann. of Math. (2)*, 122(1):41–85, 1985. (Cited on page 143.)

[275] Frances Kirwan. Moduli spaces of degree d hypersurfaces in $\mathbf{P}_n$. *Duke Math. J.*, 58(1):39–78, 1989. (Cited on pages 143 and 147.)

[276] Felix Klein. Ueber Flächen dritter Ordnung. *Math. Ann.*, 6(4):551–581, 1873. (Cited on page 174.)

[277] Felix Klein. The Evanston Colloquium lectures on mathematics, delivered at Northwestern University Aug. 28 to Sept. 1893. Reported by *Alexander Ziwet*. 2nd ed. New York: Amer. Math. Soc. XI + 109 S. 8° (1894)., 1894. (Cited on page 183.)

[278] Anthony Knapp. *Elliptic curves*, Vol. 40 of *Mathematical Notes*. Princeton, NJ: Princeton University Press, 1992. (Cited on page 21.)

[279] Kunihiko Kodaira and Donald Spencer. On deformations of complex analytic structures. I, II. *Ann. of Math. (2)*, 67:328–466, 1958. (Cited on page 32.)

[280] Kenji Koike. Moduli space of Hessian K3 surfaces and arithmetic quotients. *arXiv:1002.2854*. (Cited on page 185.)

[281] János Kollár. *Rational curves on algebraic varieties*, volume 32 of *Ergebnisse der Mathematik und ihrer Grenzgebiete. 3. Folge. A Series of Modern Surveys in Mathematics*. Berlin: Springer-Verlag, 1996. (Cited on pages 81, 84, and 197.)

[282] János Kollár. Fundamental groups of rationally connected varieties. *Michigan Math. J.*, 48:359–368, 2000. Dedicated to William Fulton on the occasion of his 60th birthday. (Cited on page 89.)

[283] János Kollár. Unirationality of cubic hypersurfaces. *J. Inst. Math. Jussieu*, 1(3):467–476, 2002. (Cited on page 87.)

[284] János Kollár, Karen Smith, and Alessio Corti. *Rational and nearly rational varieties*, Vol. 92 of *Cambridge Studies in Advanced Mathematics*. Cambridge: Cambridge University Press, 2004. (Cited on pages 156 and 158.)

[285] Maxim Kontsevich and Yuri Tschinkel. Specialization of birational types. *Invent. Math.*, 217(2):415–432, 2019. (Cited on pages 239, 271, 289, and 337.)

[286] Hanspeter Kraft, Peter Slodowy, and Tonny Springer (eds.). *Algebraische Transformationsgruppen und Invariantentheorie*, Vol. 13 of *DMV Seminar*. Basel: Birkhäuser Verlag, 1989. (Cited on page 141.)

[287] Thomas Krämer. Summands of theta divisors on Jacobians. *Compositio Math.*, 156(7):1457–1475, 2020. (Cited on page 227.)

[288] Andreas Krug, David Ploog, and Pawel Sosna. Derived categories of resolutions of cyclic quotient singularities. *Q. J. Math.*, 69(2):509–548, 2018. (Cited on page 398.)

[289] Stephen Kudla and Michael Rapoport. On occult period maps. *Pacific J. Math.*, 260(2):565–581, 2012. (Cited on pages 186, 246, and 249.)

[290] Ravindra Kulkarni and John Wood. Topology of nonsingular complex hypersurfaces. *Adv. in Math.*, 35(3):239–263, 1980. (Cited on pages 15 and 24.)

[291] Ernst Kunz. *Residues and duality for projective algebraic varieties*, volume 47 of *University Lecture Series*. Providence, RI: Amer. Math. Soc. 2008. With the assistance of and contributions by David A. Cox and Alicia Dickenstein. (Cited on page 44.)

[292] Alexander Kuznetsov. Hochschild homology and semiorthogonal decompositions. *arXiv:0904.4330*. (Cited on pages 361, 369, and 384.)

[293] Alexander Kuznetsov. Homological projective duality for Grassmannians of lines. *arXiv:math/0610957*. (Cited on page 389.)

[294] Alexander Kuznetsov. Derived categories of quadric fibrations and intersections of quadrics. *Adv. Math.*, 218(5):1340–1369, 2008. (Cited on pages 379 and 388.)

[295] Alexander Kuznetsov. Lefschetz decompositions and categorical resolutions of singularities. *Selecta Math. (N.S.)*, 13(4):661–696, 2008. (Cited on page 392.)

[296] Alexander Kuznetsov. Derived categories of cubic fourfolds. In *Cohomological and geometric approaches to rationality problems*, Vol. 282 of *Progr. Math.*, pp. 219–243. Boston, MA: Birkhäuser Boston, 2010. (Cited on pages 263, 384, 385, 386, 391, and 392.)

[297] Alexander Kuznetsov. Base change for semiorthogonal decompositions. *Compositio Math.*, 147(3):852–876, 2011. (Cited on pages 367 and 369.)

[298] Alexander Kuznetsov. Derived categories view on rationality problems. In *Rationality problems in algebraic geometry*, Vol. 2172 of *Lecture Notes in Math.*, pp. 67–104. Cham: Springer, 2016. (Cited on pages 262, 384, and 386.)

[299] Alexander Kuznetsov. Calabi–Yau and fractional Calabi–Yau categories. *J. Reine Angew. Math.*, 753:239–267, 2019. (Cited on page 366.)

[300] Alexander Kuznetsov and Dimitri Markushevich. Symplectic structures on moduli spaces of sheaves via the Atiyah class. *J. Geom. Phys.*, 59(7):843–860, 2009. (Cited on pages 365 and 397.)

[301] Alexander Kuznetsov and Alexander Perry. Homological projective duality for quadrics. *J. Algebraic Geom.*, 30(3):457–476, 2021. (Cited on page 364.)

[302] Martí Lahoz, Manfred Lehn, Emanuele Macrì, and Paolo Stellari. Generalized twisted cubics on a cubic fourfold as a moduli space of stable objects. *J. Math. Pures Appl. (9)*, 114:85–117, 2018. (Cited on page 396.)

[303] Martí Lahoz, Juan Carlos Naranjo, and Andrés Rojas. Geometry of Prym semicanonical pencils and an application to cubic threefolds. *arXiv:2106.08683*. (Cited on page 199.)

[304] Kuan-Wen Lai. New cubic fourfolds with odd-degree unirational parametrizations. *Algebr Number Theory*, 11(7):1597–1626, 2017. (Cited on pages 148, 338, 349, and 402.)

[305] Herbert Lange and Christina Birkenhake. *Complex abelian varieties*, Vol. 302 of *Grundlehren der Mathematischen Wissenschaften*. Berlin: Springer-Verlag, 1992. (Cited on pages 216, 221, 222, and 226.)

[306] Michael Larsen and Valery Lunts. Motivic measures and stable birational geometry. *Mosc. Math. J.*, 3(1):85–95, 259, 2003. (Cited on pages 111 and 240.)

[307] Robert Laterveer. A remark on the motive of the Fano variety of lines of a cubic. *Ann. Math. Qué.*, 41(1):141–154, 2017. (Cited on pages 113, 306, 404, and 405.)

[308] Robert Laterveer. A family of cubic fourfolds with finite-dimensional motive. *J. Math. Soc. Japan*, 70(4):1453–1473, 2018. (Cited on pages 306 and 405.)

[309] Robert Laterveer. Lagrangian subvarieties in the Chow ring of some hyperkähler varieties. *Mosc. Math. J.*, 18(4):693–719, 2018. (Cited on page 321.)

[310] Kristin Lauter. Geometric methods for improving the upper bounds on the number of rational points on algebraic curves over finite fields. *J. Algebraic Geom.*, 10(1):19–36, 2001. With an appendix in French by J.-P. Serre. (Cited on page 236.)

[311] Radu Laza. The moduli space of cubic fourfolds. *J. Algebraic Geom.*, 18(3):511–545, 2009. (Cited on pages 350 and 352.)

[312] Radu Laza. The moduli space of cubic fourfolds via the period map. *Ann. of Math. (2)*, 172(1):673–711, 2010. (Cited on pages 307 and 349.)

[313] Radu Laza. Maximally algebraic potentially irrational cubic fourfolds. *Proc. Amer. Math. Soc.*, 149(8):3209–3220, 2021. (Cited on page 310.)

[314] Radu Laza, Giulia Saccà, and Claire Voisin. A hyper-Kähler compactification of the intermediate Jacobian fibration associated with a cubic 4-fold. *Acta Math.*, 218(1):55–135, 2017. (Cited on page 298.)

[315] Radu Laza and Zhiwei Zheng. Automorphisms and periods of cubic fourfolds. *Math. Z.*, 300(2):1455–1507, 2022. (Cited on page 40.)

[316] Robert Lazarsfeld. *Positivity in algebraic geometry. II*, Vol. 49 of *Ergebnisse der Mathematik und ihrer Grenzgebiete. 3. Folge. A Series of Modern Surveys in Mathematics*. (Cited on page 103.)

[317] Joseph Le Potier. *Lectures on vector bundles*, Vol. 54 of *Cambridge Studies in Advanced Mathematics*. Cambridge: Cambridge University Press, 1997. Translated by A. Maciocia. (Cited on pages 135 and 136.)

[318] Christian Lehn. Twisted cubics on singular cubic fourfolds–on Starr's fibration. *Math. Z.*, 290(1–2):379–388, 2018. (Cited on page 269.)

[319] Christian Lehn, Manfred Lehn, Christoph Sorger, and Duco van Straten. Twisted cubics on cubic fourfolds. *J. Reine Angew. Math.*, 731:87–128, 2017. (Cited on page 342.)

[320] Chunyi Li, Laura Pertusi, and Xiaolei Zhao. Derived categories of hearts on Kuznetsov components. *arXiv:2203.13864*. (Cited on page 394.)

[321] Anatoly Libgober. On the fundamental group of the space of cubic surfaces. *Math. Z.*, 162(1):63–67, 1978. (Cited on page 147.)

[322] Anatoly Libgober and John Wood. On the topological structure of even-dimensional complete intersections. *Trans. Amer. Math. Soc.*, 267(2):637–660, 1981. (Cited on page 14.)

[323] Yuchen Liu. K-stability of cubic fourfolds. *J. Reine Angew. Math.*, 786:55–77, 2022. (Cited on page 353.)

[324] Yuchen Liu and Chenyang Xu. K-stability of cubic threefolds. *Duke Math. J.*, 168(11):2029–2073, 2019. (Cited on page 244.)

[325] Michael Lönne. Fundamental groups of projective discriminant complements. *Duke Math. J.*, 150(2):357–405, 2009. (Cited on page 147.)

[326] Eduard Looijenga. *Isolated singular points on complete intersections*, Vol. 77 of *London Mathematical Society Lecture Note Series*. Cambridge: Cambridge University Press, 1984. (Cited on page 29.)

[327] Eduard Looijenga. The period map for cubic fourfolds. *Invent. Math.*, 177(1):213–233, 2009. (Cited on pages 300, 308, 347, and 349.)

[328] Eduard Looijenga. Teichmüller spaces and Torelli theorems for hyperkähler manifolds. *Math. Z.*, 298(1–2):261–279, 2021. (Cited on pages 295 and 348.)

[329] Eduard Looijenga and Rogier Swierstra. The period map for cubic threefolds. *Compositio Math.*, 143(4):1037–1049, 2007. (Cited on pages 246 and 248.)

[330] Shouhei Ma. On the Kodaira dimension of orthogonal modular varieties. *Invent. Math.*, 212(3):859–911, 2018. (Cited on page 349.)

[331] Ian Macdonald. The Poincaré polynomial of a symmetric product. *Proc. Cambridge Philos. Soc.*, 58:563–568, 1962. (Cited on page 114.)

[332] Emanuele Macrì and Paolo Stellari. Fano varieties of cubic fourfolds containing a plane. *Math. Ann.*, 354(3):1147–1176, 2012. (Cited on page 341.)

[333] Emanuele Macrì and Paolo Stellari. Lectures on non-commutative K3 surfaces, Bridgeland stability, and moduli spaces. In Andreas Hochenegger, Manfred Lehn, and Paolo Stellari (eds.), *Birational geometry of hypersurfaces*, Vol. 26 of *Lect. Notes Unione Mat. Ital.*, pp. 199–265. Cham: Springer, 2019. (Cited on page 309.)

[334] Yuri Manin. Correspondences, motifs and monoidal transformations. *Mat. Sb. (N.S.)*, 77 (119):475–507, 1968. (Cited on page 229.)

[335] Yuri Manin. *Cubic forms*, Vol. 4 of *North-Holland Mathematical Library*. Amsterdam: North-Holland Publishing Co., 2nd ed. 1986. Algebra, geometry, arithmetic, Translated from the Russian by M. Hazewinkel. (Cited on pages 156, 158, 159, and 170.)

[336] Laurent Manivel and Emilia Mezzetti. On linear spaces of skew-symmetric matrices of constant rank. *Manuscripta Math.*, 117(3):319–331, 2005. (Cited on page 276.)

[337] Eyal Markman. A survey of Torelli and monodromy results for holomorphic-symplectic varieties. In *Complex and differential geometry*, Vol. 8 of *Springer Proc. Math.*, pp. 257–322. Heidelberg: Springer, 2011. (Cited on pages 295, 308, 309, 341, and 348.)

[338] Eyal Markman. Prime exceptional divisors on holomorphic symplectic varieties and monodromy reflections. *Kyoto J. Math.*, 53(2):345–403, 2013. (Cited on pages 308 and 326.)

[339] Eyal Markman. Lagrangian fibrations of holomorphic-symplectic varieties of $K3^{[n]}$-type. In *Algebraic and complex geometry*, Vol. 71 of *Springer Proc. Math. Stat.*, pp. 241–283. Cham: Springer, 2014. (Cited on page 342.)

[340] Dimitri Markushevich and Xavier Roulleau. Irrationality of generic cubic threefold via Weil's conjectures. *Commun. Contemp. Math.*, 20(7):1750078, 12, 2018. (Cited on page 238.)

[341] Dimitri Markushevich and Alexander Tikhomirov. The Abel–Jacobi map of a moduli component of vector bundles on the cubic threefold. *J. Algebraic Geom.*, 10(1):37–62, 2001. (Cited on page 220.)

[342] John Mather and Stephen S. T. Yau. Classification of isolated hypersurface singularities by their moduli algebras. *Invent. Math.*, 69(2):243–251, 1982. (Cited on page 46.)

[343] Hideyuki Matsumura. *Commutative algebra*, Vol. 56 of *Mathematics Lecture Note Series*. Reading, MA: Benjamin/Cummings Publishing Co., Inc., 2nd ed. 1980. (Cited on page 31.)

[344] Hideyuki Matsumura and Paul Monsky. On the automorphisms of hypersurfaces. *J. Math. Kyoto Univ.*, 3:347–361, 1963/1964. (Cited on pages 34 and 38.)

[345] Teruhisa Matsusaka. On a characterization of a Jacobian variety. *Mem. Coll. Sci. Univ. Kyoto Ser. A. Math.*, 32:1–19, 1959. (Cited on page 227.)

[346] Teruhisa Matsusaka and David Mumford. Two fundamental theorems on deformations of polarized varieties. *Amer. J. Math.*, 86:668–684, 1964. (Cited on page 35.)

[347] Laurentiu Maxim and Jörg Schürmann. Twisted genera of symmetric products. *Selecta Math. (N.S.)*, 18(1):283–317, 2012. (Cited on page 121.)

[348] Evgeny Mayanskiy. Intersection lattices of cubic fourfolds. *arXiv:1112.0806*. (Cited on page 332.)

[349] René Mboro. Remarks on the CH_2 of cubic hypersurfaces. *Geom. Dedicata*, 200:1–25, 2019. (Cited on page 401.)

[350] Stephen McKean. Rational lines on smooth cubic surfaces. *arXiv:2101.08217*. (Cited on page 183.)

[351] Ciaran Meachan. A note on spherical functors. *Bull. Lond. Math. Soc.*, 53(3):956–962, 2021. (Cited on page 398.)

[352] James Milne. Jacobian varieties. In *Arithmetic geometry (Storrs, Conn., 1984)*, pp. 167–212. New York: Springer, 1986. (Cited on page 236.)

[353] Rick Miranda. Triple covers in algebraic geometry. *Amer. J. Math.*, 107(5):1123–1158, 1985. (Cited on page 66.)

[354] Giovanni Mongardi and John Christian Ottem. Curve classes on irreducible holomorphic symplectic varieties. *Commun. Contemp. Math.*, 22(7):1950078, 15, 2020. (Cited on pages 309, 310, and 330.)

[355] Riccardo Moschetti. The derived category of a non generic cubic fourfold containing a plane. *Math. Res. Lett.*, 25(5):1525–1545, 2018. (Cited on page 385.)

[356] Shigeru Mukai. *An introduction to invariants and moduli*, Vol. 81 of *Cambridge Studies in Advanced Mathematics*. Cambridge: Cambridge University Press, 2003. Translated from the 1998 and 2000 Japanese editions by W. M. Oxbury. (Cited on pages 135, 139, 147, 185, and 186.)

[357] Shigeru Mukai and Hirokazu Nasu. Obstructions to deforming curves on a 3-fold. I. A generalization of Mumford's example and an application to Hom schemes. *J. Algebraic Geom.*, 18(4):691–709, 2009. (Cited on page 70.)

[358] David Mumford. Theta characteristics of an algebraic curve. *Ann. Sci. École Norm. Sup. (4)*, 4:181–192, 1971. (Cited on page 180.)

[359] David Mumford. Prym varieties. I. In *Contributions to analysis (a collection of papers dedicated to Lipman Bers)*, pp. 325–350. New York: Academic Press, 1974. (Cited on pages 221 and 238.)

[360] David Mumford. Hilbert's fourteenth problem–the finite generation of subrings such as rings of invariants. In *Mathematical developments arising from Hilbert problems (Proc. Sympos. Pure Math., Vol. XXVIII, Northern Illinois Univ., De Kalb, Ill., 1974)*, pp. 431–444. Providence, RI: Amer. Math. Soc., 1976. (Cited on page 135.)

[361] David Mumford. Stability of projective varieties. *Enseign. Math. (2)*, 23(1–2):39–110, 1977. (Cited on pages 147 and 185.)

[362] David Mumford, John Fogarty, and Frances Kirwan. *Geometric invariant theory*, Vol. 34 of *Ergebnisse der Mathematik und ihrer Grenzgebiete (2)*. Berlin: Springer-Verlag, 3rd ed., 1994. (Cited on pages 135, 136, 137, and 139.)

[363] Daniel Murfet. Residues and duality for singularity categories of isolated Gorenstein singularities. *Compositio Math.*, 149(12):2071–2100, 2013. (Cited on pages 372 and 374.)

[364] Jacob Murre. Algebraic equivalence modulo rational equivalence on a cubic threefold. *Compositio Math.*, 25:161–206, 1972. (Cited on pages 193, 197, 198, 228, and 229.)

[365] Jacob Murre. Reduction of the proof of the non-rationality of a non-singular cubic threefold to a result of Mumford. *Compositio Math.*, 27:63–82, 1973. (Cited on pages 193, 221, and 238.)

[366] Jacob Murre. Some results on cubic threefolds. In *Classification of algebraic varieties and compact complex manifolds*, Vol. 412 of *Lecture Notes in Math.*, pp. 140–160. Berlin: Springer, 1974. (Cited on page 228.)

[367] Jacob Murre, Jan Nagel, and Chris Peters. *Lectures on the theory of pure motives*, Vol. 61 of *University Lecture Series*. Providence, RI: Amer. Math. Soc. 2013. (Cited on pages 4 and 112.)

[368] Jan Nagel and Morihiko Saito. Relative Chow–Künneth decompositions for conic bundles and Prym varieties. *Int. Math. Res. Not. IMRN*, (16):2978–3001, 2009. (Cited on page 229.)

[369] Iku Nakamura. Planar cubic curves, from Hesse to Mumford. Vol. 17, pp. 73–101. 2004. Sugaku Expositions. (Cited on page 147.)

[370] Isao Naruki. Cross ratio variety as a moduli space of cubic surfaces. *Proc. London Math. Soc. (3)*, 45(1):1–30, 1982. With an appendix by Eduard Looijenga. (Cited on page 185.)

[371] Johannes Nicaise and Evgeny Shinder. The motivic nearby fiber and degeneration of stable rationality. *Invent. Math.*, 217(2):377–413, 2019. (Cited on pages 239, 271, 289, and 337.)

[372] Howard Nuer. Unirationality of moduli spaces of special cubic fourfolds and K3 surfaces. *Algebr. Geom.*, 4(3):281–289, 2017. (Cited on page 348.)

[373] Georg Oberdieck, Junliang Shen, and Qizheng Yin. Rational curves in holomorphic symplectic varieties and Gromov–Witten invariants. *Adv. Math.*, 357:106829, 8, 2019. (Cited on pages 318, 319, and 329.)

[374] Kieran G. O'Grady. Desingularized moduli spaces of sheaves on a K3. *J. Reine Angew. Math.*, 512:49–117, 1999. (Cited on page 298.)

[375] Peter Orlik and Louis Solomon. Singularities. II. Automorphisms of forms. *Math. Ann.*, 231(3):229–240, 1977/78. (Cited on pages 34 and 46.)

[376] Dimitri Orlov. Triangulated categories of singularities and D-branes in Landau–Ginzburg models. *Tr. Mat. Inst. Steklova*, 246:240–262, 2004. (Cited on page 372.)

[377] Dmitri Orlov. Derived categories of coherent sheaves and equivalences between them. *Uspekhi Mat. Nauk*, 58(3(351)):89–172, 2003. (Cited on page 355.)

[378] Dmitri Orlov. Derived categories of coherent sheaves and triangulated categories of singularities. In *Algebra, arithmetic, and geometry: in honor of Yu. I. Manin. Vol. II*, Vol. 270 of *Progr. Math.*, pp. 503–531. Boston, MA: Birkhäuser Boston, 2009. (Cited on pages 372 and 373.)

[379] David Oscari. Cubic surfaces as Pfaffians. *arXiv:1911.09754*. (Cited on page 173.)

[380] Giorgio Ottaviani. Spinor bundles on quadrics. *Trans. Amer. Math. Soc.*, 307(1):301–316, 1988. (Cited on page 364.)

[381] John Christian Ottem. Nef cycles on some hyperkähler fourfolds. In *Facets of algebraic geometry. Vol. II*, Vol. 473 of *London Math. Soc. Lecture Note Ser.*, pp. 228–237. Cambridge: Cambridge University Press, 2022. (Cited on page 317.)

[382] Genki Ouchi. Lagrangian embeddings of cubic fourfolds containing a plane. *Compositio Math.*, 153(5):947–972, 2017. (Cited on page 397.)

[383] Xuanyu Pan. Automorphism and cohomology I: Fano varieties of lines and cubics. *Algebr. Geom.*, 7(1):1–29, 2020. (Cited on pages 39 and 215.)

[384] Kapil H. Paranjape. Cohomological and cycle-theoretic connectivity. *Ann. of Math. (2)*, 139(3):641–660, 1994. (Cited on pages 5 and 400.)

[385] Alexander Perry. The integral Hodge conjecture for two-dimensional Calabi–Yau categories. *Compositio Math.*, 158(2):287–333, 2022. (Cited on page 383.)

[386] Laura Pertusi. Fourier–Mukai partners for very general special cubic fourfolds. *Math. Res. Lett.*, 28(1):213–243, 2021. (Cited on page 395.)

[387] Chris Peters. On a motivic interpretation of primitive, variable and fixed cohomology. *Math. Nachr.*, 292(2):402–408, 2019. (Cited on pages 5 and 403.)

[388] Chris Peters and Joseph Steenbrink. *Degeneration of the Leray spectral sequence for certain geometric quotients*, *Moscow Math. Journal*, 2(3): 1085–1095, 2003. Dedicated to Vladimir Igorevich Arnold on the occasion of his 65th birthday. (Cited on pages 144, 145, and 146.)

[389] Chris Peters and Joseph Steenbrink. *Monodromy of variations of Hodge structure*, *Acta Appl. Math.*, 75:183–194, 2003. Monodromy and differential equations (Moscow, 2001). (Cited on page 25.)

[390] Chris Peters and Joseph Steenbrink. *Mixed Hodge structures*, Vol. 52 of *Ergebnisse der Mathematik und ihrer Grenzgebiete. 3. Folge. A Series of Modern Surveys in Mathematics*. Berlin: Springer-Verlag, 2008. (Cited on page 117.)

[391] Alexander Polishchuk and Arkady Vaintrob. Chern characters and Hirzebruch–Riemann–Roch formula for matrix factorizations. *Duke Math. J.*, 161(10):1863–1926, 2012. (Cited on pages 372 and 374.)

[392] Irene Polo-Blanco and Jaap Top. A remark on parameterizing nonsingular cubic surfaces. *Comput. Aided Geom. Design*, 26(8):842–849, 2009. (Cited on page 166.)

[393] Bjorn Poonen. Varieties without extra automorphisms. III. Hypersurfaces. *Finite Fields Appl.*, 11(2):230–268, 2005. (Cited on pages 38 and 40.)

[394] Pavel Popov. Twisted cubics and quadruples of points on cubic surfaces. *arXiv:1810.04563*. (Cited on pages 177, 375, and 399.)

[395] Ziv Ran. On subvarieties of abelian varieties. *Invent. Math.*, 62(3):459–479, 1981. (Cited on page 227.)

[396] Kristian Ranestad and Claire Voisin. Variety of power sums and divisors in the moduli space of cubic fourfolds. *Doc. Math.*, 22:455–504, 2017. (Cited on page 349.)

[397] Michael Rapoport. Complément à l'article de P. Deligne "La conjecture de Weil pour les surfaces K3". *Invent. Math.*, 15:227–236, 1972. (Cited on page 11.)

[398] Miles Reid. The complete intersection of two or more quadrics. PhD thesis, University of Cambridge, 1972. (Cited on page 229.)

[399] Emanuel Reinecke. Moduli space of cubic surfaces. `www.math.uni-bonn.de/people/huybrech/Reineckefinal.pdf`. Bachelor thesis, University of Bonn, 2012. (Cited on pages 183 and 185.)

[400] Max Rempel. Positivité des cycles dans les variétés algébriques. PhD thesis, University of Paris, 2012. (Cited on page 317.)

[401] Ulrike Rieß. On the non-divisorial base locus of big and nef line bundles on $K3^{[2]}$-type varieties. *Proc. Roy. Soc. Edinburgh Sect. A*, 151(1):52–78, 2021. (Cited on pages 308 and 309.)

[402] Xavier Roulleau. Elliptic curve configurations on Fano surfaces. *Manuscripta Math.*, 129(3):381–399, 2009. (Cited on pages 198 and 215.)

[403] Xavier Roulleau. The Fano surface of the Klein cubic threefold. *J. Math. Kyoto Univ.*, 49(1):113–129, 2009. (Cited on pages 216 and 244.)

[404] Xavier Roulleau. The Fano surface of the Fermat cubic threefold, the del Pezzo surface of degree 5 and a ball quotient. *Proc. Amer. Math. Soc.*, 139(10):3405–3412, 2011. (Cited on page 198.)

[405] Xavier Roulleau. Fano surfaces with 12 or 30 elliptic curves. *Michigan Math. J.*, 60(2):313–329, 2011. (Cited on page 216.)

[406] Xavier Roulleau. Quotients of Fano surfaces. *Atti Accad. Naz. Lincei Rend. Lincei Mat. Appl.*, 23(3):325–349, 2012. (Cited on page 215.)

[407] Xavier Roulleau. On the Tate conjecture for the Fano surfaces of cubic threefolds. *J. Number Theory*, 133(7):2320–2323, 2013. (Cited on page 217.)

[408] Francesco Russo and Giovanni Staglianò. Congruences of 5-secant conics and the rationality of some admissible cubic fourfolds. *Duke Math. J.*, 168(5):849–865, 2019. (Cited on page 337.)

[409] Francesco Russo and Giovanni Staglianò. Trisecant flops, their associated K3 surfaces and the rationality of some cubic fourfolds. JEMS (2022), to appear. arXiv:1909.01263. (Cited on page 337.)

[410] Sergey Rybakov and Andrey Trepalin. Minimal cubic surfaces over finite fields. *Mat. Sb.*, 208(9):148–170, 2017. (Cited on page 159.)

[411] Giulia Saccà. Birational geometry of the intermediate Jacobian fibration of a cubic fourfold, with an appendix by C. Voisin. *arXiv:2002.01420*. (Cited on pages 270 and 298.)

[412] Igor Šafarevič. *Basic algebraic geometry. 1.* Heidelberg: Springer, 3rd ed. 2013. Varieties in projective space. (Cited on page 159.)

[413] Kyoji Saito. Einfach-elliptische Singularitäten. *Invent. Math.*, 23:289–325, 1974. (Cited on page 45.)

[414] George Salmon. On the triple tangent planes of surfaces of the third order. *Cambridge and Dublin Math. Journal*, IV:252–260, 1849. (Cited on page 175.)

[415] George Salmon. On quaternary cubics. *Phil. Trans. R. Soc.*, 150:229–239, 1860. (Cited on pages 21 and 184.)

[416] George Salmon. *A treatise on the analytic geometry of three dimensions.* Dublin: Hodges, Figgis, 1865. (Cited on page 183.)

[417] Ludwig Schläfli. An attempt to determine the twenty-seven lines upon a surfaces of the third order, and to divide such surfaces into species in reference to the reality of the lines upon the surface. *Quart. J. Math.*, 2:110–120, 1858. (Cited on page 183.)

[418] Chad Schoen. Varieties dominated by product varieties. *Internat. J. Math.*, 7(4):541–571, 1996. (Cited on page 238.)

[419] Stefan Schreieder. Theta divisors with curve summands and the Schottky problem. *Math. Ann.*, 365(3–4):1017–1039, 2016. (Cited on page 237.)

[420] Beniamino Segre. Le rette delle superficie cubiche nei corpi commutativi. *Boll. Un. Mat. Ital. (3)*, 4:223–228, 1949. (Cited on page 183.)

[421] Edoardo Sernesi. *Deformations of algebraic schemes*, Vol. 334 of *Grundlehren der Mathematischen Wissenschaften*. Berlin: Springer-Verlag, 2006. (Cited on pages 36, 37, and 81.)

[422] Jean-Pierre Serre. *A course in arithmetic*. Vol. 7 of *Graduate Texts in Mathematics*. New York, Heidelberg: Springer-Verlag. (Cited on pages 12 and 14.)

[423] Jean-Pierre Serre. On the fundamental group of a unirational variety. *J. London Math. Soc.*, 34:481–484, 1959. (Cited on page 88.)

[424] Jean-Pierre Serre. *Local algebra*. Springer Monographs in Mathematics. Berlin: Springer-Verlag, 2000. Translated from the French by CheeWhye Chin and revised by the author. (Cited on page 43.)

[425] Junliang Shen and Qizheng Yin. K3 categories, one-cycles on cubic fourfolds, and the Beauville–Voisin filtration. *J. Inst. Math. Jussieu*, 19(5):1601–1627, 2020. (Cited on page 401.)

[426] Mingmin Shen. Surfaces with involution and Prym constructions. *arXiv:1209.5457*. (Cited on pages 105 and 106.)

[427] Mingmin Shen. On relations among 1-cycles on cubic hypersurfaces. *J. Algebraic Geom.*, 23(3):539–569, 2014. (Cited on pages 5, 229, and 400.)

[428] Mingmin Shen. Hyperkähler manifolds of Jacobian type. *J. Reine Angew. Math.*, 712:189–223, 2016. (Cited on pages 264, 294, 295, 311, and 313.)

[429] Mingmin Shen and Charles Vial. The Fourier transform for certain hyperkähler fourfolds. *Mem. Amer. Math. Soc.*, 240(1139):vii+163, 2016. (Cited on pages 306, 329, and 401.)

[430] Bernard Shiffman and Andrew Sommese. *Vanishing theorems on complex manifolds*, Vol. 56 of *Progress in Mathematics*. Boston, MA: Birkhäuser Boston, Inc., 1985. (Cited on page 319.)

[431] Ichiro Shimada. On the cylinder isomorphism associated to the family of lines on a hypersurface. *J. Fac. Sci. Univ. Tokyo Sect. IA Math.*, 37(3):703–719, 1990. (Cited on pages 127 and 132.)

[432] Evgeny Shinder. Torsion in the cohomology of Fano varieties of lines. *mathoverflow question 434409*. (Cited on page 119.)

[433] Tetsuji Shioda. An example of unirational surfaces in characteristic p. *Math. Ann.*, 211:233–236, 1974. (Cited on page 89.)

[434] Tetsuji Shioda. On unirationality of supersingular surfaces. *Math. Ann.*, 225(2):155–159, 1977. (Cited on page 89.)

[435] Tetsuji Shioda. Some remarks on Abelian varieties. *J. Fac. Sci. Univ. Tokyo Sect. IA Math.*, 24(1):11–21, 1977. (Cited on page 117.)

[436] Tetsuji Shioda. The Hodge conjecture for Fermat varieties. *Math. Ann.*, 245(2):175–184, 1979. (Cited on page 72.)

[437] Tetsuji Shioda and Toshiyuki Katsura. On Fermat varieties. *Tohoku Math. J. (2)*, 31(1):97–115, 1979. (Cited on page 72.)

[438] Fei Si. Cohomology of moduli space of cubic fourfolds I. *arXiv:2103.04282*. (Cited on page 353.)

[439] Roy Smith and Robert Varley. A Riemann singularities theorem for Prym theta divisors, with applications. *Pacific J. Math.*, 201(2):479–509, 2001. (Cited on pages 235 and 238.)

[440] Andrew Sommese. Complex subspaces of homogeneous complex manifolds. II. Homotopy results. *Nagoya Math. J.*, 86:101–129, 1982. (Cited on page 122.)

[441] Samuel Stark. Deformations of the Fano scheme of a cubic hypersurface. *arXiv:2207.08762*. (Cited on page 109.)

[442] Peter Swinnerton-Dyer. Cubic surfaces over finite fields. *Math. Proc. Cambridge Philos. Soc.*, 149(3):385–388, 2010. (Cited on page 159.)

[443] Sho Tanimoto and Anthony Várilly-Alvarado. Kodaira dimension of moduli of special cubic fourfolds. *J. Reine Angew. Math.*, 752:265–300, 2019. (Cited on page 349.)

[444] Fabio Tanturri. Pfaffian representations of cubic surfaces. *Geom. Dedicata*, 168:69–86, 2014. (Cited on page 173.)

[445] Jenia Tevelev. *Projective duality and homogeneous spaces*, Vol. 133 of *Encyclopaedia of Mathematical Sciences*. Berlin: Springer-Verlag, 2005. Invariant Theory and Algebraic Transformation Groups, IV. (Cited on page 80.)

[446] Andreĭ Tjurin. The Fano surface of a nonsingular cubic in $\mathbb{P}^4$. *Izv. Akad. Nauk SSSR Ser. Mat.*, 34:1200–1208, 1970. (Cited on pages 193, 210, 215, 225, 230, and 234.)

[447] Andreĭ Tjurin. The geometry of the Fano surface of a nonsingular cubic $F \subset \mathbb{P}^4$, and Torelli's theorems for Fano surfaces and cubics. *Izv. Akad. Nauk SSSR Ser. Mat.*, 35:498–529, 1971. (Cited on pages 207 and 236.)

[448] Andreĭ Tjurin. Five lectures on three-dimensional varieties. *Uspehi Mat. Nauk*, 27(5):(167), 3–50, 1972. (Cited on page 193.)

[449] Andreĭ Tjurin. The intermediate Jacobian of three-dimensional varieties. In *Current problems in mathematics*, Vol. 12 (Russian), pp. 5–57, 239 (loose errata). Moscow: VINITI, 1979. (Cited on page 234.)

[450] Eugenio Togliatti. Una notevole superficie de 5° ordine con soli punti doppi isolati. *Vierteljschr. Naturforsch. Ges. Zürich*, 85(Beibl, Beiblatt (Festschrift Rudolf Fueter)):127–132, 1940. (Cited on page 323.)

[451] Orsola Tommasi. Stable cohomology of spaces of non-singular hypersurfaces. *Adv. Math.*, 265:428–440, 2014. (Cited on page 146.)

[452] Burt Totaro. The integral cohomology of the Hilbert scheme of two points. *Forum Math. Sigma*, 4:e8, 20, 2016. (Cited on pages 119 and 294.)

[453] Semion Tregub. Three constructions of rationality of a cubic fourfold. *Vestnik Moskov. Univ. Ser. I Mat. Mekh.*, (3):8–14, 1984. (Cited on page 289.)

[454] Semion Tregub. Two remarks on four-dimensional cubics. *Uspekhi Mat. Nauk*, 48(2(290)):201–202, 1993. (Cited on pages 276, 287, and 337.)

[455] Nguyen Chanh Tu. Non-singular cubic surfaces with star points. *Vietnam J. Math.*, 29(3):287–292, 2001. (Cited on page 182.)

[456] Nguyen Chanh Tu. On semi-stable, singular cubic surfaces. In *Singularités Franco-Japonaises*, Vol. 10 of *Sémin. Congr.*, pp. 373–389. Paris: Soc. Math. France, 2005. (Cited on page 186.)

[457] Ravi Vakil and Melanie Matchett Wood. Discriminants in the Grothendieck ring. *Duke Math. J.*, 164(6):1139–1185, 2015. (Cited on page 146.)

[458] Bart van den Dries. Degenerations of cubic fourfolds and holomorphic symplectic geometry. PhD thesis, Utrecht University, 2012. (Cited on page 350.)

[459] Gerard van der Geer. On the geometry of a Siegel modular threefold. *Math. Ann.*, 260(3):317–350, 1982. (Cited on page 245.)

[460] Gerard van der Geer and Alexis Kouvidakis. A note on Fano surfaces of nodal cubic threefolds. In *Algebraic and arithmetic structures of moduli spaces (Sapporo 2007)*, Vol. 58 of *Adv. Stud. Pure Math.*, pp. 27–45. Tokyo: Math. Soc. Japan, 2010. (Cited on pages 239 and 241.)

[461] Gerard van der Geer and Alexis Kouvidakis. The rank-one limit of the Fourier–Mukai transform. *Doc. Math.*, 15:747–763, 2010. (Cited on page 228.)

[462] Bartel van der Waerden. *A history of algebra.* Berlin: Springer-Verlag, 1985. From al-Khwārizmī to Emmy Noether. (Cited on pages 156 and 181.)

[463] Bert van Geemen. Half twists of Hodge structures of CM-type. *J. Math. Soc. Japan*, 53(4):813–833, 2001. (Cited on page 249.)

[464] Bert van Geemen and Elham Izadi. Half twists and the cohomology of hypersurfaces. *Math. Z.*, 242(2):279–301, 2002. (Cited on page 72.)

[465] Michael van Opstall and Răzvan Veliche. Variation of hyperplane sections. In *Algebra, geometry and their interactions*, Vol. 448 of *Contemp. Math.*, pp. 255–260. Providence, RI: Amer. Math. Soc., 2007. (Cited on page 69.)

[466] Victor Vasil'ev. How to calculate the homology of spaces of nonsingular algebraic projective hypersurfaces. *Tr. Mat. Inst. Steklova*, 225(Solitony Geom. Topol. na Perekrest.):132–152, 1999. (Cited on page 146.)

[467] Misha Verbitsky. Mapping class group and a global Torelli theorem for hyperkähler manifolds. *Duke Math. J.*, 162(15):2929–2986, 2013. Appendix A by E. Markman. (Cited on pages 295 and 348.)

[468] Misha Verbitsky. Errata for "Mapping class group and a global Torelli theorem for hyperkähler manifolds" by Misha Verbitsky. *Duke Math. J.*, 169(5):1037–1038, 2020. (Cited on pages 295 and 348.)

[469] Charles Vial. Projectors on the intermediate algebraic Jacobians. *New York J. Math.*, 19:793–822, 2013. (Cited on page 404.)

[470] Eckart Viehweg and Kang Zuo. Complex multiplication, Griffiths–Yukawa couplings, and rigidity for families of hypersurfaces. *J. Algebraic Geom.*, 14(3):481–528, 2005. (Cited on page 49.)

[471] Claire Voisin. Birational invariants and decomposition of the diagonal. In Andreas Hochenegger, Manfred Lehn, and Paolo Stellari (eds.), *Birational geometry of hypersurfaces*, Vol. 26 of *Lect. Notes Unione Mat. Ital.*, pp. 3–71. Cham: Springer, 2019. (Cited on page 239.)

[472] Claire Voisin. Théorème de Torelli pour les cubiques de $\mathbb{P}^5$. *Invent. Math.*, 86(3):577–601, 1986. (Cited on pages 59, 255, 256, 257, 259, 263, 292, 300, 312, 323, 324, 325, 326, and 347.)

[473] Claire Voisin. Sur la stabilité des sous-variétés lagrangiennes des variétés symplectiques holomorphes. In *Complex projective geometry (Trieste, 1989/Bergen, 1989)*, Vol. 179 of *London Math. Soc. Lecture Note Ser.*, pp. 294–303. Cambridge: Cambridge University Press, 1992. (Cited on page 316.)

[474] Claire Voisin. *Théorie de Hodge et géométrie algébrique complexe*, Vol. 10 of *Cours Spécialisés*. Paris: Société Mathématique de France, 2002. (Cited on pages 1, 24, 27, 28, 29, 45, 47, 49, 50, 52, 57, 116, 118, 133, 153, 266, 324, and 399.)

[475] Claire Voisin. Intrinsic pseudo-volume forms and K-correspondences. In *The Fano Conference*, pp. 761–792. Turin: Univ. Torino, 2004. (Cited on page 327.)

[476] Claire Voisin. Some aspects of the Hodge conjecture. *Jpn. J. Math.*, 2(2):261–296, 2007. (Cited on pages 309 and 400.)

[477] Claire Voisin. Erratum: “A Torelli theorem for cubics in $\mathbb{P}^5$”. *Invent. Math.*, 172(2):455–458, 2008. (Cited on pages 255, 300, and 347.)

[478] Claire Voisin. On the Chow ring of certain algebraic hyper-Kähler manifolds. *Pure Appl. Math. Q.*, 4(3, Special issue: In honor of Fedor Bogomolov. Part 2):613–649, 2008. (Cited on pages 401 and 404.)

[479] Claire Voisin. Coniveau 2 complete intersections and effective cones. *Geom. Funct. Anal.*, 19(5):1494–1513, 2010. (Cited on pages 214 and 317.)

[480] Claire Voisin. Abel–Jacobi map, integral Hodge classes and decomposition of the diagonal. *J. Algebraic Geom.*, 22(1):141–174, 2013. (Cited on pages 309, 400, and 402.)

[481] Claire Voisin. Unirational threefolds with no universal codimension 2 cycle. *Invent. Math.*, 201(1):207–237, 2015. (Cited on pages 220 and 402.)

[482] Claire Voisin. Stable birational invariants and the Lüroth problem. In *Surveys in differential geometry 2016. Advances in geometry and mathematical physics*, Vol. 21 of *Surv. Differ. Geom.*, pp. 313–342. Somerville, MA: Int. Press, 2016. (Cited on page 239.)

[483] Claire Voisin. On the universal CH_0 group of cubic hypersurfaces. *J. Eur. Math. Soc.*, 19(6):1619–1653, 2017. (Cited on pages 26, 112, 239, 402, and 403.)

[484] Claire Voisin. Hyper-Kähler compactification of the intermediate Jacobian fibration of a cubic fourfold: the twisted case. In *Local and global methods in algebraic geometry*, Vol. 712 of *Contemp. Math.*, pp. 341–355. Providence, RI: Amer. Math. Soc. 2018. (Cited on page 298.)

[485] Claire Voisin. Schiffer variations and the generic Torelli theorem for hypersurfaces. *Compositio Math.*, 158(1):89–122, 2022. (Cited on page 154.)

[486] A. M. Šermenev. The motif of a cubic hypersurface. *Izv. Akad. Nauk SSSR Ser. Mat.*, 34:515–522, 1970. (Cited on page 229.)

[487] C. T. C. Wall. On the orthogonal groups of unimodular quadratic forms. *Math. Ann.*, 147:328–338, 1962. (Cited on page 14.)

[488] C. T. C. Wall. Diffeomorphisms of 4-manifolds. *J. London Math. Soc.*, 39:131–140, 1964. (Cited on page 30.)

[489] Joachim Wehler. Deformation of varieties defined by sections in homogeneous vector bundles. *Math. Ann.*, 268(4):519–532, 1984. (Cited on page 216.)

[490] Li Wei and Xun Yu. Automorphism groups of smooth cubic threefolds. *J. Math. Soc. Japan*, 72(4):1327–1343, 2020. (Cited on page 40.)

[491] Peter Weibel. *Negative space: Trajectories of sculpture in the 20th and 21st centuries*. Cambridge, MA: MIT Press, 2021. (Cited on page 182.)

[492] André Weil. Abstract versus classical algebraic geometry. In *Proceedings of the International Congress of Mathematicians, 1954, Amsterdam, Vol. III*, pp. 550–558. Amsterdam: North-Holland Publishing Co., 1956. (Cited on pages 16 and 159.)

[493] André Weil. Sur le théorème de Torelli. In *Séminaire Bourbaki, Vol. 4*, pp. Exp. No. 151, 207–211. Paris: Soc. Math. France, 1995. (Cited on page 236.)

[494] Song Yang and Xun Yu. Rational cubic fourfolds in Hassett divisors. *C. R. Math. Acad. Sci. Paris*, 358(2):129–137, 2020. (Cited on pages 337 and 349.)

[495] Mutsumi Yokoyama. Stability of cubic 3-folds. *Tokyo J. Math.*, 25(1):85–105, 2002. (Cited on page 243.)

[496] Mutsumi Yokoyama. Stability of cubic hypersurfaces of dimension 4. In *Higher dimensional algebraic varieties and vector bundles*, RIMS Kôkyûroku Bessatsu, B9, pp. 189–204. Res. Kyoto: Inst. Math. Sci. (RIMS), 2008. (Cited on pages 350 and 352.)

[497] Yuri Zarhin. Cubic surfaces and cubic threefolds, Jacobians and intermediate Jacobians. In *Algebra, arithmetic, and geometry: in honor of Yu. I. Manin. Vol. II*, Vol. 270 of *Progr. Math.*, pp. 687–691. Boston, MA: Birkhäuser Boston, 2009. (Cited on page 238.)

[498] Jun Zhang. Geometric compactification of moduli space of cubic surfaces and Kirwan blow-up. PhD thesis, Rice University, 2005. (Cited on page 191.)

[499] Zhiwei Zheng. Orbifold aspects of certain occult period maps. *Nagoya Math. J.*, 243:137–156, 2021. (Cited on pages 186, 190, 210, 236, 248, and 300.)

[500] Jian Zhou. Calculations of the Hirzebruch χ_y-genera of symmetric products by the holomorphic Lefschetz formula. *arXiv:math/9910029*. (Cited on page 121.)

[501] Steven Zucker. The Hodge conjecture for cubic fourfolds. *Compositio Math.*, 34(2):199–209, 1977. (Cited on pages 254 and 309.)

Index

Abel–Jacobi map, 133, 217, 228, 243, 399, 400, 404
  injective, 404
Abel–Prym map, 218, 223, 225
adjoint functor, 357, 362
admissible subcategory, 356
ample cone, 162
Arnold trinity, 180
automorphism
  cubic threefold, 209
  generic hypersurface, 37, 142
  Grassmann variety, 108
  infinitesimal, 31
  order, 40
  Pfaffian, 277
  polarized, 33, 46, 138
  symplectic, 40
  triple cover, 188, 190
automorphism group
  on cohomology, 39, 153, 154
  cubic surface, 160, 170
  cubic threefold, 236
  curve, 236
  family, 35, 138
  finite, 34, 46, 138, 151
  hyperkähler manifold, 296
  intermediate Jacobian, 236, 244
  Jacobian, 236
  Klein cubic, 238, 245
  line configuration, 164, 179
  versus monodromy group, 152

Beauville–Bogomolov–Fujiki form, 294, 295, 300
  cohomology class, 313
  Fano variety, 252, 264, 296, 303, 304, 314
  Hilbert scheme, 280, 294
Bertini involution, 170
Betti numbers, 5, 73
  cubic fourfold, 251
  cubic threefold, 193
  Fano variety, 120
  hypersurface, 3
  primitive, 5, 6
bitangent line, 170, 180
Bloch–Srinivas, 228, 254, 400, 402
Bondal–Orlov theorem, 356
Bott vanishing, 3, 4, 9, 56, 103, 363, 387
Brauer–Severi variety, 261, 379, 385, 386

Calabi–Yau category, 363, 374
  fractional, 363, 367
  matrix factorization, 372
canonical bundle, 356
  cubic fourfold, 251
  cubic surface, 157
  cubic threefold, 193
  Fano surface, 102, 194, 240
  Fano variety, 101, 122, 194, 252, 292, 401
  Fano variety F_2, 318
  hypersurface, 3, 7
Cayley surface, 68, 172, 174, 186
Cayley–Salmon equation, 175
Chern class of cubic, 5
Chow dimension, 403
Chow group
  cubic fivefold, 400
  cubic fourfold, 254, 306, 399, 401
  cubic hypersurface, 5
  cubic threefold, 195, 228, 229, 242, 399, 401
  Fano correspondence, 132
  Fano variety, 125, 306, 400
  moduli space, 185
Chow motive
  blow-up formula, 112
  category of, 4, 110, 285

Grothendieck group, 113
cubic fourfold, 254, 404–406
cubic threefold, 195, 404
Fano variety, 195, 254, 306, 325
finite-dimensional, 113, 306, 403, 405, 406
hypersurface, 4, 403
K3 surface, 264
transcendental, 406
Clebsch surface, 172–174, 182
Clifford algebra, 260, 338, 378, 388
Cohen–Macaulay module, 370
cohomology
equivariant, 145
integral
algebraic, 2
generator, 2
torsion free, 2
moduli space, 144
primitive, 5, 149
conic fibration, 63, 159, 204, 229, 323
cubic curve
stable, 139
cubic fivefold, 123, 317, 400, 401, 405
cubic fourfold
Betti numbers, 251
Chow group, 399
diagonal, 403
Fano correspondence, 253, 258, 263, 300, 301, 304, 310, 315, 326, 340, 397
Hodge classes, 254
Hodge conjecture, 254
Hodge numbers, 251
Kuznetsov component, 367, 384, 393, 406
lattice theory, 255
moduli space, 148, 151
motive, 306, 404–406
nodal, 265–268, 270, 385, 390
Fano variety, 269, 270
Pfaffian
see Pfaffian cubic fourfold
plane, 252, 253, 385
associated K3 surface, 259
dimension of space, 256
Fano correspondence, 263
Kuznetsov component, 385, 386
motive, 264
two, 256
quadric fibration, 257, 385
rationality, 262, 402, 403
special, 254, 273, 306, 334, 335, 344, 399
cubic surface
Chow group, 399
Kuznetsov component, 367, 405
moduli space, 146, 147, 183
cohomology, 146
monodromy group, 29
period map, 189
semi-stable, 184, 185
stable, 185
cubic threefold
Betti numbers, 193
Chow group, 399
diagonal, 403
Euler number, 193
Fano correspondence, 195, 217, 224, 243, 316
Hodge classes, 195
Hodge numbers, 193
intermediate Jacobian, 403, 406
Kuznetsov component, 367, 381, 405, 406
moduli space, 151
motive, 404
nodal, 253
simply connected, 196
Torelli, 189

decomposition
connected sum, 30
diagonal, 401–403
semi-orthogonal, 355, 360, 361
blow-up, 355
Brauer–Severi variety, 355
hypersurface, 364
projective bundle, 355
quadric, 364
deformation, 35
again hypersurface, 37
complex, 393
cubic surface, 168
Fano variety, 109, 129, 216, 297
first order, 58, 69, 83, 98, 109, 153, 198, 211, 256, 396, 397
global, 37
infinitesimal, 36
line, 81, 83, 200, 207
second type, 99
through point, 98
polarized, 35
universal, 36, 141, 143, 150, 152, 194
degree shift functor, 366, 368, 372, 373
diagonal
blow-up, 110, 293
decomposition, 401, 403
cubic, 403

Hassett divisor, 402
representable, 142
discriminant, 20, 137, 144
elliptic curve, 21, 47
lattice, 12, 13, 23, 24, 30, 227, 255, 307, 313, 333, 334
quadric fibration, 62, 63
discriminant divisor, 16–18, 20, 41, 61, 137, 147
cubic surface, 21, 179, 184
dual variety, 19
quadric fibration, 61, 126, 196, 204, 257, 259, 260, 322, 385, 387

Eckardt point
Clebsch surface, 174
cubic surface, 177, 181, 182
number of, 182
cubic threefold, 197, 198, 201, 214
effective cone, 162
Eisenstein integers, 187, 247
elliptic curve, 9, 31, 47, 73, 139, 147, 198, 216, 227
moduli space, 147
Euler number, 5, 251, 253
Euler sequence, 3, 8
exceptional collection, 360
$\mathbb{P}^n$, 360
exceptional object, 359

Fano correspondence
Chow group, 125, 132, 306, 400
cohomological, 124, 125, 127, 128, 254
cubic fourfold, 253, 258, 263, 300, 301, 304, 315, 326, 340, 397
cubic threefold, 195, 217, 224, 243, 316
derived category, 374, 386, 397
dual, 132, 254, 309, 310
geometric, 78, 122, 257
Hodge structure, 128, 263, 301, 340
nodal cubic, 267
pairing, 128, 219, 301, 333, 340
Plücker polarization, 125
quadratic, 130
quadric, 80, 365
Fano functor, 76
Fano surface, 194, 195, 201, 208, 212, 213, 215, 225, 239, 240, 244
Fano variety
automorphism, 108
Betti numbers, 120, 123
canonical bundle, 101
Chern character, 102
cohomology, 103, 118, 119, 122, 304, 311, 375
connected, 103
construction, 75, 76
cubic fivefold
cohomology, 123
cubic fourfold, 252
canonical bundle, 102, 252
cohomology, 123
dimension, 252
Euler number, 253
Hodge numbers, 123, 253
Plücker polarization, 252, 258
cubic threefold, 194
Betti numbers, 194
canonical bundle, 102, 194
cohomology, 123
Hodge numbers, 123, 194
Hodge structure, 195
Plücker polarization, 194
deformation, 109, 129, 215, 296
degree, 113, 115, 194
derived category, 374, 375
dimension, 80, 86, 89
Euler number, 113
Hirzebruch genus, 120
Hodge conjecture, 120
Hodge numbers, 120, 123
Hodge structure, 118, 122
hyperkähler manifold, 296
Kuznetsov component, 396
lines of the second type, 89, 98, 196, 252, 312, 318, 327
universal, 95
motive, 110, 113, 116, 306, 375
no vector fields, 108
not empty, 80
Picard group, 101, 103, 121, 291, 307
Plücker embedding, 78, 113
quadric, 80
relative, 146, 258
simply connected, 103, 107, 122
smooth, 80, 86
tangent bundle, 86
tangent space, 83, 396
universal, 85
Fano visitor, 376, 393
Fermat cubic, 58, 73, 86
automorphisms, 40
curve, 73
fourfold
plane, 256, 258
hyperplane section, 69, 70
Jacobian ring, 43

lines in, 86, 90
quadric fibration, 62
rational, 65
surface, 182
threefold, 198
Fermat hypersurface, 17
Fermat pencil, 22
Fermat quartic, 47
field of definition, 141
field of moduli, 141
Fourier–Mukai functor, 355, 390
convolution, 367
mutation, 364
Fujiki constant, 294, 295, 297, 314

Gauss map, 70, 71, 89, 93, 97, 233, 235
Geiser involution, 170
Grassmann functor, 76
Grassmann variety, 76
automorphism, 108
Bott vanishing, 103
isotropic, 80
universal line, 106
Grothendieck group
Hodge structure, 117
Kuznetsov component, 405
motive, 113
thick subcategory, 357
Grothendieck ring, 110, 113, 146, 270, 285, 289, 375
Gysin sequence, 54, 146, 316

Hassett conjecture, 337, 384, 392
Hassett divisor, 256, 334–339, 342, 344, 345, 348, 349, 385, 393, 402, 405
cubics with a plane, 256
diagonal, 402, 403
unirational, 148
Hilbert scheme, 76, 81, 83, 111
ample cone, 309
K3 surface, 251, 253, 268, 277, 279, 293, 295, 296, 307, 309, 310, 312, 315, 341, 342
cohomology, 280, 295
Hilbert square, 110, 112, 384, 402
cohomology, 119
derived category, 375
Fano correspondence, 130
Hodge structure, 117
Kuznetsov component, 397
Hilbert's 14th problem, 135
Hilbert–Mumford criterion, 139, 186
Hirzebruch genus, 7, 8, 120, 121
Hirzebruch signature theorem, 10
Hirzebruch surface, 37, 165, 286
Hirzebruch–Riemann–Roch, 8, 14, 157, 216, 313
Hodge conjecture
cubic fourfold, 251, 254, 309, 310, 400
Fano variety, 120, 310
hypersurface, 26
Hodge filtration, 53, 56, 148
Hodge isometry, 153, 187, 230, 267, 290, 300, 304, 383, 394
Hodge numbers, 7
cubic fourfold, 251
cubic threefold, 193
F_2, 318
in family, 150
Fano variety, 120, 123, 253
hypersurface, 3, 10, 149
Hodge structure, 148
associated K3 surface, 337
blow-up, 116, 119, 266
cubic fourfold, 149, 251, 271, 302, 331, 338, 340
cubic hypersurface, 11, 25
cubic threefold, 149
Fano correspondence, 125, 128, 131, 253, 263, 301, 302, 309
Fano variety, 118, 122, 280, 312
Hilbert scheme, 117
hyperkähler manifold, 295
hypersurface, 8
K3 surface, 264, 330
Pfaffian cubic, 290
polarization, 116, 149
pure, 116
Tate twist, 11, 117
triple cover, 72
Hodge–de Rham spectral sequence, 15, 53
hyperkähler manifold, 102, 107, 109, 251–253, 270, 271, 287, 289, 292, 294–296, 300, 303, 306–310, 313, 315, 316, 326, 341, 342
cohomology ring, 297
Jacobian type, 295

intermediate Jacobian, 11, 133, 190, 217, 382
automorphism, 190, 244
blow-up, 220
categorical, 377
cubic threefold, 217, 219, 220, 225, 227, 229, 231, 235, 236, 238, 383
irreducible, 191, 235
isogeny, 403
Klein cubic, 238, 245
minimal cohomology classes, 227
modular, 231

nodal cubic, 241
reduction, 238
specialization, 239
theta divisor, 225, 235, 384
under blow-up, 237
versus Albanese, 217
versus Prym, 221
intersection form, 13–15, 24, 127, 132, 148
cubic fourfold, 330, 331
disc = 14, 334, 335
disc = 2, 335
disc = 3, 302
disc = 6, 335
disc = 8, 264, 334, 337
discriminant, 333
Fano variety, 313
nodal, 267
Pfaffian, 272, 287, 307
plane, 255, 306
quadric, 256
Veronese, 272
cubic surface, 157, 159
cubic threefold
Chow, 228
Kuznetsov component, 381
curve, 230
Fano surface, 227
Fano variety, 195
K3 surface, 294, 330
invariant cycle theorem, 24, 253, 324
invariant ring, 184
invariants
cubic fourfold, 251
cubic threefold, 193
hypersurface, 1
irrationality, 66
Jacobian ring, 41, 367
cubic surface, 46
determining hypersurface, 51
dimension, 42
Gorenstein, 42
Lefschetz property, 46
matrix factorization, 372
multiplication, 49, 57
Poincaré polynomial, 42, 46
primitive cohomology, 52
K3 surface
ample cone, 162
associated, 263, 268, 270, 272, 281, 282, 336, 337, 339, 342, 346, 350, 351, 386, 389, 390, 393
automorphism, 34, 143
Brauer group, 261, 385, 388
degree 14, 275, 277, 279, 283, 284, 290, 303
degree 2, 260, 271, 313, 333
degree 6, 265, 266, 268, 270, 308, 346
derived category, 291, 339, 385
double plane, 171, 192
global Torelli, 155, 267, 295, 346, 347
as Hessian, 185
Hilbert scheme, 253, 269, 270, 277, 279, 293, 295–297, 307, 310, 312, 315, 341, 342
cohomology, 280
Hodge structure, 330
as hyperkähler manifold, 293
moduli space, 141, 349, 350
moduli stack, 142
motive, 264
Mukai lattice, 338
period, 343, 345, 350
Plücker polarization, 280
quartic, 31, 47
quasi-polarized, 351
twisted, 264, 339, 350, 351, 393
universal family, 141
Klein cubic, 216, 238, 239, 245
Kodaira vanishing, 4, 15, 37, 103, 107, 109, 218, 315
Koszul complex, 42, 50, 83, 103, 107, 319, 368
Kuznetsov component, 364, 367
cubic fourfold, 384, 394, 395, 397, 406
Pfaffian, 291, 389
plane, 385
cubic surface, 378, 405
cubic threefold, 377, 378, 380, 381, 395, 405, 406
geometric, 367, 386
Grothendieck group, 405
matrix factorization, 373
motive, 405, 406
quadric, 364
quadric fibration, 386
Serre functor, 366, 367
symmetric square, 398
Kuznetsov conjecture, 384, 392
Lüroth problem, 88, 236, 403
Lagrangian fibration, 270, 340, 342
Lagrangian subvariety, 270, 287, 289, 291, 315–318, 321
lattice
cubic fourfold, 331
K3 surface, 298, 330
Lefschetz hyperplane theorem, 1, 3, 4, 275

Lefschetz pencil, 16, 21, 27, 71, 158, 324
  cubic surface, 172
Lefschetz property, 46
line
  contained in plane, 257
  first type, 86, 89, 91, 93, 98, 196, 198–200, 202, 204, 205, 252, 258, 322
  normal bundle, 84
  second type, 86, 89, 93, 94, 98, 196–200, 202, 205, 252, 268, 300, 320, 329
  through a point, 198, 253
linear subspace
  contained in hypersurface, 2
LLSvS eightfold, 342, 396

MacDonald formula, 114, 121
matrix factorization, 369
  graded, 371
  Kuznetsov component, 373
moduli space
  of abelian varieties, 151, 245
  coarse, 140, 147
  of cubic fourfolds, 148, 252, 256, 272, 273, 277, 334, 347–350, 352
  of cubic surfaces, 146, 147, 183, 185
    cohomology, 146
  of cubic threefolds, 194, 216, 243, 244
  of elliptic curves, 147
  GIT, 147
  of K3 surfaces, 343, 346, 347, 349–351
  of sheaves, 220, 270, 308, 341
  tangent space, 143
  unirational, 147
moduli stack, 142, 151, 244, 347
  Deligne–Mumford, 142, 334
  tangent space, 143
monodromy group, 16, 23, 24, 27, 150
  27 lines, 146
  algebraic, 25
  bitangents, 181
  cubic fourfolds, 347
  cubic surfaces, 29, 163, 179
  cubic threefolds, 152, 202, 225
  generators, 27
  versus diffeomorphisms, 29, 164
monodromy operator, 27
Mumford–Tate group, 26
mutation, 358
  braid relations, 361
  as Fourier–Mukai functor, 364

Nakai–Moishezon criterion, 161, 167
nodal cubic, 66, 228, 241, 243, 245, 253, 266–269, 308, 346, 377, 385
  Fano correspondence, 267
  Fano variety, 242, 350
  global Torelli, 241, 267
  intermediate Jacobian, 241
  Kuznetsov component, 390
  rational, 67, 240
nodal hypersurface, 66
node
  see ordinary double point, 66
Noether–Lefschetz divisor, *see* Hassett divisor
normal bundle, 83
  $F(Y)$ in Fano variety, 317
  F_2 in Fano variety, 318
  F_L in Fano variety, 324
  Fano variety in Grassmann variety, 102
  K3 surface in $\mathbb{P}^4$, 265
  line in cubic fourfold, 252, 299
  line in cubic hypersurface, 85, 86
  line in cubic threefold, 194
  line in projective space, 84
  scroll in cubic fourfold, 286
  universal line, 105
normal bundle sequence, 3, 5, 7, 8, 32, 36, 86, 102, 210, 256, 286, 289, 317

orbit closure, 147
ordinary double point, 19, 21, 27, 62, 63, 66, 67, 139, 199, 205, 206, 239, 241, 265, 267–270, 308, 318, 323, 335
  cubic surface, 185, 186
  maximal number, 67, 245

period domain, 150
  compact dual, 150
  cubic fourfold, 342, 344, 347
  K3 surface, 344
period map
  cubic fourfold, 152, 342
  cubic surface, 189
  cubic threefold, 152, 248
Pfaffian cubic fourfold, 273
  Fano variety, 279, 291, 296, 306, 399
  Hassett divisor, 334, 335
  Hodge structure, 290
  K3 surface, 275, 281, 284, 303, 342
  Kuznetsov component, 291, 385, 388, 389, 395
  motive, 285, 289
  plane, 257, 385
  quartic scroll, 272, 285, 291
  quintic del Pezzo, 276, 287–289
  rational, 288, 289, 337

very general, 287, 291, 307
Pfaffian cubic surface, 172, 245, 246
Pfaffian cubic threefold, 244–246
Picard group
cubic fourfold, 251
cubic surface, 157, 159
cubic threefold, 215
Fano variety, 101, 121, 264, 291, 306, 307, 320
Hilbert scheme, 309
hypersurface, 3, 4
K3 surface, 264
Picard number
Fano variety, 195, 216, 238, 245, 310, 325
Picard–Lefschetz formula, 28
Plücker coordinates, 107
Plücker polarization, 79, 100, 107, 126, 129, 130, 194, 258, 275, 280, 297, 301, 309, 311, 316, 320, 325, 329, 340, 387
BBF square, 304
degree, 194, 252, 304
Fano correspondence, 125
K3 surface, 280
Pfaffian, 279
spinor varieties, 81
Voisin endomorphism, 327
polar, 69, 129, 169
pole filtration, 56
Porteous formula, 98, 283, 312, 323
Prym variety, 221, 231
dimension, 222
not a Jacobian, 238
as polarized abelian variety, 223

quadric, 105, 108, 130, 241, 265, 266, 268, 270, 280
cubic fourfold, 271, 272
Fano variety, 80
Kuznetsov component, 364
spinor bundle, 364
quadric fibration, 58, 60
Chow group, 401
cubic fourfold, 257, 262
cubic surface, 192
cubic threefold, 63, 204
derived category, 388
discriminant divisor, 61–63, 257
smooth, 257, 259
Fermat cubic, 62
section, 64, 262, 386
unirationality, 64
quotient, 135
categorical, 135, 140
geometric, 136, 140
GIT, 136, 144, 183
good, 136, 137

rational, 2, 16, 64, 65
cubic fourfold, 65, 256, 262, 337, 338, 384
cubic surface, 66
cubic threefold, 237, 239
nodal cubic, 67, 240
Pfaffian cubic fourfold, 288, 289
stably, 239
residual category, *see* Kuznetsov component
residual conic, 62, 171, 204, 205
residual line, 199–202, 208, 242, 252
residual quadric, 61, 62, 256–258, 263
resultant, 20
Roitman's theorem, 401

Schläfli double six, 179
Schläfli graph, 178
scroll, 271, 272, 285, 289, 291, 335, 337, 389
Segre cubic, 68, 244, 245
Serre functor, 361
hypersurface, 363
Kuznetsov component, 366, 369
orthogonal complement, 362
spinor variety, 81
stable, 135
(co)tangent bundle, 4
bundle, 352
cubic fourfold, 350, 352, 353
cubic surface, 185
cubic threefold, 240, 243
hypersurface, 137
semi-, 137, 243, 350, 352
sheaf, 270, 341
subcategory
admissible, 360, 362
orthogonal, 357, 360, 362
thick, 357, 358
Sylvester form, 173, 184, 245
symmetrizer lemma, 49
symplectic structure, 292, 296, 300, 331, 396

tangent bundle
Fano surface, 209
stable, 4
tangent space
projective, 91
tangent to a line, 89–92, 98, 199, 252, 327, 328
Tate motive, 5, 113
Tate twist, 116
Torelli
categorical, 380

general, 154, 155
generic, 154, 155
cubic threefold, 155
geometric, 107, 215, 301, 382, 395
global, 154
cubic fourfold, 300, 326, 347, 394
cubic threefold, 189, 190, 231, 382
curve, 230, 234
derived, 377, 380, 394
hyperkähler manifold, 292, 295, 300, 309
K3 surface, 155, 267, 295, 346, 347
nodal cubic, 240, 267
infinitesimal, 52, 57, 129, 153–155, 190
local, 153, 155
local versus infinitesimal, 152
strong
cubic curve, 236
cubic fourfold, 300
cubic threefold, 236
variational, 58, 154, 155, 232
transcendental lattice, 304
triple cover, 58, 71, 73, 171, 187, 190, 405
tritangent plane, 176, 177, 180, 181
Tschirnhaus bundle, 66
unirational, 2, 4, 63–65, 87, 89, 245, 338
Hassett divisor, 148, 348, 349
Lüroth problem, 88, 236
moduli space, 147
universal coefficient theorem, 2, 204
universal family, 134, 140
étale locally, 141
cubic surface, 181
K3 surface, 141
line
through point, 197
non-existence, 141
over open subsets, 141
quartic curves, 181
universal hypersurface, 14, 16, 79, 146, 183
universal line, 103–106, 111, 115, 122, 194, 196, 253, 259
universal Pfaffian, 273

vanishing class, 28, 29
vanishing cohomology, 28
vanishing lattice, 29
vanishing sphere, 28
Veronese surface, 271, 272, 335
Voisin endomorphism, 327, 329

Weil conjectures, 16, 124, 159, 193, 238

Printed in the United States
by Baker & Taylor Publisher Services